1 MONTH OF
FREE
READING

at
www.ForgottenBooks.com

By purchasing this book you are eligible for one month membership to ForgottenBooks.com, giving you unlimited access to our entire collection of over 1,000,000 titles via our web site and mobile apps.

To claim your free month visit:
www.forgottenbooks.com/free1248984

ISBN 978-0-332-77394-0
PIBN 11248984

This book is a reproduction of an important historical work. Forgotten Books uses
state-of-the-art technology to digitally reconstruct the work, preserving the original format
whilst repairing imperfections present in the aged copy. In rare cases, an imperfection in
the original, such as a blemish or missing page, may be replicated in our edition. We do,
however, repair the vast majority of imperfections successfully; any imperfections that
remain are intentionally left to preserve the state of such historical works.

TABLE OF CONTENTS

PREFACE

The Sea Island Coastal Region of
South Carolina and Georgia is rich in
natural resources, including moderate
climate, dramatic scenic qualities,
fertile soils, water, fish, wildlife and
minerals. Those resources are valuable
for a variety of often competitive uses,
including active and passive recreation,
transportation, agriculture, commercial
fisheries, industrial development, preservation, and so forth.

A significant trend in the management and development of coastal resources is the growing realization that
rational decisions and final judgements
can be made only when all available information on local environmental conditions is considered. This trend
recognizes the need for a holistic
approach and has promoted the ecosystem concept in natural resource
management.

Recognition of the need for an ecological approach in managing coastal
resources has developed from increasing
evidence that man's utilization of this
environment has brought about major,
yet often subtle, changes in the func-
tioning of ecosystems. In order to
perpetuate the economic, aesthetic, and
biological values of coastal ecosystems,
we must understand their functional
relationships. As expressed by Odum
(1964), our modern ecology must be a
"systems ecology," or a hybridization of
both ecology and systems methodology.
The theory behind this approach embodies
an important ecological principle: an
ecological system is comprised of many
components, no one of which can be
altered without affecting the total system since no one part functions inde-
pendently. By including a full assessment
of the total ecosystem, management
efforts - at both the field and administrative levels - can be designed to
maximize the economic, social, and biological benefits derived from natural
resources. Recognizing this, the U.S.
Fish and Wildlife Service is employing
the ecosystem concept as a holistic
mechanism for managing natural resources
and is developing ecological characterization as one basic tool for this application.

An ecological characterization is
a synthesis of existing information and
data structured in a manner which
identifies functional relationships
between natural processes and the
various components of an ecosystem
(Preface Fig. 1). Specifically,
objectives of the Sea Island
Ecological Characterization were to:
1) assemble, review, and synthesize
existing biological, physical, and

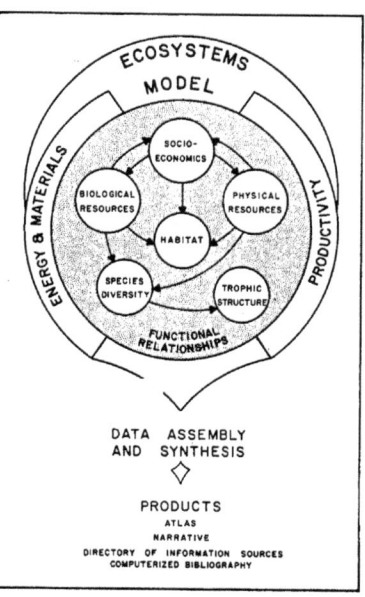

Preface Figure 1. Components and final
products of an Ecological Characterization of the Sea Island
Coastal Region.

socioeconomic information and establish
a sound information base for decision-
making; 2) identify and describe
various components (subsystems, habitats,
communities, and key species) in this
coastal ecosystem; 3) describe major
physical, biological, and socioeconomic
components and interactions; 4) describe
known and potential ecosystem responses
to man-induced changes; and 5) identify
major information deficiencies for
further study and decision-making needs.

Ecological characterizations are
designed primarily to assist coastal
resource managers engaged in comprehensive planning efforts such as
assessment of the environmental impacts
of development in the coastal zone.
Other applications include the preparation of mitigation procedures and
development alternatives. Characterization also provides an immediate data
base for specific action programs
(offshore oil and gas development, coastal
construction permit reviews, etc.) and
guidance in selecting parameters that need
study in further defining coastal ecological systems.

Detailed discussions of the natural coastal ecosystem characterization effort can be found in Tait (1977), Barclay (1978), Johnston (1978), and Palmisano (1978).

SEA ISLAND ECOLOGICAL CHARACTERIZATION

In February 1977, the U.S. Fish and Wildlife Service contracted with the Marine Resources Division of the South Carolina Wildlife and Marine Resources Department to develop an ecological characterization for the Sea Island Coastal Region of South Carolina and Georgia. The project area includes the coastal tier of counties between the Georgetown/Horry county line in northern South Carolina south to the St. Marys River on the Georgia/Florida border, and the three lowland counties of Dorchester, Berkeley, and Effingham (Preface Fig: 2).

The Sea Island Ecological Characterization is designed to yield products that will assist decision makers in evaluating and predicting impacts of man-induced perturbations (e.g., oil and gas development, dredging and filling, water resource projects), and in general coastal zone planning. The study identifies critical habitats and sensitive life history stages of important species, addresses functional interactions at the habitat level, and provides socioeconomic information relative to the coastal environment.

Data assimilated for this project are partitioned into three segments for descriptive purposes: physical features (i.e., geology and hydrology), socioeconomic features (i.e., demographic characteristics and industrial development), and biological features (i.e., an ecological treatment of animals, plants, and their habitats).

The overall framework for the preparation of ecological characterization materials was provided by conceptual models. These conceptual models have been modified for inclusion in the final products to facilitate understanding of ecosystem functions. To accommodate the broadest range of potential users, a three-tier model presentation was used and includes the following elements for each ecosystem: 1) a technical energese model demonstrating energy flow into and within the subject ecosystem, functional relationships among representative components of the system, and flow of energy in various forms from the system; 2) a less technical pictorial model of the same ecosystem illustrating representative flora and fauna; and 3) a representative food web indicating trophodynamics within the subject ecosystem.

Organization of Final Products

Several products are being developed from the Sea Island Ecological Characterization effort, as follows:

1) Characterization Atlas - the Atlas is an oversized document (28 x 42 in) that presents data in condensed form in several series at scales ranging from 1:24,000 to 1:1,000,000. The Physiographic Series (1:100,000) describes wetlands, physiographic features, ecological habitats, and land use. The Geology Series presents stratigraphic, structural, and geophysical information about the characterization area at several scales. Two topographic series at 1:250,000 and 1:100,000 depict various wildlife, archeological and recreational resources, military and educational institutions, water quality, spoil disposal, utilities, railroads and airports. Enlargements of the five major urban areas give more detailed information on industires, point source discharge, power plants, etc. All maps are printed in color.

2) Narrative Volumes - Detailed narrative treatment is provided for the three major ecosystem components; the physical, socioeconomic, and biological features of the Sea Island Coastal Region. Because conceptual models are particularly valuable in identifying ecosystem components and in relating their functional significance and regulatory processes, appropriate sections of the narrative text are prefaced by exemplary models. These models serve as a tool to promote understanding of the functional relationships within and between systems, and the impacts of various impingements and perturbations on their components. Narrative materials are arranged as follows:

a) Physical features section - Detailed treatment is provided for topical areas such as climate, physiography, geologic history and structure, coastal and nearshore erosion and deposition, hydrology, and descriptions of individual coastal islands of the study area.

b) Socioeconomic features section - Data are presented on population, labor force characteristics and trends, transportation, industrial development, agricultural practices, public utilities, energy resources, fish and wildlife conservation and utilization, and recreational resources.

c) Biological features - This section describes biotic components along ecological lines. This approach facilitates the treatment of major community or habitat types, and generally deals with organisms at the population level. Functional relationships and areas of ecological sensitivity are stressed.

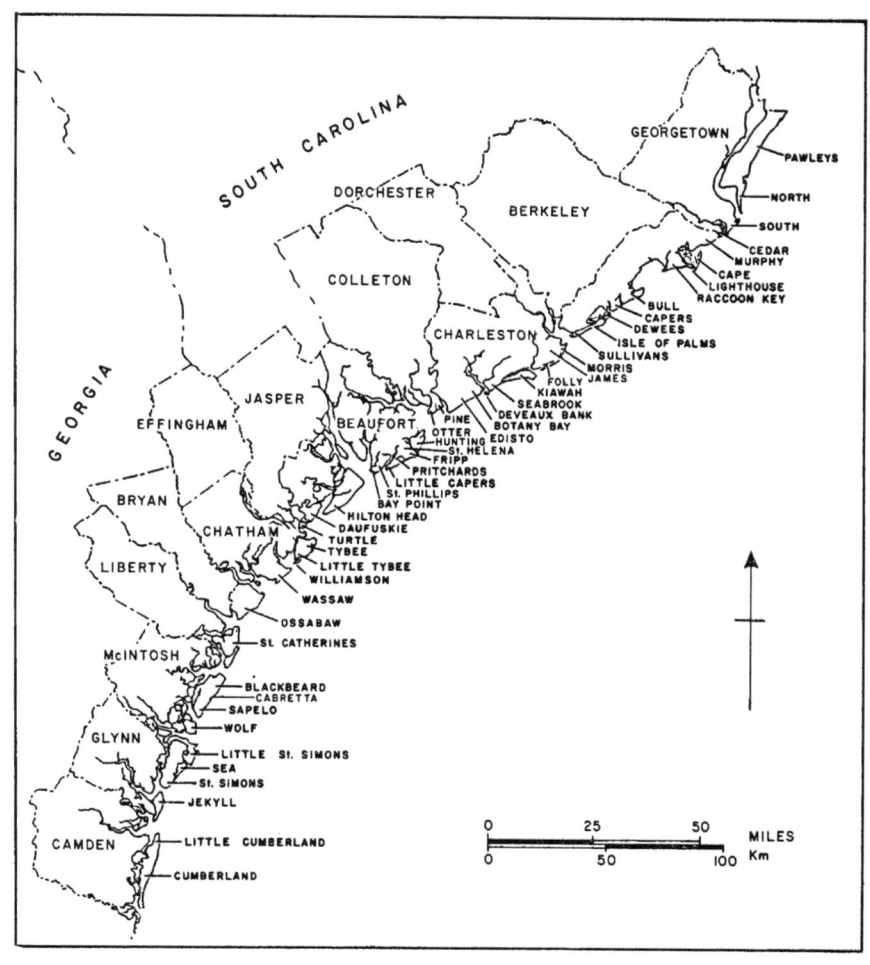

Preface Figure 2. Study area of the Sea Island Coastal Region.

3). Directory of Information
Sources - This document identifies and
describes major data sources relevant to
the ecological characterization of coastal
South Carolina and Georgia. The main purpose of the Directory is to guide users
to known sources of data pertinent to
specific subject areas. It is intended
to serve as a referral service between
groups or organizations with differing
needs.

4) Bibliography - A computerized
bibliography of over 8,000 references has
been assembled as a central component of
the Sea Island Characterization. The
system is designed for periodic updating,
and all entries can be retrieved in a
variety of ways, including key work and
author searches.

CONTRIBUTORS

Barbara S. Anderson.......................................Species list, editorial staff, index
William D. Anderson.......................................Oysters
Donna J. Barber...Species list
Charles M. Bearden..Estuarine-marine fishes
James M. Bishop...Amphibians and reptiles
Victor G. Burrell, Jr.....................................Zooplankton
Dale R. Calder..Editor, benthic, nektonic and intertidal
                                                          invertebrates, food webs
W. David Chamberlain......................................Birds
Peggy Ann Dana..Scientific illustrations
Jane S. Davis...Scientific illustrations
Patricia J. DuPree..Manuscript typing
Peter J. Eldridge...Clams
Lacey L. Gaddy..Vascular flora
Patricia M. Griffin.......................................Editorial staff
Donald L. Hammond...Freshwater fishes
James S. Hart, Jr...Scientific illustrations
Sally R. Hopkins..Alligators and marine mammals
Norris B. Jeffrey...Anadromous and freshwater fish
Myra H. Jones...Manuscript typing
Edwin B. Joseph...Birds, nonmarine mammals
John J. Manzi...Editor, modelling, phytoplankton, bacteria
                                                          and fungi, food webs
Donald E. Marchette.......................................Anadromous fishes
Michael D. McKenzie.......................................Birds, nonmarine mammals
John V. Miglarese...Editor, vascular flora, fishes, food webs,
                                                          species list, index
Lois E. Mishoe..Manuscript typing
James M. Monck..Scientific illustrations
Elizabeth C. Roland.......................................Editorial staff
Paul A. Sandifer..Editor, wetland and rice field inverte-
                                                          brates, crabs, shrimp, food webs,
                                                          endangered species, ecology, benthos
Emily S. Schroeder..Manuscript typing
Malcolm H. Shealy, Jr.....................................Insects, estuarine-marine fishes
Amelia R. Smith...Scientific illustrations
Karen R. Swanson..Scientific illustrations
Ralph W. Tiner, Jr..Wetland vascular plants
Glenn F. Ulrich...Marine reptiles, anadromous fishes

     We wish to express special acknowledgement to Patricia J. DuPree and Lois E. Mishoe for
typing the final camera-ready copy of this volume.

LIST OF FIGURES

LIST OF FIGURES (Continued)

LIST OF TABLES

CHAPTER ONE

INTRODUCTION

I. ECOSYSTEM DEFINITION

Odum (1971) defined an _ecosystem_ as
"any unit that includes all the organisms
(i.e., the 'community') in a given area
interacting with the physical environment
so that a flow of energy leads to clearly
defined trophic structure, biotic diversity, and material cycles (i.e., exchange of materials between living and
nonliving parts) within the system." He
further recognized six major components
of ecosystems: 1) inorganic substances
such as water, carbon dioxide, and nitrogen; 2) organic compounds such as proteins
and carbohydrates; 3) climate regime (the
physical "forcing factors" which dictate
the kind of biotic community that can
develop); 4) autotrophic producers (primarily plants); 5) macroconsumers; and
6) microconsumers and decomposers. These
components are involved in the ecosystem
processes of 1) energy fixation and flow,
2) material cycling, 3) development of
spatial and temporal diversity patterns,
4) succession, 5) reaction (to short-and
long-term variations and perturbations),
and 6) interactions. This Ecological
Characterization of the Sea Island Coastal
Region of South Carolina and Georgia is
at the "macroecosystem" (i.e., large system) level and is organized along both
structural (i.e., component) and functional (i.e., process) lines.

II. MAJOR PRINCIPLES
PERTAINING TO ECOSYSTEMS

(based primarily on Odum 1971 and
Gosselink 1978 unless otherwise stated)

A. PRINCIPLES PERTAINING TO ENERGY

"Life is kept going by the continuous inflow of sun energy from outside"
(Odum 1971). According to the First Law
of Thermodynamics, energy can be transferred from one form to another (e.g.,
from potential to kinetic, or from chemical to heat or light), but can neither be
created nor destroyed. However, no spontaneous energy transformation or conversion is 100% efficient (Second Law of
Thermodynamics). During such changes,
some of the energy is always dispersed as
unavailable heat; thus, the total amount
of directly utilizable energy decreases
with each use or conversion within a system. Since energy flows through ecosystems and is not recycled, all energy
entering a system can be accounted for as
energy which is stored in or dispersed
from that system.

B. PRINCIPLES PERTAINING TO MATTER,
SPACE, AND TIME

Energy flows through a system only
once, but materials circulate or cycle
within ecosystems, being used over and
over again (Law of Conservation of Matter).
Stated another way, energy is used once
by a given organism, becomes converted
into heat, and is eventually dispersed
from the system. On the other hand, materials such as carbon, nitrogen, and
water may circulate repeatedly between
biotic and abiotic components of ecosystems and among different compartments
(i.e., different organisms or physical
states) within each of these components
(for examples, see biogeochemical cycles,
pp. 3 - 7).

Productivity is a measure of energy
fixation or food production. Primary
production refers to photosynthesis, the
conversion of light energy and simple inorganic materials (e.g., water, carbon
dioxide) into complex organic food molecules by autotrophic (i.e., "self-
nourishing") plants. Gross primary production is defined as "the total rate of
photosynthesis, including the organic matter used up in respiration, during the
measurement period" (Odum 1971). Net primary production is the amount of organic
matter incorporated into plant tissues in
excess of respiratory utilization. That
is, net primary production refers to the
amount of energy (organic matter) available for use by heterotrophs (consumers).

After its production by plants, food
energy is transferred through a series of
consuming organisms, each eating and being
eaten in turn, called a food chain. Within a food chain, energy conversion efficiencies are generally quite low (on the
order of 10% - 20% at each step); thus,
at each step of being eaten, 80% - 90% of
the potential energy available at the previous level is lost as heat (Odum 1971).
Losses of this magnitude limit most food
chains to no more than four or five steps
and, obviously, the shorter the food
chain, the more energy there is available.

Odum (1971) defined two basic types
of food chains: 1) a _grazing food chain_
in which the primary production of plants
is grazed directly by herbivores (plant-
eaters), which in turn are consumed by
carnivores (animal-eaters); and 2) a
detritus food chain in which much of the
plant production is allowed to die, whereupon it is acted upon by microorganisms
and then consumed by detritivores (detritus-eaters), which are in turn eaten by
predators, etc. Both types of food chains
are represented in the ecosystems described
in this report.

CHAPTER ONE

INTRODUCTION

I. ECOSYSTEM DEFINITION

Odum (1971) defined an ecosystem as "any unit that includes all the organisms (i.e., the 'community') in a given area interacting with the physical environment so that a flow of energy leads to clearly defined trophic structure, biotic diversity, and material cycles (i.e., exchange of materials between living and nonliving parts) within the system." He further recognized six major components of ecosystems: 1) inorganic substances such as water, carbon dioxide, and nitrogen; 2) organic compounds such as proteins and carbohydrates; 3) climate regime (the physical "forcing factors" which dictate the kind of biotic community that can develop); 4) autotrophic producers (primarily plants); 5) macroconsumers; and 6) microconsumers and decomposers. These components are involved in the ecosystem processes of 1) energy fixation and flow, 2) material cycling, 3) development of spatial and temporal diversity patterns, 4) succession, 5) reaction (to short-and long-term variations and perturbations), and 6) interactions. This Ecological Characterization of the Sea Island Coastal Region of South Carolina and Georgia is at the "macroecosystem" (i.e., large system) level and is organized along both structural (i.e., component) and functional (i.e., process) lines.

II. MAJOR PRINCIPLES PERTAINING TO ECOSYSTEMS

(based primarily on Odum 1971 and Gosselink 1978 unless otherwise stated)

A. PRINCIPLES PERTAINING TO ENERGY

"Life is kept going by the continuous inflow of sun energy from outside" (Odum 1971). According to the First Law of Thermodynamics, energy can be transferred from one form to another (e.g., from potential to kinetic, or from chemical to heat or light), but can neither be created nor destroyed. However, no spontaneous energy transformation or conversion is 100% efficient (Second Law of Thermodynamics). During such changes, some of the energy is always dispersed as unavailable heat; thus, the total amount of directly utilizable energy decreases with each use or conversion within a system. Since energy flows through ecosystems and is not recycled, all energy entering a system can be accounted for as energy which is stored in or dispersed from that system.

B. PRINCIPLES PERTAINING TO MATTER, SPACE, AND TIME

Energy flows through a system only once, but materials circulate or cycle within ecosystems, being used over and over again (Law of Conservation of Matter). Stated another way, energy is used once by a given organism, becomes converted into heat, and is eventually dispersed from the system. On the other hand, materials such as carbon, nitrogen, and water may circulate repeatedly between biotic and abiotic components of ecosystems and among different compartments (i.e., different organisms or physical states) within each of these components (for examples, see biogeochemical cycles, pp. 3 - 7).

Productivity is a measure of energy fixation or food production. Primary production refers to photosynthesis, the conversion of light energy and simple inorganic materials (e.g., water, carbon dioxide) into complex organic food molecules by autotrophic (i.e., "self-nourishing") plants. Gross primary production is defined as "the total rate of photosynthesis, including the organic matter used up in respiration, during the measurement period" (Odum 1971). Net primary production is the amount of organic matter incorporated into plant tissues in excess of respiratory utilization. That is, net primary production refers to the amount of energy (organic matter) available for use by heterotrophs (consumers).

After its production by plants, food energy is transferred through a series of consuming organisms, each eating and being eaten in turn, called a food chain. Within a food chain, energy conversion efficiencies are generally quite low (on the order of 10% - 20% at each step); thus, at each step of being eaten, 80% - 90% of the potential energy available at the previous level is lost as heat (Odum 1971). Losses of this magnitude limit most food chains to no more than four or five steps and, obviously, the shorter the food chain, the more energy there is available.

Odum (1971) defined two basic types of food chains: 1) a grazing food chain in which the primary production of plants is grazed directly by herbivores (plant-eaters), which in turn are consumed by carnivores (animal-eaters); and 2) a detritus food chain in which much of the plant production is allowed to die, whereupon it is acted upon by microorganisms and then consumed by detritivores (detritus-eaters), which are in turn eaten by predators, etc. Both types of food chains are represented in the ecosystems described in this report.

Food chains are not isolated from each other, but generally are interconnected by numerous links, forming complex food webs. Each step through which food energy is transferred in a food chain is called a trophic level. The first trophic level is the primary production or plant level, the next is the primary consumer or herbivore level, the next is the secondary consumer or herbivore-eater level, and so on.

For organisms to grow and reproduce, it is necessary that required materials (e.g., water, nutrients, etc.) be available at the appropriate time and place in the appropriate form and concentration. Hence, living organisms depend upon the mechanisms that control resource availability, and the productivity of a given food chain is determined to a large degree by the rates at which required materials are recycled through the system. Too, the efficiency with which energy and materials can be utilized by an individual organism and its population depends in part upon the spatial distribution of the organisms.

The distribution of an organism or population in space and time is determined in part by resource availability and in part by the limits of its tolerances to various factors. With regard to resources, that material which is necessary for growth and reproduction and which is available in amounts most closely approaching the critical minimum level required by that organism will tend to be the factor which limits growth (Liebig's "Law" of the Minimum). For each physical factor, an organism has a range of tolerance bounded by minimum and maximum levels (Shelford's "Law" of Tolerance). Odum (1971) included several corollaries of this "Law": 1) a given organism may have a wide tolerance range for one factor and a narrow range for another; 2) generally, eurytolerant organisms (those with wide tolerance ranges for all or most factors) are likely to be most widely distributed; 3) suboptimal conditions of one factor may cause an organism's tolerance of other factors to be reduced; 4) frequently, organisms do not occur or are not most abundant within their optimal range for a particular physical factor; in these cases other factors, often biological (e.g., predation, competition, parasitism, etc.), are more important; and 5) environmental factors are most likely to be critically limiting for a given organism during its reproductive period. Thus, the parts of an ecosystem which are inhabited by any given species at any given time are determined by the availability of required resources, the ranges of tolerance for all pertinent physical factors, and biotic interactions.

C. PRINCIPLES PERTAINING TO DIVERSITY

Diversity refers to the number of different kinds (i.e., species) of organisms found in a given ecosystem, community, or trophic level. Typically, a community or trophic group is composed of a relatively small proportion of dominant species (i.e., those with many individuals, large biomass, or high productivity) and a large proportion of less "important" species that are relatively uncommon or rare. Because of their numerical and biomass dominance, the common species account for most of the energy flow through the community or trophic level (Odum 1971). However, since there generally are many more rare than common species, the rare species largely determine the species diversity of the group. Calculation of various ratios between the number of species and numbers of individuals (or biomass) yield so-called species diversity indices, which simply allow mathematical expression of the relative diversities of different communities or groups. Such indices, especially the Shannon-Weaver formula, have been used widely in ecological studies (Goodman 1975). Recently, however, Peet (1975) presented evidence that these relativized diversity indices are mathematically inappropriate for most ecological studies. Thus, considerable care must be exercised in the use of diversity indices, especially in evaluating the effects of environmental alterations.

Over the last few decades, the idea that high species diversity is directly correlated with community or trophic system stability has become widespread and perhaps too generally accepted. This is the so-called "diversity-stability hypothesis." Odum (1971) provides a good example of the reasoning behind this hypothesis:

"Higher diversity, then, means longer food chains and more cases of symbiosis (mutualism, parasitism, commensalism, and so forth), and greater possibilities for negative feedback control, which reduces oscillations and hence increases stability. . . . Consequently, communities in stable environments such as the tropical rain forest, have higher species diversities than communities subjected to seasonal or periodic perturbations by man or nature. What has not yet been measured is the extent to which an increase in community diversity can, in itself, increase the stability of the ecosystem in the face of external oscillations in the physical habitat."

In contrast, Goodman (1975) critically reviewed the biological and mathematical bases for the diversity-stability
hypothesis and concluded that diversity
and stability in ecosystems are not directly related. A similar conclusion was
reached by Leffler (1978), based on data
from laboratory microcosm experiments.
With regard to the often-cited example of
the highly diverse and stable tropical
rain forest, Goodman (1975) suggested that
its apparent stability may well be an il-
lusion fostered by insufficient study of
exceedingly complex communities. As support for his argument, he noted that such
rain forests are being found to be pain-
fully vulnerable to man-made perturbations.
Goodman (1975) summarized his conclusions
as follows:

> "The expectations of the di-
> versity-stability hypothesis
> are borne out neither by ex-
> periments, by observations,
> nor by models; its theoretical
> formulations have no necessary
> connection with secure scien-
> tific law, and its preconcep-
> tions are inconsistent with an
> evolutionary perspective.
> Clearly, the belief that more
> diverse communities are more
> stable is without support. It
> is not so clear what sort of
> relation we should expect be-
> tween diversity and stability."

More recently, Huston (1979) noted
that high diversity has been related to
both intense and reduced competition and
correlated both positively and negatively
with productivity. With this confusing
situation as a background, he presented
a new hypothesis of diversity based on
"differences in the rates at which populations of competing species approach
competitive equilibrium."

From the above discussion, it is
clear that species diversity and its relationships to community or ecosystem
stability and productivity are poorly
understood. Much additional theoretical
and experimental research is needed in
this area.

D. PRINCIPLES PERTAINING TO ECOSYSTEM
    STABILITY AND RESILIENCE

According to Odum (1971), "The
habitat of an organism is the place where
it lives, or the place where one would go
to find it." The term ecological niche,
on the other hand, "includes not only the
physical space occupied by an organism,
but also its functional role in the community (as, for example, its trophic
position) and its positions in environmental gradients of temperature, moisture,
pH, soil, and other conditions of existence." Gosselink (1978) points out that
the species occupying a given niche
typically are more efficient users of the

resources of that niche than are foreign
species which might attempt to replace
them. This idea leads to the hypothesis
that a natural community is the optimum
adaptation to a given environment; consequently, any disturbance which changes
the natural community reduces the overall
efficiency of the system.

Ecosystems develop with time through
a process called ecological succession.
This is an orderly, reasonably directional,
and therefore predictable sequence of
changes in species composition and community processes through time; it occurs because the community changes the physical
environment (Odum 1971). Succession cul-
minates in a stabilized climax community
"in which maximum biomass (or high information content) and symbiotic function between organisms are maintained per unit of
available energy flow" (Odum 1971).

The stability principle states that
any natural ecosystem tends to change
until a stable situation, with homeostatic
(self-regulating) mechanisms, is developed
(Odum 1971). These mechanisms tend to return the system to its stable state following short-term changes caused by outside factors. Natural and man-made
perturbations are such factors. The resilience of a given system (i.e., its
ability to return to its stable state following external perturbations) depends on
many factors and is very difficult to predict. It depends on the characteristics
of the particular system involved; the
type, magnitude, and frequency of the perturbations; and other factors. Because of
differences in systems and perturbations,
some effects may be immediate (e.g., death
of part or all of a population), while
others may not appear until much later
(e.g., bioaccumulation of pesticides, genetic damage). Further, because of the in-
terlocking nature of the populations and
communities of ecosystems, perturbations in
one part of a system may result in perturbations in other parts. In some cases,
these latter effects may be more severe
than those caused directly by the original
perturbation. Also, the effects of a particular perturbation may be greater when
it is applied to a system repeatedly.

III. BIOGEOCHEMICAL CYCLES

A. GENERAL FEATURES

A major functional component of an
ecosystem is the exchange of materials,
chemicals, between living and nonliving
structural components of the system. This
continuous, more or less circular exchange
of the elements and inorganic compounds
necessary for life is known as nutrient
or biogeochemical cycling (Odum 1971).
Likens et al. (1977) consider ecosystems
to have four physical compartments among
which biogeochemicals cycle. These are:

1) organic matter, 2) available nutrients in soil or dissolved in water, 3) the minerals of soils and rocks, and 4) the atmosphere. The physical movement of these chemicals within or among ecosystems may result from meteorologic forces (e.g., wind, rain, currents), geologic forces (e.g., drainage, gravitational settling, vulcanism), or biological forces (e.g., animal movement, plant seed dispersal).

Biogeochemical cycles themselves may be divided into two compartments: a relatively small exchange pool that is cycled rapidly between organisms and their immediate environment, and a large, slow-moving reservoir pool in the sediment, atmosphere, or hydrosphere (oceans). Biogeochemical cycles based upon sedimentary reservoirs (such as phosphorus and calcium) are termed sedimentary types, while those based on atmospheric or hydrospheric reservoirs (e.g., CO_2 and N) are called gaseous types (Odum 1971).

The general features of several of the more important biogeochemical cycles are similar for most macroecosystems. Therefore, the cycles for water, carbon, nitrogen, and phosphorus are treated here in general outline to prevent redundancy.

B. THE HYDROLOGIC (H_2O) CYCLE

Figure 1-1 illustrates the major aspects of the global water cycle. Several important features of this cycle are readily apparent but deserve emphasis here. First, while the global reservoir of water is large, most of this water (~94%) is tied up in the lithosphere (earth's crust) and circulates very slowly. In contrast, the actively cycling atmospheric pool is quite small and vulnerable to perturbations (Odum 1971). Second, more water evaporates from the sea than is returned via precipitation, while the reverse holds true for the land. Thus, some of the water that supports land systems originates from the sea. Third, the rate of groundwater recharge determines how much water will be available in subsurface aquifers for man's use and, in coastal areas, the extent of saltwater intrusion into aquifers. The rate of recharge is determined by the amount of precipitation, the runoff rate, and the porosity of the substrates. A common effect of many of man's activities, such as agriculture and construction, is to increase the runoff rate, which has a secondary and potentially disastrous effect of reducing the rate of groundwater recharge. Fourth, the rate, volume, and seasonal variations of surface runoff govern 1) the distribution of salinity in estuaries and coastal waters, 2) the circulation patterns in coastal waters, and 3) the amount of dissolved and suspended sediments, nutrients, organic materials,

and contaminants contributed to estuarine and coastal waters (Clark 1974).

C. THE CARBON CYCLE

Basic features of the carbon cycle are diagrammed in Figure 1-2. The carbon cycle is characterized by a relatively small, but very actively circulating, atmospheric pool in the form of carbon dioxide (CO_2). Incorporation of atmospheric CO_2 through biological fixation (photosynthesis) provides the base upon which nearly all biological productivity depends. The system is buffered by the tremendous dissolving power of the oceans, which maintain a major carbon reservoir primarily in the form of carbonates. However, man's extensive burning of fossil fuels, intensive agriculture (which releases soil CO_2), and deforestation apparently have exceeded the rate at which the ocean can take up CO_2, resulting in an increase in the amount of atmospheric CO_2 (Odum 1971). This, in turn, may increase the amount of heat held within the atmosphere and eventually may have a major effect on the climate and weather (i.e., the so-called "greenhouse" effect).

D. THE NITROGEN CYCLE

The nitrogen cycle is a good example of a complex gaseous cycle (Fig. 1-3). The primary reservoir is the atmosphere, which consists of 80% nitrogen. Nitrogen is a major element necessary for protein synthesis, and thus its availability over the long term may influence the abundance of organisms. The nitrogen cycle is based on the activity of two general groups of microorganisms: 1) the denitrifying bacteria which degrade complex nitrogenous compounds and return elemental nitrogen to the air, and 2) the nitrogen-fixing bacteria and algae which take elemental nitrogen and transform it into forms (nitrates, nitrites, ammonia) useful to organisms in the synthesis of amino acids and other biological molecules. Some nitrogen is also fixed by electrification and photochemical reactions, but this amount is relatively small. The nitrogen cycle is basically the same in terrestrial, freshwater, and oceanic environments (Fig. 1-3), although the species of microorganisms involved, the relative importance of algae and bacteria, and the cycle dynamics may vary appreciably. Such differences also will be true even within major environments. For example, while nitrogen-fixing algae are very important in the ocean's photic zone, they appear to be of little significance in the aphotic zone.

E. THE PHOSPHOROUS CYCLE

The phosphorous cycle appears quite a bit simpler than the nitrogen cycle, as indicated by Figure 1-4. Like nitrogen, phosphorus is required in the synthesis

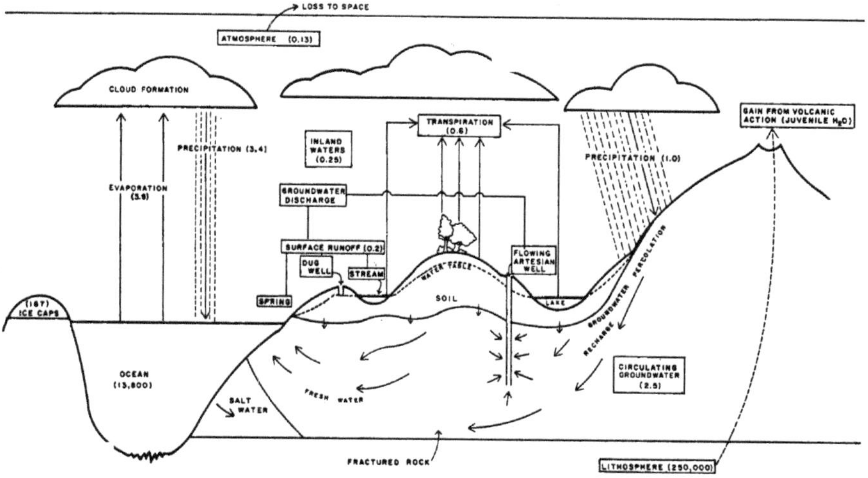

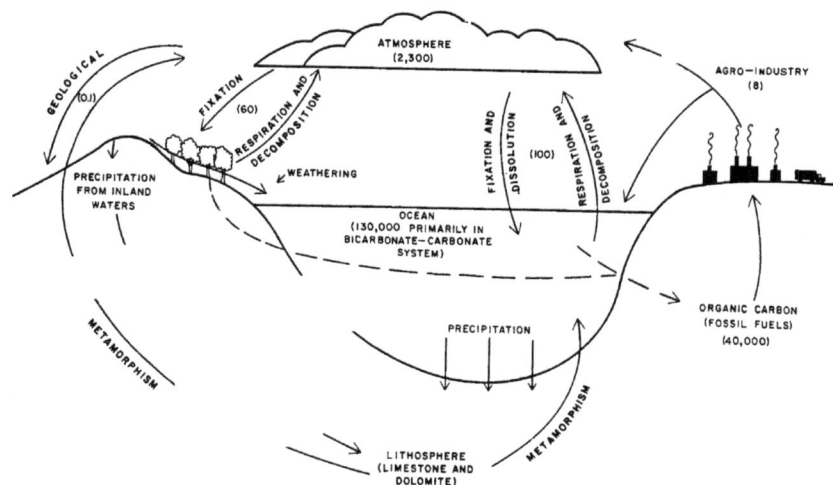

Figure 1-1. The global hydrologic (H_2O) cycle [diagram modified from Energy Resources Co., Inc. 1978 adaptation of a figure from Caswell 1977, with additional information from Odum 1971 and Clark 1974; numbers are from Odum 1971 and indicate geograms (10^{20} grams) H_2O in the major compartments of the biosphere].

Figure 1-2. The global carbon cycle (adapted from Odum 1971; numbers are 10^9 tons of CO_2 and are taken from Odum 1971).

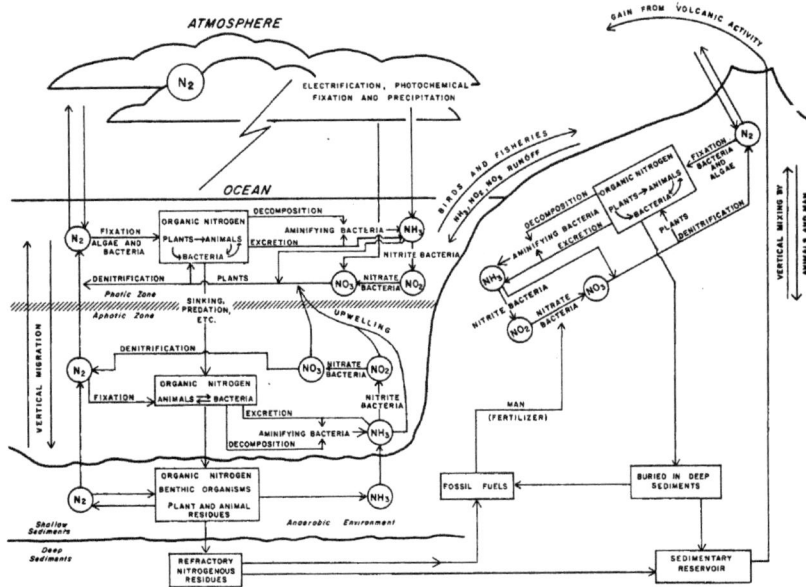

Figure 1-3. The global nitrogen cycle (adapted from Odum 1971).

of biological molecules and often can be more limiting to biological productivity than the availability of nitrogen. This limiting effect may be related to its relatively low availability (according to Odum 1971, it is 23 times less abundant than nitrogen in natural waters) and to the fact that it occurs in only one biologically useful inorganic form, phosphate. However, phosphates appear to be regenerated rather rapidly (Fig. 1-4). In contrast to the situation for nitrogen, the phosphorous reservoir is located primarily in the sediments, and release is accomplished primarily by erosion (plus man's mining efforts). Although marine birds, in particular, play a major role in recycling sedimented phosphates into easily weatherable guano deposits, much phosphate is buried in marine sediments and essentially lost to the cycle. However, many such deposits in coastal areas can be mined and the phosphates recycled.

F. THE SULFUR CYCLE

Major elements of the sulfur cycle are illustrated in Figure 1-5. Sulfur is essential for the manufacture of certain

amino acids, and sulfate (SO_4) is the principal form utilized by plants and incorporated into proteins. Organisms require much less sulfur than nitrogen or phosphorus, so sulfur is much less frequently limiting to plant growth. Nonetheless, sulfates are necessary, particularly since the sulfur cycle is inextricably coupled with the carbon cycle in anoxic environments and provides a pathway by which reduced carbon compounds can be recycled to the atmospheric pool.

The sulfur cycle is characterized by the following major features: 1) a large sedimentary reservoir pool; 2) a much smaller atmospheric pool; 3) a rapidly fluxing pool of several forms of sulfur, with the transformations mediated by specialized microorganisms; 4) "microbial recovery" of sulfur from deep sediments by bacterial transformation of SO_4 in the sediments to H_2S gas, which then rises to the photic zone where it can be utilized by other microorganisms; and 5) the interaction and interdependence of air, water, and soil pools and geochemical, meteorological, and biological processes (Odum 1971). Also, Odum (1971) points out

6

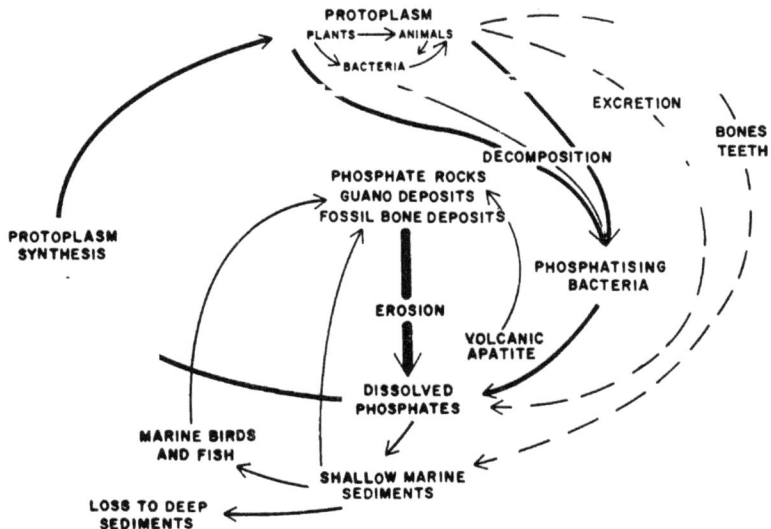

PROTOPLASM
PLANTS ⟶ ANIMALS
BACTERIA

EXCRETION

BONES
TEETH

DECOMPOSITION

PHOSPHATE ROCKS
GUANO DEPOSITS
FOSSIL BONE DEPOSITS

PROTOPLASM
SYNTHESIS

PHOSPHATISING
BACTERIA

EROSION

VOLCANIC
APATITE

DISSOLVED
PHOSPHATES

MARINE BIRDS
AND FISH

SHALLOW MARINE
SEDIMENTS

LOSS TO DEEP
SEDIMENTS

Figure 1-4. The global phosphorous cycle (adapted from Odum 1971 with very minor modifications).

that the formation of iron sulfides in the sediments results in the conversion of phosphorus from insoluble to soluble form, thus releasing it for biological use.

IV. ECOSYSTEMS OF THE SEA ISLAND COASTAL REGION

A. STUDY AREA

The study area includes the coastal tier of counties in South Carolina and Georgia and the adjacent lowland counties of Dorchester, Berkeley, and Effingham. To the north, the study area is delimited just below the broad crescent of the "Grand Strand" by the Horry/Georgetown County line in South Carolina and to the south by the St. Marys River on the Georgia/Florida border. The east-west boundaries are the seaward 3-mi (4.8 km) territorial limit and the inland county lines, respectively (see Preface Fig. 2). This coastal study area extends almost 480 km (300 mi) and is characterized by numerous islands, inlets, and sounds.

On the basis of the wetland classification scheme of Cowardin et al. (1977), the study area was divided into the following seven interrelated "macroecosystems"

for characterization: 1) coastal marine, 2) maritime, 3) estuarine, 4) riverine, 5) palustrine, 6) lacustrine, and 7) upland. These ecosystems are defined briefly below, and the remainder of this volume is devoted to a detailed characterization of each system (systems 4 - 6 are treated together in one chapter on Freshwater Ecosystems). The general distribution and areal extent of the major systems in the study area are shown in Atlas plate 1 (frontispiece).

B. SYSTEM DEFINITIONS (based on Cowardin et al. 1977)

1. Coastal Marine Ecosystem

The coastal marine ecosystem may be considered an edge system where land and water have unobstructed access to the open ocean. Water regimes and chemistry are determined primarily by tidal action. Oceanic water is only minimally diluted except opposite mouths of estuaries, and salinities generally exceed 30°/oo. Some coastal marine systems (e.g., open ocean beach) are considered high energy environments in terms of wave and current energy.

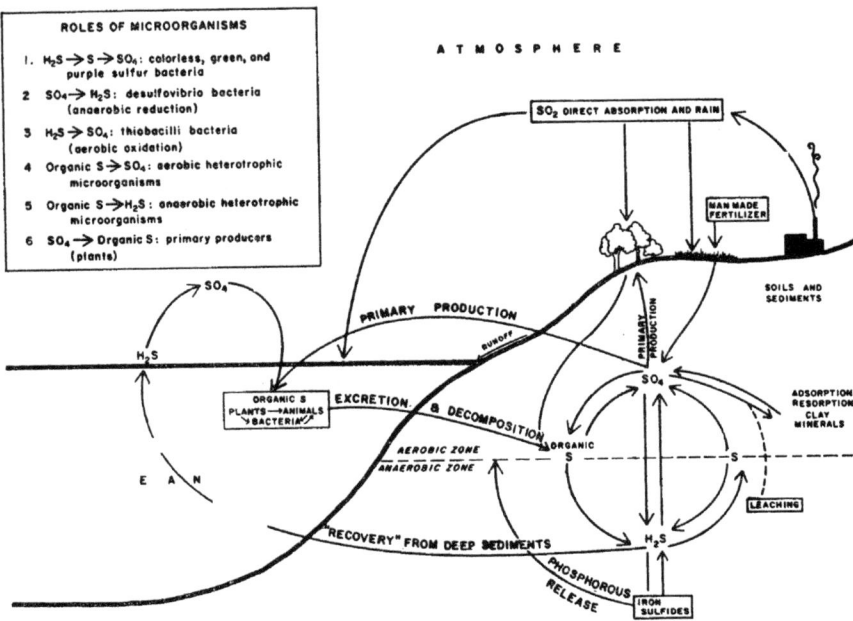

Figure 1-5. The global sulfur cycle (adapted with minor modification from Odum 1971).

For this characterization, the seaward limit of the coastal marine ecosystem was defined as the outer boundary of the 3-mi (4.8 km) territorial sea. However, the more realistic seaward limit is the outer edge of the continental shelf (Cowardin et al. 1977). The inland boundary of the coastal marine ecosystem was defined by Cowardin et al. (1977) as "1) the landward limit of tidal inundation, including the splash zone from breaking waves; 2) the seaward limit of wetland emergents, trees, or shrubs where they extend into open ocean waters; or 3) the seaward limit of the Estuarine System where this limit is determined by factors other than vegetation."

The coastal marine ecosystem is divided into two subsystems: the subtidal and the intertidal. The subtidal subsystem consists of coastal waters extending seaward of extreme low spring tide level and with salinities consistently exceeding 30°/oo. The intertidal subsystem consists of tidal beaches and bars contiguous to coastal waters.

2. Maritime Ecosystem

The maritime ecosystem is defined as all upland forest areas and dunes located on barrier islands (islands bordered on one side by tidal marshes, creeks, and/or rivers and on another side by an active ocean-formed beach). Components of this system are influenced to varying degrees by salt water.

3. Estuarine Ecosystem

The estuarine system "consists of deep-water tidal habitats and adjacent tidal wetlands which are usually semi-enclosed by land, but have open, partially obstructed or sporadic access to the open ocean and in which ocean water is at least occasionally diluted by fresh water runoff from the land" (Cowardin et al. 1977). At times, evaporation may increase the salinity above that of the open ocean. Estuaries are influenced more by terrestrial processes than is the contiguous coastal marine system.

8

According to Cowardin et al. (1977), "Estuaries extend upstream and landward to the place where ocean-derived salts measure less than 0.5⁰/oo during the period of average annual low flow. The seaward limit of the Estuarine System is: 1) a line closing the mouth of a river, bay, or sound; 2) a line enclosing an off-shore area of diluted sea-water with typical estuarine flora and fauna; or 3) the seaward limit of wetland emergents, shrubs, or trees where these plants grow seaward of the line closing the mouth of a river, bay, or sound."

The estuarine system is divided into two subsystems, the subtidal and the intertidal. Open water sounds, bays, and tidal rivers and streams having salinities greater than 0.5⁰/oo make up the subtidal subsystem. The intertidal subsystem consists of beaches, bars, flats, oyster rocks, marshes, etc., exposed by tidal action. Impounded wetlands with average salinities greater than 0.5⁰/oo also are included in the intertidal subsystem, although technically they are not intertidal.

4. Riverine Ecosystem

The riverine ecosystem includes "all wetlands and deep water habitats contained within a channel, except: 1) wetlands dominated by trees, shrubs, persistent emergents, nonaquatic mosses or lichens, and 2) habitats with waters containing ocean-derived salts in excess of 0.5⁰/oo" (Cowardin et al. 1977). A channel, as defined here, may be natural or artificial, but must contain either periodically or continually moving water (intermittent stream channels are included in the riverine system), or the channel may be a link between two bodies of standing water.

The emergent and forested wetlands of flood plains are here considered part of the palustrine ecological system. Where large flood plains are present and the river channel is not easily defined, the riverine system ends where persistent emergents (e.g., cat-tails and cordgrasses) become dominant. The riverine ecological system is, therefore, bordered inland by the palustrine, lacustrine (where a river channel empties into or from a lake), or upland systems, and is bordered by the estuarine system (where salinities are greater than 0.5⁰/oo) on the seaward side.

The riverine system is divided into four subsystems: the tidal, the lower perennial, the upper perennial, and the intermittent. The tidal subsystem is characterized by water velocity fluctuating under tidal influence, a low gradient, a streambed composed mainly of mud, occasional oxygen deficits, and a well-developed flood plain. The lower perennial subsystem has nontidal flowing water throughout the year, low flow velocities,

a substrate of sand and mud, occasional oxygen deficits, a fauna dominated by still-water and planktonic forms, a low gradient, a well-developed flood plain, and warm temperatures [average monthly water temperatures are >20⁰C (>68⁰F)]. The upper perennial subsystem is characterized by fast-flowing water throughout the year; a substrate predominantly of rocks, cobbles, or gravel; high oxygen concentration; fauna characteristic of running water with few, if any, planktonic forms; high gradient; and little floodplain development. Finally, the intermittent subsystem has flowing water only part of the year; the remainder of the time, channels may be dry or have water present in isolated pools. The riverine ecosystem in the Sea Island Coastal Region is dominated by the tidal subsystem, with the lower perennial subsystem represented upstream in major drainage systems. The intermittent subsystem is well represented in flood plains, but the upper perennial subsystem does not occur in the Sea Island Coastal Region.

5. Palustrine Ecosystem

The palustrine ecosystem "includes all nontidal wetlands dominated by trees, shrubs, persistent emergents, nonaquatic mosses or lichens, and all such wetlands that occur in tidal areas where salinity due to ocean-derived salts is below 0.5⁰/oo" (Cowardin et al. 1977). Wetlands which lack the vegetation described above, but which are less than 8 ha (20 acres) in size, lack a wave-formed or bedrock shoreline, have a maximum depth less than 2 m (6.5 ft) at low water, and have a salinity <0.5⁰/oo are also included in the palustrine system.

The palustrine ecosystem includes swamps, marshes, flood plains, savannahs, and other similar, extensively vegetated wetlands. In fact, all wetlands that do not fall into the marine, estuarine, riverine, or lacustrine systems are considered palustrine. Such environments often occur adjacent to lacustrine, riverine, or estuarine areas and may appear to grade into these systems. The palustrine ecosystem has no subsystems.

6. Lacustrine Ecosystem

According to Cowardin et al. (1977), the lacustrine ecosystem "includes wetlands and deep-water habitats with all of the following characteristics: 1) situated in a topographic depression or a dammed river channel; 2) lacking trees, shrubs, persistent emergents, nonaquatic mosses or lichens with greater than 30 percent areal coverage; and 3) greater than 8 hectares (20 acres) in size." Wetlands smaller than 8 ha (20 acres) may be included in the lacustrine system if they have an active wave-formed or bedrock shoreline and if maximum water depth at low water is greater than 2 m (6.5 ft).

It is possible to have a tidal lacustrine body, but the salinity must be less than 0.5⁰/oo.

By the above definition, the lacustrine system includes permanently flooded, tidal, oxbow, and intermittent lakes and reservoirs. This type of ecosystem is represented best in the Sea Island Coastal Region by reservoirs and oxbow lakes. Such systems typically exhibit large areas of deep water and much wave action, and frequently may encompass islands of palustrine wetlands.

The lacustrine ecosystem is composed of two subsystems, the littoral and the limnetic. The littoral subsystem extends from the shoreward boundary of the lacustrine body to a depth of 2 m (6.5 ft) (at low water), or to the maximum extent of non-persistent emergents (if they are beyond the 2 m or 6.5 ft depth point). Thus, all vegetated areas are considered littoral. If the lake has no vascular vegetation present, the littoral zone is determined by the 2 m (6.5 ft) depth. The deep, open-water zone is, consequently, the limnetic subsystem.

7. Upland Ecosystem

In general, uplands include all lands that are not part of the five previously defined wetlands or aquatic systems (i.e., lacustrine, palustrine, riverine, estuarine, and marine). For this ecological characterization, we have divided these "uplands" into two distinct ecosystems. These are 1) an upland ecosystem affected by fresh water and 2) a maritime ecosystem defined as all upland forest areas and dunes located on barrier islands. Therefore, the upland ecosystem comprises all non-maritime uplands.

Following Cowardin et al. (1977), one might define upland or terrestrial ecological systems as those areas not classified as wetlands or aquatic systems and characterized by the water table not being at, near, or above the land surface for sufficient time each year to promote the formation of hydric soils and the growth of hydrophytes as the dominant plant type. Soils of upland areas, then, would be predominantly non-hydric, and the vegetation would be predominantly mesophytic or xerophytic rather than hydrophytic. In addition, uplands should be characterized further as lands that are never flooded during years of normal precipitation.

V. ECOSYSTEM MODELS

Any system, particularly one as complex as the Sea Island Coastal Region of South Carolina and Georgia, can be considered to be composed of an infinite number of subsystems. Our choice of systems to model was dictated primarily by the wetland classification scheme of Cowardin et al. (1977), as previously indicated. Thus, independent models were constructed for each of the seven major ecosystems within the characterization area.

Each ecosystem model is characterized by four basic elements: compartments, flows between compartments, major inputs or external driving forces, and major outputs or products. These four elements are connected, following the compartmentalized approach of Odum (1967), Odum (1971), and Patten (1971), to produce abstract models which represent the structural and functional components of the ecosystems. In particular, the models attempt to show the pathways through which energy flows and materials cycle in the various ecosystems, and thus they are termed "energese" models. A number of standard symbols are used in the models to convey the concepts of distribution, utilization, storage, and loss or resistance. These symbols are listed below (see Odum 1971).

Circles indicate external energy sources, either potential or kinetic (e.g., the sun). These normally indicate energy input to the system (i.e., the driving forces for the system).

"Silos" represent passive storage. These are normally used to indicate internal storage (i.e., nutrient pools) and provide the mechanism for nutrient regeneration.

"Bullets" represent primary producers (plant populations) or autotrophic activity.

Hexagons indicate consumers or heterotrophic activity. The trophic level is indicated by a number within the hexagon (i.e., 1⁰ means primary consumer, etc.).

10

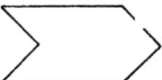

Pointed blocks represent "work gates," where one flow of energy assists another to pass over some energy barrier. These are used here also with the following symbol to indicate energy loss in all transfers of energy.

Arrows with ground indicate sinks (i.e., loss or dispersion from the system). The use of this symbol is limited almost entirely to work gates in the characterization documents.

Arrows without ground indicate flows between compartments.

As a whole, these models are largely speculative. Energy flows and material cycling in most ecosystems are generally not well defined. Therefore, the energese diagrams presented here do not treat ecosystems in a quantitative manner, but rather indicate relative dominance of energy sources and flows and probable relationships among compartments. Hopefully, by organizing the components and flows of a given system, the models will help the reader develop an understanding of how each ecosystem likely works. Such an understanding is the first step in realistic management.

Each energese model is also accompanied by a simplified food web featuring representative organisms and by a pictorial which illustrates the major biotic and physical attributes of the system and their interrelationships. The pictorials present realistic cross sections of the ecosystems, displaying representative organisms as they might appear in their natural habitats. As independent units, the pictorials were prepared to familiarize the reader with the various habitats encompassed by the ecosystem definitions and display these habitats in a form which creates easily identifiable composites. The food webs demonstrate trophic dependencies and act as links between the models and pictorials by providing illustrations of representative flora and fauna which are tagged with appropriate model symbols. Together, the energese models, food webs, and pictorials

form a tripartite presentation which provides the reader with options for understanding the attributes of the various ecosystems under consideration. Because of the physical size of these tripartite presentations, they are presented in the atlas volume. Only the energese models and a simple block diagram of food webs are presented here in the text. Thus, the reader should consult the Atlas for the full ecosystem model presentations (Atlas plates 2 - 8).

VI. ENDANGERED SPECIES

A. BACKGROUND INFORMATION

Over a period of many years, people throughout the world have come to realize that a growing number of species of wildlife and plants are likely to become extinct soon unless man takes steps to protect them. In the United States, this growing realization led to the passage of the Endangered Species Preservation Act of 1966, which was amended in 1969 to the Endangered Species Conservation Act. Later, a stronger law, the Endangered Species Act of 1973, was passed by the 93rd Congress (Public Law 93-205). The major purposes of this law were: 1) "to provide a means whereby the ecosystems upon which endangered species and threatened species depend may be conserved," and 2) "to provide a program for conservation of endangered species and threatened species . . ."

Central to such legislation, of course, are the concepts of "endangered," "threatened," and "rare" species and what may be termed the "conservation ethic." The ecological and philosophical background for the development of the endangered species concept is discussed aptly by Smith (1976), but perhaps the best definitions of endangered and threatened species are those given by Rayner et al. (1979):

> "Endangered - A taxon [organism] whose population level is naturally low or which has become reduced in numbers throughout all or a significant portion of its range, or whose natural habitat has been altered and/or reduced to the extent that reproductive populations are small and vulnerable to extirpation. Without protection and management throughout its range, a taxon in this category could well be extirpated from its natural habitats within the immediate future."

> "Threatened - A taxon which is not in immediate danger of extirpation, but one whose populations have been depleted or

are decreasing at a significant rate and/or whose natural
habitat has been altered or
destroyed throughout much of
its original range. Without
protection and management,
these plants [organisms] are
likely to become endangered
within the foreseeable future
throughout all or a significant portion of their range."

The most recent list of animals and plants
officially designated as endangered or
threatened under the 1973 Act appeared in
the Federal Register on 17 January 1979
(U.S. Department of Interior, Fish and
Wildlife Service 1979a).

Besides establishing a national effort to conserve and preserve endangered
and threatened organisms, the Endangered
Species Act of 1973 encouraged States to
develop their own legislation to protect
endangered species. Georgia's State leg-
islature took action rapidly, and in 1973
passed an Endangered Wildlife Act and a
Wildflower Preservation Act. These laws
required the Georgia Department of Natural
Resources to identify and protect those
species of wildlife and plants in need of
protection in Georgia. The first lists
of protected wildlife and plant species
were approved by the Georgia Board of
Natural Resources in 1975, and have since
been revised (McCollum and Ettman 1977,
Odom et al. 1977). The following year,
the South Carolina General Assembly passed the South Carolina Nongame and Endangered Species Conservation Act of 1976.
This law required the South Carolina
Wildlife and Marine Resources Commission
to develop a list of those species of
nongame wildlife endangered within the
State, and to conduct programs for the
management of nongame and endangered
species. Twenty-five species of animals
were then declared to be "Endangered
Wildlife Species of South Carolina" under
the rules and regulations provision of
the Act (South Carolina Legislative
Council of the General Assembly 1977).
Plants were not covered under this law.

In 1978, the Federal Endangered
Species Act of 1973 was amended (Public
Law 96-632) to direct the Secretary of
the Interior and the Secretary of Commerce
to develop "recovery plans" for the conservation and survival of federally listed
endangered or threatened species. Re-
sponsibility for development of recovery
plans for birds and terrestrial species
was delegated to the Fish and Wildlife
Service (Department of the Interior),
while that for marine species except birds
and sea turtles was given to the National
Marine Fisheries Service (Department of
Commerce). Development of recovery plans
for sea turtles was to be a joint respon-
sibility of the two agencies.

In response to the 1978 law, Recovery
Teams were established to develop recovery
plans for a number of endangered species.
These teams usually have from three to
seven members, and include employees of
Federal and State agencies, plus profes-
sionals from academic or conservation organizations (U.S. Department of Interior,
Fish and Wildlife Service 1979b). Most
or all team members are engaged actively
in research or management efforts on the
subject species. Each recovery team is
charged to develop a recovery plan that
". . .justifies, delineates, and schedules
those actions required for restoring and
securing an E/T [endangered/threatened]
species as a viable self-sustaining member
of its ecosystem" (U.S. Department of
Interior, Fish and Wildlife Service
1979b). At present, recovery teams for
the following endangered or threatened
animals of the Sea Island Coastal Region
have been established: American alligator,
bald eagle, brown pelican, Florida manatee,
Florida panther, Kirtland's warbler,
peregrine falcon, red-cockaded woodpecker,
sea turtles, and shortnose sturgeon.
Information on the recovery team leaders
and the status of recovery plans is given
in the brief species summaries presented
later in this section.

B.   ENDANGERED AND THREATENED VASCULAR
     PLANTS

Vascular plants presently considered
to be in need of protection in coastal
South Carolina and Georgia are listed in
Tables 1-1 and 1-2, respectively. Although only one species from this region
is included in the official "List of Endangered and Threatened Wildlife and
Plants" (U.S. Department of Interior,
Fish and Wildlife Service 1979a), a total
of 71 species is considered to be in
jeopardy by State authorities.

While South Carolina does not yet
have a State law governing endangered
plants, an ad hoc advisory committee has
made great progress in determining what
species are in need of protection in the
State. This committee produced the first
list of rare and endangered plants of
South Carolina (Rodgers and Clark 1977),
and its "unofficial" list was updated and
published recently (Rayner et al. 1979).
This new list is now the primary reference
on rare and endangered plants of South
Carolina.

Rayner et al. (1979) grouped plants
believed to be endangered or threatened
in South Carolina into four categories:
Of National Concern; Of Regional Concern;
Of Statewide Concern; and Of Concern,
Status Unresolved. These categories are
defined below (adapted from Rayner et al.
1979). Readers should keep in mind that
none of the species listed in Table 1-1
from Rayner et al. (1979) are considered

Table 1-1. Endangered and threatened vascular plants of South Carolina's coastal region (Rayner et al. 1979).

SPECIES[a]	STATUS[b]	REASON(S) FOR LISTING/COMMENT	PREFERRED HABITATS
Agrimonia incisa (Incised groovebur)	Threatened Statewide		Dry pine woods
Aletris obovata (White colic root)	Endangered Statewide	One of two South Carolina populations destroyed by urban development	Savannahs, pinelands, moist open areas
Amaranthus pumilus (Coast pigweed)	Threatened Statewide		Sand dunes, coastal beaches
Amphicarpum muhlenbergianum (Muhlenberg's amphicarpum)	Threatened Statewide		Low pinelands; moist, sandy roadsides; moist, sandy margins of lime-sinks
Anthaenantia rufa (Purple silkyscale)	Of Concern Status Unresolved		Flatwoods and sand hills
Arenaria godfreyi (Godfreys sandwort)	Threatened Nationally Endangered in S.C.	Endemic to coastal plain of N.C., S.C., Ga., and Fla.; endangered in S.C.; restricted habitat	Wooded seepage slopes of marl soils
Asplenium heteroresiliens (Carolina spleenwort fern)	Endangered Nationally	Endemic to coastal plain of N.C., S.C., Ga., Fla.; few very small populations in few localities	Marl outcrops
Asplenium resiliens (Black-stemmed spleenwort)	Endangered Statewide	Only four small populations known in S.C.	Marl outcrops
Aster spectabilis (Low showy aster)	Threatened Statewide		Pine barrens and wood-land borders
Balduina uniflora (Single-flowered balduina)	Of Concern Status Unresolved		Savannahs and pine flat-woods
Calamovilfa brevipilis (Riverbank sandreed)	Endangered Regionally	Far disjunct from main population (N.J.); habitat dependent on fire in management	Savannahs and bogs
Canna flaccida (Golden canna lily)	Threatened Statewide		Pine savannahs and marshes
Carex chapmanii (Chapman's sedge)	Threatened Regionally		Dry sandy woods and road-sides
Carya myristicaeformis (Nutmeg hickory)	Threatened Regionally Endangered in S.C.	No recent collections from S.C.	Swamp forest and river-banks
Chrysobalanus oblongifolius	Threatened Statewide	Known from only two counties in S.C.	Dry pinelands and sand-

13

Table 1-1. Continued

SPECIES[a]	STATUS			
Cliftonia monophylla (Buckwheat tree)	Of Concern Status Unresolved	Thought to be extinct in S.C., but one S.C. population reported but not verified	Non-alluvial swamps and bogs	P
Crotonopsis linearis (Narrow-leaved rushfoil)	Of Concern Status Unresolved		Roadsides and thinly woody sandy habitats	U
Dicerandra odoratissima (Rose dicerandra)	Threatened Regionally	Rare in S.C. and Fla.; not rare in Ga.	Thin scrub oak woods of the coastal plain	U
Dionaea muscipula (Venus' fly trap)	Threatened Nationally Endangered in S.C.	Exploited commercially and by collectors; habitat being eliminated by development; eliminated from two of three S.C. counties where originally found	Open, sandy bog margins	P
Echinodorus parvulus (Little burhead)	Threatened Statewide		Sandy shores or mud bottom in shallow water	P
Eryngium aquaticum var. ravenelii (Ravenel's buttonroot)	Threatened Statewide		Coastal plain pinelands	U
Habenaria integra (Yellow fringeless orchid)	Threatened Statewide		Moist pine flatwoods and savannahs	P U
Habenaria lacera (Green fringeless orchid)	Threatened Statewide		Bogs, marshes, wet meadows	P
Ilex amelanchier (Sarvis holly)	Threatened Nationally	Habitat destruction by stream channelization, flood control projects, drainage of swamps	Swamps, river banks, floodplains	P
Ipomoea macrorhiza (Large-rooted morning glory)	Endangered Statewide	Known from only two small populations in S.C.	Sandy clearings and beaches	U
Kalmia cuneata (White wicky)	Threatened Nationally Endangered in S.C.	Endemic to coastal plain of N.C. and S.C.; highly susceptible to succession and disturbance	Pocosin ecotones and bay margins	P
Lachnocaulon beyrichianum (Southern bog-buttons)	Of Concern Status Unresolved		Sandy pine-palmetto flats	U
Lepuropetalon spathulatum (Southern lepuropetalon)	Threatened Statewide		Sandy ditches and wet sandy soil	P
Lindera melissaefolium (Jove's fruit or Southern spicebush)	Endangered Nationally	No recently verified collections	Margins of limesink ponds	P
Litsea aestivalis (Pond spice)	Threatened Regionally	Endemic to N.C., S.C., Ga., and Fla.; S.C. probably has more populations than elsewhere	Pond and swamp margins and low, wet woodlands	P L U

14

Table 1-1. Continued

SPECIES[a]	STATUS[b]	REASON(S) FOR LISTING/COMMENT	PREFERRED HABITAT(S)
Lygodium palmatum (American climbing fern)	Threatened Statewide		Moist thickets and wood-lands on acid soils
Magnolia macrophylla (Umbrella tree)	Endangered Statewide	Known from old collections; not verified recently	Alluvial woods and sheltered valleys in coastal plai
Monotropis odorata (Pigmy-pipes)	Threatened Nationally Endangered in S.C.	Known from four populations in S.C.; none recently verified	Mixed deciduous forests
Myriophyllum laxum (Loose watermilfoil)	Threatened Nationally	Endemic to coastal plain of N.C., S.C., Ga., Fla.; known from few, dispersed locations; habitat threatened by development	Ponds and sinks
Narthecium americanum (Yellow asphodel)	Threatened Nationally Endangered in S.C.	Known from single location in S.C.; not verified recently	Bogs and pocosins
Nyssa ogeche (Ogeechee plum)	Endangered Statewide	Endemic to coastal plain of S.C., Ga., Fla., and Ala.	Swamps and bog forests
Parnassia caroliniana (Carolina grass-of-parnassus)	Endangered Statewide	Endemic to coastal plain of N.C., S.C., and Ala.; has been extirpated from nearly 1/2 its original range	Savannahs
Peltandra sagittaefolia (White arrow-arum)	Threatened Regionally		Bogs, swamps, muddy ponds and margins of sluggish streams
Pieris phillyreifolia (Climbing fetterbush)	Threatened Regionally	Only known population in S.C. on U.S. Forest Service land	Climbs trunks of pond cypress
Pinckneya pubens (Georgia fever-bark)	Threatened Statewide	Endemic to S.C., Ga., and Fla.	Sloping swamp borders
Polygala nana (Low milkwort)	Of Concern Status Unresolved		Wet, open, sandy areas of coastal plain
Potamogeton foliosus (Leafy pondweed)	Of Concern Status Unresolved		Fresh and brackish streams and ponds
Rhapidophyllum hystrix (Needle palm)	Threatened Regionally Endangered in S.C.	Rare in S.C., Fla., Ala.; has reproductive problems; is cultivated commonly	Coastal plain swamps and hammocks
Rhexia aristosa (Sun-petaled meadow-beauty)	Threatened Regionally	Rare in southern part of its range	Coastal plain savannahs and low pinelands
Rudbeckia mollis (Soft-haired cornflower)	Threatened Regionally	Rare in northern 3/4 of range (includes Ga. and S.C.)	Open sandy woods in coastal plain

15

Species	Status	Distribution	Habitat	
Sageretia minutiflora (Small-flowered buckthorn)	Threatened Regionally	Rare throughout range except in Fla.	Shell mounds, calcareous hammocks, rocky bluffs of coastal plain	U M
Sarracenia rubra (Sweet pitcher-plant)	Threatened Statewide		Acid shrub bogs and savannahs	P
Schisandra glabra (Bay starvine)	Threatened Regionally Endangered in S.C.		Rich woods	P U
Schwalbea americana (Chaff-seed)	Threatened Statewide		Coastal plain savannahs and pine flatwoods	P U
Scirpus erismanae (Georgia bullrush)	Threatened Regionally Endangered in S.C.	Endemic to Fla., Ga., S.C.; rare in Ga. and S.C.; habitat threatened by natural succession	Open sandy pond margins	P
Scirpus subterminalis (Swaying bullrush)	Of Concern Status Unresolved		Sluggish streams in coastal plain	P
Scleria baldwinii (Baldwin's nutrush)	Threatened Statewide		Pond cypress savannahs, moist pine woods, low open areas	P
Solidago verna (Spring-flowered goldenrod)	Threatened Nationally	Endemic to coastal plain of S.C. and N.C.; very restricted range	Savannahs, bogs, pine barrens	P U
Spiranthes laciniata (Lace-lip ladies tresses)	Threatened Statewide		Wet cypress savannahs and pinelands, marshes coastal plain	P
Spiranthes longrilabris (Giant spiral-orchid)	Threatened Statewide	Known in S.C. from two populations	Swamps, marshes, wet savannahs and meadows	P
Sporobolus teretifolius (Wire-leaved dropseed)	Threatened Regionally	Endemic to coastal plain of N.C., S.C., and Ga.; rare in N.C. and S.C.	Savannahs and moist pine woods	U P
Trillium pusillum var. pusillum (Dwarf or Carolina trillium)	Endangered Nationally	Endemic to N.C. and S.C.	Moist woods, bog and savannah margins	P
Triphore trianthophora (Three-birds orchid)	Threatened Statewide		Damp woods and thickets of coastal plain and mountains	U
Utricularia floridana (Florida bladderwort)	Endangered Statewide	Endemic to coastal plain of S.C., Ga., and Fla.; known in S.C. from single locality	Old pond	L
Zephranthes simpsonii (Rain lily)	Threatened Nationally	Endemic to coastal plain of S.C., Ga., and Fla.; exploited by collectors	Low, sandy pinelands	U

a. Species listed in alphabetical order by scientific name following convention of Rayner et al. (1979).
b. See text for explanation of status categories used by Rayner et al. (1979).

Table 1-2. Endangered and threatened vascular plants of Georgia's coastal region (McCollum and Ettman 1977).

SPECIES[a]	STATUS	REASON(S) FOR LISTING/COMMENT	PREFERRED HABITAT(S)	ECOSYSTEM(S)
Asplenium heteroresiliens (Spleenwort)	Threatened (Ga.)	Rare in Georgia; Georgia populations widely separated from main populations	Marl outcrops	Upland
Baptisia arachnifera (Hairy wild-indigo)	Endangered[b] (Ga. and U.S.)	Rare throughout its natural range	Sandy pine woods	Upland
Elliottia racemosa (Georgia plume)	Endangered (Ga.)	Rare throughout its natural range	Various sandy soil habitats	Upland
Fothergilla gardenii (Dwarf witch alder)	Threatened (Ga.)	Rare throughout its natural range	Savannahs and swamps	Palustrine
Hartwrightia floridana (Hartwrightia)	Threatened (Ga.)	Rare throughout its natural range	Swamps, marshes and wet grasslands	Palustrine
Hymenocallis coronaria (Spider-lily)	Endangered (Ga.)	Rare throughout its natural range	Swamp margins and stream banks	Palustrine
Litsea aestivalis (Pond spice)	Threatened (Ga.)	Rare throughout its natural range	Swamp margins, limestone ponds and wet woodlands	Palustrine
Myriophyllum laxum (Water milfoil)	Threatened (Ga.)	Rare throughout its natural range	Sinks and shallow pools	Maritime Palustrine Lacustrine
Sarracenia flava (Fly-catcher)	Threatened (Ga.)	Suffering from commercial exploitation and rapid habitat loss	Bogs, savannahs and low pine woods	Palustrine
Sarracenia minor (Hooded pitcher-plant)	Threatened (Ga.)	Suffering from commercial exploitation and rapid habitat loss	Bogs, savannahs and low pine woods	Palustrine
Sarracenia psittacina (Parrot pitcher-plant)	Threatened (Ga.)	Suffering from commercial exploitation and rapid habitat loss	Bogs, savannahs and low pine woods	Palustrine

a. Species listed in alphabetical order by scientific name; common names used in this table are from McCollum and Ettman 1977.

b. This is the only species found in coastal Georgia or South Carolina that occurs on the official U.S. "List of Endangered and Threatened Wildlife and Plants" (U.S. Department of Interior, Fish and Wildlife Service 1979a).

endangered or threatened at the Federal
level.

1. Of National Concern - species
endangered or threatened over their entire range.

2. Of Regional Concern - species
endangered or threatened throughout significant portions of their ranges, including South Carolina and other States
of the region. Such species may be of no
special concern outside the region, but
their rarity and definite threats to
their populations make them of concern in
this part of the country. Included here
also are South Carolina populations that
are far removed from the main populations
of certain species.

3. Of Statewide Concern - species
which are considered endangered or
threatened only in South Carolina.

4. Of Concern, Status Unresolved -
species believed to be rare and likely
endangered or threatened, but for which
too little information is available to
allow a determination of status.

Under provisions of the Georgia
Wildflower Preservation Act of 1973, the
Georgia Department of Natural Resources
was required to determine those species
of plants in need of protection in
Georgia. A first list of 99 species was
approved by the Board of Natural Resources
in 1975. Subsequent work led to the removal of 41 species from the list, and at
present 58 species are protected by law
(McCollum and Ettman 1977). McCollum and
Ettman (1977) designated species as endangered or threatened in Georgia based
on consideration of the following five
criteria:

1. Rare Throughout - The species is
rare throughout its entire natural range
regardless of political boundaries.

2. Rare Disjunct - The species is
rare in Georgia and is separated from the
major body of the population - which may
be located outside Georgia - by a considerable distance.

3. Rare peripheral - The species is
rare in Georgia and represents a range
extremity. The species may be much more
plentiful in other parts of its range.

4. Exploited - The species has been
subjected to commercial exploitation.
The species is not necessarily rare, but
the natural population cannot support an
extensive commercial market and remain
stable.

5. Rapid Habitat Loss - The species
requires a habitat which is being altered
at an especially rapid rate when compared
to the alteration of Georgia's natural
environments over all.

For each species listed, McCollum and
Ettman (1977) give a brief description of
the plant, its habitat requirements, and
what is known of its distribution in
Georgia.

C. ENDANGERED AND THREATENED ANIMALS

A total of 30 species of animals are
considered endangered or threatened in
the Sea Island Coastal Region (Table 1-3).
Unlike the situation for plants, all but
three of these species are already on the
official Federal list of endangered and
threatened species (U.S. Department of
Interior, Fish and Wildlife Service 1979a).
All of the species on the South Carolina
State list (South Carolina Legislative
Council of the General Assembly 1977) are
considered endangered; the 1976 State law
did not include a threatened category.
In contrast, Georgia recognizes four possible classes of protected species - endangered, threatened, rare, and unusual
(Odom et al. 1977).

A brief discussion of each species of
endangered or threatened wildlife listed
in Table 1-3, plus some other species of
concern in the region, follows. The
reader also should consult the appropriate
ecosystem chapters and Atlas plates 31 -
40.

1. Mammals

a. Eastern Cougar. Historically,
this species ranged from Canada into South
Carolina, but today it is known from only
a few scattered locations at best. No
specific habitat requirements are known,
except that it needs a large wilderness
area with an adequate food supply (usually
deer). Cougars were probably common
predators in coastal environments at one
time, but they are virtually unknown now
in the Sea Island Coastal Region. No recovery team has been appointed to date
for the cougar, but it is likely that many
of the recommendations developed for the
Florida panther (another subspecies) can
be applied to the cougar. For further
information, see the section on mammals
of maritime forests in Chapter Three.

b. Pocket Gophers. Two species of
pocket gophers are considered endangered
in Georgia, but are not included on the
Federal list and do not occur in South
Carolina. The colonial pocket gopher's
entire range in North America is restricted to about 12 mi^2 (31 km^2) in southeastern Georgia (Camden County). Presently, it is known from a single 200 ha
(494 acre) area near Scotchville, Georgia
(Winchester et al. 1978). Similarly,
Sherman's pocket gopher was known in
Georgia from only a single population near
Savannah, but recent attempts to find
specimens were unsuccessful (Odom et al.
1977).

Table 1-3. Endangered and threatened animals of the Sea Island Coastal Region of South Carolina and Georgia (South Carolina Legislative Council of the General Assembly 1977; Odom et al. 1977; and U.S. Department of Interior, Fish and Wildlife Service 1979a. Asterisk indicates Recovery Plan available from U.S. Fish and Wildlife Service).

SPECIES[a]	STATUS[b]	REASON(S) FOR LISTING/COMMENT	PREFERRED HABIT.
MAMMALS			
Cougar, eastern (Felis concolor cougar)	Fed. Endangered S.C. Endangered Ga. Endangered	Habitat loss; human disturbance and deliberate shootings; reduction of deer herds in past years	Upl
Gopher, colonial pocket (Geomys colonus)	Fed. Not Listed S.C. Not Listed Ga. Endangered	Modification of habitat primarily by timber industry; only one population known (Camden County, Ga.)	Upl
Gopher, Sherman's pocket (Geomys fontanelus)	Fed. Not Listed S.C. Not Listed Ga. Endangered	Reasons for population decline unknown; only one colony known (Chatham County)	Upl
*Manatee, West Indian (Florida) (Trichechus manatus)	Fed. Endangered S.C. Endangered Ga. Endangered	Poaching in past years; habitat destruction from dredging and water pollution; injuries and death from propellers of power boats	Mar Est Riv
*Panther, Florida (Felis concolor coryi)	Fed. Endangered S.C. Not Listed Ga. Endangered	Habitat loss; human disturbance and deliberate shootings; reduction of deer herds in past years	Pal Upl
Whale, Atlantic right (Eubalaena glacialis)	Fed. Endangered S.C. Endangered Ga. Endangered	Over-harvesting	Mar
Whale, blue (Balaenoptera musculus)	Fed. Endangered S.C. Endangered Ga. Not Listed	Over-harvesting	Mar
Whale, bowhead (Balaena mysticetus)	Fed. Endangered S.C. Endangered Ga. Not Listed	Over-harvesting	Mar
Whale, finback (Balaenoptera physalus)	Fed Endangered S.C. Endangered Ga. Not Listed	Over-harvesting	Mar
Whale, humpback (Megaptera novaeangliae)	Fed. Endangered S.C. Endangered Ga. Endangered	Over-harvesting	Mar
Whale, sei	Fed. Endangered	Over-harvesting	Mar

Table 1-3. Continued

SPECIES[a]	STATUS[b]	REASON(S) FOR LISTING[c]		
Whale, sperm (Physeter catodon)	Fed. Endangered S.C. Endangered Ga. Not Listed	Over-harvesting	Open ocean	Marine
BIRDS				
Curlew, Eskimo (Numenius borealis)	Fed. Endangered S.C. Endangered Ga. Not Listed	Excessive shooting; other factors unknown; rare migrant on S.C. coast	Marshes and tundra	Estuari
Eagle, bald (southern bald) (Haliaeetus leucocephalus leucocephalus)	Fed. Endangered S.C. Endangered Ga. Endangered	Loss of habitat; lowered reproduction from chlorinated hydrocarbons; illegal shooting; human disturbance; siltation in rivers.	Undisturbed wetland areas with large trees for nesting and open water for feeding	Estuari Riverin Upland Palustr Lacustr
Falcon, American peregrine (Falco peregrinus anatum)	Fed. Endangered S.C. Endangered Ga. Endangered	Lowered reproduction from chlorinated hydrocarbons; illegal shooting	Rocky cliffs, tall trees for nesting; migrant along S.C. and Ga. barrier islands	Upland Maritim
Falcon, Arctic peregrine (Falco peregrinus tundrius)	Fed. Endangered S.C. Endangered Ga. Endangered	Lowered reproduction from chlorinated hydrocarbons; illegal shooting	Rocky cliffs, tall trees for nesting; migrant along S.C. and Ga. barrier islands	Upland Maritim
*Pelican, brown (eastern brown) (Pelecanus occidentalis carolinensis)	Fed. Endangered S.C. Endangered Ga. Endangered	Lowered reproduction from DDT and its metabolites and other chlorinated hydrocarbons; entanglement in fishing lines and fish hooks, illegal shooting	Nearshore marine waters, sounds and estuaries	Marine Estuari
Warbler (wood), Bachman's (Vermivora bachmanii)	Fed. Endangered S.C. Endangered Ga. Endangered	Unclear; possibly related to modification or loss of habitat from cutting mature southern swamp and from cane farming in its winter range in Cuba	Not positively known; associated with swamps	Palustr
Warbler (wood), Kirtland's (Dendroica kirtlandii)	Fed. Endangered S.C. Endangered Ga. Endangered	Very specific habitat requirements; loss of habitat and parasitism from cowbirds	Nests in jack pine stands and winters in Bahamas; migrant on S.C. and Ga. coasts	Upland
Woodpecker, ivory-billed (Campephilus principalis)	Fed. Endangered S.C. Endangered Ga. Endangered	Loss of habitat from cutting of mature hardwood forests and cypress swamps; possibly other factors and the inability to adapt to changing habitat	Over-mature hardwood, fresh and cypress swamps	Upland Palustr

20

Table 1-3. Continued

SPECIES[a]	STATUS[b]	REASON(S) FOR LISTING/COMMENT	PREFERRED HABITAT(S)	ECOSYSTEM(S)
*Woodpecker, red-cockaded (Picoides borealis)	Fed. Endangered S.C. Endangered Ga. Endangered	Loss of habitat from forest management practices which reduce amount of timber 60 years or older	Over-mature pine forest with under-story < 1.5 m in height; red heart disease associated with cavity trees	Upland

REPTILES

SPECIES[a]	STATUS[b]	REASON(S) FOR LISTING/COMMENT	PREFERRED HABITAT(S)	ECOSYSTEM(S)
*Alligator, American (Alligator mississippiensis)	Fed. Endangered/Threatened[c] S.C. Special Regulations[d] Ga. Endangered	Over-harvesting, poaching, loss of habitat; indiscriminate killing	River swamps, lakes, ponds, marshes, and impoundments	Estuarine Riverine Palustrine Lacustrine
Snake, eastern indigo (Drymarchon corais couperi)	Fed. Threatened S.C. Not Listed Ga. Threatened	Over exploitation for pet trade; habitat loss; killed incidentally to "rattlesnake roundups"; possible detrimental effects of mirex and dieldrin	Longleaf pine-turkey oak-wire grass community; burrows of gopher tortoise	Upland
Tortoise, gopher (Gopherus polyphemus)	Fed. Not Listed S.C. Endangered Ga. Not Listed	Loss of habitat; over collection for pet trade; killed incidentally to "rattlesnake roundups"	Sand ridges in lob-lolly pine-turkey oak-wire grass communities	Upland
Turtle, green sea (Chelonia mydas)	Fed. Threatened[e] S.C. Not Listed Ga. Not Listed	Over-harvesting; loss of nesting beaches; once much more abundant along Southeastern coast than at present	Shallow tropical waters	Marine
Turtle, Atlantic hawksbill (Eretmochelys imbricata imbricata)	Fed. Endangered S.C. Not Listed Ga. Endangered	Shells taken for jewelry; adults and eggs taken as food; predation on eggs and young; loss of nesting beaches	Shallow coastal waters of Atlantic, Caribbean, and Gulf of Mexico	Marine
Turtle, Kemp's ridley (Lepidochelys kempii)	Fed. Endangered S.C. Endangered Ga. Endangered	Adults and eggs taken as food; predation of eggs and young; loss of nesting habitat; incidental catch	Shallow tropical waters	Marine
Turtle, leatherback (Dermochelys coriacea)	Fed. Endangered S.C. Endangered Ga. Endangered	Loss of nesting beaches; adults and eggs taken as food	Open ocean	Marine
Turtle, Atlantic loggerhead (Caretta caretta caretta)	Fed. Threatened S.C. Special Regulations[d] Ga. Not Listed	Predation of eggs and young; eggs taken as food; loss of nesting habitat; incidental catch; the only marine turtle which nests in S.C. and Ga.	Inshore waters and open ocean; nest on S.C. and Ga. barrier islands	Marine

Table 1-3. Concluded

SPECIES	STATUS		Major river drain-ages on eastern seaboard	Marine Estuari Riverin

FISHES

Shortnose sturgeon (*Acipenser brevirostrum*) — Fed. Endangered, S.C. Endangered, Ga. Endangered — Not well known; probably over-fishing, pollution and damming of rivers

a. Listed in alphabetical order by common name within major groups, following convention of Federal list (U.S. Department of Interior, Fish and Wildlife Service 1975a).

b. See text for explanation of status categories.

c. Populations in three coastal parishes of Louisiana were removed from Federal list in 1977; populations of several other coastal areas, including those of South Carolina and Georgia, were reclassified as threatened.

d. Not included on State list but protected by special regulations of the South Carolina Wildlife and Marine Resources Department (S. R. Hopkins, 1979, South Carolina Wildlife and Marine Resources Department, Charleston, pers. comm.).

e. Endangered in breeding areas (Florida and Pacific coast of Mexico).

22

c. West Indian (Florida) Manatee. This species was probably once more than twice as abundant as presently. In the U.S., it occurs primarily in Florida, but during the warmer months individuals disperse as far north as North Carolina along the Atlantic coast. Generally, manatees frequent fresh, brackish, or salt waters at least 1.5 m (4.9 ft) deep with an adequate food supply (aquatic vegetation). Manatee populations were impacted initially by overharvesting. Hunting has been outlawed, but collision with boats and loss of habitat have resulted in heavy mortality. A recovery team has been appointed with Mr. John C. Oberhen as Team Leader (U.S. Fish and Wildlife Service, Jacksonville Area Office, 900 San Marco Boulevard, Jacksonville, Florida, 32207). A draft recovery plan has been prepared (West Indian (Florida) Manatee Recovery Team 1979). Objectives of this plan are to minimize injuries and mortalities to manatees caused by humans, minimize reduction of manatee habitat, monitor manatee populations, and reduce harassment of manatees. For more information on the manatee in the Sea Island Coastal Region, see the section on mammals of the Subtidal Coastal Marine Subsystem in Chapter Two.

d. Florida Panther. This species once occurred in Arkansas, Louisiana, Mississippi, Alabama, Georgia, Tennessee, and South Carolina. It survives today, but probably only in small, scattered populations, especially in South Florida. Documentation of its continuing presence in the characterization area is limited to relatively few sightings. Although some of the sightings were made by qualified individuals, hard evidence of its presence is generally lacking (Odom et al. 1977). Like the more northerly ranging eastern cougar, the Florida panther requires large wilderness areas for hunting. Such areas are rapidly disappearing in the region. Overhunting also has contributed to its decline. A recovery team has been appointed, with Mr. Robert C. Belden as Team Leader (Florida Game and Fresh Water Fish Commission, Wildlife Research Projects Office, 4005 S. Main St., Gainesville, Florida, 32601). A draft recovery plan was completed in 1978 (Florida Panther Recovery Team 1978). The recovery plan noted that a primary factor limiting efforts to preserve the Florida panther is an extreme dearth of information on its present status and occurrence. As an initial step toward determining areas of active panther populations for further investigation, the plan recommends that each State within the panther's historic range establish a sighting clearinghouse, with a central clearinghouse being located in Florida. The draft further recommends that necessary habitat and range be set aside for any existing panther populations located, that attempts be made to reestablish stocks through carefully controlled captive breeding and restocking efforts, and that more emphasis be placed on public education programs.

e. Whales. Seven species of whales which occasionally occur in ocean waters off South Carolina and Georgia are considered endangered throughout their range worldwide (U.S. Department of Interior, Fish and Wildlife Service 1979a). These are the Atlantic right, blue, bowhead, finback, humpback, sei, and sperm whales (Table 1-3). All of these whales are in real danger of extinction for the same reason, over-harvesting for meat and oil. Although international conservation efforts have resulted in protection of these species by a number of nations, a complete international moratorium on whaling has not yet materialized.

2. Birds

Over the years, much information has been amassed on birds through a national network of State and Federal researchers and volunteer field observers. Results of such observations are published through local and national ornithological periodicals. Journals like The Chat, The Oriole, Wilson Bulletin, and American Birds have been used extensively for compiling information on the birds of the Sea Island Coastal Region. Major sources of information include Georgia Birds (Burleigh 1958) and South Carolina Birdlife (Sprunt and Chamberlain 1970).

Nine species or subspecies of birds are endangered in the Sea Island Coastal Region (Table 1-3).

a. Eskimo Curlew. In all of North America, there has been only one convincing sighting of this species in the last few years. Two birds were seen together on Martha's Vineyard, Massachusetts, in August 1972. This species was never more than a casual transient in coastal South Carolina, and Burleigh (1958) states that more definite evidence would be required for it to be included in the Georgia list. Its populations have suffered from overharvesting and perhaps other unknown factors.

b. Bald Eagle. The bald eagle is considered endangered throughout the continental United States except for the populations in Michigan, Minnesota, Oregon, Washington, and Wisconsin, which are classified as threatened (U.S. Department of Interior, Fish and Wildlife Service 1979a). Causes of population declines include lowered reproductive success through effects of accumulated pesticides; loss of feeding and nesting areas; illegal shooting; human disturbance; and, more recently, lead poisoning from ingestion of lead shot occurring in crippled or dead waterfowl.

The primary concern for this species throughout the conterminous States is its marked decline in reproductive success. In contrast, populations in Canada and Alaska appear to be in good shape. This is certainly not the case, however, in the Southeastern United States. As little as 20 years ago, nesting reports of bald eagles in the Southeast were commonplace; but in more recent years, nesting records have become alarmingly scarce outside of Florida. Although bald eagles are observed every winter in the Sea Island Coastal Region, only 19 active eagle territories are known presently within the characterization area, 18 in South Carolina and one in Georgia (T. M. Murphy, Jr., 1979, South Carolina Wildlife and Marine Resources Department, Green Pond, pers. comm.). In contrast, South Carolina alone apparently had 42 active eagle territories during the 1960's and early 1970's (Murphy and Coker 1978). Despite this gloomy picture, data on reproduction provided by Murphy and Coker (1978) indicate that eagles in South Carolina are at least holding their own for the present. In South Carolina, active eagle territories are frequently found in association with areas of impounded marsh. This led Murphy and Coker (1978) to speculate that South Carolina's extensive areas of impounded wetlands may be an important reason why many more eagles nest in South Carolina than in North Carolina or Georgia.

A bald eagle recovery team has been appointed, with Mr. Thomas M. Murphy as its leader (South Carolina Wildlife and Marine Resources Department, Division of Wildlife and Freshwater Fisheries, Poco Sabo Plantation, Green Pond, South Carolina, 29446). A recovery plan is in preparation. For further information on eagles, see the section on birds of non-forested wetlands in Chapter Five and Atlas plates 3 - 8.

c. American Peregrine Falcon. This raptor has experienced a dramatic decline within the continental United States, possibly as a result of lowered reproductive success due to accumulated pesticides and illegal shooting. At present no American peregrines breed in the wild in the Southeastern United States (Edwards 1978), and the meager population in the Rocky Mountain States has serious reproductive problems. Undoubtedly, the species did at one time breed fairly regularly in the mountains of northwestern South Carolina and Georgia, and plans to reintroduce birds to this area from captive breeding stock at Cornell University's Laboratory of Ornithology are under consideration (Edwards 1978). A recovery team for peregrine falcons has been appointed, with Rene M. Bollengier as Team Leader (U.S. Fish and Wildlife Service, P.O. Box 1518, Concord, New Hampshire, 03301). A draft recovery plan was completed in 1976 (Eastern Peregrine Falcon Recovery Team

1976). This plan recommends intensified protection, public education programs, and further reductions in the levels of persistent biocides in the environment.

d. Arctic Peregrine Falcon. Although populations of this falcon have also declined, the species continues to be reported regularly in the Sea Island Coastal Region. The arctic peregrine falcon does not breed in the U.S., and it is most often seen in South Carolina and Georgia on barrier island beaches. Although pesticides affected this subspecies in essentially the same manner as the American peregrine falcon, mark-recapture (banding) studies indicate that substantial numbers of these birds still migrate along the eastern seaboard (Edwards 1978). The arctic peregrine falcon is included in the recovery plan mentioned above for the American peregrine falcon.

e. Eastern Brown Pelican. Although this species suffered serious population declines over much of its range, especially in Louisiana and Texas, its status in the Southeastern United States appears to be improving. The primary cause of its decreased abundance was apparently impaired reproduction caused by ingestion of pesticide residues in the fish it fed upon. The major chemicals involved were DDT and its metabolites, polychlorinated biphenyls, dieldrin, and endrin. Such pesticides are blamed for the drastic reduction in the Texas population and the extirpation of the original Louisiana stock.

Brown pelicans feed on fishes in estuarine and coastal waters. They frequently utilize sand spits and bars for roosting or "loafing" areas, and nest on small coastal islands. In South Carolina, nesting areas are located in the Cape Romain National Wildlife Refuge and on Deveaux Bank, now a National Audubon Society refuge. The status of breeding populations in these locations seems to be improving (Blus et al. 1979). Pelicans do not nest in Georgia presently, and apparently did not do so historically.

A recovery team led by Mr. Lovett Williams (Florida Game and Fresh Water Fish Commission, 4005 S. Main St., Gainesville, Florida, 32601) has prepared a recovery plan for the eastern brown pelican (Williams 1978). The major objectives of this plan are 1) to restock the species in vacant nesting habitat and thus restore its wide distribution, and 2) to maintain natural and restocked colonies through natural reproduction. A wide range of recommendations accompanies these objectives.

For additional information on eastern brown pelicans in the characterization area, see the section on birds of bird key and bank communities in Chapter Three, and Atlas plate 5.

f. Bachman's Warbler. This species is the rarest and least known of North American warblers, and it appears to be continuing its decline toward extinction. No confirmed nest observations have been made in the United States since 1937. Most of the prior records were from I'on Swamp in the Francis Marion National Forest, Charleston County, South Carolina. The species continues to be reported sporadically from this area, but recent systematic searches have failed to produce even one sighting (Hamel and Hooper 1978). However, if the species still exists in the United States, it is most likely to occur in the palustrine forested wetlands of I'on Swamp. Specific factors responsible for the decline of this species are not known, but loss of habitat is suspected. No recovery team has been appointed for this species.

g. Kirtland's Warbler. This species of warbler is the second rarest in North America and occurs in the study area only as a rare transient during migration between its breeding grounds in Michigan and the Bahama Islands. There have been few records of the Kirtland's warbler in the study area; population levels of this species are extremely low, and singing male counts during "Kirtland's Warbler Censuses" show about 200 males over the last several years.

A recovery team led by Mr. John Byelich (Michigan Department of Natural Resources, Wildlife Division, Mason Building, Lansing, Michigan, 48926) completed a draft recovery plan in 1976 (Kirtland's Warbler Recovery Team 1976). Major recommendations of this plan include maintenance of suitable nesting areas, reintroduction of birds to their former breeding grounds, protection for the warbler during migration and wintering, reduction of factors detrimental to its reproduction and survival, and monitoring of breeding populations.

h. Ivory-billed Woodpecker. This species formerly ranged throughout the Southeastern and Gulf States and occurred in the Mississippi Valley northward to southern Illinois and Indiana. According to the most recent information, the ivory-billed woodpecker is probably extinct in the United States. No indisputable records to the contrary have been obtained during the last few years. Unconfirmed reports continue to be received, but none have been verified by competent observers. The decline of the species was probably related to the cutting of low-land hardwood forests. No recovery team has been appointed.

i. Red-cockaded Woodpecker. This woodpecker generally inhabits old (60+ years) pines, especially those trees infected with the red-heart disease fungus. The decline in populations of this species appears to have accompanied the cutting

of most pine forests with 60+-year old trees. However, populations within the Sea Island Coastal Region appear to be stable. In fact, in South Carolina, the red-cockaded woodpecker is receiving much attention because the resident clans there are among the densest in the species' range (Jackson 1978). The U.S. Forest Service's recent interest in research on this woodpecker in the Francis Marion National Forest (South Carolina) is also encouraging.

A recovery team led by Jerome A. Jackson (Mississippi State University, Department of Zoology, P.O. Box Z, Mississippi State, Mississippi, 39762) developed a recovery plan for this species (Red-cockaded Woodpecker Recovery Team 1979). The recovery plan contains very detailed recommendations, which may be summarized as follows: 1) identify existing populations of red-cockaded woodpeckers and maintain a file of this information; 2) protect existing populations and develop and implement management techniques on private and public property; 3) reestablish populations in areas formerly occupied where suitable habitat exists; 4) link extant populations with protected "habitat corridors" that could allow expansion and eventual merging of scattered populations; and 5) continue protection for the species.

j. Other Species. In addition to the nine endangered species discussed above, an ad hoc South Carolina Bird Committee proposed listing 20 additional species, two as endangered, four as threatened, 12 as "of special concern" and two as "peripheral" (Gauthreaux et al. 1979), as follows:

(1) Proposed Endangered. Swallow-tailed Kite: The South Carolina Bird Committee felt that this species should be considered endangered in the study area because of its declining numbers. Individual swallow-tailed kites are observed every spring and summer in South Carolina, and the bird breeds sparingly in the swamplands of the lower coastal plain. Some observers feel that this species has declined dramatically throughout its range and should be of national concern. Whereas the swallow-tailed kite formerly bred over an extensive area of the central and eastern United States, its breeding now is limited to the southern States. For more information on this species, see the sections on birds of nonforested and forested wetlands in Chapter Five.

Ipswich Sparrow: This distinct race of the Savannah sparrow winters along the barrier beach dunes of the South Carolina-Georgia coast. It has been rare historically, but because of its highly specific habitat requirements - it occurs only in dune fields - the Committee felt it should be considered endangered within South Carolina. Extensive destruction of the

25

dune habitat might eliminate this species completely from its wintering grounds in South Carolina and Georgia. In turn, such loss of wintering grounds might adversely affect the breeding status of this sparrow in Nova Scotia. For more information on the Ipswich sparrow, see the section on dune bird communities, Chapter Three.

(2) Proposed Threatened.
American Osprey: Although this species still is observed frequently in the Sea Island Coastal Region, it is definitely not as common as it once was. Ospreys continue to breed in the characterization area, but the birds have suffered major setbacks in reproduction in past years over the entire Atlantic Coastal Plain. As was the case with several other species, persistent pesticides, accumulated through their food source (fish), appears to have been a major cause of the osprey's reproductive difficulties. According to the "Blue List" of American Birds (Arbib 1975), 88% of all respondents strongly supported inclusion of the osprey on the official "List of Endangered and Threatened Wildlife and Plants." Recent studies in South Carolina indicate that reproduction is adequate to maintain the species in the State and that the crisis of the pesticide era may be a thing of the past (S. R. Hopkins, 1979, South Carolina Wildlife and Marine Resources Department, Charleston, pers. comm.). Nevertheless, protection is still needed to assure the species' comeback. For additional information on the osprey, see the sections on birds of forested wetlands and birds of the riverine ecosystem, Chapter Five.

Wood Stork: Because it nests in tall cypress trees in remote places, its status in the coastal plain is becoming threatened as more forests are cleared through draining and logging (Kahl 1964). Wood storks have been observed during summer, and there are recent records of individuals nesting in the characterization area. Such areas should be protected from disturbance to encourage the establishment of nesting colonies.

Least Tern: This species is in jeopardy because of its nesting habits. It tends to nest on low areas of beaches where the eggs and young can be washed by high, spring tides. It also nests on beaches where there is high human use, and often people will walk through a colony without knowing that they are stepping on the eggs. Increased human use and development of barrier island beaches will magnify the problem for these birds. Recently, however, it has been noted that least terns are starting to nest on the tops of buildings in large shopping centers, and this is encouraging. This practice has been reported from Charleston and North Charleston in South Carolina, and also from other large cities in the coastal plains of the Southeastern and Gulf coasts. Additional studies of this behavior should be undertaken to see if this practice is a possible solution to the declining population numbers. For additional information on least terns, see the section on birds of the intertidal marine subsystem in Chapter Two and the sections on birds of bird key and bank communities and dune bird communities in Chapter Three.

Cooper's Hawk: In the recent "Blue List" (Arbib 1975), this species received the second highest total of votes for inclusion on the official threatened list. The occurrence of this species in the Sea Island Coastal Region is on the decline, and the resident populations within the region are quite small. There is a definite need for additional information on the status of populations of Cooper's hawks in the Southeastern United States. For additional information, refer to sections on birds of forested wetlands in Chapter Five and birds of the upland ecosystem in Chapter Six.

(3) Species "Of Special Concern." The species placed in this category are those which, although not considered threatened or endangered, show probable cause for concern because of their rarity, exploitability, vulnerability to specific pressures, or other reasons identifiable by experienced investigators. This category also includes species of undetermined status where data are insufficient for precise assessment.

The species proposed as "of special concern" by the South Carolina Bird Committee are listed, along with the reasons for concern, in Table 1-4.

(4) "Peripheral" Species. Only two species are included in this category: the golden eagle and the long-billed curlew. The golden eagle breeds mainly to the north of the Sea Island Coastal Region in the Appalachian Mountains, but is seen on the Southeastern Coastal Plain during winter. Although once declining, it appears to be making a comeback as habitat is managed to encourage breeding. The long-billed curlew occurs east of the Mississippi River only rarely. It occurs in South Carolina and Georgia as a rare visitor, mainly during winter.

3. Reptiles

a. American Alligator. The American alligator is the only crocodilian occurring naturally in South Carolina and Georgia. The species occurs in wetlands from North Carolina through the Southeastern and Gulf States to Texas, and also occurs in southeastern Oklahoma and southern Arkansas. Alligators long have been prized for their hides and, somewhat more recently, have been slaughtered for their meat. Also, as a large predator, the alligator has been

Table 1-4. Species proposed as "of special concern" by the South Carolina
Bird Committee (Gauthreaux et al. 1979).

Species	Reason for Concern	Probable Cause
Barn owl	Apparent population decline; status not well determined	?; Perhaps same as for other raptors
Bewick's wren	Tremendous population declines	
Canvasback	Population declines	Hunting pressure
Great horned owl	Apparent population decline; status not well determined	?; Perhaps same as for other raptors
Ground dove	Population decline	Habitat destruction
Hairy woodpecker	Possible population decline; not well documented	Habitat destruction?
Loggerhead shrike	Population decline	?; Perhaps same as for other raptors
Merlin	Definite population decline	?; Perhaps same as for other raptors
Mississippi kite	Population decline?	Habitat destruction
Red-headed woodpecker	Population decline in Southeastern U.S.	Habitat destruction + competition
Swainson's warbler		Possible habitat destruction
Wilson's plover	Population decline	Habitat destruction

subjected to widespread indiscriminate
killing. It also suffers loss of habitat
through man's activities. In response to
these pressures, alligator population
levels declined significantly throughout
much of its range, reaching a low point
in the late 1950's to mid 1960's. In
recognition of the precarious position of
the species, the U.S. Secretary of the
Interior listed it as "endangered" in
1967. This action was followed by legal
protection in all States of its occurrence
under the Federal Endangered Species Act
of 1973.

Through the imposition of protective
measures, the alligator's population decline has been reversed, to the point
that in some localities alligators are
once again numerous enough to become nui-
sances to man. Because of the dramatic
increase in the numbers of animals, the
alligator populations of Cameron,
Calcasieu, and Vermilion parishes of
Louisiana were removed from the Federal
list of endangered and threatened species
in 1977. Additional populations of
coastal areas, including those of South
Carolina and Georgia, were reclassified as
"threatened" in 1977; this reclassifica-
tion was confirmed in 1979 (U.S. Department of Interior, Fish and Wildlife Service
1979a).

A recovery team led by Mr. Ted Joanen
(Louisiana Wildlife and Fisheries Commission, Route 1, Box 20-B, Grand Chenier,
Louisiana, 70643) was appointed and has
recently completed a draft recovery plan
(American Alligator Recovery Team 1979).
The team concluded that the long-range
survival of the alligator hinges on three
critical factors: "(1) the maintenance
and proper management of essential wetlands throughout its range, (2) the con-
tinuation and enforcement of laws regulating the harvest and movement of skins and
other parts through commerce, and (3) the
creation of a high level of tolerance toward the species by citizens who must live

in frequent contact with the species"
(American Alligator Recovery Team 1979).

For additional information on alligators, see the sections on reptiles in Chapter Five.

b. Eastern Indigo Snake. The indigo snake is a large (average length, 1.8 m (5.9 ft), Odom et al. 1977), docile snake that has become quite popular in the pet trade. Its primary habitat is the "sandhill" community (typically a turkey oak-longleaf pine-three awn grass community), which is also the habitat of the gopher tortoise (Odom et al. 1977, Speake et al. 1978). The indigo snake uses the burrows of gopher tortoises as wintering sites, as often do rattlesnakes (Speake et al. 1978). This makes them quite vulnerable to gassing and death during "rattlesnake roundups." This snake occurs throughout coastal Georgia but is not documented from South Carolina.

Although overexploitation and indiscriminant killing have been important factors in the decline of indigo snakes, probably the most significant factor has been the loss of habitat through farming, construction, forestry, and other development practices. No recovery team has been appointed for this species, but it has been recommended that recovery efforts include acquisition of habitat, encouragement of forest management practices that produce desirable indigo snake habitat, increased public education efforts, study of limiting factors, and possibly captive breeding and restocking.

c. Gopher Tortoise. This species is not considered endangered by Federal authorities and is fairly common in Georgia. However, since South Carolina represents its northernmost range and it occurs in only two South Carolina counties, it is considered endangered in the State. Further, in South Carolina, the species lives in a few loose colonies which are vulnerable to many agricultural and forestry practices.

d. Marine Turtles. The Atlantic leatherback, green sea, Atlantic hawksbill, and Kemp's ridley turtles are all considered transient species at present in the Sea Island Coastal Region. None of these species are known to nest in South Carolina or Georgia, but all four are observed with some regularity (although uncommonly) in the area. All of these species have suffered population declines through overexploitation of adults and eggs for food (and, in some cases, jewelry), loss of nesting beaches through resort development, heavy predation of nests and young, and incidental catch and drowning by commercial fishermen. The green sea and the ridley turtles especially were once much more abundant along the Southeastern United States than at present.

The Atlantic loggerhead turtle is the only species of marine turtle which nests in the Sea Island Coastal Region, and it is considered a resident species. The barrier island beaches of South Carolina and Georgia provide some of the most important loggerhead nesting sites in the United States (see Atlas plates 31 - 40). In South Carolina, the most significant nesting area is in the Cape Romain National Wildlife Refuge, especially Cape Island. In Georgia, Little Cumberland, Cumberland, and Jekyll islands provide the most important loggerhead nesting beaches. As is the case for the other marine turtles, major problems include the depredation of nests by such organisms as human poachers, raccoons, and ghost crabs; human development of nesting beaches; overexploitation; and incidental catch. Despite these difficulties, the loggerhead populations appear to be in somewhat better shape than those of the other marine turtles.

A recovery team for marine turtles has been appointed, with Ms. Sally Hopkins (South Carolina Wildlife and Marine Resources Department, Non-game and Endangered Species Section, P.O. Box 12559, Charleston, South Carolina, 29412) and Dr. Peter C. H. Pritchard (Florida Audubon Society, P.O. Drawer 7, Maitland, Florida, 32751) as co-team leaders. A recovery plan is in preparation.

For additional information on marine turtles, refer to sections on reptiles of the subtidal and intertidal marine subsystems, Chapter Two.

4. Fishes

Shortnose Sturgeon. This is the only species of fish considered endangered in the Sea Island Coastal Region. The shortnose sturgeon apparently has become rare throughout its range (Atlantic seaboard rivers from New Brunswick to northern Florida), and hence is considered to be endangered. In the past, no distinction was made between Atlantic and shortnose sturgeon in published commercial landings, and actually very little is known about the historical or present abundance of shortnose sturgeon. However, it is likely that its populations have suffered serious declines, along with those of the Atlantic sturgeon. For the Atlantic sturgeon, overfishing and the destruction of river spawning sites through damming and deterioration of water quality are considered the probable causes of declining abundance (Murawski and Pacheco 1977).

Until recently, the shortnose sturgeon was known in South Carolina from a single collection near Charleston, and, although undocumented, it was thought to occur in the Altamaha and Savannah rivers of Georgia (Odom et al. 1977). A number of specimens have now been taken in South Carolina during studies on the Atlantic

sturgeon (T. I. J. Smith and D. E.
Marchette, 1979, South Carolina Marine
Resources Division, Charleston, pers.
comm.). Collection records include the
Edisto, Pee Dee, Santee, and Waccamaw
rivers, plus Charleston Harbor and Winyah
Bay. Further, Heidt and Gilbert (1978)
recently reported that shortnose sturgeon
have been collected in the Altamaha,
Ogeechee, and Savannah rivers of Georgia.

A shortnose sturgeon recovery team
has been appointed, with James Hoff
(Southeast Massachusetts University,
Biology Department, North Dartmouth,
Massachusetts, 02747) as leader.

CHAPTER TWO

COASTAL MARINE ECOSYSTEM

I. MAJOR CHARACTERISTICS

A. DEFINITION (abbreviated; see Chapter
One for details)

The coastal marine ecosystem is an
edge system where land and water have unobstructed access to the open ocean. As
such, it has two components: the intertidal and the subtidal. The intertidal
subsystem consists of tidal ocean beaches
(virtually all on barrier islands in the
Sea Island Coastal Region) and bars contiguous to coastal waters. The subtidal
subsystem consists of coastal waters extending seaward of extreme low spring
tide level and with salinities consistently exceeding 30°/oo.

B. GENERAL DESCRIPTION AND MODEL

For the purposes of this characterization, the coastal marine ecosystem may
be considered as beginning just seaward
of the beach dunes and continuing to the
3-mi (4.8 km) territorial limit (Fig.
2-1). It is a continuous system forming
the seaward border of the characterization
area (Preface Fig. 2). The substrate is
typically sandy from the beach to well
beyond the arbitrary 3-mi (4.8 km) limit
imposed here.

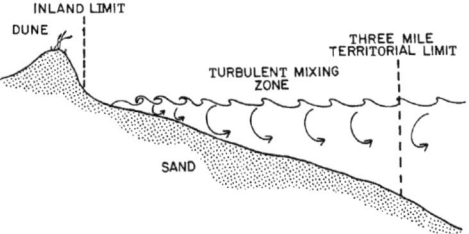

Figure 2-1. Diagrammatic representation
of the boundaries of the
coastal marine ecosystem of
the Sea Island Coastal
Region.

The sandy bottom region in coastal
waters of 20 m (65.6 ft) or less was referred to by Day et al. (1971) in North
Carolina as the turbulent zone. Because
of the generally high energy nature of
the characterization area coastline and
its resultant turbulent mixing characteristics, the entire coastal marine ecosystem within the study area could properly
be termed the turbulent zone. However,
Tanner (1960) has reported that the
Georgia coast has the lowest energy levels
along the Southeastern United States.

The primary energy source for the
marine ecosystem, as for all ecosystems,
is the sun (Fig. 2-2). The sun provides
the basic energy for the synthesis of
organic carbon molecules from CO_2 by autotrophic organisms (plants) through photosynthesis. In turn, the organic carbon
produced by the autotrophs provides the
basic energy for nearly all other life
forms (the heterotrophs). The primary
producers (primarily phytoplankton in the
coastal marine system of the study area)
are consumed by first level heterotrophs
(generally herbivorous zooplankton, such
as many of the copepods), which in turn
are consumed by second level heterotrophs
(e.g., fish larvae, chaetognaths), which in
turn are eaten by third level consumers,
and so forth. Excretion and death at each
level result in the release of nutrients
either directly into the water or through
the action of decomposers (bacteria and
fungi). The decomposers themselves may
release nutrients directly to the water
or may be grazed by heterotrophs, which in
turn excrete and decompose. Nutrients may
also be regenerated through various
physical processes, including current
movement, upwelling, vulcanism, etc.
(See Chapter One for general information
on important biogeochemical cycles.)
Regeneration of nutrients provides stimuli
for additional primary productivity, which
in turn may be consumed by heterotrophs.
Any primary production not utilized within
the area may be exported through the
action of currents. Thus, a phytoplankton
bloom that is not consumed completely might
be exported out of or imported into the
local system.

An additional source of organic input
is detritus, i.e., dead organic material
with associated decomposer flora and epifauna. Much of the detrital input is produced elsewhere (e.g., estuarine wetlands)
and imported into the system by water
movements. Some detritus may be consumed
directly by herbivores and omnivores, with
much of the remainder acted on by decomposers. A portion of the detritus is
often exported from the system via currents, while another portion may be
essentially lost to the system through
sedimentation.

The physical forcing factors which
shape the coastal marine ecosystem include
primarily the tide, wind, waves, currents,
precipitation, runoff, and estuarine
(river) flow. Tides are a result of the
gravitational pull of the moon and sun
acting on the ocean. The other factors,
like primary productivity, are related to
insolation (which drives atmospheric circulation, evaporation, etc.). Together,
these forcing factors determine the major
physical characteristics of the ecosystem
and, therefore, the kinds of biota which
populate it.

The wave climate, which is determined
by local winds, storms, and bottom

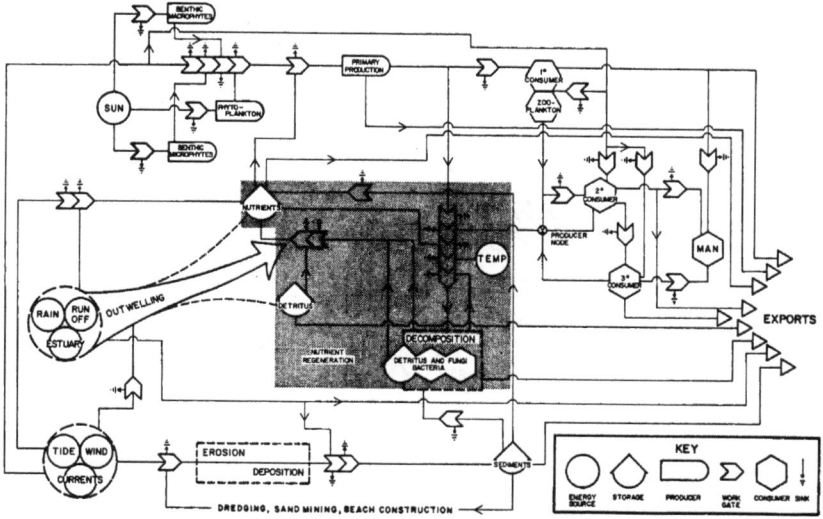

Figure 2-2. A generalized energese model of the coastal marine ecosystem of the Sea Island Coastal Region.

topography, controls to a large degree the amount of turbulent mixing and hence the erosion, transport, and deposition of near-shore bottom and beach sediments. The height and frequency of tides determines many physical characteristics of the intertidal zone (such as duration of periods of submergence and emergence and the amount of interstitial water and oxygen). Further, tides transport sus-pended and dissolved materials and create inshore currents. Currents also suspend, transport, and deposit materials and organisms. Longshore and rip currents, which are caused by tidal currents and waves interacting with beach fronts, affect the distribution of materials and organisms in shallow water and are pri-marily of local significance. Coastal currents are major forces in the shaping of beaches through erosion and deposition.

The Gulf Stream, a major ocean cur-rent situated well beyond the 3-mi (4.8 km) boundary of the characterization area, has major effects throughout the area. These effects are primarily of three types: 1) the transport of colonizing stages of organisms from other areas, principally from the south, into the characterization area, 2) nutrient regeneration, and 3) ameliora-tion of temperature and coastal climate. Frictional drag, storms, and other physical factors result in the formation of gyres and eddies along the western (landward) boundary of the Gulf Stream. These gyres

and eddies break off in the "resident" water of the area, carrying colonizing stages of organisms, nutrients, and other materials (e.g., Sargassum spp. seaweed). This type of transport may have a major influence on species composition and re-population of certain faunal elements of offshore live bottom areas, and it almost certainly accounts for the occasional occurrence of juveniles of spiny lobsters (Panulinus argus) and other Caribbean species in inshore waters of South Carolina (P. A. Sandifer, 1978, South Carolina Marine Resources Division, Charleston, pers. comm.). Equally, if not more important, is the role of the Gulf Stream as a source of nutrients for shelf and coastal waters. Subsurface intrusions of deep, nutrient-rich Gulf Stream waters, coupled with the eastward deflection of the Gulf Stream off the characterization area (which allows upwelling of deep ocean water from beyond the shelf break), appar-ently supply significant amounts of nitrates and phosphates necessary to sus-tain a high level of primary productivity in shelf and coastal waters (U.S. Depart-ment of Interior, Bureau of Land Management 1977; Mathews and Pashuk 1977) (see Fig. 2-3). Finally, the warm and generally stable temperature regime of the Gulf Stream ameliorates the temperature of the surrounding waters, atmosphere, and coastal land mass. (See Volume I, Chapter Four, Regional Climatic Trends, for details.)

31

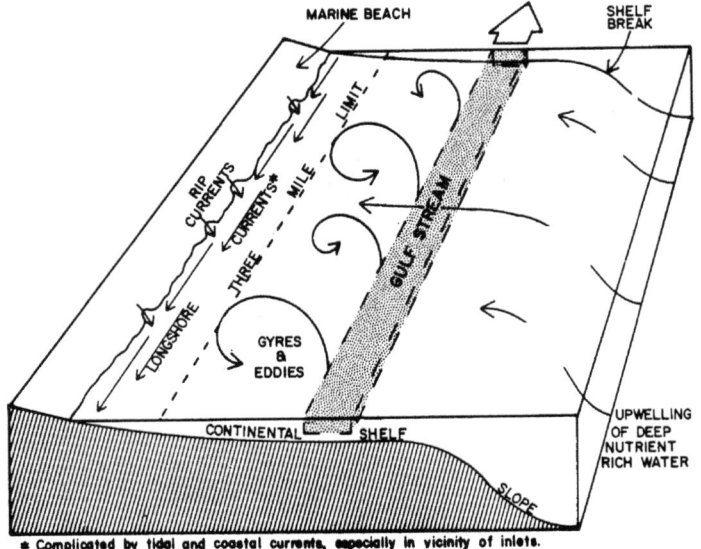

MARINE BEACH

SHELF BREAK

GULF STREAM

RIP CURRENTS

CURRENTS*

THREE MILE LIMIT

LONGSHORE

GYRES & EDDIES

CONTINENTAL SHELF

SLOPE

UPWELLING OF DEEP NUTRIENT RICH WATER

* Complicated by tidal and coastal currents, especially in vicinity of inlets.

Figure 2-3.   Major currents affecting the coastal marine ecosystem of the Sea Island Coastal
Region.

The effects of precipitation, surface runoff, and estuarine (river) flow on the coastal marine ecosystem may be considered jointly as "outwelling" effects. These may be considered to be of two primary types: 1) freshwater input and 2) contribution of materials and organisms. The volume of fresh water and its variation with time (season and other) regulates the salinity characteristics of nearshore marine waters and markedly affects circulation patterns. The salinity regime, in turn, may effect the distribution and abundance of organisms. The volume and seasonal variation of river flow and runoff also determine the amount of suspended particulate matter (including sediments). detritus, nutrients, contaminants, and estuarine organisms carried into the nearshore marine waters. The magnitude of these outwelling contributions will, in turn, affect the occurrence, standing stock levels, and trophic dynamics of biological populations in the coastal marine system.

Thus, there are three primary sources of nutrients and materials for the coastal marine ecosystem (Fig. 2-4): 1) inwelling of materials from oceanic waters by current, wind, and wave action; 2) upwelling

from deep ocean waters; and 3) outwelling from estuarine and terrestrial environments. These three input sources combine to make the coastal marine system highly productive, especially from a fisheries standpoint.

C.   GENERALIZED FOOD WEB AND RELATIONSHIPS
     (Fig. 2-5)

As Steele (1974) has pointed out, the ability of herbivores in the ocean to graze nearly all the plant production is a major difference between marine and terrestrial systems. In natural terrestrial systems, usually <10% of the plant material is consumed alive; thus, most of the primary production is transferred to higher trophic levels via the decomposer pathway. In contrast, nearly all primary production in the ocean is accomplished by phytoplankters, which are consumed almost as rapidly as they are produced. Here, the transfer of plant energy to higher trophic levels occurs primarily through the herbivore pathway. Of course, phytoplankton production may outstrip herbivore grazing for a short time on a local or seasonal basis so that some phytoplankters may settle out and contribute to the food of the bottom fauna. However, grazers

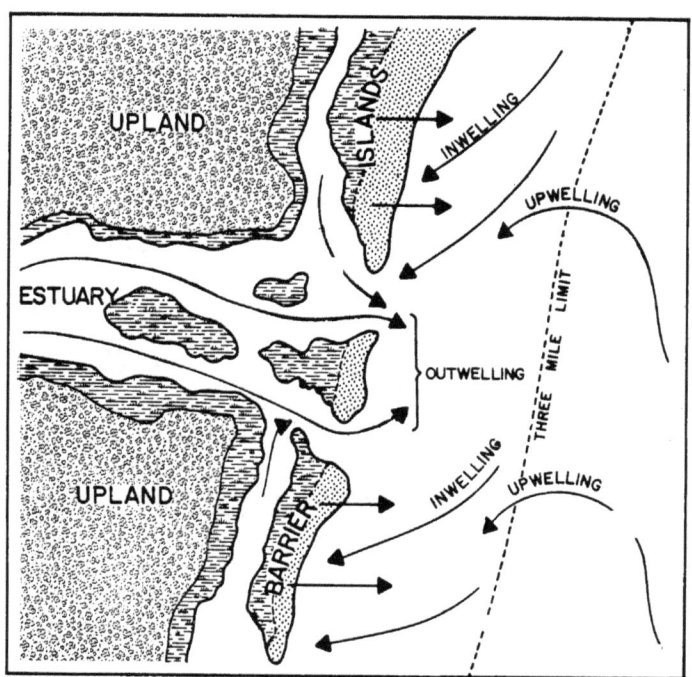

Figure 2-4. Diagrammatic representation of major sources of nutrients and materials for the coastal marine ecosystem of the Sea Island Coastal Region.

generally tend to overtake such blooms rapidly so that their direct contribution to the benthic food budget is generally minor.

Because phytoplankton populations are grazed almost as fast as they are produced, there is no accumulation of surplus primary production that can be utilized by the pelagic (planktic) herbivores to "even out" variations in phytoplankton productivity. Thus, the herbivores are re-source-limited (Steele 1974); that is, at any given time the magnitude of their populations depends directly upon the levels of phytoplankton available for them to feed on. Further, since most of the primary productivity is cycled through herbivores, benthic organisms must depend in large part upon the feces (plus dis-solved organics and corpses) of herbivores and other planktic or pelagic animals for their food. The benthic production then

provides a food source for a variety of larger pelagic organisms. Hence, as far as the general food situation in coastal subtidal waters is concerned, there are two compartments of special significance: 1) primary production, and 2) feces. These two compartments provide the basis for animal production in the nearshore marine environment.

The trophic situation in the marine intertidal environment can be considered as merely a special case of the repre-sentative food web presented in Fig. 2-5 for the entire coastal marine ecosystem. However, trophic dynamics in the marine intertidal zone differ in several important respects from those discussed above for subtidal waters.

There is apparently very little _in situ_ primary production in marine beaches, although some diatoms and blue-green algae

33

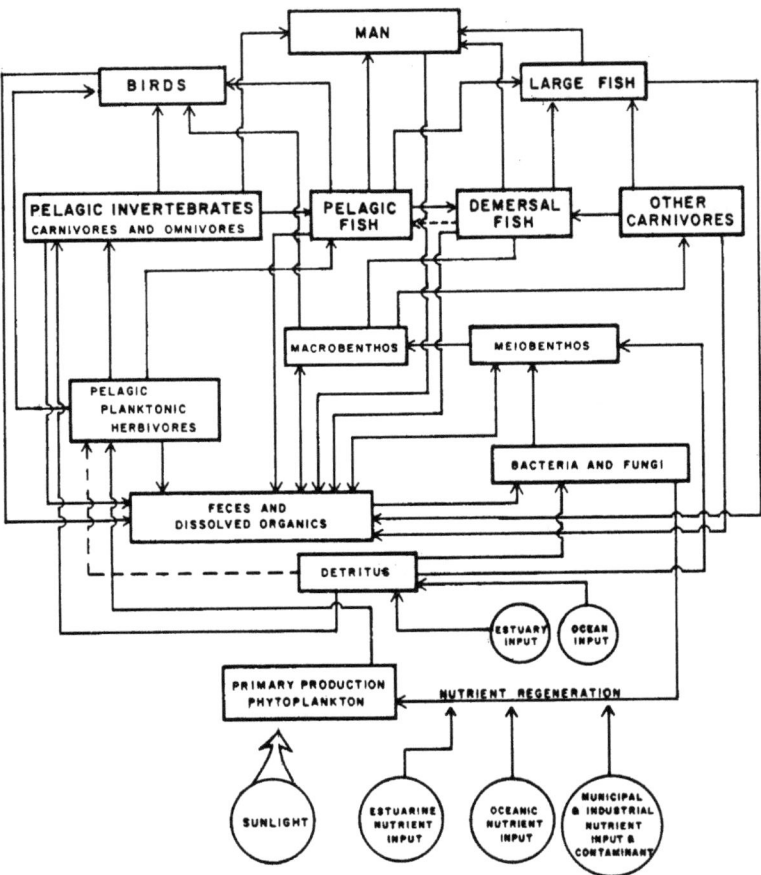

Figure 2-5. A generalized food web for the coastal marine ecosystem of the Sea Island Coastal Region (modified substantially from Steele 1974).

are generally present and occasionally numerous (Riedl and McMahan 1974). The turbulence and continually shifting substrate severely restrict occurrence and production of plants in this environment. In the absence of significant primary production, imported detritus apparently provides the basis for most secondary (heterotrophic) production on beaches (Riedl and McMahan 1974). However, dissolved organic materials in seawater may also be important to beach community energetics. Riedl (in Riedl and McMahan 1974) suggested that a beach may function as a "bubble filter," utilizing wave energy to strip organic matter from seawater, concentrate it, and convert it to a particulate form which can be used as a direct food source. Recent evidence (Scully 1978) suggests that such organic foam can be used directly as food by some beach-dwelling macrofauna.

Food webs in the coastal marine environment are highly complex, with many branches, connections, and overlaps. Figure 2-5 illustrates in a very generalized fashion some of the trophic interrelationships of major groups of organisms (e.g., pelagic fish, macrobenthos) represented in the characterization area. As implied by the numerous connecting links in the diagram, these interrelationships are far from simple. Few large groupings of organisms, other than the primary producers, can be limited to a single trophic level. Much more commonly, representatives of a given group, whether defined taxonomically or ecologically, feed at more than one trophic level, resulting in multiple "horizontal links" in the food web.

Within a food web, the magnitude of production at each trophic level is determined by three primary factors: 1) trophic efficiency, 2) transfer rate, and 3) number of horizontal links. Trophic efficiency is the efficiency with which a population transfers energy to the next step (horizontal or vertical) in the food web. This efficiency is frequently taken to be a more or less "constant" of one order of magnitude for each trophic level (Odum 1971). However, while this 10% figure may be approximately correct on the average, it undoubtedly does not hold for every trophic level or population. For example, Steele (1974) states that ". . . transfer efficiencies around 20 percent appear to be required of the pelagic herbivores and also, possibly of the benthic infauna that feed on fecal material." Similarly, Ryther (1969) gave 15% as the average ecological efficiency in a coastal marine environment. Such differences between the actual transfer efficiency of a population or trophic level and a theoretical average can become quite important when one begins to consider such things as the potential size of fish stocks, for example. The second factor, transfer rate, refers to the rate

at which energy is transferred from one step in the food web to the next through grazing, predation, waste, or death. It is, in part, determined by the generation time or "turnover rate" of the population and partly by the grazing or predation rate. Finally, the number of horizontal links is important because the more links present, the greater the degradation of energy within a given trophic level and the less energy there is available for transfer to the next trophic level. Steele (1974) illustrated this concept very effectively by constructing two hypothetical trophic relationships, one a straight chain and the other a web with one branch, among phytoplanktic producers, zooplanktic herbivores, and primary carnivores (see Fig. 2-6).

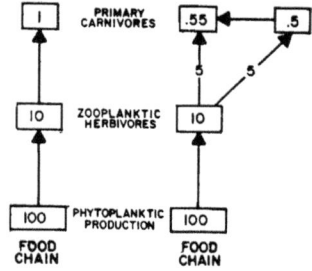

Figure 2-6. A simple example of the effect of branching in a food chain on the energy output, assuming 10% trophic efficiency at each level (redrawn with minor modifications from Steele 1974; numbers refer to units of energy).

Despite the preceding summary, very little is really known about the trophic dynamics of coastal marine waters, especially along the South Carolina-Georgia coast. As Riedl and McMahan (1974) concluded: "The energy input and output relationships of the beach system -- all the correlations with neighboring environments -- are largely unresolved." A similar statement could be made about the entire coastal marine ecosystem of the Sea Island Coastal Region.

II. SUBTIDAL COASTAL MARINE SUBSYSTEM

A. DESCRIPTION

The subtidal coastal marine subsystem, which extends seaward from the extreme low spring tide mark, has been termed the "Coastal Plankton" system by Odum et al. (1974) because its food webs are predominantly phytoplankton-based (see Fig. 1-1 for example). It also might be termed an edge or buffer system, lying between and influenced by the ocean to seaward and the estuaries to landward. In the case of the South Carolina-Georgia shelf, this subsystem might best be described as that area of coastal marine waters and associated sea bottom lying between the beach and the landward edge of the Gulf Stream.

Because of the effects of tidal oscillations, runoff, mass migrations of organisms, and outfalls, the subtidal coastal marine subsystem along the South Carolina-Georgia coast is probably more influenced by estuaries than by the deep ocean. In some areas, such as off the Santee River in South Carolina and the Altamaha River in Georgia, estuarine conditions occasionally may penetrate well down the coastal marine system, particularly during times of flooding or high river flow. Water in the coastal plankton zone is generally highly turbid, nutrient-rich, and less saline compared to oceanic waters.

In addition to the plankton populations, diverse and abundant benthic and epibenthic fauna are representative of the coastal subtidal marine subsystem. These benthic assemblages are responsible for a considerable fraction of the total respiration and nutrient cycling within the system (Odum et al. 1974) and support sizable fish populations. These coastal waters form perhaps the single most important marine environment worldwide, at least in terms of useful biological production, since they are the principal location of commercial fisheries (Ryther 1969).

A number of major factors determine the dominant characteristics of coastal subtidal waters in the characterization area. For example, circulation patterns and stratification are affected by long-shore drift of various water masses, current gyres, runoff and low salinity waters, and upwelling upwelling of oceanic water through action of currents and storms. Other important factors include seasonal cycles in temperature and precipitation, primary production by phytoplankton, and the significant storage capacity of the bottom fauna (Odum et al. 1974).

The substrate of the continental shelf along the South Carolina-Georgia coast typically is composed of shifting sandy sediments that provide relatively poor fish habitat (Barans and Burrell 1976). Occasional outcrops of rock, relict worm tube reefs, or other material provide some vertical relief and substrate for attachment of a variety of sessile benthic invertebrates, which in turn attract motile invertebrates and concentrations of fishes. The extensive assemblages of filter-feeding and detritivorous invertebrates on such substrates may concentrate the potential energy of the plankton and detritus in one area, thus increasing the efficiency of energy transfer to higher trophic levels. These highly productive areas, known locally as "live bottoms" or "reefs," are not uncommon but generally occur well beyond the 5-mi (4.8 km) territorial sea (U.S. Department of Interior, Bureau of Land Management 1977; J. A. Barans and V. G. Burrell, Jr., 1976, South Carolina Marine Resources Division, Charleston, pers. comm.). Although these "reefs" occur seaward of the study area, they are briefly described here since they are such important features of the shelf.

At least five types of live bottom 'reefs' already have been identified in the Georgia Embayment. These include: 1) the shelf break (Eddy et al. 1967, MacIntyre and Milliman 1970), 2) a relict lithothamnion reef in North Carolina (Menzies et al. 1966), 3) coral outcroppings (Huntsman and MacIntyre 1970), 4) black rock reefs (Pearse and Williams 1951), and 5) Gray's Reef in Georgia (Hunt 1974). The definition of a "reef" given by Hunt 1974 is generally applicable to all of these: ". . . an outcrop of a body of rock on an otherwise sandy bottom which expresses relief above the surrounding bottom and supports accumulation of sessile benthos." However, as yet little is known of the areal extent of these types of habitats, their geographical distribution, or the biological communities which inhabit them.

The continental shelf off South Carolina and Georgia is believed to have three distinct, although longitudinally discontinuous, zones of live bottoms. The first zone is in relatively shallow water approximately 15 - 30 m or 49 - 98 ft near shore. This zone contains the features locally termed "Blackfish Banks" (Bearden and McKenzie 1972 in South Carolina and the widely known Gray's Reef in Georgia. Further offshore, in waters of about 15 - 55 m (81 - 180 ft, occur the so-called 'Snapper Banks' and other highly localized, discontinuous 'reefs.' The third zone of live bottoms is found at the shelf edge in depths of approximately 55 - 100 m (180 - 656 ft). This shelf edge 'reef' is more or less continuous along the entire shelf off South Carolina and Georgia and is fished extensively by recreational and commercial fishermen (Bearden and McKenzie 1972). This deeper zone apparently contains the largest concentration of live bottoms on the South Carolina-Georgia shelf.

Very little is known about most of the live bottom areas in the Georgia Embayment, either geologically or biologically. Even for Gray's Reef, which has been studied in more detail than any other live bottom on the South Carolina-Georgia shelf (Hunt 1974), few quantitative data exist on community structure, sizes of populations, or species composition for many groups. Nevertheless, it is well documented that live bottom habitats are important to the fisheries of the region (Menzies et al. 1966; Struhsaker 1969; Bearden and McKenzie 1971; Sekavec and Huntsman 1973; MARMAP Program, 1978, South Carolina Marine Resources Division, Charleston, unpubl. data). Commercial fisheries are beginning to develop in the live bottom areas off South Carolina and Georgia, while well-established and still expanding recreational fisheries have existed for many years in the same locales. (See Volume II, Chapters Seven and Nine, for details.) The economic impact of these latter fisheries is hard to quantify, but conservatively it might be estimated to be on the order of $25 million yearly in South Carolina and Georgia (D. M. Cupka, 1978, South Carolina Marine Resources Division, Charleston, pers. comm.). Additional information concerning these habitats may be found in papers by Porter and Jenner (1967), MacIntyre and Pilkey (1969), MacIntyre (1970), Cain (1972), Shoemaker (1972), Roberts (1974a), and a recent environmental impact statement review (U.S. Department of Interior, Bureau of Land Management 1977).

B. PRIMARY PRODUCERS

1. Phytoplankton

Primary production in the coastal marine environment is almost entirely a function of the phytoplankton. By strict definition, phytoplankton consists of microscopic autotrophs drifting in the aquatic environment and transported primarily by currents and wind. In deep waters, phytoplankton populations are composed almost completely of holoplanktonic (i.e., planktonic throughout their lives) forms. In contrast, neritic (i.e., nearshore, shallow water) phytoplankton populations are decidedly more meroplanktonic (i.e., planktonic for a portion of their life cycle) and tychoplanktonic (i.e., normally living on bottom surfaces but carried into the water column by turbulence). In general, the more obvious members of the coastal marine phytoplankton are the diatoms and the dinoflagellates.

Extremely important, and as yet not well studied, components of marine phytoplankton communities are the nano- (5 - 60 μm in size) and ultraplankton (<5 pm). These populations often comprise more than 50% of the standing phytoplankton biomass, and they are particularly important in neritic waters. Aside from a small number of diatoms and dinoflagellates, these populations include genera of the Cyanophyta, Euglenophyta, Prasinophyta, Xanthophyta, Chrysophyta, Cryptophyta, and Chlorophyta, most of which are single cell flagellates. These organisms are so ubiquitous that many researchers have termed them collectively the phytoflagellates as a major subdivision of the phytoplankton.

In general, the microscopic algae comprising the phytoplankton are segregated not only by size but also by position in the water column. The phytoplankton found within the water column are termed pelagic flora. Algae associated with substrates are benthic flora, and phytoplankton found at or near the water/air interface (in the surface microlayer) are called phytoneuston. Recent studies have indicated that the phytoneuston is an important component of the total phytoplankton, particularly in neritic marine and estuarine waters (Manzi et al. 1977a).

The nearshore marine subtidal systems of the Southeastern United States, particularly those of South Carolina and Georgia, have been little studied. This is especially true for the neritic flora, for which there is poor documentation of both temporal and spatial relationships. Nevertheless, some pertinent data exist for the Sea Island Coastal Region and adjacent areas.

Stenohaline marine phytoplankters of water off the Southeastern United States have been listed in several papers, principally those of Hulburt (1963a, b, 1967), Hulburt and Rodman (1963), Marshall (1966, 1968, 1969a, b, 1971, 1976, 1978) and Hulburt and MacKenzie (1971). These accounts, however, deal only with algal populations offshore of the characterization area.

Of the above studies, Marshall's recent work (1976) is probably most pertinent. He reported results of collections made from a total of 542 stations over 22 cruises (1964 - 1974). A total of 609 species was identified from these collections, including 277 diatoms, 247 dinoflagellates, 54 coccolithophores, 9 silicoflagellates, 6 cyanophytes and 16 other forms from the Chlorophyta, Euglenophyta, and Cryptophyta. Unfortunately, Marshall's work included very little data on the temporal and spatial distribution or dominance of the various species. Nonetheless, his studies do indicate some general trends in the distribution of phytoplankton populations off the coast of the Southeastern United States. The largest numbers of species and highest total concentrations of algae normally occurred in neritic waters within 50 mi (80 km) from shore. The number of diatom species generally decreased seaward, as did the number of species of dinoflagellates and silicoflagellates. On the other hand, more species of coccolithophores were found in waters beyond the continental shelf, while the blue-green algae were well represented only within 100 mi (161 km) of the coast.

predominantly phytoplankton-based (see Fig. 2-5 for example). It also might be termed an edge or buffer system, lying between and influenced by the ocean to seaward and the estuaries to landward. In the case of the South Carolina-Georgia shelf, this subsystem might best be described as that area of coastal marine waters and associated sea bottom lying between the beach and the landward edge of the Gulf Stream.

Because of the effects of tidal oscillations, runoff, mass migrations of organisms, and outfalls, the subtidal coastal marine subsystem along the South Carolina-Georgia coast is probably more influenced by estuaries than by the deep ocean. In some areas, such as off the Santee River in South Carolina and the Altamaha River in Georgia, estuarine conditions occasionally may penetrate well into the coastal marine system, particularly during times of flooding or high river flow. Water in the coastal plankton zone is generally highly turbid, nutrient-rich, and less saline compared to oceanic waters.

In addition to the plankton populations, diverse and abundant benthic and epibenthic fauna are representative of the coastal subtidal marine subsystem. These benthic assemblages are responsible for a considerable fraction of the total respiration and nutrient cycling within the system (Odum et al. 1974) and support sizable fish populations. These coastal waters form perhaps the single most important marine environment worldwide, at least in terms of useful biological production, since they are the principal location of commercial fisheries (Ryther 1969).

A number of major factors determine the dominant characteristics of coastal subtidal waters in the characterization area. For example, circulation patterns and stratification are affected by long-shore drift of various water masses, current gyres, runoff and low salinity waters, and inwelling/upwelling of oceanic water through action of currents and storms. Other important factors include seasonal cycles in temperature and precipitation, primary production by phytoplankton, and the significant storage capacity of the bottom fauna (Odum et al. 1974).

The substrate of the continental shelf along the South Carolina-Georgia coast typically is composed of shifting sandy sediments that provide relatively poor fish habitat (Barans and Burrell 1976). Occasional outcrops of rock, relict worm tube reefs, or other material provide some vertical relief and substrate for attachment of a variety of sessile benthic invertebrates, which in turn attract motile invertebrates and concentrations of fishes. The extensive assemblages of filter-feeding and detritivorous invertebrates on such substrates may concentrate the potential energy of the plankton and detritus in one area, thus increasing the efficiency of energy transfer to higher trophic levels. These highly productive areas, known locally as "live bottoms" or "reefs," are not uncommon but generally occur well beyond the 3-mi (4.8 km) territorial sea (U.S. Department of Interior, Bureau of Land Managment 1977; C. A. Barans and V. G. Burrell, Jr., 1978, South Carolina Marine Resources Division, Charleston, pers. comm.). Although these "reefs" occur seaward of the study area, they are briefly described here since they are such important features of the shelf.

At least five types of live bottom "reefs" already have been identified in the Georgia Embayment. These include: 1) the shelf break (Eddy et al. 1967, MacIntyre and Milliman 1970), 2) a relict Lithothamnion reef in North Carolina (Menzies et al. 1966), 3) coral outcroppings (Huntsman and MacIntyre 1971), 4) black rock reefs (Pearse and Williams 1951), and 5) Gray's Reef in Georgia (Hunt 1974). The definition of a "reef" given by Hunt (1974) is generally applicable to all of these: ". . . an outcrop of a body of rock on an otherwise sandy bottom which expresses relief above the surrounding bottom and supports accumulation of sessile benthos." However, as yet little is known of the areal extent of these types of habitats, their geographical distribution, or the biological communities which inhabit them.

The continental shelf off South Carolina and Georgia is believed to have three distinct, although longitudinally discontinuous, zones of live bottoms. The first zone is in relatively shallow water (approximately 15 - 30 m or 49 - 98 ft) near shore. This zone contains the features locally termed "Blackfish Banks" (Bearden and McKenzie 1971) in South Carolina and the widely known Gray's Reef in Georgia. Further offshore, in waters of about 25 - 55 m (82 - 180 ft) occur the so-called "Snapper Banks" and other highly localized, discontinuous "reefs." The third zone of live bottoms is found at the shelf edge in depths of approximately 55 - 200 m (180 - 656 ft). This shelf edge "reef" is more or less continuous along the entire shelf off South Carolina and Georgia and is fished extensively by recreational and commercial fishermen (Bearden and McKenzie 1971). This deeper zone apparently contains the largest concentration of live bottoms on the South Carolina-Georgia shelf.

Very little is known about most of the live bottom areas in the Georgia Embayment, either geologically or biologically. Even for Gray's Reef, which has been studied in more detail than any other live bottom on the South Carolina-Georgia shelf (Hunt 1974), few quantitative data exist on community structure, sizes of populations, or species composition for many groups. Nevertheless, it is well documented that live bottom habitats are important to the fisheries of the region (Menzies et al. 1966; Struhsaker 1969; Bearden and McKenzie 1971; Sekavec and Huntsman 1973; MARMAP Program, 1978, South Carolina Marine Resources Division, Charleston, unpubl. data). Commercial fisheries are beginning to develop in the live bottom areas off South Carolina and Georgia, while well-established and still expanding recreational fisheries have existed for many years in the same locales. (See Volume II, Chapters Seven and Nine, for details.) The economic impact of these latter fisheries is hard to quantify, but conservatively it might be estimated to be on the order of $25 million yearly in South Carolina and Georgia (D. M. Cupka, 1978, South Carolina Marine Resources Division, Charleston, pers. comm.). Additional information concerning these habitats may be found in papers by Porter and Jenner (1967), MacIntyre and Pilkey (1969), MacIntyre (1970), Cain (1972), Shoemaker (1972), Roberts (1974a), and a recent environmental impact statement review (U.S. Department of Interior, Bureau of Land Management 1977).

B. PRIMARY PRODUCERS

1. Phytoplankton

Primary production in the coastal marine environment is almost entirely a function of the phytoplankton. By strict definition, phytoplankton consists of microscopic autotrophs drifting in the aquatic environment and transported primarily by currents and wind. In deep waters, phytoplankton populations are composed almost completely of holoplanktonic (i.e., planktonic throughout their lives) forms. In contrast, neritic (i.e., nearshore, shallow water) phytoplankton populations are decidedly more meroplanktonic (i.e., planktonic for a portion of their life cycle) and tychoplanktonic (i.e., normally living on bottom surfaces but carried into the water column by turbulence). In general, the more obvious members of the coastal marine phytoplankton are the diatoms and the dinoflagellates.

Extremely important, and as yet not well studied, components of marine phytoplankton communities are the nano- (5 - 60 μm in size) and ultraplankton (<5 μm). These populations often comprise more than 50% of the standing phytoplankton biomass, and they are particularly important in neritic waters. Aside from a small number of diatoms and dinoflagellates, these populations include genera of the Cyanophyta, Euglenophyta, Prasinophyta, Xanthophyta, Chrysophyta, Cryptophyta, and Chlorophyta, most of which are single cell flagellates. These organisms are so ubiquitous that many researchers have termed them collectively the phytoflagellates as a major subdivision of the phytoplankton.

In general, the microscopic algae comprising the phytoplankton are segregated not only by size but also by position in the water column. The phytoplankton found within the water column are termed pelagic flora. Algae associated with substrates are benthic flora, and phytoplankton found at or near the water/air interface (in the surface microlayer) are called phytoneuston. Recent studies have indicated that the phytoneuston is an important component of the total phytoplankton, particularly in neritic marine and estuarine waters (Manzi et al. 1977a).

The nearshore marine subtidal systems of the Southeastern United States, particularly those of South Carolina and Georgia, have been little studied. This is especially true for the neritic flora, for which there is poor documentation of both temporal and spatial relationships. Nevertheless, some pertinent data exist for the Sea Island Coastal Region and adjacent areas.

Stenohaline marine phytoplankters of water off the Southeastern United States have been listed in several papers, principally those of Hulburt (1963a, b, 1967), Hulburt and Rodman (1963), Marshall (1966, 1968, 1969a, b, 1971, 1976, 1978) and Hulburt and MacKenzie (1971). These accounts, however, deal only with algal populations offshore of the characterization area.

Of the above studies, Marshall's recent work (1976) is probably most pertinent. He reported results of collections made from a total of 542 stations over 22 cruises (1964 - 1974). A total of 609 species was identified from these collections, including 277 diatoms, 247 dinoflagellates, 54 coccolithophores, 9 silicoflagellates, 6 cyanophytes and 16 other forms from the Chlorophyta, Euglenophyta, and Cryptophyta. Unfortunately, Marshall's work included very little data on the temporal and spatial distribution or dominance of the various species. Nonetheless, his studies do indicate some general trends in the distribution of phytoplankton populations off the coast of the Southeastern United States. The largest numbers of species and highest total concentrations of algae normally occurred in neritic waters within 50 mi (80 km) from shore. The number of diatom species generally decreased seaward, as did the number of species of dinoflagellates and silicoflagellates. On the other hand, more species of coccolithophores were found in waters beyond the continental shelf, while the blue-green algae were well represented only within 100 mi (161 km) of the coast.

Within the characterization area proper, Zingmark (1975) reported results of net phytoplankton sampling in neritic ocean waters off Kiawah Island, South Carolina. Three diatom species (Skeletonema costatum, Asterionella glacialis, and Rhizosolenia alata) dominated his collections. All three species occurred in the plankton year-round, but S. costatum was markedly dominant during summer and fall, A. glacialis was dominant during midwinter, and R. alata was co-dominant with S. costatum during early summer but was relatively unimportant during the remainder of the year (Fig. 2-7). These diatoms are ubiquitous along the Georgia-South Carolina coast and, together with a few additional species (e.g., Chaetoceros curvisetus, Nitzschia pungens, N. delicatissima), dominate the net phytoplankton in southeastern neritic waters.

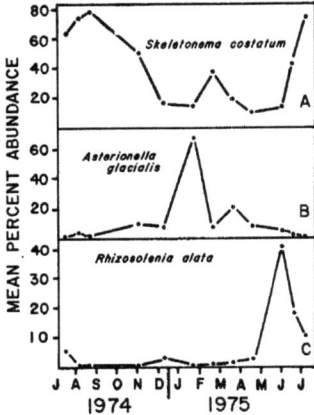

Figure 2-7. Mean monthly abundance of Skeletonema costatum, Asterionella glacialis, and Rhizosolenia alata from oceanic waters off Kiawah Island, South Carolina (Zingmark 1975). This is a good example of temporal segregation in use of a habitat by several species.

In addition to the work of Zingmark (1975), records from random collections from several neritic locations on the South Carolina coast are included in the Annotated Checklist of the Biota of the Coastal Zone of South Carolina (Zingmark 1978). All known records of marine subtidal phytoplankton are included in the phytoplankton section (Manzi and Zingmark 1978) of the checklist, and the marine

subtidal macroscopic algae are covered in the benthic marine algae section (Wiseman 1978). No comparable data are available for Georgia.

Besides the diatoms and dinoflagellates, at least one genus of filamentous blue-green algae, Oscillatoria (Trichodesmium), is important in the Georgia Bight. These algae are of special interest for two reasons. First, they are large filamentous forms (generally 300 - 440 um long), and it is unknown how they are utilized by heterotrophs. Secondly, they are known to possess the ability to fix atmospheric nitrogen (Dunstan and Hosford 1977).

The only detailed investigation of Oscillatoria accomplished in the Georgia Bight was conducted offshore of the characterization area by Dunstan and Hosford (1977). These investigators sampled from 12.5 km (4.8 mi) offshore to the shelf break along six transects between Charleston, S.C., and Jacksonville, Fla., during fall (September 1973), winter (December 1973), spring (April 1974), and summer (July 1974). They found Oscillatoria relatively abundant and widely distributed throughout the year, but filaments were most numerous in summer. During the fall cruise, concentrations of Oscillatoria were encountered most frequently near shore, with localized blooms apparent off the Cooper River in South Carolina and just north and south of the Altamaha River in Georgia. A similar situation was observed during the winter cruise. Spring samples showed increased occurrence of filaments over fall and winter cruises, with the largest blooms inshore just off the St. Johns River and up the Georgia coast about midway between the St. Johns and Altamaha rivers. During summer, Oscillatoria filaments were so prevalent as to constitute virtually a "shelf-wide bloom," with concentrations of 2,000 - 5,600 filaments/ liter at most stations and a maximum concentration of 36,000 filaments/liter at mid-shelf.

Dunstan and Hosford (1977) were unable to clearly demonstrate a relationship between Oscillatoria blooms and hydrographic conditions. However, they found higher concentrations of filaments in shelf waters than in Gulf Stream waters, and Oscillatoria blooms did not appear to be linked to intrusions of Gulf Stream water.

Although calculations by Dunstan and Atkinson (1976) indicate that no more than 5% of the "new" nitrogen required for primary production in the Georgia Bight could be the result of fixation, Oscillatoria must be significant in shelf waters, at least in terms of biomass. As Dunstan and Hosford (1977) stated, "During blooms the levels of carbon represented by Oscillatoria are 50% or more of the Particulate Organic Carbon in surface waters (100 - 150 μg Carbon Liter^{-1})." These

authors further assert that during
Oscillatoria blooms, the abundance of
300 - 400 µm long clusters of Oscillatoria
cells, in contrast to the usual dominance
by diatoms and coccolithophores 5 - 60 µm
in size, must significantly affect secondary production (i.e., the herbivores and
their predators) in shelf waters.

Concentrations of chlorophyll a, long
used to indicate phytoplankton abundance,
have been studied in the Georgia Bight
(although offshore of the Sea Island
Coastal Region) for the last several years.
Several different studies (Haines 1974,
Haines and Dunstan 1975, Dunstan and
Hosford 1977, University of Georgia
Institute of Ecology 1978) have shown
that chlorophyll a concentrations generally decrease with distance from shore
and with increasing salinity. Normally,
the chlorophyll a concentrations exceed
0.5 mg/m^3 inshore of the 20 m (65.6 ft)
depth contour while ranging from 0.1 -
0.5 mg/m^3 in deeper waters. Concentrations
of <0.1 mg/m^3 were the rule beyond the
200 m (656.2 ft) contour. Nearshore waters
between the Savannah and St. Marys rivers
frequently exhibit concentrations exceeding 1.0 mg/m^3. However, in one summer
cruise (June 1975) concentrations to
29.72 mg/m^3 were found in the coastal region south of Savannah (University of
Georgia Institute of Ecology 1978).

Turner et al. (1979) measured primary production along four transects per-
pendicular to the coast between Charleston,
South Carolina, and Jacksonville, Florida,
during 3 months (June, August, and
November). They found that primary production decreased dramatically within
10 km (6.2 mi) of shore, but followed
similar seasonal patterns in nearshore
(<20 m) and offshore (20 - 200 m or 65.6 -
656.2 ft) waters. Peak production occurred
during late summer. However, large increases in photosynthetic activity may
occur offshore during winter in response
to storms.

Although phytoplankton dynamics in
the Georgia Bight exhibit seasonal
heterogeneity, relatively short-term
phenomena such as storms and intrusions
of Gulf Stream waters are more important
in influencing algal production (Haines
and Dunstan 1975, Dunstan and Atkinson
1976). Nevertheless, seasonal climatological factors - in particular solar
radiation, temperature, wind, and water
transparency - directly influence phytoplankton growth and, together with both
horizontal and vertical currents, are
prime factors in determining phytoplankton
distribution. The numerous estuaries and
extensive wetlands along the South
Carolina-Georgia coast provide a constant
outwelling of nutrients and suspended
organic and inorganic material. This
outwelling, coupled with vertical mixing,
helps set growth limits for shelf phytoplankton populations. The coupling

between estuarine and shelf plankton
communities was perhaps described best
by Turner et al. (1979). They noted that
estuarine phytoplankton are flushed from
the turbid estuarine waters into the
clearer, nearshore ocean waters by the net
seaward flow typical of estuaries. The
greater availability of light promotes as
much as a six-fold increase in photosynthesis here, which, in turn, results
in nutrient depletion. Primary production
falls off in the nutrient-poor water
further seaward.

Although the relative importance of
vertical mixing along the broad shallow
shelf of the South Carolina-Georgia coast
is poorly known, it undoubtedly influences
phytoplankton production in at least two
ways: by upwelling nutrients and by reducing water transparency. Vertical mixing
and stratification are a result of the
major temperature changes occurring in the
surface waters. Water is a poor conductor
of heat, and temperature changes result
in density changes. During spring and
summer, temperatures of surface waters
increase rapidly, resulting in the establishment of a discontinuity layer or
thermocline. This layer provides resistance to vertical mixing and will persist
as long as surface heating continues
(Fig. 2-8). Eventually, surface waters
will approach oligotrophic conditions and
phytoplankton blooms, which are prevalent
in spring and early summer, will dissipate
and become spatially and temporally
sporadic. With the decrease in solar
radiation and air temperature and the increase in wind velocity accompanying the
onset of fall and winter, the discontinuity
layer loses integrity and extensive vertical mixing will occur throughout the
cooler months (Fig. 2-8). Thus, deeper
nutrient-rich water is transported to the
surface.

The second mode of phytoplankton
growth regulation by vertical mixing is
its influence on the penetration of solar
radiation. Phytoplankton populations can
become light-limited when vertical mixing
increases the suspended solids load of
surface waters. Thus, a critical mixing
depth can be established for particular
locations (Gran and Braarud 1935, Sverdrup
1953) based on such parameters as light
penetration, transparency of the water
column, and phytoplankton photosynthesis/
respiration ratios. Critical mixing depths
are not known along the coast of South
Carolina and Georgia. However, because
of the generally high turbidity of these
waters, it would be reasonable to estimate
a shallow critical depth, perhaps an
average of 5 - 10 m (16.4 - 32.8 ft).

Although, as pointed out earlier,
short-term phenomena are probably more
important than seasonal factors in controlling algal production in the Georgia
Bight, seasonal effects are still fairly
distinct. Figure 2-9 illustrates in a

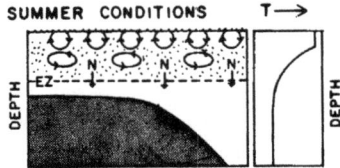

SUMMER CONDITIONS T→

DEPTH

WINTER CONDITIONS T→

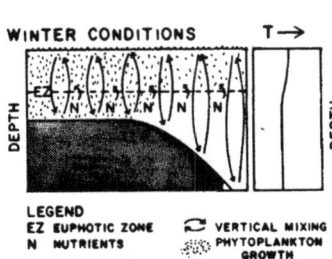

DEPTH

LEGEND
EZ EUPHOTIC ZONE   ⌇ VERTICAL MIXING
N NUTRIENTS   ⣿ PHYTOPLANKTON GROWTH

Figure 2-8. Diagrammatic illustrations of thermal factors partially regulating phytoplankton populations in neritic temperate waters (Yentsch 1971, 1977).

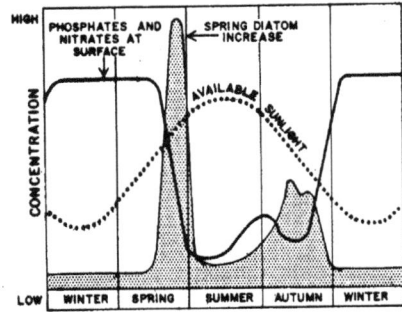

Figure 2-9. A stylized sequence of diatom population growth and the seasonal features influencing growth and primary production (redrawn from Russell-Hunter 1970).

simplified fashion the interacting effects of seasonal changes in light and nutrients (vertical mixing) on a diatom population. Nutrient concentrations in surface waters are greatest in winter. These decrease in spring and remain low throughout summer and early fall. In contrast, solar radiation is lowest in winter and increases in spring, remaining high throughout summer and early fall. Thus, diatom populations are generally light-limited in winter and nutrient-limited in warmer months.

The activities of man can greatly influence nonvascular floral populations, although such effects are often ameliorated in the marine subtidal environment by the great dilution potential of oceans. Neritic areas are especially affected by the outwelling of contaminants as well as nutrients from coastal rivers. Interactions between the products resulting from man's activities and nonvascular plant populations have been studied to some extent, but our knowledge of the effects of most environmental perturbations on subtidal flora is far from adequate. Some basic documentation is available concerning effects of oil spills (Schramm 1973), oil-spill emulsifiers (Boney 1970, 1975), pesticides (Cox 1971, Kapoor et al. 1972, Kemp et al. 1973, Scura and Theilacker 1977), heavy metals (Hannan and Patouillet 1972a, b, c), offshore construction (Rounsefell 1972), radioactive contaminants (Ancellin et al. 1973) and thermal pollution (Steidinger and Van Breedveld 1971, Levin et al. 1972). However, very little is known specifically about the effects of man's activities on the marine nonvascular flora of the Sea Island Coastal Region, and this can be considered a major data gap for the area.

2. Benthic Macrophytes

The macroscopic, benthic marine subtidal algae of South Carolina colonize a variety of natural and artificial substrates (e.g., remnant, subtidally exposed geological formations; shells and shell fragments; wrecks; man-made fishing reefs; harbor jetties). Schneider (1976) surveyed the deep-water algal flora of the Carolinas and reported 150 taxa of marine algae (1 blue-green, 24 greens, 21 browns, 104 reds). In South Carolina, his sampling was confined to Long Bay. Schneider reported 40 taxa from South Carolina, including two (Apoglossum ruscifolium and Spermothamnion investiens var. cidaricola) which were collected only in South Carolina. Earlier, Williams (1951) had investigated the marine algal flora of a shallow water, submerged "black rock" formation off Little River, South Carolina. Schneider (1976) found that 35 of the 104 taxa reported by Williams (1951) from Little River, South Carolina, and New River, North Carolina, were not represented in his deepwater collections. As Schneider (1976) pointed out: "That fact is, however, not surprising, in that 29 of the Williams total were found in water less than 6.5 m deep. . .it is obvious that Williams was sampling a component of the Carolina flora which I did not." In reference to Williams' list of 104 species, Wiseman and Schneider (1976) have noted that, with the exception of the 'Carolinas only' notation under which he listed 11 taxa, the format used by Williams has made it impossible to ascertain with certainty records for South Carolina.

The following species were prominent in collections from Sapelo Island, Georgia (Chapman 1971), and are also expected to occur along the South Carolina coast: Enteromorpha lingulata, Ulva lactuca var. latissima, Chaetomorpha minima, Codium isthmocladium subsp. clavatum, Sargassum fluitans, Lithothamnion spp., Halymenia spp., Gracilaria foliifera, Chrysymenia sp., Botryocladia occidentalis, Rhodymenia pseudopalmata, Lomentaria baileyana, Callithamnion sp., Ceramium leptozonum, Grinnellia americana, Dasya pedicellata, Polysiphonia subtilissima and Chondria littoralis. Further, J. J. Manzi (1979, South Carolina Marine Resources Division, Charleston, pers. comm.) recorded Enteromorpha lingulata, Ulva lactuca, Bryopsis plumosa, Codium isthmocladium, Dictyota dichotoma, Sargassum fluitans, Bangia fuscopurpurea, Porphyra leucosticta, Gracilaria sp., Grinnellia americana, Polysiphonia denudata and Polysiphonia sp. commonly in various places along the South Carolina coast.

Schneider and Searles (1978) described two distinct algal assemblages from North Carolina, the "jetty" and "shelf" floras. Of the 280 taxa listed (162 reds, 61 greens, 57 browns), 116 are restricted to the "jetty" flora, 101 to the "shelf" flora, and 63 are shared by both floras. Within the "jetty" flora, 16% of the taxa reach their southern limit of distribution in North Carolina, 38% their northern limit, and 40% have no boundary in the Carolinas. Within the "shelf" flora, 3% reach their southern limit in the Carolinas, 72% their northern limit, and 20% have no boundary in the Carolinas according to these authors. Many of the taxa reported from the "shelf" and "jetty" assemblages of North Carolina have not been reported from South Carolina; nevertheless, most probably will be recorded there eventually. For example, the tropical green alga, Arainvillea longicaulis, cited by Schneider (1976) from North Carolina recently has been collected offshore of South Carolina and the red alga, Halymenia floridana, reported from North Carolina has been collected in abundance from a coquina outcrop at Myrtle Beach (R. L. Wiseman, 1978, College of Charleston, Charleston, unpubl. data). The subtidal algae of the latter habitat have never been surveyed completely. Preliminary collections from the coquina outcrop have revealed that the dominant subtidal species of the surf zone are Corallina officinalis and Gracilaria foliifera (R. L. Wiseman, 1978, College of Charleston, Charleston, unpubl. data). Most of the shallow water (less than 6 m or 19.7 ft) species reported by Williams (1951) from Little River, South Carolina, probably inhabit this extensive outcrop at Myrtle Beach. The submerged, subtidal parts of the jetties now being constructed at Murrells Inlet should be colonized by a variety of species, some perhaps new to South Carolina.

3. Vascular Flora

Marine flowering plants are generally absent from the waters of the Sea Island Coastal Region. Although Small (1933) described the range of eel grass as "Florida to Newfoundland" (as have others), Radford et al. (1968) reported no marine flowering plants for South Carolina. However, Curtis (1972) listed eel grass from the Santee and Cooper rivers, but this record has not been confirmed. Radford et al. (1968) reported a herbarium record of eel grass from Georgia, but J. R. Bozeman and J. Phillips (1979, Georgia Department of Natural Resources, Brunswick, per. comm.) have confirmed that eel grass does not occur in Georgia.

Strandings of non-native flora along southeastern beaches are common and may account for some of the discrepancies in the presence or absence of herbarium records. For example, A. E. Radford (1978, University of North Carolina, Chapel Hill, pers. comm.) reported a specimen of turtle grass from the North Carolina coast, although turtle grass does not grow north of Florida. The absence of beds of marine flowering plants from the Sea Island Coastal Region is probably due to the fact that no protected shallow sounds (such as those in northeastern North Carolina) are present.

Widgeon grass has been considered a marine flowering plant by some authors. However, in the Sea Island Coastal Region, it is found in brackish waters and is dealt with under the Estuarine Ecosystem (see Chapter Four).

4. Detritus

Detritus is discussed here because it is basically surplus primary production produced elsewhere and enriched with a variety of heterotrophic/decomposer micro- and macro-organisms. Such surplus production arises principally in the vast estuarine wetlands of this region and is "imported" into the nearshore waters via the numerous estuaries and inlets which cut the South Carolina-Georgia coast. The importance of detritus in marine food webs decreases rapidly with distance from shore, but it is probably quite significant, especially for benthic organisms, within the 3-mi (4.8 km) zone considered here. C. D. Harris (1979,

Georgia Department of Natural Resources, Brunswick, pers. comm.) documented the presence of Spartina alterniflora detritus (stems) on live bottom areas off Georgia.

For convenience sake, we may also include with detritus the complex organic aggregates commonly found in near-surface neritic waters and known as "marine snow." Such aggregates have not been studied in the Georgia Bight. Off the California coast, these aggregates are generally >0.5 mm in size and typically contain highly enriched populations of diatoms, dinoflagellates, microzooplankters, fecal pellets, and other waste materials compared to the surrounding water (Silver et al. 1978). Such aggregates tend to concentrate micro food resources such as fecal pellets and other waste material (e.g., crustacean exoskeletons) into larger particles, and to increase the retention time of such particles in near-surface waters and thus their availability to potential pelagic grazers (Silver et al. 1978). In so doing, they may also reduce the amounts and rates of food sinking from the euphotic zone to the benthos.

C. CONSUMERS

1. Zooplankton

The zooplankton community of near-shore marine waters is quite diverse compared to other marine habitats. Zooplankton abundance tends to be greater in nearshore, shallow waters and to decrease with increasing depth across the shelf (Turner et al. 1979). Grice and Hart (1962) indicated that relative ratios of mean numbers of zooplankters in western Atlantic habitats approximate the following: 30 (shelf): 6 (Gulf Stream): 2 (slope): 1 (Sargasso Sea). Numbers of zooplankters reported from nearshore waters off South Carolina and Georgia are nearly identical to those found in similar habitats off New York by Grice and Hart (1962) (see Anderson et al. 1956a, b; and Anderson and Gehringer 1957a, b, 1958a, b, 1959a, b, c). Nutrients washed (outwelled) from adjacent lands and estuaries and upwelled from deep ocean layers support abundant phytoplankton populations, which form the base of the food chain. The zooplankton includes herbivores, detritivores, and carnivores, linking producers to pelagic secondary and tertiary consumers. Zooplankton metabolic wastes are decomposed and utilized by phytoplankton; fecal pellets and carcasses of zooplankters sink to the bottom where they are utilized by benthic forms.

In contrast to oceanic and freshwater zooplankton, that of neritic waters contains a large meroplanktonic component, consisting of those forms that occur in the plankton for only a portion of their life cycles (e.g., egg and larval stages). The meroplankton is most abundant

relatively near shore (Reeve 1975), and it is highly diverse, consisting chiefly of the larval stages of benthic invertebrates such as crustaceans, echinoderms, mollusks, and polychaetes, the eggs and larvae of fishes, and hydrozoan medusae.

The species composition of coastal-marine zooplankton may reflect strong influences from contiguous estuaries (Turner et al. 1979). For example, many of the meroplankters are larvae of species (e.g., blue crab, penaeid shrimp, sciaenid fishes) commonly found in estuaries during other phases of their life cycles. Also, estuarine holoplankters may be significant contributors to the biomass of coastal marine habitats in the vicinity of inlets when runoff is great (Bowman 1971).

Calanoid copepods are the most characteristic group of marine holo-plankton in general, and they also dominate coastal marine habitats. Chaetognaths, siphonophores, salps, ctenophores, pteropods, scyphozoans, radiolarians, and dinoflagellates on occasion may dominate the zooplankton in nearshore waters (Raymont 1963). (Dinoflagellates may be considered as members of both phyto- and zooplankton.)

The coastal marine zooplankton of the Sea Island Coastal Region is best known from samples taken during eight cruises by the U.S. Fish and Wildlife Service's M/V Theodore N. Gill (Anderson et al. 1956a, b; Anderson and Gehringer 1957a, b, 1958a, b, 1959a, b, c). Collections were made quarterly during 1953 and 1954 at a number of regular stations in continental shelf waters off South Carolina, Georgia, and Florida. Seven of these stations (no. 23, 34, 35, 36, 44, 45, and 46) were relatively nearshore (6 - 57 km or 3.7 - 35.4 mi) between Georgetown, South Carolina, and Jacksonville, Florida. In addition, Environmental Science and Engineering, Inc., studied the zooplankton of Kings Bay, Georgia (St. Marys River - Cumberland Sound area). This firm sampled six sites just beyond the mouth of the St. Marys River quarterly in 1976 (U.S. Naval Facilities Engineering Command 1977). Zooplankton samples have also been taken offshore beyond the territorial limits of South Carolina and Georgia since 1972 by the Marine Resources Research Institute, South Carolina Wildlife and Marine Resources Department, as part of its continuing MARMAP Program (Powles and Stender 1978). More recently, zooplankton of the Georgia Bight was sampled during a baseline survey supported by the Bureau of Land Management (Texas Instruments, Inc. 1978).

Zooplankters in the Gill collections were generally most numerous nearshore and decreased in abundance with distance offshore. Seasonally, total zooplankton concentrations were generally greatest in summer and least in winter over the entire shelf and at the nearshore stations (Table 2-1; see also Turner et al. 1979). Samples from the St. Marys River mouth also indicated a zooplankton minimum in winter, but maximum numbers were recorded in fall rather than summer (U.S. Naval Facilities Engineering Command 1977). Zooplankton concentrations in the mouth of the St. Marys River generally exceeded those in the <u>Gill</u> samples except in winter.

Copepods were generally the most abundant organisms in samples from the <u>Gill</u> cruises. Only in summer and fall 1954 were they (apparently) exceeded by protozoans. Copepods were generally abundant throughout the year at the nearshore stations sampled by the <u>Gill</u>, but they were somewhat more numerous in summer (Table 2-2). Copepods accounted for more than 50%, in one case 97%, of the total zooplankters in samples taken from the mouth of the St. Marys River (U.S. Naval

Facilities Engineering Command 1977). Planktonic copepods are either herbivorous filter feeders, combine filter feeding and predatory feeding habits, or are strictly predatory. The mouth parts of these copepods are modified according to the mode of feeding. There are six naupliar and five copepodid larval stages which, like the adult, are planktonic.

Bowman (1971) described three major copepod associations from the 1953 <u>Gill</u> cruises: 1) a coastal association of the eurytolerant estuarine species <u>Acartia tonsa</u> and the coastal species <u>Labidocera aestiva</u>; 2) a shelf association of four species (<u>Paracalanus parvus</u>, <u>Centropages furcatus</u>, <u>Eucalanus pileatus</u>, and <u>Temora turbinata</u>); and 3) an oceanic association of an additional seven species (Table 2-3). Common nearshore species included <u>Acartia tonsa</u>, <u>Labidocera aestiva</u>, <u>Paracalanus crassirostris</u>, <u>Paracalanus indicus</u>, <u>P. quasimodo</u>, <u>P. aculeatus</u>, <u>Temora stylifera</u>, <u>T. turbinata</u>, <u>Centropages hamatus</u>, and <u>C. furcatus</u>.

Table 2-1. Total numbers of zooplankters/m^3 taken in 0.5 m plankton nets by M/V <u>Theodore N. Gill</u> at seven inshore stations in the Georgia Bight during 1953 and 1954 (compiled from Anderson et al. 1956a, b; Anderson and Gehringer 1957a, b, 1958a, b, 1959a, b, c).

SEASONS

Winter 1953 (Feb. - Mar.)	1,215	1,084	502	414	217	413	776	660
Spring 1953 (Apr. - May)	360	789	1,689	217	515	255	343	595
Summer 1953 (July - Aug.)	4,422	1,956	1,440	1,529	2,218	2,538	3,606	2,530
Fall 1953 (Oct. - Nov.)	514	504	1,523	2,115	ND[a]	1,331	1,489	1,246
Winter 1954 (Jan. - Feb.)	1,238	1,333	199	653	17	ND	1,221	777
Spring 1954	ND	ND	ND	ND	ND	ND	ND	ND
Summer 1954 (June - July)	1,503	1,387[b]	1,022[b]	2,328	383	4,333	3,501	2,065
Fall 1954 (Nov. - Dec.)	1,029	ND	195[b]	1,867[b]	303[b]	777[b]	808	830

a. ND = no data available from 0.5 m net
b. Numerous radiolaria not counted

Table 2-2. Seasonal abundance of copepods (No./m^3) taken in 0.5 m plankton nets by M/V Theodore N. Gill at seven inshore stations in the Georgia Bight during 1953 and 1954 (compiled from Anderson et al. 1956a, b; Anderson and Gehringer 1957a, b, 1958a, b, 1959a, b, c).

| SEASONS | T. N. GILL NEARSHORE STATIONS | | | | | | | |
	23	34	35	36	44	45	46	MEAN
Winter 1953 (Feb. - Mar.)	1,198	717	448	295	179	397	761	571
Spring 1953 (Apr. - May)	15	125	663	118	168	82	187	194
Summer 1953 (July - Aug.)	526	533	650	753	1,147	1,028	1,731	910
Fall 1953 (Oct. - Nov.)	394	157	650	564	ND[a]	478	790	506
Winter 1954 (Jan. - Feb.)	334	478	58	336	8	ND	741	326
Spring 1954	ND	ND	ND	ND	ND	ND	ND	ND
Summer 1954 (June - July)	326	524	337	640	234	1,060	1,442	652
Fall 1954 (Nov. - Dec.)	853	ND	154	195	88	328	86	284

a. ND = no data available from 0.5 m net.

Acartia tonsa was the most abundant species in the mouth of the St. Marys River (U.S. Naval Facilities Engineering Command 1977). Most of these species are present year around and have also been recorded from the Florida Current (Owre and Foyo 1967); thus, it appears that the calanoid copepod component of the coastal zooplankton has a distinct southern affinity.

In contrast to the situation for calanoid copepods, very little information is available concerning the cyclopoid copepods of the characterization area. In a study of nearshore waters off Beaufort, North Carolina, Sutcliffe (1950) found the following cyclopoids: Oithona spp., Coryaceus spp., and Sapphirina nigromaculata. Several of these probably would be found among the copepod fauna of inshore South Carolina and Georgia waters as well. Atkinson (1975) found Corycella carinata at an inshore station off Charleston, South Carolina. Samples from the mouth of St. Marys River contained Oithona sp. and harpacticoids of the genus Euterpina.

Pierce and Wass (1962) reported the chaetognaths of the Gill collections. Five of 12 species found in these samples were present at nearshore stations.

Sagitta tenuis was most abundant, followed by S. helenae, S. inflata, S. hispida and Krohnitta pacifica. Sagitta hispida, considered by Pierce and Wass to be the best indicator of coastal water, occurred at five of the nearshore stations. Chaetognaths are major predators on other zooplankton forms.

Williams (1965) reported some 220 species of marine decapod crustaceans from the Carolinas, and recent work has increased to 272 the number of decapod species "known or presumed to occur in South Carolina waters from the intertidal zone to the 200 m contour" (Young 1978). The planktonic larval stages of these species contribute significantly to the diverse meroplankton of the characterization area and were generally common in Gill samples except in winter. However, of the meroplanktonic forms collected by the Gill, only the crab larvae, especially of the genus Callinectes, have been studied in any detail (Nichols and Keney 1963). (Some investigators question the reliability of Nichols and Keney's generic identification of Callinectes larvae in view of the poor state of knowledge of larvae of the Portunidae.) Early larval stages of Callinectes were found nearest the coast from May to December, with late

44

Table 2-3. Copepod-water mass associations in waters off the Southeastern.United States (Bowman 1971).

WATER MASS	COPEPOD SPECIES
Coastal	Acartia tonsa Labidocera aestiva
Shelf	Paracalanus parvus Centropages furcatus Eucalanus pileatus Temora tubinata
Oceanic	Lucicutia flavicornis Temora stylifera Paracalanus aculeatus Clausocalanus furcatus Calanus minor Undinula vulgaris Euchaeta marina

stage zoeae collected 20 - 40 mi (32 - 64 km) offshore from July to September. Other crab larvae identified by Nichols and Keney (1963) were Polyonyx sp., Emerita sp., Hepatus sp., Portunus sp., Panopeus sp., Eurypanopeus sp., Neopanope sp., Menippe sp., Rhithropanopeus sp., Pinnotheres sp., Sesarma sp., Uca sp., a representative of the family Leucosiidae, plus a genus resembling Ethusozoea and unidentified specimens. The seasonality and distribution of these larvae were not reported.

More recently, Sandifer and Eldridge (1976) studied the distribution and abundance of penaeid shrimp and scyllarid lobster larvae taken during MARMAP cruises in continental shelf waters between Cape Fear, North Carolina, and Cape Canaveral, Florida. They found Penaeus larvae at relatively few stations and in generally low concentrations during two winter cruises (Table 2-4). These larvae were assumed to be P. a. aztecus on the basis of the time of their occurrence, since peak recruitment of P. a. aztecus postlarvae into estuaries occurs during February and March in this region (Bearden 1961, Williams 1965, Hoese 1973). Distributional data indicated that winter spawning of Penaeus is probably concentrated in the southern half of the shelf area sampled, with one area of high concentration of larvae (656/1000 m³) just north of Cape Canaveral. Larval Penaeus were most numerous (to a maximum of ∿ 1300/1000 m³) and widespread during 1973 in May. These larvae were attributed to P. setiferus since white shrimp spawn during spring and summer along this coast (Calder et al. 1974a). During fall, Penaeus larvae, thought to be P. a. aztecus, were widespread in mid-shelf, especially south

of 32° N Latitude. No summer samples were worked. Overall, Sandifer and Eldridge (1976) reported that all large concentrations of Penaeus larvae (750/1000 m³) were collected in water <70 m (230 ft) deep, and most were taken within 20 km (12.4 mi) of shore and in relatively less saline, shallow (<25 m or 82 ft) waters. Catches of Penaeus larvae also were positively correlated with silicate concentration, but the significance of this was not known. Larvae of other penaeid shrimp genera and of spiny (palinurid and scyllarid) lobsters were widely distributed throughout the study area during all seasons sampled (Sandifer and Eldridge 1976).

Besides contributing heavily to the meroplankton, some decapods are holoplanktonic. Species which occur in the Sea Island Coastal Region include the caridean Leptochela serratorbita (Williams 1965, Young 1978) and two common sergestid shrimp, Acetes americanus and Lucifer faxoni, both of which are abundant in late summer and fall (Williams 1965, Bowman and McCain 1967).

Protozoan groups important in the zooplankton include the dinoflagellates, radiolarians and tintinnids. Some members of this phylum contain photosynthetic pigments and are autotrophic; however, many are predators. Reproduction is often by fission, but complex processes comparable to sexual reproduction frequently occur in the same species (MacGinitie and MacGinitie 1968).

The Gill cruise reports provide the only data available on the coastal planktonic protozoa of the Sea Island Coastal Region. Data given for protozoan plankters in these reports presumably represent minimal estimates of actual numerical abundance

45

Table 2-4. Relative abundance of Penaeus larvae and postlarvae during five cruises in continental shelf waters between Cape Fear, North Carolina, and Cape Canaveral, Florida (P. A. Sandifer and P. J. Eldridge, 1976, South Carolina Marine Resources Division, Charleston, unpubl. data; data subject to revision).

CRUISE/SEASON	TOTAL NO. STATIONS	STATIONS WITH PENAEUS		MEAN NO. PENAEUS/1000 m^3 [a]
		No.	(%)	
WINTER				
Jan. - Feb. 1975	28	3	(10.7)	2
Feb. - Mar. 1973	67	8	(11.9)	11
SPRING				
April 1974	42	14	(33.3)	8
May 1973	42	21	(47.7)	60
FALL				
Oct. - Nov. 1973	27	10	(37.0)	20

a. Calculated as sum of concentrations (No./1000 m^3) for all stations on cruise ÷ number of stations occupied on cruise.

because of the large mesh aperture (0.417 mm) of the nets used and the tendency of very soft bodied species to be extruded. Protozoa were listed in nearly all the Gill collections. Maximum numbers occurred in fall 1954, and in summer 1954 radiolarians were recorded as too numerous to be counted accurately.

Cnidarian medusae (jellyfish) are perhaps the most conspicuous zooplankters. They are relatively large, have venomous stinging cells, and some species (e.g., Stomolophus meleagris) are so overwhelmingly abundant at times that they interfere with commercial fishing activities. The Cnidaria are almost exclusively marine or estuarine, with few freshwater species. All marine planktonic cnidarians belong to the orders Hydrozoa and Scyphozoa. Species of these groups typically exhibit metagenesis, that is, two major life stages. These two stages are an attached polyp, which reproduces asexually, and the sexually reproducing planktonic medusa. Most of the larger, more conspicuous jellyfish are scyphozoans. However, the siphonophoran hydrozoans (which include the Portuguese Man O'War) may measure 30 cm (11.8 in) across the medusa bell.

Calder and Hester (1978) list 44 species of cnidarian medusae from euhaline waters of coastal South Carolina. Kraeuter and Setzler (1975) reported on the seasonal cycle of Scyphozoa and Cubozoa at three locations in coastal Georgia, one of which was in coastal

marine waters off Sapelo Island. Several other medusae were reported by Allwein (1967) from Beaufort, North Carolina; some or all of these may also occur in South Carolina waters, although they have yet to be reported (Calder and Hester 1978). Keiser (1976) found the cabbage head medusa, Stomolophus meleagris, so abundant at times that it clogged nets and reduced fishing time of shrimp boats. Among the larger jellyfish, Chrysaora quinquecirrha, Cyanea capillata and Chiropsalmus quadrumanus are also often abundant in coastal waters of South Carolina and Georgia.

Ctenophores (comb jellies are exclusively marine or estuarine, and, except for the aberrant genus Platyctena, all are planktonic. All species are hermaphroditic and carnivorous. Members of the Class Tentaculata (e.g., Mnemiopsis leidyi) are voracious predators of small crustaceans. Such major zooplankton groups as copepods and cladocerans may be virtually eradicated in areas where these comb jellies occur (Burrell and Van Engel 1976). Beroe, the only genus of Class Nuda, feeds chiefly on the tentaculate ctenophores and may at times be responsible for an increase in the abundance of crustacean zooplankters through its reduction of the tentaculate predator population (Burrell and Van Engel 1976). Mnemiopsis leidyi, M. macradyi and Beroe ovata are present and sometimes abundant in coastal waters of South Carolina (Calder and Burrell 1978).

46

Annelid worms contribute to the zooplankton primarily through their meroplanktonic larvae. In addition, some species are holoplanktic, and some benthic species may be temporarily abundant in the plankton when they move to the surface to spawn. This latter phenomenon has been witnessed in South Carolina coastal waters (V. G. Burrell, Jr., 1978, South Carolina Marine Resources Division, Charleston, pers. comm.). Worms were present throughout the year at coastal Gill stations, and were most abundant in winter and fall 1954.

Planktic mollusks include the holoplanktic heteropods, pteropods, pelagic Janthina, and the tropical nudibranch Phyllistioe, plus meroplanktonic larvae of gastropods and bivalves and the early stages of the benthopelagic cephalopods. The food resources exploited by these planktic mollusks are varied, even within the same group. For example, the shelled pteropods apparently feed on diatoms and protozoans, while the naked forms are highly predacious on other zooplankters. Pteropods were recorded from Gill samples every season of the year, and were most abundant in fall 1953.

Cladocerans are primarily freshwater organisms; however, members of the genera Podon, Evadne and Penilia occur in marine and estuarine environments and frequently are important components of the zooplankton. While none were reported from Gill stations, they are presumably common in coastal waters of the area. Cladocera are chiefly primary and secondary consumers, feeding on algae and protozoa. Eggs are carried in a brood chamber, and development is direct in marine forms. Some species exhibit predictable morphological variations in response to environmental changes.

Planktonic amphipods for the most part belong to the suborder Hyperiidae and frequently are a conspicuous component of continental shelf zooplankton. Some (e.g., Parathemisto sp.) prey on other zooplankters, while a number of species live in close association with cnidarian medusae and salps and apparently feed on them (Bowman and Gruner 1973). Eggs are carried in a brood chamber, and production of several broods a year is common among marine species (Barnes 1963). Amphipods were taken at Gill stations in all cruises and were most abundant during the fall (October - November) cruise.

Only two genera of isopods (Munnopsis and Eurydice) are commonly holoplanktonic, and both are probably predacious. Frequently, however, benthic isopods are swept into the water column and become part of the plankton for a short period. Also, early stages of some parasitic forms are commonly meroplanktonic for a brief period until a suitable host is located. Eggs of marine isopods are brooded and the hatching stage is a postlarva (Davis 1955). Isopods were recorded irregularly at Gill stations in all seasons, but they were never abundant.

Echinoderms were abundant at coastal Gill stations in spring and summer 1954. These probably were the meroplanktonic larvae of coastal forms.

The Urochordata (a subphylum of the Chordata) is divided into three classes of tunicates. Two, the Thaliacea and Larvacea, are strictly marine and planktonic while the third, the Ascidiacea, are sessile as adults but have meroplanktonic "tadpole" larvae. All tunicates are filter feeders. Common genera of the Thaliacea are the solitary forms Salpa and Doliolum and the colonial Pyrosoma. Larvacea are all solitary, and genera likely to be represented in the coastal plankton of the Southeastern United States are Oikopleura, Appendicularia and Kowalewskaria (Barnes 1963). Tunicates were present at inshore Gill stations during all seasons sampled, but were most abundant in spring.

Fish eggs and larvae are seasonally abundant meroplankters. Sciaenids, clupeids, and gadids are among the prominent fish groups found in southeastern coastal waters. Fish eggs and larvae were present in nearly all Gill samples, but were abundant only in spring and summer. This seasonality was very apparent at the inshore stations sampled by the Gill (Table 2-5; see also Turner et al. 1979).

Powles and Stender (1976) reported on ichthyoplankton collected during three MARMAP cruises (winter, spring, and fall 1973) in shelf waters off the Southeastern United States. Samples from both neuston (surface) and obliquely-towed bongo nets were analyzed. Some differences in the species composition of the catches from the two habitats sampled by these devices were noted. Powles and Stender (1976) reported that larval and juvenile fish were more abundant in the neuston during winter, and least numerous in fall. In contrast, the bongo net collections indicated greatest abundance of ichthyoplankton in the spring and least in winter; the summer season was not sampled. These latter results are in good agreement with those obtained from the Gill's 0.5 m net collections.

Powles and Stender (1976) distinguished three basic types of distribution patterns among the ichthyoplankton: shelf, shelf/slope, and slope. Families of fish exhibiting each of these patterns are listed in Table 2-6. These investigators also found that the distributions of larval and juvenile fish in continental shelf waters off South Carolina and

47

Table 2-5. Seasonal abundance of fish eggs and larvae combined (No./m^3) taken in 0.5 m plankton nets by M/V Theodore N. Gill at seven inshore stations in the Georgia Bight during 1953 and 1954 (compiled from Anderson et al. 1956a, b; Anderson and Gehringer 1957a, b, 1958a, b, 1959a, b, c).

	T. N. GILL NEARSHORE STATIONS							
SEASONS	23	34	35	36	44	45	46	MEAN
Winter 1953 (Feb. - Mar.)	2.2	0.1	0.6	0.8	0.3	T[a]	T	0.6
Spring 1953 (Apr. - May)	84.1	11.7	38.7	25.4	20.1	9.9	25.5	30.6
Summer 1953 (July - Aug.)	5.6	49.2	15.1	48.3	35.2	169.1	28.2	50.1
Fall 1953 (Oct. - Nov.)	0.2	0.2	2.9	2.7	ND[b]	0.7	0.5	1.2
Winter 1954 (Jan. - Feb.)	0.2	0.1	0.1	0.6	0.5	ND	0.1	0.3
Spring 1954	ND	ND	ND	ND	ND	ND	ND	ND
Summer 1954 (June - July)	3.8	40.9	5.1	14.4	4.1	36.7	24.0	18.4
Fall 1954 (Nov. - Dec.)	T	ND	0.5	1.4	0.1	0.2	0.3	0.4

a. T = trace; <0.1/m^3.
b. ND = no data available from 0.5 m nets.

Table 2-6. Families of fishes exhibiting various distribution patterns in the ichthyoplankton off the Southeastern United States (adapted from Powles and Stender 1976).

SHELF PATTERN	SHELF/SLOPE PATTERN	SLOPE PATTERN
Sciaenidae	Bothidae	Myctophidae
Gadidae	Carangidae	Scombridae (in spring)
Triglidae	Mugilidae	Labridae (in winter)
Clupeidae	Pomatomidae	
	Serranidae	
	Mullidae (in winter)	
	Gobiidae (in winter)	
	Monacanthidae (in fall)	

Georgia generally exhibited no discernible north-south trends. However, some onshore-offshore patterns were apparent, probably related to the major offshore currents which generally parallel the coastline in this area. Powles and Stender (1976) concluded that "young of some fishes of sizes sampled effectively by the bongo net [e.g., Bothidae, Carangidae, Clupeidae, Sciaenidae, Serranidae] are rarer in the plankton of inshore, runoff-influenced waters than in deeper waters of the shelf."

The coastal marine environment is perhaps the most important zooplankton habitat of the Sea Island Coastal Region. Nearly all species of fish and crustaceans

which support commercial and recreational
fisheries within the area spawn either in
nearshore waters or further offshore, and
then move through the zone (as larvae or
postlarvae) enroute to estuarine nursery
areas (Burrell 1975). Coastal marine zooplankton is also characterized by its
vulnerability to environmental perturbations that occur on land and at sea.
Beaches and estuaries provide broad avenues
and physical transport mechanisms to
communicate man's "additions" to the aquatic
environment's nearshore zone, while prevailing wind and currents help transmit
effects of offshore disasters (e.g., oil
spills) to the highly productive coastal
waters. In addition, estuarine species
contribute heavily to the inshore marine
zooplankton, thereby directly trans-
ferring impacts from man's upland activities to the coastal fauna.

In spite of the importance of coastal
marine zooplankton and its vulnerability,
little is known about the dynamics of the
populations or the effects of perturbations
on these populations. Modest beginnings
have been made in a few of the more
obvious problem areas, although for the
most part outside the characterization
area. An exhaustive review of this
literature is not attempted here, but
several of the most pertinent studies are
discussed.

Perhaps the most ambitious attempt to
determine effects of pollutants on plankton
populations in coastal marine waters is
the Controlled Ecosystem Pollution
Experiment (CEPEX). This interdisciplinary
study utilizing large, transparent,
flexible bags to contain experimental
populations within a surrounding water
body was begun in 1973 in Canadian
(British Columbia) waters. The CEPEX
study design permits the investigation of
the effects of various contaminants on
several trophic levels simultaneously under
nearly natural conditions (Reeve et al.
1976). Because of the nature of the
experiments, the results have been very
complex and often difficult to interpret.
Nevertheless, Reeve et al. (1976) inferred
that increasing concentrations of copper
(initially 10 and 50 ppb) resulted in increased mortality of major zooplankton
groups, but this effect was overshadowed
by influence of zooplankton predators
(primarily jellyfish). However, direct
impact of copper on the jellyfish was
readily apparent. The number of jellyfish
in the 10ppb treatment decreased steadily
after addition of copper, reaching zero
by day 25, while in the 50ppb treatment,
the jellyfish population declined by 89%
immediately following addition and reached
the zero level by day 17. Similar results
were reported by Gibson and Grice (1977).
Laboratory studies with copepods also
demonstrated detrimental effects of copper
at these concentrations (Reeve et al.
1976). Grice et al. (1977) reported no
lasting effect of mercury (as HgCl$_2$) at

initial concentrations of 1 and 5 ppb on
the standing crop or production of
phytoplankton in a CEPEX study. However,
the effect on bacterial heterotrophic
activity was similar to that reported by
Vaccaro et al. (1972) for copper, ". . .
namely an immediate 10-fold reduction in
activity followed by rapid recovery
within five days." They further reported
that calanoid copepods, particularly of
the genus _Pseudocalanus_, were the most
important macrozooplankters in the study.
The higher mercury level resulted in a
reduced concentration of _Pseudocalanus_,
especially of early stages, and failure
of females to molt and produce eggs.

Toxic effects of pollutants dispersed
in open coastal waters have been even
more difficult to demonstrate and evaluate
than in the CEPEX program. Vaccaro et al.
(1972) and Grice et al. (1973) could not
demonstrate a decline in zooplankton
standing crop as a result of acid-iron
waste disposal in New York Bight, even
though laboratory tests showed that acid-
iron waste would kill copepods. The
buffering action of seawater probably reduced toxicity of the acid solution very
rapidly as it was dumped into the sea.
Yet, Longwell (1976) found that at least
80% of mackerel eggs (_Scomber scombrus_)
collected in the New York Bight showed
developmental abnormalities. In Guam,
Tsuda and Grosenbaugh (1977) found decreased numbers of zooplankton in the
vicinity of the outfall of a coastal
secondary treatment sewage plant, in
contrast to a control area nearby. Lee
and Nicol (1977) found coastal zooplankters
less vulnerable to toxic effects of the
water soluble fractions of No. 2 fuel
oil than oceanic species. However, the
response of the coastal species to light
was modified whereas the oceanic species
were apparently unaffected. Conover
(1971) found that spilled Bunker C oil
from the wrecked tanker _Arrow_ had no
deleterious effect on planktonic copepods.
He surmised that copepods may play a
significant role in dispersal and degradation of spilled oil.

Some attention also has been focused
on uptake and retention of agricultural
and industrial compounds by planktonic
organisms in coastal waters. Concentrations of chlorinated hydrocarbons in
coastal zooplankton have been investigated
in the Gulf of Mexico by Giam et al.
(1973). These investigators found DDT and
PCB levels in zooplankton comparable to
those in small whole fish or muscle
tissue of larger fish, with the majority
of nearshore samples containing about
100 μg (wet weight) PCB's. Williams and
Holden (1973) found a gradient in levels
of PCBs, DDT, and dieldrin with distance
from the mouth of a polluted estuary and
reaffirmed the general assumption that
these residues are progressively concentrated in marine food chains. Fowler
et al. (1976a, b, c) conducted a series

of investigations on the concentrations of trace metals, plutonium, and mercury in zooplankton off the coast of Europe. They noted that food chain amplification of heavy metal concentrations generally did not occur in their samples, although a carnivorous amphipod (**Phrosina semilunata**) did concentrate cadmium. Further, they suggested that crustacean plankton may play an important role in downward transport of plutonium in the ocean via the sinking of shed exoskeletons. They further found that methylmercury was retained in zooplankton to a much greater extent (biological half-life of 450 days) than inorganic mercury ($HgCl_2$) (biological half-life of 10 days).

Little has been done to study pollutants other than oil spills that may move from offshore areas into coastal waters. Studies of coastal currents along the Georgia and South Carolina coasts do provide a rough means of predicting movements of pollutants moving inshore (see Volume I, Chapter Five), but the impacts on zooplankton have not been assessed. This is essentially a data gap for the area.

At present, the best information on the effects of environmental alterations result from fishery-related studies. For the most part, these studies concern natural perturbations. Low water temperatures have been implicated in poor shrimp production in coastal waters of South Carolina during some years (C. M. Bearden, 1978, South Carolina Marine Resources Division, Charleston, pers. comm.). During 1977, water temperatures remained below 47°F (8.3°C) for more than 4 consecutive weeks in the Charleston Harbor area, resulting in high mortalities of overwintering white shrimp (Farmer and Whitaker 1978). Salinity is important in the success of some species with meroplanktonic larvae, such as penaeid shrimp (Gunter et al. 1964). Thus, excessive rainfall or drought may affect the populations of such species. Ekman transport has been correlated with harvests of menhaden (W. R. Nelson, 1978, National Marine Fisheries Service, Beaufort, pers. comm.); it appears that a favorable wind regime is responsible for most menhaden larvae being recruited to inshore populations; otherwise they are carried out to sea and lost.

2. Insects

To live successfully in the sea, insects have had to overcome the physical constraints of buoyancy and surface tension, and the physiological problems of respiration and osmoregulation (Cheng 1976). Possibly as a result of these contraints, diversity in the truly oceanic insect community is extremely limited. Only five species of insects, all in the genus **Halobates**, the ocean skaters (Hemiptera), have been successful in

colonizing the open ocean. Much of the following information is based on the recent review of water-striding insects by Anderson and Polhemus (1976).

The ocean skaters form part of the pleuston, the animal community at the air/water interface. The organisms associated with **Halobates** in this community, as well as the chemical and physical characteristics peculiar to the sea surface, have been reviewed by David (1965) and Cheng (1975).

Of the five ocean skaters, four are typically tropical in distribution. Only one species, **Halobates micans**, is likely to occur in the Sea Island Coastal Region. **Halobates micans** is wingless and lives successfully on the surface of the open ocean throughout its life cycle. Elsewhere in the Atlantic Ocean, Cheng (1973) was unable to detect any distinct seasonality in the occurrence of the species.

Where abundant, **H. micans** may play an important role in near-surface marine food chains. All species of **Halobates** are predacious fluid feeders, but food preferences of individual marine waterstrider species are not well known. Cheng (1974) offered various organisms collected from surface waters to **Halobates**, and found them to feed on planktonic crustaceans and fish larvae trapped on the surface film. Ashmole and Ashmole (1967) analyzed stomach samples of various seabirds and found **Halobates** in 68% of the stomach samples, at times constituting as much as 16% by number (7% by volume) of the food items. Also, the high lipid content of ocean skaters suggests that they might be a rich source of food for surface-feeding fishes (Lee and Cheng 1974).

Aeolian (wind-driven) transport and offshore deposition of insects from adjacent landmasses have been described elsewhere as being quite important at times (Bowden and Johnson 1976). Insects are abundant along the South Carolina-Georgia coast, and those blown out to sea may provide appreciable amounts of organic matter to marine food webs in surface and near-surface waters. A number of insects obviously blown seaward from adjacent coastal lands have been observed in neuston samples collected near the Sea Island Coastal Region in recent years (C. A. Barans and B. W. Stender, 1978, South Carolina Marine Resources Division, Charleston, pers. comm.). From a number of accounts summarized by Bowden and Johnson (1976), predominant orders contributing to this nutrient source include Homoptera, Diptera, Hymenoptera, Neuroptera, Coleoptera, Lepidoptera, and Heteroptera.

3. Benthic Invertebrates

As mentioned previously, the coastal marine ecosystem of the characterization area falls entirely within a region characterized as the "turbulent zone" by Day et al. (1971). Sediments consist largely of sand and ground shell, and to a depth of approximately 20 m (65.6 ft) the bottom is ripple-marked due to wave action. According to Day et al. (1971), species characteristic of inshore areas of this zone in North Carolina include the sand dollar Mellita quinquesperforata, the polychaete Magelona papillicornis, the decapod Dissodactylus mellitae, the mollusks Olivella mutica and Spisula raveneli, and the amphipod Platyischnopus sp.

Nearshore benthic communities have been studied in some detail off the coast of Georgia (Frankenberg 1965, 1971, Smith 1971, 1973, Dörjes 1972, 1977, Leiper 1973, Frankenberg and Leiper 1977). Frankenberg (1971) and Frankenberg and Leiper (1977) found significant spatial and temporal variations in the density of benthic invertebrates at stations located in fine sand within 3 mi (4.8 km) of the beach at Sapelo Island, Georgia. During 1973 and 1974, the polychaete Spiophanes bombyx, the bivalve Tellina texana?, and the cumacean Oxyurostylis smithi accounted for over 60% of the animals collected in February and March, but made up less than 10% of those collected from June through November. In contrast, the decapod Pinnixa chaetopterana, the ophiuroid Hemipholis elongata, the polychaete Magelona sp., and the decapod Callianassa sp. made up over 40% of the collection from June through November, but less than 10% from December through May (Frankenberg 1971). Somewhat different patterns were observed in 1969 and 1970. Wide variations in numbers from one sampling period to another were observed by Frankenberg and Leiper (1977); the density of Spiophanes bombyx varied from 1 to 1,464 individuals/m² at a nearshore station. These findings, along with other evidence of temporal changes in community structure (see Boesch et al. 1976a), emphasize the limited utility of short-term surveys as environmental impact or baseline characterizations. Contrary to the belief of some, the community structure of shallow water benthic communities is far from static in temperate areas. Seasonal sampling, preferably over a several-year interval, is necessary for such surveys to be of value in assessing environmental change. Boesch et al. (1976a) found significant variations in benthic community structure at one station sampled over an 11-year period in the York River, Virginia. Large fluctuations were observed in the abundance of constituent species, although the qualitative composition was more persistent. The inconstant nature of the communities they studied led Boesch et al. (1976 a) to emphasize the importance of understanding community dynamics in assessing the causes for observed changes.

The reasons for such variations in benthic communities are not fully understood.

Frankenberg and Leiper (1977) suggested that some of the differences they observed may have been attributable to such factors as breeding aggregation, patchy larval settlement, predation, simple longevity, and seasonal migrations, as well as the difficulty of precisely relocating the sampling sites. Boesch et al. (1976a) observed that physicochemical variables such as salinity, temperature, dissolved oxygen, substrate changes, and pollution are important, as are biological factors affecting reproduction, recruitment, and survival.

In South Carolina, D. M. Knott, D. R. Calder, R. F. Van Dolah (1978, South Carolina Marine Resources Division, unpubl. data) conducted a seasonal investigation of the nearshore macrobenthos in depths from 1 to 5 m (3.3 to 16.4 ft) at Murrells Inlet. An assemblage of 205 taxa, dominated by polychaetes in terms of both species (83) and numbers of individuals, was observed. Community structure varied seasonally as the numerically dominant species underwent wide fluctuations in abundance. The ten most abundant species (Spiophanes bombyx, Protohaustorius nr. deichmannae, Scolelepis squamata, Acanthohaustorius millsi, Tellina sp., Donax variabilis, Ensis directus, Amphipoda–undescribed species, Maldanidae–undetermined, and Bathyporeia parkeri) accounted for 80.2% of the fauna. Animal densities, which ranged from 297 to 35,162 individuals/m², were highest during winter and spring due to heavy recruitment of polychaetes (Spiophanes bombyx, Scolelepis squamata), bivalves (Tellina sp., Donax variabilis, Ensis directus), and certain haustoriid amphipods. Population levels were progressively reduced during summer and autumn.

D. R. Calder, R. F. Van Dolah, and D. M. Knott (1979, South Carolina Marine Resources Division, unpubl. data) conducted an assessment of the macrofauna at 40 stations ranging between 4 - 19 km (2.5 - 11.8 mi) from shore off Charleston during summer 1978. Fine nearshore sands gave way to coarser sediments offshore, with an accompanying change in community composition. Assemblages in the study area were faunistically richer, and were characterized by higher species diversity than expected from previous studies of shelf benthos in the Southeast. The infauna, comprised of more than 500 species, were dominated numerically by polychaetes (37.5%), the cephalochordate Branchiostoma caribaeum (19.6%), amphipods (10.0%), and pelecypods (7.0%). Although B. caribaeum was the single most abundant species, it is a migratory, seasonally abundant organism whose numbers reflected the August sampling period. In terms of species, the fauna was dominated by polychaetes (42.8%), pelecypods (10.8%), gastropods (9.9%), and decapods (9.9%). Species diversity was high, with H' ranging from 3.46 to 6.13. The distribution of individuals among the

51

various species was remarkably even; the 20 most abundant species made up only 56.4% of the total fauna. Faunal densities ranged from 208 to 7,932 individuals/m^2. Although 171 epifaunal or partly epifaunal species were identified in qualitative dredge samples, the sessile biota was typically sparse because of shifting sandy sediments in the area. A few localized patches of large octocorals (Tidanideum frauenfeldii), sponges, and other epibenthic taxa were encountered, but none of the sampling sites constituted "live bottom" areas.

While most studies of macrobenthic community structure are based on counts of individual animals, numerical dominance clearly cannot always be equated with either biomass or functional dominance (Table 2-7). In a study off the Georgia coast, Smith (1971, 1973) observed that numerically dominant crustaceans and polychaetes were less important in overall biomass, which was dominated by echinoderms and mollusks. Dominance ranking based on respiration varied seasonally. Total community respiration was positively correlated with water temperature, and was estimated at 676.6 liters of oxygen/m^2/year. The macrofauna accounted for only 5% - 25% of this amount; the meiofauna and microbenthos (fauna and flora) accounted for 25% - 58%, while bacterial respiration amounted to 30% - 60% of the total. This relative importance of the meiofauna and flora has become recognized widely only during the last decade or so. Collectively, meiofauna are characterized as having relatively low biomass but rapid turnover rates. Gerlach (1971) has estimated that yearly production of meiobenthos in the North Sea is an order of magnitude greater than its biomass. Nevertheless, while a general appreciation of the importance of meiobenthos in the marine environment is now widespread, little is known of the amount of available energy meiobenthic organisms consume or the amount of their production utilized by macrobenthos (Steele 1974). This is definitely an important data gap for the Sea Island Coastal Region, where even descriptive information on the meiobenthos of nearshore shelf waters is lacking.

Coastal waters of South Carolina and Georgia, particularly near mouths of rivers, are characteristically turbid due to transport of organic and inorganic materials out of the estuaries (Odum and de la Cruz 1967, Howard et al. 1972, Dörjes 1977). Feeding types of benthic invertebrates in nearshore waters of Georgia were believed by Leiper (1973) to be correlated with the availability of detritus. The proportion of deposit feeders was highest inshore where detrital material was relatively abundant, while the percentage of suspension feeders increased offshore. Seasonal differences in the relative numbers of these two feeding types were also believed to be correlated with seasonal variations in the export of detritus from saltmarsh areas.

While nearshore infauna is taxonomically rich, Tenore et al. (1978) report that macrofaunal biomass is low over much of the Georgia shelf due to an unfavorable sediment regime and low nutrient levels. They observed that nutrients outwelling into coastal waters from marshes and rivers are inhibited from moving across the shelf by a turbidity front about 20 km (12.4 mi) from shore, and that periodic Gulf Stream intrusions chiefly influence the shelf break. Neither process appears to affect the middle shelf to any appreciable degree.

In a different type of study, McCloskey (1970) investigated the fauna associated with the coral Oculina arbuscula on the north jetty off Charleston Harbor. A total of 63 species was found among coral heads from Charleston (Table 2-8), a low number in comparison with samples from areas studied in North Carolina (Shark Shoal Jetty, Cape Lookout Jetty, Lookout Outcrop). Species diversity was also lowest in the samples from Charleston. This was correlated with the environmental characteristics of the area, including a relatively high degree of pollution compared with the other sampling sites. Community structure of fauna from coral heads was found to be quite stable and predictable.

The Florida-Hatteras shelf is predominantly sandy with varying amounts of calcium carbonate, and it is marked by several long, sinuous shoals including one off Cape Romain, South Carolina (Emery and Uchupi 1972, Milliman et al. 1972). While shelf channels and submarine canyons such as those off the Hatteras-Cape Cod shelf are absent, several types of outcrops known as reefs, hard banks, or "live bottom" areas occur along the Florida-Hatteras shelf. Most of these hard banks occur beyond the 3 mi (4.8 km) limit and are, therefore, outside the scope of this study. These biologically productive areas have been described briefly on pp. 36 - 38.

The Cape Hatteras area has long been recognized as an area of faunal transition between the Virginian region to the north and the Carolinian region to the South. The complexity of this zone as a zoogeographical barrier was demonstrated by Cerame-Vivas and Gray (1966). As expected, they found a Virginian fauna above and a Carolinian fauna below the cape, but they also encountered a Caribbean assemblage offshore. Differences in the extent of the Virginian and Carolinian zones were observed from summer to winter. Off the coasts of South Carolina and Georgia, the fauna is Carolinian inshore, with distinct Caribbean affinities in offshore areas and particularly on hard banks or "live

Table 2-7. Key invertebrate species in subtidal benthic communities off Sapelo Island, Georgia.

	SOURCE	
Frankenberg (1971)	Smith (1973)[a]	Dörjes (1977)
Oxyurostylis smithi	JULY	Callianassa biformis
Spiophanes bombyx	Motomastus sp.	Scoloplos fragilis
Tellina texana (?)	Mellita quinquesperforata	Spiophanes bombyx
Pinnixa chaetopterana	Glycera disbranchiata	Saccoglossus kowalevskii
Hemipholis elongata	Haliactus sp.	Tellina cf. texana
Abra aequalis	Nephtys picta	Hemipholis elongata
Paraphoxus epistomus	OCTOBER	Pectinaria gouldii
Magelona sp.	Callianassa sp.	Oxyurostylis smithi
Callianassa sp.	Pinnoxa chaetopterana	Abra aequalis
Tellina iris	Abra aequalis	Glycera americana
Scoloplos fragilis	Micropholis atra	Notomastus latericeus
Nephtys picta	Magelona sp.	Owenia fusiformis
Solen viridis	JANUARY	Spiochaetopterus oculatus
	Abra aequalis	Edotea montosa
	Notomastus sp.	Capitomastus cf. aciculatus
	Callianassa sp.	Magelona sp.
	Micropholis atra	Arabella iricolor
	Pinnixa chaetopterana	Prionospio cf. cirrifera
	APRIL	Nucula proxima
	Abra aequalis	
	Sabellides oculata	
	Solen viridis	
	Micropholis atra	

a. Based on respiration studies.

bottom" areas. The estuarine fauna, particularly in lower salinities, bears a striking qualitative resemblance to that of the mid-Atlantic region.

4. Nektonic Invertebrates

The invertebrate component of the marine nekton is limited largely to a few species of squids, and to crabs of the family Portunidae. Several species of large scyphozoan jellyfishes are seasonally abundant in waters of the study area, but these are more planktonic than nektonic. Species such as the commercially important penaeid shrimps (see also Chapter Four of Volume III, and Chapter Seven of Volume II) are more benthic

53

Table 2-8. Animals associated with heads of the coral
            Oculina arbuscula from a jetty at the entrance
            of Charleston Harbor, South Carolina (McCloskey
            1970).

SPECIES	TOTAL NUMBERS
Phylum Rhynchocoela	
Nemertean D	4
Nemertean E	1
Phylum Nematoda	
Enoplus communis	8
Prooncholaimus sp.	8
Pseudocella panamaense	37
Anticoma columba	2
Oncholaimus dujardinii	9
Paracanthoncus sp.	2
Polygastrophora sp.	1
Nematoda (undet.)	4
Eurystomina americana	24
Phanoderma sp.	1
Euchromadora sp.	3
Phylum Mollusca	
Odostomia dianthophila	1
Odostomia seminuda	1
Gastropod (undet.)	1
Brachidontes exustus	5
Diplothyra smithi	3
Musculus lateralis	2
Ostrea equestris	9
Phylum Annelida	
Autolytus prolifer	1
Brania sp.	3
Dodecaceria concharum	12
Hydroides dianthus	7
Exogone sp.	22
Hypsicomus phaeotaenia	1
Lepidonotus variabilis	1
Nereis succinea	2
Nereis occidentalis	1
Odontosyllis sp.	14
Polydora sp.	63
Sabella microphthalma	3
Sabellaria vulgaris	4
Syllis gracilis	33
Syllid (undet.)	20
Syllid (undet.)	2
Thelepus setosus	1
Phylum Sipunculida	
Dendrostoma alutaceum	2
Golfingia pellucida	2
Phylum Arthropoda	
Tanystylum orbiculare	1
Ostracod (undet.)	2
Balanus trigonus	1
Balanus venustus	2
Cumacean (undet.)	18
Tanaid (undet.)	1
Jaeropsis coralicola	13
Paracereis tomentosa	1
Caprella geometrica	51
Corophium sp.	164
Eurystheus sp.	3
Lembos smithi	41
Leucothoe sp.	8
Melita sp.	6
Photis pugnator	10

Table 2-8. Concluded

SPECIES	TOTAL NUMBERS
Phylum Arthropoda (Cont.)	
Stenothoe sp.	14
Menippe mercenaria	2
Neopanope sayi	11
Pilumnus sayi	2
Synalpheus fritzmuelleri	2
Phylum Echinodermata	
Axiognathus squamata	10
Phylum Chordata	
Styela plicata	1
Hypsoblennius hentzi	1

than nektonic, but will be discussed briefly here because they are infrequently included in studies of quantitative benthic ecology.

Whitaker (1978) investigated various aspects of the biology of two species of squid, Loligo pealei and L. plei, in shelf waters of the Southeastern United States. L. pealei, having a center of distribution north of Cape Hatteras, was most frequent in waters of 8° - 22°C (46.4° - 71.6° F). This species was generally collected at depths exceeding 100m (328 ft) except in winter and spring, when it occurred in shallower waters. L. plei, a species with more southern affinities, was most abundant in waters above 22°C (71.6°F). It was collected each of the four seasons in waters of less than 50m (164 ft) depth. Diets of the two species consisted principally of crustaceans and chaetognaths for small squids, while larger individuals preyed largely on other squids and fishes. Spawning in L. pealei occurred during winter, while L. plei was believed to spawn in autumn. A third species, Lolliguncula brevis, occurs all year in shallow coastal waters and in estuaries.

Penaeid shrimps (Penaeus setiferus, P. aztecus) constitute the principal fishery resource of Georgia and South Carolina (Calder et al. 1974b). Spawning in white shrimp (P. setiferus) occurs in nearshore waters along the coast. It begins in April (Georgia) or May (South Carolina) and continues into September (Lindner and Anderson 1956). Little is known about the months or sites of spawning in brown shrimp (P. aztecus aztecus), although postlarvae enter the estuaries in greatest numbers during winter. Larvae and early postlarvae of both species are planktonic and migrate from coastal waters into the estuarine nursery grounds. Once in the estuaries, they congregate in shallow waters and grow rapidly. Juveniles eventually move into

deeper waters of the estuary before returning to the ocean. Penaeids are known to undergo migrations along the coast in addition to their inshore and offshore movements. Brown shrimp disappear completely from the coasts of South Carolina and Georgia during autumn. In contrast, most of the spawning stock of P. setiferus consists of overwintering, nonmigratory shrimp. This coastal system is thus linked to the estuaries by migrating subsystems (Odum et al. 1974), such as shrimp and other species of invertebrates as well as fishes that use the estuaries as nursery grounds. A third species of penaeid, the pink shrimp (P. duorarum), is frequent in the study area but is of major commercial importance on this coast only in North Carolina. In addition, other species of penaeid shrimps, principally Sicyonia spp., Trachypenaeus constrictus, and Xiphopenaeus kroyeri, are probably also important in the characterization area (P. A. Sandifer, 1978, South Carolina Marine Resources Division, pers. comm.).

In studies of the incidental catch of the shrimp fishery in coastal waters of South Carolina, Keiser (1976) reported that benthic and nektonic invertebrates, other than shrimp, comprised 4% by weight of trawl samples. Of these other invertebrates, miscellaneous crustaceans comprised about 40% by weight. According to Keiser, blue crabs (Callinectes sapidus) comprised a significant portion of the crustacean catch, but accurate estimates of their quantity were impossible because many active specimens escaped as the trawls were emptied on deck. This was believed to have greatly decreased the total invertebrate component of the samples taken for study. In addition to C. sapidus, other common nektonic decapods recorded by Keiser from trawl catches included the portunids Arenaeus cribrarius, C. similis and C. ornatus, Ovalipes ocellatus, Portunus gibbesii, and P. spinimanus. Two species of squid, Loligo pealei and Lolliguncula brevis, were reported by Hoese (1973) to be abundant off the beach at Sapelo Island, Georgia, while _Loligo plei_ was the most common squid in offshore waters. The same portunids reported in catches from South Carolina by Keiser (1976) appeared to be common in Georgia as well. The scyphomedusa _Stomolophus meleagris_ accounted for 15% by weight of Keiser's invertebrate catch. These medusae are sporadic in occurrence, but were particularly abundant in 1974, the first year of Keiser's 2-year study.

Anderson et al. (1977) reported a number of swimming invertebrates from seine catches in the surf zone off Folly Beach, South Carolina. Of these, the crab _Arenaeus cribrarius_ was most abundant. Also collected were _Penaeus setiferus_, _Ovalipes ocellatus_, _Callinectes sapidus_, _Palaemonetes pugio_, _Portunus anceps_, _Acetes americanus carolinae_, _Emerita talpoida_, _Nereis succinea_, _Squilla empusa_, and an unidentified scyphozoan. Several of these may be regarded as more benthic or planktonic than nektonic. Eight other invertebrate species were represented in the samples by a single specimen.

Hoese (1973) observed that striking changes occur in the fauna of inshore and offshore areas. The inshore fauna is warm-temperate while the immediate offshore fauna is somewhat subtropical, particularly during warmer periods of the year. Pelagic species were generally more widely distributed over these areas than benthic forms, and their distribution appears to be correlated primarily with water masses.

5. Fishes

The coastal marine waters of the Sea Island Coastal Region provide a diversity of habitat types for a variety of fishes. The majority of fishes found in the study area are of two general categories: 1) resident species which spend their entire lives in nearshore or estuarine environments (e.g., killifishes, silversides, and bay anchovy), and 2) seasonal migrants which utilize the area during only a part of their life cycle (e.g., mullets, menhaden, and many sciaenids). The first group of fishes is restricted to nearshore and estuarine waters, while the latter (which is the predominant group in terms of numbers of species and individuals) generally spawns offshore, moving into the coastal zone as larvae or postlarvae.

Estuaries provide valuable habitat, nursery areas rich in food, and refuge from predators. Vast numbers of young-of-the-year, motile species are found in the estuarine zone, moving seaward in response to physiological and environmental changes, especially during the summer and fall. Like penaeid shrimp, such fishes are "migrating subsystems"

(Odum et al. 1974) linking the marine and estuarine environments. The major types of habitat supporting marine and estuarine fishes within the Sea Island Coastal Region are the following: subtidal marine waters; intertidal marine surf zone; subtidal estuarine waters (tidal rivers, channels, creeks); intertidal estuarine bottoms including flats, marshes, and oyster reefs; and salt or brackish water impoundments. Most common marine fish species are not confined to one habitat type and may occur in several habitats on a seasonal basis or at different stages in their life cycles. For further information on estuarine species, see Chapter Four.

Those species which have adapted to the rigors of highly variable salinities, temperatures, and habitat types are generally the most successful in coastal marine waters. Some euryhaline species, such as several sciaenids, menhaden, mullet, and others are found up-river in fresh water at times, especially as juveniles. Although no all-inclusive list of fish species occurring in marine waters of South Carolina and Georgia is given here, Table 2-9 presents the characteristic habitats of some of the more important species.

Many of the previous and current surveys of marine and estuarine fishes in the Sea Island Coastal Region have been based on sampling by otter trawl, a biased sampling device which primarily takes demersal species. A considerable amount of future work, including sampling with different gear types and within various habitats, will be required to provide a more thorough understanding of the distribution and occurrence, diversity, abundance, biology, and trophic relationships of the marine and estuarine fishes. The following discussion of the fishes of the subtidal marine system, however, is based primarily on data from trawl studies, since such data are the most abundant and comparable. Predominant fishes taken by trawling in the marine subtidal habitat are listed in Table 2-10.

According to Keiser (1976), sciaenids (especially spot, kingfish, Atlantic croaker, and star drum), Atlantic menhaden, anchovies, and gadids were by far the predominant bottom fishes caught by shrimp trawlers in coastal marine waters of the Sea Island Coastal Region (Table 2-11, Fig. 2-10). Similarly, the predominant species taken in trawls off the Georgia coast by Mahood et al. (1974a, b, c, d) were star drum, Atlantic croaker, spot, sea catfish, weakfish, Atlantic menhaden, blackcheek tonguefish, southern kingfish, and silver perch.

Most of the fishes comprising the subtidal marine community are transient or migratory species which spawn offshore, migrate to estuaries as juveniles,

56

Table 2-9. Characteristic habitats of some of the more important marine and estuarine fishes occurring in South Carolina and Georgia. -Rare, *Uncommon, **Common, ***Abundant

HABITAT TYPE

SPECIES	SUBTIDAL MARINE	SURF ZONE	SUBTIDAL ESTUARINE	INTERTIDAL ESTUARINE	IMPOUNDMENTS
Carcharhinus spp.	***	**	**	*	-
Sphyrna spp.	**	*	**	*	-
Spiny dogfish	***	*	**		
Clearnose skate	**	-	*	-	
Atlantic stingray	*	*	**	**	-
Atlantic sturgeon	*	-	**	**	***
Longnose gar	*	-	**	**	-
Ladyfish	**	*	**	**	**
American eel	**	*	**	*	-
Atlantic menhaden	***	*	***	**	***
Dorosoma spp.	*	-	***	**	*
Atlantic thread herring	***	*	***	**	*
Alosa spp.	**	*	**	*	
Carp	-		*	**	**
Bay anchovy	**	***	***	***	***
Mummichog	*	*	**	***	***
Striped killifish	-	***	*	**	***
Sailfin molly	-	*	*	**	**
Sheepshead minnow	-	**	*	***	**
Mosquitofish	-	*	*	*	
White catfish	-	-	***	*	***
Sea catfish	***	*	***	***	-
Oyster toadfish	**	*	**	*	-
Urophycis spp.	***	*	***		
Menidia spp.	*	***	*	**	**
Morone spp.	-	*	**	**	*
Bluefish	**	*	**	**	
Spanish mackerel	***	*	*	*	-
Atlantic bumper	**	*	*	*	*
Trachinotus spp.	**	***	**	**	**
Caranx spp.	**	**	**	*	*
Eucinostomus spp.	*	*	*	**	**
Pigfish	**	-	**	**	*
Pinfish	**	*	**	**	**
Silver perch	**	*	**	**	**
Sheepshead	**	*	***	***	***
Weakfish	*	*	**	***	***
Banded drum	***	*	***	**	**
Sp ѣ	**	*	**	*	-
Southern kingfish	***	**	***	**	***
Gulf kingfish	***	*	*	**	-
Atlantic croaker	*	*	**	-	-
Black drum	***	**	**	**	**
Star drum	***	*	***	*	-

57

Table 2-9. Concluded

| | HABITAT TYPE | | | | |
SPECIES	SUBTIDAL MARINE	SURF ZONE	SUBTIDAL ESTUARINE	INTERTIDAL ESTUARINE	IMPOUNDMENTS
Red drum	**	**	**	**	**
Mugil spp.	**	***	***	***	***
Southern stargazer	*	*	**	*	-
Hypsoblennius spp.	*	-	**	*	*
Gobionellus spp.	*	-	**	**	*
Gobiosoma spp.	***	-	**	-	-
Atlantic cutlassfish	**	*	**	-	-
Peprilus spp.	***	*	**	*	-
Prionotus spp.	**	*	**	-	-
Windowpane	***	-	***	*	*
Citharichthys spp.	***	*	***	*	*
Fringed flounder	**	*	***	***	**
Paralichthys spp.	***	*	***	*	*
Hogchoker	***	*	***	*	*
Blackcheek tonguefish	***	*	***	-	*
Planehead filefish	***	*	***	-	-
Northern puffer	**	*	**	*	-
Striped burrfish	**	*	**	**	*

58

Table 2-10. Common fishes taken by trawling in nearshore marine waters of South Carolina and Georgia.

FAMILY AND SCIENTIFIC NAME	COMMON NAME
Carcharhinidae	
Mustelus canis	Smooth dogfish
Rhizoprionodon terraenovae	Atlantic sharpnose shark
Squalidae	
Squalus acanthias	Spiny dogfish
Rajidae	
Raja eglanteria	Clearnose skate
Dasyatidae	
Dasyatis sabina	Atlantic stingray
Gymnura micrura	Smooth butterfly ray
Clupeidae	
Brevoortia tyrannus	Atlantic menhaden
Opisthonema oglinum	Atlantic thread herring
Engraulidae	
Anchoa mitchilli	Bay anchovy
Anchoa hepsetus	Striped anchovy
Synodontidae	
Synodus foetens	Inshore lizardfish
Ariidae	
Arius felis	Sea catfish
Bagre marinus	Gafftopsail catfish
Gadidae	
Urophycis regius	Spotted hake
Urophycis floridanus	Southern hake
Urophycis earlli	Carolina hake
Carangidae	
Chloroscombrus chrysurus	Atlantic bumper
Selene vomer	Lookdown
Caranx hippos	Crevalle jack
Sciaenidae	
Stellifer lanceolatus	Star drum
Leiostomus xanthurus	Spot
Micropogonias undulatus	Atlantic croaker
Menticirrhus spp.	Kingfishes
Cynoscion spp.	Seatrouts, weakfishes
Larimus faciatus	Banded drum
Scombridae	
Scomberomorus maculatus	Spanish mackerel
Scomberomorus cavalla	King mackerel

Table 2-10. Concluded

FAMILY AND SCIENTIFIC NAME	COMMON NAME
Stromatéidae	
Peprilus alepidotus	Harvestfish
Peprilus triacanthus	Butterfish
Triglidae	
Prionotus spp.	Searobins
Bothidae	
Paralichthys spp.	Flounders
Citharichthys spilopterus	Bay whiff
Etropus crossotus	Fringed flounder
Soleidae	
Trinectes maculatus	Hogchoker
Cynoglossidae	
Symphurus plagiusa	Blackcheek tonguefish

Table 2-11. Comparison of the 10 most abundant fish species (by weight) of trawl samples off South Carolina and Georgia (Keiser 1976).

Fish species composition of South Carolina shrimp trawl samples May to December 1974 and May to mid-August 1975

RANK	SPECIES	PERCENT BY WEIGHT OF DISCARD
1	Spot	40.2
2	Atlantic menhaden	10.3
3	Atlantic croaker	8.8
4	Star drum	5.0
5	Southern kingfish	5.0
6	Sea catfish	3.4
7	Weakfish	3.0
8	Cownose ray	2.2
9	Spanish mackerel	2.1
10	Banded drum	1.9
	TOTAL PERCENT	81.9

Fish species composition of Georgia shrimp trawl from July 1969 to June 1971 (Knowlton 1972)

1	Spot	28.0
2	Atlantic croaker	20.9
3	Whitings (kingfish)	8.9
4	Atlantic menhaden	7.0
5	Weakfish	6.9

Table 2-11. Concluded

RANK	SPECIES	PERCENT BY WEIGHT OF DISCARD
6	Star drum	4.6
7	Stingrays	3.6
8	Sea catfish	3.3
9	Banded drum	3.2
10	Atlantic cutlassfish	2.8
	TOTAL PERCENT	89.2

and return to oceanic waters as sub-adults or adults. Some, such as the mackerels and pompanos, are coastal migrants found here only during the warmer months. Only a few species (e.g., the Gulf kingfish) are found in the nearshore marine habitat during most or all of their life cycle.

Previous trawl investigations have indicated that the biomass of fishes within coastal sounds and estuaries is much higher than in the nearshore subtidal marine area (Hoese 1973, Mahood et al. 1974a, b, c, d). In a 3-year study along the coast of Georgia, Mahood et al. (1974a, b, c, d) reported that only 16% of all fish taken in trawl samples were caught in ocean waters, whereas 84% of the total catch came from within the sounds and tidal creeks. It should be pointed out, however, that otter trawls collect slower moving demersal species primarily. Thus, the trawl surveys may not give a true indication of the actual biomass of fishes. For example, in gill net sampling along the Georgia coast, Mahood (1974a, b, c, d) found that 52% of fishes collected were taken in the inshore ocean waters, versus 24% for sounds and 24% in the tidal creeks. Unfortunately, almost no other investigations of fishes of the subtidal marine habitat using collecting gear other than otter trawl have been conducted. Data on populations of coastal migratory species such as Spanish mackerel, bluefish, sharks, menhaden, thread herring, and smaller schooling fishes (e.g., anchovies) commonly occurring in the subtidal marine habitat are lacking, as is information on the abundance, etc., of species such as adult spot and mullet, which travel in huge schools close to the beaches at certain times of the year.

Although much of the marine subtidal habitat within the Sea Island Coastal Region is typified by a relatively flat bottom of sand or sand-mud mixtures, specialized habitats in the form of jetties and groins, shipwrecks, rocks, buoys, and man-made artificial reefs exist at some locations. Structures such as jetties and artificial fishing reefs provide valuable additional habitat for marine fishes, and are important both for food production and cover. The sunken materials become encrusted with organisms such as barnacles, algae, sponges, and bryozoans and are eventually populated by a variety of resident and transient fish species. In the nearshore marine area, fishes commonly associated with these types of habitat include black sea bass, sheepshead, oyster toadfish, bluefish, flounders, blennies, gobies, juvenile snappers, groupers, Atlantic spadefish, cobia, tripletail, and sciaenids such as black drum and spotted seatrout. No detailed investigations of the fish assemblages of these habitats and their trophic relationships are known to have been conducted along the South Carolina and Georgia coasts. However, some information is available for artificial reef habitats further offshore in both States (South Carolina Marine Resources Division, 1978, Charleston, unpubl. data; Georgia Department of Natural Resources, 1978, Brunswick, unpubl. data).

Little information is available on the trophic relationships of fishes specifically in the subtidal marine habitat of the Sea Island Coastal Region. Nevertheless, various investigators have studied the food habits of a number of demersal fishes common to both estuaries and coastal marine waters (Dawson 1958, Bearden 1963, Sikora et al. 1972, Stickney et al. 1974, 1975, Heard 1975, Kjelson and Johnson 1976, Stickney 1976). Results of these studies are summarized in Table 2-12, and are discussed in more detail in Chapter Four, Estuarine Ecosystem. Although most of these studies have dealt with specimens collected from within the sounds and tidal creeks, it is likely the food habits of the same life history stage would be similar in nearshore ocean waters.

The trophic relationships of the larger motile fishes occurring in the subtidal marine habitat are little known or understood. These fishes are mostly top carnivores (sharks, mackerels, bluefish, jack crevalle, seatrouts, drums, etc.). Sharks are known to feed largely on demersal fishes and

61

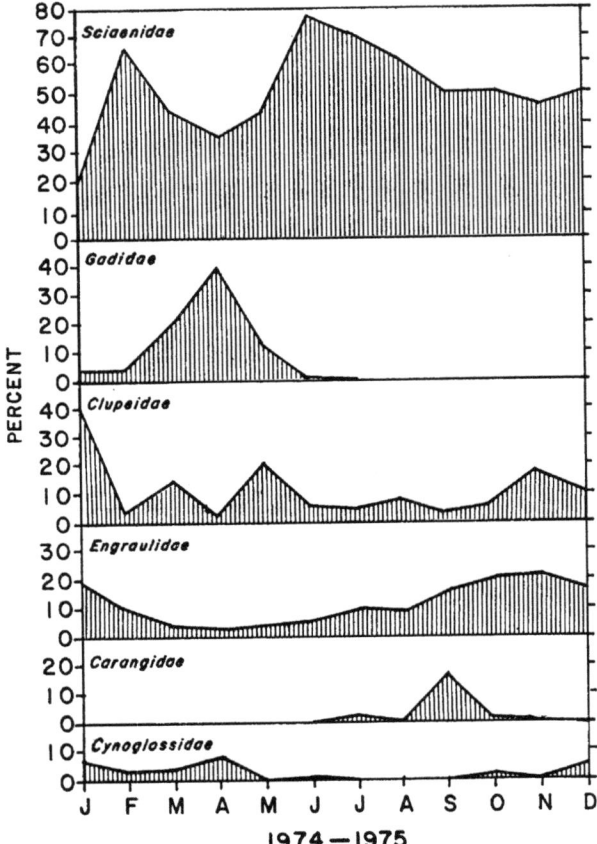

Figure 2-10. Percent contribution by numbers of six fish families in trawler catch samples in South Carolina (January to April data based on R/V Carolina Pride catches; May to December data based on shrimp trawler catches) (Keiser 1976).

Table 2-12. Trophic levels of some of the more important marine and estuarine fishes occurring within the Sea Island Coastal Region.

SPECIES	MAJOR FOOD ITEMS							
	ORGANIC DETRITUS	VASCULAR PLANTS	ALGAE	PHYTOPLANKTON	ZOOPLANKTON	BENTHIC INVERTEBRATES	INSECTS	FISH
I. Herbivores (predominantly)								
Striped mullet	x	-	x	x	-	x	-	-
Atlantic menhaden	x	-	x	x	x	-	-	-
Sheepshead minnow	x	x	x	-	x	-	x	-
Sailfin molly	x	-	x	-	x	-	x	-
II. Omnivores								
Atlantic sturgeon	-	x	x	-	-	x	x	x
Carp	-	x	x	-	-	x	x	x
Dorosoma spp.	x	-	-	-	x	x	-	x
White catfish	x	x	-	-	x	x	x	x
III. Primary Carnivores								
Bay anchovy	-	-	-	-	x	-	-	-
Atlantic silverside	-	-	-	-	x	-	-	-
Mosquitofish	-	-	-	-	-	-	x	-
IV. Mid Carnivores								
Atlantic stingray	-	-	-	-	-	x	-	x
American eel	-	-	-	-	-	x	x	x
Sea catfish	-	-	-	-	-	x	-	x
Oyster toadfish	-	-	-	-	-	x	-	x
Urophycis spp.	x	-	-	-	x	x	-	x
Mummichog	x	-	x	-	x	x	x	x
Striped killifish	-	-	-	-	x	x	-	x
Atlantic bumper	-	-	-	-	x	-	-	x
Trachinotus spp.	-	-	-	-	x	x	-	x
Peprilus spp.	-	-	-	-	x	x	-	x
Pinfish	-	-	-	-	x	x	-	x
Silver perch	-	-	-	-	x	x	-	x
Banded drum	-	-	-	-	x	x	-	x
Southern kingfish	-	-	-	-	x	x	-	x
Spot	-	-	-	-	x	x	-	x
Atlantic croaker	-	-	-	-	x	x	-	x
Star drum	-	-	-	-	x	x	-	x
Gobionellus spp.	-	-	-	-	x	x	-	x
Hypsoblennius spp.	-	-	-	-	x	x	-	x
Prionotus spp.	-	-	-	-	x	x	-	x
Bay whiff	-	-	-	-	x	x	-	x
Fringed flounder	-	-	-	-	x	x	-	-
Hogchoker	-	-	-	-	=	x	-	-
Blackcheek tonguefish	-	-	-	-		x	-	-

Table 2-12. Concluded

SPECIES	ORGANIC DETRITUS	VASCULAR PLANTS	ALGAE	PHYTOPLANKTON	ZOOPLANKTON	BENTHIC INVERTEBRATES	INSECTS	FISH
V. Top Carnivores[a]								
Carcharhinus spp.	-	-	-	-	-	x	-	x
Longnose gar	-	-	-	-	-	x	-	x
Atlantic needlefish	-	-	-	-	-	x	-	x
Striped bass	-	-	-	-	-	x	-	x
White perch	-	-	-	-	-	x	-	x
Ladyfish	-	-	-	-	-	x	-	x
Bluefish	-	-	-	-	-	x	-	x
Spanish mackerel	-	-	-	-	-	x	-	x
Cynoscion spp.	-	-	-	-	-	x	-	x
Red drum	-	-	-	-	-	x	-	x
Paralichthys spp.	-	-	-	-	-	x	-	x

a. Larval and juvenile stages may function as mid carnivores.

invertebrates in the nearshore marine area, principally the species taken by shrimp trawl (Keiser 1976). The pelagic mackerels, bluefish, and crevalle jack, which migrate in large schools along the coast, feed primarily on midwater and surface forage fishes (e.g., anchovy, thread herring, and menhaden). Larger bottom fishes, such as red drum and spotted seatrout, feed primarily on small fishes, crabs, shrimp, and other benthic invertebrates in the subtidal marine habitat.

The nearshore subtidal marine habitat is subject to influence by freshwater coastal rivers insofar as salinity and turbidity are concerned. Within the Sea Island Coastal Region, distribution of fishes is correlated with water masses affected by river discharge, as well as with bottom sediment types (Hoese 1973). Water temperature has a strong effect on species and diversity within this zone. Many migratory fishes move offshore or southward during the colder months and are replaced by migrants from more northerly waters, including gadids and clupeids. Most elasmobranch fishes leave nearshore waters during the fall and do not return until the following spring. During this period in past years, intrusions of large numbers of spiny dogfish have occurred (Bearden 1965).

The effects of man-induced perturbations on fishes of subtidal marine waters in the Sea Island Coastal Region are poorly known. Investigations of the impact of environmental perturbations on marine and estuarine fishes in South Carolina and Georgia have been limited primarily to studies on pesticide (and occasionally heavy metal) residues (U.S. Environmental Protection Agency 1971, Markin et al. 1974, Reimold and Shealy 1976), and to the effects of dredged sediments on fishes (Stickney 1972, Hoss et al. 1973.). The U.S. Department of Health, Education, and Welfare (1965) conducted investigations of fish kills in Charleston Harbor caused by organophosphorous pesticides and other pollutants. Physical alterations, including dredging of channels and overboard disposal of dredge spoil, take place near major ports, but their possible effects on marine fishes are not known. Commercial fishing, especially shrimp trawling, is carried on intensively in the nearshore marine area during much of the year. Most of the fishes taken during trawling are small demersal forms and are discarded at sea along with the incidental catch of invertebrates. Previous investigations do not indicate that shrimp trawling has had any long-range adverse environmental impact on marine fish populations, either by direct mortalities or by agitation of the bottom and its resulting increases in turbidity. The discarding of large quantities of small fishes and invertebrates would appear likely to have a direct impact on the food chain in the nearshore marine area, and this practice may be quite important trophically for larger carnivores, such as sharks,

bluefish and others.

6. Reptiles

The marine turtles are the only representatives of the Class Reptilia which have been documented as occurring in the coastal marine waters of South Carolina and Georgia (see also reptiles of the subtidal estuarine ecosystem in Chapter Four). These animals are of particular interest because of the alarming decreases in their abundance. The Atlantic leatherback, Kemp's ridley, and Atlantic hawksbill turtles, all considered to be transient species in the Sea Island Coastal Region, are listed as endangered under the 1973 U.S. Endangered Species Act (see Chapter One, Section VI). Caribbean and Florida populations of the Atlantic green turtle, another transient species in South Carolina and Georgia waters, also are listed as endangered. The resident Atlantic loggerhead turtle is considered threatened (U.S. Department of Interior, Fish and Wildlife Service 1979a).

Sea turtles are associated with marine coastal waters during at least three periods of their life cycles (Richardson and Hillestad 1978). Mature females mate in coastal waters (Caldwell 1959) and then move on to sandy beaches where they nest. Later, hatchling turtles emerge from the beach nesting sites and move out into the sea, apparently leaving coastal waters. The turtles then typically reappear in the coastal (feeding) areas as juveniles and subadults during the warmer months.

In spite of the interest in, and concern for, the survival of marine turtles, large gaps exist in our knowledge of their basic biology and ecology. Most studies of the life histories of marine turtles have been conducted on nesting beaches (see for example Caldwell 1962, Bell and Richardson 1978, Hopkins et al. 1978, Richardson and Hillestad 1978, J. Richardson et al. 1978, T. Richardson et al. 1978); unfortunately, such studies provide data only on adult females. In contrast, few data are available on the nektonic phase of female sea turtles' lives, or on adult males and juveniles. Information on early life history (first 1-2 years) and migratory or dispersal mechanisms of sub-adults is particularly scant. Most of the occurrence records for marine turtles in the Sea Island Coastal Region are the result of incidental capture by commercial fishery vessels or strandings of dead animals on beaches (Hillestad et al. 1978, Ulrich 1978).

The period between the time when the hatchling turtle enters the ocean and when it returns to inshore feeding areas is often referred to as the "lost year" (Carr 1967). The work of several investigators has indicated that hatchling turtles move directly offshore from their natal beaches into the influence of offshore

65

current systems (Frick 1976, Fletemeyer 1978). Pen-reared, yearling green turtles released in Florida also exhibited the same movement pattern (Witham 1976). On the basis of tag returns and the acceptability of pelagic coelenterates, ctenophores, and salps as food for marine turtles, Witham (1976) postulated that turtle hatchlings from Florida are transported around the North Atlantic gyre, entering inshore habitats after a year or more of pelagic existence. Bell and Richardson (1978) suggest, on the basis of tagging data from Atlantic loggerhead turtles, that young turtles may migrate from their natal beaches in Georgia northward to Cape Hatteras using the Gulf Stream, and later in the fall move to warmer mid-Atlantic waters (the Sargasso Sea). Hatchling loggerhead turtles in offshore waters are reported to be associated with floating mats of Sargassum weed, which may provide both protection from predators and a source of food (Caldwell 1968, Fletemeyer 1978).

Of the five species of marine turtles considered here, only the Atlantic loggerhead can be considered a resident species in South Carolina and Georgia, and its occurrence in inshore waters is seasonal. Loggerhead turtles are not observed in inshore waters during winter months, but there is evidence that some turtles overwinter in offshore waters of the area. South Carolina snapper fishermen report sighting loggerhead turtles in 20 - 30 fathoms (36.6 - 54.9 m) of water in January and February. Similar sightings during winter in offshore waters of Georgia by divers and commercial fishermen are also reported (J. Williamson, 1978, Savannah Science Museum, Savannah, Georgia, pers. comm.). Recently, loggerhead turtles were found in a "dormant" state, partially buried in the mud of Florida's Canaveral ship channel during winter (L. Ogren, 1977, National Marine Fisheries Service, Panama City, Fla., pers. comm.). It is possible that this type of wintering behavior also may occur in suitable areas of Georgia and South Carolina. However, no evidence of such "dormant" turtles occurring in channels within the Sea Island Coastal Region was found in recent surveys sponsored by the National Marine Fisheries Service.

Atlantic loggerhead turtles are first seen in inshore waters of the characterization area during March and April when mating pairs are sighted in tidal creeks behind the barrier islands (Caldwell 1959). Nesting commences in mid-May and continues through mid-August. The Atlantic loggerhead turtle is the only marine turtle nesting on South Carolina and Georgia beaches, and it utilizes most of the barrier island beaches to a greater or lesser degree. Presently available data on nesting are presented in the Marine Intertidal section of this chapter. In South Carolina inshore waters, Atlantic loggerheads are taken incidentally

by shrimp trawlers from mid-May (normal opening of shrimp season) until at least the end of October (Ulrich 1978). Hillestad et al. (1978) reported incidental capture of Atlantic loggerheads from June through early October in Georgia.

Juvenile Atlantic green turtles (carapace length approximately 30 cm or 11.8 in) have been observed stranded on South Carolina beaches, and specimens have been taken incidentally by shrimp trawlers in this State (G. F. Ulrich, 1978, South Carolina Marine Resources Division, Charleston, pers. comm.). Not over three such specimens have been reported in any one season. Interviews with Georgia shrimpers indicated that small numbers of juvenile green turtles are also captured incidentally by trawlers there (Hillestad et al. 1978). Presently a rare species, the Atlantic green turtle was reported to have been considerably more abundant in past years. Coker (1906) states that the Atlantic green turtle was common at Beaufort, North Carolina, in former years, but by 1906 the capture of an Atlantic green was a rare occurrence. Coker (1906) further stated, "the exhaustive fishery in more southern waters and the despoiling of the nests for eggs doubtless accounts for their present scarcity."

Coker (1906) reported that the Kemp's ridley was common in the Beaufort, North Carolina, area during the warmer months of the early 1900's. It is likely that this turtle was equally common in South Carolina and Georgia waters during this period. It is now considered very rare in South Carolina, understandably so in light of the cataclysmic decline in its population noted during the last 30 years on the breeding grounds near Taumalipas, Mexico.

The Charleston Museum has four specimens of Kemp's ridley, all taken in the vicinity of Charleston Harbor. The most recent of these specimens was obtained in 1954. Acquisition data for a specimen taken in a trawl in 1935 indicated that these turtles were considered reasonably common. According to carapace measurements, all of the Charleston Museum specimens were juveniles. In Georgia, Knepton (1956) reported the occurrence of Kemp's ridley in Chatham, Glynn, and McIntosh counties. Recent occurrences of Kemp's ridley have been documented in Georgia by Hillestad et al. (1978), who state that it is commonly encountered by shrimp trawlers in Georgia's inshore marine and estuarine waters.

The Atlantic leatherback turtle is not abundant in the coastal waters of South Carolina and Georgia. However, data from interviews with shrimp fishermen in these States indicate that the species is observed or captured incidentally with some regularity during spring and summer (Hillestad et al. 1978, Ulrich 1978). The

66

Atlantic leatherback has the most temperate and pelagic distribution of the marine turtles, but its nesting distribution is tropical. In Georgia, Hillestad et al. (1978) reported that 24 Atlantic leatherback turtles were caught by shrimp trawlers during 1976, primarily 3 - 5 km (1.9 - 3.1 mi) offshore during spring. Although considered to be pelagic in nature, an 875 lb (398 kg) female was captured inside Calibogue Sound in September 1975. Interviews with South Carolina shrimpers indicated that at least three Atlantic leatherbacks were caught during the 1976 season, all of which were reportedly released unharmed.

In general, the ecological relationships of marine turtles are poorly understood. The majority of the available information concerns the loggerhead turtle and the following material deals only with this species.

The Atlantic loggerhead turtle is an opportunistic carnivore. In the coastal and estuarine waters occupied by this species in South Carolina and Georgia, it is unlikely that food supply would ever be a limiting factor. Stomachs from dead loggerheads stranded on South Carolina beaches contained anemones, crabs, small sciaenid fishes, penaeid shrimp, and marine gastropods.

After entering the sea, hatchling marine turtles are preyed upon by sea birds and predatory fishes, although direct documentation of this predation is not common. According to a note from the Charleston Museum's files, a young loggerhead turtle was taken from the stomach of a black sea bass captured in 14 fathoms (26 m) offshore of Charleston, South Carolina. Brongersma (1972) reported two loggerhead turtle hatchlings taken from the stomach of a deep water shark, Carcharhinus longimanus, and Witham (1974) reported the recovery of one hatchling Atlantic green and eight Atlantic loggerhead turtles from the stomach of a dolphin, Coryphaena hippurus, 19 km (11.8 mi) east of St. Lucie Inlet, Florida. Predation rates on juvenile and adult loggerhead turtles are unknown, but common observations of turtles with missing or partially missing limbs suggest that large sharks may be a significant mortality source.

Man's greatest impact on marine turtles in the subtidal marine system is undoubtedly the incidental capture and associated mortality caused by commercial fishing. Bullis and Drummond (1978) analyzed the incidental captures of turtles by research vessels operating in the Southeastern United States and Caribbean region from 1950 - 1976. Seven marine turtles (six Atlantic loggerhead and one Atlantic hawksbill) were captured during this period in Faunal Zone II, which includes the offshore areas south of Cape Hatteras to Brunswick, Georgia.

A study initiated in July 1976 in South Carolina collected data on the incidental catch of marine turtles by commercial fishermen (Ulrich 1978). The observed locations of incidental captures of Atlantic loggerhead turtles by shrimp trawlers from July - October 1976 and from 15 June - 30 September 1977 are shown in Figure 2-11. The number of turtles caught per trawling hour averaged 0.037 and 0.040, respectively, for the 1976 and 1977 seasons. Extrapolation of these CPUE (catch per unit effort) values to total estimated trawling effort in South Carolina yields estimated incidental captures of 4,480 turtles in 1976 and 3,199 in 1977. Mortality associated with these captures may have been as high as 806 turtles in 1976 and 1,375 in 1977, based on the maximum observed mortality rate (18% and 43%) in the respective years (Ulrich 1978).

Hillestad et al. (1978) used an interview approach to conduct an incidental turtle catch assessment in Georgia. They obtained catch rates similar to those reported by Ulrich (1978), but their mortality estimate was considerably lower (7.8%). This study determined an average capture of 30.7 turtles per vessel per year. Expansion of this capture rate to a 321-vessel shrimp fleet yields an estimated incidental catch of 9,855 turtles in Georgia in 1976. Based on the 7.9% average mortality rate reported by the vessel captains, Hillestad et al. (1978) estimated minimum mortality due to incidental capture at 778 turtles. These investigators also documented the dramatic correlation between opening of the inshore waters to shrimping and the occurrence of dead turtles on Georgia beaches.

Incidental capture of Atlantic loggerhead turtles by shrimpers is often thought to pose the most serious threat to breeding females. Yet, the majority (88%) of the incidentally captured turtles in Georgia were juveniles (Hillestad et al. 1978). Similarly, Ulrich (1978) reported that adult females made up only 18% and 10% of turtles observed during sampling on-board shrimp vessels in 1976 and 1977, respectively, in South Carolina. Thus, available data indicate that contact by shrimpers with the adult segment of the Atlantic loggerhead turtle populations is quite small. Likewise, hatchling turtles also appear to be excluded from the incidental catch (Hillestad et al. 1978). A gill net fishery for Atlantic sturgeon in the Winyah Bay area of South Carolina also results in a relatively small number of turtle drownings related to incidental capture (Ulrich 1978). Nevertheless, the impact of incidental catch mortality on the continued survival of marine turtles in the Southeastern United States is not presently known because of our lack of information on turtle population sizes, natural survival rates, and annual recruitment.

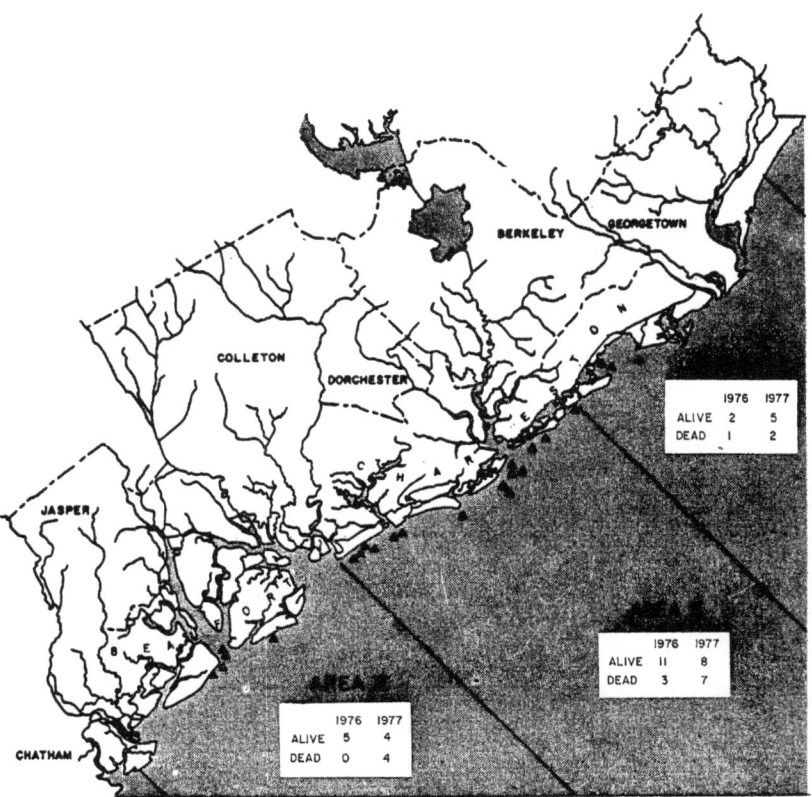

Figure 2-11. Sites of incidental captures of Atlantic loggerhead turtles by shrimp trawlers in South Carolina during 1976 (July – October) and 1977 (June – September) (G. F. Ulrich, 1978, South Carolina Marine Resources Division, Charleston, unpubl. data).

7. Birds

a. Overview. Nearshore waters teem with marine life and provide food for approximately 30 commonly occurring species of marine and pelagic birds, plus a number of rarer species (Audubon Field Notes 1967 - 1970, American Birds 1971 - 1977). Most species utilizing this environment are piscivorous, although omnivorous scavengers and benthic feeders are also widely represented.

Ten species of birds are permanent residents of the coastal marine habitat (Table 2-13). Of these, seven are dominant and characteristic: the double-crested cormorant, eastern brown pelican, Forster's tern, royal tern, herring gull, laughing gull, and ring-billed gull. Although all seven species are common along the coasts of both South Carolina and Georgia, only the royal tern breeds in Georgia, while the brown pelican, the laughing gull, and the royal tern breed in South Carolina (Burleigh 1958, Sprunt and Chamberlain 1970, Forsythe 1978).

Five species -- Audubon's shearwater, the gannet, the greater shearwater, the northern phalarope, and Wilson's petrel -- are primarily pelagic, occurring well offshore much of the time. Concentrations of these species normally are found well seaward of the 3-mi (4.8 km) territorial sea, and their distributions have been linked to location of the Gulf Stream. The abundance of squid and certain fishes associated with the Gulf Stream is thought to have great influence on distribution of Audubon's and greater shearwaters (Murphy 1967, Ashmole 1971). Plankton concentrations associated with the Gulf Stream convergent front also have been suggested as an explanation for the variable distribution of the northern phalarope (Ashmole 1971). These concentrations of zooplankton are normally encountered some 40 - 60 mi (64 - 97 km) from shore and are highly variable. At times, the Gulf Stream front can be found as close as 15 mi (24 km) to the shoreline or as far away as 90 mi (145 km) (MARMAP data records, 1973 - 1976, South Carolina Marine Resources Division, Charleston, unpubl. data). Wass (1974) summarized available data pertaining to pelagic bird species in the South Atlantic. Although a significant number of these birds do occur offshore of the Sea Island Coastal Region, they are outside boundaries of the characterization area and thus will not be considered further here.

As noted previously, 10 species are regarded as permanent residents of sub-tidal marine waters in the characterization area (Table 2-13). However, populations of five of these (the black skimmer, brown pelican, common tern, laughing gull, and royal tern) decrease during winter, with most of the birds moving southward. Correspondingly, populations of the five

remaining species (the caspian tern, double-crested cormorant, Forster's tern, herring gull, and ring-billed gull) increase as additional birds move down the Atlantic coast to winter in South Carolina and Georgia (Burleigh 1958, Sprunt and Chamberlain 1970, Forsythe 1978).

Seasonal variation in the total number of bird species utilizing this habitat is not marked; 22 species overwinter here while 18 occur during summer. Twelve species, including six species of waterfowl, are considered winter residents (Table 2-13). Summer residents (eight species) include three terns which breed in coastal South Carolina and Georgia (Cabot's tern, the gull-billed tern, and the least tern). Altogether, nine species of marine birds breed in coastal South Carolina and Georgia, but they require other habitats to do so.

Little information exists concerning the number of birds that utilize marine coastal waters in the characterization area. Chamberlain and Chamberlain (1975) reported that more than 42,000 individual birds utilized the water off Kiawah Island, South Carolina, during summer. Teal (1959a) recorded several thousand lesser scaup scattered over the water off Doboy Sound. Large concentrations of sea ducks, often numbering in the thousands, are frequently encountered in these waters during winter (Audubon Field Notes 1967 - 1970, American Birds 1971 - 1977). In a study confined to gull species, Forsythe (1973) recorded more than 9,000 individuals patrolling the waters around Charleston in the course of a year. However, to date no serious attempt has been made to quantify marine avian populations off the coasts of South Carolina and Georgia. This should be considered an important data gap.

b. Trophic Relationships. The trophic habits of birds of marine coastal waters are highly varied (Fig. 2-12). Two pelagic species, Wilson's petrel and the northern phalarope, feed on the abundant zooplankton resource of nearshore marine waters (Sprunt and Chamberlain 1949, Van Tyne and Berger 1959, Ashmole 1971), while diving ducks (e.g., the lesser scaup, surf and black scoters, and the canvasback) consume benthic organisms such as mollusks, crustaceans, and worms (Cottam 1939, Sprunt and Chamberlain 1949, Bent 1962a, Johnsgard 1975). However, most birds that utilize coastal marine waters are piscivorous. Fish are taken by aerially diving birds (the gannet, brown pelican, and terns), swimming birds (loons, double-crested cormorant, and diving ducks), and surface skimming birds (the black skimmer, gull-billed tern, and occasionally gulls). At the top of this avian trophic pyramid are the scavengers, represented in the coastal marine habitat primarily by the gulls (Fig. 2-12). These birds consume various dead animal matter, including

69

Table 2-13. Birds of marine subtidal waters of the Sea Island Coastal Region (compiled from Sprunt and Chamberlain 1949, 1970, Audubon Field Notes 1967 - 1970, Chamberlain 1968, American Birds 1971 - 1977, Forsythe 1978).

DOMINANT OR CHARACTERISTIC			
Eastern brown pelican	C	PR	
Double-crested cormorant	C	PR	
Herring gull	C	PR	
Ring-billed gull	C	PR	
Laughing gull	C	PR	
Royal tern	C	PR	
Lesser scaup	C	WR	Oct. - April
Surf scoter	C	WR	Oct. - June
Black scoter	C	WR	Oct. - May
Forster's tern	C	PR	

MODERATE			
Common loon	C	WR	Oct. - April
Gull billed tern	FC	SR	April - Oct.
Common tern	FC	PR	
Least tern	C	SR	March - Oct.
Caspian tern	FC	PR	
Bonaparte's gull	C	WR	Oct. - May
Horned grebe	C	WR	Oct. - March
Gannet	FC	WR	Pelagic
Black tern	FC	SR	May - Sept.

MINOR			
Red-throated loon	FC	WR	Oct. - April
Canvasback	FC	WR	Nov. - March
American goldeneye	FC	WR	Nov. - March
Cabot's tern	FC	SR	April - Oct.
Black skimmer	C	PR	
Ruddy duck	C	WR	Oct. - April
Wilson's petrel	FC	SR	Pelagic
Great black-backed gull	U	WR	Nov. - May
Northern phalarope	FC	SR	Pelagic
Audubon's shearwater	U	SR	Pelagic
Greater shearwater	FC	SR	Pelagic

Note: Dominance indicates relative importance of the species as a group in the community. This concept is not based necessarily on taxonomic relationships but rather on numbers, size, and trophic dynamics.

KEY: C - Common, seen in good numbers
       FC - Fairly common, moderate numbers
       U - Uncommon, small numbers irregularly
       PR - Permanent resident, present year around
       WR - Winter resident
       SR - Summer resident

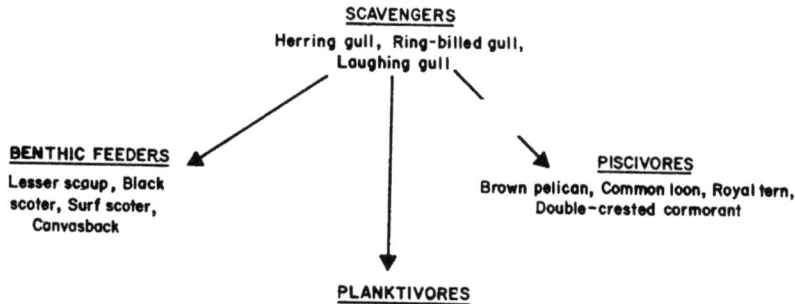

SCAVENGERS
Herring gull, Ring-billed gull,
Laughing gull

BENTHIC FEEDERS
Lesser scaup, Black
scoter, Surf scoter,
Canvasback

PISCIVORES
Brown pelican, Common loon, Royal tern,
Double-crested cormorant

PLANKTIVORES
Wilson's petrel, Northern phalarope

Figure 2-12.   Generalized trophic relationships of representative birds of marine coastal waters
of the Sea Island Coastal Region.

the carrion of other marine birds. Gulls
also may pirate the prey of other birds,
and can themselves be highly predatory in
marine coastal waters. In coastal South
Carolina and Georgia, the laughing gull
preys heavily on young brown pelicans and
the eggs of other marine species (Bent
1963a, Chamberlain and Chamberlain 1975).

     c. Representative Species. Because
of their abundance and important roles in
the coastal marine habitat, brown pelicans,
laughing gulls, royal terns, and black
scoters have been selected for more detailed discussion.

          (1) Brown Pelican. The brown
pelican is a common, broadly distributed
permanent resident of the Sea Island
Coastal Region (Sprunt and Chamberlain
1949, Burleigh 1958, Forsythe 1978). It
frequents most inshore waters and is regularly found in tidal rivers, creeks,
and saline impoundments. Because of its
impressive size and appearance, it is
perhaps the most widely recognized and
best known bird of the marine environment.

     Brown pelicans have not always been
as common along the coasts of South
Carolina and Georgia as they are today.
At the turn of the century, brown pelican
populations suffered heavily from the
commercial collection of its eggs (Wayne
1910). Then, in the post-World War II
period, the species fell victim to the
metabolites of insecticides (DDE and
dieldrin) and PCB's (polychlorinated
biphenyls) which were magnified through
its food web (Schreiber and Risebrough
1972). Reproductive success in South
Carolina dropped to its lowest point

in 1965 (Beckett 1966). This decline was
attributed to eggshell thinning, which
was linked directly to the bird's absorption of DDE (Blus et al. 1974a, b, 1977,
1979). Today, brown pelican population
levels appear to be improving a bit (see
Chapter Three), but the species is still
listed as endangered nationally (U.S.
Department of Interior, Fish and Wildlife
Service 1973, 1979a).

     In feeding habits, the brown pelican
is exclusively piscivorous, relying
heavily on silversides and menhaden in the
study area (Sprunt and Chamberlain 1949,
Burleigh 1958). In other parts of its
range, the species takes small amounts of
commercially valuable fish, but in no way
threatens commercial or recreational
fisheries (Bent 1963a).

     Brown pelicans nest in only two lo-
cations in South Carolina, Deveaux Bank
in the mouth of the North Edisto River
and Marsh Island in the Cape Romain
National Wildlife Refuge. No breeding
populations are known from Georgia. For
details regarding breeding populations of
this species, see Chapter Three (Maritime
Ecosystem, Section VII); additionally,
see Atlas plate 5 for nesting localities
of brown pelicans.

          (2) Laughing Gull. The laugh-
ing gull is a common permanent resident
of the Sea Island Coastal Region, occurring not only in coastal marine waters,
but also in estuaries, tidal rivers and
creeks, and agricultural areas (Sprunt
and Chamberlain 1949, Burleigh 1958,
Forsythe 1978). The species is common
in both South Carolina and Georgia, but

71

no total population figures are available for the area. The breeding population in the Cape Romain National Wildlife Refuge is estimated to be approximately 1,600 pairs (Blus 1977), while recent studies of the Charleston area indicate a breeding population of 3,200 to 6,000 pairs (Forsythe 1973, Chamberlain and Chamberlain 1975). Although present throughout the year, the laughing gull's population is much reduced during winter. Forsythe (1973) found the population low point to occur in February, when as few as 5 - 10 individuals could be observed regularly in the entire Charleston area. There is only one published record of this species occurring in Georgia during the month of January (Burleigh 1958).

The laughing gull consumes a variety of animal matter including rough fish, crustaceans, earthworms, insects, and the young of other birds (Bent 1963a). Food is obtained by scavenging, particularly behind commercial fishing vessels, and through piracy of other birds' prey, especially that of the brown pelican. When the pelican dives for fish, the laughing gull will harass and peck the diver until it releases its prey, which is then consumed by the gull (Bent 1963a, Sprunt and Chamberlain 1970). In other habitats, the laughing gull is insectivorous. It is often seen following plows, where it finds an abundance of grubs, beetles, and grasshoppers (Sprunt and Chamberlain 1970). On the South Carolina breeding grounds, the laughing gull displays a close association with the brown pelican and frequently consumes the eggs and young of the larger bird (Chamberlain and Chamberlain 1975).

The laughing gull is a common breeding bird along the South Carolina coast today, but it was not regarded as such prior to the 1930's (Wayne 1910, Sprunt and Chamberlain 1949). Nesting takes place colonially on selected keys and banks (see Chapter Three, Section VII). According to Burleigh (1958), this species has never been known to breed in Georgia.

(3) Royal Tern. The most abundant sea bird in coastal South Carolina, and a permanent resident, is the royal tern. It is found throughout the South Carolina coastal area, and occasionally ventures inland (Sprunt and Chamberlain 1949). Its population is drastically reduced during winter, but in summer the breeding population is the largest of all South Carolina maritime species (Sprunt and Chamberlain 1949). Breeding occurs on selected keys and banks in the Cape Romain National Wildlife Refuge and south of Charleston on Deveaux Bank. (See Chapter Three, Section VII, and also Atlas plate 5.)

On the Georgia coast, the royal tern is more commonly found in winter and is an uncommon summer resident (Burleigh 1958). However, Teal (1959a) reported this species as common along the beach year around. Breeding in the Georgia population takes place on Blackbeard Island, Oysterbed Island, and Little Egg Island (Burleigh 1958, Kale and Teal 1958).

The royal tern is predominantly piscivorus and obtains food by aerial dives, often from considerable height. Preferred prey organisms include menhaden, silversides, perch, bluefish, shrimp, and squid (Burleigh 1958, Sprunt and Chamberlain 1970).

(4) Black Scoter. The black scoter is a common winter resident in all coastal counties of South Carolina, but this has not always been the case (Forsythe 1978). The first record of its occurrence in South Carolina was not obtained until 1884 (Sprunt and Chamberlain 1949), and Wayne (1910) regarded the species as very rare south of New Jersey. Conversely, the surf scoter was once very abundant in South Carolina, but its population has since declined drastically. Evidently, there has been a winter range shift between these two species (Sprunt and Chamberlain 1949). The black scoter is much less common in Georgia coastal waters, and its occurrences there have been termed accidental (Burleigh 1958).

The black scoter is a benthic feeder, consuming primarily pelecypods, often from a depth of up to 12 m (39.4 ft) (Cottam 1939). Mussels are the most common mollusk eaten, and they appear to be a favorite in South Carolina (Wayne 1910, Cottam 1939, Sprunt and Chamberlain 1970). Other food items include gastropods (typically periwinkles), assorted crustaceans, and small fish (Cottam 1939).

The black scoter is found primarily in large flocks off beaches. In periods of adverse weather, it will also utilize large bays and sounds. While few population estimates are available for the study area, flocks of up to 1,025 individuals are commonly encountered in the Cape Romain area (Audubon Field Notes 1970), and flocks of approximately 12,000 birds have been reported in Bulls Bay, South Carolina. In the southern part of South Carolina, near the Beaufort/Savannah area, flocks with approximately 10,000 birds have been reported in open marine waters (Sprunt and Chamberlain 1970).

d. Impacts. Oil pollution of the sea probably poses man's greatest adverse impact on birds in subtidal marine waters. Such impact has been aggravated in recent years through increases in the fleet of oil tankers and the amount of oil transported by ships. World-wide concern over oil tanker catastrophies has been expressed during the last 25 years (Erickson 1963, Hartung 1967, Goethe 1968).

Although the coast of South Carolina and Georgia has not yet experienced major oil spills, the trend toward exploitation of hydrocarbon resources in outer continental shelf areas, coupled with increased tanker traffic, suggest that it is only a matter of time. Oil pollution can cause severe effects on populations of sea and coastal birds, as well as on their food organisms (e.g., small fish, crustaceans, etc.). The greatest impact on birds is likely to come from accidental discharges of oil during transportation (tanker disasters), drilling, and production; from pipelines; and by intentional discharge from tankers. The probability of one or.more spills greater than 1,000 barrels impacting on pelagic bird rookeries in the South Atlantic has been estimated as approximately 8% - 9% over the production life of any oil leases developed offshore (Table 2-14).

Mass fatalities of aquatic birds are often observed after spills of crude and heavy fuel oils. Bird fatalities were estimated at 3,686 for the Santa Barbara, California, blowout (Straughan 1972), and at 7,000 for the San Francisco spill of Bunker C fuel oil (Chan 1972, Boesch et al. 1974). An estimated 10,000 birds, primarily ducks, died as a result of seepage of a heavy, tar-like oil from a sunken barge in the Chesapeake Bay.

The immediate effect of oil on birds is the fouling of their feathers. Clark (1969) reports that feathers become matted together, and the water repellant and insulation properties are lost, as well as their buoyancy. Some birds become so soaked that they drown (Tuck 1960); others lose the ability to fly (Erickson 1963) or dive for food (Chubb 1954). Hartung (1967) reports that heavily oiled ducks lose more than twice the normal amount of body heat due to the breakdown of the insulating properties of their plummage. To compensate for the additional loss, they develop a very high metabolic rate, which often leads to accelerated starvation. Ingested oil is usually toxic to birds and often results in inflammation of the digestive tract or disturbance of physiological processes (Boesch et al. 1974). One study showed that seawater-adapted mallard ducklings which ingested 12.5 ppm (lowest level tested) of crude oil demonstrated a diminution of intestinal mucosal transfer rate of sodium ions and water, thus diminishing the amount of water available to the organism (Crocker et al. 1974). The study further postulated that dehydration resulting from impairment of the mucosal transfer mechanism may be an important factor contributing to high mortalities among oil-contaminated seabirds.

Not all birds are equally vulnerable to oil slicks. Clark (1971) reports that in Western Europe, auks, puffins, razorbills, murres, and sea ducks suffer the most fatalities as a consequence of spills. Boesch et al. (1974) report that loons and grebes, which accounted for 7% - 10% of the total bird population, suffered 64% of the mortalities resulting from the Santa Barbara spills. Generally, the most vulnerable species are diving birds and others that are attracted by slicks and consequently land on them. Bourne (1968) suggested that oil attracts birds by calming the water or by resembling food, tide rips, or shoaling fish. He also observed that swimming birds do not notice slicks until they come in contact with them and are consequently trapped. While some species (e.g., gulls) fly away, others (e.g., murres) dive beneath the surface. Surfacing in a slick can be fatal, as observed for alcids in the _Torrey Canyon_ spill (Goethe 1968).

The treating of oil-contamined birds has met with very limited success. According to Boesch et al. (1974), most oiled birds perish soon after capture, while others do not survive the cleaning procedures or the following recuperation period. Improper feeding and handling of the treated birds is thought to have been responsible for the failure of early attempts at treatment. Boesch et al. (1974), however, report that established British centers with trained personnel achieve nearly complete survival of treated birds.

The time of day, season, and sea conditions all play important roles in the species of birds that occur in a specific area, and in the probability that they could be impacted by an oil spill. Spills occurring during migration and wintering periods would probably affect the greatest numbers and diversity of birds in the Sea Island Coastal Region. The birds most likely to be impacted are shorebirds, waterfowl, and predatory birds, although other birds may also be affected. The migratory route of the already endangered arctic peregrine falcon extends along the entire length of the Eastern Atlantic Coast. It has been estimated that there would be at least an 89% probability that a major oil spill occurring within the Georgia Bight could impact, at some point, the migration path of this species (Table 2-14). In contrast, the probability that an oil spill in the potential outer continental shelf production area off the Southeastern United States would adversely affect brown pelican rookeries, bald eagle nesting areas, or the dusky seaside sparrow habitat is $\leq$4% (Table 2-14). However, estimates were not given for the probability of such an oil spill impacting the feeding territories of brown pelicans, bald eagles, or other coastal bird species.

Table 2-14. Probabilities of one or more oil spills greater than 1,000 barrels occurring, and impacting biological resources and recreational areas in t South Atlantic area over the production life of the entire lease area. Also, the size of a major spill in relation to the extent of expose resource (U.S. Department of Interior, Bureau of Land Management 1977).

Event/Resource	PROBABILITY (percent)				Ratio o mean spil to extent o expos resources[a]
	Based on pipeline transport to Jacksonville and Savannah	Based on pipeline transport to Charleston and Savannah	Based on tanker transport to Jacksonville and Savannah	Based on tanker transport to Charleston and Savannah	
Probability of coming ashore	93	92	96	95	.04
Beaches	93	92	95	95	.05
Recreation areas (State and Federal)	40	39	45	43	.24
Wildlife refuges	8	8	9	9	.53
Historical sites	29	32	33	33	1.00
Marsh and wetlands	15	16	17	14	.10
Areas of high sedimenta- tion rate	55	55	61	62	.12
Brown pelican rookeries	3	2	4	3	3.33
Coastal or pelagic bird rookeries	8	8	9	9	.37
Bald eagle nesting sites	b	b	b	b	10.00
Dusky seaside sparrow habitat	b	b	b	b	25.00
Arctic peregrine falcon migration routes	89	89	93	93	.04
White, brown, and pink shrimp	62	62	69	69	.07
Royal red shrimp	?	?	2	2	2.50
Commercial fishing grounds	95	95	97	97	.02

Table 2-14. Concluded

Event/Resource	PROBABILITY (percent) Based on pipeline transport to Jacksonville and Savannah	Based on pipeline transport to Charleston and Savannah	Based on tanker transport to Jacksonville and Savannah	Based on tanker transport to Charleston and Savannah	Ratio of mean spills to extent of expose resources[a]
Sport fishing area	4	5	5	5	.71
Commercial scallop grounds	43	44	47	49	.22
Crabs and oysters	53	58	59	65	.06
Bay scallops	6	7	7	7	.53
Sea turtle nesting sites	69	69	75	75	.05

a. Onshore resources compared on the basis of length and offshore resources on the basis of area.
b. Less than 0.5%.

Migratory waterfowl (especially the common loon, canvasback, redhead, whistling swan, mallard, and ruddy duck) are probably among the most susceptible to oil pollution due to their flocking habits and the large number of birds using the Atlantic flyway. Large concentrations of several of these birds are often found in relatively small areas. Thus, if a spill were to affect an area where the birds were congregated, high mortalities might be expected. Entire flocks of these waterfowl could be impacted during the spring and fall migrations. Also, many waterfowl winter in the Sea Island Coastal Region, and they would be susceptible during the winter. If oil should impact these areas, populations could be reduced through loss of habitat or direct fouling. Fortunately, however, most of the major bird refugia are not liable to direct impacts from spilled oil during drilling, production, or transportation activities, as most of them are situated within inlets or are protected by barrier islands.

If a major spill should occur as a result of outer continental shelf development, habitats utilized by shorebirds in the South Atlantic could be impacted. As Erickson (1963) reports, both habitats and food supply are greatly reduced by oil deposits. Many shorebirds rely on intertidal areas which could be affected as a result of a spill.

Despite the massive bird kills often associated with large oil spills, such disasters may actually do less long-term damage to birds than may chronic exposure to low levels of petroleum products. This may become a very significant problem in the future, considering the amounts of petroleum released into the oceans during normal industrial operations. All hydrocarbons do not degrade readily in water, and Murphy (1971) has suggested that the carcinogenic ones may be the more persistent in the marine environment. Further, Stickel and Dieten (1979) have shown clearly that low levels of oil contamination may have a variety of detrimental effects on aquatic birds, especially young stages. They found that very small quantities of crude and fuel oil (5 - 20 µl of oil per egg) caused significant mortalities of mallards, common eider, great black-backed gulls, laughing gulls, Louisiana herons, and sandwich terns; no embryos tested (mallards) survived exposure to 50 µl of oil. Embryos which survived treatment with oil showed an increased incidence of abnormalities when compared with controls, and this effect was intensified by addition of metals (vanadium, nickel, mercury) found in oil. Weathering of the oil for 2 - 4 weeks reduced, but did not eliminate, its toxic effects on embryos. Stickel and Dieten (1979) further considered the possibility that

small quantities of oil might come into contact with the eggs of aquatic birds by transfer from the plumage of adults exposed to oil. This was tested in the laboratory with laughing gulls. The feathers covering the brood patch of each test bird were treated with No. 2 fuel oil; control birds were treated with water. Mortality of embryos from the oiled birds was 41%, compared with 2% for the controls. A similar effect was demonstrated with mallards, where breeding adults were allowed to swim in clean (control) or oil-contaminated water prior to laying eggs. In other experiments, Stickel and Dieten (1979) found that adult waterfowl are able to tolerate relatively high concentrations of oil in their diet as long as they are not otherwise stressed. However, inclusion of oil as 2.5% - 3% of the diet of mallard hens for several months resulted in a 50% - 100% decrease in egg laying. The eggs produced by hens fed oil hatched as well as did control eggs, and the hatchlings weighed as much as did the controls. However, young birds fed oil as 2.5% or 5% of their diet did not develop flight feathers.

Another important impact on birds in the coastal marine waters is the exposure to and accumulation of biocides. This topic is dealt with in detail for brown pelicans in Chapter Three, Section VII. Also, Stickel and Dieten (1979) report that ducks can accumulate petroleum hydrocarbons through food chains.

8. Mammals (see also Chapter One, Section VI)

The marine mammals of the Sea Island Coastal Region are poorly known. What little information exists is derived primarily from strandings of animals in coastal areas and from a relatively few sightings at sea. Since stranding data do not necessarily correlate with species abundance in adjacent waters (Brown 1975), virtually nothing is known about marine mammal population levels or movements in the characterization area. This situation is true for the common bottle-nosed dolphin as well as for the less frequently encountered species.

A total of 25 species of cetaceans, two pinnipeds, and one siren have been recorded from or are likely to occur in the coastal waters of South Carolina and Georgia (Table 2-15). However, most of the cetaceans are known only from very few stranding records (Tables 2-16 and 2-17).

The Atlantic bottle-nosed dolphin is the most common, and only resident, marine mammal in the Sea Island Coastal Region (Johnson et al. 1974, Neuhauser and Ruckdeschel 1978, Sanders 1978). Its known range extends from New England southward to Florida, westward in the

76

Table 2-15. Checklist of marine mammals known, or likely, to occur in coastal waters of South Carolina and Georgia (compiled from Caldwell and Caldwell 1974, Neuhauser and Ruckdeschel 1978, Sanders 1978).

	OCCURRENCE		
SPECIES	SOUTHEASTERN REGION[a]	S.C.	GA.
Cetaceans			
Bottle-nosed dolphin	+		
Bridled dolphin	+	–	
Common (saddleback) dolphin	+	?	
Long-beaked (spinner) dolphin	+	+	–
Rough-toothed dolphin	+	–	+
Spotted dolphin	+	?	+
Striped dolphin	+	+	
Grampus	+	+	
Harbor porpoise	+		
Antillean beaked whale	+	–	
Atlantic (dense-) beaked whale	+	+	
Atlantic right whale	+	+	+
Bryde's whale	+	–	+
Dwarf sperm whale	+	+	+
False killer whale	+	+	+
Fin-backed whale	+	?	–
Goose-beaked whale	+	+	+
Humpback whale	+	+	+
Killer whale	+	+	
Minke whale	+	+	–
Pygmy sperm whale	+	+	+
Sei whale	+	+	–
Short-finned whale	+	+	+
Sperm whale	+	+	
True's beaked whale	+	?	
Pinnipeds			
California sea lion	+	+	+
Harbor seal	+	+	?

Siren
West Indian (Florida) manatee

a. Region between Cape Hatteras, North Carolina, and Cape Canaveral, Florida.

Table 2-16. Occurrence of stranded marine mammals in South Carolina (compiled from Sanders 1978).

Bottle-nosed dolphin	Beaufort	3
	Charleston	13
Common dolphin	Charleston	1
Long-beaked dolphin	Charleston	
Spotted dolphin	Beaufort	1
	Charleston	1

77

Table 2-16. Concluded

SPECIES	STRANDING LOCATION (COUNTY)	NUMBER OF STRANDING EVENTS
Striped dolphin	Charleston	1
Grampus	Charleston	
Atlantic beaked whale	Charleston	1
Atlantic right whale	Charleston	2
	Horry	1
Dwarf sperm whale	Charleston	1
False killer whale	Unknown	1
Fin-backed whale	Georgetown	1 (?)
Goose-beaked whale	Charleston	6
Humpback whale	Charleston	1
Killer whale	Charleston	1
Minke whale	Georgetown	1
Pygmy sperm whale	Beaufort	3
	Charleston	10
	Georgetown	3
	Jasper	1
Sei whale	Charleston	1
Short-finned pilot whale	Beaufort	1
	Charleston	6
	Horry	1
Sperm whale	Beaufort	2
True's beaked whale	Charleston	-

Gulf of Mexico and throughout the West Indies and Caribbean to Venezuela. In the southern portion of the United States, dolphins are found near shore and often enter bays, inlets, and estuaries (Leatherwood et al. 1976), where they are frequently observed feeding in pairs or small bands. Principal food items are shrimp, eels, catfish, menhaden, mullet, and miscellaneous other fishes. Perhaps more is known about the biology of this marine mammal than any other because of its long use in commercial oceanaria. Despite its abundance, however, virtually nothing is known about the actual size of its population, local movements, seasonality, recruitment, etc., in South Carolina and Georgia waters. This is a major data gap for the Sea Island Coastal Region. For details of its status and habits elsewhere, see Lowery (1974) and Odell (1975).

The pygmy sperm whale is perhaps the second most abundant cetacean in the Sea Island Coastal Region, judging from stranding records (Tables 2-16 and 2-17). This apparent relative abundance may simply be due to the fact that it tends to travel closer to shore than do most other cetaceans. The normal range for this species is unknown, since specimens are rarely observed at sea. However, stranded specimens have been found on beaches from Halifax, Nova Scotia, to Cuba, and as far west as Texas in the Gulf of Mexico. This small whale feeds primarily on squid, but it will also take fish (Leatherwood et al. 1976). Stranded specimens have been recorded throughout the year on South Carolina and Georgia beaches (Neuhauser and Ruckdeschel 1978, Sanders 1978). Mortality of this species appears to be greatest during summer (June - August) and least in winter

78

Table 2-17. Occurrence of stranded marine mammals in Georgia (compiled
from Neuhauser and Ruckdeschel 1978).

SPECIES	STRANDING LOCATION (COUNTY)	NUMBER OF STRANDING EVENTS
Bottle-nosed dolphin	Camden	21
	Chatham	10
	Glynn	2
	Liberty	2
	McIntosh	6
Rough-toothed dolphin	Camden	1
Antillean beaked whale	Chatham	1
Atlantic (dense-) beaked whale	Camden	1
Atlantic right whale	Chatham	1
Bryde's whale	Chatham	1
Dwarf sperm whale	Camden	6
	Chatham	1
False killer whale	Chatham	1
Goose-beaked whale	Camden	2
	Chatham	1
	Glynn	1
	Liberty	1
Humpback whale	McIntosh	1
Pygmy sperm whale	Camden	5
	Chatham	6
	Glynn	5
	Liberty	1
	McIntosh	7
Short-finned pilot whale	Camden	2
	Chatham	5
	Glynn	7
	Liberty	1
	McIntosh	1
	?	1

(December - February) (Neuhauser and
Ruckdeschel 1978). No population estimates
are available, and nothing is known about
its status in the characterization area.

The largest cetacean to strand in
numbers in the Sea Island Coastal Region
is the short-finned pilot whale. This
species travels in large schools and,
although there have been only eight recorded strandings in South Carolina,
these have involved about 100 individuals.
A single stranding on Bull Island in
March 1944 involved 65 animals (Sanders
1978). Another mass stranding of some
35 pilot whales occurred on Kiawah Island
in October 1973. In Georgia, mass strandings of pilot whales have been recorded on
several occasions (unknown locations,
1939, ∿24 whales; Richardson

Creek, 1955, 24 - 50 whales; St. Simons
Island, 1962, 15 - 25 whales; Little St.
Simons Island, 1968, 53 whales; Cumberland
Island, 1977, 15 whales) (Neuhauser and
Ruckdeschel 1978). There is as yet no
fully accepted explanation for these
mass strandings. Autopsies of some
of the pilot whales stranded on Kiawah
Island, South Carolina, in 1973 revealed massive nematode infestations in
the middle ear. Whether these infestations
played any causal role in the strandings
is not known. There does appear to be
some seasonality to the strandings,
however, with mass strandings occurring more
frequently during fall and winter, and
individual or small group strandings
occurring more often during spring
(Neuhauser and Ruckdeschel 1978).

79

The goose-beaked whale is known from six strandings in South Carolina and five in Georgia (Table 2-16 and 2-17). However, there is essentially no other information available on this species in the characterization area.

The dwarf sperm whale is known from only a single stranding record in South Carolina, but seven specimens have stranded on Cumberland and Little Cumberland islands, Georgia, since 1971 (Neuhauser and Ruckdeschel 1978). Stranding records are more common in cool weather, but little else is known about the species in Georgia or South Carolina waters.

Spotted dolphins are known from two strandings in South Carolina (Table 2-14). This species apparently prefers deeper offshore waters. It has not been reported stranded in Georgia, but it has been sighted offshore in April, May, and June (Neuhauser and Ruckdeschel 1978). No other information is available concerning the occurrence of this species in the Sea Island Coastal Region.

Several other species of cetaceans have been reported stranded on the South Carolina or Georgia coast on one to a few occasions (Tables 2-16 and 2-17). Virtually nothing else is known about these species in the characterization area.

Management of cetaceans is virtually impossible at the State level, so it has been delegated to national and international agencies such as the National Oceanic and Atmospheric Administration and the International Whaling Commission. Man's over-exploitation of whales for meat and oil is well known. Effects of other factors such as pollutants in the ocean need much further study. As one step to gain more information, the National Marine Fisheries Service established the Southeast Region Marine Mammal Stranding Network in 1978. The purpose of this network is to accumulate reports of strandings and to acquire as much information as possible concerning seasonal distribution, food habits, reproduction, diseases, parasites, accumulation of pollutants, etc., through rapid-response examination of the stranded animals. Additional work, particularly population surveys of healthy animals, is urgently needed.

Two pinnipeds, the California sea lion and the harbor seal, are seen occasionally along the coasts of South Carolina and Georgia. The sea lions were likely released accidentally or intentionally from oceanaria, but the harbor seals are regular, although uncommon, winter residents of the Sea Island Coastal Region (Caldwell et al. 1971, Sanders 1978). No estimates of population size are available for either species anywhere

in the southeast.

Although common in certain areas of coastal Florida, the endangered West Indian (Florida) manatee is seen only rarely in Georgia and South Carolina. Manatees migrate northward through sheltered marine sounds and rivers during the warmer months (Colley 1962). One individual moved up the Santee River, through the locks of Pinopolis Dam and into Lake Moultrie, South Carolina, in June 1965. More recently, a large male was caught in a shrimp trawl off Hilton Head Island, South Carolina, during August 1977. Unfortunately, the specimen drowned before it could be removed from the net. Several days later, a cow and calf were observed for several hours feeding on smooth cordgrass in the marina at Parris Island Marine Depot. The size of the calf indicated that it was probably born in South Carolina waters. Manatees are also occasionally sighted or collected on the Georgia coast (Tomkins 1956, 1958). Cooperative aerial surveys now being conducted annually by the U.S. Fish and Wildlife Service Manatee Recovery Team and wildlife biologists in both States should produce additional information on abundance, distribution, and movement of these marine mammals in Georgia and South Carolina. Some detailed information is already available on the manatee in Florida's Crystal River (Hartman 1969, 1971). For additional information see Chapter One, section VI.

D. DECOMPOSERS AND NUTRIENT REGENERATION

1. Overview

Bacterial and fungi are present in virtually all accumulations of water, from the interstitial waters of sandy beaches and mud flats to the open waters of rivers, lakes, and oceans. Almost all aquatic bacteria and fungi are heterotrophic organisms, complementing aquatic primary producers by the reduction of organic matter and the subsequent recycling of inorganic nutrients. Most aquatic bacteria are saprophytic, a small number are either photoautrophs (i.e., fix carbon in the presence of light; examples, green and purple bacteria) or chemoautotrophs (fix carbon through light - independent chemical reaction; examples, nitro-sulfur and iron bacteria), and only a very few are true parasites. Morphologically, most marine bacteria are cocci or bacilli, but filamentous, band-shaped, and stalked forms are also found (Rheinheimer 1974). Some aquatic bacteria tend to form aggregates or colonies with specific morphology, commonly spheres or stars and ribbons or sheets. True aquatic bacteria are normally characterized by their ability to use nutrients available in extremely small concentrations (Wright and Hobbie 1966), and are either free-living in the water column or attached to a substratum (e.g., detritus).

Bacteria in the coastal marine environment may be either endemic or derived from soil, air, plants, and animals (including man). Many of the introduced forms are able to live for only a short time in ocean waters, and are unable to proliferate there. True marine bacteria have an obligate growth requirement for NaCl and a specific requirement for Na+, which precludes a simple osmotic explanation. Pratt (1974) and Pratt and Tedder (1974) found that in seawater samples from areas free of terrestrial contamination, 90% of the bacteria present required seawater nutrient media. Genera which are extremely common in coastal and oceanic waters include Pseudomonas, Vibrio, Spirillum, Achromobacter, Flavobacterium, and Bacillus. The bulk of marine bacteria belong to the orders Pseudomonadates and Eubacteriates, but other orders also are often represented.

Marine and brackish water environments are distinguished by the presence of the majority of luminous bacteria. These may be free-living or symbiotic on mollusca (cephalopods) and bony fish. Pigmentation is also common in halophilic microphytes, and Zobell (1946) concluded that more than 50% of marine bacteria are, to some degree, pigmented.

The vast majority of aquatic bacteria and fungi are true heterotrophs, requiring presynthesized organic molecules for their metabolism. Thus, organic compounds originating from primary production and subsequent consumption are converted back to lower-energy organics and are finally remineralized by the aquatic microflora.

In addition to the heterotrophic bacteria, photoautotrophic and chemoautotrophic bacteria are present in neritic and oceanic waters. Bacteria capable of photosynthesis (photoautotrophs) have rather specific habitat requirements (e.g., anaerobic or microaerobic conditions, light, sufficient hydrogen donors [H_2S or organic acids], etc.). Normally, these conditions do not occur in open systems, but where they do occur, as in eutrophic lakes and some lagoons and pools of coastal regions, photosynthetic bacteria can be major producers of organic matter. Chemoautotrophic bacteria enjoy a much broader distribution and are particularly numerous in coastal waters where sulphur bacteria oxidize sulphur, nitrifying bacteria oxidize ammonia to nitrite or nitrate, and iron and manganese bacteria oxidize Fe++ and Mn++ to Fe+++ and Mn+++, respectively (Watson 1963, Rheinheimer 1967). The contribution of chemoautotrophs to total production of organic matter is not known. The role that bacteria play in primary production is only speculative. Zobell (1963) conservatively estimated that

annual bacterial production in an aquatic system was about 7.3 mg carbon/m^3; many would argue that this estimate is too conservative.

Aquatic fungi, like bacteria, are ubiquitous in most waters, and populations are composed of true aquatics as well as fungi more typical of soils. All aquatic fungi are heterotrophic organisms, and they exist as saprophytes or as parasites on a large variety of plants and animals. Most aquatic fungi are obligate aerobes requiring free O_2 to break down pectins, hemicellulose, cellulose, lignin and chitin, as well as common proteins, sugars, starches, and fats. Although capable of existing at extremes of both pH and temperature, aquatic fungi are almost singularly mesophilic. In contrast to aquatic bacteria, fungi have larger cells, are eukaryotic, and display a greatly varied morphology, particularly in fruiting bodies.

Representatives from all four fungal classes have been found in marine habitats and, aside from true halophilic fungi, many salt-tolerant forms of limnetic and terrestrial origin have been identified (Rheinheimer 1974). The Phycomycetes are generally considered the primary aquatic fungi, and in marine habitats are represented by saprophytic forms and a large number of parasites.

The distributions of bacteria and fungi in marine environments have not been well defined and in the Sea Island Coastal Region are basically unknown. The paucity of data on microfloral distribution is partially the result of the sampling difficulties involved in any ecologically oriented microbial study. This is a major data gap for the entire region.

2. Function

According to Wood (1967), bacteria play two major roles in the marine environment: 1) conversion of dissolved organic material into particulate form (the bacteria themselves), and 2) the breakdown of complex substances. As shown in Chapter One, Section III, bacteria play pivotal roles in the biogeochemical cycles through which nutrients are regenerated. In addition to these cycles discussed in Chapter One (i.e., water, carbon, nitrogen, phosphorus, and sulfur cycles), bacteria also function in the iron and manganese cycles of the hydrosphere. A large group of bacteria can oxidize ferrous to ferric compounds and manganous to manganic compounds.

In addition to the activities of microorganisms, nutrient regeneration is also affected by element residence times in the ocean. Cherry et al. (1978) defined such residence time as "the average time an element spends in ocean water between introduction into the ocean and incorporation into the sediments." They further reported that residence times of a number of elements were controlled primarily by the sinking rates of zooplankton fecal pellets. However, as pointed out previously (see section on detritus in this chapter), pellet sinking rates can be markedly slowed, and thus element residence times and probabilities of reincorporation into living organisms increased, through formulation of the organic aggregates known as "marine snow" (Silver et al. 1978).

Bacteria are involved in the trophic dynamics of marine ecosystems from the water's surface into the deep sediments. Sieburth (1976) has pointed out that the surface layer of ocean water exhibits a much higher concentration of dissolved organic matter (on the order of 2,000 mg C/1) than do sub-surface waters, and that this high organic medium may support bacterial populations of at least 10^5 cells/ml. Maximum subsurface bacterial populations are at least an order of magnitude lower. This "bacterioneuston" was assumed to be an important source of food for neustonic (i.e., surface-living) protozoans.

Bacteria are the first microorganisms to colonize solid materials introduced into the marine environment (Sieburth 1976). Such colonization leads to the formation of bacterial films, which may be important for the settling of other fouling organisms. Bacterial fouling occurs on living and dead macroorganisms, as well as on abiotic materials, and the bacterial populations provide a rich food source for protozoans and other small consumers. In fact, it has been shown that the growth of microbiota on detritus (e.g., marsh grass) is a major source of the food value of the detritus to a consumer (Sieburth 1976).

Some workers have suggested that bacteria occur primarily on substrates in ocean waters. However, Sieburth (1976) presented evidence indicating that marine bacteria in the water column are probably themselves planktic and not attached to other particulates. Such bacterioplankton populations in euphotic waters may be on the order of 10^4 cells/ml, and they may serve as an important source of food for zooplankton and attached filter feeders (Sorokin 1971, Sieburth 1976).

The consumers in marine waters convert about 25% of this food into feces, which, along with the carcasses of dead organisms, cast exoskeletons, and the like, from what Sieburth (1976) terms the "fecal-sestonic" ecosystem. As pointed out previously (see Section I

of this chapter), this fecal-sestonic material is the energy input that supports the benthic biota. Bacteria and other microorganisms colonize this organic debris, enrich it through their own production, and break it down into simpler forms, ultimately releasing inorganic nutrients. In the sediments, bacteria continue a dual role as decomposers of complex substances and as direct food for deposit feeders. Further, the interactions of microorganisms with sediments are closely related to the formation of certain resources. Bacteria and fungi have contributed to the formation of peat, coal, petroleum, sulphur, and certain ore deposits (iron, manganese, etc.).

III. INTERTIDAL MARINE SUBSYSTEM

A. DESCRIPTION

The marine intertidal subsystem consists of tidal beaches and bars contiguous to coastal marine waters. Virtually all marine beaches of the Sea Island Coastal Region are on barrier islands. These barrier beaches form a narrow fringe or chain which extends from the northern to the southern limits of the study area, and which is interrupted at frequent intervals by inlets, estuaries, harbors, bays, and sounds. This intertidal subsystem extends from the landward limit of the coastal marine ecosystem (extreme high water of spring tides) to the extreme low spring tide mark, and it is characterized by frequent and regular exposure of its substrate to air by tides. The Sea Island Coastal Region includes a total of some 293 km (183 mi) of marine intertidal beaches, excluding the 96 km (60 mi) long Grand Strand area of South Carolina.

Major factors which determine the physical characteristics of beaches include wave action, storms, tides and tidal range, beach inclination, amount of groundwater input, and the activities of man. Proximity to inlets and estuaries may also markedly influence the amount of outwelled material that is eventually deposited on beaches (see Fig. 2-4). All of these factors affect the processes of sedimentation, erosion, and sand grain sorting which shape beaches. Because of the high wave energy forces which continually sort the sediments of beach fronts, the substrate can be described as essentially fluid and continually shifting. Nevertheless, many open interstitial spaces are maintained between sand grains due to the inability of most fine particles to settle out in this turbid environment (Riedl and McMahan 1974). Thus, the marine intertidal zone of the South Carolina-Georgia coast may be characterized as a high energy, highly turbid environment with a continually

shifting substrate of sand containing
very little fines (mud, clay, or silt).
These characteristics, coupled with its
regular and frequent exposure to the air,
make it a harsh environment indeed.

Within the beach substrate itself,
physical factors such as temperature,
water saturation, salinity, oxygen concentration, levels of free CO_2, water hard-
ness, light, and concentration of organic
materials vary markedly (Riedl and
McMahan 1974). These factors generally
exhibit rhythmic variations with tidal
day-night and seasonal cycles also. For
example, the amount and characteristics
of interstitial water are determined by
interactions of the ocean (tides),
evaporation and precipitation, and seasonal variations in groundwater input.
Most of these factors are important in
controlling the distribution of organisms
in the marine intertidal subsystem, but
perhaps the most important factors are
degree of desiccation, salinity, and
sediment characteristics.

The organisms of sandy beaches may
be divided into three groups on the
basis of their distribution in relation
to the substrate (Riedl and McMahan 1974).
These groups are 1) the epipsammon, or
surface-dwelling, generally motile macrofauna; 2) the endopsammon, or burrowing
species that are too large to live in
interstitial spaces; and 3) the
mesopsammon, or intersititial fauna. The
epipsammon is limited to a few groups such
as fish or birds which are strong or speedy
enough to exploit this turbulent environment. The endopsammon is also limited
to a relatively few taxa, mainly the
hardy and highly mobile crustaceans (e.g.,
certain amphipods, crabs, and shrimp),
some mollusks, and a few echinoderms and
polychaetes. The mesopsammon, on the
other hand, is quite diverse, ". . .each
beach system containing more than a
thousand species, compared with several
dozen in the epi- and endopsammon together" (Riedl and McMahan 1974).
Further, while the epipsammon consists
primarily of predators and the endopsammon of a mixture of predators,
filterers, and scavengers, the mesopsammon
appears to be markedly dominated by
detritus-feeders. However, as Riedl and
McMahan (1974) point out, ". . .'detritus'
and 'detritus feeders' are vague con-
ceptions and often a group is labeled as
detritus feeding only because of the
absence of formed food particles in the
gut." Besides these groups, the decomposers are believed to be very important
in beach systems, and may themselves serve
as rich food sources.

B. PRIMARY PRODUCERS

Primary production in high energy
beaches such as those on barrier islands
in the characterization area is generally

quite low., Benthic diatoms are probably
the most important producers in this
regard, followed by blue-green algae,
but in reality very little is known about
these groups in marine beaches. Also,
because light cannot penetrate the sediments, primary production must be limited
to plants occurring in the upper few
millimeters of the beach substrate.
Nevertheless, Riedl (in Riedl and McMahan
1974) indicated that in the deeper layers
blue-green algae may reach a density of
10,000 cells/l of sediment, but noted that
their trophic importance was unknown.

Members of several algal classes can
be found growing attached to solid objects
such as rocks, shells, pilings, tree
stumps, decaying marsh grass, Sargassum,
etc., and in sand in the marine intertidal zone of the Sea Island Coastal
Region. As yet, however, these algae have
not been studied intensively. Chapman
(1971) conducted a qualitative survey of
macroalgae along the Georgia coast,
principally in the area of Sapelo and
Cabretta islands. He recorded 69 taxa,
of which more than half (36 species) were
red algae. However, 14 of these species
were taken only in dredge collections
offshore from the beach and were not recorded from the intertidal zone itself. Also
included in his survey were 11 species of
blue-green algae, 17 species of green
algae, and 5 species of brown (including
2 species of Sargassum) algae. Overall,
Chapman recorded 25 species of algae
specifically from objects in the intertidal zone.

R. L. Wiseman (1978, College of
Charleston, Charleston, South Carolina,
unpubl. data) has studied the algae of
a coquina outcrop at Myrtle Beach, South
Carolina. The dominant species on these
rocks throughout the year are Gracilaria
foliifera, Gelidium crinale, Gymnogongrus
griffithesiae, Chondria bailevana (all
red algae), and a number of species belonging to the green algal genera Ulva
and Enteromorpha. Conspicuously absent
are brown algal species so typical of
rocky intertidal areas throughout the
world. In spring, however, the brown
alga, Leathesia difformis, forms pro-
tuberant masses on the surfaces of the
blades of Gracilaria foliifera, and various
other species occur as epiphytes on some
of the dominant reds. The tide pools
created in the coquina rock present a
spectacular contrast to the miles upon
miles of strand line devoid of such
habitats. In Georgia, Chapman (1971)
reported the green alga, Monostroma
oxyspermum, as common in beach tidal
pools.

The algal flora of the intertidal
sand of marine beaches is a rich and in-
triguing one dominated by a variety of
diatoms and euglenoids. Blooms of the
euglenoid, Eutrepia sp., often color the

83

sand surfaces with a deep green film. These sand-dwelling algae provide an abundant harvest for the sand-dwelling meiofauna. The seawalls and beach jetties are characteristically colonized by members of the green algal genera, Ulva and Enteromorpha, by a variety of blue-greens, and, during the colder months, by the rhodophytes, Bangia atropurpurea and Porphyra spp. Throughout the year, the dominant red alga is the turf-forming Bostrychia radicans, which is also found in salt marshes and growing upon hard substrates in harbors. This species also is seen commonly in similar habitats in Georgia (Chapman 1971).

As the preceding short discussion indicates, very little qualitative and almost no quantitative information is available concerning the primary producers of marine beaches in the Sea Island Coastal Region. The significance of these plants in beach trophodynamics is also unknown. This area, then, can properly be termed a major data gap for the region.

C. CONSUMERS

1. Zooplankton

The zooplankton of the surf zone has not been studied in the Sea Island Coastal Region or elsewhere nearby. Nevertheless, we would expect its general composition to be similar to that of the contiguous nearshore marine waters (see Section II of this Chapter). It is also likely to include numerous small members of the benthic fauna which occasionally become suspended in the water column by wave turbulence, or which periodically (e.g., at night or at high tide) move into the water column to feed. Larval stages of ghost crabs, burrowing (mole) crabs, and other forms are released in the surf zone, and postlarvae of these species must be recruited to the beach via the surf zone. Unfortunately, nothing is known of the larval dynamics of any of these species. Knowledge of the beach zooplankton, then, is an important data gap for this region.

2. Benthic Invertebrates

Environmental conditions on sandy beaches present a rigorous challenge for invertebrate animals. To occur in such a habitat, a species must withstand strong wave and current action, tidal rise and fall, shifting substrata, heavy predation, and wide variations in salinity and temperature. Given such conditions, the fauna is specialized and highly adapted for survival.

As mentioned previously, beach-dwelling organisms have been divided into several categories based upon where they occur in relation to the substrate (Riedl and McMahan 1974). Species living on the surface, including certain insects, belong to the epipsammon. Burrowing species

typically living in the sediment, such as amphipods, pelecypods, and gastropods, constitute the endopsammon. The interstitial fauna, or mesopsammon, includes various meiobenthic taxa--gastrotrichs, gnathostomulids, turbellarians, tardigrades, and harpacticoid copepods. The richest assemblage of species occurs in the mesopsammon, but the groups represented are generally rather poorly known and will not be discussed further here.

As expected in a high-stress environment, few macrobenthic invertebrate species occur on an open beach. However, a few of those represented occur in large numbers; in a study of the fauna inhabiting a North Carolina beach, Dexter (1969) found both low diversity and low density, with dominance by only a few species. Most are rapid burrowers and some, such as the coquina clam Donax variabilis and the mole crab Emerita talpoida, are known to migrate up and down the beach as the tides rise and recede (Pearse et al. 1942, Turner and Belding 1957, Roberts 1974a). Filter feeders dominate the benthos since little organic detritus accumulates intertidally in the sediments of exposed beaches. As a result, the system acts as an extensive food-filtering system (Riedl and McMahan 1974). Among the most abundant macroinvertebrates of sandy beaches are the amphipods. Filter-feeding haustoriids are abundant intertidally, while various talitrids or "beach hoppers" occur near the high tide area where they feed on detritus in beach drift. Various species of burrowing decapods are represented near or below the low tide zone, including Callianassa, Lepidopa, and Emerita. Also present are burrowing polychaete worms such as Scolepis squamata.

Perhaps the most commonly recognized macroinvertebrate of marine beaches is the abundant ghost crab, Ocypode quadrata. This crab burrows in the beach from the upper intertidal region to the dune backshore, with the youngest and smallest crabs generally tending to burrow nearer the water and the older, adult crabs farther into the dunes (up to 0.4 km or .25 mi from the ocean) (Williams 1965). It is the most terrestrial of the decapod crustaceans inhabiting the coastal region of South Carolina and Georgia, only entering the water at intervals to moisten its gills and aerate egg masses or release larvae. Ghost crabs generally will not run into water even when disturbed (Williams 1965). The crabs can actively take up moisture from damp sand (Wolcott 1976a, b), thus further reducing their dependence on the sea.

Ocypode quadrata is nocturnal in its feeding habits, feeding primarily as a predator on Emerita and Donax populations in the intertidal zone (Wolcott 1978) and also as a scavenger along the drift line (Williams 1965). They are occasionally

cannibalistic as well, and frequently
prey on eggs of loggerhead turtles
(Hopkins et al. 1978). In turn, the
ghost crab is preyed upon by shorebirds
and probably by raccoons and other animals
that feed in the dunes and along the
foreshore (Roberts 1974b). Thus, although the ghost crab does not actually
enter the sea to feed, it acts as a direct
trophic link between the marine and
terrestrial environments.

Burrowing activities are generally
carried out during the day, with burrows
of larger crabs reaching 4 ft (1.2 m) in
depth (Williams 1965). Ghost crabs burrow
and feed actively during most of the year
(late March to mid-December in Texas; Haley
1972), but apparently seal their burrows
and lie dormant during winter (Milne and
Milne 1946, Haley 1972). A similar seasonality probably occurs along the South
Carolina-Georgia coast.

Ghost crabs attain sexual maturity
at a carapace width of about 24 mm for
males and 26 mm for females (Haley 1969).
Haley (1972) studied reproduction of O.
quadrata in Texas and reported cyclic
copulatory activity with peaks in April
and September. He presented evidence for
two maturations per female per year, one
very synchronous one involving virtually
all postpuberty females in April and a
second more diffuse one in summer.
Similarly, the frequency curve for ovi-
gerous females was bimodal, with a small
peak in April and a large one in late
summer (especially August). Williams
(1965) reported the egg-laying season
for this species to be from April to July
in the Carolinas. Haley (1972) estimated
the life span of female O. quadrata to be
about 3 years.

Larval development in the ghost crab
consists of five zoeal stages and a
megalopa (Diaz and Costlow 1972). The
larvae are apparently released directly
into the surf at night, and presumably
development takes place at sea. Crane
(1940) suggested that the basic morphology
of the megalopa (stout and globose in
appearance and structured so that the legs
can be drawn tightly against the body,
increasing its ball-like appearance) is
an adaptation for recruitment through the
rough surf zone to the beach. Presumably
O. quadrata reaches the beach as a megalopa, metamorphoses there to the crab
stage, and then gradually moves further
away from the water as it grows.

Characteristic general features of
the ghost crab are its burrows, nocturnal
feeding habits (it tends to clean up
organic refuse on the beach), light and
variable color which makes it difficult
to detect against the beach sand, extreme
running speed and agility, large retract-
able eyes, and sound production. This
latter feature has been the subject of

several behavioral studies, and it has now
been demonstrated that O. quadrata can
perceive as well as produce sound (see for
example Horch and Salmon 1969). Ghost
crab populations tend to decrease as human
use of beaches increases and as more and
more dunes are leveled to make way for
human habitations.

Beaches of the Sea Island Coastal
Region typically receive relatively moderate wave energy, and can be considered as
only medium instead of high energy environments. Roberts (1974a) reported that such
beaches have a more diverse fauna than
that of high energy areas. He noted that
Donax and Emerita are of lesser importance
in such places than on exposed beaches,
and that polychaetes are better represented,
although haustoriid amphipods are numerically dominant. Two species of haustoriids,
Neohaustorius schmitzi and Acanthohaustorius
millsi, accounted for nearly 90% of the
fauna on the partially sheltered beach
studied by Dexter (1969) in North Carolina.

Another detritus-feeder common on
many low/moderate energy beaches is the
familiar hermit crab, Pagurus longicarpus
(Roberts 1968). Recent observations by
Scully (1978) have shown that this species
is able to use the organic foam generated
on marine beaches as a direct food source.
Perhaps other species are able to do this
as well, as suggested by Riedl (in Riedl
and McMahan 1974).

The fauna of two beach-related tidal
flats on Sapelo Island, Georgia, were
studied by Howard and Dörjes (1972) and Dörjes
(1977). Fifty species each were found on
Nannygoat and Cabretta flats, but only 27
species were common to the two areas. While the
haustoriid amphipods Acanthohaustorius
sp. and Bathyporeia sp. were the two most
abundant macroinvertebrate species at both
locations, they were much more numerous
on Cabretta Flat. On Nannygoat Flat, at
the entrance of Doboy Sound, polychaetes
accounted for 38% of the fauna and crustaceans 36%. Among the characteristic
species at this site were Heteromastus
filiformis and Diopatra cuprea (polychaetes),
Mulinia lateralis (pelecypod), and
Callianassa sp. (decapod). Water rich in
organic material was observed over this
flat at high tide, providing an abundant
source of food for deposit and suspension
feeders. Mud comprised 5% - 16% of the
sediment at this site. On Cabretta Flat,
polychaetes and crustaceans accounted for
28% and 40% of the fauna, respectively.
Species characteristic of this flat, besides haustoriids, included Oliva sayana
(gastropod), and Callianassa sp. and
Ogyrides alphaerostris (decapods). Although this flat is protected from the
ocean by an intertidal shoal, the flat is
open at either end and fine sediments are
carried away by currents.

D. M. Knott, D. R. Calder, and R. F. Van Dolah (1979, South Carolina Marine Resources Division, Charleston, unpubl. data) conducted seasonal quantitative sampling of the beach macrofauna at Murrells Inlet during construction of a rubble jetty. Although 88 species were found intertidally during the study, over 78% of the animals collected belonged to three species, the polychaete Scolelepis squamata, the amphipod Neohaustorius schmitzi, and the bivalve Donax variabilis. Macroinvertebrate assemblages were depauperate at high tide, but species numbers, diversity, and densities of individuals were higher at mid-tide and low tide levels of the beach. Community structure varied widely from season to season, reflecting the population dynamics of the numerically dominant species. Faunal densities were highest during periods of peak recruitment in winter and spring, and were lowest in autumn. Construction was accompanied by increases in species numbers and faunal densities on the sheltered side of the jetty. This was attributed to the direct and indirect effects of reduced wave energy.

Stephenson and Stephenson (1952) studied the intertidal biota of rock riprap used to form jetties and breakwaters in the vicinity of Charleston, South Carolina. A number of different zones were evident intertidally on jetties at the entrance of the harbor. In the supralittoral fringe, the barnacle Chthamalus fragilis was present in a black band of blue-green algae and lichens. An abundance of C. fragilis was also evident high in the intertidal zone. Dominant invertebrates in the middle intertidal zone included oysters (Crassostrea virginica), mussels (Brachidontes exustus), and barnacles (Balanus eburneus, B. improvisus). While barnacles and mussels were also well-represented in the lower intertidal zone, oysters were much less abundant. Other invertebrates reported from this zone included Urosalpinx cinerea (gastropod), Oculina arbuscula (coral), Molgula manhattensis (ascidian), Asterias forbesi (asteroid), Arbacia punctulata (echinoid), Tubularia crocea (hydroid), Anguinella palmata (bryozoan), Bunodosoma cavernata (actiniarian), and a red sponge believed to be Hymeniacidon heliophila.

Within Charleston Harbor, a number of changes in the intertidal invertebrate assemblage were noted. Brachidontes exustus was less abundant, the mussel Geukensia demissa was more frequent, oysters were much more numerous and changed in general shape from flat to erect, Balanus eburneus was more abundant, and Molgula manhattensis replaced many of the plants and animals in the lower intertidal. Littorina irrorata (gastropod) was observed in the supralittoral fringe, and the isopod Ligia exotica was well represented.

The fauna of sandy beaches is exposed to a number of perturbations beyond the normal stresses of the environment, including severe storms, oil spills, domestic and industrial pollution, construction activities including "beach nourishment" projects, and the recreational activities of humans. Of these, only the effects of severe storms have been investigated in the study area. Croker (1968) found little impact on the abundance and distribution of haustoriid amphipods at Sapelo Island following passage of two hurricanes in 1964. The effects of extensive freshwater runoff were believed to represent a greater threat to these organisms than wind and wave action. Similar observations were made following a hurricane at Panama City Beach, Florida, by Soloman and Naughton (1977). Although beach erosion and property damage were severe, the storm was not accompanied by particularly heavy rainfall and its impact on the macrobenthos was relatively minor. Numbers of individuals remained virtually unchanged before and after passage of the hurricane. Species numbers actually increased temporarily following the storm, due to the introduction of organisms from other environments.

No studies on the effects of beach nourishment have been undertaken in this area, but investigations conducted elsewhere offer insight into the possible effects of such operations. Although changes in beach profile may alter the extent of habitat available to the fauna, the long-term effects of beach nourishment appear to be relatively minor if sediments used resemble those of the original beach (Thompson 1973). Sediments of a different type or grain size could alter community structure significantly. Faunal diversity and density on a beach are generally lowest at and immediately above the high tide mark, and deposition of sediments in this area of a beach would appear to have the least impact on the system. The fauna of beaches is made up of species having high resilience to natural perturbations, and recolonization of properly nourished beaches should be rapid. The impacts of oil spills and pollution, although little studied, would be much more deleterious. The effects of human use of beaches for recreation on the macroinvertebrate fauna have not been investigated.

Table 2-18 lists some of the most typical benthic macroinvertebrates of medium-energy sand beaches along the South Carolina-Georgia coast. Additional information on epibenthic species that may be encountered in the surf zone is presented in the preceding section on benthic invertebrates of subtidal marine waters. Also, the common horseshoe crab, Limulus polyphemus, may be regarded as benthopelagic and may be considered an

Table 2-18. Species of macroinvertebrates typical of medium-energy sand beaches in the Carolinian Province (from Pearse et al. 1942, Croker 1967, 1968, Dexter 1967, 1969, Howard and Dörjes 1972, Shealy et al. 1975, Calder et al. 1976, 1977a, Dörjes 1977).

Phylum Annelida
Glycera dibranchiata
Haploscoloplos fragilis
Nerinides sp.
Onuphis sp.
Scolelepis squamata

Phylum Mollusca
Polinices duplicatus
Donax variabilis
Mulinia lateralis

Phylum Arthropoda
Chiridotea sp.
Acanthohaustorius millsi
Amphiporeia virginiana
Bathyporeia sp.
Lepidactylus dytiscus
Monoculodes sp.
Neohaustorius schmitzi
Orchestia spp.
Parahaustorius longimerus
Talorchestia spp.
Callianassa sp.
Emerita talpoida
Lepidopa websteri
Ocypode quadrata

occasional and relatively common component of the surf fauna. Specimens of L. polyphemus are frequently stranded in the intertidal zone of the beaches.

3. Insects

A variety of nuisance insects can be found on marine beaches of the Sea Island Coastal Region. These include the non-biting seaweed and beach flies (Dobson 1976, Simpson 1976) and biting species such as several mosquitoes, sand flies (gnats), horse flies, and deer flies (Axtell 1974). On occasion, hordes of these insects may descend upon the intertidal marine beaches, rendering them essentially unfit for man's recreational use. Predacious tiger beetles are commonly seen foraging on various invertebrates in the detrital debris stranded on the beaches during spring and summer months. These and other species commonly found on coastal islands are discussed in more detail in Chapter Three.

4. Fishes

Several surveys of the fishes occurring in marine intertidal waters have been conducted in the Sea Island Coastal Region. Nevertheless, little work has been done on the ecological relationships or biology of fishes occupying this zone in South Carolina and Georgia. Investigations of surf zone fishes in South

Carolina and Georgia have been limited primarily to small beach seine surveys (Miller and Jorgenson 1969, Cupka 1972, Dahlberg 1972, Anderson et al. 1977). These studies have provided information on the species composition, diversity, seasonal occurrence, relative abundance, and length-frequencies of smaller fishes in the surf zone habitat. In addition, Fields (1962) conducted seining studies at three permanent sites along the Georgia coast and presented data on recruitment, size, and food habits of juvenile pompano occurring in the surf zone. Results of these studies show that the extensive surf zone of the Sea Island Coastal Region is significant as a habitat for both resident fishes (e.g., the striped killifish, silversides, and anchovies) and migratory species (e.g., Gulf kingfish, pompanos, and mullets).

Cupka (1972) sampled six regular surf-zone stations along the South Carolina coast monthly for one year (1971) (Fig. 2-13). He collected a total of 39 species of fish, representing 18 families. On the basis of their occurrence in his samples, these species could be grouped into three categories: resident species (5 species), seasonal migrants (17 species), and strays and occasionals (17 species) (Table 2-19). Species diversity and biomass were greatest in summer and least in winter, while the number of individual fish captured was highest

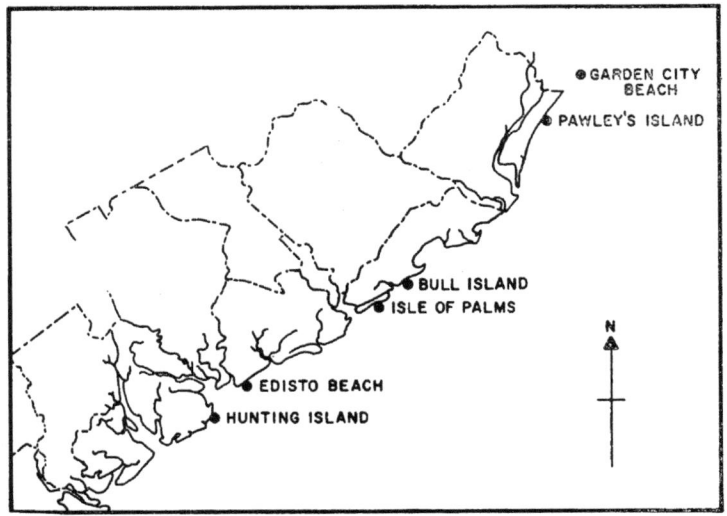

Figure 2-13. Locations of seining survey stations along the coast of South Carolina during 1971 (adapted from Cupka 1972).

in spring and lowest in fall (Table 2-20). In an effort to identify the most important fish species in the beach surf zone, Cupka (1972) calculated an "importance rank" based on the numbers, biomass, and occurrences of each species collected. Results of this ranking are summarized in Table 2-21. Not surprisingly, five of the top seven species are also resident species (Table 2-19).

Anderson et al. (1977) took seine samples from the surf and a large tidal pool at Folly Beach, South Carolina, at approximately biweekly intervals from October 1969 to October 1971. They collected a total of 41 species during the 2-year study. Specimens of all 41 species were collected from the surf, but the tidal pool yielded only 16 species. In the surf zone, species diversity was greatest in summer and least in winter; however, the number of individuals and biomass captured were greatest in winter (Table 2-22). Species diversity was also lowest during winter in the tidal pool, but so were the numbers of individuals and biomass. Here, species diversity, numbers of individuals, and biomass were greatest in fall and spring (Table 2-22). The six most important species of fish taken on Folly Beach are listed by rank in Table 2-23. Note the similarity

of this list to that given by Cupka (1972) (Table 2-21). Note also that the Atlantic silverside was the most important species in both the surf zone and the tidal pool, while the bay anchovy and Gulf kingfish were captured almost exclusively in the surf zone and the striped killifish was taken primarily in the tidal pool.

Miller and Jorgenson (1969) conducted a seine survey of the fishes of the surf zone on St. Simons Island, Georgia. They sampled at biweekly intervals over an 8-year period (March 1953 – May 1961), and recorded a total of 98 species of fish from this habitat. This is greater than twice the diversity for surf-zone fish populations found in South Carolina during the 1- and 2-year studies of Cupka (1972) and Anderson et al. (1977), respectively. However, the top 10 most abundant species made up 95% of the total number of specimens collected by Miller and Jorgenson (1969) (Table 2-24). Further, despite the greater species diversity, the surf zone in Georgia was dominated for the most part by the same species that dominated on South Carolina beaches. In fact, eight of the 10 most abundant species captured by Miller and Jorgenson (1969) were among the 10 top-ranked species of Cupka (1972), and four

Table 2-19. Species of fish captured in the surf zone of South
Carolina, grouped by occurrence pattern (compiled from
Cupka 1972).

RESIDENT SPECIES	SEASONAL MIGRANTS	STRAYS AND OCCASIONALS
Atlantic silverside	Striped anchovy	Blueback herring
Bay anchovy	Crevalle jack	Finetooth shark
Striped killifish	Atlantic bumper	Southern stargazer
Gulf kingfish	Spot	Silver perch
Striped mullet	Rough silverside	Atlantic menhaden
	Northern kingfish	Horse-eye jack
	White mullet	Striped burrfish
	Planehead filefish	Sheepshead minnow
	Broad flounder	Gizzard shad
	Bluefish	Lancer stargazer
	Red drum	Tidewater silverfish
	Northern puffer	Southern kingfish
	Atlantic needlefish	Summer flounder
	Florida pompano	Southern flounder
	Permit	Northern searobin
	Palometa	Lookdown
	Bluntnose stingray	Northern pipefish

Table 2-20. Number of species, total number of individuals, and total weight of fish captured
by season during a 1971 beach seining survey in South Carolina (Cupka 1972).

SEASON	NO. OF SPECIES	PERCENT OF TOTAL	NO. OF SPECIMENS	PERCENT OF TOTAL	WEIGHT[a] (g)	PERCENT OF TOTAL
Winter (Jan. - Mar.)	8	20.5	1,217	18.7	4,161.5	18.7
Spring (Apr. - June)	21	53.8	2,598	39.8	6,103.7	27.4
Summer (July - Sept.)	26	66.7	1,673	25.7	7,768.4	34.8
Fall (Oct. - Dec.)	17	43.6	1,028	15.8	4,272.3	19.1

a. 1 g = 0.035274 oz

were included in the top six species of
Anderson et al. (1977) (see Tables 2-21,
2-23, 2-24).

Of the 10 most abundant species
captured by Miller and Jorgenson (1969),
four could be termed permanent residents
of the surf zone, while the other six
species were common there only seasonally
(Table 2-24). The bay anchovy was the
most abundant species encountered; it
dominated the surf fauna from May through
October and was a fairly common resident
species during the remainder of the year.
The striped mullet also could be termed

a resident species, but it was most abundant from November through April and
occurred in only low numbers during the
warmer months. Both species of silversides also were captured year around, but
they exhibited somewhat different patterns
of seasonal abundance. Low population
levels were noted from September through
November for the Atlantic silverside, and
from March through May and October through
December for the rough silverside. Spot
were quite numerous in winter and early
spring but almost disappeared from the
beaches from July through December. In
contrast, the striped anchovy, Florida

Table 2-21. Numerical rank, weight rank, appearance rank, and importance rank of the 16 most important species collected at six surf zone stations during a 1971 seining survey in South Carolina (Cupka 1972).

SPECIES	NUMERICAL RANK	WEIGHT RANK	APPEARANCE RANK	SUM OF RANKS	AVERAGE	IMPORTANCE RANK
Atlantic silverside	1	1	1	-	1.0	1
Bay anchovy	2	2	2	6	2.0	2
Florida pompano		4	5	12	4.0	
Gulf kingfish	4	8	2	14	4.7	4
Striped mullet	8	3	3	14	4.7	4
Rough silverside	6	11	4	21	7.0	5
Striped killifish	·7	10	5	22	7.3	6
Striped anchovy	5	16	5	26	8.7	
Spot	18	9	8	35	11.7	8
Permit	13	17	6	36	12.0	9
Southern flounder	22	5	10	37	12.3	10
Finetooth shark	22	6	10	38	12.7	11
Bluntnose stingray	21		10	38	12.7	11
Atlantic bumper	10	23	8	41	13.7	12
Planehead filefish	9	24	9	42	14.0	13
Crevalle jack	15	25		47	15.7	14

pompano, and Gulf kingfish were most numerous in summer, with few specimens captured during the cooler months. The Atlantic menhaden was most numerous in spring and early summer, while the dusky anchovy was taken only during mid-summer (July).

Dahlberg (1972) sampled the fish populations of nine coastal habitats in Georgia, including Sapelo Island beach, seven estuarine habitats, and a fresh-water creek. He sampled the beach surf zone a total of about 20 times during all seasons from April 1967 through February 1970. The beach habitat had the highest diversity of fish (114 species) of the habitats sampled, and this total did not include 38 of the species recorded by Miller and Jorgenson (1969) from St. Simons Island beach. Again, however, the assemblage of important species was similar to those reported from other beaches in the Sea Island Coastal Region: the rough silverside, Atlantic silverside, striped killifish, bay anchovy, and striped mullet were common throughout much of the year, while the Atlantic bumper, Florida pompano, white mullet, southern kingfish, and Gulf kingfish were abundant only during the warmer months.

Hammond and Cupka (1977) conducted a survey of the ocean pier fishery located along the northern South Carolina coast, including the catch composition of fishes. This fishery is located in the nearshore marine habitat and includes species which frequent the intertidal zone. During the study period (April – November 1974), 8,109 fish representing 28 families and 58 species were observed. Four families accounted for over 96% of the total number of fish taken from the piers: sciaenids (85.8%); jacks (4.6%); bluefish (3.9%); and sea catfishes (1.7%). The predominant species observed were spot, Atlantic croaker, and kingfishes (Menticirrhus spp.). Silver perch, Florida pompano, bluefish, sea catfish, and seatrout were other common species. Numbers of fish harvested were highest during April, July, and October, when almost 61% of the catch was made. Most of the fish taken by the selective

Table 2-22. Numbers of species, specimens, and total weight of fishes collected by seining on Folly Beach, South Carolina, from October 1969 to October 1971 (data from both years combined) (Anderson et al. 1977).

Season	SURF			TIDAL POOL			SURF + TIDAL POOL		
	Number Species	Number Specimens	Total Weight(g)[a]	Number Species	Number Specimens	Total Weight(g)	Number Species	Number Specimens	Total Weight(g)
Fall	20	561	2,037	11	502	2,100	20	1,063	4,137
Winter	11	1,206	4,359	4	32	125	12	1,238	4,484
Spring	19	330	2,758	10	1,509	1,657	20	1,839	4,415
Summer	24	650	3,712	6	305	570	26	955	4,282
TOTAL	41	2,747	12,866	16	2,348	4,452	41	5,095	17,318

a. 1 g = 0.035274 oz

Table 2-23. Ranks of the six most important species of fish collected by seining on Folly Beach, South Carolina. Ranks are in parentheses. Appearance equals number of days collected (because of this, the number of days collected in surf and tidal pool combined does not equal number of days collected in surf + number of days collected in tidal pool) (Anderson et al. 1977).

Location	Species	Number of Specimens	Number of Appearances	Total Weight(g)	Seasonal Index[a]	Sum of Ranks	Importance
Surf	Atlantic silverside	1,485(1)	36(1)	4,025(1)	109.7(1)	4	1
	Gulf kingfish	291(3)	34(2)	831(4)	72.9(2)	11	2
	Bay anchovy	362(2)	29(3.5)	350(5)	63.7(3)	13.5	3
	Florida pompano	266(4)	29(3.5)	1,018(3)	62.0(4)	14.5	4
	White mullet	121(5)	9(5)	2,591(2)	11.7(5)	17	5
Tidal pool	Atlantic silverside	630(2)	18(1)	2,201(1)	31.2(1)	5	1
	Florida pompano	710(1)	13(3)	501(3)	21.8(2)	9	2
	Striped killifish	341(4)	15(2)	734(2)	20.9(3)	11	3
	White mullet	545(3)	7(4)	118(4)	8.5(4)	15	4
Surf and tidal pool combined	Atlantic silverside	2,115(1)	40(1)	6,226(1)	126.9(1)	4	1
	Florida pompano	976(2)	29(4)	1,519(3)	76.2(2)	11	2
	Gulf kingfish	294(6)	34(2)	843(4)	73.0(3)	15	3
	White mullet	666(3)	14(6)	2,709(2)	29.2(5)	16	4
	Bay anchovy	372(4)	30(3)	365(6)	66.5(4)	17	5
	Striped killifish	344(5)	16(5)	744(5)	23.0(6)	21	6

a. Seasonal index $= 1 + \log N \cdot 4 \sum_{i=1}^{\epsilon} \frac{D_i C_i}{M_i}$

Where N = total number collected
D_i = days collected in ith season
C_i = months collected in ith season
M_i = number of months in ith season

Table 2-24. Monthly relative abundance of the 10 most common fishes taken at St. Simons Island beach from March 1953 to May 1961 (Miller and Jorgensen 1969) (upper figure = number of fish; middle figure = percent of total of all species for the month; lower figure = average number of fish per sampling day; * <0.05%).

SPECIES	JANUARY	FEBRUARY	MARCH	APRIL	MAY	JUNE	JULY
Bay anchovy	353	529	3,827	2,011	7,176	31,691	21,029
	3.3	2.6	12.4	29.3	57.0	65.5	43.5
	22.1	27.8	191.4	77.3	341.7	1,760.6	876.2
Spot	1,228	9,800	18,883	1,077	287	7,311	45
	11.6	47.6	61.3	15.7	2.3	15.1	0.1
	76.8	515.8	944.2	41.4	13.7	406.2	1.9
Striped anchovy		1	22	85	510	903	15,786
		*	0.1	1.2	4.1	1.9	32.7
		0.1	1.1	3.3	24.3	50.2	657.7
Striped mullet	6,941	7,098	780	827	198	51	154
	65.4	34.5	2.5	12.0	1.6	0.1	0.3
	433.8	373.6	39.1	31.8	9.4	2.8	6.4
Atlantic menhaden	68	794	5,991	613	219	3,648	78
	0.6	3.8	19.4	8.9	1.7	7.5	0.2
	4.2	41.8	299.6	23.6	10.4	202.7	3.2
Atlantic silverside	1,092	1,260	978	1,601	581	388	1,218
	10.3	6.1	3.2	23.3	4.6	0.8	2.5
	68.2	66.3	48.9	61.6	27.7	21.6	50.7
Rough silverside	800	812	96	104	164	1,179	2,828
	7.5	3.9	0.3	1.5	1.3	2.4	5.8
	50.0	42.7	4.8	4.0	7.8	65.5	117.8
Florida pompano			1	55	2,254	1,386	883
			*	0.8	17.9	2.9	1.8
			0.1	2.1	107.3	77.0	36.8
Dusky anchovy	1						2,804
	*						5.8
	0.1						116.8
Gulf kingfish	2	2	2	8	54	227	422
	*	*	*	0.1	0.4	0.5	0.9
	0.1	0.1	0.1	0.3	2.0	12.6	17.6

93

Table 2-24. Concluded

SPECIES	AUGUST	SEPTEMBER	OCTOBER	NOVEMBER	DECEMBER	TOTAL BY SPECIES	CUMULATIVE PERCENT
Bay anchovy	4,856 49.6 231.2	16,941 85.2 891.6	3,602 58.2 211.9	1,132 40.2 70.8	2,770 63.4 173.1	95,917 43.4 411.7	43.4
Spot	8 0.1 0.4	13 0.1 0.7	1 * 0.1		21 0.5 1.3	38,674 17.5 166.0	60.9
Striped anchovy	1,017 10.4 48.4	176 0.9 9.3	65 1.1 3.8	4 0.1 0.2	2 * 0.1	18,571 8.4 79.7	69.3
Striped mullet	42 0.4 2.0	20 0.1 1.1	8 0.1 0.5	901 32.0 56.3	451 10.3 28.2	17,471 7.9 75.0	77.2
Atlantic menhaden	9 0.1 0.4	20 0.1 1.1		15 0.5 0.9	4 0.1 0.2	11,459 5.2 49.2	82.4
Atlantic silverside	929 9.5 44.2	147 0.7 7.7	239 3.9 14.1	203 7.2 12.7	706 16.1 44.1	9,342 4.2 40.1	86.6
Rough silverside	492 5.0 23.4	430 2.2 22.6	184 3.0 10.8	64 2.3 4.0	160 3.6 10.0	7,313 3.3 31.4	89.9
Florida pompano	794 8.1 37.8	326 1.6 17.2	344 5.6 20.2	100 3.6 6.2	5 0.1 0.3	6,148 2.8 26.4	92.7
Dusky anchovy						2,805 1.3 12.0	94.0
Gulf kingfish	334 3.4 15.9	364 1.8 19.2	195 3.2 11.5	59 2.1 3.7	37 0.8 2.3	1,706 0.8 7.3	94.8

gear (hook and line) were larger
specimens.

South Carolina also has a beach haul
seine fishery. This fishery is pursued
chiefly from September through December,
when large schools of adult spot and
mullet move close to shore along the
South Carolina coast. Not surprisingly,
these species dominate the catch.

Although the pier and haul seine
fisheries in South Carolina include
species which overlap between the intertidal surf zone and the subtidal marine
habitat, most fishes taken occur in
very shallow water within a few hundred
yards of the beach. Of the larger fishes,
those which are definitely known to occur
within the intertidal (surf) zone include
stingrays (Dasyatis spp.), carcharhinid
sharks, Florida pompano, sea catfish, Gulf
kingfish, southern kingfish, spot, red
drum, black drum, and under certain conditions, spotted seatrout, bluefish, sheepshead, and flounders. Other species,
including most of those taken in the pier
and haul seine fisheries, may invade this
zone on higher stages of the tide.

Species inhabiting the surf zone
must tolerate wide ranges in turbidity,
turbulence, current velocity, and bottom
characteristics, as well as variations in
temperature and salinity (Anderson et al.
1977). It is not surprising, then, that
even among the most common species inhabiting this zone, there are substantial
variations in occupancy and salinity
tolerance (Table 2-25). Of these common
species, three (striped mullet, spot, and
red drum) can be classified as euryhaline,
being found from marine to freshwater
habitats. Strictly marine or stenohaline
species include striped killifish, pompano,
and Gulf kingfish. Species tolerant of
moderate salinity variations include
Atlantic silverside, bay anchovy, and
southern kingfish.

Smaller fishes, such as striped killifish, silversides, juvenile mullets, and
pompano are often abundant in tidal pools
and sloughs which retain sea water after
the tide has receded. Tidal pools may
provide protection for these fishes from
larger predators (Anderson et al. 1977).
However, some of the larger, more permanent tidal pools have been found to
contain large numbers of Atlantic needlefish, and tidal sloughs have been found in
some instances to contain large flounders
(C. M. Bearden, 1978, South Carolina
Marine Resources Division, Charleston,
unpubl. data).

Studies of the food habits and trophic relationships of fishes inhabiting
the surf zone along the South Carolina
and Georgia coasts are quite limited.
Fields (1962) reported that juvenile
pompano feed chiefly on small mollusks

and crustaceans (amphipods, copepods,
isopods, and crabs), polychaetes, and
other invertebrates, as well as larval
fishes. Bearden (1963) reported that
juvenile Gulf kingfish feed on amphipods,
while stomachs of adults contained fish
remains, mole crabs (Emerita sp.) and
stomatopods. Odum (1970a) reported that
mullet from Sapelo Island beach, Georgia,
fed largely on macroplant detritus.
From available data, it appears that most
of the predominant small fishes of the
surf zone feed largely on zooplankters,
small mollusks, crustaceans, and organic
detritus. Trophic levels and other information for the more important species
occurring in the intertidal marine habitat
of the Sea Island Coastal Region are
summarized in Table 2-25.

Although Anderson et al. (1977)
investigated the occurrence of fishes
in the beach zone in relation to environmental variables such as temperature and
salinity, little is known concerning the
effects of man-induced alterations on
the fishes of this habitat. Groins,
jetties, and other structures for beach
erosion control can serve as habitat and
feeding areas for certain species. These
structures often are covered with sessile
marine invertebrates (barnacles, bryozoans,
mollusks, etc.) and provide habitat for
numerous crustaceans and small fishes,
thereby attracting larger predators (e.g.,
sheepshead, flounder, seatrout, red drum,
and oyster toadfish) during favorable
tidal stages.

Mortalities of surf-zone fishes along
the South Carolina coast have occurred occasionally as a result of mosquito abatement programs involving the aerial application of
pesticides along coastal residential and
resort areas. One such kill occurred in
1977, apparently as the result of aerial
application of malathion at Sullivans
Island, South Carolina. Tissues of
striped killifish, juvenile pompano, and
other fishes were found to contain 135 ppb
of malathion (S.C. Department of Health
and Environmental Control, 1977, Charleston,
unpubl. data). Presumably, fishes frequenting the shallow beach zone, including
tidal pools and sloughs, are quite likely
to be affected by pesticides aimed at
nuisance insects. Oil spills could also
have immediate toxic or cumulative effects.

Future investigations of this important
habitat should incorporate multiple gear
sampling (small mesh beach seine, haul
seine, variable mesh gill net) on a seasonal
basis to provide a better understanding of
the fish assemblages present. Studies of
the ecological relationships, food habits,
and life histories of the surf-zone fishes,
and of the effects of environmental perturbations on these fishes, are needed.
Previous beach seine surveys have provided much information on the smaller,
slower moving fishes present, but little

Table 2-25. Trophic levels, seasonal and life-stage occurrence, and salinity tolerances of some of the more important fishes occurring in the intertidal marine habitat of the Sea Island Coastal Region.

	LIFE STAGES PRESENT	SEASONAL OCCURRENCE	SALINITY TOLERANCE
I. Herbivores			
Striped mullet[a]	Postlarvae→Adult	Year-round	Euryhaline
White mullet	Postlarvae→Adult	Spring, Summer	Marine
II. Primary Carnivores			
Striped anchovy	All Stages	Spring-Fall	Mesohaline
Bay anchovy	All Stages	Year-round	Mesohaline
Atlantic silverside	All Stages	Year-round	Mesohaline
Rough silverside	All Stages	Year-round	Marine
III. Mid Carnivores			
Striped killifish	All Stages	Year-round	Marine
Florida pompano	Juvenile→Adult	Spring, Fall	Marine
Gulf kingfish	All Stages	Year-round	Marine
Southern kingfish	Juvenile→Adult	Spring-Fall	Mesohaline
Spot	Postlarvae→Adult	Spring-Fall	Euryhaline
Sea catfish	Juvenile→Adult	Spring, Summer	Mesohaline
IV. Top Carnivores			
Sharks	Juvenile→Adult	Spring, Summer	Marine
Atlantic needlefish	Juvenile→Adult	Spring, Summer	Euryhaline
Bluefish	Juvenile→Adult	Spring-Fall	Mesohaline
Seatrouts	Subadult→Adult	Spring-Fall	Mesohaline
Red drum/channel bass	Young of the Year→Adult	Spring-Fall	Euryhaline
Black drum	Juvenile→Adult	Spring-Fall	Mesohaline
Flounders	Subadult→Adult	Spring-Fall	Euryhaline

a. Although the striped mullet is generally considered a herbivore and detritus feeder as an adult, recent data (Bishop and Miglarese 1978) have shown that adults are carnivorous on occasion.

is known of the ecology of the larger species (e.g., adult mullet, red drum, adult kingfishes, and several species of sharks and rays) occurring in this zone.

For a detailed discussion of the commercial and recreational fisheries of the Sea Island Coastal Region, refer to Volume II, Chapter Seven.

5. Reptiles

The Atlantic loggerhead turtle is the only marine reptile nesting on the barrier island beaches of South Carolina and Georgia. These barrier beaches constitute some of the most important loggerhead nesting habitat in the Southern United States.

The nesting season in this area generally extends from mid-May through mid-August, with peak nesting activity normally occurring in late June and early

July. Female loggerhead turtles may nest as many as five times during a nesting season (Davis and Whiting 1977), emerging from the sea at approximately 2-week intervals to deposit eggs. Average clutch size for turtles nesting on Cape Island, South Carolina, was 124 eggs (Caldwell 1959), and for Cumberland Island, Georgia, was 115 eggs (Richardson 1978). Age at sexual maturity is not known.

The general concensus among turtle biologists is that mature females return to their natal beaches to nest, but this belief has not been substantiated. There is high probability, however, that turtles will return to the same beach where they nested previously. Bell and Richardson (1978) and T. Richardson et al. (1978) have observed that there is a 49% probability that a turtle which nested on Cumberland or Little Cumberland Island, Georgia, will return to nest a second time. After the second nesting, the probability of remigration increases to 70%. Return migrations to nesting beaches occur at intervals of 1 to 4 years. In Georgia, the majority (56%) of re-migrations recorded on Cumberland and Little Cumberland islands occurred at 2-year intervals, while 31% remigrated in 3 years, and only 3% returned annually (Richardson and Hillestad 1978, T. Richardson et al. 1978). Whether some of the turtles which do not remigrate shift to new nesting grounds is not known. However, LeBuff (1974) noted that a logger-head turtle originally tagged on Florida's lower Gulf coast was observed nesting on the Atlantic coast (Melbourne Beach, Florida) over 4 years later. The available literature suggests variability in nest site tenacity among populations. Such variability and migration patterns for South Carolina and Georgia populations need further study.

Although the reproductive potential of marine turtles is high, survival of eggs and hatchlings is low due to a number of biological and physical factors. Egg predation appears to be a serious problem on many nesting beaches, and it has increased on Cape Island, South Carolina, since Baldwin and Lofton conducted their study in 1939 (Caldwell 1959). Baldwin and Lofton reported that 44% of the nests hatched without being disturbed by pre-dators. Ghost crabs (Ocypode quadrata) were the most serious predators noted at this time, entering 41% of nests; however, some of these nests may have produced some hatchlings. Raccoons destroyed only 6% of the nests, but this was after a winter of serious raccoon control measures. In contrast, recent data for four South Carolina barrier islands, including Cape Island, indicate that raccoons are now by far the most serious predator on loggerhead turtle nests, while crabs destroy relatively few (Table 2-26). Where they occur, feral hogs have been noted as important pre-dators on loggerhead turtle nests (Johnson et al. 1974). Although pro-hibited by law, human predation (poach-ing) of loggerhead eggs has been noted frequently on some South Carolina beaches, and was the major cause of nest destruc-tion on Sand Island during 1977 (Table 2-26).

Physical factors are also responsible for some nest destruction. Such factors are usually much less significant than biological factors, but on occasion may be quite important (e.g., see Table 2-26, Sand Island 1978 and 1979). High spring tides accompanied by storm winds may erode nesting beaches and destroy turtle nests. Abnormal amounts of rainfall, usually associated with a tropical storm or hurricane, also may drown nests by raising an island's water table and in-undating the eggs (Ragotzkie 1959a). Dean and Talbert (1975) reported the loss of 90% of the 1974 year class of turtles on Kiawah Island, South Carolina, due to excessive rainfall.

Adult turtles usually are not con-sidered prone to predation except by man. (See Section II of this chapter for dis-cussion of impact of fishing activities on adult loggerhead turtles.) Baldwin and Lofton (in Caldwell 1959) reported that two hound dogs were implicated in killing two female loggerhead turtles on Edisto Island in 1929. Before their ex-tinction on the coastal islands, large predators such as cougars and wolves may have killed occasional turtles but over-all were probably beneficial to sea tur-tles by reducing the raccoon populations. Adult turtles may also be attacked by large sharks, particularly when returning to the sea from a nesting emergence. At such times, turtles are exhausted and less able to escape or fend off an attack.

Hatchlings emerging from nests may be eaten by ghost crabs, raccoons, gulls, and, to a lesser extent, crows. Most hatchling emergences occur at night, which provides protection from predatory birds. Unfortunately, essentially nothing is known about what happens to hatchlings from the time they enter the ocean until they are observed again in coastal waters as juveniles of 50 cm carapace length or larger. Further, it is not known whether the juveniles which appear along the coasts of South Carolina and Georgia actually hatched there (Richardson and Hillestad 1978). This question of re-cruitment is a critical one, but so far few data are available. J. Richardson et al. (1978) and Richardson and Hillestad (1978) observed no evidence of increased recruitment to the Little Cumberland Island population, although a hatchery on that island has produced and released 6,000 - 10,000 hatchlings yearly since 1965. A major data gap here is that age to maturity is not known for Atlantic loggerheads, although it is believed to be in the 12 - 20+ years range (Richardson and Hillestad 1978, J. Richardson et al. 1978). Thus, there may be several ex-planations for the lack of observable effect of the hatchery on recruitment: the hatchlings may suffer extremely high mortality in the ocean, they may not re-turn to their natal beach, or none may have yet reached sexual maturity (J. Richardson et al. 1978). Nevertheless, Richardson and Hillestad (1978) predict that the hatchery program eventually may increase recruitment by a factor of 2.3 - 7.0.

A Sea Grant funded project to deter-mine the amount of loggerhead turtle nesting on the coastal islands of South Carolina was initiated in 1976. Aerial reconnaissance of beaches was conducted during 1976, 1977, and 1978. However, data from this survey have not yet been completely analyzed and are not available (O. R. Talbert, 1979, University of South

97

Table 2-26. Fates (as percent destroyed or hatched) of 1,590 loggerhead turtle nests on four South Carolina barrier islands during three nesting seasons (1977 - 1979) (data from Hopkins et al. 1978 and S. R. Hopkins, 1979, South Carolina Wildlife and Marine Resources Department, Charleston, pers. comm.).

	NORTH ISLAND		SAND ISLAND			SOUTH ISLAND			CAPE ISLAND		
	%		%			%			%		
	1977 N=79	1978 --	1977 N=165	1978 N=217	1979 N=229	1977 N=88	1978 N=156	1979 N=121	1977 N=134	1978 N=192	1979 N=209
Abiotic Factors											
Erosion & inundation	5.2	--	19.6	49.2	10.0	3.4	3.9	5.7	20.1	3.1	8.6
Hurricane David	--	--	--	--	26.2	--	--	1.7	--	--	23.9
Biotic Factors											
Humans	2.5	--	47.5	10.1	2.2	0.0	0.6	0.8	0.0	0.0	0.0
Ghost Crabs	2.5	--	3.2	0.5	3.1	4.6	0.6	3.3	0.0	0.0	2.4
Crabs/Raccoons	8.9	--	6.3	0.5	2.2	1.1	3.2	1.7	1.5	1.1	0.0
Raccoons	69.5	--	16.4	15.7	33.6	86.3	87.2	86.8	75.4	95.8	59.3
Fox/Dog	--	--	--	--	13.1	--	--	--	--	--	--
Hatched	11.4	--	7.0	24.0	9.6	4.6	4.5	0.0	3.0	0.0	5.8
TOTAL	100.0	--	100.0	100.0	100.0	100.0	100.0	100.0	100.0	100.0	100.0

Carolina, Columbia, pers. comm.). The
following assessment of relative nesting
activity is based upon communications from
O. R. Talbert of the University of South
Carolina and S. R. Hopkins of the Non-
game and Endangered Species Section, South
Carolina Wildlife and Marine Resources
Department.

The coast from North Inlet to Bull
Island contains the most significant
loggerhead turtle rookeries in South
Carolina, with the majority of the nesting
activity occurring within the Cape Romain
National Wildlife Refuge. Approximately
half of the nesting activity in the State
is on Cape Island. Raccoon Key, Sand,
North, Cedar, and South islands rank second
through sixth in nests or emergences per
mile of beach. The beaches of the Grand
Strand (from Pawleys Island to Little
River) and the Charleston area (from
Capers Island to Stono Inlet) support
only sporadic nesting. Nesting on the
southern coast of the State is of low to
moderate density; the more important nesting beaches in this area are Dolphin Head,
Fripp, Harbor, Otter, and Bay Point islands.
Kiawah Island had reduced nesting activity
in 1976 and 1977 compared to the years
1972 - 1975, when approximately 200 nests
per season were constructed. (Dean and
Talbert 1975).

In Georgia, Hillestad et al. (1977,
see also Richardson and Hillestad 1978)
conducted an aerial survey of loggerhead
nesting activity on Georgia beaches during
mid-summer 1977. Their results indicate
that nearly all beaches with suitable
dune habitat support some nesting
activity. The greatest numbers of nests
were observed on the Ossabaw-Raccoon
Key-Wassaw complex of islands, on
Blackbeard Island, and on the Cumberland-
Little Cumberland-Jekyll Island complex
(Table 2-27). Fewest nests were observed
on St. Simons, Tybee, Pine, Little Wassaw,
and Sea islands.

A number of projects to collect data
on the nesting ecology of loggerhead
turtles or provide protection for the eggs
and hatchlings have been conducted or are
still in operation in the Sea Island
Coastal Region. Unfortunately, little of
the data collected has been analyzed or
published to date. Brief status reports
for these projects are given in Table 2-
28.

Alteration of coastal habitats by
construction of beachfront dwellings and
removal of dunes and natural vegetation,
with the associated beach lighting and
vehicular and pedestrian traffic, may
affect the utilization of beaches by nesting turtles and reduce survival of hatchlings. Caldwell (1962) noted the rapid
abandonment of the Jekyll Island beach by
nesting loggerheads after significant
ʰeachfront development took place. In this

Table 2-27. Number of loggerhead turtle
nests observed by aerial
surveys of Georgia beaches
during the period 26 June -
31 July 1977 (data from
Hillestad et al. 1977).

ISLAND	NO. OF NESTS
Tybee	1
Little Tybee	11
Wassaw	32
Pine and Little Wassaw	7
Raccoon Key	36
Ossabaw	80
St. Catherines	29
Blackbeard	65
Sapelo	16
Wolf	18
Little St. Simons	23
Sea	8
St. Simons	0
Jekyll	24
Little Cumberland	39
Cumberland	52
TOTAL	441

case, the nesting turtles shifted their
activities to the undeveloped and strictly
controlled beaches of Little Cumberland
and Cumberland islands. Dean and Talbert
(1975) have postulated that the increase
in nesting on Cape Island resulted from
loggerhead turtles abandoning the developed
beaches of the Grand Strand and Charleston
areas for the undisturbed beaches of the
refuge. However, Mann (1977, 1978) found
that in Southeastern Florida, female
loggerheads did not avoid lighted, developed
beaches in favor of "natural" beaches.

Even in areas where turtles continue
to nest on developed beaches, the disorientation of hatchlings by artificial
lighting may seriously interfere with the
reproductive potential of the beach
(McFarlane 1963, Mann 1977, 1978). In
some cases, hatchlings emerge from nests
at night and move toward highway or
building lights instead of toward the
sea, which is the brightest horizon in a
natural situation. The disoriented turtles often are crushed by passing automobiles on roads adjacent to the beaches,
or wander inland where they die of deby-
dration. Also, heavy pedestrian traffic
and the use of heavy equipment on beaches
have been shown to increase mortality of
hatchlings before they leave their nests
(Mann 1977, 1978). Beach nourishment
projects may affect hatchling-emergence if
the amount of sand placed above existing
nests is too great. Ideally, projects
of this nature should be conducted before nesting begins or after all hatchlings
have emerged.

Table 2-28. Summary of research projects dealing with nesting of loggerhead turtles in South Carolina and Georgia.

LOCATION	TYPE OF ACTIVITY	AGENCY/INVESTIGATOR	DURATION	NUMBER OF NESTING FEMALES OR NESTS	AVAILABILITY OF DATA
SOUTH CAROLINA					
1. Cape Island, Cape Romain National Wildlife Refuge	Daylight beach patrols by jeep to count emergences and nests and document predation of nests.	U.S. Fish and Wildlife Service, National Wildlife Refuge	1938-present	An increase in nesting activity sin e 1939 has b ea n tedo This may be the result of a real increase in the Cape Island nesting population or a shift b Cape Island of females hat formerly used other beaches which are now highly developed. Baldwin and Lofton found 400 nests/season on C pa Island during the years 1938 - 1940. Recent censuses have indicated that 2,000 or more nests are laid on the island annually (G. Garris, 1978, Cape Romain NWR, pers. mm.).	Baldwin's and Lofton's observations for 1938 - 1940 were published in Caldwell (1959). Records of recent beach censuses are available at the refuge office, Awendaw, S.C.; contact George Garris, Refuge Manager.
2. Kiawah Island	Nightly beach patrols, tagging of nesting females, indoor hatchery opera- tion b combat raccoon preda- tion.	Kiawah Beach Company	1972-present	The annual nesting population on Kiawah Island for the years 1972 - 1975 was about 70 - 75 females that produced an average of 200 nests/season (Dean and Talbert 1975).	An analysis of data from 1972 to 1975 is provided in the report, "Environmental Inventory of Kiawah Island" (Campbell et al. 1975). Subsequent data are held by the Kiawah Beach Company.
3. South, Sand, North, and portion of Cape islands	Daylight be a patrols b monitor emer- gences, nests, nest predation, and hatching success; sonic and radio track- ing of interest- ing movements of fem les off- shore.	Non-Game and Endangered Species Sec- tion, South Carolina Wildlife and Marine Resources Department	1977-present	In 1977, 88 nests (30 n ts/km of be h) w ra noted on South Island, while in 1978 and 1979, 156 and 121 nests respectively, wera counted. Similarly, approximately 165 n ts (55 nests/km) were observed on Sand Island in 1977, 217 n ts in 1978, and 229 nests in 1979. North Island was surveyed only in 1977, when approximately 79 nests were observed. On the portion of Cape Island survey- ed, 134 nests wera observed in 1977, 192 nests in 1978, and 209 nests in 1979 (S. Hopkins, 1979, S.C. Wildlife and Marine Resources Department, Charleston, pers. mm.)o	Report published (Hopkins et al. 1978) on biotic and abiotic factors affecting nesting success. Contact T. Kholsatt, S. Hopkins, or T. Murphy, Non- game and Endangered Species Section, S.C. Wildlife and Marine Resources Department, for additional data.

Table 2-28. Continued

	TYPE OF ACTIVITY	AGENCY/INVESTIGATOR	DURATION	NUMBER OF NESTING FEMALES OR NESTS	AVAILABILITY OF DATA
4. Cedar and Cape Islands and entire coast	Coast-wide aerial survey with daylight patrols on Cedar and C p a islands b pr e vide ground truth for fly-over surveys. D a taken on nesting and nonnesting emergences and predation rates, with other effort to evaluate relocation of nests as d em-rent to raccoon predation.	Dr. Steven Stancyk, Belle W. Baruch Institute for Marine Biology and Coastal Research, University of South Carolina, Principal Investigator	1976-present	No data reports presently available	No published reports available. For information contact Dr. Steven Stancyk, Baruch Institute, University of South Carolina.
GEORGIA (based on compilation of Hillestad et al. 1977).					
1. Wassaw Island	Nightly beach patrols, tagging of nesting females, hatchery operations.	Cooperative venture of Savannah Science Museum and U.S. Fish and Wildlife Service	1972-present	Average of 40 nesting females/year.	D ta not presently analyzed or published; information entered in University of Georgia cooperative tagging project data processing system; contact Jerry Williamson, Savannah Science Museum, for present status of project and data.
2. Ossabaw Island	Nightly beach patrol, tagging of females and hatchery operation.	Georgia Governor's Intern Program, New York Zoological Society and private donation	1972-present except for 1976	Number of nesting females tagged ranged from 14 to 45.	No published data available.

101

Table 2-28. Concluded

LOCATION	TYPE OF ACTIVITY	AGENCY/INVESTIGATOR	DURATION	NUMBER OF NESTING FEMALES OR NESTS	AVAILABILITY OF DATA
3. Jekyll Island	Nightly beach patrol, tagging of females.	Cooperative project involving Georgia Game and Fish Div., Jekyll Island Authority, Brunswick Junior College, Coastal Georgia Audubon Society, Governor's Emergency Fund, and University of Georgia	1958, 1972–present	Average of 41 nesting turtles observed annually.	Data entered in Univeristy of Georgia data processing system and presently being analyzed as p rt of a study of the St. Andrews Sound turtle nesting area (includes Cumberland and Little Cumberland islands).
4. Little Cumberland Island	Nightly beach patrol, tagging of nesting females, hatchery.	J. I. Richardson, University of Georgia, Principal Investigator	1964–present (This is the largest continuous ne ting beach project in the U.S.)	Average of 82 nesting turtles (range 44–125) utilize the beach annually.	See Bell and Richardson 1978, Richardson 1978, J. Richardson et al. 1978, T. Richardson et al. 1978. Contact J. I. or T. H. Richardson for project details.
5. Cumberland Island	Nightly beach patrol, tagging of nesting females, hatchery.	National Park Service and owners of Candler estate	1972–present	An average of 55 turtles nest on this island annually.	See above.

6. Birds

a. **Overview and Trophic Relationships.**
Excluding rare or unusual species, approximately 36 species of birds regularly utilize the marine intertidal habitat. The majority of these are scavengers or piscivores that are also prominent inhabitants of marine subtidal waters. There are, however, several species that feed primarily on macrobenthic invertebrates found only in the surf zone. These birds are largely confined to the intertidal beach habitat.

The avifauna of intertidal beaches in South Carolina and Georgia is composed primarily of permanent residents (25 species), 14 species of which are dominants (Table 2-29). In addition, one winter resident (the knot) and one summer resident (the least tern) also are considered dominant. Other inhabitants of the intertidal zone include an additional five winter and four summer residents (Table 2-29). No serious attempt has yet been made to quantify the use of the beach habitat by various bird species in the Sea Island Coastal Region. Existing data are, at best, spotty due to infrequent censusing.

Many of the birds that utilize the beach habitat are primarily marine species. With the exception of the insectivorous black and gull-billed terns, these birds feed on nektonic organisms from the adjacent subtidal waters (Fig. 2-14; see also Section II of this chapter). These marine birds, then, use the beach primarily for "loafing" or breeding rather than for feeding.

A variety of birds commonly use intertidal beaches of coastal islands as feeding grounds. The base trophic level of this habitat utilized directly by the avifauna is the macrobenthos, which includes annelids, crustaceans, and mollusks. These macrobenthic organisms are consumed by American oystercatchers, scaup, ruddy turnstones, willet, knot, sanderlings, plovers, and sandpipers. Piscivorous species include such dominants as the brown pelican, royal tern, and black skimmer. Beach scavengers include the gulls plus five terrestrial species (the fish crow, common crow, boat-tailed grackle, turkey vulture, and black vulture). All of these birds rely heavily on dead animal matter as a source of food. Finally, the top avian predator on the beach is the peregrine falcon, which preys on all the birds found in the intertidal zone. Unfortunately, the peregrine falcon is uncommon and endangered (U.S. Department of Interior, Fish and Wildlife Service 1973, Forsythe 1978, Gauthreaux et al. 1979; see also Chapter One, Section VI).

b. **Representative Species.** Since most birds that utilize intertidal beaches are also common residents of other habitats, only four species which are limited to intertidal beaches have been selected for further discussion. These are sanderling, knot, piping plover, and Wilson's plover. Other species are treated in detail, where appropriate, in other chapters.

The sanderling is a common permanent resident, but does not breed in the Sea Island Coastal Region (Tomkins 1936, Burleigh 1958, Forsythe 1978). Those individuals present in midsummer are thought to be immature (Wayne 1910); adults breed in the Arctic at this time. The sanderling is nearly cosmopolitan, since its range includes the Pacific and the southern hemisphere to southern Chile and Peru as well as the Atlantic coast of North America (Bent 1962b). Throughout this extensive range, the species is virtually confined to the beach environment (Bent 1962b). In South Carolina, Wayne (1910) never recorded it anywhere but on a sand beach. However, in Georgia, Burleigh (1958) cites four records of this species away from the coast. The sanderling's decided preference for beaches is supported by its choice of food. It probes the sand in the surf swash for sand fleas, shrimp, small crustaceans, small mollusks, marine worms, and small insects (Bent 1962b). The sanderling favors high energy beaches and is the most common and well known beach sandpiper.

The knot is a fairly common winter resident in the characterization area (Burleigh 1958, Forsythe 1978). It is most numerous during migration and was previously believed to be only a transient (Wayne 1910, Sprunt and Chamberlain 1949). Like the sanderling, the knot breeds in the Arctic and winters as far south as the extreme tip of South America (Hall 1960). The knot is also limited to the beach environment, and is seldom recorded in other localities (Sprunt and Chamberlain 1949). According to Burleigh (1958), the knot is so restricted to outer beaches in its seasonal movements that it is one of the less familiar shorebirds in the study area. Food of this species includes minute mollusks and crustaceans, young horseshoe crabs, and marine worms (Bent 1962b). In South Carolina and Georgia, the knot feeds extensively on the butterfly shell clam and other mollusks (Wayne 1910, Burleigh 1958). Although still found in seemingly large flocks, the knot was once a prized game species during the days of shorebird hunting. As such, it was shot by the thousands when decoyed, and its numbers were dramatically reduced by the turn of the century (Hall 1960).

The piping plover is a winter resident of beaches along the coast of South Carolina and Georgia from August through May (Forsythe 1978). It is fairly common during winter, but is more numerous during migration (mid-March to mid-May and August to October) (Sprunt and Chamberlain 1949, Burleigh 1958). Wayne (1910) regarded the piping plover as a rare transient, extremely shy and confined to sandy

Table 2-29. Birds of intertidal beaches in South Carolina and Georgia (compiled from Sprunt and Chamberlain 1949, 1970, Audubon Field Notes 1967 – 1970, Chamberlain 1968, American Birds 1971 – 1977, Forsythe 1978).

DOMINANT

Species			
Brown pelican	C	PR	
Black-bellied plover	C	PR	
Ruddy turnstone	C	PR	
Willet	C	PR	
Knot	FC	WR	
Least sandpiper	C	PR	
Dunlin	C	PR	
Semipalmated sandpiper	C	PR	
Western sandpiper	C	PR	
Sanderling	C	PR	
Herring gull	C	PR	
Ring-billed gull	C	PR	
Laughing gull	C	PR	
Least tern	C	SR	
Royal tern	C	PR	
Black skimmer	C	PR	

MODERATE

Species			
American oystercatcher	C	PR	
Wilson's plover	C	SR	Mar.-Oct.
Dowitcher	C	PR	
Bonaparte's gull	C	WR	Oct.-May
Gull-billed tern	FC	SR	Apr.-Oct.
Forster's tern	C	PR	
Common tern	C	PR	
Sandwich tern	FC	SR	Apr.-Oct.
Caspian tern	FC	PR	
Black tern	FC	SR	May-Sept.
Fish crow	C	PR	
Boat-tailed grackle	C	PR	
Piping plover	FC	WR	Aug.-May

MINOR

Species			
Snowy egret	C	PR	
Scaup	C	WR	Oct.-Apr.
Turkey vulture	C	PR	
Black vulture	C	PR	
Peregrine falcon	U	WR	Sept.-Apr.
Great black-backed gull	U	WR	Nov.-May
Common crow	C	PR	

KEY: C – Common, seen in good numbers
FC – Fairly common, moderate numbers
U – Uncommon, small numbers irregularly
PR – Permanent resident, present year around
WR – Winter resident
SR – Summer resident

NOTE: Dominance indicates relative importance of the species as a group in the community. This concept is not based necessarily on taxonomic relationships but rather on numbers, size, and trophic dynamics.

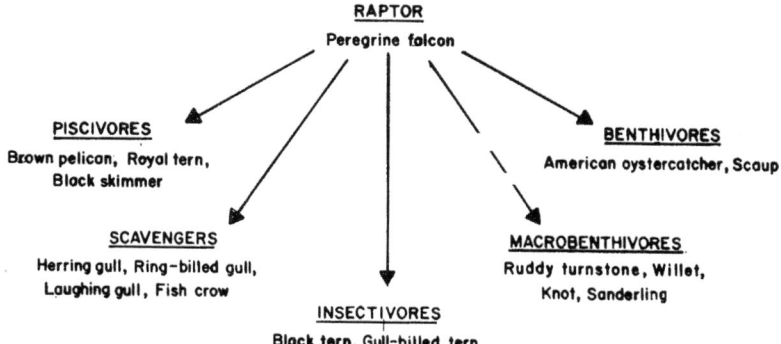

RAPTOR
Peregrine falcon

PISCIVORES
Brown pelican, Royal tern,
Black skimmer

BENTHIVORES
American oystercatcher, Scaup

SCAVENGERS
Herring gull, Ring-billed gull,
Laughing gull, Fish crow

MACROBENTHIVORES
Ruddy turnstone, Willet,
Knot, Sanderling

INSECTIVORES
Black tern, Gull-billed tern,
Least sandpiper

Figure 2-14. Generalized feeding groups (with representative species) of birds on intertidal ocean beaches of the Sea Island Coastal Region.

beaches. This species is still not numerically abundant, but can be observed regularly on beaches in winter. The piping plover feeds on marine worms, insects, crustaceans, and mollusks, especially the butterfly shell clam (Sprunt and Chamberlain 1949, Bent 1962a). Like many shorebirds, this species was previously much more numerous, but was reduced almost to the point of extinction during the late 1800's (Bent 1962a).

The Wilson's plover is a common summer resident from March to mid-October. It is also found predominately on beaches, and favors inlets between barrier islands (Sprunt and Chamberlain 1949). This bird also frequents mud flats and spoil banks, but prefers beaches where it nests in the dune field among the foredunes (Wayne 1910, Sprunt and Chamberlain 1949, Burleigh 1958, Chamberlain and Chamberlain 1975). Recently, this species has become the subject of special concern due to the reduction of its breeding habitat through resort development (Gauthreaux et al. 1979). The Wilson's plover feeds on fiddler crabs, other small crabs, shrimp, mollusks, and insects such as beetles, flies, ants, and spiders (Sprunt and Chamberlain 1949, Hall 1960, Bent 1962a).

Man's impact on the intertidal beach habitat, and its subsequent effects on birds, is closely related to alterations of the maritime dune and beach communities (see Chapter Three). The potential impact of oil pollution on birds of coastal

marine waters has already been discussed (see Section II of this Chapter).

7. Mammals

The intertidal beaches of the Sea Island Coastal Region of South Carolina and Georgia have not been studied explicitly as mammal habitat. Nevertheless, a number of scattered observations have been recorded, and inferences can be drawn from surveys of the mammals on several islands. (For more detailed information on the mammals of coastal islands, see Chapter Three, Maritime Ecosystem.)

The intertidal marine beach cannot be considered permanent habitat for any mammal, but a number of species resident in adjacent dunes and forests will enter this environment to forage or to escape insects. Several such species may have a considerable impact on the intertidal zone. Scavengers and predators that can be considered regular nocturnal visitors to the beaches include the Virginia opossum, the eastern mole, the eastern wood rat, the cotton rat, the raccoon, the feral hog, man, and occasionally the bobcat and the nine-banded armadillo (Johnson et al. 1974, Sanders 1978). In addition, white-tailed deer and cattle also may come onto the beach.

In general, the raccoon (especially), feral hog, and man are the most important mammals occurring on marine intertidal beaches in South Carolina and Georgia.

105

Raccoons make nightly feeding forays on many partially developed and undeveloped beaches. Recent research on the nesting success of the Atlantic loggerhead turtle in South Carolina indicated that raccoons destroyed 16% - 34% of the turtle nests laid on Sand Island and 59% - 96% of those laid on three other islands (Table 2-26). Depredation of loggerhead nests by raccoons also has been noted at Hutchinson Island, Florida (Gallagher et al. 1972); Kiawah Island, South Carolina (Dean and Talbert 1975); and Cape Island, South Carolina (Caldwell 1959). Raccoons also feed on a variety of fish, crustaceans, and mollusks deposited in the drift line or captured on beaches.

In Georgia, feral hogs are a major problem on Ossabaw Island and probably would take all turtle eggs laid if these eggs were not moved to a hatchery. The herds of feral hogs have been reduced on Cumberland and Little Cumberland islands, and no longer pose a great threat to turtle nests (J. I. Richardson, 1978, Southeastern Wildlife Services, Inc., Athens, Georgia, pers. comm.). Feral hogs also occur on many undeveloped South Carolina islands. They often scavenge along the beaches and root extensively along the transition area where dune vegetation spreads onto the upper beach.

Deer tracks are common on intertidal beaches of islands where deer populations exist. Deer are thought to graze to some extent on the dune vegetation which spreads onto the upper beach, but the intertidal habitat is obviously not a principal feeding habitat. Rather, the deer are thought to come to the ocean beaches at night in an attempt to escape the biting insects of the forest and marsh (P. Wilkinson, 1978, South Carolina Wildlife and Marine Resources Department, Charleston, pers. comm.). They may also use the beaches as a source of salt, by licking salt left by evaporation on exposed objects. Like other mammals which penetrate this habitat, deer are almost entirely nocturnal.

Pelton (1975) has recorded bobcat tracks on the beach at Kiawah Island, and many of the coastal islands have bobcat populations (see Chapter Three for more information). Thus, it is reasonable to assume that bobcats patrol the beaches at night. The many seabirds which roost on the beaches at night constitute a likely source of prey.

Most of the small rodents and insectivores which occupy the dry sandy dune areas may be expected to enter the beach habitat at night. For example, the old-field mouse is also known as the beach mouse. It is recorded from some, but not all, of the coastal islands of Georgia (Johnson et al. 1974, Hillestad et al. 1975). The coastal range of this species in the Sea Island Coastal Region is apparently restricted to Georgia, as it is yet unreported from the coastal islands of South Carolina (Golley 1966).

Rabbit tracks and fecal pellets are not unusual on the upper beach adjacent to dunes. Since both eastern cottontail rabbits and marsh rabbits are known to occupy the dunes, it is likely that both penetrate the upper beach. The only food sources here are the succulent plants of the upper beach-dune transition zone.

The Virginia opossum was once abundant on Cumberland Island, but has since been extirpated there and on Little Cumberland Island (Hillestad et al. 1975). In contrast, the nine-banded armadillo invaded Cumberland Island in 1973 and was seen on Little Cumberland Island in 1974 (Hillestad et al. 1975). Armadillos are now regularly seen on Little Cumberland Island (J. I. Richardson, 1978, Southeastern Wildlife Services, Inc., Athens, Georgia, pers. comm.).

In some cases, activities of man have had a dramatic impact on the beaches as mammal habitat, both directly and indirectly. The most dramatic direct effects have resulted from residental and resort development. Several kinds of impacts can be recognized easily. The most obvious is destruction of adjacent habitat, especially the dunes and seaward portions of maritime forests from which most of the transient mammals enter the beach zone. On some highly developed beaches (e.g., Folly Beach), the dune zone has been largely destroyed and extensive bulkheading has taken place. Even where partial dunes remain, bulkhead construction completely alters the ectonal zone between dune and upper beach. Examples where greater or lesser alteration have occurred include the Isle of Palms and Sullivans Island near Charleston, Tybee Island near Savannah, and St. Simons and Jekyll islands near Brunswick.

As the amount of human activity on the beaches increases, especially at night, use of the beaches by transient mammals is inhibited even if adjacent habitat has not been destroyed. The use of off-road vehicles on beaches at night is clearly detrimental to the use of this habitat by nocturnal mammals.

Beach erosion resulting from both sea level changes and man-made coastal alterations has dramatically altered many beaches as habitat for mammals. In extreme cases, the upper beach, which would normally be covered only on spring and storm tides, has been eliminated completely by erosion.

D. DECOMPOSERS AND NUTRIENT CYCLING

Undoubtedly, most of the nutrient input to intertidal marine beaches comes

from the sea, either in the form of
macro-detritus, small particulate matter,
or dissolved organics. Significant
amounts of these materials are probably
imported from nearby estuaries and wetlands, and their importance as direct
food sources for beach fauna has already
been mentioned. In addition, the detritus
and particulate and dissolved organics
provide substrates for the growth of a
fairly dense bacterial flora, which in
turn releases inorganic nutrients and
serves as a food resource for meiobenthic
nematodes, turbellarians, and copepods
(Sieburth 1976). Humm (in Pearse et
al. 1942) reported some 200,000 bacterial
cells/g of sediment on a North Carolina
beach. These interstitial bacteria were
primarily involved in the nitrogen cycle
and chitin digestion. Sulfur bacteria
are undoubtedly also important in the
deeper black layers of marine beaches.
Steele (1974) noted that even though decomposers play a key role in the detritus-
based trophic dynamics of marine beaches,
virtually nothing is known about bacterial
community structure or function in marine
intertidal sediments. This is a major
data gap which should receive attention
at the earliest possible opportunity.
Intertidal marine beaches are particularly
susceptible to catastrophic damage, both
ecological and economic, by oil spills.
Yet, we know virtually nothing about the
occurrence and dynamics of the important
petroleum-degrading microflora of beaches
of the Sea Island Coastal Region.

CHAPTER THREE

MARITIME ECOSYSTEM

I. GENERAL DESCRIPTION

A. DEFINITION

Upland ecological systems include those areas not classified as wetland or aquatic systems by Cowardin et al. (1977). Further, uplands are never flooded during years of normal precipitation. For this ecological characterization, we have divided uplands into two distinct ecosystems: 1) an upland ecosystem (see Chapter Six), occurring only on the mainland, and 2) a maritime ecosystem, which is defined simply as all upland areas located on barrier islands. Colquhoun and Pierce (1971) described a generalized barrier island, such as those of the Sea Island Coastal Region, as follows:

"Recent barrier islands, in plain view, are gently arcuate-shaped sand bodies developed along shore lines. Their seaward margins are curvilinear in plain view, their landward margins, irregular, resulting from washover fans, tidal deltas, and marshes . . . True barrier islands are separated from the mainland by sounds or bays and the mainland is expressed as a submerged shoreline. Where separated from the mainland, barrier islands are relatively narrow features, usually less than a mile in width." (For more geological information on barrier islands, see Volume I.)

The maritime ecosystem, then, is limited on its ocean side by the extreme high spring tide mark (i.e., the landward limit of the marine intertidal subsystem) and on the landward side by tidal marshes, creeks, or rivers.

B. SUBSYSTEMS

We recognize four subsystems within the maritime ecosystem, each differentiated from the other primarily by physical features and vegetative communities. One of these, the bird key and bank subsystem, is a special unit physically separated from the others by water. The other three subsystems (the dune, transition shrub, and maritime forest units) are contiguous and grade into one another (Fig. 3-1).

1. Bird Key and Bank Subsystem

Bird keys and banks (also called sand spits and sand bars) are small, isolated islands usually found in tidal inlets and broad bays. Following the terminology of Hayes and Kana (1976), these features would generally be classified as ebb or flood tidal deltaic islands or as swash bars (see Volume I, section VI. D.). All exhibit low topographic profiles and are frequently subject to overwash by spring tides and storm action. Additionally, these small islands are known to be geologically unstable and to migrate in response to inlet morphology (Hayes et al. 1975).

Sand keys and banks (bars) are found throughout the Sea Island Coastal Region and generally are less than one square mile (2.6 km^2) in size. They are found commonly near river mouths but also may be associated with harbor and inlet jetties. These bars are characterized by erratic fluctuations in size, shape, and vegetative cover from year to year as the sand shifts in response to storms and other physical forces. A good example of such changes is Deveaux Bank, which lies near the mouth of the North Edisto River, South Carolina. In 1975, it was composed of 35% vegetated area, 55% mud flats, and 10% bare sand; by 1977 these proportions had changed dramatically to 15% vegetated areas, 35% mud flats, and 50% bare sand (Gaddy 1977).

2. Dune Subsystem

A variety of dunes, ranging from low, sprawling dune fields to high, well-developed dunes, are found in the maritime ecosystem. All of the dunes discussed here are "open dunes;" "closed dunes," or former dunes covered with woody vegetation, are discussed under the transition shrub and maritime forest subsystems.

Sand, wind, and vegetation interact to form coastal dunes. Windblown sand accumulates behind wrack (dead grass culms, usually smooth cordgrass washed up on the beach), which provides seedbed for plants such as beach hogwort, sea oats, Russian thistle, seabeach orach, sea rocket, sea beach panic grass, and creeping spurge, all of which aid in the initial stages of dune-building (Hosier 1975). Sea oats and sea beach panic grass, however, are of longer-term significance in dune formation (Johnson et al. 1974). Because these two perennials tolerate intense salt spray, as well as develop extensive lateral root systems, they are instrumental in the stabilization and growth of newly formed dunes (Fig. 3-2).

3. Transition Shrub Subsystem

The transition shrub community, also called the "maritime shrub thicket," is a dense but generally narrow ecotonal band between the maritime forest and the dune community. Its location depends on salt spray intensity. The community is characterized by low diversity of plant species, extremely dense structure with little or no understory, and a total

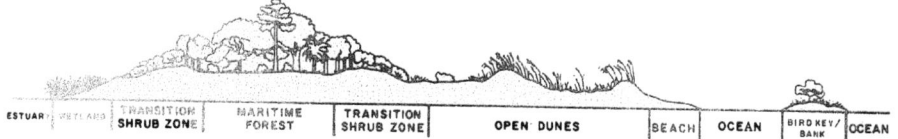

| ESTUARY WETLAND | TRANSITION SHRUB ZONE | MARITIME FOREST | TRANSITION SHRUB ZONE | OPEN DUNES | BEACH | OCEAN | BIRD KEY/ BANK | OCEAN |

Figure 3-1. Generalized cross section of the maritime ecosystem of the Sea Island Coastal Region.

height of up to 13 ft (3.9 m) (Hosier 1975). Typically, the bank may be only a few yards in width but, in unusual cases, may be several hundred yards in depth. Wells and Shunk (1938) reported the occurrence of transition shrub communities from 0.1 mi (0.16 km) (Myrtle Beach, South Carolina) to 1.25 mi (2 km) (Kitty Hawk, North Carolina) from the mean high water mark. The precise location and width of the transition shrub zone is a function of the angle of the beach in respect to the prevailing winds and the height of the fore and back dunes (i.e., the amount of protection the zone has from salt spray).

4. Maritime Forest Subsystem

Precise limits of the maritime forest are extremely difficult, if not impossible, to define. Bourdeau and Costing (1959) noted that dominance by live oak seemed to be a good indicator of the extent of the maritime forest. Wells and Shunk (1938) and Oosting (1954) used loblolly pine and turkey oak as indicators of the beginning of non-maritime forests. Rayner (1974) termed maritime vegetation "all vegetation within a distinct zone extending from the surf zone of the outer beach inland to the beginning of the southeastern oak-hickory association." Other researchers have generally defined maritime vegetation in terms of the presence of salt spray (Oosting and Billings 1942, Boyce 1954). The zone of maritime forest vegetation has been called the "maritime strand," "the salt spray zone," and, more simply, "maritime forest" (Wells 1939, Oosting 1954). Here, the maritime forest is considered that zone of forest vegetation between the transition shrub communities on the seaward and landward sides of barrier islands (see Fig. 3-1).

C. MODEL

The distinct zonation of plants is the characteristic feature of the maritime ecosystem. Salt spray, which is the result of tide, wind, and wave interaction, is the major causative factor contributing to the characteristic plant distribution and zonation of the maritime ecosystem. Details of the physical and

climatic features of barrier island maritime systems are given in Volume I, Physical Features.

In the maritime ecosystem, as in all others, the primary source of energy is the sun (Fig. 3-3). The sun provides the basic energy for the synthesis of organic molecules from CO_2 by autotrophic organisms (plants) through photosynthesis. In turn, the organic carbon produced by the autotrophs provides the basic energy for nearly all other life forms (the heterotrophs). In the maritime ecosystem, as in other upland systems, the trees and shrubs account for nearly all primary production; algae and phytoplankton are generally insignificant in this regard. These primary producers are grazed extensively by first level heterotrophs (e.g., many insects, birds, amphibians, herbivorous and omnivorous mammals) that feed on leaves, stems, twigs, fruits, seeds, roots, etc. Many primary heterotrophs then become food for secondary heterotrophs (e.g., insects, reptiles), which in turn are eaten by tertiary consumers (e.g., hawks, bobcats, and man). The size of predator/parasite populations tends to regulate the size of prey populations and vice versa. Excretion and death at each level result in the release of nutrients either directly into the soil or through the action of decomposers (bacteria and fungi). The decomposers themselves may release nutrients directly to the soil or may be grazed by heterotrophs, which in turn excrete and decompose. An important source of nutrient input is the salt spray, which is characteristic of the maritime ecosystem (see Maritime Forest Subsystem and Figure 3-3 for more details). Other avenues of nutrient renewal are discussed in Chapter One under Biogeochemical Cycles.

Upland systems such as the maritime ecosystem are characterized by having a large amount of primary production which is not directly grazed by heterotrophs (Steele 1974). Thus, much of the total primary production eventually falls to the maritime forest or shrub floor as litter (dead leaves, twigs, stems, seeds, flowers, tree trunks, etc.), where it may be browsed directly by a complex litter

109

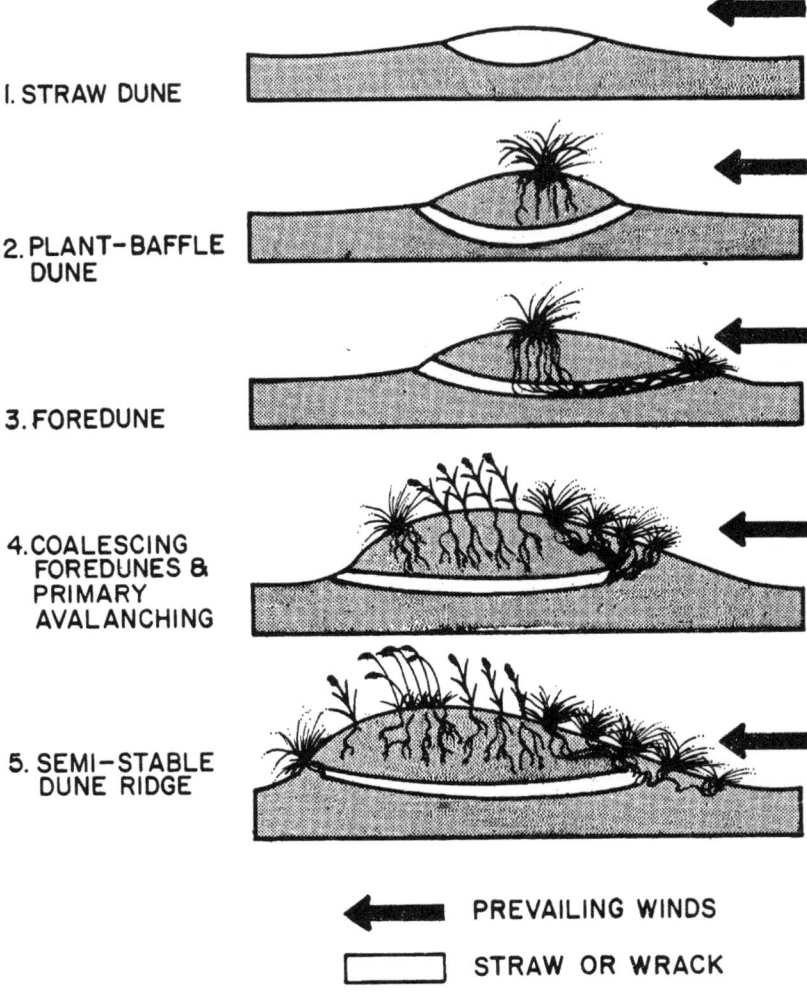

1. STRAW DUNE

2. PLANT-BAFFLE DUNE

3. FOREDUNE

4. COALESCING FOREDUNES & PRIMARY AVALANCHING

5. SEMI-STABLE DUNE RIDGE

PREVAILING WINDS

STRAW OR WRACK

Figure 3-2. Successional stages of dune formation (adapted from Oertel and Larsen 1976).

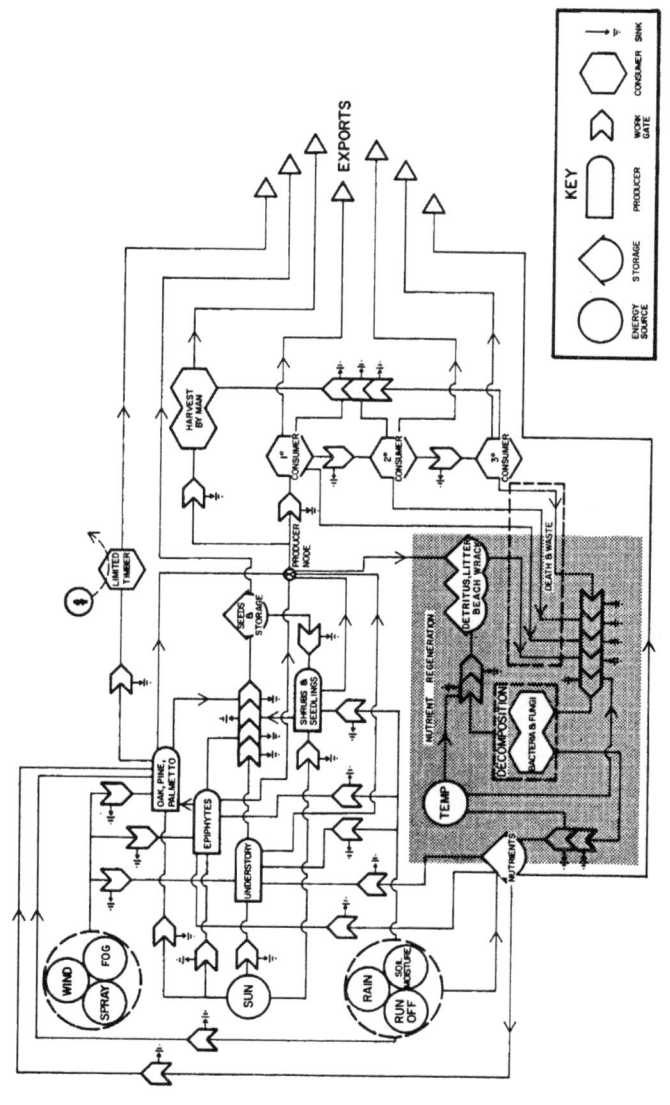

Figure 3-3. A generalized energese model of the maritime ecosystem of the Sea Island Coastal Region. (Shading = nutrient regeneration.)

111

fauna or, more commonly, be decomposed. In addition, the maritime ecosystem receives inputs of detritus on its seaward-facing beach and on its landward marsh side. Although most of the detritus originates in estuarine wetlands and becomes deposited on the margins of the maritime system by tides, that of the ocean beach often contains quite a bit of Sargassum, Ulva, and other more oceanic material, in addition to dead cordgrass culms. Both types of detritus are utilized extensively by a wide variety of browsers and decomposers.

Not all of the maritime ecosystem's production is utilized in situ; considerable amounts are exported and such exports, like the imports of detritus, serve to connect this system to others. For example, many of the motile animals (rabbits, raccoons, deer, bobcats, birds, etc.) which feed extensively in maritime habitats are not restricted to these habitats, although there may be resident populations. Such feeding by nonresident species results in a net export of organic matter. However, the reverse situation also holds true; for example, maritime resident populations of raccoon and deer may feed extensively in adjacent beach and wetland systems, respectively, thereby serving as mechanisms of import. Other kinds of export include harvest of timber resources and game populations by man, runoff into oceanic or estuarine environments, island erosion, etc.

Throughout the ecosystem, physical factors such as temperature, nutrient concentrations, degree of soil moisture, quantity of salt spray, wind, etc., determine limits for the various kinds and rates of biological production. Populations of a given kind of organism flourish only within certain limits of environmental conditions; when those conditions change, the populations are reduced or eliminated and other organisms are favored (see Chapter One). Similarly, growth and reproduction of many organisms vary seasonally with such factors as temperature and day length.

D. GENERALIZED FOOD WEB

Food webs in the maritime ecosystem are highly complex, with many branches, connections, and overlaps. Figure 3-4 illustrates in a very generalized fashion some of the trophic interrelationships of major groups of organisms represented in the Sea Island Coastal Region. As implied by the numerous connecting links in the diagram, these interrelationships are far from simple. Few large groupings of organisms, other than the producers, can be confined to a single trophic level. Much more commonly, representatives of a given group feed at more than one trophic level, resulting in many "horizontal links" in the food web.

Within a food web, the magnitude of production at each trophic level is determined by three primary factors: trophic efficiency, transfer rate, and number of horizontal links. Trophic efficiency is the efficiency (usually taken to be 10%, Odum 1971) with which a population transfers energy to the next step (horizontal or vertical) in the food web. Transfer rate refers to the rapidity with which energy is transformed from one step in the food web to the next through grazing, predation, waste, or death. The more horizontal links present in the food web, the greater the degradation of energy within a given trophic level, and the less energy available for transfer to the next level. These factors are discussed in more detail in Chapter Two, Marine Ecosystem.

The outstanding feature of almost any terrestrial community, including the maritime community, is the presence of large, rooted green plants, which are the chief primary producers. This situation is in distinct contrast to the marine environment, where single cell phytoplankters account for nearly all the primary production (see Chapter Two). Further, usually <10% of the plant material produced in terrestrial ecosystems is consumed alive, despite the presence of large, active herbivores (e.g., rodents, birds, deer) (Steele 1974). Yet, in the oceans, herbivores graze down the phytoplankton almost as fast as it is produced. Thus, herbivores in the maritime ecosystem generally are not as food resource-limited as are those in the marine environment (Steele 1974).

Because relatively little of the maritime ecosystem's primary production is consumed directly, much of it falls to the forest floor as litter (leaves, stems, fruits, seeds, tree trunks, etc.). This litter buildup decays, forming a large organic reserve on and within the forest soil, and provides nutrition for a complex assemblage of small consumers known as the litter fauna. It is this mass of dead and decaying plant matter, seeds, and decomposer organisms that forms the major food compartment in the maritime ecosystem. Thus, the energy fixed by the trees enters the maritime food web primarily by way of decomposers (especially fungi), plus invertebrate consumers of wood, foliage, and other plant parts, a few grazers and browsers, and many fruit- and mast-eating birds and mammals (Johnson et al. 1974). Within the maritime forest, it is likely that live oak acorns comprise the single most important food source for many species of wildlife including deer, raccoons, feral hogs, wild turkeys, and other birds (Elton 1968, Johnson et al. 1974).

Secondary and tertiary consumers include various insects, reptiles, birds, and mammals. There are relatively few

112

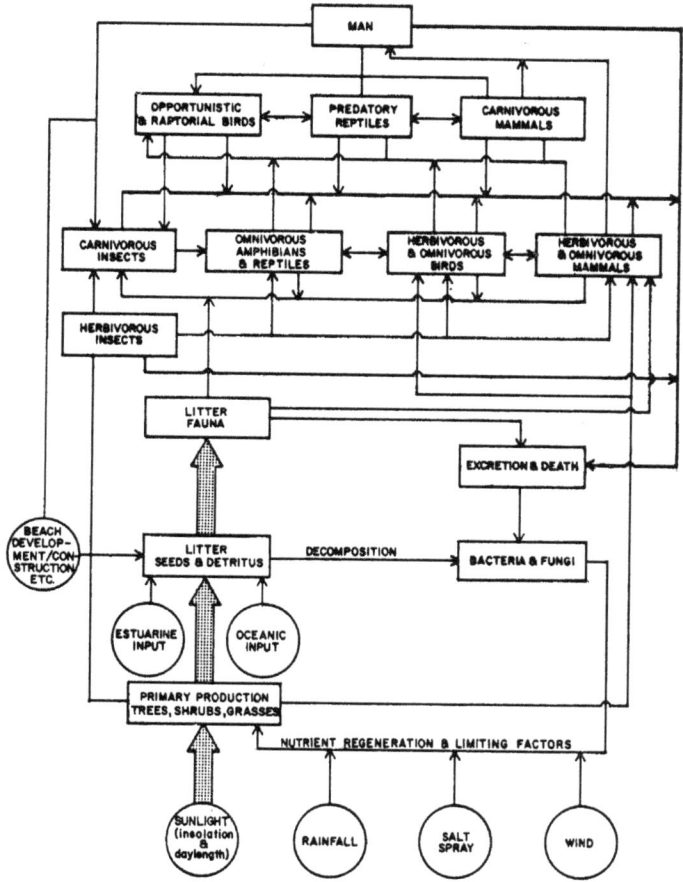

Figure 3-4. A generalized food web for the maritime ecosystem of
the Sea Island Coastal Region.

mammalian carnivores on the islands, most
of those formerly present (e.g., foxes,
wolves, pumas, bears) having been extermi-
nated. The terminal carnivore niches are
filled mainly by predatory birds, the
Eastern diamondback rattlesnake, the bobcat, and relatively inefficient mammalian
predators/omnivores such as raccoons and
opossums (Johnson et al. 1974).

II. REPRESENTATIVE BIOTA
AND INTERACTIONS

The four subsystems of the maritime
ecosystem are occupied by distinctly different vegetative communities. On the
other hand, the communities of motile consumer species are often quite similar,
especially in the contiguous subsystems,

although there may be marked differences in the relative importance of various species. Thus, we have described the plant communities separately for each subsystem and have lumped discussions of animal populations, taking care to point out major differences in the roles of particular animal species or groups in the various subsystems.

A. PRODUCERS

1. Bird Key and Bank Communities

Zonation of vegetation in coastal environments is determined by two major factors: 1) wind as a carrier of salt spray, and 2) tolerance of plants to salt-water immersion (Wagner 1964). Not surprisingly, then, vegetation of sand keys and banks is limited to only pioneer, salt-tolerant plants. Two studies of the plant communities on sand keys and banks have been conducted in South Carolina (Pinson 1973, Gaddy 1977), and their results are summarized below. No data specifically pertinent to this system are available for Georgia, but considerable information on the similar dune subsystem is available and is summarized in the next section.

Pinson (1973) described the vegetation of Bird Bank, a 3.2 km (2 mi) long by 100 - 200 m (100 - 200 yds) wide sand key located across the south jetty of Winyah Bay, South Carolina. This bank consists of "low back dunes" dominated by saltmeadow cordgrass, and "higher foredunes" dominated by panic grass with scattered clumps of sea oats and beach elder. Saltmeadow cordgrass and panic grass were the overwhelming dominants; beach elder, a co-dominant, was the only woody plant found (Table 3-1).

Gaddy (1977) listed seven plant communities from Deveaux Bank, a 1.2 km (0.75 mi) long sand bank in the North Edisto River, South Carolina. These were: 1) a glasswort community, 2) a smooth cordgrass community, 3) a saltmeadow cordgrass community, 4) a saltmeadow cordgrass-panic grass association, 5) a mixed shrub-forb-grass association (dominated by dog fennel, camphorweed, beach elder, sea myrtle, and saltmeadow cordgrass), 6) a panic grass community, and 7) a mixed smooth cordgrass-sea purslane-glasswort association. Communities 1, 2, and 7 are marsh communities, while 3, 4, 5, and 6 are dune communities. Figure 3-5 illustrates the zonation of these communities on Deveaux Bank (because of its rarity, community 7 is not included in the figure). Panic grass dominates high foredunes because of its tolerance to intensive salt spray, and saltmeadow cordgrass is the low dune dominant because of its tolerance to immersion (Gaddy 1977). The relative rarity of sea oats on Deveaux Bank, Bird Bank, and in several other South Carolina dune communities is discussed in the following section.

2. Dune Communities

In South Carolina, dune vegetation has been described in most detail by Pinson (1973) (for Debidue, North, South, Cedar, Murphy, Cape, Bull, Capers, Dewees, Isle of Palms, Kiawah, Seabrook, Edisto, Hunting, Fripp, and Hilton Head islands), Stalter (1974a) (for Huntington Beach, Isle of Palms, and Hunting islands), and Hosier (1975) (for Kiawah Island). Figure 3-6 presents a generalized profile of the zonation of dune plants commonly encountered in South Carolina, based primarily on the work of these authors.

Pinson (1973) recognized five floristic zones in the dunes of the islands he studied: 1) foreslope of foredunes, 2) back slope of foredunes, 3) foreslope of back dunes, 4) back slope of back dunes, and 5) interdune. He listed a total of 68 species from these zones, including four not previously reported from South Carolina (golden aster, seashore paspalum, beach grass, and sweet grass). Plants most commonly encountered by Pinson (1973) were sea oats, saltmeadow cordgrass, sand grass, panic grass, camphorweed, beach elder, euphorbia, horseweed, evening primrose, and beach pennywort. From calculations of community "coefficients of similarity," he concluded that three general floristic patterns or communities occurred in his study sites: 1) a sea oats-panic grass-beach elder pattern (peak density on the foreslopes of foredunes), 2) a sand grass-camphorweed-evening primrose pattern (peak density on the slopes of back dunes), and 3) a saltmeadow cordgrass-horseweed-beach pennywort pattern (peak density in interdune areas).

In general, Pinson (1973) found that dune plant communities were quite similar and were dominated by sea oats on all the islands he studied except those (South, Cedar, and Murphy) in the Santee Delta. There, panic grass was more important (Fig. 3-7). He speculated that the lower salinity of the Santee River, the presence of Piedmont clays, or a high nitrogen content (see Wagner 1964) may be responsible for the dominance of panic grass on these islands.

Stalter (1974a) concluded, based on sample quadrants, that sea oats most commonly dominate on tops of the foredunes and back dunes. Beach elder favors the front and top of foredunes, although it may also be dominant in other zones. Saltmeadow cordgrass was most commonly found in the depressions, as were sandspurs, euphorbia, beach pennywort, and diodia. Horseweed and camphorweed were found "well away from the ocean," frequently in spur areas on old dunes.

Hosier (1975) described six vegetation zones (strand line, foredune, dunefield, reardune, mesic slack, and xeric slack) from the dunes of Kiawah Island, South Carolina. The strand line is

114

Table 3-1. Frequency[a] and density of dune vegetation on Bird Bank, South Carolina, based upon 186 samples of m^2 plots (Pinson 1973).

Species	Frequency	Density
Sea oats	9.1	0.67
Saltmeadow cordgrass	43.0	7.13
Panic grass	32.3	6.38
Beach elder	16.1	1.23
Euphorbia	7.5	0.14
Horseweed	2.1	0.04
Sandspurs	3.7	0.13
Sea rocket	3.7	0.12
Russian thistle	4.3	0.16
Sea purslane	1.1	0.02
Finger grass	0.5	0.01

a. Frequency is percent total plots containing a given species. Density is average number of individual plants per unit area (Wagner 1964).

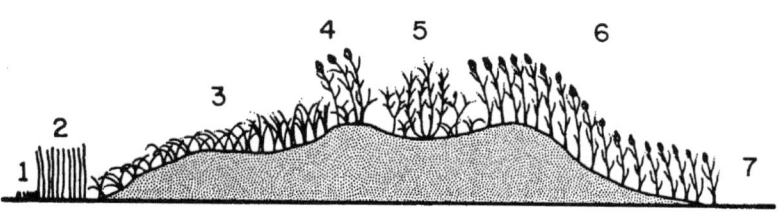

1-GLASSWORT FLAT   3- SALTMEADOW   5-MIXED SHRUB/
                        CORDGRASS   FORB/GRASS
2-SMOOTH CORDGRASS   4- SALTMEADOW   6-PANIC GRASS
                        CORDGRASS/   7-BEACH
                        PANIC GRASS

Figure 3-5. Zonation of major vegetative communities on Deveaux Bank, South Carolina (based on Gaddy 1977).

115

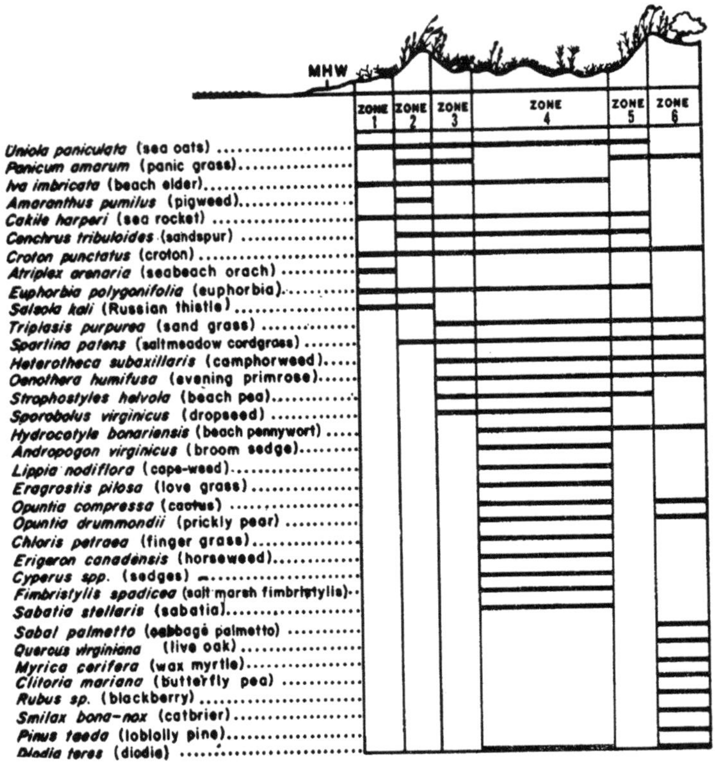

Figure 3-6. Generalized zonation of common South Carolina dune plants, based on Coker (1905), Pinson (1973), Stalter (1974a), and Hosier (1975). (MHW = mean high water; zone 1 = strand line; zone 2 = foreslope of foredunes; zone 3 = back slope of foredunes; zone 4 = interdune area; zone 5 = foreslope of back dunes; zone 6 = back slope of back dunes; solid black bar indicates presence of species.)

dominated by beach elder, with beach hogwort (croton), sea oats, and several annuals present. The dominant foredune plant on Kiawah Island is sea oats, comprising over 50% of the plant cover, with beach hogwort, beach elder, and panic grass also common. Dune fields (old dune ridges that have become poorly defined due to wind action) occur between fore and rear dunes. Here again, sea oats are dominant, with camphorweed, sand grass, beach pennywort, evening primrose, broom sedge, and horseweed also present. Rear dunes are initially vegetated with catbrier and butterfly pea (Clitoria mariana); however, wax myrtle, live oak, red bay,

and yaupon holly are the succeeding dominants (see Transition Shrub section). Hosier (1975) defined slacks as "low depressions formed during dune development or by blowouts in the dune field." Mesic (moist) slacks are dominated by beach pennywort, salt marsh fimbristylis, little bluestem, and wax myrtle. In xeric (dry) slacks on Kiawah Island, beach pennywort, salt marsh fimbristylis, marsh-gentian, and camphorweed are dominant. Shrub "thickets" are also present in slacks on Kiawah Island. They are interspersed among xeric and mesic slacks, and are dominated by yaupon holly and wax myrtle. These thickets are very similar in species

116

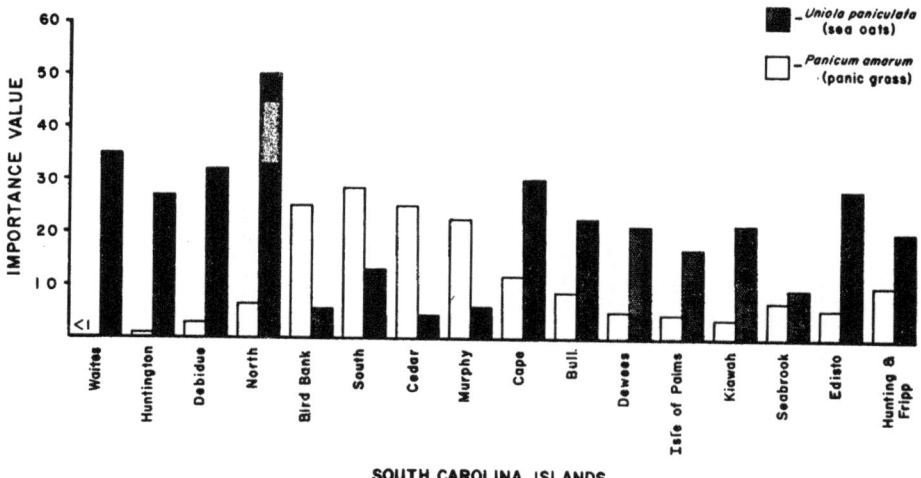

Figure 3-7. Importance of sea oats and panic grass on South Carolina dunes (adapted from Pinson 1973).

composition to transition shrub communities.

In Georgia, several authors have reported on the vegetation of maritime islands. Duncan (1955), Francisco et al. (1970) and Worthington (1972) listed the plants of Sapelo, Wassaw, and Cumberland islands, respectively, but did not characterize separate communities. However, Somes and Ashbaugh (1973) described the vegetation of St. Catherines Island, including two distinct dune communities: 1) a sea oats grassland type, and 2) a beach grass type. The sea oats grassland was dominated by sea oats, with beach pennywort, sandspur, beach hogwort (croton), cape-weed, euphorbia, and prickly pear also common. The beach grass community included salt grass, sandspur, Bermuda grass, beach hogwort, cape-weed, beach pennywort, euphorbia, and Russian thistle. Bozeman (1975) described six dune communities from Cumberland Island. The species composition and occurrence of these communities are listed in Table 3-2. The fore and rear dunes of Cumberland Island seem to be vegetatively similar to those of South Carolina's barrier islands. However, the presence of extensive interdune woody plant communities suggests that the dune field and interdune communities of Cumberland Island are at a more advanced successional stage than are those of South Carolina interdune communities. (See Maritime Forest Communities for information on succession on barrier islands.)

Plant zonation in dune communities results from the considerable differences in plant tolerance to salt spray (Wells and Shunk 1937). Boyce (1954) explained that droplets from breaking waves are carried inland by wind. Intensity of this "salt spray" is directly proportional to distance from the ocean. Vegetation of foreslopes of the front dunes, therefore, is limited to only the most salt-tolerant plants (sea oats, beach elder, etc.). Plants less tolerant to salt spray may occur on the leeward side of front dunes (foredune backslope in Pinson's (1973) terminology), because the salt spray is much decreased there. However, foreslopes of back dunes are unprotected from wind and spray and thus cannot harbor the same plants as the foredune backslope. Oosting (1945) determined that seabeach orach, saltmeadow cordgrass, sea oats, beach elder, finger grass (Chloris petraea), and beach hogwort (croton) were among the plants most tolerant of salt spray (in the order listed). As one moves into zones of lesser salt spray intensities, non-tolerant and mildly tolerant species dominate.

Generally, woody vegetation occurs only in very restricted areas in dune communities of the Southeastern United States. However, where extremely high foredunes or extensive interdune areas are present, communities of mildly salt-tolerant woody species such as wax myrtle, yaupon holly, live oak, and red bay may exist (see

117

Table 3-2. Dune plant communities of Cumberland Island, Georgia (Bozeman 1975).

COMMUNITY	SPECIES COMPOSITION	OCCURRENCE
Dune grass - forb	Seashore paspalum, beach pennywort, beach hogwort, cape-weed, euphorbia, sandspur, dropseed, Russian thistle, railroad vine, and sea oats.	Foredunes and low interdunes
Dune shrub thicket	Wax myrtle, saw palmetto, Spanish bayonet, tough buckthorn, yaupon holly, bamboo brier, pepper vine, cross vine, red bay, live oak, and Drummond's prickly pear.	Upper half of foreslope of rear dunes.
Dune oak-buckthorn scrub forest	(See following section "Transition Shrub Communities" for a discussion of the species here).	Top and rear slopes of rear dunes.
Interdune grass-sedge meadow	Evening primrose, beach elder, wax myrtle, moundlily yucca, cape-weed, beach pennywort, seashore paspalum, horseweed, centipede grass, ground cherry, muhly grass, salt marsh fimbristylis, sedges (Cyperus spp.), toad rush, marsh-gentian, coastal love grass, and star-rush.	High and low interdune areas.
Interdune shrub thicket	Shrubs and Vines Wax myrtle, cabbage palmettos, live oak, hackberry, slash pine, dahoon, willows, Southern elderberry, Hercules' club, saw palmetto, pepper vine, muscadine grape, Virginia creeper, and bamboo brier. Herbs Sedges (Cyperus spp.), dog fennel, cape-weed, saltmeadow cord-grass, marsh fleabane, marsh pennywort, climbing hempweed, salt marsh fimbristylis, fringe-leaved paspalum, rush, star-rush, broom sedge, and false nettle.	Interdunes protected from salt spray and marginal areas of interdune ponds and sloughs.
Interdune pine-mixed	Canopy Slash pine, loblolly pine, cabbage palmetto, hackberry, live oak, southern red cedar, sweet bay, swamp tupelo. Shrubs and Vines Wax myrtle, saw palmetto, button bush, muscadine grape, laurel greenbrier, pepper vine. Herbs False nettle, spike grass, lizard's tail, beak rush, water pimpernel, Chinaman's shield.	Broad interdune flats and depressions.

Table 3-2). The limit of "open" dune communities is usually determined as that area where salt spray intensity has decreased to such a degree that a transitional shrub thicket is present. Woody plants such as live oak, red bay, yaupon holly, and southern red cedar are permanently established here, with an "espalier" canopy being the obvious effect of salt spray (see next section).

Dune communities are especially susceptible to destruction by the natural forces of wind and water. Beach erosion, of course, is a natural process, but it may be accelerated by human activity (road construction, dune removal, plant removal) or by grazing. Oosting and Billings (1942) and Bozeman (1975) have pointed out that intense grazing eventually destroys dune plants (which stabilize the dunes), ultimately creating wildly shifting dunes. These unvegetated dunes intrude into maritime forests, burying acres of trees. During storms, such dunes may be completely destroyed, leaving blowouts that are easily flooded.

3. Transition Shrub Communities

Shrub communities are commonly found in three habitats on barrier islands: 1) in the transition zone between beach front dunes and maritime forest communities; 2) in the transition zone between high marsh and maritime forest communities (normally on the landward side of barrier islands and around the perimeter of marsh islands or hammocks); and 3) in interdune slacks or depressions, as discussed in the preceding section. The communities to be discussed below (primarily numbers 1 and 2) are collectively referred to as transition shrub communities or "thickets." The distinction between shrub communities and transitional communities occurring between shrub and forest communities is not clear. Bozeman (1975) separated these two community types by referring to one as "shrub" community and the other as "scrub" community. As evident in Figure 3-1, no sudden change from one community to the next is obvious; instead, there is a gradual gradation from shrub to forest. Shrub communities alone will be dealt with here; for information on scrub and low forest communities, see Maritime Forest Communities.

The shrub zone between the front dunes and the seaward margin of the maritime forest is perhaps the best-known and documented of the barrier island shrub communities. This dune-forest community is noted for its characteristic sheared or "espaliered" canopy. The height of the canopy increases with distance inland, ranging from ground level to the point at which the shrub community blends into maritime forest (see Fig. 3-1). Sharitz (1975) reported that on Kiawah Island, South Carolina, the dune-forest transition shrub community averaged from 4 to 6 ft

(1.5 to 2 m) in height with a cover of 90% - 95%. Rayner (1974) described shrub communities on three South Carolina barrier islands (Bull, Kiawah, and Daufuskie) as ranging from ground level to 10 ft (3 m).

It was originally thought that the shape of this transition shrub community was due to wind intensity, as is the case in alpine wind-formed canopies (Wells 1932). Wind-blown sand was also suggested as a major factor in shaping the espalier canopy. Later, Wells and Shunk (1938) reported that intensity of salt spray, not wind per se, was primarily responsible for the shape of the shrub zone. Doutt (1941) proposed a theory combining the drying effect of wind, wind-blown sand, and salt spray. Wells and Shunk's (1938) theory was later confirmed by Oosting and Billings (1942) and Boyce (1954), who added that wind and wind-blown sand are of secondary importance in the determination of the shape and extent of the "salt spray" shrub community.

Rayner (1974) studied transition shrub communities on Bull, Kiawah, and Daufuskie islands (among others) in South Carolina. The most frequently encountered plants were wax myrtle and yaupon holly; however, live oak, eastern red cedar, red bay, and various vines such as catbrier, pepper vine, trumpet vine, and Virginia creeper were also common. Sharitz (1975) pointed out that a thick shrub layer is present near the edge of the maritime forest on Kiawah Island in "areas of more recent dune development and . . . extreme exposure to salt spray and wind." Here wax myrtle dominates, with live oak, French mulberry, cabbage palmetto, and sea myrtle also abundant. On Kiawah Island, this community was covered by a dense growth of pepper vine. In Georgia, Bozeman (1975) described a "dune shrub thicket" from Cumberland Island. This community was formed on the foreslopes of the rear dunes and was "highly unstable and mobile." Wax myrtle, yaupon holly, red bay, and live oak were common, as also were saw palmetto and tough buckthorn, both of which are absent from South Carolina shrub communities.

In areas of well-developed dunes, small shrub communities or thickets may be present in dune slacks (interdune areas). Rayner (1974), Bozeman (1975), Hosier (1975), and Sharitz (1975) all noted the presence of these thickets. Species composition of these communities is generally the same as the dune-forest transitional shrub communities (wax myrtle and yaupon holly are dominants). The major difference between these communities is that slack communities are interdune and isolated, whereas the transition community grades into maritime forest.

The marsh-forest transitional shrub community on the landward side of barrier

119

islands probably owes its shrubby nature
to periodic invasion by extremely high
(spring and storm) tides or to sterile
soil conditions (Hosier 1975). Because
of the absence of salt spray here, this
community grades to forest more abruptly
than do beach front shrub communities.
It also does not have the characteristic
"espaliered" slope of the beach front
dune-forest community. Here the transition zone is narrower and more abrupt
(Fig. 3-1). A transitional shrub community may be present around entire marsh
islands or hammocks.

Hosier (1975) termed the marsh-forest
transition community of Kiawah Island,
South Carolina, a "marsh thicket." He
noted that this "thicket" occurs on marsh
islands as well as barrier islands. The
dominants are sea myrtle, marsh elder,
sea ox-eye, black needlerush, salt marsh
fimbristylis, and sea lavender. Hosier
(1975) also indicated that the width of
the transition shrub zone is dependent on
the slope from high ground to marsh.
Elsewhere in South Carolina, Tiner (1977)
listed sea myrtles, orach, switchgrass,
wax myrtle, broom sedges, and seaside
goldenrod as the possible dominants of the
community (see also Chapter Four, Estuarine Ecosystem for additional information).

In Georgia, Bozeman (1975) characterized the marsh-forest transitional
shrub community as a narrow zone "infrequently or rarely" flooded. Marsh elder,
sea ox-eye, sea myrtles, wax myrtle,
yaupon holly, southern red cedar, and live
oak are present here, along with Florida
privet, which is absent from South
Carolina.

Other sources of information on
coastal island maritime shrub communities
include the following: for South Carolina,
Coker (1905), Stalter (1971, 1972a, 1973a,
b, 1974a, b, c, d, e, 1975a, b, c), Pinson
(1973), and Radford (1976); for Georgia,
Duncan (1955), Francisco et al. (1970),
Worthington (1972), Somes and Ashbaugh
(1973), and Oertel and Larsen (1976).

Destruction of the transition shrub
communities often results in the invasion
of the maritime forest by dunes (in the
case of beach front transition communities), or in severe tidal erosion (in the
case of the marsh-forest transition community). Sharitz (1975) ranked the shrub
communities of Kiawah Island, South
Carolina, as the most critical areas on
the island. She noted that destruction of
these zones would allow sand and salt
spray to invade the forest proper, de-
stroying inland species that are more
sensitive to salt spray than those of
shrub communities.

4. Maritime Forest Communities

For the purpose of this study, maritime forests have been defined loosely as
"those forests found on barrier islands."
Most maritime forests are distinct from
inland forest communities; however, on
barrier islands of Pleistocene origin
(e.g., Hilton Head, Jekyll, Ossabaw, and
Cumberland), communities may be present
that are also found in inland forests.
These communities are discussed both here
and under the Upland Ecosystem (Chapter
Six).

As noted previously, maritime zonation is determined by a combination of:
1) salt spray (especially), 2) the drying
effect of wind, and 3) wind-blown sand
(Wells 1932, Wells and Shunk 1938, Doutt
1941, Oosting and Billings 1942, Boyce
1954). Few trees can withstand the rigors
of a climate created by salt spray and
wind. Live oak and cabbage palmetto are
more tolerant of salt spray than any other
trees. Live oaks occur within the rear
dunes (of the beach front), often becoming
part of the transition shrub community and
almost always dominating the low forest
that exists between the shrub community
and the maritime forest proper. Cabbage
palmettos often form a supercanopy above
the level of the transition shrub community, facing the direct impact of the
salt spray.

As one moves inland away from intense
salt spray, both species and structural
diversity increase (Rayner 1974). Gradually, the effect of salt spray lessens,
the species number increases, and finally,
all signs of salt spray disappear (in the
Sea Island Coastal Region, this usually
does not occur until one has left the barrier islands and moved inland).

Sharitz (1975) provided the most detailed description of a South Carolina
maritime forest in her work on Kiawah Island. She distinguished the following
five major types of maritime forest communities:

1) Oak-pine (see Fig. 3-8) -
This community is dominated by
laurel oak, with loblolly and
longleaf pines forming a supercanopy. Red bay, hickories,
cabbage palmetto, and sweet gum
are present in lesser numbers.
Sapling and shrub layers are
dominated by yaupon holly and
red bay, along with American
holly and blueberries.

2) Oak-palmetto-pine (see
Fig. 3-8) - This community is
dominated by laurel oak, cabbage palmetto, and pines (again

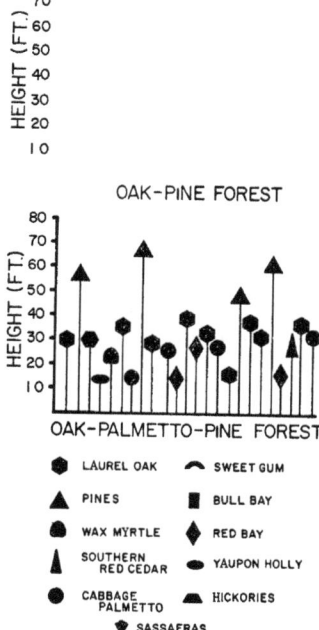

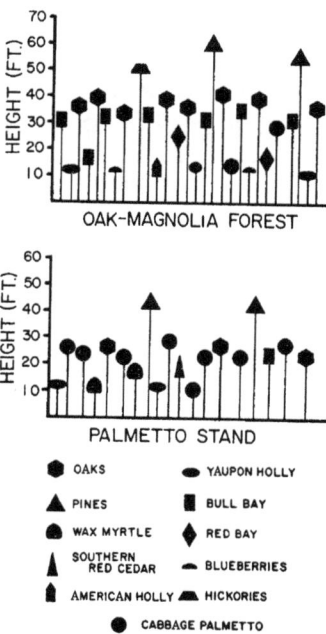

Figure 3-8. Profiles of oak-pine and oak-palmetto-pine forest types (adapted from Sharitz 1975).

Figure 3-9. Profiles of oak-magnolia and palmetto forest types (adapted from Sharitz 1975).

forming a supercanopy). Live oak and southern red cedar are important subcanopy trees. This community occurs adjacent to transition shrub communities where salt spray intensity is great. Longleaf pines and live oak are more common near shrub communities, but as one moves inland, loblolly pine and laurel oak become increasingly dominant. Yaupon holly and red bay again dominate the understory. Elsewhere, a slash pine-saw palmetto forest has been reported in flatwoods of Pleistocene origin on Hilton Head Island, South Carolina. This community is nearly identical to the slash pine community of upland forests (see Chapter Six) (National Wetlands Inventory 1978).

3) Oak-magnolia or oak-bay (see Fig. 3-9) - In these

communities, which also may be called mixed oak-hardwood communities, oaks (laurel or live) dominate, with bull bay (magnolia) or red bay as co-dominants. Pines are still present in the supercanopy, but in lesser densities. Yaupon holly and red bay dominate the understory.

4) Palmetto (Fig. 3-9) - The palmetto forest community occurs along edges of ponds on Kiawah Island. Cabbage palmetto is dominant, with laurel oak also present in high densities. A pine supercanopy contains both loblolly and longleaf pines. Wax myrtles, southern red cedar, and bull bay (magnolia) are also found here, but in lower proportions. Yaupon holly dominates the understory.

5) Low oak woods (Fig. 3-10) - This community is usually found

121

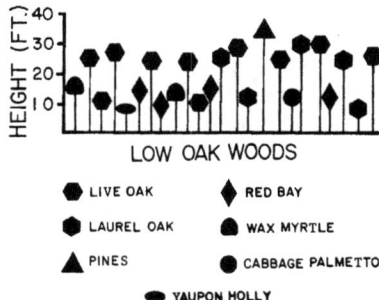

HEIGHT (FT.)

LOW OAK WOODS

● LIVE OAK   ◆ RED BAY

⬡ LAUREL OAK   ⬡ WAX MYRTLE

▲ PINES   ● CABBAGE PALMETTO

● YAUPON HOLLY

Figure 3-10. Profile of low oak woods
              forest type (adapted from
              Sharitz 1975).

in a band immediately behind
the transition shrub community.
Pines form a sparse supercanopy
with live oak and some laurel
oaks dominating the sub-canopy.
Red bay and wax myrtle are
abundant under the sub-canopy.
This community is somewhat
transitional in nature, grad-
ing into the oak-palmetto-pine
community on the inland side,
and into the transition shrub
community on the seaward side.

Another significant report on South
Carolina maritime forests is that of
Rayner (1974), who studied the vegetation
of several South Carolina barrier islands.
He concluded that the various maritime
forest communities are different stages
of maturity of the same basic forest type.
However, his transects covered only the
obvious salt spray zones and did not extend completely across each island. According to Rayner's (1974) findings,
transition shrub communities are the most
immature of the maritime wooded communities, and the live oak-palmetto-bull bay
communities are the most mature. Pine is
a supercanopy tree only in Rayner's (1974)
mature maritime forest, which is higher
in basal area and diversity than the more
immature communities. Rayner (1974)
pointed out that in areas where salt spray
intensity is low, laurel oak (which sometimes dominates), sweet gum, water oak,
and pignut hickory may be present along
with cabbage palmetto, pines, and live
oak.

On Hunting Island, South Carolina,
Radford (1976) reported that slash pine,
live oak, laurel oak, cabbage palmetto,
and red bay dominated the canopy and sub-
canopy. He also noted that as one moves
down the coast from North Carolina to
Georgia, slash pine replaces loblolly

pine; palmettos gradually appear; and,
finally, saw palmetto replaces dwarf
palmetto in the understory.

Bozeman (1975) provided the most detailed analysis of a Georgia maritime
forest in his work on Cumberland Island.
He described the following eight vegetative
communities:

1) Dune oak-buckthorn scrub
community - This community is
similar to Sharitz's (1975)
"low oak woods." It occurs on
the tops and backslopes of rear
dunes, and is transitional between the transition shrub community (Bozeman's "dune shrub
thicket") and more inland plant
associations.

2) Oak-juniper-palm (palmetto)
community - This community is
also essentially transitional.
It occurs adjacent to the marsh-
forest transition shrub community and grades into a taller
canopy. It is also found on
islands ("hammocks") in the
marsh and on peninsulas.

3) Lowland mixed hardwood
community - This community is
found in drains and depressions,
and is dominated by evergreen
trees and shrubs (see also
Chapter Five, Palustrine
Forests).

4) Pine-oak scrub community -
This association occupies moderate to poorly drained soils,
and is dominated by pond and
slash pine.

5) Oak scrub community - This
community is "characterized by a
dense scrubby growth of broad-
leaved evergreens and scattered
pines."

6) Oak-palmetto community -
Here live oak, red bay, and cabbage palmetto are dominants.

7) Oak-pine forest - This is a
secondary community, growing on
land that was formerly cultivated.

8) Mixed oak-hardwood forest -
This community represents the
most advanced successional stage
on Cumberland Island, and is less
widespread than the oak-pine
forest.

In an earlier study of sand ridge
vegetation in the coastal plain of Georgia,
Bozeman (1971) sampled forest communities
on several barrier islands (Jekyll,
Blackbeard, Wassaw, Sapelo, and Skidaway).
He described three forest associations
from these islands. Two of these, the

122

live oak-yaupon holly and the myrtle oak-Chapman oak association, are true forest communities, while his live oak-yaupon holly-sea oats association is a composite of dune, shrub, and dwarf forest communities. Other dominants of the live oak-yaupon holly association are catbrier, cabbage palmetto, and tough buckthorn. Bozeman (1971) noted that, although this community is confined to barrier islands, the same species may also be found in inland sand ridge communities. The myrtle oak-Chapman oak community includes rusty lyonia and giant-seeded beak rush as other dominants. This community is especially interesting in that it is found on barrier island sand ridges of Pleistocene origin and on inland sand ridges (see also Chapter Six, mixed Hardwood Forest Community).

More recently, Wharton (1978) described four generalized maritime forest types from Georgia. His "lowland maritime forest" corresponds to Bozeman's (1975) "lowland hardwood" community. What Bozeman (1975) had termed oak-juniper-palm is called "maritime strand forest" by Wharton (they agree on species content). Wharton's "upland forest" corresponds to Bozeman's "mixed oak-hardwood" community and is the climax for barrier islands (Wharton 1978). Finally, Wharton has combined Bozeman's "dune oak-buckthorn scrub" and "dune shrub thicket" communities [corresponding to Sharitz's (1975) "low oak woods" and "wax myrtle thicket"] into one community, which he calls the "dune oak-evergreen shrub" type. Wharton's (1978) species information for each community or forest type is similar to Bozeman's (1975), with minor modifications.

From the preceding descriptions, it is clear that maritime forests in South Carolina and Georgia are quite similar. Live oak, cabbage palmetto, and bull bay (magnolia) seem to be significant dominants in South Carolina and Georgia maritime forests, along with laurel oak and various pines. However, there are some important floristic differences. Tough buckthorn is much more important on Georgia barrier islands than on South Carolina islands. Saw palmetto is present in some South Carolina maritime forest communities, but it is not found as extensively as it is in Georgia and it does not occur in interdune communities in South Carolina. Some floristic elements of the Georgia islands, such as rusty lyonia, tarflower, Florida privet, dwarf blueberry, milk pea, myrtle oak, and paw-paw are completely absent from South Carolina barrier islands.

Further information on the floristic characteristics of maritime forests may be obtained from the following sources: 1) general works – Wells (1939), Boyce (1954), Oosting (1954), Brown (1959), Art (1976), Godfrey and Godfrey (1976).

2) South Carolina works – Coker (1905), Stalter (1971, 1972a, b, 1973a, b, 1974b, c, d, e, 1975a, b, c), and Radford (1976); and 3) Georgia works – Duncan (1955), Francisco et al. (1970), Worthington (1972), Somes and Ashbaugh (1973), and Oertel and Larsen (1976).

Wells and Shunk (1938) described the zones of maritime vegetation as polyclimaxes (i.e., the zones are climaxes in their own right, though maintained by salt spray, and are not advancing toward a more advanced stage). Wells (1939) coined the term "salt spray climax" and pointed out that, in general, maritime zonation proceeded according to polyclimax theory. Oosting and Billings (1942) termed maritime communities "salt-spray maintained sub-climaxes." Oosting (1954) agreed with Wells (1939) somewhat, but continued to advocate a modified mono-climax theory. Rayner (1974) avoided the term "climax" and described maritime zonation in terms of stages of maturity; earlier stages of development (e.g., transition shrub communities) are held static by the influence of salt spray, whereas more advanced stages of maturity (including the appearance and dominance of laurel oak) occur where the influence of salt spray is lessened. Art (1976) explained that labels such as "pioneer" and "climax" are meaningless in environments that are continually disturbed (salt spray zones). He felt that the terms "succession" and "climax" are more useful for describing directional changes in inland communities.

Despite the differences of opinion on succession and climax in maritime communities, it is nonetheless clear that succession does occur in these communities. Probably the best available information on vegetative succession on barrier islands is that provided by Bozeman (1975) for Cumberland Island, Georgia. The successional trends on this island are summarized in Figure 3-11. Bozeman (1975) noted that succession in dune and transition shrub communities occurs "only as a result of the growth of the island seaward, thus reducing the effect of the salt spray." In the forests, Bozeman indicates that the pine-oak scrub and oak-shrub communities are successional stages of the more advanced oak-palmetto community. Also, the oak-pine community on Cumberland Island is a secondary stage of old field succession and is relatively young. The most advanced successional stage in the forests of Cumberland Island is the mixed hardwood community (Fig. 3-11).

The relationships between vegetative community development and nutrient cycles, sources, and limits have not been studied in maritime environments of the Southeastern United States. However, information is available from two studies (Art et al. 1974, Art 1976) of an ecosystem on Long Island, New York, that is comparable to barrier island forests. Art et al. (1974)

123

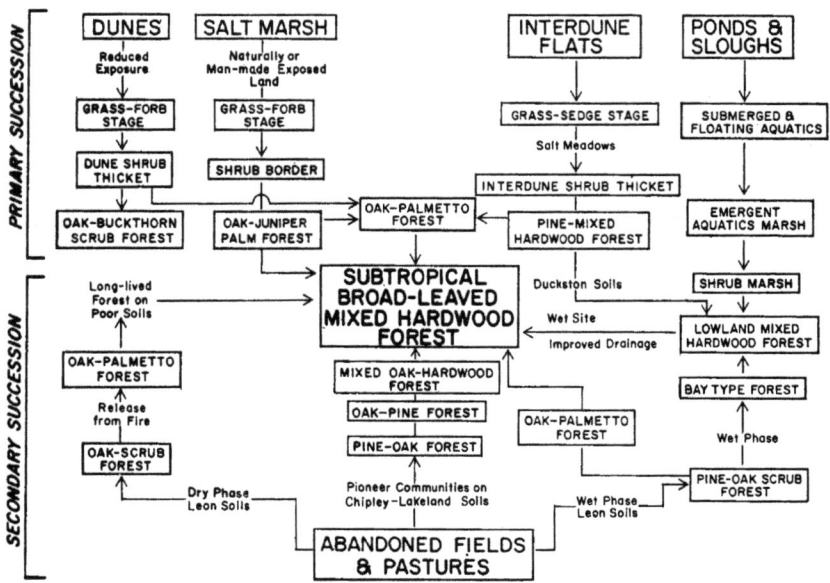

Figure 3-11. Successional relationships of plant communities on Cumberland Island, Georgia (redrawn from Bozeman 1975).

pointed out that in the Sunken Forest (a dwarf dune forest strongly influenced by salt spray), the primary source of nutrients is meteorological in the form of aerosols from salt spray. The leaves of the trees are coated with minerals from the spray, and these minerals are subsequently washed down to the roots of the trees by rainfall (Fig. 3-12). Art (1976) further noted that in some forests (including barrier island forests on sterile sands), the sole source of nutrients may be meteorological. In maritime forests (especially in the salt spray zone), meteorological input is enhanced by a large leaf area facing seaward, many finely divided twigs to capture minerals present in salt spray, and an extensively developed root system for the absorption of nutrients. In inland forested ecosystems, soil nutrients are generally supplied by weathering, which may produce as much as five times the nutrient flow that meteorological forces produce.

Monk (1966) suggested that evergreen plants may play a major role in nutrient

cycling and conservation. In evergreen wetlands (e.g., Cumberland Island) and in predominantly evergreen forests (e.g., maritime forests), leaves are dropped year around. The evergreen habitat thus may have evolved on poorer soils (sterile dune sands) where year-round leaf litter provides a major source of soil nutrients. Monk's theory has not been tested, but appears plausible in light of the ideas of Art (1976) (Fig. 3-12).

As Sharitz (1975) pointed out, transition shrub communities are the most critical vegetated areas on barrier islands. They essentially act as a buffer between the salt spray and other wind-driven forces and the forest. If they are destroyed, then the forest becomes vulnerable to invasion and destruction by dunes, salt water, and erosion.

Within maritime forests, fire is probably the most destructive force other than man. However, fire can perpetuate as well as destroy certain forest communities (Bozeman 1975) (for a detailed discussion

124

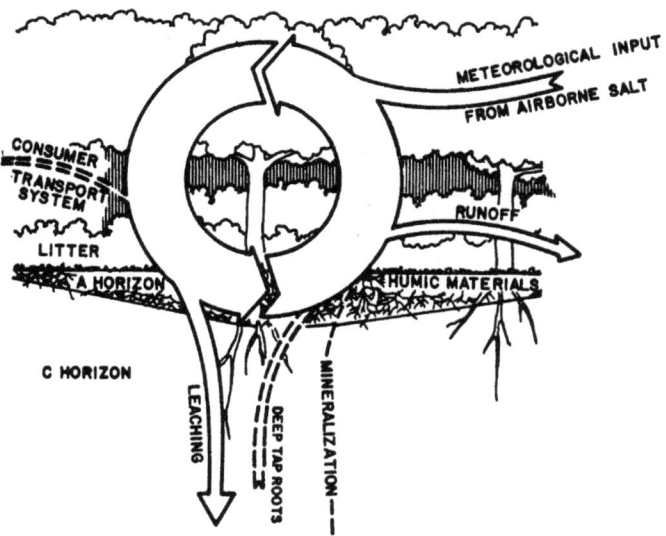

METEOROLOGICAL INPUT
FROM AIRBORNE SALT

CONSUMER TRANSPORT SYSTEM

RUNOFF

LITTER

A HORIZON

HUMIC MATERIALS

C HORIZON

LEACHING

DEEP TAP ROOTS

MINERALIZATION

Figure 3-12. A hypothetical outline of sedimentary nutrient flow on a sandy barrier island. Arrow width represents the proportion of flow relative to other sections of the model. Solid lines represent measurements from a Long Island, New York barrier island (Art et al. 1974); dotted lines represent hypothetical flows which have not been measured (adapted from Richardson and Worthington 1975).

of the effects of fire in upland forests, see Chapters Five and Six). Cabbage palmettos are fire-resistant when mature, but extremely susceptible to destruction by fire when young (Sharitz 1975). Live oak is tolerant of "cool" fires, but where live oak is mixed with other hardwood species, these trees may be destroyed by moderately hot fires.

The most obvious perturbation or disturbance on barrier islands is clearing of land for development. All unprotected (not owned by State, Federal, private conservation agencies) barrier islands are vulnerable to this form of alteration.

B. CONSUMERS

1. Invertebrates

Invertebrate communities of maritime environments are composed primarily of numerous insects and spiders, plus a heterogeneous soil (litter) fauna and one species of crab. Information on forest insect communities and soil fauna is presented in detail in Chapter Six (Upland Ecosystem), and will not be repeated here.

The invertebrate population of dune communities is limited almost entirely to insects. Absent are animals with moist skins, such as worms. The only macrocrustacean inhabiting the dune environment of high energy sandy beaches is the common ghost crab, Ocypode quadrata. This crab burrows in the beach from the upper intertidal region to the dune backshore, with the youngest and smallest crabs generally tending to burrow nearer the water, while some older, adult crabs burrow farther into the dunes (up to 0.25 mi or 0.40 km from the ocean) (Williams 1965). It is the most terrestrial of the decapod crustaceans inhabiting the coastal region of South Carolina and Georgia, only entering the water at intervals to moisten its gills and to aerate egg masses or release larvae. For a more complete discussion of the common ghost crab, see Chapter Two (Intertidal Marine Subsystem).

Many insect species cannot be placed conveniently into single habitats. Numerous species traverse many of the habitats being categorized in this study. Indeed, for some species of flying insects under consideration, individuals are

125

capable of flying to, or over, all of the habitats involved. Further, different life stages of a single species of insect often require markedly different habitats. For example, the immature naiad stage of a dragonfly requires an aquatic habitat and would be found in a lacustrine, palustrine, or riverine setting, while after emergence the aerial adult can fly with ease to a number of other habitat types, such as salt marsh, dune, transition shrub community, or mixed hardwood forest. Treatment of this group is also complicated by the presence of many transients, vagrants, and other species whose capacity to range widely often places them in locations other than those of which they are generally characteristic. Because of these complications, this study will consider primarily the insect communities typically found in particular habitats.

·Due to the mobility of maritime dune and transition shrub insects, particularly aerial species, these communities are basically interchangeable and are, therefore, treated together here. Primary factors affecting distribution of dune and transition shrub insects (and other invertebrates in these habitats) include: mechanical damage by airborne sand particles; physical injury by desiccation and salt content of onshore wind; shelter; and food (Heerdt and Bruyns 1960).

In other areas, Ranwell (1972) reported that the Coleoptera, Hymenoptera, and Diptera were the insect groups best represented in dune and shrub habitats. Ardo (1957) found that the dipterous fauna of sand dunes was "overwhelmingly rich."

Hillestad et al. (1975) surveyed insects of interdune flats (dry site) on Cumberland Island, Georgia. They found a rich fauna of orthopterans (short-horned, long-horned, and wingless long-horned grasshoppers, crickets, and pygmy mole crickets), hemipterans (assassin and lygaeid bugs), hemopterans (leaf hoppers), dermapterans (earwigs), coleopterans (click beetles, ground beetles, darkling beetles, scarab beetles, rove beetles, tumbling flower beetles, sap beetles, hister beetles, snout beetles, and pediliid beetles), hymenopterans (spider wasps and ants), and numerous dipterans. Also known to be present in maritime dune and transition shrub habitats are various butterflies and moths (lepidopterans), and dragonflies and damselflies (odonatans).

Davis (1978) provided a checklist of insect species known or expected to occur in coastal South Carolina, including those of dunes, the sometimes moist depressions between dunes, and other maritime habitats. Reproduction of this extensive list (nearly 750 species) is not practical here.

A number of nuisance insects, predominantly dipterans, inhabit the dune and transition shrub zone and venture onto adjacent beaches, often in overwhelming numbers. The annoyance caused by these insects, especially during summer, may adversely affect tourism and interfere with man in recreational activities. At times, beaches can be rendered almost completely unsuitable for recreation due either to large numbers of biting insects or non-biting dipterans such as seaweed flies (Dobson 1976) or beach flies (Simpson 1976). The more important biting insects include several mosquitoes (*Anopheles*, *Aedes*, and *Culex*), *Culicoides* sand flies (also known commonly as "no-see-ums," "punkies," "biting gnats," or "biting midges"), and tabanid flies (horse flies and deer flies) (Axtell 1974). All of these biting insects attack man and other warm-blooded animals in search of blood meals. However, while biting adult stages are abundant in transition shrub zones, maritime dunes, and the adjacent beaches, the wetter marsh and upland habitats are primarily responsible for the production and early development of these species (see Chapters Four and Six, Invertebrates).

Insects important in the maritime forest environment include flies, mosquitoes, midges, grasshoppers, katydids, crickets, cockroaches, mantids, walking sticks, and fleas. A more detailed treatment of this community is given in Chapter Six, Upland Ecosystem.

2. Amphibians and Reptiles

a. Bird Key and Dune Communities. Normally, bird keys and banks support no resident populations of amphibians or reptiles. Occasionally, Carolina diamondback terrapins may utilize bird keys for basking and perhaps egg laying, but most nesting activity takes place on high ground along tidal creeks (Johnson et al. 1974). The Atlantic loggerhead turtle may crawl up on bird keys seeking nesting sites, but rarely nests here (G. F. Ulrich, 1977, South Carolina Marine Resources Division, Charleston, pers. comm.). Any other reptiles found on bird keys are stragglers.

The rigors of maritime dune habitats are not unlike many experienced in arid desert-like conditions. Animals inhabiting coastal dunes are subject to salt spray, limited vegetation, wind, shifting sand, drought, occasional flooding, full light intensity, high evaporation, high temperatures, and predation (Johnson et al. 1974, Hillestad et al. 1975). In addition to the problems of physiological adaptations, food supply is both quite limited in quantity and restricted in type (Parnell and Adams 1971). Possibly indicative of the similarity between coastal dune communities and the arid southwestern United States is the establishment of a feral population of Texas horned lizards on Sullivans Island and the adjacent Isle of Palms, South Carolina. This species is an introduction and appears to be limited to the dune communities (Gibbons 1978).

The six-lined racerunner is probably the most characteristic reptile species of maritime dunes. It has been recorded in this specific habitat on the Isle of Palms and Kiawah Island, South Carolina (Poer 1967, Gibbons and Harrison 1975), and St. Catherines Island, Georgia (Zweifel and Cole 1974). This lizard is also known from a number of other islands in the Sea Island Coastal Region (Hillestad et al. 1975, Ringler 1977). Its presence among dunes on North Carolina as well as Florida islands (Engels 1942, 1952, Blaney 1971, Parnell and Adams 1971) indicates habitation of this community where available. Engels (1952) and Poer (1967) observed six-lined racerunners to retreat into burrows of ghost crabs for refuge when pursued.

Other reptiles characteristically encountered in dunes include the island glass lizard and the eastern coachwhip. Little is known about the habitats of the island glass lizard, but its distribution in Georgia and South Carolina is limited to coastal areas. This lizard is considered rare in coastal South Carolina (Gibbons 1978), but Neill (1948) found it to be "astonishingly abundant" in the tidal wrack area in front of insular dunes in Georgia. Gibbons and Harrison (1975) observed eastern coachwhips most often among dunes on Kiawah Island but did not consider it an abundant species.

Specific records of other species occurring in dune habitats within the characterization area include three species of toads, one lizard species, three species of snakes, and one species of turtle. After periods of rain, Gibbons and Harrison (1975) found southern toads, eastern spadefoot toads, and eastern narrowmouth toads in dunes on Kiawah Island, and Poer (1967) captured southern toads and eastern spadefoot toads in dunes on the Isle of Palms. The eastern glass lizard, southern black racer, northern scarlet snake, and eastern diamondback rattlesnake also have been recorded from dune areas (Zweifel and Cole 1974, Gibbons and Harrison 1975, Hillestad et al. 1975, Ringler 1977). Insular dunes are the prime nesting habitat of the Atlantic loggerhead turtle (see Chapter Two, Marine Intertidal Subsystem).

The number of amphibian and reptile species that may occur in maritime dunes is undoubtedly more extensive than listed here, but specific records are absent. The locomotory abilities of herpetofauna would allow almost any of the insular species to wander into dunes, if only as transients. Frequency of transients might be expected to increase in cool or wet periods (Gibbons and Harrison 1975). Locally occurring species reported from dune areas on islands off North Carolina and in the Gulf of Mexico include the southern leopard frog, eastern box turtle, cottonmouth, southern black racer, and the eastern glass lizard (Engels 1942, Blaney 1971, Parnell and Adams 1971). Specific food habits of herpetofauna in dune habitats are unknown, but probably do not differ from those of mainland species except for the limited supply of species and numbers of prey. Maritime dunes are particularly vulnerable to development and human encroachment, and these impacts are probably more detrimental to herpetofauna than to other vertebrate groups. A review of amphibians and reptiles recorded from the islands along the Georgia and South Carolina coast is provided by Gibbons and Coker (1978).

b. Transition Shrub and Maritime Forest Communities. Greater numbers of individuals and species of amphibians and reptiles are found in maritime shrub habitats than among adjacent dunes. Shrub habitats, located immediately inland from dunes, are not as demanding physiologically as dune areas, and they offer greater protection and diversity of vegetation.

With the exception of the island glass lizard, all species recorded in dune habitats on South Carolina islands also have been observed in transition shrub zones (Poer 1967, Gibbons and Harrison 1975, J. W. Gibbons, 1977, Savannah River Ecology Laboratory, Aiken, pers. comm.). Gibbons and Harrison (1975) did not discuss the transitional shrub zone separately, but they captured the following additional species in this habitat: slimy salamanders, green treefrogs, squirrel treefrogs, southern leopard frogs, green anole, ground skink, broadhead skink, southeastern five-lined skink, rough green snake, and southeastern crowned snake (J. W. Gibbons, 1977, Savannah River Ecology Laboratory, Aiken, pers. comm.).

In Georgia, Martof (1963), Johnson et al. (1974), Zweifel and Cole (1974), Hillestad et al. (1975), Anderson et al. (1976), and Ringler (1977) mention a number of species as being distributed throughout an island, but only the eastern diamondback rattlesnake was specifically noted as a common inhabitant of transition shrub areas on Cumberland Island (Hillestad et al. 1975). Engels (1942, 1952), Blaney (1971), and Parnell and Adams (1971) observed several species common to the Sea Island Coastal Region from transition shrub habitats on islands off Florida and North Carolina. The mobility of herpetofauna would allow almost any of the island species to traverse the shrub habitats. Perturbations and depredations upon herpetofauna in transition shrub communities are similar to those experienced in dune habitats.

All non-marine herpetofauna found on the coastal islands utilize maritime forests. It is here that the most protected and "stable" freshwater habitats occur, that food is most abundant, and that physiological stresses are reduced. Although no herpetofauna are endemic to the coastal islands within the study area, subspeciation has occurred on islands and adjacent mainlands north and south (for examples, see Carr and Goin 1942, Conant and Lazell 1973). Gibbons and Coker (1978) attributed the absence of endemism on South Carolina and Georgia barrier islands to their geologically recent origin and a high potential for gene-flow from the mainland.

Within the Sea Island Coastal Region, however, distribution of the island glass lizard appears to be limited to coastal islands and the immediate mainland. Collecting sites for this species in Georgia and South Carolina are generally barrier islands (Neill 1948, McConkey 1954, Martof 1963, Johnson et al. 1974, Hillestad et al. 1975, Gibbons 1978), but Neill (1948) reported specimens from a short distance inland (i.e., Bluffton, South Carolina). In peninsular Florida, this species occurs in sand-pine scrub and adjacent flatwoods of the interior and differs in coloration. The insular population in Georgia and South Carolina may warrant taxonomic recognition (McConkey 1954), but the species is poorly known and considered rare (Gibbons 1978).

Gibbons and Coker (1978) attempted to correlate selected variables with the number of species recorded from eight barrier islands located from Virginia to Georgia. They obtained significant positive correlations between the number of insular reptile species and the area of maritime forests, and between number of insular amphibian and reptile species and number of mainland species plus area of maritime forest.

Relatively complete lists of herpetofauna of eight barrier islands within the Sea Island Coastal Region are available (four were included in Gibbons and Coker 1978): Isle of Palms (Poer 1967, Harrison 1978), Kiawah (Gibbons and Harrison 1975, Gibbons and Coker 1978), Ossabaw (Johnson et al. 1974, Zweifel and Cole 1974, Ringler 1977), St. Catherines (Johnson et al. 1974, Zweifel and Cole 1974), Sapelo (Martof 1963, Teal and Teal 1964, Johnson et al. 1974, Anderson et al. 1976), Little Cumberland (Johnson et al. 1974, Hillestad et al. 1975), Cumberland (Johnson et al. 1974, Hillestad et al. 1975), and Jekyll (Johnson et al. 1974, Anderson et al. 1976). Occurrences of species on other islands are mentioned by Johnson et al. (1974), Gibbons (1978), and Harrison (1978), but these records refer to particular species rather than island check-lists. Poer (1967), Zweifel and Cole (1974), and Gibbons and Coker (1978)

listed insular herpetofauna as well as species occurring on the adjacent mainland (i.e., potential colonizers).

Available lists for the aforementioned eight islands in the characterization area were combined for comparison (Table 3-3). Several shortcomings are inherent in such lists. Some of the island surveys are not comprehensive (D. B. Means, 1977, Tall Timbers Research Station, Tallahassee, Fla., pers. comm.), and are therefore conservative for numbers of species. In addition, the exact distributional limits of certain species are not well known on the adjacent mainland and their classification as potential colonizers according to Conant's (1975) distribution maps is questionable (G. K. Williamson, 1977, Savannah Science Museum, Savannah, pers. comm.). County records of Georgia herpetofauna in the Savannah Science Museum collection were also used to judge mainland distributions (Williamson and Moulis 1979).

Despite numerous data gaps, colonization trends of specific herpetofauna groups are obvious, with salamanders and turtles being most successful and lizards most successful (Gibbons and Coker 1978) (Fig. 3-13). Another striking feature of colonizers and potential colonizers is the repeated occurrence or absence of particular species on the islands (Table 3-3).

The paucity of salamander species on the islands undoubtedly results from their inability to osmoregulate in hypertonic solutions, their intolerance to dry conditions, and their limited ambulatory abilities. Cumberland Island exhibited the greatest amount of colonization with four (central newt, mole salamander, southern dusky salamander, dwarf salamander) of a possible 15 species represented; one salamander was stated to occur on adjacent Little Cumberland, but Hillestad et al. (1975) failed to mention the species. No salamanders were found by Poer (1967) on the Isle of Palms, South Carolina.

Frogs and toads are better represented on barrier islands than are salamanders. Again, Cumberland Island exhibited the largest number of anuran colonizers with 14 of 21 possibilities, while the Isle of Palms exhibited the fewest with only 6 of 23 possibilities. Overall, 37.6% of the frogs and toads known from the region have successfully colonized barrier islands (Fig. 3-13). Species common to most of the eight islands include the southern toad, eastern spadefoot toad, eastern narrowmouth toad, southern leopard frog, green treefrog, and squirrel treefrog. Obvious unsuccessful colonizers on most or all islands include the spring peeper, southern chorus frogs, bronze frog, and bullfrog. Gibbons and Coker (1978) attributed the absence of the latter group to their dependence on standing water during winter for breeding or

128

Table 3-3. Successful and potential herpetofauna colonizers of eight islands along the Georgia and South Carolina coast. Potential colonizers are those established on the adjacent mainland according to Conant (1975) and Williamson and Moulis (1979). Typical marine or introduced species are not included. Blank spaces indicate that the species has not been recorded on the island or the adjacent mainland because the area is outside the limits of the current known range of the species. Information sources are listed in text. Plural forms represent more than one subspecies.

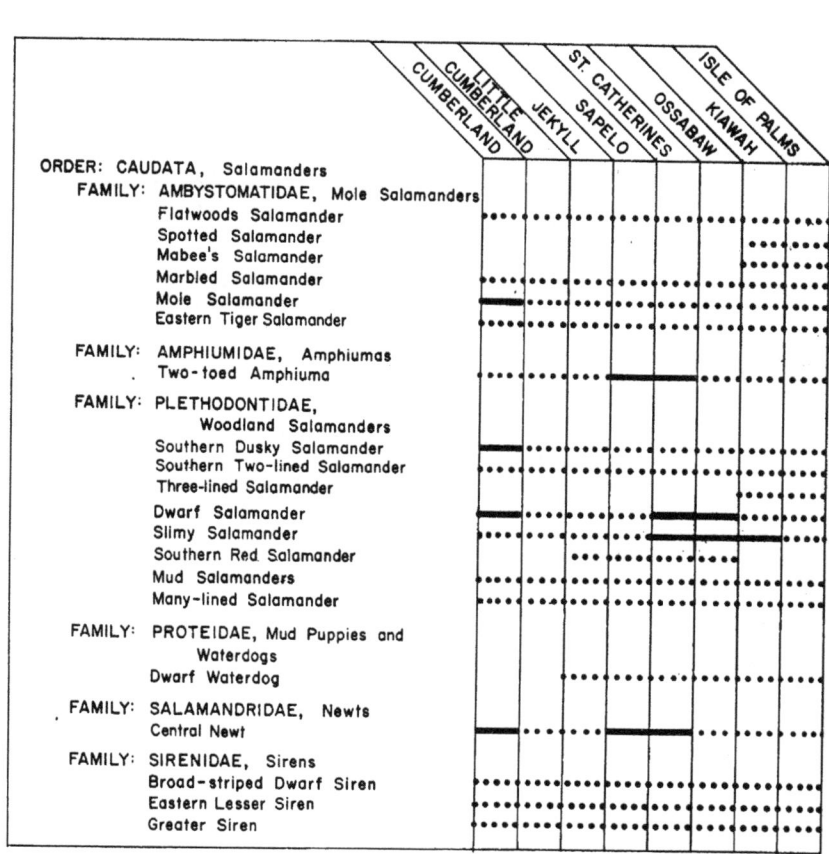

PRESENT

●●●●● POTENTIALLY OCCURRING

(Blank) ABSENT

Table 3-3. Continued

	CUMBERLAND	LITTLE CUMBERLAND	JEKYLL	SAPELO	ST. CATHERINES	OSSABAW	KIAWAH	ISLE OF PALMS
ORDER: ANURA, Frogs and Toads								
FAMILY: BUFONIDAE, Toads								
Southern Toad	━	━	━	━	━	━	━	━
Oak Toad	━	•••	•••	•••	━	•••	•••	•••
FAMILY: HYLIDAE, Tree, Cricket and Chorus Frogs								
Southern Cricket Frogs	━	•••	•••	•••	━	━	•••	•••
Green Treefrog	━	━	━	━	━	━	━	━
Spring Peepers	•••	•••	•••	•••	━	━	•••	•••
Pine Woods Treefrog	━	•••			━	━	•••	•••
Barking Treefrog	━				━	━		
Squirrel Treefrog	━	━	━	━	━	━	━	━
Gray Treefrogs	━	•••	•••	•••	━	━	•••	•••
Little Grass Frog	━	•••			━		•••	•••
Southern Chorus Frog	━	•••	•••	━		━	•••	•••
Ornate Chorus Frog	•••	•••	•••	•••	━	•••	•••	•••
Brimley's Chorus Frog						•••	•••	•••
FAMILY: MICROHYLIDAE, Narrowmouth Toads								
Eastern Narrowmouth Toad	━	━	━	━	━	━	━	━
FAMILY: PELOBATIDAE, Spadefoot Toads								
Eastern Spadefoot Toad				•••			•••	
FAMILY: RANIDAE, True Frogs								
Crawfish Frogs	•••	•••	•••	•••	•••	•••	•••	•••
Bullfrog	•••	•••	━	•••	•••	•••	•••	•••
Bronze Frog	•••	•••	•••	•••	•••	•••	•••	•••
Pig Frog	━	•••	•••	•••	━	━	•••	•••
River Frog	•••	•••	•••	•••	•••	━	•••	•••
Pickerel Frog						•••	•••	•••
Southern Leopard Frog	━	━	━	━	━	•••	━	━
Carpenter Frog	•••	•••	•••	•••	•••	•••	•••	•••

━━ PRESENT

••••• POTENTIALLY OCCURRING

(Blank) ABSENT

Table 3-3. Continued

	Cumberland	Little Cumberland	Jekyll	St. Catherines	Sapelo	Ossabaw	Kiawah	Isle of Palms
ORDER: CROCODILIA, Crocodilians								
FAMILY: ALLIGATORIDAE, Alligators								
American Alligator	—	—	—	•	•			
ORDER: SQUAMATA								
SUBORDER: LACERTILIA, Lizards								
FAMILY: ANGUIDAE, Lateral-fold Lizards								
Eastern Glass Lizard	—	—	—	—	—	—	—	—
Eastern Slender Glass Lizard	•	•	•	•	•	•	•	•
Island Glass Lizard	—	—	—	•	•	•	•	•
FAMILY: IGUANIDAE, Iguanid Lizards								
Green Anole ("Chameleon")	—	—	—	—	—	—	•	•
Southern Fence Lizard	—	•	•	—	—	—	—	•
FAMILY: SCINCIDAE, Skinks								
Ground Skink	—	—	•	•	—	—	—	—
Five-lined Skink	—	•	•	•	•	—	—	•
Broadhead Skink	—	—	—	—	—	—	•	•
Southeastern Five-lined Skink	—	•	•	—	—	—	—	•
Northern Mole Skink	•	•	•	•	•	•	•	
FAMILY: TEIDAE, Whiptails								
Six-lined Racerunner	—	—	•	•	—	—	—	—
SUBORDER: SERPENTES, Snakes								
FAMILY: COLUBRIDAE								
Banded Water Snake	—	—	—	—	—	—	•	•
Redbelly Water Snake	•	•	•	•	•	•	•	•
Florida Green Water Snake	•	•	•	•	•	•	•	•
Brown Water Snake	•	•	•	•	•	•	•	•
Glossy Crayfish Snake	•	•	•	•	•	•	•	•
Black Swamp Snakes	•	•	•	•	•	•	•	•
Eastern Garter Snake	—	—	—	—	—	—	—	•
Eastern Ribbon Snakes	—	—	—	—	—	—	—	•
Eastern Earth Snake	•	•	•	•	•	•	•	•
Rough Earth Snake	•	•	•	•	•	•	•	•

CONTINUED

— PRESENT

••••• POTENTIALLY OCCURRING

(Blank) ABSENT

Table 3-3. Continued

ORDER: SQUAMATA
  SUBORDER: SERPENTES, Snakes
    FAMILY: COLUBRIDAE (continued)

Columns (left to right): CUMBERLAND, LITTLE CUMBERLAND, JEKYLL, ST. CATHERINES, SAPELO, OSSABAW, KIAWAH, ISLE OF PALMS

Species	Cumberland	Little Cumberland	Jekyll	St. Catherines	Sapelo	Ossabaw	Kiawah	Isle of Palms
Pine Woods Snake	•••	•••	•••	•••	•••	•••	•••	•••
Redbelly Snakes	•••	•••	•••	•••	•••	•••	•••	•••
Brown Snakes	•••	•••	•••	•••	•••	•••	•••	•••
Eastern Hognose Snake	•••	•••	•••	•••	•••	•••	•••	•••
Southern Hognose Snake	•••	•••	•••	•••	•••	•••		
Eastern Worm Snake							•••	•••
Southern Ringneck Snake	•••	•••	•••	•••	•••	•••	•••	•••
Rough Green Snake			•••					
Eastern Mud Snake			•••					
Rainbow Snake	•••	•••						
Southern Black Racer	▬▬	▬▬	▬▬	▬▬	▬▬	▬▬	▬▬	▬▬
Eastern Coachwhip			•••			▬▬		
Eastern Indigo Snake	•••	•••	•••	•••	•••	•••		
Pine Snakes	▬▬		•••					
Yellow Rat Snake	▬▬	▬▬	▬▬	▬▬	▬▬	▬▬	▬▬	▬▬
Corn Snake	▬▬	▬▬	▬▬	▬▬	▬▬	▬▬	•••	•••
Northern Scarlet Snake	▬▬							
Scarlet Kingsnake			•••				•••	•••
Mole Kingsnake	•••	•••	•••	•••	•••	•••	•••	•••
Eastern Kingsnake	▬▬		•••					•••
Southeastern Crowned Snake							▬▬	•••

    FAMILY: VIPERIDAE, Vipers

Species	Cumberland	Little Cumberland	Jekyll	St. Catherines	Sapelo	Ossabaw	Kiawah	Isle of Palms
Cottonmouths	▬▬		•••					
Southern Copperhead			•••	•••	•••	▬▬		•••
Pygmy Rattlesnakes	•••	•••	•••	•••	•••	•••	•••	•••
Canebrake Rattlesnake	▬▬	•••	•••		▬▬	•••		
Eastern Diamondback Rattlesnake	▬▬		•••				•••	•••

    FAMILY: ELAPIDAE, Coral Snakes, Cobras

Species	Cumberland	Little Cumberland	Jekyll	St. Catherines	Sapelo	Ossabaw	Kiawah	Isle of Palms
Eastern Coral Snake	•••	•••	•••	•••	•••	•••	•••	•••

▬▬▬ PRESENT
••••• POTENTIALLY OCCURRING
(Blank) ABSENT

larval development. Fresh water on barrier islands is often reduced or absent during winter months (Gibbons and Coker 1978).

Alligators have been recorded in the literature from each of the eight islands except Jekyll, and breeding populations exist on Kiawah, Ossabaw, St. Catherines, Sapelo, and Cumberland islands (Martof 1963, Zweifel and Cole 1974, Gibbons and Harrison 1975, Hillestad et al. 1975). More recently, alligators also have been reported from Jekyll Island (Georgia Department of Natural Resources, Brunswick, unpubl. data). Although saline environments are not prime alligator habitat (Chabreck 1972a, Joanen et al. 1972), records of alligators in marine or estuarine situations are not uncommon (Neill 1958). In fact, alligators probably occur on virtually all barrier islands along the South Carolina-Georgia coast and breed on many.

Table 3-3. Concluded

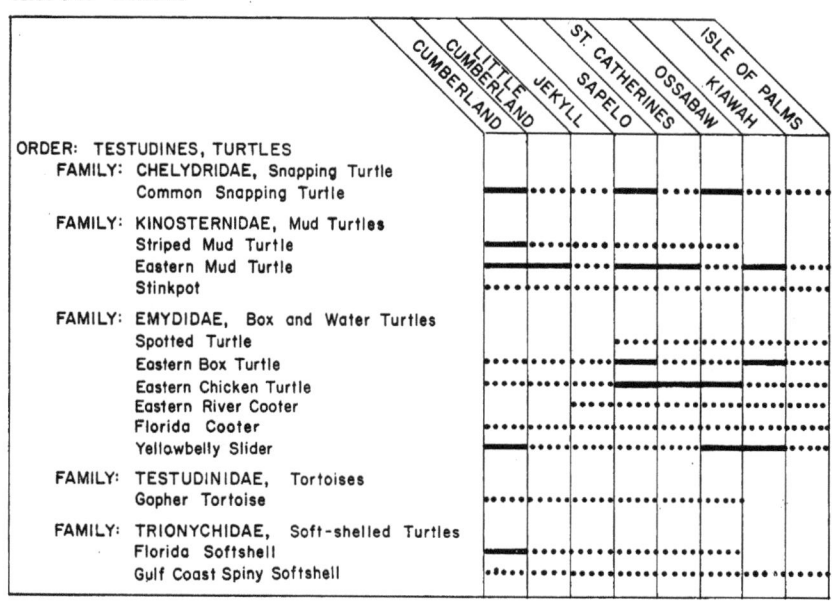

Columns (left to right): CUMBERLAND, LITTLE CUMBERLAND, JEKYLL, SAPELO, ST. CATHERINES, OSSABAW, KIAWAH, ISLE OF PALMS

ORDER: TESTUDINES, TURTLES
  FAMILY: CHELYDRIDAE, Snapping Turtle
    Common Snapping Turtle

  FAMILY: KINOSTERNIDAE, Mud Turtles
    Striped Mud Turtle
    Eastern Mud Turtle
    Stinkpot

  FAMILY: EMYDIDAE, Box and Water Turtles
    Spotted Turtle
    Eastern Box Turtle
    Eastern Chicken Turtle
    Eastern River Cooter
    Florida Cooter
    Yellowbelly Slider

  FAMILY: TESTUDINIDAE, Tortoises
    Gopher Tortoise

  FAMILY: TRIONYCHIDAE, Soft-shelled Turtles
    Florida Softshell
    Gulf Coast Spiny Softshell

▬▬▬ PRESENT

••••• POTENTIALLY OCCURRING

(Blank) ABSENT

As a group, lizards are the most successful barrier island colonizers (Table 3-3, Fig. 3-13). Their small body size, frequency of association with rafting material, insectivorous diet, and tolerance of xeric and saline conditions probably account for their success in this regard (Gibbons and Coker 1978). An average of 55.8% of mainland lizard species were recorded on the eight islands; again, Cumberland Island showed the highest colonization level (9 of 11 possible species), and the Isle of Palms the lowest (3 of 10 possible species) (Fig. 3-13). The eastern glass lizard is known from all eight of the islands; the green anole, ground skink, and broadhead skink are known from all but one of the eight; and the six-lined racerunner is recorded from all but two of the listed islands (Table 3-3).

Serpents are the third most successful group of island colonizers (average 30.8%), based on the eight island herpetofaunal surveys available. They exhibit an interesting pattern of island representation. Ten species were present on six or more of the islands; these include the banded water snake, eastern garter snake,

eastern ribbon snake, rough green snake, southern black racer, yellow rat snake, corn snake, northern scarlet snake, eastern kingsnake, and cottonmouth. The repeated presence of these species attests to their mobility and adaptability. The eastern coachwhip is known from four islands. Canebrake rattlesnakes have been recorded from Cumberland, St. Catherines, and Kiawah islands. Kiawah is the only island of the eight reviewed that supports populations of southern copperheads and southeastern crowned snakes (Gibbons and Harrison 1975); pine snakes and scarlet kingsnakes were found only on Cumberland Island (Hillestad et al. 1975). Again, 44.1% of the potential colonizers were established on Cumberland Island, while only about a third as many were found on Isle of Palms (Fig. 3-13).

Most species of aquatic and small, cryptic snakes are surprisingly poor colonizers. Reasons for their absences may be behavioral or availability of chance transportation. Pettus (1958) found that the saltmarsh subspecies of the banded watersnake (the Gulf salt marsh snake) would not drink saltwater of 30°/oo,

133

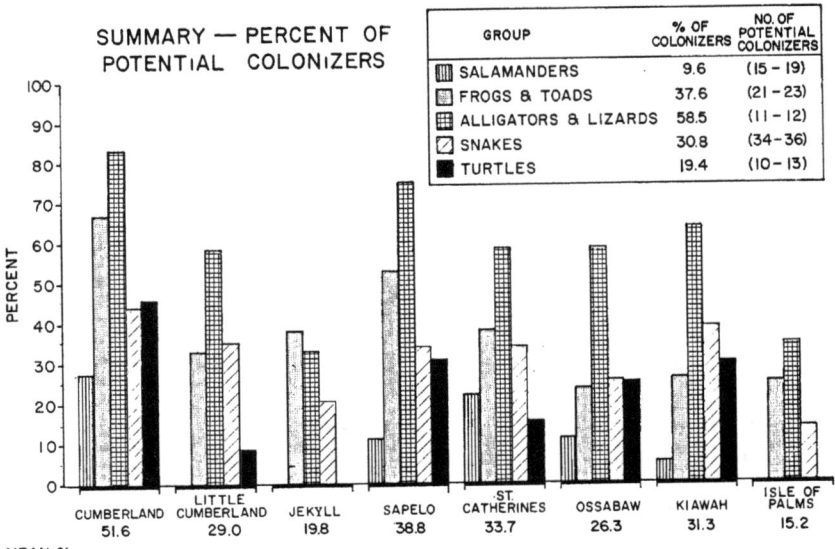

GROUP	% OF COLONIZERS	NO. OF POTENTIAL COLONIZERS
SALAMANDERS	9.6	(15 – 19)
FROGS & TOADS	37.6	(21 – 23)
ALLIGATORS & LIZARDS	58.5	(11 – 12)
SNAKES	30.8	(34 – 36)
TURTLES	19.4	(10 – 13)

	CUMBERLAND	LITTLE CUMBERLAND	JEKYLL	SAPELO	ST. CATHERINES	OSSABAW	KIAWAH	ISLE OF PALMS
MEAN %	51.6	29.0	19.8	38.8	33.7	26.3	31.3	15.2

Figure 3-13. Summary of percent of potential herpetofauna colonizers on selected South Carolina and Georgia coastal islands (box shows overall percent of successful colonizers by herpetofauna group) (information sources listed in text).

whereas the freshwater subspecies (the broad-banded water snake) would drink salt water and subsequently succumb to its effects. Different species of mainland water snakes in the study area may exhibit drinking behavior similar to that of the broad-banded water snake in saltwater areas, and thus experience the same fate. Absences of two other highly aquatic species, the eastern mud snake and the rainbow snake, may result not only from an inability to osmoregulate in hypertonic salinities, but also from their highly restricted diet of eel-like salamanders (Amphiuma) and eels, respectively. The small size of ground snakes (Storeria, Tantilla, Virginia, Diadophis, Carphophis, Micrurus) would greatly decrease their ability to survive inhospitable conditions in the marshes and estuaries which must be crossed to reach barrier islands. Also, these species are burrowers in litter, decayed logs, dense ground cover, etc., which probably reduces their chances for recruitment to insular habitats. The absence of the eastern and southern hognose snakes is an enigma.

Non-marine turtles are generally absent from barrier islands. No species are recorded from two of the eight islands surveyed (Jekyll and Isle of Palms), but five of a possible 11 mainland species are established on Cumberland Island. The gopher tortoise is established on Cumberland Island, but its presence may have resulted from an introduction by man (Johnson et al. 1974); thus, it was not listed as a colonizer (Table 3-3). Overall, an average of 19.4% of the 10 - 12 mainland species (excluding the gopher tortoise) were represented on the islands (Table 3-3). Gibbons and Coker (1978) attributed the diminutive colonization success of non-marine turtles to the possible absence of fresh water during severe droughts or saltwater flooding. Raccoons are the most significant predator of marine turtle eggs in the Sea Island Coastal Region (Johnson et al. 1974, Hopkins et al. 1978), and they may also prey heavily on eggs of freshwater turtles.

The eastern mud turtle is the most widespread insular non-marine species of turtle, with viable populations on five of the eight islands. Its success probably results from its tolerance of brackish water and desiccation (Zweifel and Cole 1974, Gibbons and Coker 1978). Freshwater ponds on Sapelo, St. Catherines, and Ossabaw islands support populations of eastern chicken turtles, and yellowbelly sliders are found on Cumberland, Ossabaw, and Kiawah islands (Martof 1963, Zweifel and Cole 1974, Gibbons and Harrison 1975, Ringler 1977). It is not known why such widely distributed and abundant species as the common snapping turtle, stinkpot, and eastern box turtle are not more successful in colonizing barrier islands (Gibbons and Coker 1978).

Except for the cottonmouth, thorough studies of behavior, biology, and ecology of the herpetofauna of coastal islands have not been made. Wharton (1969) investigated this venomous snake on Sea Horse Key, Florida, and found that it aggregates beneath bird rookeries in spring and extensively utilizes carrion dropped by adult and nestling birds. Similar behavior for the cottonmouth has been noted near heron and ibis roosts on the Isle of Palms and Capers Island, South Carolina (Sass 1926, M. D. McKenzie, 1977, South Carolina Marine Resources Division, Charleston, pers. comm.).

Differences in colonization of barrier islands by herpetofaunal species may be related to the relative ages of the islands. Most barrier islands in the Sea Island Coastal Region are considered to be of Holocene age, but several islands (Cumberland, Jekyll, Sapelo, St. Catherines) are combinations of Pleistocene and Holocene deposits (Hoyt 1968). Thus, time for colonization is far greater for such islands than for those of only Holocene age (e.g., Little Cumberland, Ossabaw, Kiawah, Isle of Palms). It should not be surprising, then, that the three islands (Cumberland, Sapelo, St. Catherines) with the largest percentage of mainland species (Fig. 3-13) are of Pleistocene and Holocene age.

Human habitation may also influence the occurrence of herpetofauna on islands. Jekyll Island is relatively heavily populated, and records of poisonous snakes are absent. These species may well have been present historically, but were extirpated because of their undesirability in urban situations. The apparent low overall percentage (15.2%) of colonization on the Isle of Palms may be related to the intensity and duration of human habitation, but it is also likely a function of the fact that the herpetofauna of this island has been very incompletely surveyed to date. Typical island species such as the southern leopard frog, green anole, broadhead skink, eastern garter snake, eastern ribbon snake, corn snake, and rough green snake probably occur on the island, but simply were not observed by Poer (1967).

3. Birds

a. Bird Key and Bank Communities. Within the characterization area, there are three major bird key and bank communities that support the overwhelming

majority of the area's breeding marine birds (see Atlas Plates 31 - 40). The most significant of these is in the Cape Romain National Wildlife Refuge, South Carolina, where seven small islands are regarded as one unit. The second major area is Deveaux Bank at the mouth of the North Edisto River, South Carolina (Beckett 1966); this bank is under lease to the National Audubon Society as the Alex Sprunt, Jr., Sanctuary. The third major bird bank community is on Little Egg Island, Georgia. This island served as a major bird bank community during the mid-1950's, and was the subject of ornithological observations by Kale and Teal (1958), Kale et al. (1965), and Kale (1967). Previous observations showed nesting colonies of black skimmers; royal, least, and gull-billed terns; American oystercatchers; Wilson's plovers; willets; and boat-tailed grackles. However, Shanholtzer (1974a) made seven trips to Little Egg Island from mid-April to the second week of August 1970 and observed no evidence of nesting by any of these species. Shanholtzer (1974a) also stated that the abandoned black skimmer colony of Little Egg Island appears to have relocated at a pelican spit near Little St. Simons Island. He recorded approximately 110 skimmers from the air and, subsequent to the nesting season, observed a small population of young skimmers. No evidence of terns or other species typical of nesting on South Carolina bank communities was reported.

In addition to these three major bird banks, there are several smaller communities of some importance. Bird Key near Folly Beach, South Carolina, provides suitable habitat for nesting marine species. Hillestad et al. (1975) noted that approximately 200 pairs of black skimmers nested near the mouth of Christmas Creek on Cumberland Island, Georgia, during 1973. Dopson and Richardson (1968) reported 15 downy young and 400 adult and immature skimmers on a small bird key and bank community in St. Andrews Sound, Georgia.

A total of 18 bird species regularly breed on the keys and banks of the characterization area (Table 3-4). Dominant species include the royal tern, laughing gull, brown pelican, Louisiana heron, and snowy egret. These birds comprise the bulk of colonial activity, particularly on the larger islands. Several other species (see Table 3-4) nest on these islands in moderate to low numbers. One of these, the least tern, suffers from heavy predation by laughing gulls and, consequently, prefers to site its colonies well away from larger species. Several other species (e.g., Wilson's plover, common tern, black skimmer, American oystercatcher) are compatible with the least tern, however, and can be found nesting around the edges of its colonies.

Table 3-4. Bird species that breed on maritime keys and banks in the Sea Island Coastal Region (compiled from Sprunt and Chamberlain 1949, 1970, Burleigh 1938, Audubon Field Notes 1967 – 1970, Chamberlain 1968, American Birds 1971 – 1977, Shanholtzer 1974b, Forsythe 1978).

DOMINANT

Species		
Royal tern	C	PR
Laughing gull	C	PR
Brown pelican	C	PR
Louisiana heron	C	PR
Snowy egret	C	PR

MODERATE

Species			
Black skimmer	C	PR	
Gull-billed tern	FC	SR	mid-March – mid-Sept.
Sandwich tern	C	SR	April-Nov.
American oystercatcher	C	PR	
Great egret	C	PR	

MINOR

Species			
Least tern	C	SR	mid-March – late Oct.
Common tern	FC	PR	
Caspian tern	FC	PR	
Wilson's plover	FC	SR	March-Oct.
Cattle egret	C	PR	
Glossy ibis	C	PR	
Little blue heron	C	PR	
Willet	C	PR	

Note: Dominance indicates relative importance of the species as a group in the community. This concept is not based necessarily on taxonomic relationships but rather on numbers, size, and trophic dynamics.

KEY: C – Common, seen in good numbers

     FC – Fairly common, moderate numbers

     PR – Permanent resident, present year around

     SR – Summer resident

Reproductive success of individual species varies from year to year (e.g., see Table 3-5), with weather (principally tidal overwash) the main physical factor affecting reproduction. Species such as the least tern, American oystercatcher, black skimmer, and gull-billed tern are most often affected by overwash. Winter storms also produce problems through reduction of dunes. This eliminates the preferred nesting sites for the brown pelican, causing yearly fluctuations in breeding populations. To partially compensate for this factor, the brown pelican may relocate breeding colonies (Chamberlain and Chamberlain 1975), but smaller dune-nesting species often suffer seasons with very low reproductive success.

Brown pelicans are perhaps the most widely known and recognized birds that breed on keys and banks. However, relatively few islands within the Sea Island Coastal Region provide acceptable breeding sites for brown pelicans. Brown pelican nesting islands apparently must meet two principal requirements: 1) isolation from predators such as raccoons, and 2) sufficient elevation to preclude widescale flooding of the nests (Blus et al. 1974a). A list of brown pelican breeding islands in South Carolina is given in Table 3-6. No brown pelican nesting activities have been reported from Georgia key and bank communities. Brown pelicans sighted in Georgia are non-nesting adults or young produced at colonies in the Cape Romain, South Carolina, area or at St. Augustine, Florida (Hillestad et al. 1975).

Historically, few data are available on the numbers of pelicans occurring in South Carolina; however, Blus et al. (1974a) have provided reliable records on the Cape Romain colony since the 1930's. Brown pelicans have been reported to nest at various times on the Refuge at Marsh Island (Vessel Reef), Bird Bank, White Banks, Cape Island, and Raccoon Key (Sandy Point). Blus et al. (1974a) also reported that as many as three colony sites on the refuge have been occupied simultaneously.

Brown pelicans have used the Cape Romain National Wildlife Refuge as a breeding area for many years, but the number of nesting birds has varied considerably from year to year (Blus et al. 1974a, 1979). Sprunt and Chamberlain (1949) reported a maximum of 800 breeding pairs in 1946. However, Blus et al. (1974a) stated that E. Milby Burton (Charleston Museum) recorded over 1,000 nests in one colony on the refuge in the 1930's. Estimates of the number of nesting pairs on the refuge ranged from 500 to 650 from 1962 through 1968; but over 1,000 nests were actually counted in 1969 (Blus et al. 1974a).

Available data indicate distinct annual fluctuations in brown pelican reproductive success (Tables 3-7 and 3-8).

According to Blus et al. (1979), reproductive success for pelicans has been generally higher (except for 1969) on Deveaux Bank than at Cape Romain, although more breeding pairs utilize the Cape Romain sites (Table 3-8). They attributed this difference to tidal flooding of nests, which occurs frequently at Cape Romain but uncommonly on Deveaux Bank. They also noted that in any year reproductive success in one colony paralleled that in the other.

Despite the fact that the size of the brown pelican breeding population in South Carolina is growing steadily, reproductive success has often been below the recruitment standard of 1.2 - 1.5 fledglings per nest per year believed necessary to maintain a stable population (Beckett 1966, Henry 1972, Blus et al. 1979) (Table 3-8). Nevertheless, successful reproduction in 1973, 1976, and 1977 (see Table 3-8 and Blus et al. 1979) suggests that the situation may be improving.

Within the primary nesting colonies of marine birds in the characterization area, the brown pelican tends to occupy the higher ground, particularly areas where there is ample vegetation handy for nest construction. The laughing gull constructs its nests on the fringes of the brown pelican nesting area to take advantage of the food source provided by the pelican eggs and young (Beckett 1966). The laughing gull also preys heavily on the eggs and young of terns, particularly the royal tern (Bent 1963a, Hatch 1970). However, the density of royal tern nests is extremely high, and this offers good protection against predation. Only those nests on the edge of the royal tern colony are regularly predated (Buckley and Buckley 1977). Other species such as the Wilson's plover, black skimmer, and willet are scattered, and have no special associations.

Little information exists on the use of keys and banks by non-marine species. Shorebirds such as the knot and sanderling can be expected on the beach regularly, but, due to the general paucity of vegetation, other species are uncommon. Those noted, however, include the nighthawk [found breeding on Bird Bank in Bulls Bay by Sprunt (1925)], and the blue-winged teal [found roosting in large numbers during migration on Bird Key in the Stono River by Chamberlain and Chamberlain (1975)]. The bald eagle was observed preying on inhabitants of key breeding colonies as early as 1831 by Audubon and as recently as 1972 by Beckett (Herrick 1968, Chamberlain and Chamberlain 1975).

Man's impact on bird keys and bank communities has been significant during recent years. Shanholtzer (1974b) reported effects of man on the Little Egg Island skimmer and tern colony in Georgia. Although this community once served as a

Table 3-5. Breeding success of marine birds at Cape Romain, South Carolina, from 1973 through 1975 (L. J. Blus, U.S. Fish and Wildlife Service, Patuxent, Maryland, unpubl. data).

	1973		1974		1975	
	NUMBER OF NESTS	NUMBER OF YOUNG	NUMBER OF NESTS	NUMBER OF YOUNG	NUMBER OF NESTS	NUMBER OF YOUNG
Brown pelican	836	1,082	920	825	900	500
Royal tern	UNKNOWN	UNKNOWN	5,034	1,990	4,813	2,003
Laughing gull	1,636	>863	1,025	>434	1,090	UNKNOWN
Louisiana heron (COMBINED)	UNKNOWN	UNKNOWN	UNKNOWN	UNKNOWN	UNKNOWN	UNKNOWN
Snowy egret	564	>343	688	>409	1,128	950
Black skimmer	607	396	159	>21	544	280
Gull-billed tern	196	>51	125	>33	130	17
Sandwich tern	UNKNOWN	UNKNOWN	129	UNKNOWN	387	UNKNOWN
American oystercatcher	55	UNKNOWN	18	UNKNOWN	59	>79
Great egret	10	>7	8	>5	26	26
Least tern	23	UNKNOWN	177	>3	281	48
Common tern	UNKNOWN	UNKNOWN	UNKNOWN	UNKNOWN	1	UNKNOWN
Caspian tern	UNKNOWN	UNKNOWN	2	UNKNOWN	UNKNOWN	UNKNOWN
Wilson's plover	UNKNOWN	UNKNOWN	UNKNOWN	UNKNOWN	2	UNKNOWN
Cattle egret	UNKNOWN	UNKNOWN	5	UNKNOWN	2	2
Glossy ibis	1	3	1	UNKNOWN	9	2
Little blue heron	5	>4	1	UNKNOWN	3	UNKNOWN
Willet	UNKNOWN	UNKNOWN	UNKNOWN	UNKNOWN	5	>1

Table 3-6. Islands used for nesting by the brown pelican in South Carolina (Blus et al. 1974a).

Island	Year
Unidentified, Georgetown County	1934
Raccoon Key (Sandy Point)	1931; 1943 - 1946; 1952
Cape Island	1939 - 1942
White Banks	1956; 1958 - 1960; 1963; 1965
Marsh Island (Vessel Reef)	1947 - 1948; 1951; 1955 - 1959; 1962; 1964 - 1965; 1967 through 1972
Bird Bank	1901; 1915; 1926; 1928; 1951; 1957 - 1961; 1963; 1965 - 1966
Unidentified, on beach near Charleston	1901
Bird Key (mouth of Stono River)	Unknown
Deveaux Bank	1947 through 1972
Egg Bank (near Beaufort, also called Bird Bank)	1904; 1943
Unidentified, 18 mi east of Beaufort	1943
Bay Point (large colony)	1901

Table 3-7. Number of young brown pelicans fledged each year from 1949 to 1970 in the Cape Romain National Wildlife Refuge, South Carolina (Schreiber and Risebrough 1972).

Years	Estimated Numbers of Birds Fledged Annually
1949 - 1953	500 - 900
1954 - 1956	250 - 500
1957 - 1960	1200 - 1500
1961	1800
1962 - 1963	500
1964	800
1965	2000
1966 - 1968	500
1969	900
1970	500 - 600

Table 3-8. Reproductive success of brown pelicans in South Carolina, 1969 - 1976 (Blus et al. 1979).

Year	Colony	No. of Nests	No. of Young Fledged	Young Fledged Per Nest
1969	Cape Romain	1016	900[a]	0.82[a]
	Deveaux Bank	250[a]	80	0.32[a]
	Both Colonies	1266	980	0.78
1970	Cape Romain	637	500[a]	0.78[a]
	Deveaux Bank	479	445	0.93
	Both Colonies	1116	945	0.85
1971	Cape Romain	1094	949	0.87
	Deveaux Bank	375	400	1.07
	Both Colonies	1469	1349	0.92
1972	Cape Romain	763	514	0.67
	Deveaux Bank	652	456	0.70
	Both Colonies	1415	970	0.69
1973	Cape Romain	836	1082	1.29
	Deveaux Bank	810	1644	2.03
	Both Colonies	1646	2726	1.66
1974	Cape Romain	920	825	0.90
	Deveaux Bank	750	800	1.07
	Both Colonies	1670	1625	0.97
1975	Cape Romain	900	500	0.56
	Deveaux Bank	1500	1300	0.87
	Both Colonies	2400	1800	0.75
1976	Cape Romain	1440	1399	0.97
	Deveaux Bank	1100[a]	1738[a]	1.58[a]
	Both Colonies	2540	3137	1.23

a. Estimated numbers - all other figures are based on actual counts.

major colonial nesting area, recreational activities (boating, picnicking, etc.) have eliminated or reduced nesting colonies of skimmers, terns, American oyster-catchers, and plovers. Gauthreaux et al. (1979) reported that the Wilson's plover and least tern populations have shown signs of decreasing within South Carolina's coastal area because of beach habitat modification and destruction.

Perhaps man's greatest impact on the community structure of bird keys and banks has been through indiscriminate use of pesticides. Beckett (1966) related this problem to the Deveaux Bank nesting colony. Blus et al. (1974b, 1977, 1979) found that eggs collected from brown pelican nests in South Carolina after 1969 had 10% - 17% thinner shells than those collected prior to 1947. Such eggshell thinning is related to increased mortality of embryos. Residues of 10 - 13 organochlorine compounds (including DDE, dieldrin and PCB's), plus mercury and other heavy metals, were routinely encountered in brown pelican eggs in South Carolina (Blus et al. 1974a, b, 1977, 1979) However, DDE and PCB's made up the majority of the organochlorine residues identified (see Table 3-9). DDE is known as the organochlorine which exerts the most detrimental influence on pelican reproduction (Blus et al. 1974a, b, 1977, 1979), and it is highly encouraging that levels of this contaminant and other organochlorines except PCB's appear to be decreasing in pelican eggs (see Table 3-9). Nevertheless, considerable potential for deleterious effects of these chemicals remain. In South Carolina, breeding brown pelicans feed almost exclusively upon young-of-the-year Atlantic menhaden, which tend to accumulate organochlorines during their residence in estuaries (Blus et al. 1977, 1979). Biomagnification from such fish to pelican eggs may be as great as 31 times for DDE and 23 times for total organochlorine (Blus et al. 1977).

b. Dune Bird Communities. The maritime dune field community is a harsh

Table 3-9. Trends for organochlorine residues in brown pelican eggs, Deveaux Bank and Cape Romain (Marsh Island), South Carolina, 1969 – 1975 (Blus et al. 1979).

Year	Sample Size	Residues, μg/g Fresh Wet Weight					
		DDE	TDE	DDT	Σ DDT	Dieldrin	PCBs
1969	15	5.45[a],[b] A (4.44-6.70)	1.65 A (1.30-2.10)	0.45 A (0.15-0.83)	7.81 A (6.48-9.40)	1.16 A (1.03-1.52)	6.11 AB (5.00-7.45)
1970	13	3.58 B (2.23-5.72)	0.79 B (0.53-1.20)	0.55 A (0.42-0.69)	5.27 B (3.49-7.77)	0.82 B (0.52-1.32)	5.25 AB (3.92-7.04)
1971	65	2.48 C (2.27-2.71)	0.48 C (0.43-0.53)	0.17 B (0.13-0.21)	3.20 D (2.94-3.48)	0.46 C (0.40-0.52)	6.49 A (5.44-7.73)
1972	72	3.03 B (2.70-3.40)	0.36 C (0.31-0.42)	0.18 B (0.15-0.21)	3.69 C (3.31-4.12)	0.45 C (0.39-0.52)	7.51 A (6.68-8.46)
1973	104	2.09 D (1.91-2.29)	0.19 D (0.17-0.22)	0.17 B (0.15-0.20)	2.56 E (2.35-2.78)	0.45 C (0.41-0.50)	4.75 B (4.26-5.31)
1974	119	2.22 CD (2.03-2.43)	0.49 C (0.44-0.54)	0.02 C (0.01-0.04)	2.72 E (2.49-2.96)	0.58 C (0.53-0.64)	7.63 A (6.80-8.55)
1975	102	1.40 E (1.27-1.54)	0.41 C (0.37-0.46)	0.004 C (0.002-0.007)	1.80 F (1.64-1.97)	0.40 C (0.36-0.43)	6.45 A (5.75-7.24)

a. Geometric mean: 95 percent confidence limits are in parentheses.

b. A significant difference among thickness means (P<0.05) is indicated for those means in each column not sharing a common letter.

environment with such limiting stress factors as blowing sand, high summer temperature, low available water, and sparse vegetative cover. The dominant plant in the dunes is sea oats (see previous sections), which may compose 50% of the total plant cover (Hosier 1975) and is the major seed producer for granivorous bird species utilizing the dune field. That dunes are a relatively seed-rich environment is reflected in the species composition of the bird populations, which are largely small seed-eating forms.

Of a total of approximately 33 species utilizing this habitat, 11 may be considered dominant (Table 3-10). The kestrel or sparrow hawk, mourning dove, ground dove, fish crow, boat-tailed grackle, and red-winged blackbird are permanent residents. The least tern and nighthawk are common summer residents. In winter, the tree swallow, yellow-rumped warbler, and Savannah sparrow are also dominant.

Seasonal balance is about equal overall; 16 species are permanent residents, and an additional eight species each occur here during summer and winter, respectively (Table 3-10). One additional species, the bobolink, is a migrant occurring from April to May and from July to December. Of the 16 permanent residents (Table 3-10), the sharp-shinned hawk is rare during the summer, and the field sparrow is associated with other habitats during the same period (Sprunt and Chamberlain 1949, Burleigh 1958, Shanholtzer 1974b, Forsythe 1978).

Most species which utilize the dune habitat are also found in other terrestrial habitats. Sea birds are frequently observed flying over dunes, and the gull-billed tern, least tern and Wilson's plover are closely associated with dunes and breed in the foredune area.

A generalized diagram of trophic relationships among birds of the dune community is given in Figure 3-14. At the most fundamental level in the dune habitat are the granivorous species that rely heavily on sea oat seeds. This large group includes the doves, blackbirds, cardinal, painted bunting, American goldfinch, and sparrows. Insectivorous species, such as the nighthawk, swallows, chimmney swift, warblers, and the gull-billed tern comprise the second largest group. As might be suspected, granivorous species occasionally eat insects, and some insectivores may consume small amounts of vegetable matter. In addition, there are a few species that can be characterized best as omnivores. Most notable are the boat-tailed grackle and the common grackle. Both species consume about 40% animal matter and 60% vegetable, but animal matter includes crabs, shrimp, and minnows in addition to insects (Wayne 1910, Sprunt and Chamberlain 1949).

The role of avian scavenger in the dune habitat is filled largely by the fish crow. This species consumes predominantly marine organisms picked up along the beach, as well as various berries and the eggs of other birds. Scavenged material includes shellfish, minnows, fiddler crabs, and carrion (Sprunt and Chamberlain 1949, Bent 1963a).

The most common, although not the most voracious, raptor of the dune field is the sparrow hawk or kestrel. It is considered a permanent resident, but it is rare during summer (Wayne 1910, Sprunt and Chamberlain 1949, Burleigh 1958). In the study area, the sparrow hawk is predominantly insectivorous, preying especially on large grasshoppers, but it also consumes rodents, reptiles, amphibians, and small birds (Wayne 1910, Sprunt and Chamberlain 1949, Burleigh 1958).

The most voracious raptor of the dunes is the great horned owl. Its food includes mammals as large as rabbits and birds as big as large hawks (Bent 1961). Large concentrations of great horned owls have been observed in the dune field at night, and prey has been noted to include nesting colonial bird species (Sprunt and Chamberlain 1949, Chamberlain and Chamberlain 1975).

The least tern is a common piscivore of the dune habitat. It is a summer resident, occurring from mid-March to late October, and is observed only rarely in winter (Forsythe 1978). This species suffered great depredations by plume hunters around the turn of the century, but it had recovered by the 1930's (Wayne 1910, Sprunt and Chamberlain 1949, Tomkins 1959). Today, the least tern is again in trouble, primarily because of poor reproductive success and dwindling breeding habitat (Gauthreaux et al. 1979). The least tern commonly feeds in coastal marshes and waters, but relies on dunes for breeding habitat. Nesting occurs colonially in the foredune area where vegetation is sparse. Favorite localities appear to be adjacent to inlets on barrier beaches, specifically recurved spits, although other beach locations are also known (Chamberlain and Chamberlain 1975). Poor reproductive success can be attributed to a number of factors including reduction in amount of suitable habitat available, natural disasters, and high predation rates. Colonies on Kiawah Island were observed to suffer predation by hogs, cotton rats, people, snakes, ghost crabs, and the great horned owl (Chamberlain and Chamberlain 1975). Tomkins (1959) reported that the nesting success of terns in the Savannah River area was reduced by daily forays of fish crows, which eventually caused abandonment of the Oysterbed Island nesting site by terns.

Table 3-10. Birds of the maritime dune subsystem of the Sea Island Coastal Region (Sprunt and Chamberlain 1949, 1970, Burleigh 1958, Audubon Field Notes 1967 - 1970, Chamberlain 1968, American Birds 1971 - 1977, Shanholtzer 1974b, Forsythe 1978).

DOMINANT				MODERATE				MINOR			
Sparrow hawk	C	PR		American oystercatcher	FC	PR		Sharp-shinned hawk	FC	PR	
Least tern	'C	SR	March-Oct.	Wilson's plover	C	SR	March-Oct.	Great horned owl	FC	PR	
Mourning dove	C	PR		Willet	C	PR		Rough-winged swallow	C	SR	March-Sept.
Ground dove	FC	PR		Gull-billed tern	FC	SR	March-Sept.	Purple martin	C	SR	Feb.-Nov.
Nighthawk	C	SR	April-Oct.	Chimney swift	C	SR	March-Nov.	Bobolink	C	M	April-May July-Dec.
Tree swallow	C	WR	July-June	Barn swallow	C	PR		Ipswich sparrow	R	WR	
Fish crow	C	PR		Comm crow	C	PR		Dark-eyed junco	C	WR	Oct.-April
Yellow-rumped warbler	C	WR	Oct.-April	Palm warbler	C	WR	Sept.-April	Field sparrow	C	PR	
Red-winged blackbird	C	PR		Eastern meadowlark	C	PR					
Boat-tailed grackle	C	PR		Common grackle	C	PR					
Savannah sparrow	C	WR	Oct.-April	Cardinal	C	PR					
				Painted bunting	C	SR	March-Nov.				
				American goldfinch	C	WR	Oct.-May				
				Song sparrow	C	WR	Sept.-April				

Note: Dominance indicates relative importance of the species as a group in the community. This concept is not based necessarily on taxonomic relationships but rather on numbers, size, and trophic dynamics.

KEY: C - Common, seen in good numbers   SR - Summer resident

　　　 FC - Fairly common, moderate numbers   M - Migrant

　　　 R - Rare

　　　 PR - Permanent resident, present year around

　　　 WR - Winter resident

143

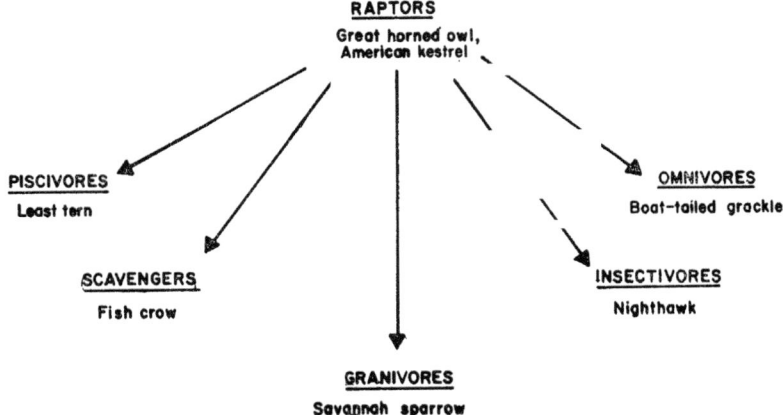

RAPTORS
Great horned owl,
American kestrel

PISCIVORES
Least tern

OMNIVORES
Boat-tailed grackle

SCAVENGERS
Fish crow

INSECTIVORES
Nighthawk

GRANIVORES
Savannah sparrow

Figure 3-14. Generalized trophic relationships of representative birds of the maritime dune community.

The Savannah sparrow is a common granivore and winter resident, occurring from late September to mid-May (Shanholtzer 1974b, Forsythe 1978). This species is by far the most common sparrow of the dune field. It is an elusive bird, running through the ground cover or flushing for short, erratic flights only to dive again into the vegetation. The Savannah sparrow is not highly social and does not flock in the true sense. Winter density of this species has been reported to be about 4 to 5 birds per acre (Norris 1963). The Savannah sparrow's diet is mostly vegetable matter, with seeds accounting for an average of 97%; the remaining 3% was found to be insects (Quay 1947).

Unlike the Savannah sparrow, which occurs in other habitats as well as dunes, the Ipswich sparrow race is restricted to the dune field. This bird was formerly regarded as a separate species (American Ornithologists' Union 1973). Because of its restricted habitat, the Ipswich sparrow has been proposed for endangered status in South Carolina (Gauthreaux et al. 1979). (See also Chapter One, Section VI.) The Ipswich sparrow is a large, pale sparrow that virtually blends in with the dunes. Historically, it has been rare in the Sea Island Coastal Region, with the first Georgia and South Carolina specimens described by Worthington (1890) and Wayne (1902), respectively. Since its discovery,

the status of this race has changed little. The Ipswich sparrow breeds only on Sable Island off Nova Scotia, where the total population ranges from 1,000 to 4,000 individuals (Burleigh 1958, Chamberlain 1974). Since the winter range includes the immediate coast from New Jersey to Cumberland Island, Georgia (Bent 1968), it is understandable that this sparrow is rare. The Ipswich sparrow's reputed limitation to the dune habitat is fully supported by the field work of Wayne, who found it on the mainland only three times in 46 years (Sprunt and Chamberlain 1949). Tomkins recorded this species only four times in Georgia. Each of his observations were on Oysterbed Island (Burleigh 1958). Food on the winter range is limited to sea oats (Wayne 1910).

The ground dove, another granivorous permanent resident, occurs commonly in the dune field except during winter (Sprunt and Chamberlain 1949). It displays a definite preference for sandy areas and breeds in the dune field. Nests are located typically in wax myrtle shrubs, found in mesic slacks between dune ridges. Food of the ground dove consists almost wholly of seeds from such plants as crab grass, wire grass, foxtail grass, purslane, ragweed, amaranth, and sedges (Sprunt and Chamberlain 1949).

The status of the ground dove has changed drastically since the turn of the century. Wayne (1910) reported this species to be "exceedingly abundant" prior to the cold wave of 1899. By 1949, it was less common, a fact attributed to the reduction of suitable breeding areas. This decline continues today due to general habitat loss, and the ground dove is now listed as a species of special concern for South Carolina (Gauthreaux et al. 1979).

Man's interference with the natural maritime dune system has destroyed many nesting colonies of birds along the South Carolina and Georgia coast. The fragile nature of dunes and the importance of dune stability to the overall ecology of the beachfront have long been recognized by scientists. However, many land developers have ignored the importance of dunes. Consequently, some of South Carolina's and Georgia's most attractive beaches and dunes have been modified drastically by development. Most of the dune complex has been destroyed by development of North Myrtle Beach, Myrtle Beach, Surfside Beach, Cherry Grove Beach, Savannah Beach, and St. Simons and Jekyll islands. Dunes at many other beaches and islands within the study area also have been altered to a great extent.

Obviously, birds such as gulls, terns, pelicans, plovers, etc., which require the beach and dune areas for nesting habitat, are forced to relocate when the system is altered and man encroaches upon the area. Most dune and beach nesting species are highly aggravated by the presence of man and seek isolated nesting sites away from human activity.

Loss of dune vegetation from overgrazing also contributes to destabilization and erosion of the dune complex, and indirectly affects the nesting success of breeding colonies. Free-ranging livestock has seriously depleted and removed dune plants from some of the barrier islands (e.g., see Johnson et al. 1974 for information on reduction of dune vegetation on Ossabaw, Little St. Simons, and Cumberland islands, Georgia, by overgrazing).

The use of vehicles on the beach and dune complex has introduced a new element into the management of barrier islands. All-terrain vehicles are transported to many of the islands by boat, and are used to traverse the dune landscape. The environmental consequences are serious when such traffic is directed over dune vegetation and through bird nesting areas. Vehicles on the beach can also play havoc with shorebirds. Heavily used beaches and dune fields have few nesting birds, since these birds require a certain degree of solitude to raise young. Vehicles and beachcombers have a tendency to frighten adult birds from their nests, exposing the eggs and young to the intense insolation and heat on the beach.

The mortality rate of young chicks is high when adults are not present to shield the sun's rays (Hillestad et al. 1975).

Birds most affected by recreational traffic in the beach and dune habitat are the gull-billed and least terns. These birds have had little breeding success in recent years on many of the barrier beaches because of uncontrolled traffic through nesting colonies. The least tern, however, has been reported recently breeding on tops of buildings in large shopping centers in Charleston and other large cities in the Sea Island Coastal Region (Gauthreaux et al. 1979).

c. Transition Shrub Bird Communities.

(1) Overview. Because of its lack of understory, generally low plant density, and low plant height, the maritime shrub community provides much less habitat for birds than does the forest, and its bird community is correspondingly smaller. A total of approximately 24 bird species regularly utilize the transition shrub community, of which nine species can be regarded as dominants (Table 3-11). Permanent residents are represented by the ground dove, mockingbird, yellowthroat, and the red-winged blackbird. All are common breeding species in this habitat. Other common breeders are the three summer residents, the eastern kingbird, yellow-breasted chat, and the painted bunting. Dominant winter residents are the abundant tree swallow and the yellow-rumped warbler.

Most species that utilize maritime shrub communities are residents of other habitats, particularly the adjacent dune and maritime forest communities. Several species breed in this shrub community, but feed extensively in adjacent habitats. This group includes the ground dove, red-winged blackbird, and painted bunting. Two species, the tree swallow and yellow-rumped warbler, are closely associated with the abundant wax myrtle bush, and utilize its fruit. These birds are both predominantly insectivorous throughout most of the year, but in winter consume large quantities of wax myrtle berries (Sprunt and Chamberlain 1970).

Twelve species found in the maritime shrub community are permanent residents (Table 3-11). Of these, the sharp-shinned hawk and catbird are much more common in winter. Their distribution in this area in summer is so scattered and irregular that they are regarded as rare during this season (Sprunt and Chamberlain 1949, Forsythe 1978). During the winter, particularly severe ones, populations of the insectivorous barn swallow and prairie warbler are drastically reduced (Sprunt and Chamberlain 1949).

Winter residents are represented by only four species, and summer residents

Table 3-11. Birds of the maritime transition shrub communities of the Sea Island Coastal Region (Sprunt and Chamberlain 1949, 1970, Burleigh 1958, Audubon Field Notes 1967 – 1970, Chamberlain 1968, American Birds 1971 – 1977, Shanholtzer 1974b, Forsythe 1978).

DOMINANT				MODERATE				MINOR			
Ground dove	C	PR		Sharp-shinned hawk	C	PR		Brown thrasher	C	PR	
Eastern kingbird	C	SR	Mar.-Oct.	Sparrow hawk	C	WR	Aug.-May	Bobolink	C	T	April-May July-Oct.
Tree swallow	C	WR	July-June	Barn swallow	C	PR		Blue grosbeak	FC	SR	April-Oct.
Mockingbird	C	PR		Chimney swift	C	SR	Late Mar. to Oct.	Indigo bunting	C	SR	April-Oct.
Yellow-rumped warbler	C	WR	Oct.-April	Catbird	FC	PR					
Yellowthroat	C	PR		American redstart	C	T	Apr.-May July-Oct.				
Yellow-breasted chat	C	SR	April-Sept.	Prairie warbler	C	PR					
Red-winged blackbird	C	PR		Palm warbler	C	WR	Aug.-April				
Painted bunting	C	SR	April-Oct.	Boat-tailed grackle	C	PR					
				Rufous-sided towhee	C	PR					
				Cardinal	C	PR					

Note: Dominance indicates relative importance of the species as a group in the community. This concept is not based necessarily on taxonomic relationships but rather on numbers, size, and trophic dynamics.

KEY: C – Common, seen in good numbers

FC – Fairly common, moderate numbers

PR – Permanent resident, present year around

WR – Winter resident

SR – Summer resident

T – Transient resident

146

by six (Table 3-11). There are two common migrants, the American redstart and the bobolink.

Granivorous species such as the ground dove and cardinal eat seeds, making them the most basic consumers in this habitat (Fig. 3-15). Since seeds and fruits are extremely limited, many species eat both insects and fruit, depending on the abundance of either food. Most consistent in this practice are omnivores such as the mockingbird and boat-tailed grackle. Some species, such as the tree swallow, are seasonally omnivorous (see following discussion). Insects are consumed by the bulk of species utilizing maritime shrub communities, as illustrated by the eastern kingbird and the yellow-throat. The sparrow hawk also consumes quantities of insects but, by including an occasional small bird or mammal in its diet, joins the sharp-shinned hawk as a top predator.

(2) Representative Species. Based on their abundance and roles in the maritime transition shrub community, two species, the tree swallow and the yellow-breasted chat, have been selected for more detailed discussion.

The tree swallow is an abundant winter resident of the South Carolina and Georgia coasts (Sprunt and Chamberlain 1949, Burleigh 1958). This species is present almost year around (July to June), but does not breed here. The southern limit of its breeding range is southeastern Virginia, which perhaps explains the tree swallow's absence for only a brief period. Although generally abundant in South Carolina and Georgia, particularly in fall, the tree swallow also has been known to be irregular in its winter occurrence. Wayne (1910) attributed this not to adverse weather, but to an inadequate food supply. Through most of the year, the tree swallow is insectivorous, with approximately 80% of its diet composed of various flies, beetles, ants, leafhoppers, and dragonflies (Sprunt and Chamberlain 1949). In fall and winter, a dietary shift is made to vegetable matter, which accounts for about 20% of the total diet. Virtually all of the vegetable material eaten is seeds of the wax myrtle. Tree swallows often concentrate when feeding on wax myrtle berries, and flocks numbering in the hundreds are commonplace. At these times, feeding is chaotically intense, and the birds are so oblivious to their surroundings that they can be captured by hand (Wayne 1910).

The yellow-breasted chat is the largest of the North American warblers, and is a common summer resident of the Sea Island Coastal Region from March through November (Sprunt and Chamberlain 1949). Although numerically common, this species is not well known due to its secretive

habits and preference for virtually impenetrable thickets. In the South Carolina and Georgia coastal plain, the chat is most common in overgrown fields with dense brush thickets; however, it also occurs on barrier islands in dense wax myrtle thickets (Chamberlain and Chamberlain 1975). The yellow-breasted chat breeds commonly in transition shrub zones, but the nest site is always confined to extremely dense thicket vegetation. For this reason, some writers have discussed the apparent scarcity of nests (Wayne 1910), but recently this myth has been dispelled by more active field work (Sprunt and Chamberlain 1970). This species is predominantly insectivorous, feeding on weevils, beetles, bees, ants, caterpillars, moths, and mayflies. Some vegetable material is also eaten, including strawberry bush, blackberries, and wild grapes (Sprunt and Chamberlain 1949).

Man's impact on the maritime transition shrub community is closely associated with his activities in maritime dune and forest systems. As these areas are exploited through development activities, the nesting habitat for ground doves, red-winged blackbirds, painted buntings, etc., is destroyed.

d. Maritime Forest Bird Communities.

(1) Overview. Relatively little information exists on avian diversity in maritime forests of the characterization area. However, some observations have been recorded for Bull, Capers, Kiawah, Turtle, Wassaw, Sapelo, Cumberland, Wolf, Tybee, and Blackbeard islands (Sprunt 1936, Folk 1939, Teal 1958a, U.S. Department of Interior, Fish and Wildlife Service 1971, Chamberlain and Chamberlain 1975, Chamberlain 1979). Apparently, the maritime forest provides essential resting and feeding habitat for many passerine migrants along the coasts of South Carolina and Georgia. Maritime forest plant associations are becoming increasingly important for many songbird species as live oaks on the mainland are replaced by pine plantations. Generally, the maritime forest is in an advanced stage of succession and supports most of the non-marine birds on barrier islands. These birds are primarily insectivores.

Excluding several rare and accidental species, the total number of species inhabiting maritime forests in the study area can be placed at 83 (Table 3-12). This number appears to be significantly lower than for mainland forests, and there is a corresponding paucity of individuals (for comparison, see sections on birds of upland forests, Chapter Six). This situation can be attributed to a number of factors. The dense vegetation characteristic of undisturbed maritime forests undoubtedly restricts bird mobility. This has been verified in tropical communities

147

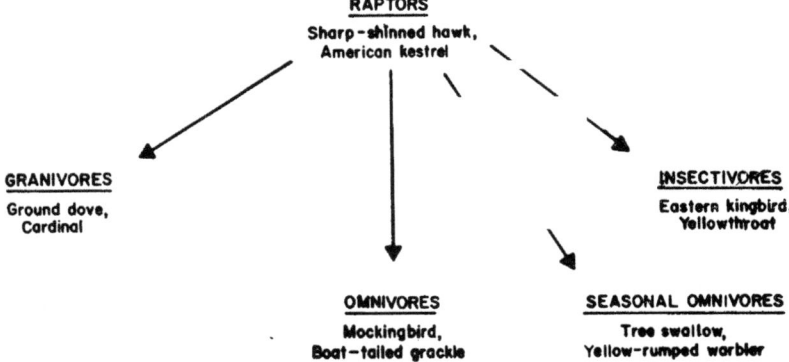

RAPTORS
Sharp-shinned hawk,
American kestrel

GRANIVORES
Ground dove,
Cardinal

INSECTIVORES
Eastern kingbird,
Yellowthroat

OMNIVORES
Mockingbird,
Boat-tailed grackle

SEASONAL OMNIVORES
Tree swallow,
Yellow-rumped warbler

Figure 3-15. Generalized trophic relationships of representative birds of maritime transition shrub communities.

and monotypic pine stands (Karr and Roth 1971). In homogeneous habitats, a decrease in the number of resident species also has been explained by patchy occurrence (MacArthur and Pianka 1966). Further, tropical species have been found to exhibit larger home ranges in homogeneous habitats (Karr and Roth 1971). The proportion of species with large home ranges favors tropical areas dramatically, 51% to 3%, over a temperate locale (Illinois) (Karr and Roth 1971). Maritime forests in South Carolina and Georgia can be considered more or less semi-tropical, perhaps to the extent that avian home ranges are sufficiently large to reduce the number of individuals present, as well as the species diversity.

The maritime forest represents the first major terrestrial habitat inland from the marine environment, and it is dominated by passerine bird species. Most species are insectivorous, and leaf-gleaners of both the understory and canopy are well represented. There are no forms unique to the maritime forest, however, and most species can be found in several terrestrial habitats.

The maritime forest can be divided into five general vertical zones or niches as far as use by birds is concerned: ground, understory, mid-story, canopy, and aerial (adapted from Dunlavy 1935). Most bird species favor a particular vertical zone, although individuals frequently deviate from this zone for escape, migration, etc. Vertical distribution is

more pronounced, however, in forests with greater canopy height than the maritime forest. Canopy height is typically only 80 ft (24 m) or less in Southeastern maritime forests (Sharitz 1975), and vertical distribution of birds is correspondingly compressed. This situation often results in species occupying more than one vertical zone (e.g., the yellow-rumped warbler).

A total of 39 species are dominant residents of the maritime forest, including 23 permanent residents, eight winter residents, seven summer residents, and one migrant (Table 3-12). Altogether, permanent residents include 40 species. Of these, populations of the red-shouldered hawk, which preys heavily on herpetofauna, are much reduced during winter. Populations of three small insectivores, the blue-gray gnatcatcher, yellow-throated warbler, and the prairie warbler, are similarly reduced, depending on the severity of the winter (Sprunt and Chamberlain 1949). Conversely, populations of the red-tailed hawk, sharp-shinned hawk, and catbird increase during winter. Summer residents are represented by a total of 16 species. These birds are primarily warblers, although swallows, summer tanager, orchard oriole, and the two buntings also must be included. Twenty-three species of winter residents, primarily wrens, hawks, and sparrows, replace a number of summer insectivores. This predominance of winter residents is offset slightly by four migrant species. The American redstart is a common transient visitor, and the

Table 3-12. Birds of maritime forest communities of the Sea Island Coastal Region (compiled from Sprunt and Chamberlain 1949, 1970, Burleigh 1958, Audubon Field Notes 1967 – 1970, Chamberlain 1968, American Birds 1971 – 1977, Shanholtzer 1974b, Forsythe 1978).

DOMINANT			MODERATE			MINOR		
Red-tailed hawk	C	PR	Black vulture	C	PR	Cooper's hawk	U	PR
Red-shouldered hawk	C	PR	Turkey vulture	C	PR	Turkey	U	PR
Mourning dove	C	PR	Sharp-shinned hawk	C	PR	Red-headed woodpecker	FC	PR
Yellow-billed cuckoo	C	SR	April-Nov.	Bobwhite	C	PR	Purple martin	C	SR	Feb.-Oct.
Great horned owl	FC	PR	Screech owl	C	PR	Starling	C	PR
Chuck-will's-widow	C	SR	March-Sept.	Barred owl	C	PR	Prothonotary warbler	FC	SR	April-Sept.
Chimney swift	C	SR	March-Oct.	Whip-poor-will	FC	WR	Sept.-April	Indigo bunting	C	SR	April-Oct.
Ruby-throated hummingbird	C	SR	March-Oct.	Nighthawk	C	SR	April-Oct.	Fox sparrow	FC	WR	Nov.-March
Common flicker	C	PR	Yellow-bellied sapsucker	FC	WR	Oct.-March
Pileated woodpecker	FC	PR	Hairy woodpecker	FC	PR
Red-bellied woodpecker	C	PR	Rough-winged swallow	C	SR	March-Sept.
Downy woodpecker	C	PR	Barn swallow	FC	PR
Great crested flycatcher	C	SR	April-Oct.	Blue jay	C	PR
Eastern phoebe	C	WR	Sept.-March	Tufted titmouse	C	PR
Tree swallow	C	WR	July-June	Brown creeper	FC	WR	Oct.-March
Common crow	C	PR	House wren	C	WR	Sept.-April
Fish crow	C	PR	Winter wren	FC	WR	Sept.-April
Carolina chickadee	C	PR	Brown thrasher	C	PR
Carolina wren	C	PR	Hermit thrush	C	WR	Oct.-April
Catbird	C	PR	Swainson's thrush	C	T	April-May Sept.-Oct.
Robin	C	WR	Oct.-April	Cedar waxwing	C	WR	Sept.-May
Blue-gray gnatcatcher	C	PR	Solitary vireo	FC	WR	Oct.-April
Ruby-crowned kinglet	C	WR	Oct.-April	Red-eyed vireo	C	SR	March-Oct.
White-eyed vireo	C	PR	Black & white warbler	FC	WR	Sept.-May

149

Table 3-12. Concluded

DOMINANT				MODERATE			
Northern parula	C	SR	March-Oct.	Orange crowned warbler	FC	WR	Oct.-April
Yellow-rumped warbler	C	WR	Oct.-April	Prairie warbler	C	PR	
Yellow-throated warbler	C	PR		Palm warbler	C	WR	Aug.-April
Pine warbler	C	PR		Hooded warbler	C	SR	March-Sept.
Yellowthroat	C	PR		Ovenbird	C	T	April-May Aug.-Oct.
American redstart	C	T	April-May July-Oct.	Northern waterthrush	C	T	March-May July-Oct.
Red-winged blackbird	C	PR		Orchard riole	C	SR	April-Aug.
Boat-tailed grackle	C	PR		Summer tanager	C	SR	April-Oct.
Common grackle	C	PR		American goldfinch	C	WR	Oct.-April
Cardinal	C	PR		Savannah sparr w	C	WR	Oct.-April
Painted bunting	C	SR	April-Oct.	Dark-eyed junco	C	WR	Oct.-April
Rufous-sided towhee	C	PR		Field sparrow	C	PR	
White-throated sparrow	C	WR	Oct.-May				
Song sparrow	C	WR	Sept.-April				
Swamp sparrow	C	WR	Oct.-May				

Note: Dominance indicates relative importance of the species as a group in the community. This concept is not based
necessarily on taxonomic relationships but rather on numbers, size, and trophic dynamics.

KEY: C - Common, seen in good numbers   SR - Summer resident

FC - Fairly common, moderate numbers   T - Transient resident

U - Uncommon, small numbers irregularly

PR - Permanent resident, present year around

WR - Winter resident

ovenbird, northern waterthrush, and
Swainson's thrush are all fairly common,
though less familiar, transients (Sprunt
and Chamberlain 1949, Shanholtzer 1974b).

While most avian forms in the maritime forest are insectivorous, the most
fundamental level is occupied by the
nectar-eating ruby-throated hummingbird.
Other species also occasionally utilize
such plant products as sap, most notably
the yellow-bellied sapsucker, ruby-
crowned kinglet, and some wood warblers.
However, these species do not rely on sap
exclusively (Sprunt and Chamberlain 1949,
Van Tyne and Berger 1959, Bent 1964).
Seeds and fruit are consumed by granivorous species such as the sparrows and
buntings, as well as by omnivores like the
robin, blackbirds, and grackles. Omnivorous birds also eat large numbers of insects, and the swallows, flycatchers,
vireos, wood warblers, wrens, thrushes,
and other small passerine species are almost totally reliant on them. Dead animal matter is eaten by two scavenging
species, the black vulture and the turkey
vulture. Raptors include the red-tailed
hawk and the sharp-shinned hawk, the latter being a major predator on small pas-
serines. Atop the predation pyramid is
the great horned owl, which takes not only
mammalian prey, but also other birds as
large as the two aforementioned hawks
(Bent 1961). Trophic relationships among
birds in this habitat are summarized in
Fig. 3-16.

The turkey was formerly native to
many of the barrier islands in South
Carolina and Georgia. Sprunt (1936) observed many wild turkeys on Cumberland
Island. South Island and Blackbeard Island provide habitat for the purest strain
of wild turkey populations remaining in
the Sea Island Coastal Region. Penreared,
semi-tame turkeys have been released over
the years on many of the islands to establish game populations, and turkeys
sighted today may not be a pure strain.

The blue jay is a common mainland
bird that has a sporadic distribution in
the maritime forest habitat. Its presence on the mainland has been reported
many times in South Carolina and Georgia
(Burleigh 1958, Sprunt and Chamberlain
1970). Tomkins (1965a) reported the absence of blue jays on some of Georgia's
barrier islands such as Sapelo, Blackbeard,
Wassaw, and Little Cumberland. However,
since that time, the blue jay has been
reported as a permanent resident on Sapelo
Island (Johnson et al. 1974). Interest-
ingly, blue jays are present in great
abundance on islands where high-density
housing development has occurred (e.g.,
Jekyll, St. Simons, etc.). Apparently,
human habitation has a positive influence
on distribution of this species (Johnson
et al. 1974).

(2) Representative Species.
The following species have been selected
for more detailed discussion on the basis
of their abundance and role in the maritime forest habitat: great horned owl,
red-bellied woodpecker, Carolina wren,
boat-tailed grackle, painted bunting, and
yellow-rumped warbler.

The great horned owl is a fairly common permanent resident throughout the
study area (Sprunt and Chamberlain 1949,
Burleigh 1958, Shanholtzer 1974b). This
is not to say that this large owl is common to the extent that smaller species are.
As a large predator, it requires a correspondingly large home range, a feature
Wayne (1910) noted by saying that it was
"by no means abundant anywhere." While
the great horned owl is known to utilize
a number of habitats from the mountains
to cypress swamps, it is particularly at
home on barrier islands where large tracts
of undisturbed maritime forest remain.
It has been noted regularly on Bull,
Capers, Kiawah, Hilton Head, Turtle,
Sapelo, Blackbeard, Cumberland, Wilmington,
and Sea islands (Sprunt and Chamberlain
1949, Burleigh 1958, Teal 1959a,
Chamberlain and Chamberlain 1975,
Laurie 1978, Chamberlain 1979). The great
horned owl eats a wide variety of food but
concentrates on small nocturnal mammals,
including various species of rats and mice,
squirrels, mink, opossums, and rabbits.
Birds taken by this large raptor include
species as large as the Canada goose and
turkey, and as small as sparrows and the
least tern (Bent 1961, Chamberlain and
Chamberlain 1975). Nesting of this
species occurs early, often in January and
February. The great horned owl usually
uses an abandoned nest of a hawk or crow,
but on occasion constructs its own nest.
On Bull Island, this species has been
known to share a nest with a bald eagle
(Sprunt and Chamberlain 1949). The great
horned owl is a retiring species, often
not detected by the casual observer.

The red-bellied woodpecker is a common permanent resident throughout South
Carolina and Georgia, and it is the most
common of the woodpeckers found there
(Sprunt and Chamberlain 1949). This
species is widely distributed, but favors
deciduous forests. It is more numerous
in winter than during the spring-summer
breeding season, a characteristic Wayne
(1910) attributed to a large home range.
The species has been observed to travel in
loose flocks on the coastal islands of
South Carolina, Georgia, and northern
Florida (Maynard 1896). The red-bellied
woodpecker is primarily insectivorous,
with ants, caterpillars, wood borers, grubs,
and grasshoppers composing more than 50%
of its diet. The remainder is vegetable
matter, including nuts, acorns, berries,
and other fruit (Sprunt and Chamberlain
1949). This woodpecker has the interesting

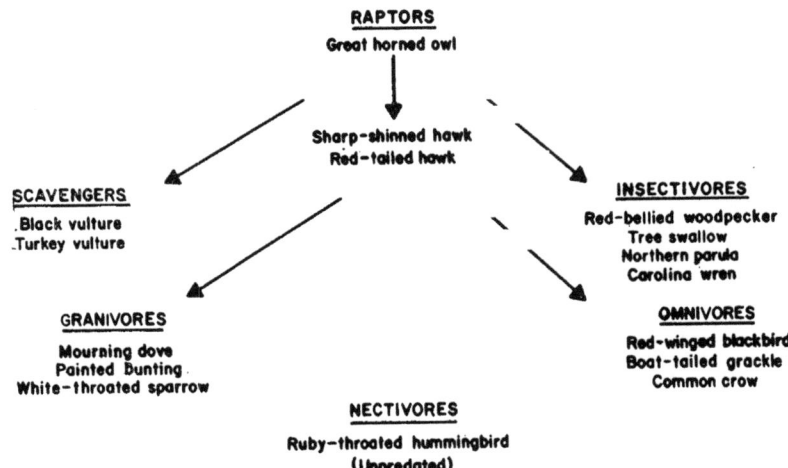

RAPTORS
Great horned owl

Sharp-shinned hawk
Red-tailed hawk

SCAVENGERS
Black vulture
Turkey vulture

INSECTIVORES
Red-bellied woodpecker
Tree swallow
Northern parula
Carolina wren

GRANIVORES
Mourning dove
Painted bunting
White-throated sparrow

OMNIVORES
Red-winged blackbird
Boat-tailed grackle
Common crow

NECTIVORES
Ruby-throated hummingbird
(Unpredated)

Figure 3-16. Generalized trophic relationships among representative birds in maritime forests of the Sea Island Coastal Region.

habit of storing food in a central location and returning to its cache to feed (Bent 1964a).

The Carolina wren is another common permanent resident throughout the coastal plain of South Carolina and Georgia (Sprunt and Chamberlain 1949). Wayne (1910) regarded this species as "exceedingly abundant," and there can be little doubt that the Carolina wren remains extremely common. This is the largest wren found in eastern North America, and when in its natural haunts, it favors dense undergrowth and brush. Here it searches for insects, which compose 95% of its diet. Food includes caterpillars, moths, beetles, grasshoppers, ants, flies, and spiders. Lizards, small frogs, and snails are also eaten, but only rarely (Sprunt and Chamberlain 1949).

The boat-tailed grackle is a common permanent resident in the coastal area of South Carolina and Georgia, but it does not occur far inland (Sprunt and Chamberlain 1949, Burleigh 1958). This species exhibits a marked preference for remaining near salt water, although it does occur up major rivers (e.g., the Santee and Savannah) to about 40 mi (64 km) inland (Wayne 1910, Bent 1965). The boat-tailed grackle does not migrate during winter, but does forage in large

flocks. These flocks are sexually segregated, and the males and females do not mix until late February or early March (Wayne 1910). Nests are located both in low shrub vegetation such as the wax myrtle and in trees as much as 80 ft (24 m) above the ground. The boat-tailed grackle is a colonial nester, and large groups often select sites at the edge of salt marsh or along the edge of a pond. This species is gregarious in feeding as well as in breeding. It is terrestrially oriented, with flocks exploring fields, beaches, and mud flats. Food is approximately 40% animal and 60% vegetable material. Animal matter includes fish, insects, shrimp, fiddler crabs, and crayfish. Vegetable matter is composed of corn, oats, and fruit of the cabbage palmetto (Wayne 1910, Sprunt and Chamberlain 1949, Bent 1965).

The painted bunting is a common summer resident of the Sea Island Coastal Region. This species is limited to the area below the fall line, and is most common on the coast (Sprunt and Chamberlain 1949, Burleigh 1958). Wayne (1910) regarded the painted bunting as "very abundant" on the coast, particularly in open country with hedgerows and scattered trees. Males typically sing from exposed perches in this open habitat and defend territories vigorously from others of the

152

same species. Nests are constructed in dense low growth or trees generally below 25 ft (8 m) in height (Bent 1968). The painted bunting is primarily granivorous, consuming 73% vegetable matter. Grass seeds are the favorite food, particularly those of foxtail grass. Insect matter taken includes beetles, grasshoppers, crickets, flies, and weevils (Sprunt and Chamberlain 1949).

Formerly known as the myrtle warbler, the yellow-rumped warbler is a common winter resident of the study area (Sprunt and Chamberlain 1949). This species is by far the most numerous warbler and justifiably can be called abundant. It occurs at just about all elevations in most terrestrial habitats, although it favors wax myrtle thickets. In South Carolina and Georgia, most food eaten by the yellow-rumped warbler is vegetable matter. The largest percentage is seeds of the wax myrtle, but red cedar, partridge pea, poison ivy and sumac seeds are also consumed. Insects taken include flies, beetles, plant lice, and insect larvae. Occasionally, the yellow-rumped warbler has been observed at fresh borings of the yellow-bellied sapsucker. It has not been established whether this warbler is utilizing the sap or the insects attracted to it (Sprunt and Chamberlain 1949, Bent 1963b).

In its natural state, the maritime forest is becoming rare in the South Carolina/Georgia barrier island complex. Man's impact on this habitat has been tremendous through development and logging operations. Maritime forests are not generally protected by land use regulations and these areas, located on the highest elevations of barrier islands, bring extremely high real estate prices. The disrupting forces of development in the maritime forest have detrimental effects on dominant species such as the warblers, sparrows, woodpeckers, flycatchers, wrens, etc. Most often, these species are dislocated and forced to seek replacement habitat elsewhere.

The most damaging developmental practice affecting avifauna of the maritime forest habitat is the wholesale reduction of understory. This practice has proven to be the major factor involved in the decline of songbird populations, particularly the Carolina wren and white-eyed vireo (Kale and Webber 1972). MacArthur and MacArthur (1961) found that niche availability for dominant ground-dwelling species varied with herbaceous and shrub vegetation density. Well-developed understory and subcanopy in a forest provides habitat diversity for low to mid-level foragers and nesters. Generally, the understory is removed first in the development of barrier islands. Construction of roads, trails, and home sites require such removal.

4. Mammals

a. Bird Key and Dune Communities. Because low-lying bird keys are subject to overwash and flooding, especially under storm conditions, they cannot be considered to have a permanent mammalian fauna. Nevertheless, these islands do become temporarily occupied by small rodents. W. David Chamberlain reports (1978, Charleston, South Carolina, pers. comm.) that cotton rats are present on Deveaux Bank and on Bird Island in Bulls Bay, South Carolina. Chamberlain also believes that an experimental introduction of a subspecies of cotton mouse (Peromyscus gossypinus anastasae) was made on Deveaux Bank by a Georgia University group. This has not been confirmed, and the status of this mouse population is not known. No published papers deal with the mammals of these transitory islands.

Although a relatively seed- and grass-rich environment, the dune community is susceptible to recurrent environmental extremes. Few mammal species permanently reside in this harsh, xeric subsystem. This dune habitat is, however, routinely traversed from forest to beach by nocturnal mammals. Specific studies of dune mammals have not been conducted in the characterization area, but papers by Golley (1962), Johnson et al. (1974), Hillestad et al. (1975), and Pelton (1975) present a considerable amount of information on the occurrence of mammals in dune environments.

Pelton (1975) found an exotic species, the house mouse, to be the numerically dominant mammal in the dune zone of Kiawah Island, South Carolina. Feral house mice feed on a wide variety of seeds and herbs (Golley 1962), and the dunes represent a seed-rich environment. Other rodents trapped in the Kiawah dunes by Pelton (1975) include the cotton mouse, the eastern wood rat, and the cotton rat, which was second in abundance only to the house mouse. In Georgia, Hillestad et al. (1975) found the cotton rat to be the most abundant small mammal on Cumberland and Little Cumberland islands. However, the cotton rat generally frequents areas of denser cover than the dune environment provides. Pelton (1975) found it more among the secondary dunes, where the typical dune grasses are mixed with wind-pruned shrubs. The cotton rat feeds largely upon seeds and herbs, but may be a serious predator on the eggs and hatchlings of birds nesting in the dunes. All of these rodents, in turn, are important food sources for birds of prey, both hawks and owls, which hunt rather extensively over the dunes.

The eastern mole can be expected among dunes as the principal insectivore. The subspecies expected to occur is Scalopus aquaticus howelli (Golley 1966). This eastern mole is known to feed largely on burrowing insects and annelids, but it also consumes plant roots. It seems likely that in the dry, sandy interdunes, roots would be relatively more important than in more moist habitats. Moles were the second most common small mammal of Cumberland Island (Hillestad et al. 1975).

Most of the rather narrow fringe of dunes in the study area is backed by maritime forest which supports deer herds, although such forest is generally not good deer habitat (see following section). Wherever this juxtaposition occurs, the deer come onto the dunes at night to graze. Pelton (1975) cites extensive grazing by deer on herbaceous plants such as the butterfly pea on the Kiawah Island dunes.

On those islands where domestic animals have been introduced, it is not unusual to find them grazing in dune areas. For example, a small flock of feral sheep (six to eight animals) still exists on Capers Island. Pelton (1975) listed feral hogs and goats grazing on Kiawah Island, and Hillestad et al. (1975) described horses, cattle, and pigs on Cumberland Island and pigs on Little Cumberland Island.

The principal carnivorous mammal in the dunes (as in most other coastal habitats) is the raccoon. Dunes are not considered a principal feeding area for raccoons, but they do traverse the dunes and seek out large insects, crabs, and eggs of nesting birds. A special case occurs during early and mid-summer months when the loggerhead sea turtle is nesting in the transition zone of upper beach and primary dunes. The raccoon is a major predator on nests of these turtles (see Chapter Two, Marine Intertidal Subsystem, for further discussion of this predatory behavior).

Both the eastern cottontail and marsh rabbits forage among the dunes at night for grasses and seeds. Their presence is readily detected by their easily recognizable fecal deposits. Although the relative numerical abundance of the two local species of rabbits has not been determined for coastal South Carolina and Georgia, the marsh rabbit is expected to be the principal species. Large owls which are reported to patrol the dunes at night are probably hunting for rabbits.

As stated previously, the primary perturbation of the dune community is a direct result of man's activities. Total destruction (levelling) or severe alteration of dunes for building sites is common. The use of all-terrain vehicles on sand dunes destroys much of the vegetation responsible for limiting erosion. Any action which reduces the already sparse dune vegetation adversely affects the use of the habitat by mammals. Rising sea level and erosion are two natural factors in loss of dune habitat; however, the rate of loss has been accelerated by man.

b. Transition Shrub Zone and Maritime Forest Communities. Maritime forests of the barrier islands along the coasts of South Carolina and Georgia exhibit a generally impoverished mammalian fauna compared to mainland habitats (Hillestad et al. 1975).

The white-tailed deer is the largest of native mammals to presently occupy maritime forests of the Sea Island Coastal Region and, ecologically, it is probably the dominant one. Although maritime forests fail to provide ideal deer habitat, deer populations of these islands are of special interest. Goldman and Kellogg (1940) described a number of subspecies of deer from the sea island coastline, based on skeletal size and pelage color. These subspecies included Odocoileus virginianus tavrinsulae from Bull Island north of Charleston; O. v. hiltonensis from Hilton Head Island south of Port Royal Sound; and O. v. nigribarbis from Sapelo, Blackbeard, and possibly other adjacent Georgia islands. The principal mainland subspecies in both States is considered to be O. v. virginianus. Whether these subspecific designations have a real base in genetic differences or were simply reflections of nutritional conditions on the several islands is open to question. Deer are known to be strong swimmers and often move freely between adjacent islands. Therefore, it seems unlikely that genetic isolation could have proceeded very far on most barrier islands. Further, over the last century there have been a number of deer introductions and transplantations, and the question of genetic isolation may now be moot.

Deer herds of varying size exist today on virtually all the islands of the Sea Island Coastal Region except for those which have been developed entirely into residential areas. South Island, near the northern limit of the characterization area, undoubtedly has the highest population density of deer of any island in the Sea Island Coastal Region. This high population level is maintained only because complete protection from hunting and substantial amounts of supplemental feed are provided. Also, a significant portion of the island has been converted into a grass-covered mall. At twilight, the herd emerges from the maritime forest to graze on the grassy areas through much of the night. In such areas, it is not unusual for observers to have 50 to 60 deer in sight at one time. This is clearly an unnatural condition and such a high density could not exist without extensive supplemental feeding. Even with the supplemental feeding, a sharp browse line is obvious in the forested areas. Undoubtedly, cessation of supplemental feeding would lead rapidly to a population collapse.

154

The three islands adjacent to South Island (North, Cedar, and Murphy) all have small deer populations. However, the size of the herd is consistent with the existing habitat, even though hunting is prohibited or highly limited. Bull Island, within the Cape Romain National Wildlife Refuge, has a fairly dense deer population. The only population control on Bull Island is through limited and highly regulated archery hunts. (Data for these hunts are presented in Volume II, Socioeconomic Features.) Both Capers and Dewees islands have limited deer populations, which currently appear to be in balance with the size of their habitat. Currently, hunting is prohibited on Capers Island and is limited to a few property owners on Dewees Island.

Near major metropolitan areas, many coastal islands have been converted into residential areas, thus eliminating deer populations. For example, Isle of Palms, Sullivans Island, and Folly Island in the vicinity of Charleston support no deer populations. Nevertheless, all of the larger islands in the southern half of South Carolina still retain deer populations, even though some (e.g., Kiawah and Hilton Head islands) have undergone residential and resort development.

In Georgia, significant deer herds exist on those islands (e.g., Cumberland) where development pressures have been minimal (Hillestad et al. 1975). Where development has largely replaced the natural habitat, the deer herds have been reduced or eliminated. Johnson et al. (1974) described the introduced European fallow deer of Jekyll, Little St. Simons, Sea, and St. Simons islands, and noted that a herd of 400 - 500 fallow deer still exists in the vicinity of Little St. Simons Island. Red deer were also introduced on Little St. Simons Island, but did not persist there.

Gray and fox squirrels are important herbivorous mammals in all forest types, and both are present in coastal maritime forests. The gray squirrel in particular exhibits interesting and presently unexplained geographic distribution patterns. The maritime forest, with its mixture of oaks, hickory, magnolia, and pines, would appear to provide ideal feeding and nesting conditions, yet gray squirrels appear to be totally absent from many islands. Early distributional studies found the gray squirrel native to Cumberland Island but not to Little Cumberland Island (Bangs 1898, Harper 1927). The same condition prevails on this pair of islands today (Hillestad et al. 1975). According to Johnson et al. (1974), the gray squirrel occurs on Cockspur, Tybee, Wassaw, Ossabaw, St. Catherines, Sapelo, Little St. Simons, St. Simons, Jekyll, and Cumberland islands in Georgia, but is missing from Oysterbed, Little Tybee, Blackbeard, Wolf, Sea, and Little Cumberland islands. Tomkins (1965b) suggested that squirrels may have

been introduced on several of these islands. The same patchy distribution occurs on South Carolina coastal islands. For example, gray squirrels are present on South, Kiawah, Hilton Head, and most of the residentially developed islands, but are absent from Bull Island and Capers Island. Published accounts are lacking for many of the islands.

The distribution of fox squirrels is also sporadic. Johnson et al. (1974) cited records from Ossabaw, St. Simons, and Cumberland islands, but not from any of the other Georgia islands. In South Carolina, Pelton (1975) recorded the fox squirrel on Kiawah Island. It is also known to occur on Bull Island, South Island, and Cat Island. It is definitely missing from Capers Island. According to Sanders (1978), a mid-western subspecies of the fox squirrel (Sciurus niger rufiventer) was introduced on Dewees Island and Capers Island. This squirrel did not survive on Capers, and its status on Dewees is unknown.

In general, it appears that those islands nearest the mainland have squirrel populations, while those more widely separated by marshes and sounds may or may not have such populations. It is difficult to accept that the small differences in geographic isolation account for this differential distribution. Whatever the reason, this phenomenon provides an excellent opportunity to study maritime forest ecology where such an important faunal element is totally missing.

Where squirrels do occur in the maritime forest, they feed predominantly on the mast of oak and hickory, although a variety of plant materials and some insects are consumed. Squirrels also feed extensively on the green stage of pine cones.

The marsh and eastern cottontail rabbits are occasionally found nesting in maritime forests, but such forests are far from ideal feeding areas for either species. The marsh rabbit is more likely to be limited to the interface between forest and wet habitats, especially salt marsh. The cottontail also could be expected in forest areas adjacent to fields or more open woods. Neither rabbit is ecologically very significant in this habitat.

Several species of rodents complete the principal assemblage of herbivorous mammals of maritime forests. However, surprisingly few small mammal studies have been conducted in this habitat on the coastal islands. Pelton (1975) found the cotton mouse, cotton rat, and eastern wood rat to be the most abundant small rodents in the maritime forests of Kiawah Island. Hillestad et al. (1975) found the cotton rat to be the most abundant small rodent on Cumberland Island. The assemblage described from Kiawah is probably typical of most islands.

The most abundant medium-sized omnivorous mammal of the maritime forests is the raccoon. However, the raccoon probably does not play a major trophic role in the maritime forest community. Rather, the forest provides required daytime cover, resting trees, and den trees. At night, most raccoons desert the forest to feed in more suitable habitat such as adjacent marshes and intertidal flats.

The opossum is a well-known omnivore occupying maritime forests. It feeds on a variety of plant and animal materials, but studies in Alabama and Georgia indicate that insects are the preferred food (Wheeler 1939). On Georgia islands, the present distribution of opossums includes generally those islands for which bridges and causeways provide avenues for invasion or reinvasion from the mainland, plus Sapelo and Little St. Simons islands, neither of which has a bridge or causeway connection. Certain islands not connected to the mainland which formerly had populations of opossums do not at present. For example, the opossums on Cumberland Island have been extirpated within the last century. Johnson et al. (1974) suggest that, in the aftermath of the Civil War, newly freed slaves may have eliminated the opossum from isolated coastal islands by using them for food. Few data are available concerning the distribution of opossums on South Carolina islands, but in general the pattern appears similar to that observed in Georgia. The opossum is apparently present on all islands connected to the mainland by causeways and bridges, and is either present or absent on islands not connected.

Feral hogs are conspicuous components of the maritime forest mammalian fauna, and on some islands may be the ecologically dominant omnivore. In some cases, the swine populations are the result of fairly recent introductions, but in other cases the feral populations are historically recorded. Lawson (1937) reported that the coastal islands north of Charleston (probably Dewees and Capers islands) were used as naturally fenced repositories of cattle and hogs at the time of his visits in the early years of the 18th century. Hanson and Karstad (1959) reviewed the history and present status of feral hogs in the coastal plains region of Georgia and South Carolina, and their implications for livestock owners and wildlife conservationists. They also studied food habits of feral swine in several Georgia coastal locations and found that the hogs consumed a variety of plant materials including acorns, roots, and tubers. In addition, feral swine have been found to feed on earthworms, insects, fiddler crabs, frogs, snakes, turtles, rodents, and the eggs and young of birds. Where deer, wild turkey, raccoons, and feral hogs exist in the same habitat, swine compete with the aforementioned for the same food items.

During the early colonial period, a number of large predators including black bear, eastern cougar, gray wolf, and bobcat occupied maritime forests of the coastal region. Of these, only the bobcat and black bear have survived, and they are limited to a few coastal sites (e.g., Francis Marion National Forest). The black bear no longer occurs on any of the coastal islands. Sanders (1978) summarizes reported visual observations which suggest that a few eastern cougars may remain in the Cape Romain and Santee Delta area.

The bobcat is the most successful large mammalian predator in maritime forests of the Sea Island Coastal Region, but its present distribution on the islands is spotty. Johnson et al. (1974) cited past records of the bobcat from Cockspur, Ossabaw, Sapelo, and Cumberland islands in Georgia, but indicated that the species had been eliminated from all four islands. Hillestad et al. (1975) reported that three bobcats were released on Cumberland Island in the early 1970's, but it is not known if they survived. In South Carolina, the bobcat is still present on several islands, particularly North, South, Cat, and Murphy islands and probably on Bull and Capers islands. All of these are seaward of the Francis Marion National Forest, which supports a substantial population of bobcats. Food habits of bobcats on coastal islands have not been investigated. However, Kight (1962) studied bobcats around the Savannah River Plant (near Aiken, South Carolina) and found that rodents and rabbits constituted the bulk of this predator's diet, although birds, reptiles, and large insects were also consumed. It seems reasonable to assume that the same general food habits would also characterize bobcats in the maritime forest, with some shifts in species composition. For example, it is likely that the marsh rabbit and rice rat would be relatively more important items of the bobcat's diet in coastal as opposed to inland locations.

Mink and river otter are present on most of the islands along the coasts of South Carolina and Georgia, and use the maritime forest to some extent for cover and nesting. However, since they are largely aquatic feeders (see Chapter Five, Freshwater Ecosystems), their trophic role is minimal in this habitat.

The short-tailed shrew, least shrew, and eastern mole have been recorded from the maritime forests of Kiawah Island by Pelton (1975) and, despite their small size, they are important predators. This same trio is likely to occur in the maritime forests of most other coastal islands, but too few detailed faunal studies have been done in this habitat to confirm their presence.

In addition to the true terrestrial mammals, some 11 or 12 species of bats are known to occur in the Sea Island Coastal Region. These include the big brown bat, silver-haired bat, red bat, hoary bat, northern yellow bat, Seminole bat, little brown myotis, evening bat, eastern pipistrelle, Rafinesque's big-eared bat, and the Brazilian free-tailed bat (Sanders 1978). While not included by Sanders (1978), the southeastern myotis has been recorded from coastal South Carolina and Georgia (including Sapelo Island), and it appears to be more common in this region than does the little brown myotis (LaVal 1967, Davis and Rippy 1968). Of these species, the most common in the characterization area are probably the red bat, Seminole bat, evening bat, eastern pipistrelle, and Brazilian free-tailed bat (Sanders 1978). Most of these common species typically roost in trees or in buildings, but the Seminole bat roosts in lumps of Spanish moss.

Bats are basically insectivorous, although some species may also consume a variety of other animal and vegetable matter. The most important source of mortality for bats is probably predation. Important predators include rat snakes, corn snakes, owls, hawks, other raptors, raccoons, opossums, striped skunks, minks, house cats, large fish, frogs, blue jays, and even cockroaches (which may consume young bats) (Barbour and Davis 1969). Another important source of mortality is accidental death through impalement on barbed wire or entanglement or collision with other man-made materials. Bats are also vulnerable to loss of roosting trees through forest reduction for resort development and to chlorinated hydrocarbon pesticides (such as DDT), which are deposited in fatty tissues. Bats may be particularly sensitive to pesticide poisoning in the spring following emergence from hibernation when their fat reserves have been depleted (Barbour and Davis 1969).

As mammalian habitat, the maritime forest is subject to dramatic impacts by human activity. The most obvious and significant impact is the use of maritime forest for residential and resort development. The islands are highly valued for these purposes, and the maritime forest, with its live oaks and palmettos and rolling forested dune ridges, makes a lovely setting for either residential or resort development. At present, the full spectrum of conditions ranging from virtually undisturbed natural maritime forest to total conversion to urban residential development can be found among the islands of the South Carolina and Georgia coastline. It is quite apparent that shifts in the mammalian fauna have and will continue to occur as development proceeds. Certain mammals will be favored by development and others will be completely eliminated. For certain mammals, limited modification may enhance the

habitat, while more intense development would eliminate that species entirely. Such situations are well known for white-tailed deer. The natural maritime forest is not ideal habitat for deer, since the upper story of vegetation is often so dense that shrubs and grass hardly exist at the ground level. When portions of the maritime forest are opened up but adjacent areas remain in the natural state, overall habitat is much improved and a larger deer herd can exist. When no natural areas remain on a given island (e.g., Isle of Palms, Sullivans Island, Folly Island and to a lesser extent the islands in the vicinity of Brunswick, Georgia), the deer will be eliminated completely except when they assume a protected and largely domesticated existence. Some mammals, like the bobcat, are not tolerant of human presence and will be eliminated at an early stage of island development. Others, such as the raccoon, adapt readily to human presence and can become pests when they abandon their natural habits and beg for food at the kitchen door or visit the garbage can. As development proceeds, the inevitable increase in domestic dogs and cats creates additional stress and competition for the native mammal populations. Also, the introduction of roads and automobiles into the maritime forest is a major mortality source for medium-sized nocturnal mammals.

The introduction of domestic animals on those islands not developed for residential or resort purposes has and will continue to have adverse impacts on the maritime forest and its natural mammalian fauna. Cattle, sheep, goats, and swine have all been introduced to some coastal islands. The negative impact on native mammals may come about through alteration of the vegetation by grazing or by direct competition for food as was described for feral hogs. Fortunately, this practice of introducing domestic livestock to the islands is diminishing. In many cases, domestic and feral animals are being actively removed or eliminated.

CHAPTER FOUR

ESTUARINE ECOSYSTEM

1. MAJOR CHARACTERISTICS

A. DEFINITION

The estuarine system was defined by Cowardin et al. (1977) as ". . .deep-water tidal habitats and adjacent tidal wetlands which are usually semi-enclosed by land, but have open, partially obstructed, or sporadic access.to the open ocean and in which ocean water is at least occasionally diluted by freshwater runoff from the land." Included under this definition are so-called hypersaline or "negative estuaries" (Perkins 1974), where evaporation may at least periodically exceed freshwater inflow. Offshore areas inhabited by organisms regarded as typically estuarine, such as oysters and mangroves, were also included in the estuarine system. This definition is a modification of the widely accepted description of an estuary by Pritchard (1955) as ". . .a semi-enclosed coastal body of water having a free connection with the open sea and within which the sea water is measurably diluted with fresh water runoff."

The limits of the estuarine system, as defined by Cowardin et al. (1977), ". . .extend upstream and landward to the place where ocean-derived salts measure less than 0.5 °/oo during the period of average annual low flow. The seaward limit of the Estuarine System is: 1) a line closing the mouth of a river, bay, or sound; 2) a line enclosing an offshore area of diluted sea-water with typical estuarine flora and fauna; or 3) the seaward limit of wetland emergents, shrubs or trees where these plants grow seaward of the line closing the mouth of a river, bay, or sound." (See the Atlas frontispiece for a graphic delineation of estuarine areas of the Sea Island Coastal Region.)

The estuarine system was subdivided by Cowardin et al. (1977) into two major subsystems, subtidal and intertidal. The subtidal subsystem includes those areas where the substrate is continuously submerged, while the intertidal subsystem consists of those areas where the substrate is periodically exposed and flooded by tides (including the associated splash zone).

B. GENERAL DESCRIPTION AND MODEL

Estuaries are generally considered to be low energy systems in terms of wave and current action, and the estuarine system is more strongly influenced by its association with land than is the marine system (Cowardin et al. 1977).

Estuarine hydrography is influenced by oceanic tides, precipitation, freshwater runoff from land, evaporation, and wind (Cowardin et al. 1977), as well as river flow, meteorological pressure centers, and the size and shape of the estuary. Estuarine areas of Georgia and South Carolina are swept by moderately strong tidal currents, the result of a 2 - 4 m (6.5 - 13.1 ft) tidal range. Rivers in the study area are typically narrow and shallow in the estuarine zone; most are less than 2 km (1.2 mi) in width and 10 m (32.8 ft) in depth. Since the tributaries are generally small, isohalines are compressed in estuaries of the Sea Island Coastal Region. Salinity, which has a significant influence on the species composition of the resident biota, fluctuates within these estuaries daily (tidally), seasonally, and as a result of freshwater runoff. The degree of these differences varies from one estuary to another. Water temperature, an important factor regulating the activities and reproductive cycles of estuarine organisms, normally ranges from about $10^{\circ}C$ ($50^{\circ}F$) in winter to $28^{\circ}C$ ($82.4^{\circ}F$) in summer. Estuaries are characteristically turbid due to the considerable sediment loads and detritus that are suspended in the water column. The resulting reduction in water transparency generally inhibits phytoplankton production (Ragotzkie 1959b) and limits the growth of macroalgae in subtidal areas. Estuarine substrates in Georgia and South Carolina vary widely. Sheltered areas behind the coastal islands are generally muddy. Bottoms of sand and shell are widespread at the mouths of rivers, sounds, and inlets. Coarse sand is common near the head of many estuaries with watersheds originating in the Piedmont region, and mud is prevalent in the middle and upper reaches. Further details on estuarine hydrography and geology are discussed in Volume I, Physical Features.

Several different types of estuaries are recognized based on water circulation patterns (see Volume I, Table 5-5). In vertically stratified estuaries, the net motion of surface and bottom layers is ideally in opposite directions, the former having a net seaward movement and the latter a net landward movement. This has important implications for the biota of estuaries, and especially for the plankton (see the section on zooplankton of the subtidal estuarine subsystem).

Estuaries also differ in terms of geomorphology. Of three types described by Perkins (1974), namely bar-built estuaries, deep-basin estuaries, and coastal plain estuaries, only deep-basin estuaries are lacking in the study area (see Volume I). Bar-built estuaries originate from the formation of a bar seaward of a low-lying coastal area. Coastal plain estuaries or drowned river valleys occur at the terminal end of river valleys and drowned river mouths.

The physiographic regions of a typical estuary, originally classified by Day (1951, 1964) and modified by Carriker (1967), are shown in Figure 4-1. According to Carriker, boundaries between zones lack precise limits because of the variety of estuarine types and the inherent variability of salinity in a given estuary.

Several classifications have been proposed which categorize different zones in an estuary based on salinity (Remane 1971). The most widely adopted of these is the so-called Venice System (Symposium on the Classification of Brackish Waters 1958):

Limnetic	< 0.5°/oo
Mixohaline	0.5°/oo – 30°/oo
Oligohaline	0.5°/oo – 5°/oo
Miooligohaline	0.5°/oo – 3°/oo
Pliooligohaline	3°/oo – 5°/oo
Mesohaline	5°/oo – 18°/oo
Miomesohaline	5°/oo – 10°/oo
Pliomesohaline	10°/oo – 18°/oo
Polyhaline	18°/oo – 30°/oo
Euhaline	30°/oo – 40°/oo
Hyperhaline	> 40°/oo

The oligohaline zone, encompassing salinities of 0.5°/oo – 5°/oo, occurs at the head of an estuary. The unidirectional flow of water characteristic of nontidal riverine environments changes to a regime of slow circulation in the oligohaline zone, and waters there are characteristically turbid (Copeland et al. 1974). Organisms inhabiting this zone must be tolerant of both low salinity and fresh water, at least for brief intervals, because of variations in saltwater intrusion and freshwater discharge. This is a region of high stress for both marine and freshwater organisms, and the number of species is low. Nevertheless, the oligohaline zone of an estuary is very important as a nursery ground for a number of species, including various fishes, shrimp, and blue crabs. After studying the oligohaline region of an estuarine system in Virginia, Van Engel and Joseph (1968) concluded that it afforded an abundant and acceptable food supply, protection from predators, and physiological suitability, all of which are essential elements of a successful nursery ground.

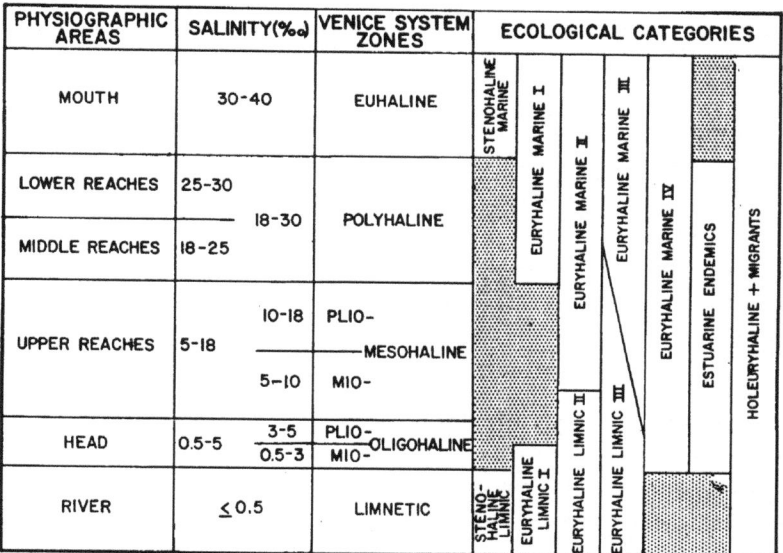

Figure 4-1. Physiographic areas (Day 1951, 1964), Venice System zones (Symposium on the Classification of Brackish Waters 1958), and ecological categories of organisms (Remane 1971) in estuaries. See text for an explanation of terms (Calder et al. 1977b).

The mesohaline zone, bounded by salinities of 5º/oo - 18º/oo, is rather restricted in estuaries of the Sea Island Coastal Region. As noted by Bellis (1974), most rivers of Georgia and South Carolina flow directly into the ocean rather than emptying into large water masses such as Chesapeake Bay, Delaware Bay, and the sounds of North Carolina, which have extensive mesohaline zones. Mesohaline areas are of considerable economic importance in most estuaries of the east coast because they support commercially exploited fisheries resources including oysters, shrimp, blue crabs, and a number of species of fishes. The species composition of the biota in this zone is very similar in both Middle Atlantic and South Atlantic States (Roberts 1974b, Watling et al. 1974).

The polyhaline zone, with salinities of 18º/oo - 30º/oo, occurs in the lower portions of many Georgia and South Carolina river systems, as well as in various creeks, bays, and sounds behind the barrier islands (Roberts 1974b). Since most rivers originating in the Sea Island Coastal Region have relatively small watersheds, freshwater discharge is typically low and salinities are generally high in the lower reaches. The entire coastal area is low, and broad expanses of marshes and mud flats occur in the estuaries. Extensive beds of the American oyster are common on these flats and along the margins of tidal creek tributaries in polyhaline areas.

The euhaline zone, bounded by salinities of 30º/oo - 40º/oo, is largely restricted in Georgia and South Carolina estuaries to inlets and bays receiving negligible freshwater discharge, and to the area near the mouths of various rivers and sounds. The number of species reaches its peak in this zone of an estuary.

Based on the magnitude of salinity oscillations, estuaries may be regarded as either fluctuating (poikilohaline) or relatively stable (homoiohaline). While useful for the definition of zones in homoiohaline estuaries, the Venice System appears to have limited applicability in poikilohaline estuaries.

Remane (1971) classified aquatic organisms according to their distribution with respect to salinity (see Ecological Categories, Figure 4-1). The fauna of estuarine areas consists largely of eurybaline marine species (Percival 1929, Gunter 1950), because freshwater animals are typically intolerant of even low salinities. Species numbers generally decline along the salinity gradient from the mouth to the head of an estuary because decreases in the number of marine species are not offset by a corresponding increase in freshwater species.

Another noteworthy classification of estuarine and coastal areas, based on

energy flow, was given by Odum et al. (1974). They recognized a total of 48 such systems in the United States.

Aspects of energy flow in the estuarine ecosystem are diagrammed in Figure 4-2. As noted in Chapter Two, the sun is the primary energy source for all ecosystems, including estuaries, in the project area. Energy input from the sun is utilized in the process of photosynthesis by autotrophic plants. Part of the light energy converted during photosynthesis is stored (production) and part is dissipated (respiration) by autotrophs. While nonvascular plants contribute to the total primary production in estuaries of Georgia and South Carolina, it is widely accepted that a greater contribution is made by the marsh grasses. Heterotrophic components of the estuarine ecosystem are directly or indirectly dependent upon the production of autotrophs for their energy, and energy is transferred to the higher trophic levels in steps from primary consumers to secondary and tertiary consumers, and so on, as outlined in the following section on food webs.

High nutrient levels provide a basis for the high productivity characteristic of most estuaries (Perkins 1974, Nelson-Smith 1977). Production generally exceeds respiration in estuarine autotrophs, and Odum (1971) speculates that nutrient and detritus are "outwelled" to coastal waters. Bound nutrients are released to the estuarine ecosystem through excretion and decomposition of organic materials, and decomposers play an important role in nutrient recycling. As noted by Odum, energy from waves and tides plays a role in nutrient recycling and in transporting plant materials to the primary consumers.

In contrast with energy, which becomes dissipated through progressive links of the food chain, substances such as pesticides, heavy metals, and radionuclides may become concentrated at the higher trophic levels. Examples of this "biological magnification" include the concentration of DDT residues in predatory birds and the high levels of mercury found in some carnivorous fishes. Estuarine filter feeders, such as oysters, may concentrate trace metals several orders of magnitude above that found in the water.

It has been a generally accepted tenet in ecology that diverse systems are more stable than ecosystems of lower diversity, and are therefore better able to resist stresses of a given magnitude. However, this belief as it applies to estuaries has recently been challenged by Cópeland (1970) and Boesch (1972, 1974), and the reverse appears to be true. Boesch observed that benthic organisms in the typically low diversity mesohaline and oligohaline zones of an estuary are more resistent to perturbations, including salinity fluctuations and multiple pollution stress, than those restricted to areas of higher salinity.

160

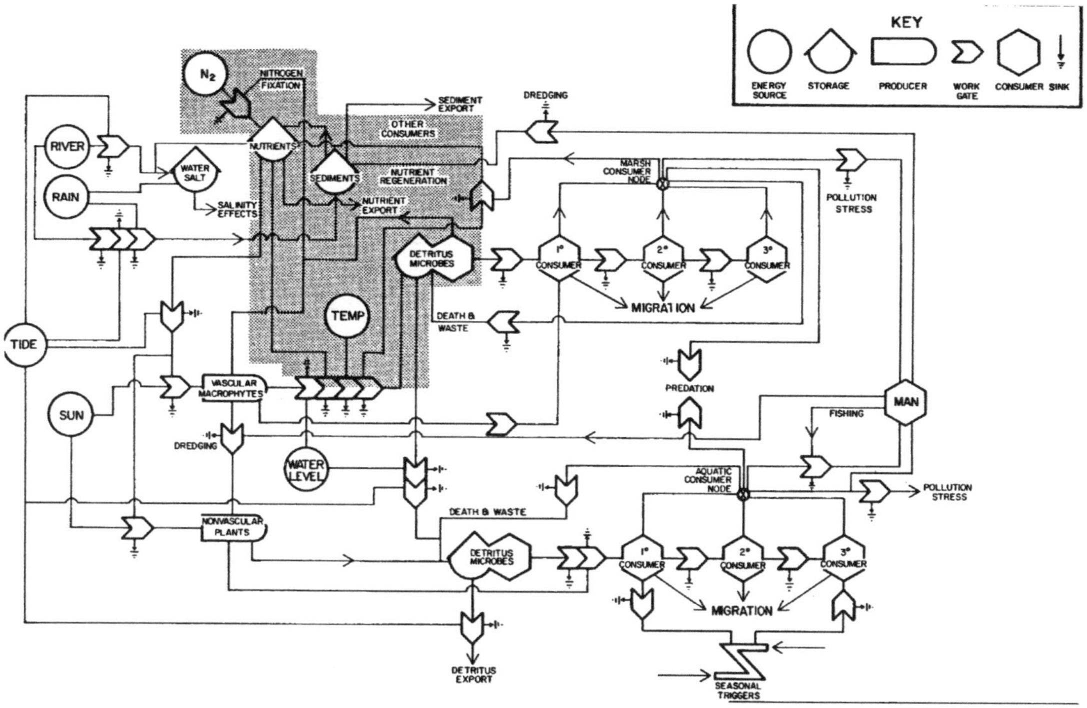

Figure 4-2. A generalized emergse model of the estuarine ecosystem of the Sea Island Coastal Region.

He also stated that species found in areas of low salinity exhibit high resilience to disturbance as a result of their typically high fecundity and rapid growth. Polluted areas in the polyhaline zone of Hampton Roads, Virginia, were found to be populated by eurytopic macrobenthic species typical of lower salinity regions (Boesch 1973). Similarly, the effects of flooding during Tropical Storm AGNES in the Chesapeake Bay system had less impact on the benthos of the upper estuarine regions than in higher salinity areas (Boesch et al. 1976b). These observations suggest that the effect of a given stress would therefore have a greater impact on estuarine biota where environmental constancy is high, where larger numbers of relatively stenotopic species (i.e., having a narrow range of adaptability) are present, and where diversity is therefore typically high. At the same time, the vulnerability of the entire estuarine habitat to human disturbance should not be underestimated (Boesch 1974). Careful and thoughtful management will be necessary if the environmental integrity of these biologically, economically, and aesthetically valuable areas is to be maintained.

C. GENERALIZED FOOD WEB AND RELATIONSHIPS (Fig. 4-3)

Food chains depict the transfer of energy from one organism to another in an ecosystem. Since a substantial percentage of the potential energy may be lost at each transfer, the number of sequential levels in most food chains is usually limited to not more than four or five (Odum 1971). Odum recognized two types of food chains: 1) a grazing food chain, involving transfer of energy from living plants to animals, and 2) a detritus food chain, involving transfer of energy from plants to animals via organic detritus. Food chains are variously interconnected, often in complex patterns, to form what is known as a food web.

Studies at the University of Georgia Marine Institute (Schelske and Odum 1962) have shown that estuaries of the Sea Island Coastal Region are among the most productive natural systems on earth. Schelske and Odum attributed the high productivity of these waters to tidal action, abundant nutrient supplies, conservation and rapid turnover of nutrients, three units of primary production (marsh grass, benthic algae, and phytoplankton), and year-round production. Much of the productivity is attributable to the extensive salt marshes behind the barrier islands, and especially to smooth cordgrass. While a small amount of the total production of this plant is consumed directly by grazers such as insects, most is reduced to detritus by bacterial and fungal decomposers. The detritus, because of its associated microbes, is believed to be more nutritional than Spartina tissue alone (Odum and de la Cruz 1967), and estuaries of Georgia and South Carolina are based to a greater degree on detritus food chains than on grazing food chains. Darnell (1967) indicated that large quantities of particulate organic detritus are ingested by estuarine consumers, whether large or small, invertebrate or vertebrate, benthic or suprabenthic. Darnell stressed the important storage, transport, and buffer aspects of organic detritus to the estuarine environment. Energy is stored in detritus and utilization occurs at different time intervals. Transport occurs through the action of tidal currents, which disperse detritus throughout the estuary. Detritus constitutes a buffer because it is available to consumers during periods when primary production is low (e.g., during winter).

Energy is transferred through the estuarine food chain both from detritus and its associated microbiota to the higher trophic levels largely through bottom-dwelling invertebrates and detritus-feeding fishes, and from phytoplankton, macrophytes, and benthic microphytes to grazers including benthic invertebrates, zooplankton, and certain fishes. Middle carnivores include larger nektonic species and zooplankton predators. Top carnivores include larger fishes, predatory birds, and mammals.

Although the number of species is typically reduced under estuarine conditions, the food web remains intricate, even in simplified terms (Fig. 4-3). The food habits of one species, the striped mullet, reflect this complexity. This fish is part of both a grazing food chain (feeding on benthic algae) and a detritus food chain (feeding on organic detritus). It is also known to shift trophic levels from primary consumer to predator, preying on polychaete worms (Bishop and Miglarese 1978). Mullet, in turn, are preyed upon by other species of fish, birds, and mammals, including man. The complexity of food relationships in estuarine nekton has been discussed by deSylva (1975).

II. SUBTIDAL ESTUARINE SUBSYSTEM

A. DESCRIPTION

The subtidal subsystem ". . .includes that part of the Estuarine System in which the substrate is continuously submerged" (Cowardin et al. 1977). Many estuaries of the Sea Island Coastal Region are small, with drainage basins restricted to the coastal plain, although the Altamaha, Ogeechee, Savannah, Edisto, Santee, and Pee Dee rivers originate within the Piedmont area or beyond (see Volume I). Georgia and South Carolina combined have an estimated 598,700 acres (242,291 ha) of shallow subtidal estuarine habitat, compared with 2,206,600 acres (892,999 ha) for North

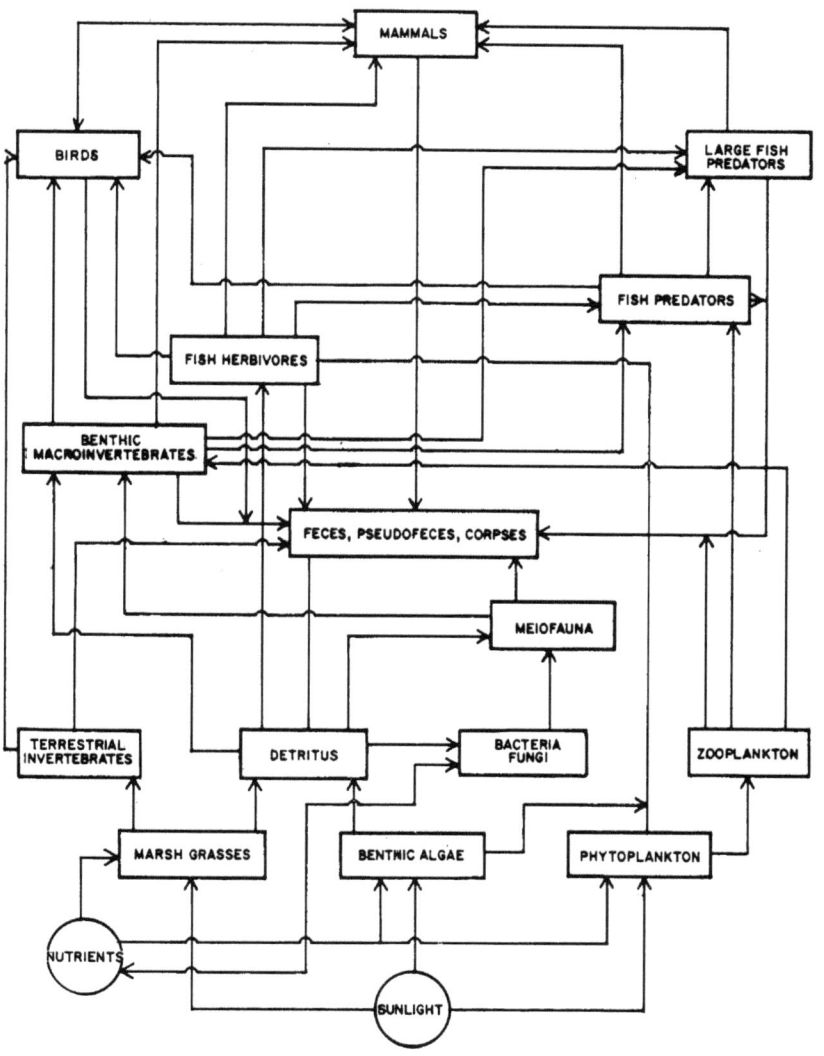

Figure 4-3. A generalized food web for the estuarine ecosystem of the Sea Island Coastal Region.

163

Carolina, 1,670,000 acres (675,840 ha) for Virginia, and 1,406,100 acres (569,041 ha) for Maryland (Spinner 1969). While the numbers of species are less than that of marine waters, densities are often high. Penaeid shrimp, oysters, blue crabs, hard clams, and certain fishes occur in sufficient abundance to support commercial and recreational fisheries in the area (see Volume II).

B. PRODUCERS

1. Phytoplankton

Although there is a paucity of information on phytoplankton populations in estuaries of South Carolina and Georgia, studies performed to date give some indication of the structure and dynamics of nonvascular flora in the myriad of habitats associated with estuarine ecosystems.

Estuaries provide a rich and varied environment for phytoplankton populations. Environmental factors in estuaries fluctuate more frequently and over wider ranges than in either fresh water or seawater. In addition, the proximity of estuarine waters to land increases the environmental variables by subjecting phytoplankton populations to urban, industrial, and agricultural discharges, as well as the normal constituents of land drainage. The dynamics of this ecosystem are thus translated to the phytoplankton, resulting in population characteristics which normally deviate greatly from either the freshwater or seawater extremes of a typical estuary. Rice and Ferguson (1975) summarized the factors influencing phytoplankton populations in estuaries (Table 4-1). As indicated in this table, a relatively large number of factors, abetted by both natural and man-imposed variation, impact upon estuarine phytoplankters. Despite these continuous perturbations, estuarine phytoplankton populations tend to be larger and thus more productive than populations in adjacent systems.

It is generally accepted that the majority of phytoplankton in lower estuarine systems are derived from the marine environment. Recent evidence, however, seems to support the concept that freshwater phytoplankton can comprise a significant portion of estuarine populations, and it has been suggested that they form an important part of the organic detritus of an estuary. A classic representation of the composition of phytoplankton populations in estuaries would show a dominance of limnetic (freshwater) forms in the oligohaline (0.5⁰/oo - 5⁰/oo) areas, an either rapid or gradual diminution of these forms in the mesohaline (5⁰/oo - 18⁰/oo) areas (dependent on tidal prism, river flow, etc.), and an increasing dominance of marine forms through the polyhaline zone (18⁰/oo - 30⁰/oo) to the sea. Few phytoplankters, either limnetic or marine, have wide environmental tolerances. True estuarine forms, if they exist, are not well represented.

Taxonomic studies of estuarine phytoplankters in South Carolina and Georgia are not numerous. A listing of all species identified or suspected from coastal South Carolina can be found in Manzi and Zingmark (1978). They listed over 360 marine/estuarine taxa; although this list is not complete, it does include all the abundant and most of the common genera associated with these aquatic systems (see Table 4-2 for a partial listing).

In terms of temporal distribution, very little is known about phytoplankton dynamics in the estuaries of South Carolina and Georgia. A compilation of abundance by season from the Santee River system (J. J. Manzi, 1977, South Carolina Marine Resources Division, Charleston, unpubl. data) is given in Table 4-3.

It is becoming more evident that the nanoplankton (phytoplankton which passes through fine-mesh plankton nets) is at least as important numerically as the net phytoplankton (phytoplankton retained by ordinary plankton nets) in estuarine waters. Data in Table 4-2, and results from several other recent studies (Campbell 1973, Van Valkenburg and Flemer 1974, Gallagher 1975), indicate that phytoplankton studies can no longer ignore the nanoplankton fraction, particularly when it may contribute up to 75% of the total phytoplankton productivity of estuarine systems (Williams and Murdoch 1966, Van Valkenburg and Flemer 1974). An additional area of high interest concerning phytoplankton distribution is the air/water interface or surface microlayer. While population densities and productivity are high in estuarine waters of South Carolina (Sellner 1973, Sellner et al. 1976, Zingmark 1977) and Georgia (Ragotzkie 1959b, Teal 1962, Pomeroy et al. 1976), it appears that population densities can be several times higher, and thus the contribution to total productivity greater, in the surface microlayer than in subsurface waters (Gallagher 1975, Manzi et al. 1977a). This is particularly important because many pollutants are also concentrated in this interface, thus magnifying the impact that pollutant additions to estuarine systems can have on microphyte productivity.

2. Macrophytes

A general absence of solid, permanent substrates, as well as turbidity and the scouring action of the tides, severely limit the growth of macroscopic algae in the estuarine environment of Georgia and South Carolina. The continuously immersed parts of sea walls, groins, oyster reefs, pilings, jetties, and subtidal substrates such as shell, man-discarded objects, and invertebrates like the gorgonian Leptogorgia

Table 4-1. Natural and man-imposed conditions which determine levels and rates of change of factors affecting abundance and succession of estuarine phytoplankton (Rice and Ferguson 1975).

Factors	Natural Conditions	Man-imposed Conditions
Salinity	Precipitation, runoff evaporation, circulation of water	Water impoundment, channelization, dredge and fill, mosquito ditching
Temperature	Latitude, season, weather, time of day, circulation of water	Heated effluent, dams, canals and waterways, stream channelization
Light Intensity		
At surface	Latitude, season, weather, time of day	Air pollution — smog
Below surface	Reflection, absorption, scattering	Dredging, waste dumping, erosion
Nutrients	Drainage, runoff, circulation of water, sediments	Sewage and industrial wastes, urban and agricultural drainage, erosion
Metabolites	Living and dead plants and animals	Sewage, urban and agricultural drainage, erosion
Toxic Substances		
Petroleum	Deposits	Leads and spills during drilling, transport, storage, use of disposal
Radionuclides	Primordial deposits, cosmic-ray produced	Fallout, nuclear power reactors, other releases
Heavy Metals	Terrestrial deposits, sediments, land drainage	Industrial and domestic wastes, mining, erosion
Synthetic Toxicants		Industrial, agricultural, domestic use

virgulata, provide places for algae to attach. Floating piers and docks offer an unusual substrate for algae, and if the salinity remains high, as it does at the marinas of Murrells Inlet, South Carolina, a diverse and luxuriant group of algae may be found. At this locality, polystyrene floats provide places of attachment for such red algal species as Grinnellia americana and Neogardhiella baileyi, and in the winter and early spring, the large, strap-shaped brown alga, Petalonia fascia, is the dominant species. Intermixed with this algal assemblage throughout the year are a variety of hydroids, bryozoans, and tunicates. This faunal and algal community offers a highly varied habitat for decapods and small fish. In less saline areas, such floats support a lower diversity of algae dominated by reds such as Polysiphonia denudata and

Ceramium strictum, chlorophytes like Enteromorpha and Ulva, ectocarpoid browns, and blue-greens.

In Charleston Harbor, very few algal species appear in trawl and dredge samples. Infrequently collected species include the red algae Gymnogongrus griffithesiae, Gracilaria verrucosa, and Chondria tenuissima; the green algae Enteromorpha and Ulva; and, during the colder months, the brown algae Ectocarpus and Giffordia. In winter and early spring, species of the sheet-like red algal genus Porphyra are a conspicuous element of shallow subtidal and intertidal areas.

The subtidal jetty flora of South Carolina is impoverished compared with that of North Carolina. Schneider and Searles (1978) listed 116 species found on North

Table 4-2. List of species of phytoplankton from estuarine stations off Kiawah Island, South Carolina (adapted from Zingmark 1975) (x = present in at least one sample; xx = predominantly found in this type of habitat; ? = present in at least one sample, but is not typically known from this type of station).

SPECIES	ABUNDANCE
Division: Bacillariophyta (Diatoms)	
Class: Centrobacillariophyceae	
Order: Eupodiscales	
Family: Coscinodiscaceae	
Actinoptychus undulatus	xx
Coscinodiscus centralis	x
C. concinnus	x
C. excentricus	x
C. excentricus var. fasciculata	xx
C. granii	x
C. lineatus	
C. radiatus	xx
C. subtilis	x
Melosira italica var. tenuissima	xx
M. moniliformis	xx
M. sulcata	x
Skeletonema costatum	
Thalassiosira sp.1	xx
T. decipiens	xx
Order: Rhizosoleniales	
Family: Corethronaceae	
Corethron hystrix	x
Family: Rhizosoleniaceae	
Guinardia flaccida	xx
Rhizosolenia alata	xx
R. alata var. indica	x
R. calcaravis	
R. castracanei	xx
R. delicatula	x
R. fragillissima	x
R. setigera	xx
R. stolterfothii	x
Order: Biddulphiales	x
Family: Bacteriastraceae	
Bacteriastrum delicatulum	x
B. elongatum	xx
B. hyalinum	xx
B. varians	x
Family: Biddulphiaceae	
Biddulphia	
B. alternans	xx
B. aurita	x

SPECIES	ABUNDANCE
B. dubia	x
B. longicruris	xx
B. longicruris var. hyalina	x
B. mobiliensis	x
B. pulchella	xx
B. rhombus	xx
B. sinensis	xx
Ceratualina bergonmi	x
Eucampia sp.1	x
Hemiaulus hauckii	xx
H. sinensis	x
Lithodesmium undulater	x
Streptotheca thamensis	x
Triceratium americanum	xx
Family: Chaetoceraceae	
Chaetoceros sp. 2	x
C. affinis	xx
C. compressus	xx
C. constrictus	x
C. convolutus	xx
C. curvisetus	xx
C. decipiens	xx
C. didymus	x
C. gracilis	x
C. hispidum	x
C. mulleri	x
C. pelagicus	x
C. pendulus	x
C. peruvianus	xx
Class: Pennatibacillariophyceae	
Order: Fragilariales	
Family: Fragilariaceae	
Asterionella glacialis	xx
Campylosira cymelliforme	xx
Plagiogramma vanherckii	x
Rhaphoneis amphiceros	x
Family: Tabellariaceae	
Synedra acus	x
Tabellaria fenestrata	x
Thalassionema nitzschioides	xx
Thalassiothrix delicatula	x
T. frauenfeldi	xx

166

Table 4-2. Continued

SPECIES	ABUNDANCE
T. longissima	x
T. mediterranea var. pacifica	x
Order: Achnanthales	
Family: Achnanthaceae	
Achnanthes	x
Order: Naviculales	
Family: Naviculaceae	
Amphiprora	xx
A. sp. 2	
A. alata	x
Diploneis didyma	x
D. elliptica	x
D. fusca var. pelagica	x
Gyrosigma	x
G. balticum	x
Mastogloia minuta	x
Navicula	x
N. sp. 2	x
N. sp. 2A	x
N. sp. 3	x
N. sp. 3A	x
N. sp. 5	x?
N. cancellata	xx
Pleurosigma sp. 1	xx
P. sp. 2	xx
P. fasciola	x
Tropidoneis sp. 1	
Order: Bacillariales	
Family: Nitzschiaceae	
Nitzschia sp. 2	
N. acicularis	x
N. delicatissima	xx
N. longissima	xx
N. obtusa	xx
N. obtusa var. scalpelliformis	xx
N. paradoxa	xx
N. pungens var. atlantica	xx
N. sigma	xx
N. sigma var. intermedia	x
N. sigmoidea	x
N. vermicularis	
Order: Surirellales	
Family: Surirellaceae	
Surirella elegans	x
S. gemma	xx

SPECIES	ABUNDANCE
Division: Pyrrhophyta	
Class: Desmophyceae	
Order: Prorocentrales	
Family: Prorocentraceae	
Exuviaella sp. 1	x
E. compressa	x
Prorocentrum micans	x
Class: Dinophyceae	
Order: Dinophysiales	
Family: Dinophysiaceae	
Dinophysis homunculus	x
Order: Gymnodiniales	
Family: Gymnodiniaceae	
Dissodinium lunula	x
(=Pyrocystis lunula)	
Order: Peridiniales	
Family: Peridiniaceae	
Diplopsalis lenticulatum f. asymetrica	x
Glenodinium sp. 1	x
Peridinium sp. 1	x
P. sp. 2	xx
P. sp. 3	x
P. conicum	xx
P. depressum	xx
P. granii	xx
P. oceanicum var. oblongum	x
P. pentagonium	x
P. quarnerense	xx
P. trochoideum	xx
Pyrophacus horologicum	
Family: Ceratiaceae	
Ceratium furca	xx
C. fusus	xx
C. atum	xx
C. macroceros	xx
C. tripos	xx
Division: Chlorophyta	
Class: Chlorophyceae	
Order: Chlorococcales	
Family: Oocystaceae	
Chotadella sp.	
Family: Micractiniaceae	
Micractinium sp.	x?
Family: Dictyosphaeriaceae	
Dictyosphaerium pulchellum	x

Table 4-2. Concluded

SPECIES	ABUNDANCE
Family: Scenedesmaceae	
Crucigenia irregularis	x?
Family: Coccomyxaceae	
Dispora sp. 1	xx
Order: Ulotrichales	xx
Family: Chaetophoraceae	
Stigeoclonium sp. 1	x?
Order: Oedogoniales	
Family: Oedongoniaceae	
Oedogonium sp. 1	x?
Order: Zygnematales	
Family: Zygnemataceae	
Mougeotia sp. 1	x?
Family: Desmidiaceae	
Cosmarium tenue	x?
Staurastrum hexacerum	x?
Division: Cyanophyta	
Class: Cyanophyceae	
Order: Chroococcales	
Family: Chroococcaceae	
Coelosphaerium sp.	x
Order: Oscillatoriales	
Family: Oscillatoriaceae	
Microcoleus sp. 1	x?
Oscillatoria sp. 1	x
Trichodesmium sp. 1	x
Division: Chrysophyta	
Class: Chrysophyceae	
Order: Dictyochales	
Family: Dictyochaceae	
Dictyocha fibula Ehrenberg	xx

Table 4-3. Seasonal abundance[a] of phytoplankton in the Santee River system, South Carolina (J. J. Manzi, 1977, South Carolina Marine Resources Division, Charleston, unpubl. data).

	Summer	Fall	Winter	Spring
Bacillariophyta	321	844	1,655	1,075
Pyrrophyta	1,251	117	201	140
Chlorophyta	427	91	30	623
Chrysophyta	918	158	164	329
Cryptophyta	63	103	83	196
Cyanophyta	123	63	86	142

a. Abundance is represented in cells/ml and is the arithmetic mean of four samples taken over a 25-hour period.

Carolina jetties. This is a mixed flora of intertidal and subtidal species, and many of the records represent species found on the Cape Lookout jetty, which has a more marine influence compared to the large harbor jetties at Winyah Bay and Charleston. However, a comparison of the algal diversity from an estuarine jetty like the one at Radio Island in Beaufort Inlet, North Carolina, with the above mentioned jetties demonstrates how impoverished the algal flora of South Carolina actually is. After surveying the Charleston Harbor jetties, Stephenson and Stephenson (1952) commented that the flora and fauna of these jetties ". . . is subnormal for a rocky marine area." The subtidal and intertidal algal flora is very depauperate due to siltation on the rock surfaces and to water turbidity. The jetties now being constructed at Murrells Inlet should provide a greater nearshore marine habitat for benthic marine algae in South Carolina.

The largest and most striking alga of the Charleston Harbor jetties is the large, membranous red alga, Grateloupia gibbesii, originally described by Harvey (1853) from specimens collected in Charleston Harbor. Rather rare species for the Western North Atlantic, e.g., Pterosiphonia pennata and Hypoglossum tenuifolium, have been reported by Wiseman and Schneider (1976) from the Charleston jetties. Oddly, these two species of red algae are found at Cape Lookout, North Carolina, but not on the jetty at Radio Island.

3. Limiting Factors

There is general agreement among marine ecologists that available NO_3 and PO_4 are major factors controlling phytoplankton abundance. The most favorable N/P (nitrogen/phosphorus) ratios usually occur between 10:1 and 30:1. In estuaries of South Carolina and Georgia, inorganic nitrogen, and more probably light, can be major limiting factors in phytoplankton abundance. A recent study by Gardner (1975) indicated that low tide runoff from intertidal marshes in South Carolina is rich in silica, phosphate, bicarbonate, and ammonia. The quantity of input of these substances into coastal waters appears to be equal to that supplied by the entire freshwater runoff in the State (Windom et al. 1975). In addition to runoff, the primary sources of nutrients in estuaries are sediments and seawater. The estuaries of South Carolina and Georgia appear to depend to a large extent on runoff (both marsh and freshwater input) and sediments for usable forms of nitrogen and phosphate. The turbid estuaries behind the barrier islands of the Sea Island Coastal Region maintain a large concentration of sediment in suspension, and there is a corresponding high rate of exchange of nutrients between sediments and water. The sediment can contain enormous amounts of nutrients, and apparently most of the phosphorus available to phytoplankton is derived from the clay sediment/water exchange (Hobbie et al. 1972, Hobbie 1976). Nutrients are also made available to estuarine waters through precipitation (Reimold and Daiber 1967, Haines 1976), groundwater (Valiela et al. 1978), microbial decomposition (Ketchum 1967, Odum and de la Cruz 1967, Reimold and Daiber 1970, Thayer 1971, 1974, Wiebe 1975, Christian and Wiebe 1978), and bacterial and algal nitrogen fixation (Whitney et al. 1975, Carpenter et al. 1978).

The classic concept of estuarine waters being the recipient of large quantities of usable organic and inorganic particulate and dissolved material from marshes is being questioned seriously. While most earlier studies seemed to indicate net exports from the marsh (Odum and de la Cruz

Family: Scenedesmus
   Crucigenia irregula
Family: Coccomyxaceae
   Dispora sp. 1
Order: Ulotrichales
Family: Chaetophoraceae
   Stigeoclonium sp. 1
Order: Oedogoniales
Family: Oedogoniaceae
   Oedogonium sp. 1
Order: Zygnematales
Family: Zygnemataceae
   Mougeotia sp. 1
Family: Desmidiaceae
   Cosmarium tenue
   Staurastrum hexacerum

Division: Cyanophyta
Class: Cyanophyceae
Order: Chroococcales
Family: Chroococcaceae
   Coelosphaerium sp.
Order: Oscillatoriales
Family: Oscillatoriaceae
   Microcoleus sp. 1
   Oscillatoria sp. 1
   Trichodesmium sp. 1

Division Chrysophyta

Table 4-3. Seasonal abundance[a] of phytoplankton in the Santee River system, South Carolina (J. J Manzi, 1977, South Carolina Marine Resources Division, Charleston, unpubl. data).

	Summer	Fall	Winter	Spring
Bacillariophyt	321	844	1,655	1,075
Pyrrophyta	1,251	117	222	140
Chlorophyta	427	92	30	623
Chrysophyta	918	159	164	329
Cryptophyta	63	223	83	196
Cyanophyta	223	63	86	142

a. Abundance i represented in cells/ml and is the arithmetic mean of four samples taken over a 25-hour period.

Carolina jettis. This is a mixed flora of intertidal ar subtidal species, and many of the recr s represent species found on the Cap Lookout jetty, which has a more marine if uence compared to the large harbor jet es at Winyah Bay and Charleston. How er, a comparison of the algal diversity r m an estuarine jetty like the one at adio Island in Beaufort Inlet, North Car ina, with the above mentioned jetties emonstrates how impoverished the algal lora of South Carolina actually is. After surveying the Charleston Harbo etties, Stephenson and Stephenson (1952 ommented that the flora and fauna of thee jetties ". . . is subnormal for a rocy marine area." The subtidal and intertial algal flora is very depauperate due t siltation on the rock surfaces and to ater turbidity. The jetties now being costructed at Murrells Inlet should provde a greater nearshore marine habitat fr benthic marine algae in South Carolin.

The largest nd most striking alga of the Charleston Hrbor jetties is the large, membranous red als, Grateloupia gibbesii, originally descried by Harvey (1853) from specimens collectd in Charleston Harbor. Rather rare specis for the Western North Atlantic, e.g., Ferosiphonia pennata and Hypoglossum tenuisfolium, have been reported by Wiseman and Schneider (1976) from the Charleste jetties. Oddly, these two species of re algae are found at Cape Lookout, North Carolina, but not on the jetty at Radio Isand.

3. Limiting Factors

Ibere is geneal agreement among marine ecologists tht available NO_3 and PO_4 are major factors controlling phytoplankton abundance The most favorable N/P (nitrogen/phosphorus) ratios usually occur between 10:1 and 30:1. In estuaries of South Carolina and Georgia, inorganic nitrogen, and more probably light, can be major limiting factors in phytoplankton abundance. A recent study by Gardner (1975) indicated that low tide runoff from intertidal marshes in South Carolina is rich in silica, phosphate, bicarbonate, and ammonia. The quantity of input of these substances into coastal waters appears to be equal to that supplied by the entire freshwater runoff in the State (Windom et al. 1975). In addition to runoff, the primary sources of nutrients in estuaries are sediments and seawater. The estuaries of South Carolina and Georgia appear to depend to a large extent on runoff (both marsh and freshwater input) and sediments for usable forms of nitrogen and phosphate. The turbid estuaries behind the barrier islands of the Sea Island Coastal Region maintain a large concentration of sediment in suspension, and there is a corresponding high rate of exchange of nutrients between sediments and water. The sediment can contain enormous amounts of nutrients, and apparently most of the phosphorus available to phytoplankton is derived from the clay sediment/water exchange (Hobbie et al. 1972, Hobbie 1976). Nutrients are also made available to estuarine waters through precipitation (Reimold and Daiber 1967, Haines 1976), groundwater (Valiela et al. 1978), microbial decomposition (Ketchum 1967, Odum and de la Cruz 1967, Reimold and Daiber 1970, Thayer 1971, 1974, Wiebe 1975, Christian and Wiebe 1978), and bacterial and algal nitrogen fixation (Whitney et al. 1975, Carpenter et al. 1978).

The classic concept of estuarine waters being the recipient of large quantities of usable organic and inorganic particulate and dissolved material from marshes is being questioned seriously. While most earlier studies seemed to indicate net exports from the marsh (Odum and de la Cruz

1967, Keefe 1972), many recent studies have had more ambiguous results. Axelrad et al. (1976), in studies of irregularly flooded brackish marshes, found a net loss of estuarine-derived phosphorus (total, dissolved organic, and orthophosphate) and nitrogen (nitrate and nitrite) to the marsh. With both nutrients, particulate forms were imported to the marsh where presumably they were mineralized and returned to the estuary as dissolved inorganic and organic forms. Heinle and Flemer (1976), in a Chesapeake marsh system similar to the one studied by Axelrad et al., found little measurable net export of particulate carbon. Further, Woodwell et al. (1977, 1979) found no annual net exchange of organic matter between salt marsh and estuary. Thayer (1974) suggested that microbial immobilization of nitrogen and phosphorus during decomposition of organic matter imported to estuarine waters from marshes may limit nutrient availability to phytoplankton. He also suggested that the annual cycles of nutrient concentration in estuarine waters may, in part, result from shifts in the equilibrium between microbial immobilization and remineralization. The above evidence, if nothing else, indicates that the exchange of nutrients between marshes and estuarine waters is not well understood and that these exchanges, as well as their relationships to phytoplankton populations, must be subjected to further intensive study. See Gardner and Kitchens (1978) for a more detailed summary of nutrient exchanges between marshes and contiguous waters.

Estuarine productivity is dependent on three separate, but interdependent, units of primary production, i.e., marshes and their resultant detritus, benthic micro- and macrophytes, and phytoplankton. Both benthic and planktonic algae are extremely important as the basis of food chains, reaching not only every habitat of the estuarine ecosystem but also habitats of marine, palustrine, and riverine systems. Estuarine microphytes are directly utilized by the zooplankton, fish, macrofaunal, and meiofaunal populations. Their importance in various food chains has been estimated to some extent (Odum 1970a, Fisher 1975, Moll 1975, Kirby-Smith 1976), but no data exist on their quantitative and qualitative significance in the estuaries of South Carolina and Georgia.

Food chains in estuaries are, in general, markedly different than in the open sea. In the past, this has been attributed to the predominance of the detritus-based food system in estuaries. It could be speculated that the difference between the contributions of benthic macro- and microalgae in estuaries and the open sea is in part responsible for this discrepancy. Another feature of estuarine food chains is that zooplankters are quantitatively less important, while

organisms which utilize detritus (or rather the microorganisms associated with detritus) and benthic algae are the dominant herbivores. Thus, a likely distribution of standing stock in a Southeastern estuary would be: benthic microheterotrophs > benthic macroheterotrophs > benthic microautotrophs > nekton > aquatic autotrophs (Thayer 1974). Many interpretations of this hierarchy are possible, but all must accept the potential of rapid autotrophic production. Since a typical effect of sublethal concentrations of heavy metals and trace contaminants is a significant reduction in photosynthesis (Reish et al. 1978), any impacts on estuarine autotrophs are likely to be magnified in higher trophic levels. Table 4-4 lists several of the more common substances toxic to phytoplankton and the concentrations necessary to either change photosynthetic rate or cause mortality (see also Zingmark and Miller 1975, Reish et al. 1978).

C. CONSUMERS

1. Zooplankton

Zooplankton biomass and numbers are usually greater in estuaries than in any other aquatic habitat, reflecting the overall high productivity of the estuarine environment. Estuarine zooplankton is predominantly of marine origin, but it also includes groups originating from freshwater and terrestrial ancestors (Green 1968a). Also, the assemblage includes forms with herbivorous, detritivorous, and carnivorous feeding habits, each of which provides an essential link in the estuarine food web. Estuarine zooplankton is of considerable trophic importance. The species of fish and shellfish responsible for over 85% by weight of the commercial fisheries landings of the Southeastern Atlantic States are estuarine or estuarine-dependent at some life stage (Burrell 1975). For many species that depend on estuaries as spawning or nursery grounds (e.g., Atlantic croaker, Atlantic menhaden, seatrout or drum, blue crab, and white shrimp), an abundant zooplanktonic population is a necessity. Small herbivores, particularly copepods and cladocerans, are essential as food for early fish larvae and for larger predacious zooplankters, which in turn are fed upon by late larval and postlarval fish and other forms. Detritivores, such as mysid shrimp and gammarid amphipods, may be the most important food chain link in estuaries bounded by extensive salt and brackish marshes, which themselves often are important fish nursery grounds (Ragotzkie 1959b, Van Engel and Joseph 1968, University of Georgia Marine Institute 1971).

Estuarine zooplankton must be able to accommodate regular salinity variations, which may fluctuate more than $12^o/oo$ during a single tidal cycle (Van Engel and Joseph 1968). Therefore, true estuarine species

170

Table 4-4. Concentrations of substances toxic to phytoplankton (Rice and Ferguson 1975).

Substance Tested		Species Tested	Range of Minimum Concentrations Producing Toxicity (ppm)			Duration of Experiment
Type	Number	Number	Low	High	Geometric Mean	Days
Chlorine	1	2	1.5	> 20	--	< 0.1
Detergents	21	32	0.1	> 1000	14.00	1 - 210
Heavy metals (Hg, Cu, Zn, Ag)	5	53	0.001	16	0.34	< 1 - 26
Hydrocarbons	2	14	0.01	100	0.52	2 - 5
Nitrogen (NO$_3$, NO$_2$, NH$_4$)	3	4	1.1	28,000	120.00	< 1 - 42
Pesticides	50	46	0.000015	> 200	0.70	1 - 30
Phosphorus (PO$_4$)	1	2	--	70,000	--	
Polychlorinated biphenyls	1	5	0.01	> 1	0.80	4

are characteristically euryhaline, with more stenohaline species restricted to either upper or lower portions of the estuary according to their affinity.

Dispersion is also a problem for zooplankters because of their limited locomotive capabilities. Tidal and river currents appear to play the major role in moving both holo- and meroplankton within estuaries. When a salt wedge is present, net transport is upstream on the bottom for some distance above the mouth, and zooplankton may be transported up the estuary in lower strata. In a thoroughly mixed estuary, zooplankton may be transported up the estuary by selectively rising into the water column on favorable tides and settling to the bottom when tidal direction is unfavorable. Such dispersal strategy has been suggested or demonstrated for estuarine or coastal zooplankton by Carriker (1951), Bousfield (1955), Sandifer (1975), and others.

Estuarine zooplankters must also have means of repopulation. Holoplanktonic forms may be recruited from the sea, by reproduction of resident stocks coupled with retention of the progeny, or from resistant eggs in the sediments. Rivers having large drainage basins originating above the fall line may on occasion be fresh throughout, even beyond the mouth (Burrell 1977). Many stenohaline marine species would not survive the dilute mixture, while nearly all would be swept out to sea. Repopulation of true estuarine holoplankters after these events would probably be

from "pocket populations"; such pockets may occur due to restricted flushing (as in bays lying at the mouths of some river systems), marsh pot holes, tidal creeks with high sills at the mouth, or from coastal waters and adjacent estuaries. Meroplanktonic forms are, for the most part, larvae of nektonic or benthic organisms, and repopulation would be a problem only if inclement conditions, such as low salinity or excessively cold temperatures, affected survival of adults or larvae, or if high runoff prevented recruitment of larval forms to adult populations.

The zooplankton of several estuaries of the Sea Island Coastal Region has been described in varying degrees of detail. Bears Bluff Laboratories, Inc. (1964, 1965) surveyed the Ashley, Cooper, Wando, and Santee rivers, and Price Inlet, South Carolina, over an annual cycle. Lonsdale and Coull (1977) described seasonal aspects of zooplankton in North Inlet, South Carolina, and the U.S. Naval Facilities Engineering Command (1977) completed a year-long zooplankton study in the St. Marys-Cumberland Sound Estuary in Georgia. Stickney and Knowles (1976) described vertical distribution of selected groups in the Skidaway River, Georgia, over three seasons. In South Carolina, the Marine Resources Research Institute has recently completed a field effort to characterize the zooplankton of the Edisto, Cooper, Wando, and Santee rivers and Winyah Bay, but only a cursory summary of information is presently available. Data from different studies are difficult to compare because mesh size, net diameter, volume strained estimates, depth sampled, and level of identifications are different for nearly every investigation.

Studies by Bears Bluff Laboratories, Inc. (1964, 1965) indicated that maximum numbers of net zooplankters occurred in each estuary in the fall, while Lonsdale and Coull (1977) and the U.S. Naval Facilities Engineering Command (1977) found maximum abundance in spring. This discrepancy appears to be due to differences in the methods of sampling employed; nets used in Bears Bluff studies had large mesh sizes which failed to capture small forms, such as copepods, in proportion to their true abundance.

Copepods, coelenterates, and crab larvae were reported as the most numerous zooplankters in all Bears Bluff studies except in the Price Inlet area, where copepods ranked ninth. These forms were most abundant in June, July, and August, and made up over 98% of all zooplankton collected in some systems studied by Bears Bluff investigators. Mysid shrimp were most abundant during fall and winter in all Bears Bluff samples except those from Price Inlet. Grass shrimp larvae (Palaemonetes spp.) were more abundant in the North Santee than in other South Carolina estuaries. They reached maximum abundance in June and July. Gammarid amphipods were very common at upriver stations, particularly in the North Santee River. They were present nearly all year and were most abundant in August. Other common zooplankters of South Carolina estuarine systems included a sergestid shrimp (Lucifer faxoni), miscellaneous crab larvae (such as xanthids), penaeid shrimp postlarvae, and fish larvae and postlarvae.

Hester (1976) recorded 20 species of hydromedusae from South Carolina. Blackfordia virginica, Nemopsis bachei, Phialidium languidum, and Phialucium carolinae were the most common. Calder and Hester (1978) listed the most common scyphomedusae in South Carolina as Cyanea capillata, Rhopilema verrilli, Stomolophus meleagris, and Chrysaora quinquecirrha, while Kraeuter and Setzler (1975) found that medusae of Chiropsalmus quadrumanus were abundant and Rhopilema and Cyanea were rare in Georgia. Calder and Burrell (1978) listed Mnemiopsis leidyi, M. mccradyi, and Beroe ovata as the most abundant ctenophores.

Lonsdale and Coull (1977) found that copepods made up between 64% - 69% of the zooplankton in North Inlet, South Carolina. The principal holoplankton species found in North Inlet were the calanoid copepods Parvocalanus crassirostris and Acartia tonsa, the cyclopoid copepod Oithona colcarva, the harpacticoid Euterpina acutifrons, and the larvacean Oikopleura sp. Barnacle nauplii were the most common meroplankters, with bivalve and polychaete larvae also important. Greatest numbers of zooplankters were present between April and July. Lonsdale and Coull (1977) considered that this fauna was most closely allied with that of Florida waters.

The Marine Resources Research Institute at Charleston has completed field work of a 2-year study of zooplankton of the Wando River, a relatively low salinity estuary near Charleston Harbor (V. G. Burrell, 1975, South Carolina Marine Resources Division, Charleston, unpubl. data). Five stations, extending from the mouth to an area 22 km (13 mi) up estuary, were occupied monthly during this study. The most abundant zooplankter each season was the copepod Acartia tonsa. The jellyfish Blackfordia virginica, barnacle larvae, and the zoeae of mud crabs and fiddler crabs were the predominant meroplanktonic elements. A preliminary listing of species from all stations (Table 4-5) indicates that many of the same species found in North Inlet by Lonsdale and Coull (1977) occur in the Wando River as well. Chief differences are the greater contribution to the total population by the copepod Parvocalanus crassirostris in North Inlet and the presence of Oikopleura sp. and pluteus larvae there. Eurytemora affinis, a polyhaline copepod, was seasonally abundant in the Wando River, but was not found in North Inlet. Distribution of these species is indicative of different salinity regimes in the two study areas, with North Inlet having the higher salinity.

Acartia tonsa was the most abundant zooplankton species in the Dauphin, Skidaway, St. Marys and Wilmington rivers and Doboy Sound in Georgia (Jacobs 1968, Stickney and Knowles 1975, 1976, U.S. Naval Facilities Engineering Command 1977). Other common plankters were the copepods Parvocalanus crassirostris and Pseudodiaptomus coronatus, tunicates, and, in certain seasons, the larvae of barnacles, gastropods, and pelecypods.

How zooplanktonic populations react to human activities, such as harvesting of commercial species, lowering of water quality, altering flow, etc., is virtually unknown for most estuaries, with those of the Sea Island Coastal Region being no exception. In some cases, alterations within a system can be inferred from the change in zooplankton distribution and abundance after such events. Most commercial species with planktic larvae, except in rare cases when the entire brood stock is virtually eliminated, have good or bad year classes as a result of poorly understood environmental conditions.

Investigations to determine the impact of industrial effluent on zooplankton have not, in most instances, documented a direct relationship. Studies in South Carolina by Bears Bluff Laboratories, Inc. (1964, 1965) in Price Inlet and North Santee River, both

Table 4-5. A preliminary list of the most common species of zooplankton found in the Wando River, August 1972 - August 1974 (V. G. Burrell, 1975, South Carolina Marine Resources Division, Charleston, unpubl. data). XXX abundant, XX common, X present.

	WINTER	SPRING	SUMMER	FALL
Arthropoda				
Copepoda				
Calanoida				
Acartia tonsa	xxx	xxx	xxx	xxx
Centropages furcatus		x		
Centropages hamatus	x			
Eucalanus sp.				x
Eurytemora affinis	xxx			x
Labidocera aestiva	x			
Parvocalanus crassirostris	x	x		x
Pseudodiaptomus coronatus	xx	x	xx	xxx
Temora longicornis			x	
Temora setacaudatus			x	
Tortanus setacaudatus				
Cyclopoida				
Oithona sp.	x			
Orthocyclops sp.	x	x	x	
Ergasilus sp.	x	x		
Harpacticoida				
Euterpina acutifrons				xx
Scottolana canadensis		x		
Harpacticoid (unidentified)	x	x	x	
Calagoida				
Unidentified sp.	x			
Other				
Cladocerans	x		x	
Ostracods	x			x
Barnacle larvae	x	xxx	xxx	xxx
Mysids	xx	xx	x	x
Cumaceans	x			
Isopods				
Caprellid amphipods				x
Gammarid amphipods		x		
Penaeus sp. larvae				
Sergestid shrimp				
Palaemonetes sp.	x			
Pagurid larvae	x			
Rhithropanopeus harrisii	x	xxx	xxx	x
Pinnotheres sp.			x	
Uca sp.		xxx	xxx	x
Decapod megalopae		x	x	
Cnidaria				
Blackfordia virginica	x	xxx	xx	xxx
Moerisia lyonsi		x	x	x
Nemopsis bachei	x		xxx	x
Lovenella gracilis	x			
Eutima mira			x	
Obelia sp.			x	
Ctenophora				
Mnemiopsis sp.	x	xx	x	xxx
Annelida				
Polychaete larvae	xxx			xx
Mollusca				
Gastropod larvae	x			
Pelecypod larvae	x			
Chordata				
Chaetognaths	x	x	x	x
Tunicate larvae			x	x
Fish eggs	x	x	x	
Fish larvae		x	x	

relatively pristine areas, and in the Cooper River, a highly industrialized area, show remarkably similar numbers of zooplankters. Conceivably, planktivorous anadromous fishes may not use a polluted estuary to as great an extent as a less impacted one, thereby masking detrimental effects of pollution on plankton populations.

Pulpwood processing is a significant activity in the Sea Island Coastal Region. Studies of kraft mill waste water showed no toxic effects on the shrimp Palaemonetes paludosus or on larvae of the mud crab Rhithropanopeus harrisii, even at 100% concentration (University of Georgia Marine Institute 1971). However, these tests were of short duration and long-term effects were not determined.

Channel dredging, a constant necessity in many areas of the Sea Island Coastal Region, may adversely affect zooplankton populations in several ways. Increased turbidity lowers primary production, thereby reducing food of herbivores (Williams and Murdoch 1966), while contributing to the fouling of substrates necessary for attachment of some benthic organisms. Metals and hydrocarbons associated with industrial processes, agricultural chemicals, and those that leach naturally into streams, are often concentrated in sediments lying near the mouths of estuaries. When these sediments are disturbed, concentrations in the water column may increase dramatically. Studies of Charleston Harbor water containing dredged sediments have shown it to be toxic to planktonic species (University of South Carolina 1973, Hoss et al. 1974, DeCoursey and Vernberg 1975).

A variety of studies have considered the effects of pesticides, PCB's, heavy metals, oil, and chlorine on estuarine zooplankters, particularly the meroplanktonic larvae of estuarine invertebrates. Some of these are listed below by way of example. For more details on the effects of pollutants in estuarine waters, the reader is referred to Reish et al. (1978) and other reviews.

The pesticides mirex, methoxychlor, malathion, insect juvenile hormone mimics, and kepone have been shown to be not only acutely toxic to crab larvae, but also to cause such sublethal effects as delayed development, increased incidence of abnormalities, and behavioral modifications (Bookhout et al. 1972, 1976, 1979, Forward and Costlow 1976, Bookhout and Monroe 1977, Christiansen et al. 1977a, b, Costlow 1977). Petroleum hydrocarbons, PCB's, and endrin have been shown to have similar effects on larval, juvenile, and adult grass shrimp, Palaemonetes pugio (Tatem 1975, Tyler-Schroeder 1976, 1979). However, Nimmo et al. (1974) found that P. pugio exhibited no mortality when exposed to PCB-contaminated sediments in open-water cages or in flowing-water laboratory systems.

Studies on the effects of heavy metals (mercury, copper, zinc) on bryozoan larvae, tube worms, bivalve mollusks, and brine shrimp by Wisely and Blick (1967) showed their toxicity to be greater at lower pH but still significant at the pH of seawater. Benijts-Claus and Benijts (1975) found that low concentrations (up to 50 ppb) of zinc and lead caused a significant delay in larval development of the estuarine mud crab, Rhithropanopeus harrisii. Shealy and Sandifer (1975) found that just a 48-hour exposure of grass shrimp (P. vulgaris) larvae to sublethal concentrations of mercury resulted in delayed effects such as reduced survival through metamorphosis, prolonged development, more larval instars, and increased incidence of deformities. Calabrese et al. (1977) exposed oyster (Crassostrea virginica) and clam (Mercenaria mercenaria) larvae to mercury, silver, copper, nickel, and zinc (clam only). The orders of toxicity were as follows: oyster, Hg > Ag > Cu > Ni; clam, Hg > Cu > Ag > Zn > Ni. Low levels of metal had no detectable influence on larval growth, but at the LC_{50} concentration (i.e., the level toxic to 50% of the test organisms), growth was markedly reduced. Paffenhoffer and Knowles (1978) found that the holoplanktonic copepod, Pseudodiaptomus coronatus, fed and grew at a normal rate in water containing 5 μg/l cadmium, but its reproductive rate was reduced by 50%.

Besides pesticides and heavy metals, chlorine-induced oxidants in estuarine waters have been shown to have detrimental effects on the survival and development of mud crab (Panopeus herbstii), hermit crab (Pagurus longicarpus), and clam (Mulinia lateralis) larvae (Roberts et al. 1979). Chlorine is used extensively as a disinfectant in sewage treatment plants and for control of fouling in power plants, and thus is frequently released into estuarine waters.

A major gap in our understanding of the effects of pollutants on estuarine organisms concerns the long-term effects of continuous or intermittent exposures of these organisms to sublethal contaminant concentrations. To date, not very much work has been done in this area, although the situation is improving. However, results from one recent study (Tyler-Schroeder 1979) serve as an example of how important considerations of life-cycle effects are (see Table 4-6). From these data, it is obvious that levels of the pesticide endrin substantially lower than the 96-hour LC_{50} "toxic" concentrations may have significant detrimental effects on life-cycle processes over a single generation.

174

Table 4-6. Toxicity of endrin to grass shrimp (Palaemonetes pugio) in a 145-day life-cycle exposure (Tyler-Schroeder 1979).[a]

Generation	Life Stage or Function	Effect	Endrin Concentrations Producing Effect (μg/1)
Parental	Juvenile adult	Death	0.38, 0.79
	Reproduction	Delay in onset of spawning Reduction in number of females spawning	0.03, 0.05, 0.11, 0.18, 0.38, 0.79
F_1	Larvae	Death	0.11, 0.18, 0.38, 0.79
	Postlarval stage	Delayed metamorphosis	
		Decrease in growth, final length	0.1, 0.18, 0.38, 0.74
		Decrease in growth, final weight	0.05, 0.11, 0.18, 0.38, 0.79

a. The acute 96-hour LC_{50} was 1.2 μg/1 for larvae, 0.35 μg/1 for juveniles, and 0.69 μg/1 for adults.

Another major gap in our understanding of the effects of contaminants on estuarine zooplankton concerns the possible synergistic interactions of pollutants with the highly variable physical conditions of estuaries and with the myriad other pollutants dumped into estuarine waters. Relatively few studies involving zooplankters have considered such synergistic effects (Cairns 1968). Much of the work that has been done concerns the effects of power plants, since one of their major impacts on estuarine areas is thought to be entrainment of zooplankton. A major concern is that power plants may entrain enough fish eggs and larvae in a given locality to adversely affect recruitment to populations of commercially or recreationally important fishes (Dahlberg 1978). The combination of temperature shock, exposure to biocides (chlorine, copper), perhaps mechanical damage resulting from passage through power plant condensers is known to kill copepods (Heinle 1969) and fish eggs and larvae. Hoss et al. (1975, 1977) demonstrated experimentally that thermal shock increased the mortality of juvenile estuarine fishes exposed to chlorine or copper as they pass through a power plant condenser. This finding is particularly significant since the need for biocides and anti-corrosion measures increases with salinity; the further seaward on an estuary that a power plant is located, then, the greater its potential impact on the zooplankton.

In other studies, DeCoursey and Vernberg (1972) and Vernberg et al. (1973, 1977) showed that temperature, salinity, and contaminant (mercury, PCB's) stress interacted synergistically to affect larvae of the fiddler crab, Uca pugilator. Similarly, Middaugh and Floyd (1978) described some synergistic effects of sublethal cadmium concentrations and suboptimal salinities on grass shrimp (P. pugio) larvae. In contrast, Benijts-Claus and Benijts (1975) reported that some combinations of lead and zinc tended to neutralize the greater toxicity of the lead to larval mud crabs (R. harrisii) and in fact to accelerate larval development. Nimmo and Bahner (1976) attempted to assess the toxic effects of a heavy metal (cadmium), a pesticide (methoxychlor), and a PCB, separately and in combination, on an estuarine shrimp (Penaeus duorarum). They observed additivity of effects, but no dramatic synergistic interactions. This general area of synergistic interactions of pollutants needs a great deal more study.

2. Benthic Meiofauna

Benthic meiofauna (defined as those bottom-dwelling animals which pass through a 500-μm seive but are retained by a 63-μm seive) have been little studied in estuarine subtidal waters of the Sea Island Coastal Region. The most pertinent data have been generated by Coull and Fleeger (1977), who studied the meiofauna of two dissimilar stations 1 km (.62 mi) apart in the North Inlet estuary, South Carolina, over a 3-year period. One station was characterized by a muddy, organic-rich substrate, while the other area consisted of medium sand exposed to continual wave action. Coull and Fleeger (1977) found that the meiobenthic copepod communities at these two stations differed markedly in species composition; similarity measurements indicated that < 5% of the fauna was shared between the stations. The copepod fauna at the sand station consisted primarily of interstitial forms, while burrowing or epipelic species predominated at the mud station. The mud station community showed distinct, regular seasonal shifts, while the sand station fauna exhibited no marked seasonality. Despite these and other differences between the two communities, they exhibited roughly equal diversity. Coull and Fleeger (1977) suggested that maintenance of equivalent species diversity was made possible by the greater microhabitat diversity and niche overlap within the sand habitat and by seasonal cycling of species suites at the mud station.

Potential effects of perturbations on estuarine subtidal meiofauna of the characterization area are essentially unknown. Vernberg and Coull (1975) examined the combined effects of temperature, salinity, and anaerobiosis on the three species of meiobenthic copepods most commonly encountered in the North Inlet samples (Thompsonula hyaenae, Pseudobradya pulchera, and Scottolana canadensis). Temperature-salinity interactions were not significant for any of these species over the ranges tested [15°, 22°, 37.5°C (59°, 71.6°, 99.5°F); 25º/oo, 30º/oo, 35º/oo]. However, there was a great difference in the sensitivity of the species to high temperature, with T. hyaenae the most sensitive. Both temperature and salinity affected resistance to anaerobiosis, and, not surprisingly, in a comparative test involving an additional six species of copepods, those typically found in muddy substrates proved more resistant to anaerobiosis than the sand-dwelling species. Thus, it is likely that major thermal additions or perturbations that produce anaerobic conditions in estuarine sediments will cause changes in the composition of local meiofaunal communities. Coull and Vernberg (1975) reported that the dominant meiobenthic copepods in the North Inlet estuary appear to remain in reproductive condition throughout the year, while less abundant species exhibited seasonal reproductive periods. Thus, short-term perturbations which interfere with copepod reproduction are likely to have a relatively greater effect on the recruitment of less abundant species.

For additional information on estuarine meiobenthos, see sections on intertidal subsystems.

3. Benthic Macroinvertebrates

a. General Ecology and Distribution. Studies of subtidal macrobenthic communities have been conducted in several South Carolina estuaries, including Little River Inlet (Calder et al. 1977a), Murrells Inlet (Calder et al. 1976), the North and South Santee rivers (Calder et al. 1977b), the estuary complex around Kiawah Island (Coull 1975), the North and South Edisto rivers (Calder et al. 1977b), and Port Royal Sound (Parrish 1972). These systems encompass a wide range of conditions from highly stable to highly variable.

Species composition of the subtidal macroinvertebrate communities of the North and South Santee rivers reflected the highly variable salinity conditions typical of these two estuaries. Fewer species were represented in both the epifauna and infauna compared with more homoiohaline (salinity-stable) areas such as the North Edisto River (Calder et al. 1977b) and Murrells Inlet (Calder et al. 1976). The fauna consisted largely of species known to be eurytopic, and the samples were dominated by euryhaline opportunists, estuarine endemics, and freshwater species. The presence of subtidal oyster reefs near the mouths of both estuaries was attributed to the wide fluctuations of salinity which limit oyster predators and competitors. There was no obvious decline in species diversity with distance from the ocean (as was observed by Boesch (1972) in the more homoiohaline Chesapeake-York-Pamunkey estuary, Virginia), because of both the lack of a stable, uniform halocline and the effects of two floods during the study. The impact of flooding appeared to be greatest on benthic communities in the lower portions of each river and least on those present near the head of the estuaries.

Despite their proximity, the North and South Edisto rivers differ significantly in hydrography and benthic community structure (Calder et al. 1977b). Salinities were low in the South Edisto, except at the mouth, because most of the fresh water from the system is discharged through this tributary. In contrast, little fresh water enters the North Edisto, and salinities were polyhaline for the most part throughout the entire river. Due to variations in freshwater flow, the South Edisto is a moderately fluctuating or poikilohaline estuary, although salinities are much less variable than in the Santee estuary. Only minor fluctuations of salinity were observed in the North Edisto, and the estuary is regarded as homoiohaline (salinity-stable).

Because of the absence of a significant freshwater source, a well defined halocline from the mouth to the head of the North Edisto River was generally lacking during this study. Differences were noted in the benthos from one location to another in this estuary, but these were largely attributable to dissimilarities in substrate type rather than to a salinity gradient. Similar circumstances were observed in Murrells Inlet by Calder et al.

176

(1976), where salinities were relatively homogeneous throughout. Major differences in the benthos among various stations in the inlet were due primarily to substrate. Likewise, a sharp break in bottom type from sand to shell along the inner channel of Little River Inlet was accompanied by a pronounced change in the benthic assemblage (Calder et al. 1977a). Similarly, Coull (1975) found that polychaetes, mollusks, and crustaceans dominated the macroinvertebrate fauna at five stations near Kiawah Island, South Carolina, and variations in benthic community structure reflected differences in the nature of the substrate (coarse sand, fine-mud sand, medium sand, fine clean sand). Likewise, Parrish (1972) found that substrate influenced the benthic communities of Port Royal Sound. Most of the organisms collected were detrital feeders, and densities were typically highest in spring. The number of species was high in bottoms consisting of silt and sand or silt, clay and sand, while fewest species were collected in medium to fine sand. Substrates of clay with silt or fine sand and detritus also harbored a reduced number of species.

A distinct salinity gradient was evident on the South Edisto River. Salinities varied from polyhaline or euhaline at the mouth to essentially limnetic at the uppermost station. As expected for a gradient estuary, the combined number of species in dredge and grab samples declined from a maximum at the highest salinity station to a minimum at the station having the lowest salinity. Dissimilarities in species composition among stations on this estuary were due partly to substrate differences, but salinity was considered to be the primary factor.

Under conditions of high and relatively uniform salinity, the number of species observed in the North Edisto River was relatively high in both dredge and grab samples. A total of 126 species of epifaunal invertebrates was identified in dredge collections from the North Edisto, compared with 71 from the South Edisto and 41 from the entire Santee estuarine area (Calder et al. 1977b). The number of species was also highest in grab samples from the North Edisto River; more than 70 taxa were identified from five of the eight stations sampled. By comparison, the greatest number of species on the South Edisto River was 32, while the maximum number at a given station on the Santee estuary was 51. The mean number of species per collection (three grabs) at stations on the North Edisto was 27, compared with 11 from the South Edisto and 13 from the Santee estuaries. Species diversity was consistently higher in collections from the North Edisto, where environmental constancy was relatively high, than in those from either the South Edisto or Santee, where stresses were

higher and environmental constancy was lower.

Several investigations of benthic macroinvertebrates have been undertaken in estuaries of Georgia. Howard and Frey (1975), citing the work of Heard and Heard (1971), listed 268 species from trawl and bucket dredge samples in Sapelo Sound, St. Catherines Sound, and the North and South Newport rivers. Most of the species captured (85%) were arthropods, mollusks, and polychaetes.

Dörjes and Howard (1975) conducted animal-sediment studies in the Ogeechee River-Ossabaw Sound estuary using both a box corer and a beam trawl. A total of 163 species, mostly polychaetes, crustaceans, and mollusks, was found in box core samples. Collections from the transition area between fresh and brackish water at the head of the estuary consisted largely of the polychaete Scolecolepides viridis and the amphipod Lepidactylus dytiscus. No epifaunal invertebrates were collected from this region of the estuary. The infauna of low salinity areas was dominated by Heteromastus filiformis (polychaete), Bostrichobranchus pilularis (ascidian), Cyathura burbancki (isopod), Nereis succinea (polychaete), and Macoma constricta (bivalve). Epifaunal samples taken by beam trawl yielded specimens of Molgula manhattensis (ascidian) and Neopanope sayi (decapod). Numbers of species increased markedly in areas of higher salinity. Numerical dominants in such regions included Spiophanes bombyx (polychaete), Solen viridis (bivalve), Oxyurostylis smithi (cumacean), and Pinnixa cf. chaetopterana (decapod). Twenty epifaunal species were collected, including 10 arthropods, 4 echinoderms, 3 mollusks, 2 cnidarians, and 1 ascidian. Salinity was believed to constitute a more important factor than sediment texture in observed patterns of faunal zonation.

Differences in the vertical and lateral distribution of species in the relatively high-salinity Doboy Sound were attributed primarily to water circulation patterns and velocities by Mayou and Howard (1975). Maximum species diversity and density were observed subtidally along channel margins; areas characterized by poor circulation and low current velocity had much lower faunal densities and species diversity. The observed fauna was dominated numerically by polychaetes, which accounted for over 60% of the individuals collected.

Five distinct communities of macroinvertebrates were recognized in the Ogeechee River, Georgia, by Dörjes (1977). Two of these were found off the mouth of the river and three were within the estuary. Of the three in the estuary, the amphipod Lepidactylus dytiscus and the polychaete Scolecolepides viridis were dominant in the oligohaline-limnetic area. Characteristic species of the mesohaline zone were the bivalve _Macoma constricta_, the polychaete _Heteromastus filiformis_, and the ascidian _Bostrichobranchus pilularis_. Burrowing or tube-dwelling species were typical of muddy polyhaline areas. In fine sands near inlet shoals, the haustoriid amphipods _Acanthohaustorius_, _Haustorius_, and _Bathyporeia_ were dominant; other common species included the polychaetes _Onuphis eremita_, _O. microcephala_, _Magelona_ sp., and _Spiophanes bombyx_, the bivalves _Solen viridis_ and _Tellina_ cf. _taxana_, the sand dollar _Mellita quinquiesperforata_, and the shrimp _Ogyrides alphaerostris_.

b. _Impacts._ Shoaling is a major problem in the Atlantic Intracoastal Waterway, harbors, and other estuarine areas of South Carolina and Georgia (see Volume I for details), and periodic dredging is necessary to maintain navigation channels. Dredging obviously disturbs the bottom and thus the benthic communities in the area dredged. Spoil from such dredging is usually placed on high land disposal areas, but in some cases, overboard disposal is necessary. Such overboard disposal of dredge spoil directly impacts benthic epifauna and infauna (especially) through smothering, exposure to pollutants which may be concentrated in certain sediments, and overall habitat modification. Reported effects of dredging and deposition of dredge spoil on benthic communities are inconsistent, ranging from very little to severe (see Van Dolah et al. 1979 for brief review).

The direct effects of dredging on estuarine benthic communities have been the subject of two studies in the Sea Island Coastal Region. Stickney and Perlmutter (1975) studied effects of intracoastal waterway dredging on muddy-bottom communities in the Ossabaw Sound area of Georgia. They observed that recolonization after dredging was rapid; within 2 months, species diversity and composition were similar in both dredged and control areas. In South Carolina, Van Dolah et al. (1979) studied effects of dredging and unconfined dredge spoil disposal on benthic communities in Sewee Bay, a euhaline neutral embayment just west of Bulls Bay. These investigators found that dredging had detectable effects on both benthic epifaunal and infaunal invertebrates, but these effects generally were short term and localized. Unconfined disposal of dredge spoil had a major effect at one station through smothering, but other stations were not severely disturbed, although some changes in community composition attributed to the dredged sediments were seen. Overall, Van Dolah et al. (1979) concluded that "the influence of dredging operations on benthic communities in Sewee Bay were short term and isolated. Furthermore, it is unlikely that dredged

material disposal had a profound effect at any station in terms of the ecological interactions of benthic communities with higher trophic levels." To mitigate effects of dredging and overboard disposal of dredge spoil on benthic communities, they recommended that: 1) dredging and disposal be conducted during late fall or early winter to minimize adverse effects on recruitment of benthic organisms, 2) dredging in any given area be conducted relatively often so that the volume of material removed and disposed is not large at any given time, and 3) when disposed overboard, dredge spoil be placed on several sites rather than concentrated on one.

Effects of pesticides, heavy metals, and other pollutants on selected species of estuarine benthic invertebrates have received some attention, but not a great deal (see Reish et al. 1978 for review of literature). Most of the species chosen for study have been relatively large, epibenthic forms such as penaeid and grass shrimp, crabs, and oysters, which also are of major commercial, recreational, and ecological significance. Generally, the contaminants studied have proven toxic to one or more estuarine organisms at some exposure level, but toxic concentrations of a given pollutant may vary widely among species, life history stages of the same species, and with experimental conditions. For example, Nimmo et al. (1977) reported that the grass shrimp (_P. vulgaris_) was several times more sensitive than _Penaeus duorarum_ to cadmium poisoning. Tyler-Schroeder (1979) showed that different life cycle stages of grass shrimp (_P. pugio_) differed in their sensitivity to the pesticide endrin (see Table 4-6). Sunda et al. (1978) demonstrated that the direct toxicity of cadmium to _P. pugio_ is dependent on the amount of free cadmium ion in solution; this, in turn, is determined by the degree of complexation of cadmium by chloride or chelating agents. Thus, a given amount of cadmium is less toxic to _P. pugio_ at high salinities or in the presence of a chelating agent than at low salinities.

A few studies have considered effects of pollutants on estuarine benthic communities, rather than individual species. Tagatz et al. (1976) exposed experimental communities of benthic invertebrates and sheepshead minnows to simulated runoff from areas treated with mirex to control fire ants. To simulate runoff, mirex was leached from fire ant bait by fresh water, which was then mixed with salt water for the experiment. The exposure concentration averaged 0.038 μg mirex/l. Significant mortalities were observed among all the crustaceans exposed to the leached mirex (grass shrimp, _P. vulgaris_; pink shrimp, _Penaeus duorarum_; mud crabs, _Panopeus herbstii_; and hermit crabs, _Clibanarius vittatus_). Mortalities of two filter-feeding mollusks, the oyster (_Crassostrea virginica_) and the ribbed mussel (_Geukensia demissa_), however, were lower in the mirex-treated tanks. Tagatz et al. (1976) attributed this to the significant reduction in crab predators in the treated tanks. All treated organisms and the sediment in the test tanks concentrated mirex from the water. Maximum concentrations ranged from up to 1,500X for sediment, to 5,500X for pink shrimp, and to 73,700X for oysters. In other studies, Tagatz et al. (1977, 1978) examined the effects of pentachlorophenol (PCP) on the development of benthic communities recruited from planktonic larvae in estuarine waters passed through control and test aquaria. Annelids, arthropods, and mollusks dominated the benthic communities that developed in the control tanks. The mollusks were the most sensitive to PCP; their numbers were reduced markedly in tanks treated with 7 or 16 µg/l. Annelids and arthropods decreased significantly at higher concentrations (76 or 161 µg/l). The PCP caused reduction in numbers of both individuals and species. Tagatz et al. (1978) concluded that, in estuarine systems, concentrations of PCP similar to those tested in the laboratory could disrupt stable ecological relationships, such as predator-prey interactions, through reduction in abundance of polychaetes and other important prey organisms, and decrease abundance of commercially important mollusks and crustaceans.

c. Selected Species of Commercial Importance. Several benthic macroinvertebrates are of commercial value in the Sea Island Coastal Region (see Volume II, Chapter Seven). Two of these, the American oyster (Crassostrea virginica) and the hard clam (Mercenaria mercenaria) have been chosen for more detailed discussion here..

(1) American Oyster. The oyster is common-to-abundant south of Massachusetts along the east coast of the United States, although small populations are located in Maine and New Hampshire. Crassostrea virginica is adapted to an existence in waters having considerable variations in salinity and temperature. Its optimum salinity range is roughly 10°/oo - 28°/oo, but it can survive periods of extended freshets. The sexes are separate and hermaphrodites are rare. Eggs and sperm are discharged into open waters, where fertilization occurs. The length of larval life is approximately 2 weeks, depending on food, temperature, and other environmental factors. Larvae can swim weakly using a highly developed velum, although they are usually transported by currents and tides. Prior to metamorphosis or setting, the larvae develop into pediveligers, characterized by eye spots and a foot. The pre-setting period is most critical, for the oyster will die unless suitable substrate is found. Oyster shells are the most common cultch (substrate) material. Food consumed by oyster larvae includes

microscopic phytoplankton, plant detritus, and bacteria (Galtsoff 1964, Loosanoff 1965, Shaw 1969).

Generally, oyster setting in South Carolina occurs from early May through early October, plus or minus 2 weeks (McNulty 1953). Slightly more than 1% of spat fall occurs at other times during the year. Two setting pulses are usually noted each season. The highest settlement occurs from early June through July, and a second and lesser peak takes place during August or early September. Considerable setting intensity may also occur before, between, and after the two pulses (McNulty 1953).

In Georgia, Furukawa and Linton (1970) found that maximum oyster spat sets occurred near the middle of June in a medium salinity sound at temperatures of about 25°C (77°F). Major concentrations of spat were found 100 - 160 cm (39.4 - 63.1 in) below the surface, but these concentrations shifted seasonally. Durant (1970) observed peak spawning in Georgia during July, August, and September within a spawning period of 7 months. Oysters were observed to begin spawning at temperatures above 23°C (73°F).

Oyster growth varies with size, temperature, quantity, and quality of food. Generally, oysters in South Carolina and Georgia grow throughout the year unless exposed to extreme temperatures or other adverse environmental circumstances. In South Carolina, approximately 95% of the oyster standing crop is intertidal (Lunz 1952).

Oysters are susceptible to predation by numerous enemies. The eggs, early embryos, and larvae are eaten by protozoans, ctenophores, jellyfishes, hydroids, worms, bivalves, barnacles, larval and adult crustaceans, and fishes (Loosanoff 1965). In South Carolina, oyster predator studies have been primarily concerned with pests found on natural beds. Lunz (1935, 1940, 1941, 1943) reported the following commonly occurring oyster pests and predators: the boring sponges Cliona celata, Cliona lobata, and Cliona truitti; the oyster drills Urosalpinx cinerea and Eupleura caudata; the knobbed whelk Busycon carica; the annelid worm Polydora sp.; and the starfish Asterias forbesii. Of these, boring sponges probably cause the greatest damage to South Carolina oysters (Lunz 1943). Cliona celata, the most common boring sponge in Georgia, is usually limited to waters having salinities greater than 20°/oo (Hoese and Durant 1970). Cliona lobata occurred over a salinity range from 10°/oo to 30°/oo, and Cliona truitti was found in salinities of 25°/oo or less. Although Cliona vastifica was found to be rare in Georgia (Hoese and Durant 1970), it is common in South Carolina (Hopkins 1956).

179

The epifauna associated with oyster beds was examined in South Carolina by Hopkins (1956), and in Georgia by Durant (1970). Beds in high salinity Georgia waters exhibited the greatest number (21) of associated species. Populations of large oyster drills, _Urosalpinx cinerea_, _Thais haemastoma_, and _Eupleura caudata_, have been reported in marshlands of Georgia by Hoese (1970).

Fouling organisms such as barnacles (_Balanus eburneus_), bryozoans (_Bugula neritina_), sea squirts (_Molgula manhattensis_), and hooked mussels (_Ischadium recurvum_) are commonly observed growing on oysters (Linton 1970). Infestations by mud worms (_Polydora_ spp.) are common in South Carolina (Lunz 1940, Grice 1951). In the North Santee River, South Carolina, the boring clam _Martesia_ sp. was reported at several stations. A widespread pathogenic sporozoan (_Perkinsia marina_), which infects oysters from Delaware Bay to Mexico, is commonly found in South Carolina and presents a problem when transplanting seed oysters (Anderson 1973); _P. marina_ infection rates in Georgia were observed to be highest in areas exhibiting the greatest net flushing (Hoese 1970). A common oyster commensal or parasite, the pea crab _Pinnotheres ostreum_, is found throughout the estuarine waters of the Sea Island Characterization Area.

South Carolina's first comprehensive investigation of its natural oyster beds was completed during the winter of 1890-91 aboard the steamer _Fish Hawk_. Characteristics of natural oyster beds and bottom areas suitable for cultivation were surveyed along the entire South Carolina coast, with the exception of the North and South Santee estuary (Battle 1892, Dean 1892). Lunz (1938a) studied mortality in six different areas of coastal South Carolina from Charleston to the Santee River and concluded that dredging operations conducted in the Intracoastal Waterway in 1936 were not detrimental to oysters; only those actually buried by dredged material were killed. Another report by Lunz (1938b) described the optimum salinity range (14°/oo - 32°/oo) for oysters in South Carolina. Smith (1949) summarized oyster cultivation progress in the State during 1939-40 and insisted that the most important factors limiting oyster production in South Carolina and Georgia were silting and an average tidal range of 7 ft (2 m). Other surveys of intertidal and subtidal oyster resources in South Carolina were completed by Bears Bluff Laboratories, Inc. (1964) in the Ashley, Cooper, Wando, and Santee rivers. Total oyster acreage was estimated and the standing crop was assessed in U.S. bushels, along with ancillary information concerning oyster condition and predators. Keith and Cochran (1968) estimated 700 acres of subtidal seed oysters in South Carolina,

and McKenzie and Badger (1969) surveyed intertidal oysters in the Savannah River Basin area. South Carolina's largest subtidal oyster beds and seed source, occurring in the Santee Delta (see Volume II, Chapter Seven), are threatened by the proposed Cooper River Rediversion Project (see Volume I, Chapter Six). The rediversion will result in increased freshwater flow and decreased salinity in the Santee River, with subsequent loss of subtidal oyster beds (U.S. Army Corps of Engineers 1975). This prediction is supported by Burrell's (1977) observations of high mortalities among subtidal and intertidal oyster populations in the Santee River during spring floods in 1975.

Earlier studies on distribution and abundance of intertidal oyster stocks in Georgia were completed by Drake (1891) and Galtsoff and Luce (1930). In 1966, another survey (Linton 1970) was initiated on the Florida/Georgia border and proceeded northward to completion at the Georgia/South Carolina boundary in 1967. Approximately 10,182 acres (4,120 ha) of intertidal oyster beds were assessed during the survey. Higher quality oysters were generally found to be more prolific in the northern portion of the State. As in South Carolina, very few subtidal oyster beds were located in Georgia (Linton 1970). According to Linton (1970), comparison of results with Galtsoff and Luce (1930) shows little change in the distribution of intertidal oysters in Georgia.

(2) _Hard Clam_. Another species of commercial importance in Georgia and South Carolina estuaries is the hard clam, _Mercenaria mercenaria_ (see Volume II, Chapter Seven). Hard clams are subjected to intense predation by blue crabs and mud crabs (chiefly _Panopeus herbstii_) in this area (Wetstone and Eversole 1978). Other predators include sting rays, whelks (_Busycon_ spp.), moon snails (_Polinices duplicatus_), oyster drills (_Urosalpinx cinerea_ and _Eupleura caudata_), and stone crabs (_Menippe mercenaria_). The combined effects of these predators apparently limit the distribution of clams to areas which have an abundance of shell in the substrate. Shelly areas suitable for hard clams are primarily limited to oyster "bars" and to areas near the low tide mark, but extend throughout the intertidal zone. An example of the relative density of clams in shelly substrate versus bottom without a shell matrix was provided by studies of clam distribution in the Santee River estuary. Results (P. J. Eldridge, 1973, South Carolina Marine Resources Division, Charleston, unpubl. data) revealed a density of about 100 clams/m² in shelly substrate compared with about 0.25 clams/m² in sandy bottom areas.

The distribution of clams is also determined by the stability of the substrate. Clams in this area are not found in bays

and sounds exposed to wave action, and
they also do not occur where strong currents periodically change bottom contours.
Instead, clams are found in the more protected small coastal creeks. Although
most clams are found around the low water
mark, some occur close to high water in
the roots of marsh vegetation.

Major intertidal beds of clams are
found along the Atlantic Intracoastal
Waterway (AIWW) throughout South Carolina
wherever salinity and substrate conditions are favorable (South Carolina Marine
Resources Division, 1978, Charleston,
unpubl. data). Clams are seldom found in
the channel of the AIWW, possibly due in
part either to dredging operations or to
the unsuitable nature of the bottom.

Growth studies using clams planted
in protected trays have shown that clams
can reach market size (45 - 50 mm total
length) within 18 months after planting
in both Georgia and South Carolina waters
(Godwin 1968, Eldridge et al. 1979).
Indications are that clams spawn or
are capable of spawning almost 9 months
of the year. This suggests that the limiting factor on clam abundance is not
lack of spawning or spawner abundance, but
a lack of suitable habitat for clams to
escape predation, especially from crabs.
If this is true, it would appear that
artificial enhancement of the substrate,
such as shelling of bottoms or the use of
plastic nets, could lead to extensive clam
mariculture operations in this region.

Although a few shells of the southern
quahog, Mercenaria campechiensis, have
been reported from South Carolina, live
specimens of this species have not been
recorded here. This does not mean that
the southern quahog is not present, but it
does suggest that the species occurs infrequently. Eldridge et al. (1976) reported M. mercenaria notata from 11 locations in South Carolina, but these
comprised only 1.2% of the clams sampled.

Clams in this area are apparently
relatively free of disease, although this
aspect of their biology is poorly known.
The greatest threat to the clam resource
in the Sea Island Coastal Region appears
to be coastal development and pollution.

4. Fishes

Fishes of subtidal estuarine habitats
such as sounds, bays, tidal rivers, and
large creeks, have been investigated more
intensively than those of any other environment within the Sea Island Coastal
Region. Much of the work on the ecology
and life histories of these fish species
has been based on sampling with the otter
trawl, which is selective for more slowly
moving, demersal species. The biological
conclusions of these studies should be
reviewed with this bias in mind. The

problems associated with sampling motile
estuarine animals have been discussed in
depth by Herke (1971).

a. Distribution and Relative Abundance. In South Carolina, Bears Bluff
Laboratories conducted regular trawl
sampling for fishes and invertebrates at
20 regular monthly stations located from
Price Inlet southward to Calibogue Sound,
South Carolina, during the period 1953 -
1964 (Bears Bluff Laboratories, Inc. 1964,
1965). Shealy (1974, 1975) and Shealy
et al. (1974, 1975) reported on bottom
trawl investigations conducted during
1972 - 1974 in estuaries of South Carolina.
The Marine Resources Division of the South
Carolina Wildlife and Marine Resources
Department continued these trawl sampling
investigations in the Cooper River, Santee
River, and Winyah Bay estuaries through
1978. These trawl investigations have
shown that, in South Carolina, the dominant benthic estuarine fish species are
sciaenids (including star drum, Atlantic
croaker, spot, weakfish, silver perch, and
kingfishes), as well as bay·anchovy, white
catfish, spotted hake, Atlantic menhaden,
blackcheek tonguefish, flounders, and
whiffs (Etropus sp. and Citharichthys sp.).
The subtidal estuarine habitat in South
Carolina varies from high salinity unstratified estuaries having considerable
"live bottom" habitat (Murrells Inlet,
North Inlet, Bulls Bay-Isle of Palms, Port
Royal Sound) to brackish water estuaries
having muddy bottoms and considerable
freshwater discharge from large rivers
(Winyah Bay, Santee estuary, Charleston
Harbor, South Edisto River). Although the
above-mentioned species have generally
been found to predominate in all areas of
the State, certain species are common in
some estuaries and uncommon or absent in
others.

Bears Bluff Laboratories, Inc. (1965)
collected a total of 23,420 fishes representing 67 species during a 12-month trawl
survey of the Price Inlet area, an estuarine zone of the South Carolina coast
characterized by highly saline, unstratified waters and a "live bottom" of octocorals, sponges, and other epibenthic invertebrates. Common estuarine species
accounted for about 70% of the total number of fishes captured (spot, Atlantic
croaker, star drum, hogchokers, tongue-
fishes, and flounders) (Table 4-7). However, black sea bass, pigfish, and oyster
toadfish, which are uncommon in intermediate to low salinity areas, made up
slightly more than 12% of the total number
of fishes collected.

In the more brackish-water areas of
South Carolina such as the Santee estuary,
species commonly collected and not present
in the higher salinity areas include white
catfish and threadfin shad (South Carolina
Marine Resources Division, 1975 - 1978,
Charleston, unpubl. data). These species

Table 4-7. Fishes taken by trawl in Price Inlet, South Carolina, from July 1964 through June 1965 (adapted from Bears Bluff Laboratories, Inc. 1965).

Spiny dogfish	5	0.02
Clearnose skate	65	0.3
Atlantic stingray	3	0.01
Southern stingray	52	0.2
Smooth butterfly ray	10	0.04
Sea catfish	19	0.08
Gafftopsail catfish	1	--
Conger eel	1	--
Alewife	7	0.02
Gizzard shad	23	0.09
Blueback herring	8	0.03
Atlantic thread herring	16	0.06
Atlantic menhaden	40	0.2
Bay anchovy	617	2.6
Inshore lizardfish	8	0.03
Northern pipefish	9	0.03
Chain pipefish	1	--
Lined seahorse	2	--
Spotted hake	309	1.3
Southern hake	1	--
Atlantic silverside	8	0.03
Striped mullet	2	--
Atlantic cutlassfish	11	0.04
Lookdown	39	0.16
Bluefish	4	0.01
Harvestfish	2	--
Butterfish	4	0.01
Black sea bass	1,238	5.3
Rock sea bass	217	0.9
Pigfish	1,249	5.3
Porgies	3	0.01
Sheepshead	1	--
Pinfish	106	0.05
Irish pompano	3	0.01
Spotted seatrout	3	0.01
Weakfish	815	--
Silver seatrout	--	3.5
Banded drum	13	0.05
Silver perch	369	1.6
Star drum	1,896	8.1
Spot	9,025	38.5
Atlantic croaker	1,296	5.5
Southern kingfish	177	0.8
Northern kingfish	58	0.2
Atlantic spadefish	66	0.3
Planehead filefish	56	0.2
Orange filefish	6	0.02
Puffers	66	0.3
Striped searobin	--	--
Northern searobin	189	0.8
Leopard searobin	--	--
Southern stargazer	2	--
Oyster toadfish	394	1.7
Atlantic midshipman	1	--
Clingfish	9	0.03
Feather blenny	93	0.4
Striped blenny	2	--
Striped cusk-eel	17	0.07
Summer flounder	746	3.2
Southern flounder	255	1.1
Ocellated flounder	133	0.6

Table 4-7. Concluded

Species	Number	Percent of Total
Spotted whiff	--	--
Bay whiff	672	2.9
Fringed flounder	--	--
Windowpane	210	0.9
Hogchoker	1,824	7.8
Blackcheek tonguefish	943	4.0

made up approximately 10% and 15% of the total numbers of fishes collected at Santee stations during 1975 and 1976, respectively. A total of 28,807 fish specimens, representing 77 species, were taken during 1975 and 1976 in the Marine Resources Division's survey trawl sampling program in the Santee estuary. This represents a greater diversity than has been recorded in previous trawl surveys in South Carolina for a single estuarine system. Sampling stations were located at distances of 1, 4, 7, and 11 mi (1.6, 6.4, 11.3, and 17.7 km) from the Atlantic Ocean in both the North and South Santee rivers.

Five species (Atlantic croaker, bay anchovy, hogchoker, white catfish, and silver perch) made up almost 70% of the total number of fishes collected at Santee trawl stations in 1975 and 1976. Atlantic croaker were largely young-of-the-year fish, and were most abundant at the 4- and 7-mile stations during the summer months. Bay anchovy were found during the entire year in the Santee area. These fishes were most commonly collected at the 1- and 4-mile stations, decreasing in numbers upriver to near zero at the two low salinity 11-mile stations. Hogchokers were much more common in the North Santee than in the South Santee, especially at the 4-, 7-, and 11-mile stations. Although these fishes were found throughout the year, they were most abundant in trawl catches during the fall and winter (November - February). White catfish, as expected, were most common at the lower salinity, upriver stations in the Santee area (miles 7 and 11). White catfish were collected during every month of the year. Silver perch were most abundant at the 1- and 4-mile stations in the North and South Santee rivers, decreasing to near zero at the upriver 11-mile stations. These fishes were most common in trawl collections during late summer and fall. Tables 4-8 and 4-9 present the seasonal abundance, total numbers, and rankings of the more common fish species collected by otter trawl at North and South Santee rivers stations during 1975 and 1976, respectively.

An intensive trawl investigation of the nearby Winyah Bay estuary, which is also strongly influenced by freshwater discharge from coastal rivers, has recently been completed by the Marine Resources Division, and analysis of data is currently underway.

Shealy et al. (1974) reported results of the first year of a 2-year trawl survey which encompassed 33 estuarine stations across the coastal zone of South Carolina, from Winyah Bay south to Calibogue Sound. All stations were sampled at least quarterly, and 17 stations in the Cooper, North Edisto, and South Edisto rivers were sampled monthly. A total of 62,684 fishes, representing 88 species from 46 families, was caught by bottom trawl in these estuaries during the 12-month sampling period. However, the vast majority of the total catch was comprised of but a few species. Star drum and bay anchovy accounted for over half of the total number of fishes caught during the year. These two species, along with Atlantic croaker and spot, made up 80.5% of the total number caught. Table 4-10 presents the numbers, relative abundance, and biomass of the fish species collected.

Contributions to the total catch in terms of weight were spread over a slightly larger number of species, with nine species constituting 80.6% of the total catch by weight. Fourteen species each contributed at least 1% of the total catch biomass. Except for white catfish, the five most important species by weight were sciaenids. In decreasing order of abundance, these sciaenids were star drum, Atlantic croaker, spot, silver perch, and weakfish. Because it is small, even as an adult, bay anchovy contributed only 3.5% of the total catch biomass. Total length, bottom salinity, temperature, and primary location at which the 88 fish species were collected by bottom trawling are presented in Table 4-11.

Of the three estuaries sampled intensively by Shealy et al. (1974), the North Edisto River exhibited the greatest diversity of benthic fishes (62 species), followed by the Cooper River (57 species), and the South Edisto (47 species). Star

183

Table 4-8. Seasonal abundance, total numbers, and ranking of fish species collected by trawl at North and South Santee stations, 1975 (South Carolina Marine Resources Division, 1975, Charleston, unpubl. data).

SPECIES	JANUARY-MARCH	APRIL-JUNE	JULY-SEPTEMBER	OCTOBER-DECEMBER	TOTAL	% TOTAL	RANK
Bay anchovy	143	853	1,770	621	3,387	24.90	1
Atlantic croaker	671	823	1,567	144	3,206	23.75	2
Silver perch	0	87	921	703	1,711	12.58	3
White catfish	419	49	395	375	1,238	9.10	4
Atlantic menhaden	673	138	11	17	839	6.17	5
Hogchoker	341	20	73	237	671	4.93	6
Weakfish	0	93	492	61	646	4.75	7
Star drum	10	6	103	473	592	4.35	8
Spot	93	83	106	53	335	2.46	9
Threadfin shad	109	6	18	27	160	1.18	10
Blackcheek tonguefish	12	0	4	92	108	0.79	11
Southern flounder	8	4	9	38	59	0.43	12
Other species	155	54	235	204	648	4.76	--
TOTALS	2,634	1,217	5,704	3,045	13,600		

Table 4-9. Seasonal abundance, total numbers, and ranking of fish species collected by trawl at North and South Santee stations, 1976 (M. H. Shealy, Jr., South Carolina Marine Resources Division, Charleston, unpubl. data).

SPECIES	JANUARY-MARCH	APRIL-JUNE	JULY-SEPTEMBER	OCTOBER-DECEMBER	TOTAL	% TOTAL	RANK
Atlantic croaker	569	3,302	717	151	4,739	30.73	1
Hogchoker	628	259	237	1,762	2,886	18.72	2
White catfish	766	370	264	217	1,617	10.49	3
Bay anchovy	632	204	181	77	1,094	7.10	4
Silver perch	151	49	464	200	864	5.60	5
Star drum	5	40	478	289	812	5.27	6
Threadfin shad	4	0	3	717	724	4.70	7
Blackcheek tonguefish	183	189	66	97	535	3.47	8
Weakfish	0	85	294	123	502	2.26	9
Spot	111	123	15	23	272	1.76	10
Atlantic menhaden	191	32	5	34	262	1.70	11
Southern flounder	95	28	18	85	226	1.47	12
Other species	280	173	195	238	886	5.75	--
TOTALS	3,615	4,854	2,937	4,013	15,419		

185

Table 4-10. Total numbers, total weights, ranking in order of abundance by number and weight, and percentage of total catch represented by 88 fish species captured by bottom trawl (all stations combined) in South Carolina estuaries from February 1973 through January 1974 (adapted from Shealy et al. 1974).

SPECIES	RELATIVE NUMBERS			RELATIVE BIOMASS		
	TOTAL NUMBER CAUGHT	NUMERICAL RANK	PERCENT OF TOTAL CATCH	TOTAL WEIGHT (kg)	BIOMASS RANK	PERCENT OF TOTAL CATCH
Star drum	23,992	1	38.3	105.6	1	19.3
Bay anchovy	12,074	2	19.3	19.4	7	3.5
Atlantic croaker	9,030	3	14.4	95.5	2	17.4
Spot	5,347	4	8.5	57.1	3	10.4
Weakfish	2,136	5	3.4	31.0	6	5.7
Silver perch	1,863	6	3.0	43.6	5	8.0
White catfish	1,732	7	2.8	54.2	4	9.9
Spotted hake	1,612	8	2.6	17.6	8	3.2
Atlantic menhaden	823	9	1.3	8.9	11	1.6
Atlantic bumper	578	10	0.9	3.4	20	0.6
Blueback herring	462	11	0.7	1.5	33	0.3
Hogchoker	407	12	0.6	3.5	19	0.6
Blackcheek tonguefish	362	13	0.6	6.0	14	1.1
Threadfin shad	327	14	0.5	0.9	36	0.2
Striped anch vy	216	15	0.3	1.7	30	0.3
Atlantic thread herring	214	16	0.3	1.6	32	0.3
Atlantic cutlassfish	189	17	0.3	7.6	13	1.4
Harvestfish	151	18	0.2	1.8	28	0.3
Sea catfish	90	19	0.1	10.0	10	1.8
Channel catfish	77	20	0.1	3.2	23	0.6
Oyster toadfish	76	21	0.1	4.1	17	0.8
Banded drum	75	22	0.1	2.1	26	0.4
Southern kingfish	75	22	0.1	1.8	27	0.3
Brown bullhead	53	23	0.1	2.6	24	0.5
Atlantic moonfish	53	24	0.1	0.2	52	< 0.1
Gafftopsail catfish	50	25	0.1	0.5	45	0.1
Atlantic spadefish	41	26	0.1	1.6	40	0.3
Silver seatrout	40	27	0.1	0.4	31	0.1
Lookdown	39	27	0.1	0.4	43	0.1
American eel	35	28	0.1	4.0	18	0.7
Feather blenny	35	28	0.1	0.3	50	0.1
Striped bass	33	29	0.1	0.2	55	< 0.1
American shad	28	30	< 0.1	0.2	53	< 0.1
Summer flounder	25	31	< 0.1	1.7	29	0.3
Southern flounder	25	31	< 0.1	4.9	15	0.9
Longnose gar	24	32	< 0.1	17.5	9	3.2
*Bighead searobin	24	32	< 0.1	0.1	67	< 0.1
Bay whiff	23	33	< 0.1	0.2	57	< 0.1
Crevalle jack	20	34	< 0.1	0.3	51	< 0.1

Table 4-10. Continued

SPECIES	RELATIVE NUMBERS			RELATIVE BIOMASS		
	TOTAL NUMBER CAUGHT	NUMERICAL RANK	PERCENT OF TOTAL CATCH	TOTAL WEIGHT (kg)	BIOMASS RANK	PERCENT OF TOTAL CATCH
Butterfish	18	35	<0.1	0.2	56	<0.1
Bluefish	17	36	<0.1	1.2	34	0.2
Spanish mackerel	17	36	<0.1	0.4	41	0.1
Fringed flounder	15	37	<0.1	0.1	61	<0.1
Atlantic sturgeon	14	38	<0.1	7.7	12	1.4
Black sea bass	14	38	<0.1	0.3	47	0.1
Southern stargazer	12	39	<0.1	0.1	63	<0.1
Windowpane	11	40	<0.1	0.3	48	<0.1
Planehead filefish	9	41	<0.1	0.7	69	<0.1
Rock sea bass	7	42	<0.1	0.1	60	<0.1
Spotted seatrout	6	43	<0.1	0.5	38	0.1
Pinfish	6	43	<0.1	0.4	42	0.1
Carolina hake	6	43	<0.1	0.2	54	<0.1
Gray snapper	5	44	<0.1	0.1	66	<0.1
Pigfish	5	44	<0.1	0.4	44	0.1
Inshore lizardfish	5	44	<0.1	0.3	49	0.1
Ocellated flounder	4	45	<0.1	0.1	70	<0.1
Smooth butterfly ray	4	45	<0.1	4.7	16	0.9
Black drum	4	45	<0.1	3.3	22	0.6
Atlantic stingray	3	46	<0.1	2.4	25	0.4
Skilletfish	3	46	<0.1	0.1	75	<0.1
Yellow bullhead	3	46	<0.1	0.3	37	0.1
Atlantic silversides	3	46	<0.1	0.1	77	<0.1
Yellow perch	3	46	<0.1	0.1	72	<0.1
*Northern searobin	3	46	<0.1	0.1	76	<0.1
Southern hake	3	46	<0.1	0.1	59	<0.1
Naked g by	2	47	<0.1	0.1	79	<0.1
Flat bullhead	2	47	<0.1	0.3	46	0.1
Smooth puffer	2	47	<0.1	0.1	65	<0.1
Striped mullet	2	47	<0.1	0.1	71	<0.1
White mullet	2	47	<0.1	0.1	73	<0.1
Atlantic sharpnose shark	2	47	<0.1	0.5	40	0.1
Striped cusk-eel	1	48	<0.1	0.1	62	<0.1
Striped burrfish	1	48	<0.1	0.1	82	<0.1
Conger eel	1	48	<0.1	0.1	58	<0.1
Gizzard shad	1	48	<0.1	0.1	78	<0.1
Darter g by	1	48	<0.1	0.1	85	<0.1
Sharptail g by	1	48	<0.1	0.1	68	<0.1
Marked g by	1	48	<0.1	0.1	74	<0.1
Seaboard g by	1	48	0.1	0.1	86	0.1
Black bullhead	1	48	0.1	0.1	64	<0.1
Redbreast sunfish	1	48	0.1	0.1	80	<0.1
*Striped searobin	1	48	<	0.1	83	<0.1

Table 4-10. Concluded

SPECIES	RELATIVE NUMBERS			RELATIVE BIOMASS		
	TOTAL NUMBER CAUGHT	NUMBERICAL RANK	PERCENT OF TOTAL CATCH	TOTAL WEIGHT (kg)	BIOMASS RANK	PERCENT OF TOTAL CATCH
C wn se ray	1	48	< 0.1	1.1	35	0.2
Guaguanche	1	48	< 0.1	< 0.1	84	< 0.1
Smooth hammerhead	1	48	< 0.1	0.5	39	0.1
Spiny dogfish	1	48	< 0.1	3.4	21	0.6
Dusky p peFish	1	48	< 0.1	< 0.1	87	< 0.1
Northern pipefish	1	48	< 0.1	< 0.1	81	< 0.1
GRAND TOTALS	62,684		100.0	547.7		100.0

*Tentative identification.

Table 4-11. Total length ranges, bottom salinity and temperature ranges, and primary locations at which 88 fish species were captured by bottom trawl in South Carolina estuaries from February 1973 through January 1974 (Shealy et al. 1974).

SPECIES	TOTAL LENGTH RANGE (mm)	BOTTOM SALINITY RANGE (°/oo)	BOTTOM TEMPERATURE RANGE (°C)	PRIMARY LOCATIONS
American eel	265 - 528	0.1 - 25.9	16.2 - 22.8	Upper Cooper River
American shad	50 - 181	0.1 - 27.3	11.5 - 29.3	Wide distribution
Atlantic bumper	38 - 132	13.7 - 32.3	18.4 - 30.5	South Edisto River
Atlantic croaker	20 - 293	< 0.1 - 34.2	9.2 - 31.0	Wide distribution
Atlantic cutlassfish	136 - 667	0.7 - 34.4	16.2 - 30.6	Wide distribution
Atlantic menhaden	35 - 243	0.1 - 30.3	12.0 - 31.3	Wide distribution
Atlantic moonfish	32 - 102	4.8 - 28.6	18.5 - 30.1	North and South Edisto, Cooper rivers
Atlantic sharpnose shark	298 - 412	28.2 - 33.2	27.3 - 28.2	North Edisto and Price Creek
Atlantic silverside	85	21.8	13.7 - 17.5	North Edisto, Charleston Region
Atlantic spadefish	37 - 137	14.2 - 33.2	26.5 - 30.1	North Edisto and Cooper rivers
Atlantic stingray	202 - 509	0.2 - 28.9	17.1 - 27.5	N rth and S uth Edisto, Cooper rivers
Atlantic sturgeon	110 - 615	< 0.1 - 33.2	16.2 - 29.9	Upper S uth Edisto River
Atlantic thread herring	40 - 174	0.1 - 34.2	12.7 - 30.1	North Edisto and Cooper rivers
Banded drum	37 - 146	12.9 - 34.2	16.0 - 30.5	N rth Edisto River
Bay anch vy	22 - 88	< 0.1 - 34.2	10.3 - 31.4	Wide distribution
Bay whiff	68 - 131	21.4 - 26.6	10.4 - 29.5	Port R yalp Charleston Region
Bighead searobin[a]	16 - 96	0.1 - 34.2	16.9 - 28.0	North Edisto, Charleston Region
Blackcheek tonguefish	53 - 156	0.1 - 34.2	8.6 - 30.5	Wide distribution
Black drum	180 - 512	3.4 - 4.8	11.2 - 16.2	Upper Cooper River
Black sea bass	60 - 198	10.9 - 33.8	10.3 - 29.2	N rth Edisto (Deveaux Bank)
Blueback herring	35 - 307	0.7 - 25.1	11.5 - 29.0	Wide distribution
Bluefish	65 - 280	0.1 - 34.4	16.2 - 30.5	N rth and South Edisto, Cooper rivers
Brown bullhead	59 - 269	0.1 - 3.4	8.7 - 28.8	Cooper River
Butterfish	50 - 143	14.2 - 26.7	21.1 - 30.5	N rth Edisto, Charleston Harbor
Carolina hake	112 - 174	25.0	16.9	North Edisto (Deveaux Bank)
Channel catfish	56 - 234	0.1 - 0.2	8.7 - 27.8	Upper South Edisto and Cooper rivers
Conger eel	450	< 0.1	17.8	Charleston Region, Ft. Johnson
Cownose ray	366	24.6	27.6	North Edisto (Point of Pines)
Crevalle jack	28 - 151	9.5 - 28.0	17.0 - 29.3	Cooper and North Edisto rivers
Darter goby	74	21.4	29.4	Charleston Harbor (Cummings Point)
Dusky pipefish	40	25.7	26.7	North Edisto (Dawho River)
Feather blenny	67 - 108	10.3 - 34.4	10.3 - 30.4	North Edisto and Cooper rivers
Flat bullhead	201 - 272	0.1	14.2	Cooper River (The Tee)
Fringed flounder	39 - 105	14.2 - 31.2	17.1 - 30.1	North and South Edisto, Cooper rivers

189

Table 4-11. Continued

SPECIES	TOTAL LENGTH RANGE (mm)	BOTTOM SALINITY RANGE (°/oo)	BOTTOM TEMPERATURE RANGE (°C)	PRIMARY LOCATIONS
Gafftopsail catfish	75 - 158	8.3 - 27.1	18.3 - 30.1	Wide distribution
Gizzard shad	112 -	9.5	17.0 -	Cooper River (Big Island)
Gray snapper	70 - 107	0.1 - 34.2	17.0 - 27.4	South Edisto, Charleston Region
Guaguanche	87	27.6	27.2	North Edisto (Point of Pines)
Harvestfish	20 - 131	0.9 - 33.2	17.4 - 30.1	Wide distribution
Hogchoker	21 - 152	< 0.1 - 32.3	6.6 - 30.5	Wide distribution
Inshore lizardfish	52 - 261	15.4 - 20.2	25.5 - 27.4	Charleston Harbor (Hog Island)
Longnose gar	314 - 1,018	0.1 - 18.1	9.1 - 29.9	Cooper River and Upper South Edisto
Lookdown	32 - 115	6.0 - 33.2	16.9 - 30.1	North and South Edisto, Charleston Region
Marked g by	62 - 63	0.2 - 24.1	29.4	Port Royal Sound, South Edisto River
Naked goby	61 - 72	15.0 - 22.8	27.6	Upper North Edisto River
Northern pipefish	235	33.2	10.4	Bulls Bay
Northern searobin	NOT AVAILABLE	16.9 - 32.1	NOT AVAILABLE	Upper Cooper River
Ocellated flounder	38 - 118	21.8 - 25.1	17.1 - 22.0	North Edisto, Charleston Region
Oyster toadfish	23 - 245	2.0 - 34.2	11.6 - 30.4	Wide distribution
Pigfish	171 - 264	27.6 - 34.2	19.6 - 24.4	Lower North Edisto, Northern Region
Pinfish	62 - 192	29.8 - 33.8	20.5 - 24.4	North Edisto and Charleston Region
Planehead filefish	16 - 76	4.2 - 32.3	9.2 - 30.6	Calibogue Sound, North Edisto
Redbreast sunfish	76 - 126	0.1	26.0	Upper Cooper River (The Tee)
R ck sea bass	35 -	4.8 - 25.9	21.0 - 30.0	Cooper River (mouth), Port Royal Sound
Seaboard g by	42 -	14.2	16.6 - 21.5	South Edisto River
Sea catfish	58 - 296	0.2 - 33.2	16.8 - 30.5	Wide distribution
Sharptail goby	254	24.6	8.6	North Edisto (Point of Pines)
Silver seatrout	78 - 180	23.8 - 30.8	17.4 - 30.5	North Edisto, Calibogue Sound
Silver perch	40 - 192	0.1 - 34.4	7.2 - 31.4	Wide distribution
Skilletfish	57 - 61	20.9 - 26.2	14.0 - 17.4	Lower North Edisto and Cooper rivers
Smooth butterfly ray	248 - 327	1.1 - 28.0	26.5 - 30.5	North Edisto and Calibogue Sound
Smooth hammerhead	468	25.4	30.5	Calibogue Sound (Marsh Island)

190

Table 4-11. Concluded

SPECIES	TOTAL LENGTH RANGE (mm)	BOTTOM SALINITY RANGE (°/oo)	BOTTOM TEMPERATURE RANGE (°C)	PRIMARY LOCATIONS
Smooth puffer	95 - 110	22.6 - 22.7	28.8 - 29.0	Lower South Edisto and Cooper rivers
Southern flounder	34 - 412	0.1 - 28.6	9.4 - 30.1	North and South Edisto, Charleston Region
Southern hake	155 - 184	25.0	16.9	North Edisto (Deveaux Bank)
Southern kingfish	43 - 271	0.9 - 34.2	9.2 - 30.1	Wide distribution
Southern stargazer	25 - 79	10.5 - 30.8	13.7 - 21.9	North Edisto River
Spanish mackerel	72 - 179	0.4 - 28.0	26.4 - 30.4	North and South Edist ọ Charleston Region
Spiny dogfish	916	20.9	8.7	North Edisto (Deveaux Bank)
Spot	22 - 188	0.1 - 34.4	11.6 - 31.4	Wide distribution
Spotted hake	57 - 190	3.4 - 29.8	8.6 - 22.6	Wide distribution
Spotted seatrout	202	26.2	12.7 - 17.4	Cooper River (mouth)
Star drum	16 - 217	0.9 - 34.4	8.6 - 30.5	Wide distribution
Striped anchovy	58 - 140	8.8 - 34.4	16.9 - 30.5	Edisto and Cooper rivers
Striped bass	47 - 142	< 0.1 - 0.3	14.2 - 28.8	Upper S utb Edisto (Snuggedy Swamp)
Striped burfish	44	25.4	30.5	Calibogue Sound
Striped cusk-eel	161	28.7	14.0 - 27.3	Lower North Edisto and Price Creek
Striped mullet	93	23.2	13.2 - 28.2	North Edisto and Charleston Harb ọ
Striped searobin[a]	NOT AVAILABLE	25.4	NOT AVAILABLE	Calibogue Sound (Marsh Island)
Summer flounder	45 - 250	3.4 - 28.7	8.7 - 30.6	North Edist ọ Cooper, Southern Regi ọ
Threadfin shad	36 - 134	0.1 - 32.3	16.6 - 29.9	Wide distribution
Weakfish	23 - 323	0.4 - 34.4	13.7 - 31.4	Wide distribution
White catfish	28 - 392	< 0.1 - 23.9	11.6 - 31.4	Upper South Edisto, Cooper and Ashepoo rivers
White mullet	98 - 111	23.2 - 25.0	28.1	Upper North Edisto (Yonges Island)
Windowpane	65 - 205	7.6 - 25.0	9.4 - 22.0	North Edisto River
Yellow bullhead	97 - 351	< 0.1 - 0.1	14.2	Upper South Edisto and Cooper rivers
Yellow perch	92 - 107	0.1	8.7 - 14.2	Upper Cooper River (The Tee)

a. Tentative identification.

drum, followed by a bay anchovy-spot-
Atlantic croaker assemblage, dominated
the North Edisto community; common species
included spotted hake, weakfish, silver
perch, and blackcheek tonguefish. Bottom
waters of the South Edisto River were also
dominated by star drum, followed by an
Atlantic croaker-white catfish-bay anchovy
assemblage. Atlantic bumper, spotted hake,
weakfish, spot, hogchoker, blackcheek
tonguefish, and silver perch also were
captured frequently. The demersal fish
fauna of the Cooper River was co-dominated
by star drum and Atlantic croaker, followed by a bay anchovy-Atlantic menhaden-
spotted hake assemblage which replaced the
supporting sciaenid assemblage of the
North Edisto River. Weakfish, spot, blueback herring, white catfish, threadfin
shad, and silver perch were common in
North Edisto trawl samples.

Fishes of estuarine subtidal waters
of South Carolina have been but poorly
investigated with gear other than bottom
trawls. Hicks (1972) conducted gill net
sampling in the channel reaches of Port
Royal Sound during April and July of 1970
(Table 4-12). Species, diversity, and
numbers were quite different than from
small mesh otter trawl collections taken
during other surveys within similar habitats (e.g., see Tables 4-9, 4-10).
Shealy (1975) found that bay anchovy comprised over 90% of the total catch taken
with a 1.5-in Cobb midwater trawl at 15
stations in the North Edisto, South Edisto,
and Cooper rivers.

In Georgia, major investigations of
subtidal estuarine fish populations have
been carried out by Miller and Jorgenson
(1969), Dahlberg and Odum (1970), Dahlberg
(1972), Hoese (1973), and Mahood et al.
(1974a, b, c, d). Dahlberg and Odum (1970)
conducted a 1-year survey using a 20 ft
(6.1 m) otter trawl [1¼ in (3.2 cm) mesh]
at 14 stations in the estuarine system of
Sapelo and St. Catherines sounds. A total
of 31,637 fishes comprising 70 species
and 37 families was collected. The most
common species in order of decreasing
abundance were star drum, blackcheek
tonguefish, weakfish, sea catfish, bay
anchovy, silver perch, spot, Atlantic
croaker, spotted hake, Atlantic menhaden,
fringed flounder, and hogchoker (Table
4-13). These 12 species made up 90% of
the total numbers of fishes collected.
The abundance and occurrence of predominant species in this study are quite
similar to what would be expected in mod-
erate-high salinity sounds in South
Carolina (Bearden and Farmer 1972, Shealy
et al. 1974).

Dahlberg and Odum (1970) found that
structure of the benthic fish communities
in the two estuaries sampled was influenced by seasonal changes in taxa,
numbers of species and individuals, and
average size of fishes. Species diversity

was studied with respect to seasonal variation in three ecological zones: sounds,
larger creeks, and smaller creeks. Four
diversity indices were compared, and seasonal variation was found with "equita-
bility" and "evenness" (indices of the
distribution of individuals among the
various species in a sample), but not with
the Shannon-Wiener index or species "richness" (index of species numbers divided by
the natural logarithm of the total number
of individuals of all species in a sample).
Figure 4-4 presents comparisons of the
four indices with respect to annual cycles.
A seasonal change in relative abundances
resulted primarily from the influx of
juveniles in late summer and autumn.

Hoese (1973) conducted a 12-month
trawl survey of fishes and invertebrates
on the Georgia inshore continental shelf
and adjacent Doboy Sound. The most characteristic fish species taken in estuarine
waters was the star drum. Other important
species included bay anchovy, sciaenids
(spot, Atlantic croaker, silver perch,
weakfish), spotted hake, and blackcheek
tonguefish. Hoese concluded that the absence of large benthic plant communities
in South Carolina and Georgia excludes the
large fauna associated with submerged grass
beds (turtle grass, Halodule, eel grass)
found to the north and south of the characterization area. Hoese also found that
a small, cool, temperate fauna invaded the
area during winter, when species such as
spiny dogfish and spotted hake became common in collections. The distribution of
many species of fishes was correlated with
freshwater influence of the Altamaha River,
temperature, and other hydrographic factors.
Up to eight times the biomass of fishes
and invertebrates was found in the estuarine
subtidal zone compared with offshore waters.

Dahlberg (1972) studied the distribution of marine and estuarine fishes in relation to ecological factors in the estuary formed by St. Catherines Sound, Newport
River, and Sapelo Sound, Georgia. His
investigation included trawl sampling in
the lower reach of the estuary as well as
the upper and middle reaches and an oligo-
baline creek. In the lower reach of the
estuary, 100 species of fishes were found
compared to 61 species in the middle reach.
Eurythermal species (i.e., those found
year around) were Atlantic stingray, bay
anchovy, silver perch, Atlantic croaker,
spot, southern kingfish, hogchoker, blackcheek tonguefish, and oyster toadfish.
Species common only during the warm months
included sea catfish, weakfish, star drum,
and Atlantic spadefish, while those taken
commonly only during the cold months included hakes, spotted seatrout, ocellated
flounder, and southern flounder. The
oligohaline section of Riceboro Creek
yielded only 40 fish species, including
21 species considered stragglers. Characteristic species included bay anchovy,
striped mullet, tidewater silversides,

Table 4-12. Summary of catch data from anchored gill net sets at 10 stations in channel reaches, Port Royal Sound, South Carolina, April and July surveys, 1970 (adapted from Hicks 1972).

	APRIL SURVEY		JULY SURVEY	
	No. of Stations Species Occurred	Individual No. Caught	No. of Stations Species Occurred	Individual No. Caught
Finetooth shark	4	8	8	119
Scalloped hammerhead	3	11	8	68
Bonnethead	3	6		
Atlantic guitarfish	5	13		
Atlantic stingray	5	7		
Atlantic sturgeon	1	2		
Longnose gar	8	77	2	2
Atlantic menhaden	10	314	9	630
Atlantic thread herring			4	25
Gafftopsail catfish			2	4
Sea catfish	4	10	7	29
Spotted hake	5	26		
Black sea bass	2	3		
Bluefish	6	24	2	4
Atlantic bumper			1	1
Silver perch	4	42	3	15
Spotted seatrout			1	1
Weakfish	1	1		
Spot	8	33	4	19
Northern kingfish			6	21
Atlantic croaker	1	1	4	4
Atlantic cutlassfish			1	1
Spanish mackerel			1	3
Northern searobin	1	2		
Butterfish	2	3		
Fringed flounder	1	2		
Summer flounder	3	6		
Southern flounder	3	6		
Hogchoker	1	1		
Blackcheek tonguefish	7	42		
Sharksucker			1	1
Striped burrfish	1	1		
TOTAL	89	641	64	947
NO. SPECIES	24		17	

193

Table 4-13. Seasonal occurrence of fish species taken by trawl in Sapelo and St. Catherines sounds, Georgia (adapted from Dahlberg and Odum 1970).

Carcharhinidae		
Blacktip shark	1	uncommon
Sphyrnidae		
Bonnethead	1	uncommon
Squalidae		
Spiny dogfish	1	colder months
Rajidae		
Clearnose skate	9	year around
Dasyatidae		
Southern stingray	2	uncommon
Atlantic stingray	72	year around
Bluntnose stingray	4	uncommon
Smooth butterfly ray	3	uncommon
Lepisosteidae		
Longnose gar	1	uncommon
Clupeidae		
Hickory shad	1	uncommon
American shad	11	colder months
Atlantic menhaden	574	year around
Scaled sardine	1	uncommon
Atlantic thread herring	11	uncommon
Engraulidae		
Striped anchovy	8	uncommon
Bay anchovy	1,513	year around
Synodontidae		
Inshore lizardfish	6	warmer months
Ariidae		
Gafftopsail catfish	30	year around
Sea catfish	1,682	year around
Ictaluridae		
White catfish	4	uncommon
Ophichthidae		
Shrimp eel	1	uncommon
Cyprinodontidae		
Mummichog	5	uncommon
Gadidae		
Southern hake	52	colder months
Spotted hake	580	colder months
Syngnathidae		
Northern pipefish	2	year around
Chain pipefish	2	year around
Serranidae		
Rock sea bass	61	year around
Black sea bass	50	year around
Pomatomidae		
Bluefish	2	uncommon

Table 4-13. Continued

Carangidae
    Atlantic bumper   1   uncommon

	Count	Occurrence
Carangidae		
Atlantic bumper	1	uncommon
Lookdown	3	uncommon
Pomadasyidae		
Pigfish	31	warmer months
Sciaenidae		
Silver perch	1,191	year around
Spotted seatrout	52	year around
Silver seatrout	70	warmer months
Weakfish	3,458	year around
Banded drum	129	year around
Spot	1,113	year around
Southern kingfish	1,527	year around
Atlantic croaker	947	year around
Black drum	1	uncommon
Star drum	15,272	year around
Chaetodipteridae		
Atlantic spadefish	256	warmer months
Trichiuridae		
Atlantic cutlassfish	8	uncommon
Gobiidae		
Highfin goby	1	uncommon
Triglidae		
Black-wing searobin	1	uncommon
Striped searobin	147	year around
Leopard searobin	87	year around
Uranoscopidae		
Southern stargazer	15	year around
Blennidae		
Crested blenny	1	uncommon
Feather blenny	37	colder months
Ophidiidae		
Striped cusk-eel	18	year around
Stromateidae		
Harvestfish	8	uncommon
Butterfish	24	colder months
Mugilidae		
Striped mullet	2	uncommon
Atherinidae		
Rough silverside	1	uncommon
Atlantic silverside	8	uncommon
Bothidae		
Ocellated flounder	70	colder months
Bay whiff	12	warmer months
Fringed flounder	516	year around
Gulf flounder	1	uncommon
Summer flounder	43	year around
Southern flounder	65	year around
Windowpane	62	colder months
Soleidae		
Hogchoker	299	year around

Table 4-13. Concluded

Family-species	Total Number Captured	Seasonal Occurrence
Cynoglossidae Blackcheek tonguefish	2,391	year around
Monacanthidae Planehead filefish		warmer months
Tetraodontidae Northern puffer	6	year around
Diodontidae Striped burrfish	17	warmer months
Batrachoididae Oyster toadfish	49	year around

mummichog, mosquitofish, longnose gar, white catfish, hogchoker, and southern flounder. American eel and northern pipefish were found in the oligohaline area and further upstream in the freshwater habitat of tidal rivers.

During the period October 1970 through September 1973, Mahood et al. (1974a, b, c, d) conducted a survey of the distribution, seasonal abundance, size composition, and life histories of marine and estuarine fishes and certain invertebrates of coastal Georgia. This investigation included sampling of fishes by 40 ft (12.2 m) otter trawl [2 in (5 cm) stretched mesh] and monofilament gill nets [2 7/8 in (7 cm) and 4 in (10 cm) stretched mesh] in tidal creeks and sounds.

Trawl samples yielded representatives of 44 families of fishes. The catches were dominated by sciaenids (spot, Atlantic croaker, star drum, and weakfish), which made up 63% of the total catch, plus Atlantic menhaden, sea catfish, white catfish, blackcheek tonguefish, and spotted hake (Mahood et al. 1974a, b, c, d). Tidal creeks produced 48.5%, sounds 38.6%, and nearshore marine waters 12.9% of the total catch of fishes taken by trawl. Seasonal trends in abundance of fishes collected were noted, with peak catches occurring during July and August (probably a result of maximum seaward migration of fishes of a size susceptible to the gear used). Low catches were observed during December and January. Seasonal differences were also noted among the southern (St. Andrews Sound - St. Simons Sound), central (Doboy Sound - Sapelo Sound), and northern (Ossabaw and Wassaw sounds) sections of the State. Largest catches of marine and estuarine fishes were taken from fall to mid-spring in the southern section; catches were greatest in the

central and northern sections during late spring and summer.

Gill net sampling in tidal creeks and sounds produced 24 families of marine and estuarine fishes, mostly the adult and sub-adult stages of sciaenids (spotted seatrout, weakfish, spot, Atlantic croaker, red drum, southern kingfish), bluefish, sea catfish, Atlantic menhaden, Spanish mackerel, striped mullet, and crevalle jack (Mahood et al. 1974a, b, c, d). Again, sciaenids were by far the most abundant group, comprising 56.3% of the total 3-year catch.

b. Recruitment and Life History. Previous zooplankton sampling by various workers (Bears Bluff Laboratories, Inc. 1964; South Carolina Marine Resources Division, 1973 - 1978, Charleston, unpubl. data) within the subtidal estuarine habitat has provided some information on the seasonal recruitment patterns, sizes, and occurrence of various species of fish larvae and postlarvae. Peak recruitment of larvae and postlarvae of species which spawn offshore (e.g., spot, Atlantic croaker, Atlantic menhaden, pinfish, eels, flounders, and mullets) occurred from January through March. Spot and Atlantic croaker postlarvae dominated plankton catches during this period (Bears Bluff Laboratories, Inc. 1964). Due to the selective nature of the sampling gear, however, the abundance of common species such as mullet and Atlantic menhaden probably was not accurately indicated. During spring and summer, larvae and postlarvae of nearshore and estuarine spawners became dominant in the subtidal estuarine habitat. Dominant species included gobies, bay anchovy, silversides, catfish, and sciaenids such as star drum. Again, due to gear selectivity, common species such as ladyfish, Atlantic thread herring,

196

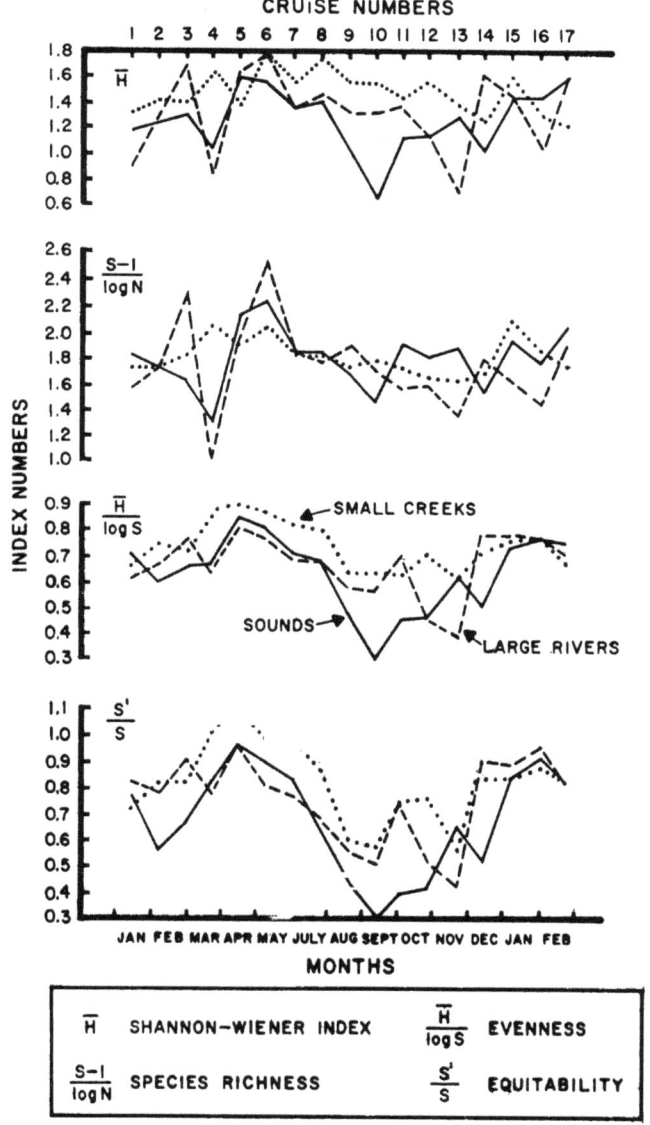

CRUISE NUMBERS

Figure 4-4. Annual cycles of species diversity indices for fishes in three Georgia estuarine zones (Dahlberg and Odum 1970).

197

hogchoker, and blackcheek tonguefish were not taken.

Powles and Stender (1976) reported on the species composition, numbers, and distribution of larval and postlarval fishes taken in ichthyoplankton survey cruises in 1973 off the South Carolina and Georgia coasts. Their data concerning species common to the estuaries, such as menhaden, hakes, mullets, bluefish, spot, and Atlantic croaker, correspond with the observed recruitment of larval and post-larval fishes into the estuarine area (Bears Bluff Laboratories, Inc. 1964; South Carolina Marine Resources Division, 1973 - 1978, Charleston, unpubl. data). Other information on the larval development and recruitment of various species, including mullets, ladyfish, pompano, and crevalle jack along the South Carolina and Georgia coasts, has been presented by several authors (Anderson 1957, 1958, Berry 1959, Gehringer 1959, Fields 1962).

Although a considerable amount of information has been amassed on the species occurrence, diversity, abundance, distribution, and life histories of the fishes occurring in the subtidal estuarine habitat, much information is lacking concerning reproduction, age and growth, population dynamics, and ecological relationships. Limited information is also available concerning the movements and migrations of those fishes in response to environmental factors. Data from previous tagging studies indicate some southerly coastal movement of several species (Atlantic croaker, southern kingfish, southern flounder) during fall and winter, while others (such as spiny dogfish) move northward in the spring (C. M. Bearden, 1971, South Carolina Marine Resources Division, Charleston, unpubl. data). Offshore movements during the fall and winter, or for spawning purposes, are known to occur in the case of many fishes, including sciaenids, mullets, Atlantic menhaden, flounders, and American eel, but very limited data exist concerning migratory patterns and offshore spawning locations. Brief life history summaries of some of the more common species are given below.

(1) Mullet. Early development, growth, and occurrence of striped mullet and white mullet have been described by Anderson (1957, 1958). Spawning of these species occurs offshore on the continental shelf near the Gulf Stream during October - February, with early development and growth up to a length of 20 - 25 mm (0.8 - 1.0 in) taking place offshore, after which they move into inshore waters (see Chapter Two). Growth is rapid, about 17 mm (0.7 in) per month, in estuarine waters of Georgia from April to November.

(2) Ladyfish. Gehringer (1959) described the early development and

metamorphosis of the ladyfish, another offshore spawner. Leptocephali of this species move into inshore waters at a length of 40 - 45 mm (1.6 - 1.8 in), where they undergo metamorphosis.

(3) Spot. Dawson (1958) and Music (1974) described the biology and life history of the spot in waters of South Carolina and Georgia. Both studies indicate that spot tolerate a wide range of salinity and temperature conditions, spawn offshore from October to March, and occur throughout the winter in estuarine waters. Length frequency data indicate that these fish reach 135 - 165 mm (5.3 - 6.5 in) by the end of their first year.

(4) Seatrout. Lunz and Schwartz (1970) and Mahood (1974) presented data on the distribution, seasonality, movements, reproduction, length frequencies, and abundance of seatrouts of the genus Cynoscion. In these fishes, spawning occurs in sounds or nearshore waters during summer. Young spotted seatrout spend their early lives in small tidal marsh creeks, whereas weakfish occupy the subtidal estuarine zone of larger streams for the most part. In many estuaries, spotted seatrout is the most common top carnivore, and thus it has little competition for an abundant food supply (Tabb 1966). Both young and adults are tolerant of rigorous environmental conditions, and the entire life cycle is spent within the estuary. Spotted seatrout are generally non-migratory, although they have been found to make seaward movements in response to freshets and low water temperatures (Tabb 1966).

(5) Kingfishes. Bearden (1963) reported that the southern kingfish, the most common species of kingfishes in the subtidal estuarine habitat, spawns offshore near the coast from May to July, and the young move into estuaries during the summer. Young-of-the-year were found to be eurythermal, whereas adults migrated to deeper water offshore during the colder months.

(6) Atlantic Croaker. The distribution and abundance of Atlantic croaker in estuarine waters of South Carolina were described by Bearden (1964). Spawning occurs well offshore primarily during winter, and then larvae and postlarvae migrate landward where they move up into tidal streams as far as salt water penetrates. Young-of-the-year croakers reach an average length of 150 mm (5.9 in) within one year. No significant decline in abundance of Atlantic croaker was noted over a 10-year period (1953 - 1962). Bearden (1964) reported that the size of Atlantic croaker varied over the estuarine salinity gradient, with the smaller fish occurring in lower salinity waters.

c. Trophic Relationships. Limited information is available concerning food habits and trophic relationships of marine and estuarine fishes found in the subtidal estuarine habitat of the characterization area. Results of several investigations in South Carolina and Georgia can be used, however, to make some generalizations along these lines (see also Table 2-12 in Chapter Two, Marine Ecosystem).

Of the more common fish species found within the subtidal estuarine zone, the predominant herbivores are mullets and menhaden. The food of striped mullet in various habitats has been described by Odum (1970a), who listed the principal items as micro-plant detritus and living algae. However, it is also known to shift trophic levels from primary consumer to predator, preying on polychaete worms (Bishop and Miglarese 1978). Menhaden postlarvae feed largely on zooplankton and function as primary carnivores when young, but undergo internal changes in the digestive system during metamorphosis. During the remainder of their lives, they are filter feeders, primarily consuming phytoplankton. Odum (1968) observed juveniles of striped mullet, white mullet, and Atlantic menhaden feeding heavily in a bloom of dinoflagellates (Kryptoperidinium) in the headwaters of the Duplin River, Georgia.

The predominant primary carnivore within the subtidal estuarine habitat is the bay anchovy, which feeds largely on copepods and other zooplankters (Odum 1970a). Most of the common young-of-the-year fishes found within the subtidal estuarine habitat (including star drum, Atlantic croaker, spot, silver perch, juvenile weakfish, flounders, hogchokers, blackcheek tonguefish, white catfish, and spotted hake) are opportunistic mid carnivores, feeding on a wide variety of planktonic and benthic organisms, including mysid shrimp, harpacticoid and calanoid copepods, amphipods, isopods, small crabs, shrimp, mollusks, polychaetes, and small fishes (Dawson 1958, Bearden 1964, Odum 1970a, Sikora et al. 1972, Stickney et al. 1974, Heard 1975, Kjelson et al. 1975, Kjelson and Johnson 1976, Stickney 1976). Top carnivores in estuarine subtidal waters include various species of sharks such as carcharinids and hammerheads (Bearden 1965, Hicks 1972), as well as longnose gar, Atlantic needlefish, striped bass, white perch, bluefish, ladyfish, Spanish mackerel, weakfish, seatrout, red drum, and flounders. The young of many of these species may function as mid carnivores or even primary carnivores as postlarvae and early juveniles. These species feed largely on smaller fishes including the herbivores, primary carnivores, and mid carnivores mentioned above, as well as on penaeid shrimp, grass shrimp (Palaemonetes spp.), crabs, squid, mollusks, and other motile and sessile invertebrates (Tabb 1966, Mahood 1974).

d. Effects of Perturbations. Investigations of the effects of environmental perturbations on fishes of the estuarine subtidal zone have dealt primarily with pesticide residue levels in tissues, the effects of pesticides on various species, and the effects of dredging and dredged materials.

In 1965, the U.S. Department of Health, Education and Welfare conducted pollution studies in Charleston Harbor. It was found that many industries, especially those on the Ashley River, were grossly polluting the area with chemical wastes. Mortalities of estuarine fishes (especially Atlantic menhaden), presumably caused by organophosphorous compounds, were a commonplace occurrence in the Ashley River system.

The U.S. Environmental Protection Agency (1971) conducted investigations on the accumulation and movement of mirex in selected estuaries of South Carolina. Mirex was found to move widely from treated areas into estuarine fishes, although no mortalities or other direct effects were noted during the study period. However, birds and other organisms which feed on fish contaminated with mirex can be affected adversely (Kaiser 1978). Markin et al. (1974) also reported on mirex and other organochlorine substances (DDT and metabolites, PCB residues) in flounder, weakfish, mullets, and Atlantic croaker samples from Savannah and Charleston. Mirex was found only in samples from Savannah, an area with a history of extensive use of this pesticide.

Reimold et al. (1973) conducted a 3-year research project to study toxaphene contamination in relation to estuarine ecology in Terry Creek and Duplin River estuary, Georgia. Toxaphene levels (ppb) were determined for muscle and liver tissue of a number of estuarine fishes, including striped mullet, American eel, mummichog, silversides, silver perch, weakfish, star drum, spot, kingfishes, menhaden, Atlantic cutlassfish, gobies, anchovies, flounders, blackcheek tonguefish, and hogchoker. Toxaphene concentrations in effluents from a plant manufacturing this chemical near Terry Creek decreased over one order of magnitude during the study period, and the toxaphene content of the flora and fauna, including estuarine fishes, decreased simultaneously. Concurrent with this was a significant increase in species diversity, as determined from otter trawl collections.

Reimold and Shealy (1976) monitored residue levels of a number of pesticides and mercury in young-of-the-year spot and silver perch from 11 estuaries representing all the major Atlantic drainage basins in South Carolina and Georgia. Chlorinated hydrocarbons for which analyses were conducted included DDT, DDE, TDE, dieldrin, endrin, polychlorinated biphenyls (PCB's), toxaphene, mirex, and chlordane. Also monitored were total mercury, phenoxy-herbicides, carbamate, and organophosphorous pesticides.

Low levels of DDT were detected in both spot and silver perch from several estuaries, but only prior to spring 1973. This disappearance may have reflected the 1972 ban on DDT use by the U.S. Environmental Protection Agency. However, the metabolities of DDT, namely DDE and TDE, were seasonally ubiquitous, albeit at relatively low concentrations, throughout the study. Residues of other chlorinated hydrocarbons (dieldrin, PCB's, toxaphene, chlordane, endrin, and mirex), phenoxy-herbicides, carbamate, and organophosphorous pesticides were not detected in concentrations above 10 µg/kg in the estuarine habitat during the 2-year study period. Mercury was detected during all seasons and in both finfish species monitored. During spring 1973, mercury levels exceeded the 500 µg/kg maximum concentration of mercury in food set by the Food and Drug Administration (Lepple 1973) by a factor of six. Mercury levels, although detectable, were lower and well within the FDA guidelines, both before and after spring 1973.

Martin (1980) investigated effects of two petroleum compounds, benzo(a)pyrene and methycholanthrene, on sheepshead minnows and channel catfish. Most of the effects on sheepshead minnows were gill-related, i.e., fungal infections or erosion and necrosis of gill lamellae. Other pathologies included hemorrhagic areas on body surfaces and hemorrhagic gills. Pathological conditions observed in channel catfish were similar to those noted in sheepshead minnows. In addition, channel catfish exhibited vertebral disorientations. Significant mortalities of both species occurred as a result of exposure to the petroleum compounds.

Further studies of impacts of man-made substances entering coastal estuarine systems and their biota are needed. Future monitoring activities should include sampling of different trophic levels at various geographic locations, particularly where significant contaminant concentrations are detected. Finfish monitoring should be continued and compared periodically to the baseline residue levels now established to determine whether these pollutants are increasing, decreasing, or maintaining a steady state in estuaries and in indicator finfish species of the Sea Island Coastal Region.

Stickney (1972) reported on the effects of intracoastal waterway dredging on ichthyofauna and benthic macroinvertebrates in Georgia. Although some short-term changes in population structure or standing crop appeared to have been associated with dredging, these effects lasted no longer than 1 or 2 months. Diversity indices failed to demonstrate any long-term effects of dredging on motile species. In South Carolina, Hoss et al. (1973) studied effects of dredged sediments on larval estuarine fishes common to Charleston Harbor. Survival and growth of the larvae were affected at the highest concentrations of extract tested, and different species were shown to have different survival rates. Larval flounder, spot, Atlantic menhaden, pinfish, and Atlantic croaker were used as test animals. Hoss et al. (1973) recommended that dredging be conducted, if possible, during periods when larval forms of more important species are not abundant in estuarine waters. Disposal of dredged materials on upland sites having impervious dikes was felt to be the least damaging type of disposal. Most previous investigations, including the aforementioned and those conducted by the South Carolina Marine Resources Division (Charleston, unpubl. data), indicate that dredging of unpolluted sediments, and the resultant increased turbidities, have little long-term effect on fishes. However, the effects of dredging sediments containing toxic materials are not well understood, and additional studies are needed to determine methods of removing and disposing of such material in a manner that will not adversely affect the biota.

Other environmental perturbations affecting estuarine fishes which are not yet fully understood and need additional investigation include the diversion of coastal streams, channelization and alteration of upland drainage patterns, marina construction, power plant siting along the coast, excavation in wetland areas, and other physical alterations in the coastal zone.

e. Anadromous Species. Six species of anadromous fishes occur in the estuarine waters of the Sea Island Coastal Region. These species are primarily transients in estuaries as they pass from the marine environment to the riverine systems during their spawning migrations. The juveniles of some species utilize estuaries as part of their nursery grounds, but most of their lives are spent in coastal marine waters.

Adult American shad traverse the estuaries in late winter/early spring, when water temperatures are approximately 10°C (50°F) to reach their spawning grounds in the coastal rivers. Juveniles are found in estuarine waters during autumn as they migrate to sea after spending approximately 6 months in fresh water.

Hickory shad enter the estuaries as adults in mid-January on their way to freshwater spawning grounds (Street 1970). Juvenile hickory shad leave the upriver spawning areas much earlier than other species of Alosa (Mansueti 1962, Pate

1972). Crochet (1976), sampling the Waccamaw and Pee Dee river systems in South Carolina, found juvenile hickory shad only in lower Winyah Bay. Godwin and Adams (1969) and Street (1970), who worked on the Altamaha River in Georgia, also found the majority of juvenile hickory shad concentrated in the lower reaches of the estuary.

Adult blueback herring migrate through the estuaries from early January through the spring as they make their spawning runs into coastal rivers (Street 1970). Juvenile blueback herring are found in the estuaries during very early summer, but migrate upstream by July and remain there until the fall when they migrate to the ocean. Street (1970) found juveniles in estuarine areas of the Altamaha River during early June (before they moved upstream), and again in November when water temperatures in the river dropped.

Striped bass adults migrate through the estuaries in late winter and early spring when water temperatures are near 14°C (57°F), enroute to spawning grounds in coastal rivers. Populations of adult striped bass south of Cape Hatteras usually return to adjacent sounds after spawning (Trent and Hassler 1968). Smith (1970) found in the Savannah, Ogeechee, and Altamaha rivers of Georgia that subadult and adult stripers utilize the estuaries during the fall, winter, and early spring months. He also found that fingerling striped bass utilize the estuarine portions of those rivers where salinities are 3.2°/oo or less. Smith (1968) collected juveniles only at estuarine stations in the Altamaha River, indicating that this was probably the major nursery area for these fish.

Adult Atlantic sturgeon enter the estuarine areas of Georgia and South Carolina coastal rivers as early as February on their way to spawning grounds in brackish and tidal freshwater regions of the rivers. This immigration coincides with Huff's (1975) report that spawning migrations begin during February in Florida, Georgia, and North Carolina.

The eggs are laid in brackish or fresh water (brackish water possibly preferred) over a hard bottom of clay, rubble, gravel, or shell in shallow running water or in water up to 30 ft (9 m) deep (Vladykov and Greeley 1963). The eggs are demersal, adhesive, and occur in stringy clusters or ribbons. Smith (1907) found North Carolina sturgeon capable of producing 1 million to 2.5 million eggs. Ryder (1890) estimated fecundity of Delaware River sturgeon to range from 0.8 million to 2.4 million eggs. Vladykov and Greeley (1963) examined a St. Lawrence River specimen that contained approximately 3.8 million eggs.

After spawning, adult sturgeon may stay in the riverine or estuarine system to feed for 2 to 3 months before returning to the ocean. However, some adults, or those in certain areas, may return to the sea very soon after completion of spawning activity. A population of the Gulf of Mexico subspecies A. oxyrhynchus desotoi in the Suwannee River, Florida, does not feed in the river, but returns to the sea shortly after spawning (Huff 1975).

After hatching, the young sturgeon will remain in the riverine and estuarine systems up to 5 years (W. Dovel, 1978, Oceanic Society, West Rectory Church, Yonkers, New York, pers. comm.). The juvenile sturgeon then migrate to the ocean, where they undergo an accelerated period of growth prior to returning to their natal stream to spawn at an age of approximately 10 - 15 years. Juveniles spend the spring and summer in fresh water, moving into estuarine areas to overwinter as water temperatures drop in the fall (Murawski and Pacheco 1977).

Data on Atlantic sturgeon populations in South Carolina and Georgia are extremely limited. Based on commercial fisheries data, the South Carolina rivers in which sturgeon are most abundant are as follows, in decreasing order of importance: Winyah Bay area (Waccamaw and Pee Dee rivers), Santee River, Edisto River, Savannah River, and Ashepoo and Combahee rivers. Little information is available on Georgia Atlantic sturgeon populations, as the fishery is inconsequential; however, the Altamaha, Ogeechee, Savannah, and St. Marys rivers are believed to be the most important sturgeon areas in the State.

Presently, the South Carolina Wildlife and Marine Resources Department's Marine Resources Division and the U.S. Fish and Wildlife Service are conducting an investigation of sturgeon populations in South Carolina (D. E. Marchette and T. I. J. Smith, 1980, South Carolina Marine Resources Division, Charleston, pers. comm.). This project will provide much needed information on the life history and commercial importance of this species.

Preliminary data from this study indicate that spawning of Atlantic sturgeon occurs as early as mid-February in the ocean off South Carolina. Ripe females can be taken in the ocean around the mouths of rivers by early to mid-March when water temperatures reach 10° - 12.6°C (50° - 55°F).

Ripe females captured to date averaged 16 years (range 11 - 27 years) and had an average total weight of 75 kg (165 lb). Ripe males averaged 12 years (range 5 - 29 years) and had an average total weight of 40.9 kg (90 lb). The average age of all fish sampled in the ocean fishery was 13.5 years (range 5 - 29 years) and mean total weight was 56.3 kg (124 lb). Average fecundity for spawning females averaged 732,500 eggs, with a range of 370,000 to 1,100,000.

Exact spawning locations are difficult to ascertain; however, the following ideal spawning sites have been identified by research personnel and commercial fishermen: the Tee of the Cooper River, the marl outcroppings at Jamestown on the Santee River, the marl hole above Givhan's Ferry on the Edisto River, Jordan Lake (an oxbow) on the Great Pee Dee, the junction of the Great and Little Pee Dee rivers, and Bull Creek, a connecting waterway between the Pee Dee and Waccamaw rivers. Investigations are currently underway to determine if any of these sites are, in fact, spawning localities for Atlantic sturgeon.

The endangered shortnose sturgeon is found in estuarine and riverine areas of South Carolina and Georgia (see Section VI in Chapter One). Apparently, this species is primarily riverine and estuarine in its distribution, and it is not known to make regular migrations to the ocean. The abundance and ecology of this species are poorly known at present. It is possible that a substantial number of the juvenile sturgeon reportedly caught in shad nets are shortnose rather than Atlantic sturgeon.

5. Reptiles

The only reptile truly characteristic of the estuarine subtidal region, and primarily restricted to this habitat, is the Carolina diamondback terrapin. Although turtles of the families Cheloniidae and Dermochelidae also occur in the estuaries, they are primarily marine forms. Certain freshwater reptiles have been noted in brackish water areas, but their incursions into the estuaries will be covered in the sections dealing with riverine, lacustrine, and palustrine habitats (Chapter Five).

Seven races of the diamondback terrapin are recognized, ranging from Massachusetts to south Texas. The race occupying waters of South Carolina and Georgia is Malaclemys terrapin centrata. DeSola and Abrams (1933) noted that other races have been found in Georgia because of the escape of individuals from commercial terrapin farms and their colonization of the surrounding marshes. It is unknown if any vestiges of the other races still exist on the Georgia coast.

Although this turtle can survive for prolonged periods in fresh water and has been noted to occur in coastal rivers above the influence of salt water (Coker 1906), it is most abundant in the creeks and estuaries of the coastal salt marsh, particularly over shell bottoms and near oyster bars. Preferred food items of Carolina diamondback terrapins in the area of Beaufort, North Carolina, are a snail (Littorina irrorata), fiddler crabs (Uca spp.), annelid worms, and smooth cordgrass (Coker 1906). Although cordgrass was found in all stomachs examined, Coker believed that it was taken incidentally by turtles eating Littorina.

Diamondback terrapins have a life span in excess of 40 years, reaching sexual maturity at an average age of 6 years (Hildebrand 1932). Nesting occurs on any sandy area rising above mean high water near salt marsh. Diamondback terrapins nest annually, laying as many as five clutches of eggs in a season (Coker 1951). Average clutch size is 8 or 9 eggs, but as many as 16 eggs have been found in a single nest. In the Beaufort area, North Carolina, hatching occurs 8 - 9 weeks after laying (Hay 1917).

In the late 1800's and early 1900's, the diamondback terrapin was hunted intensively for the gourmet food trade. Initially, the greatest demand was for northern diamondback terrapins (Malaclemys terrapin terrapin), but declining populations of this race opened the market for Carolina diamondback terrapins, which were also seriously depleted by market hunters. The market for the diamondback terrapin collapsed after World War I and never made a significant comeback. Although no published accounts of the present population status of this terrapin are available, casual observations indicate that they are relatively abundant in coastal South Carolina and Georgia.

A small scale and highly seasonal fishery for diamondback terrapins is presently conducted near Beaufort, South Carolina. Turtles are held in tidal enclosures from summer, when the fishery operates, until late fall and early winter, when demand peaks. Primary market outlets for these turtles are in Baltimore and New York.

Observations by Burger (1976) in New Jersey indicate that while man is no longer a significant predator of diamondback terrapins in most areas, predation of eggs and hatchlings by raccoons and foxes is considerable, e.g., 60% of observed nests were disturbed by these two predators. Based on experience with loggerhead turtle nests, it is likely that the major predator of Carolina diamondback eggs in the Sea Island Coastal Region is the raccoon. To date, however, no studies have been conducted on population levels, nesting activity, or predation within the characterization area.

In addition to the Carolina diamondback terrapin, estuarine waters are also frequented by the typically marine species of turtles. The extent to which marine turtles utilize estuarine waters of sounds, bays, and tidal creeks is presently unknown,

202

and our data are limited to scattered observations and incidental captures by fishermen. Atlantic loggerhead turtles are reported to ascend tidal creeks to the freshwater zone and above, and it is believed that they are abundant in estuarine waters during summer months. The taking of a mature Atlantic leatherback and a juvenile Atlantic green turtle in estuarine waters of South Carolina was noted in Chapter Two.

6. Birds

The estuarine subtidal or open water system is utilized primarily by birds for resting and feeding. In terms of trophic relationships, the primary groups occurring in this habitat are the scavengers, piscivores, benthivores, and occasional insectivores (Table 4-14; Fig. 4-5).

Dominant scavengers include the herring gull, ring-billed gull, and laughing gull. The piscivores, which should be assigned dominance status, include the brown pelican, red-breasted merganser, and royal tern. In contrast to the scavengers and fish-eating birds, benthivores including the surf scoter, black scoter, and canvasback feed on mussels, clams, algae, etc. Occasional insectivores occurring in this habitat include the laughing gull, ring-billed gull, gull-billed tern, and black tern.

Gulls are considered to occupy the upper trophic levels in estuarine subtidal open waters. Although gulls are the most conspicuous group of birds found on the coasts of South Carolina and Georgia, very little information has been gathered on their population dynamics. Forsythe (1973) studied seasonal fluctuations and numbers of gulls in the Charleston area. He found seven species, laughing, herring, ring-billed, Bonaparte's, great black-backed, black-headed, and Iceland gulls, within a 500 mi^2 (1,295 km^2) plot during the study year (1971-72). Of these, laughing, ring-billed, and Bonaparte's were the most common. About 2,200 laughing gulls were summer residents in the Charleston area, but over 3,000 were present during the fall migration. Approximately 3,600 ring-billed gulls wintered in the area, and slightly lower numbers of herring gulls were observed during the same period. Bonaparte's gulls were found from November through March, with about 400 present during February.

Gulls and terns known to nest in the study area include the laughing gull, gull-billed tern, Forster's tern, least tern, royal tern, Caspian tern, and sandwich tern. Of these, only the gull-billed tern, least tern, and royal tern have been reported as nesting in coastal Georgia (Shanholtzer 1974a). This gap in nesting occurrence for the Georgia coast is both unexplained and of great interest. Other marine birds which breed along the Atlantic

north to South Carolina but not in Georgia include the brown pelican and the double-crested cormorant.

The herring gull is the most widely distributed gull of the northern hemisphere and is probably best known as a scavenger. As such, this species renders a great service in keeping harbors and beaches relatively free from dead fish and refuse. The herring gull also captures small live fish by plunging headlong into the water. Frequently, these birds are seen in open waters feeding on schools of menhaden. Gulls feeding in this manner usually do so in flocks; Bent (1963a) gave a detailed description of such behavior.

The smaller ring-billed gull is very similar in habits to the herring gull. It feeds mostly on fish in the open water estuarine habitat, but has been observed in a variety of feeding activities. For example, it is often seen inland following plows and tractors in the field, picking insects from the ground. Also, this species feeds on field mice and other small rodents (Bent 1963a). Ring-billed gulls were observed trying to rob red-breasted mergansers of fish by Bent (1963a), who also stated that they may extensively damage eggs of other species. Meyerriecks (1965) reported observing approximately 1,500 ring-billed gulls feeding on fiddler crabs in Florida. He noted that several herring gulls and hundreds of laughing gulls were within visual and auditory range of the feeding activity but were not attracted to the food supply by the calls or behavior of the ring-billed gulls.

The laughing gull is often seen in estuarine open waters and feeds on a variety of organisms. Bent (1963a) and Sprunt and Chamberlain (1970) commented on its piratical nature, particularly directed toward the brown pelican. Bent (1963a) stated: "Wherever a number of pelicans are diving and feeding these gulls are apt to gather in large numbers and with their warning cries of 'half, half, half,' to share in the feast. As soon as the pelican appears above the surface with a pouch full of small fry one or another of the gulls attempts and often succeeds in alighting on the pelicans head and helping itself to the bountiful supply in the capacious pouch." The laughing gull, like the ring-billed gull, is an insect eater too, and also may be found in agricultural lands.

The royal tern is very common in coastal Georgia and South Carolina, especially in the estuarine open waters where it feeds entirely on fish. One of the largest nesting colonies ever found in South Carolina occurred on Cape Island, just south of the Santee River Delta (Sprunt and Chamberlain 1970). This species also nests on "Egg Bank" in St. Helena Sound, and Chamberlain (1962) reported an estimated 6,000 nests on Deveaux Bank. The only place the royal tern is

Table 4-14. Birds of the estuarine subtidal system in South Carolina (Sprunt and Chamberlain 1949, Audubon Field Notes 1967 - 1970, Chamberlain 1968, American Birds 1971 - 1977, Forsythe 1978).

DOMINANT				MODERATE				MINOR		
Brown pelican	C	PR		Common loon	C	WR	Oct. - April	Pied-billed grebe	C	PR
Double-crested cormorant	C	PR		Red-throated loon	FC	WR	Oct. - April	Redhead	U	WR	Nov. - Mar.
Lesser scaup	C	WR	Oct. - April	Horned grebe	C	WR	Oct. - March	Canvasback	FC	WR	Nov. - Mar.
Surf scoter	C	WR	Oct. - June	Ring-necked duck	C	WR	Oct. - April	Common goldeneye	FC	WR	Nov. - Mar.
Black scoter	C	WR	Oct. - May	Hooded merganser	C	WR	Nov. - April	Bufflehead	FC	WR	Nov. - April
Red-breasted merganser	C	WR	Oct. - April	Osprey	U	PR	Oct. - May	Ruddy duck	C	WR	Oct. - April
Herring gull	C	PR		Bonaparte's gull	C	WR	Oct. - May			
Ring-billed gull	C	PR		Gull-billed tern	FC	SR	April - Oct.			
Laughing gull	C	PR		Common tern	FC	PR				
Forster's tern	C	PR		Least tern	C	SR	Mar. - Oct.			
Royal tern	C	PR		Sandwich tern	FC	SR	April - Oct.			
				Caspian tern	FC	PR				
				Black tern	C	SR	May - Sept.			
				Black skimmer	C	PR				

Note: Dominance indicates relative importance of the species as a group in the community. This concept is not based necessarily on taxonomic relationships but rather on numbers, size, and trophic dynamics.

KEY: C - Common, seen in good numbers
      FC - Fairly common, moderate numbers.
      U - Uncommon, small numbers irregularly
      PR - Permanent resident, present year around
      WR - Winter resident
      SR - Summer resident

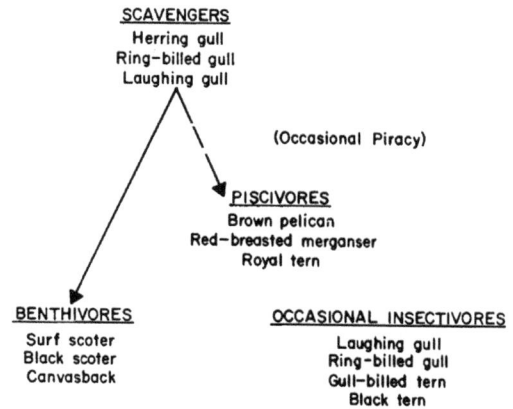

SCAVENGERS
Herring gull
Ring-billed gull
Laughing gull

(Occasional Piracy)

PISCIVORES
Brown pelican
Red-breasted merganser
Royal tern

BENTHIVORES
Surf scoter
Black scoter
Canvasback

OCCASIONAL INSECTIVORES
Laughing gull
Ring-billed gull
Gull-billed tern
Black tern

Figure 4-5. Generalized trophic relationships of representative birds
of the estuarine subtidal system.

known to nest in Georgia is on Little Egg
Island near Altamaha Sound (Kale et al.
1965). This represents the southern-most
colony of the royal tern on the Atlantic
coast. The only other accounts of this
species nesting in Georgia are Burleigh's
(1958) report of nesting on Blackbeard
Island in 1914, and the finding of a single
egg from Oysterbed Island at the Savannah
River entrance in 1933 by G. R. Rossignol
(Tomkins 1934).

Although not restricted to the estuarine habitat, the brown pelican is observed frequently in the estuarine open
waters and bays. Apparently, two breeding
colonies have sustained the brown pelican
population of the Sea Island Coastal Region,
one in the Cape Romain National Wildlife
Refuge and another on Deveaux Bank south of
Charleston. The reader is referred to the
marine subtidal system in Chapter Two and
the bird key and bank communities in Chapter Three for further discussion on this
species.

Waterfowl of estuarine open waters
and bays include the pochards or diving
ducks, dabbling ducks, and sea ducks.
Among the more common pochards known to
utilize this habitat are scaup, mergansers,
ring-necked ducks, ruddy ducks, redheads,
and canvasbacks (Table 4-14). These species generally feed in relatively deep
waters on both animal and plant matter.

The most frequently observed dabbling
duck is the baldpate or American wigeon,
which occurs in small flocks throughout
the open waters of the coastal area.

Although Sprunt and Chamberlain (1970)
stated that these birds are largely confined to freshwater situations, M. D.
McKenzie (1978, South Carolina Marine Resources Division, Charleston, unpubl. data)
has frequently observed small flocks of
baldpates feeding on what appeared to be
sea lettuce (Ulva) in subtidal open waters
of Charleston Harbor. Of the sea ducks
occurring in this habitat, the surf and
black scoters are the more common. Great
flocks of black scoters have been reported
in Bulls Bay (Sprunt and Chamberlain 1970).
These birds dive deeply and feed on mollusks and shellfish. The surf scoter, although common in estuarine open waters, is
at home on the front beaches of barrier
islands where old trees have fallen into
the surf. These birds feed on barnacles,
mussels, and other marine growth on these
fallen trees (Sprunt and Chamberlain 1970).

The osprey is the most common bird of
prey in the estuarine subtidal habitat.
However, it cannot be considered a dominant
since it prefers other habitats, as discussed elsewhere (Chapter Five).

Other species which play moderate and
minor roles in this habitat are listed in
Table 4-14. Obviously, there is considerable overlap in bird utilization between
the estuarine and marine subtidal habitats.
Both serve as feeding and resting grounds
for marine birds. Many of the large bays
and sounds of South Carolina and Georgia
serve as refuges during the waterfowl season when estuarine impoundments are hunted.
Large flocks of mixed waterfowl species

occur throughout the estuarine subtidal waters during shooting hours.

The impact of man on bird populations of the estuarine subtidal system cannot be separated from that of the estuarine intertidal emergent wetlands, and will be treated there (see also Chapters Two and Three for discussion of man's impact on marine birds).

7. Mammals

Although there is substantial mammalian activity along the edges of estuarine open waters and bays, only two mammal species, the Atlantic bottle-nosed dolphin and the river otter, are sufficiently aquatic and abundant to be considered rather consistent components of the faunal complex of this habitat.

The Atlantic bottle-nosed dolphin is the dominant mammal of estuarine bays and open waters as well as the nearshore marine habitat. Along the coastline, dolphins can be expected in every bay, inlet, and river mouth. Its feeding niche in these waters is essentially the same as in the nearshore coastal waters where dolphin prey on a variety of fishes, with striped mullet and Atlantic menhaden probably constituting the bulk of its diet. Also, as Gunter (1950) has reported, dolphins consume penaeid shrimp and probably other crustaceans. However, it is unlikely that the dolphin exerts a significant population constraint on any fish or crustacean species within the Sea Island Coastal Region (see Chapter Two for more information on dolphins).

The river otter, although not nearly as aquatic as the dolphin, does enter estuarine open waters and bays and feeds in subtidal waters. A denizen of rivers, ponds, lakes, and salt marshes, it is common in South Carolina (Sanders 1978). Otters are also fairly common on the coastal plain and in the salt marshes of Georgia, but are rare in the Piedmont (Golley 1962). Wilson (1954) reported that fish were the preferred food of otters in North Carolina marshes, appearing in 91% of the samples taken. Crustaceans were the next preferred food (39%), with insects, birds, muskrats, and clams being eaten to a lesser extent. The predator pressure exerted by otters is probably not sufficient to significantly affect prey populations. The river otter reaches sexual maturity in 1 year, and breeding probably occurs in late fall or winter. The gestation period is about 2 months, but delayed implantation may extend it up to 270 days (Lowery 1974). The litter, consisting of three to four, is born in April and there is only one litter per year.

The otter is probably not subjected to any significant predation when in this environment. Elsewhere, however, trappers may be an important cause of mortality since otter fur is highly prized (e.g., in the 1972-73 trapping season in Louisiana, an otter pelt brought $42). Otter fur is the standard by which all other furs are measured as to texture and durability. In some subtidal areas, otters may become entangled in underwater fishing gear and drown. Crab pots are numerous in the Sea Island Coastal Region, but no documentation exists that they constitute a source of mortality for the otter. Young otters may occasionally be taken by great horned owls, other birds of prey, or larger carnivores.

Several other mammals infrequently appear in this habitat, but they play no consistent role in this compartment of the ecosystem. On rare occasion, a harbor seal will appear in a given bay and remain for a few days or weeks. The West Indian manatee occasionally straggles as far north as the Santee River. When manatees appear, they are not likely to remain in the open bays since the rooted subtidal vegetation on which they subsist is generally lacking.

D. DECOMPOSERS - BACTERIA AND FUNGI

In contrast to open seas, neritic waters and estuaries are populated by bacteria and fungi which are normally classified as weakly halophilic (salt-loving) or merely halotolerant (salt tolerant) (Larsen 1962). Although there are species which are unique to brackish waters (Rheinheimer 1970), most bacteria and fungi common to neritic waters are allochthonous (derived from marine and freshwater populations).

Bacteria of estuarine and neritic waters are normally gram-negative, pleomorphic (occurring in a variety of forms) assemblages of motile (flagellated), non-spore forming facultative anaerobes (Zobell 1946, Wood 1965, 1967, Wiebe and Hendricks 1974). In addition, some terrestrial bacteria which are capable of growth in aquatic media are viable in coastal water for limited periods (Rheinheimer 1974). Soil bacteria also play an important role in the flora of river waters, including the estuarine environment. Thus, species of Azotobacter and the nitrifying bacteria Nitrosomas and Nitrobacter are common in estuaries (Brooks et al. 1971), though their principal habitat is arable soil. In addition, tidal and nontidal river systems often support large populations of the genera Vibrio, Spirilla, Thiobacilli, Micrococci, Flavobacterium, Cytophaga, Sarcinae, and Spirochaetes (Murchelano and Brown 1968, 1970, Kaneko and Colwell 1973, Cook and Goldman 1976).

As Stevenson and Erkenbrecher (1976) have pointed out, estuarine waters normally support larger bacterial populations with greater heterotrophic potential than do oceanic and some inland waters. They listed 4 major fitness traits which may contribute to the persistence of certain bacteria in estuarine environments: 1) tolerance of sea salts and associated heavy metals, 2) competitive efficiency in utilizing organic nutrients at relatively low concentrations, 3) ability to attach to and colonize particulate material, and 4) dormancy.

Relatively little is known about the natural bacterial flora of estuarine waters in the Sea Island Coastal Region. However, a few studies have been conducted in the North Inlet estuary near Georgetown, South Carolina. Sizemore et al. (1973) reported that 44% - 62% of the bacteria they isolated from the North Inlet area were proteolytic. They observed no major seasonal differences in numbers of proteolytic bacteria but found a higher percentage of proteolytic forms in the water than in the sediments. Stevenson et al. (1974) reported that the concentration of bacteria in a salt marsh creek was lowest (2 x 10^3 bacteria/ml) at maximum high tide and highest (2.2 x 10^4/ml) at low tide. They suggested that this tide-associated difference in bacterial density was due to the flushing of sediment- and detritus-associated bacteria into the water column as the tide ebbed across the marsh. Similarly, Erkenbrecher and Stevenson (1975) found that densities of aerobic, heterotrophic bacteria were greatest at two saltmarsh creek stations in the North Inlet area just before low tide. Levels of particulate organic carbon (POC) and ATP also increased during ebb tide. Later, Erkenbrecher and Stevenson (1977) studied effects of a number of variables (time, depth, temperature, current velocity, salinity, dissolved oxygen level, pH, ATP, and POC) on tidal fluctuations in bacterial populations. Flood tide waters were characterized by little relationship among the variables, while ebb tide waters exhibited significant relationships. These results appeared to support the idea that the increase in bacterial density during ebb tide was primarily due to suspension of marsh sediment bacteria. This is further supported by the fact that the number of bacteria in estuarine sediments is greatest in the surface slime and decreases with depth. This surface slime would be most readily disturbed by tidal waters.

In Florida, Ahearn et al. (1977) reported that the surface microlayer (top 10 μm) of estuarine waters contained populations of heterotrophic microorganisms (bacteria, yeasts, filamentous fungi) 1 to 2 orders of magnitude greater than the populations existing at a depth of 10 cm (3.9 in). They further found that pesticides, which often tend to become concentrated in surface slicks, inhibited the growth and metabolism of many of the surface layer microorganisms.

Although human enteric bacteria and viruses generally are not part of the normal microflora of estuaries, they are often introduced into this environment through the widespread use of estuaries for sewage disposal and through runoff. Thus, a variety of pathogenic microorganisms may occur in estuarine areas where they may pose public health and other problems. (See the Atlas for shellfish grounds which have been closed due to sewage pollution).

The density of specific enteric and pathogenic microorganisms in a given estuary depends on a variety of interacting factors, including the degree of sewage treatment, the ability of the organisms to survive in estuarine waters, and the sizes and characteristics of the human and animal populations around the estuary. Recent work by Colwell and Kaper (1978) in Chesapeake Bay have shown the incidence and survival of fecal streptococci, fecal coliforms, and such pathogenic forms as Salmonella spp. and Clostridium botulinum to be much greater than previously believed. These authors further reported that infectious human enteroviruses (such as poliovirus) could be recovered after exposure to estuarine and marine waters for 46 weeks. Temperature was the main factor affecting survival of the viruses; viral persistence was much greater during colder months. Colwell and Kaper (1978) suggested that, under conditions of environmental stress, the indigenous microflora of estuaries and coastal waters can be replaced by introduced species, many of which may be pathogenic to man.

In studies of estuarine residential canals in Galveston, Texas, Gerba and McLeod (1976) and Smith et al. (1978) demonstrated that fecal coliforms and enteric viruses survived much longer in bottom sediments than in the water column. Later, Goyal et al. (1979) showed that a significantly higher number of transferable drug-resistant bacteria, both pathogenic and non-pathogenic, occur in the canal sediments than in the overlying water. Goyal et al. (1977) suggested that the canal sediments may serve as "long-term reservoirs" of fecal bacteria; these sediments could be resuspended through the action of storms, dredging, boating, etc., and thus recontaminate the water.

In addition to introduced species, some pathogenic bacteria occur naturally in estuaries. Among the best known of these is Vibrio parahaemolyticus, which has been studied extensively in Chesapeake Bay by Colwell, Kaneko, and others (see Colwell and Kaper 1978 for references). This pathogen causes gastroenteritis, and within the bay it has been isolated from water, sediments, zooplankton, oysters, soft clams, and blue crabs. In Chesapeake Bay, V. parahaemolyticus is found only in estuarine waters; it has not been isolated from the freshwater zone nor will ocean-strength seawater support its growth. This bacterium exhibits a distinct seasonal cycle, being restricted to the sediments during winter and occurring in maximum numbers in water, zooplankton, and sediments during summer (Colwell and Kaper 1978). Although it is best documented from Chesapeake Bay, V. parahaemolyticus probably occurs in other estuaries along the Atlantic coast, including those of the Sea Island Coastal Region.

In addition to V. parahaemolyticus, Colwell and Kaper (1978) noted that the causative agent of cholera, V. cholerae, has been isolated in low numbers from Chesapeake Bay. It is found only at salinities between 5⁰/oo and 15⁰/oo, and it appears to be a ubiquitous member of the brackish-water microflora. In the bay, V. cholerae has been isolated from water, sediment, and oysters.

Fungal populations of estuarine waters in the Sea Island Coastal Region have been even less studied than bacteria. Kohlmeyer (1978) compiled a list of 14 species of marine fungi from South Carolina. Johnson and Sparrow (1961), summarizing Hohnk's work on estuarine fungi in the 1950's, reported a decrease in the number of lower fungi (the "phycomycetes") as salinity decreased, and a subsequent increase in the Ascomycetes and Fungi Imperfecti. In a study of a North Carolina estuary, Johnson (1967) found very few Phycomycetes distributed throughout the system. Only four species, a Rhizophydium, an Olpidium, and two species of Pythium occurred throughout the estuary up to salinities of 32⁰/oo. He showed no evidence for any distributional patterns in relation to pH or concentrations of dissolved oxygen, nitrates, or phosphates.

Despite our very limited knowledge of them, the microorganisms of estuaries are extremely important. The roles of bacteria in biogeochemical cycling of essential nutrients is described briefly in Chapter One and reviewed by Zobell (1973). Burchard (1972) demonstrated that a variety of autotrophic and heterotrophic bacteria were involved in the cycling and mineralization of carbon, nitrogen, sulfur, and phosphorus in the Chesapeake Bay, but he did not determine whether these bacteria were indigenous to the estuary or introduced from terrestrial sources. Coull (1973) noted that microorganisms serve as direct food for meiofauna and rapidly degrade dead meiobenthos. Bacteria also play a major role in governing the distribution of meiofauna. Most meiofauna select sands coated with a bacterial film, and many meiofaunal species can distinguish between sands inhabited by a different bacteria. Individual species are attracted to sediments inhabited by preferred bacteria. Thus, within a given sediment, the distribution of bacteria plays a major role in controlling the distribution of meiofauna.

Although some advances in knowledge of estuarine microorganisms have been made in recent years, the major data gaps remain those listed by Zobell (1973) in his review. These include the effects of estuarine microbes on pollutants such as pesticides, oils and surfactants, and quantitative data on microbial impact on nutrient cycling in estuaries, the physiochemical conditions of various estuarine environments, and the effects of estuarine microorganisms on the health and well-being of higher organisms, including man and species of commercial importance, in estuaries.

III. INTERTIDAL ESTUARINE SUBSYSTEM – EMERGENT WETLANDS

The intertidal subsystem " . . . includes that part of the Estuarine System in which the substrate is exposed and flooded by tides. It also includes the associated splash zone" (Cowardin et al. 1977). This subsystem is dominated by salt marshes and flats in the Sea Island Coastal Region of Georgia and South Carolina. Emergent wetlands are characterized by erect, rooted, herbaceous hydrophytes (Cowardin et al. 1977). In estuaries, such habitats may occur from the mouths of rivers and other tidally influenced bodies of water landward to the 0.5⁰/oo isohaline. The salt marshes of the study area are the most extensive on the Atlantic seaboard. Spinner (1969) estimated that wetlands cover over 500,000 acres (202,347 ha) in South Carolina and nearly 400,000 acres (161,878 ha) in Georgia.

The intertidal zone is typically an area of high environmental stress and low species diversity. According to Cooper (1974), salt marshes are tidally stressed environments which are subjected to rapid diurnal changes. Cooper considered the major factors limiting the occurrence of species to be salinity, drainage, and temperature. Another factor which might be added to this list is desiccation. Although environmental stresses are high in the estuarine intertidal subsystem, biological productivity is also very high. Salt marshes are believed to be among the most productive natural areas on earth (Schelske and Odum 1962). Detritus from decomposing saltmarsh plants is believed to form the primary basis of the food chain in estuaries of the area (Schelske and Odum 1962, Teal 1962). Accordingly, such wetlands are of great importance to waterfowl, invertebrates, and fishes (Spinner 1969).

A. PRODUCERS

1. Nonvascular flora

The taxonomy of estuarine microphytes and macrophytes inhabiting the benthos of

208

intertidal estuarine systems has not been well studied. The edaphic algal flora of the mud flats, marsh pannes (unvegetated sand flats), creek banks, and soils in beds of halophytic angiosperms is mainly composed of small pennate diatoms and blue-green algae. Yellowish-green films of the alga Vaucheria (Taylor 1969, Simons 1974) and euglenoids are frequently encountered. D. R. Wiseman (1978, College of Charleston, Charleston, unpubl. data) identified the following species of edaphic blue-green algae in the Leadenwah Creek area, South Carolina: Agmenellum thermale, Anacystis dimidiata, A. montana var. montana, A. marina, Gomphosphaeria appollina, Entophysalis conferta, E. deusta, Anabaena oscillarioides, Calothrix crustacea, Microcoleus lyngbyaceus, Oscillatoria lutea, O. submembranacea, Prophyrosiphon notarisii, Schizothrix arenaria, S. calcicola, and Spirulina subsalsa. Ralph (1977) suggested that the myxophycean (blue-green algae) associations found in the marsh areas may form " . . . a single ubiquitous, endemic temperate North Atlantic blue-green algal association." Comparison of the above list of blue-greens with Ralph's list from the marshes of southern Delaware confirms his hypothesis.

The other major edaphic element of the salt marsh is the pennalean (bilaterally symmetrical) diatom community. This community remains incompletely surveyed in South Carolina. Thin layers of diatoms typically occur on macroscopic algae, oyster shells, and submerged marsh vegetation. Carter (1932, 1933) described the algal flora of two salt marshes in Britain and was one of the first to recognize nonvascular floristic zones in tidal marshes. Hustedt (1955) described marine littoral diatoms from a small number of samples collected near Beaufort, North Carolina. Williams (1962) studied the ecology of diatom populations in the salt marshes near Sapelo Island, Georgia. He described 79 species and three varieties of diatoms, and speculated that intertidal diatom populations probably contain well over 400 species at certain times of the year. Williams concluded that the dark green to golden brown film which is so noticeable over much of the marsh sediments throughout most of the year consists of motile microscopic forms - primarily pennate diatoms and secondarily filamentous blue-greens and euglenoids. He listed four diatom genera, Cylindrotheca, Gyrosigma, Navicula, and Nitzschia, as most important in both Spartina marsh and bare mud intertidal areas. The principal difference in diatom populations of bare mud and vegetated areas of marsh seems to be in seasonal variability. Williams noted relatively large short-term variation on mud flats and less pronounced seasonal population dynamics. He ascribed the appreciable seasonal variability in vegetated areas to changes in grazing pressure. Table 4-15 lists the distribution of diatoms in various habitats investigated by Williams (1962). More recent studies of saltmarsh algal taxonomy have been completed in Delaware. There, Ralph (1975), Somers (1975), and Sullivan (1975, 1976) have described the diatom and blue-green components of edaphic algal communities. Sullivan (1975) encountered 104 taxa of edaphic diatoms, noting that about one-third had a general distribution over the five habitats studied (tall Spartina, dwarf Spartina, Distichlis, bare bank, and panne). D. R. Wiseman (1978, College of Charleston, Charleston, South Carolina, unpubl. data) made a preliminary survey of the edaphic diatoms of Leadenwah Creek, South Carolina. Members of the pennate genus Navicula were the most conspicuous members of the community. Members of the following pennalean genera have been identified: Amphora, Amphipleura, Amphiprora, Bacillaria, Caloneis, Denticula, Frustulia, Gyrosigma, Mastogloia, Nitzschia, Pleurosigma, Rhaphoneis, Rhopalodia, Stauroneis, Surirella, Synedra, and Scoliopheura. A number of centralean diatoms were also observed but probably represent individuals from the phytoplankton.

In general, the nonvascular marsh flora of South Carolina and Georgia seems to differ from most other areas in that it has a paucity of macroscopic forms. There also seems to be a tendency for motile microphytes (i.e., pennate diatoms, euglenoids, and blue-greens) to dominate. Williams (1962) suggested that the dominance of motile microphytes results from the high turbidity and rapid sedimentation typical of the Sea Island Coastal Region. Thus, the macroscopic Rhodophyta and Phaeophyta typical of other intertidal marshes are mostly absent from the Sea Island Coastal Region, in part because they cannot outgrow the rate of siltation. The motile microphytes can remain in the euphotic zone by migration and thus maintain a reproductive population.

The macroscopic algae colonizing estuaries of South Carolina have not been studied intensively. The mud flats, marsh pannes, creek banks, soils of halophytic angiosperm zones, oyster reefs, shell banks, pilings, and sea walls, as well as the surface films and subsurface waters that ebb and flow over the estuaries, support a diverse algal flora. Batson and Blackwelder (1974) examined the stalks of Spartina alterniflora from the Cooper and Wando rivers in the summer of 1971. They reported 15 species of algae (nine cyanophytes, three rhodophytes, and three chlorophytes). The filamentous green alga, Chaetomorpha minima, and the coccoid blue-green, Entophysalis conferta, were the most abundantly occurring species. Blackwelder (1972), in the summer of 1970, investigated the salt marshes of the Port Royal Sound area. Twenty-eight species of algae were listed (6 rhodophytes, 4 chlorophytes, and 18 cyanophytes). Most of the algal growth occurred in the wetter areas of the marshes

Table 4-15. Distribution of saltmarsh diatoms by habitat in the vicinity of Sapelo Isalnd, Georgia (Williams 1962).

WIDESPREAD OVER THE MARSH IN MUDDY AREAS	UNDER SPARTINA AND ON SHADED BARE MUD CREEK BANKS	BARE MUD FLATS	BEACH AREAS - SAND AND MUDDY SAND	SETTLED PLANKTON
Amphiprora alata*	Amphiprora conspicua	Frustulia interposita	Amphora laevis var.?	Actinoptychus senarius
Amphora acutiuscula	A. similis?	Gyrosigma angustum?	A. proteoides?	Cyclotella stylorum
Bacillaria paradoxa*	Cyclotella meneghiniana?	G. distortum var. brevis*	Diploneis litoralis?	Cymatosira belgica
Caloneis formosa	Diploneis gruendleri	Navicula groeschopfi?	Gyrosigma hummii?	C. lorenziana
Cylindrotheca gerstenbergeri*	D. smithii?	Nitzschia thermaloides*	G. subangustum?	Hemidiscus weissflogia
Gyrosigma fasciola*	Frustulia asymmetrica	N. tryblionella?	Hanzschia marina var.	Melosira nummuloides*
Nitzschia circumsuta	Gyrosigma febigeri	Scoliotropis latestriata	irregularis?	M. sulcata
N. closterium*	G. nipkowii*	Surirella inducta*	H. virgata	Pseudonitzschia australis
N. laevis*	G. scalproides var. eximia.	S. pulchra	Navicula salinicola	Raphoneis amphiceros
N. panduriformi var. minor	G. s. var. obliqua?		N. sulcifera?	Skeletonema costatum
N. punctata	G. spectabile		Nitzschia hummii	Synedra tabulata
N. sigma*	G. spencerii var.*		N. sapathulatae	S. pulchella?
Pleurosigma angulatum*	Mastogloia exigua		Pinnularia trevelyana?	
P. naviculaceum	M. pumila?		Tropidoneis lepidoptera	
Tropidoneis seriata	Navicula pygmaea?			
	N. scopulorum var.			
SANDY AREAS OF THE MARSH	trinundulata*			
	N. sapeloensis?			
Achnanthes hauckiana	Nitzschia fasciculata*			
Campylodiscus echenis?	N. granulata			
Navicula psuedony?	N. lorenziana var. subtilis			
Terpsinoe americana	N. obtusa*			
	N. o. var. scalpelliformis*			
POSSIBLY CARRIED INTO THE MARSH FROM FRESH WATER	Pleurosigma normanii			
	Rhopalodia musculus*			
Cymbella tumida	Surirella gomma			
Melosira granulata	S. litoralis?			

* Indicates that species was cultured.
? Indicates that the habitat preference is based on a few observations.

210

and along the creek banks. Dead and living Spartina stems and the marsh periwinkle, Littorina irrorata, were the main substrates. The dominant epiphytes on Spartina in the wetter areas were two species of red algae, Bostrychia radicans (reported as B. rivularis) and Caloglossa leprieurii. The Caloglossa-Bostrychia assemblage is found throughout the world's estuarine systems (Post 1936). Ulva lactuca and Chaetomorpha minima, two chlorophytes, and a variety of blue-greens shared the role of dominants with the above-mentioned reds. D. R. Wiseman (1978, College of Charleston, Charleston, South Carolina, unpubl. data) made a preliminary survey of the algae in the vicinity of Leadenwah Creek, which merges with the North Edisto River, South Carolina. In the wetter areas of the marsh, the following macroscopic algae predominate: the chlorophytes Enteromorpha spp., Ulva spp., Monostroma oxyspermum, Ulothrix flacca, Blidingia minima, Chaetomorpha minima, Bryopsis hypnoides, and Rhizoclonium riparium; the rhodophytes Bostrychia radicans, Caloglossa leprieurii, Polysiphonia denudata, and Ceramium diaphanum; and 16 species of blue-greens.

Within the estuary, oyster reefs, shell banks, pilings, and sea walls provide substrate for macroscopic algae that inhabit the intertidal zone. The intertidal zone is dominated by a number of blue-greens, greens, and reds. In the colder months, brown algae of the order Ectocarpales appear. With few exceptions, the intertidal zonation of algae in Charleston Harbor compares strikingly with the zonal patterns discerned by Earle and Humm (1964) for Beaufort Harbor, North Carolina. They described six zones, beginning with the uppermost level of algal growth and extending downward to below mean low water for the summer algal flora. The highest zone is solely inhabited by the blackish-appearing cyanophyte, Calothrix crustacea. The lowest zone, designated as the Polysiphonia zone, ranged from 4 in (10 cm) above mean low water to the mean low water level. This zone was solely colonized by P. denudata. Earle and Humm pointed out that, although there may be many obvious similarities in the intertidal zonation patterns of Beaufort and stations north and south of this locality, "There are many differences also, differences of such importance that it seems unwise to attempt to generalize or to distinguish universal or even widespread features or specific zones." In Charleston Harbor, some of the sea walls at Fort Johnson deviate somewhat from what Earle and Humm observed at Beaufort. For example, in the zone that they designated as the Enteromorpha-Lyngbya Zone [which extends from 18 to 14 in (45.7 to 35.6 cm) above mean low water], they did not report two common reds (Bostrychia radicans and Gelidium pusillum) that appear in this zone throughout the year at Fort Johnson.

Pomeroy (1959) investigated the productivity of benthic microflora in the Duplin River marshes of Georgia. He estimated that the annual gross algal production in that area was ~200 gC/m^2. Net production was estimated to be not less than 90% of the gross production. He also indicated that direct relationships exist between productivity, tide, and season, and postulated a relationship between productivity and light regime.

Among the diatoms, the edaphic pennaleans appear to contribute a significant percentage of the total primary productivity of estuarine and coastal wetlands, ranging from 5 to 246 gC/m^2/yr (Riznyk et al. 1978). An assemblage of decapods, copepods, annelids, nematodes, amphipods, and gastropods graze on diatoms. These algal cells become suspended in the water column and contribute to the food webs of the estuarine waters.

Gallagher (1975) compared the plankton communities associated with surface films and the remainder of the water column in a Georgia salt marsh. Thick films form on the tidal waters flooding the marsh surfaces. He observed that the film is mainly composed of edaphic, pennalean diatoms. At high tide, when film development appeared maximal, algal cells were about five times more abundant in the surface film than in the underlying water. As the tide ebbed, surface film counts dropped rapidly to the level of the water column. Gallagher stated that the surface film community was more strongly autotrophic in the spring than in early summer.

A summary of rates of benthic algal productivity, as reported in the literature, is presented in Table 4-16. An estimate of 685 gC/m^2/yr (1.90 gC/m^2/day) was made recently for the North Inlet estuary (Zingmark 1977). While the magnitude of this estimate does not agree well with previous work (see Table 4-16), it is only slightly higher than estimates of annual productivity in the surface microlayers of the adjacent Santee River systems, South Carolina (J. J. Manzi, 1976, South Carolina Marine Resources Division, Charleston, unpubl. data). Benthic diatoms also appear to be important sediment stabilization elements in coastal wetlands (Holland et al. 1974).

The edaphic blue-green algae are important members of the marsh community; some fix atmospheric nitrogen (Carpenter et al. 1978), others stabilize sediments, many are grazed by herbivores, and all contribute to the protein-enrichment of the sediments when they are acted upon by bacteria. Somers and Brown (1978) determined that living and dead blue-green cells have an affinity for Ca++ and perhaps contribute to calcium recycling in the marsh by exchange reactions. Moribund or dead cells could serve as temporary reservoirs of this

Table 4-16. Annual rates of benthic algal productivity for a variety of estuarine and coastal environments (adapted from Zingmark 1977).

Location	Method of Measurement	Rate of Production $gC/m^2/yr$	Reference
Georgia estuary	O_2	200	Pomeroy (1959)
South Carolina estuary	^{14}C	685	Zingmark (1977)
Puget Sound	O_2	143 – 226	Pamatmat (1968)
Danish Lake	^{14}C	143	Hunding (1971)
Western Wadden Sea	^{14}C	100 ± 40	Cadée and Hegeman (1974)
Danish Wadden Sea	^{14}C	115 – 178	Grontved (1962)
Danish fjords	^{14}C	116	Grontved (1962)
New England estuaries	^{14}C	81	Marshall et al. (1971)
Scottish estuary	^{14}C	31	Leach (1970)
Intertidal sandy beach	^{14}C	4 – 9	Steele and Baird (1968)

cation. During periods of rapid growth, the sheaths and cells of the blue-greens bind the cation and then release the calcium during their demise. Somers and Brown (1978) stated that the total mass of these algae may not be enough to make a substantial contribution to the calcium needs of a marsh community and suggested that other algae, abundant when blue-greens are not conspicuous, may play a similar role. As observed by Ralph (1977) and D. R. Wiseman (1979, College of Charleston, Charleston, South Carolina, unpubl. data), the marsh supports the largest populations of blue-greens in the summer, but significant reduced populations persist throughout the colder months in masses of macroscopic green algae, among dead culms of marsh grasses, and within marsh sediments and soils. All vegetational zones of the angiosperm halophytes are colonized by all of the blue-green species found in the saltmarsh habitat, but the density and distribution of any one species may vary between zones. The marsh at Leadenwah Creek near Bears Bluff, South Carolina, like the marshes observed by Ralph (1977) in southern Delaware, is dominated by three oscillatoriaceous species, Microcoleus lyngbyaceus, Schizothrix calcicola, and S. arenaria.

2. Vascular Flora

Marshes dominate the coastal wetlands of Georgia and South Carolina. Though a complex, nutrient-rich ecosystem, vascular plant diversity here is relatively low due to limiting factors such as salinity, drainage, temperature, and tidal influence. Estuarine emergent wetlands may be divided into two major divisions, salt and brackish marshes.

a. Salt Marsh. Salinities in salt marshes range from $10^o/oo$ to as high as $70^o/oo$ in salt pannes or salt barrens. Such marshes are composed of two zones defined by elevation. The regularly flooded zone ("low marsh") is flooded at least once but usually twice daily, while the irregularly flooded zone ("high marsh") is flooded only during spring and storm tides.

Various chemical, physical, and biological factors interact to influence the distribution of vascular plants in the estuary. Johnson et al. (1974) listed water level fluctuations, salinity, substratum type, acidity, fire, and available nutrients as major factors in determining plant composition in estuarine wetlands. Penfound (1952) included aeration, temperature, light, plant competition, animal activity, and human activities (canalizing, cutting, draining, etc.) as possible factors. Kurz and Wagner (1957) listed salt spray and soil nutrients as factors. Baden et al. (1975) pointed out that texture, organic content, and pH do not seem to be important soil-related factors in determining plant composition.

Johnson and York (1915), Wells (1928), Chapman (1938), and Hinde (1954) all agreed that tidal inundation is the most critical factor in saltmarsh zonation. Adams (1963) basically agreed with the above, but stated that the primary factor was "tide-elevation influences." Stalter (1968) indicated that saltmarsh zonation was due to the combination of salinity together with duration and depth of flooding. Johnson et al. (1974) concluded that zonation was related to "elevation as it determines frequency, depth and duration of inundation, and soil salinity." Other workers have stressed the importance of single factors such as salinity (Bourdeau and Adams 1956, Kerwin 1966) and acidity (Wherry 1920).

(1) Low Marsh. The regularly flooded marsh ("low marsh") is dominated by extensive stands of smooth cordgrass. From a distance, this zone seems to be a uniform community with no apparent zonation. Upon closer examination, however, distinctive zonation can be seen in the smooth cordgrass. Teal (1958b) described three zones in smooth cordgrass stands around Sapelo Island, Georgia (Fig. 4-6):

1) Tall Spartina Edge Marsh. This marsh type occurs along the banks of creeks. Here smooth cordgrass reaches its maximum height, ranging up to 3 m (9 ft). The substrate is characterized by low sand content (10%).

2) Medium Spartina Levee Marsh. Found atop natural levees along creek banks, smooth cordgrass here averages 1 m (3 ft) in height. Sand content of the substrate varies from 0 to 10%.

3) Short Spartina Low Marsh. Found between drainage creeks, smooth cordgrass in this zone ranges 10 - 50 cm (4 - 20 in) tall. Sand content of the substrate varies between 0 and 10%.

Stalter (1968) described two communities in the regularly flooded zone of South Carolina:

1) Low Low Marsh - dominated by tall smooth cordgrass.

2) High Low Marsh - dominated by dwarfed smooth cordgrass.

Stalter and Batson (1969) and Stalter (1974d) used the same community designations in describing other regularly flooded marshes in South Carolina.

Gallagher et al. (1975) described three types of smooth cordgrass marsh from McIntosh County, Georgia. They lumped Teal's (1958b) "tall Spartina edge" and "medium Spartina levee" marshes into "streambank" Spartina alterniflora, while describing the remainder of the regularly flooded zone as either "creekhead Spartina" or, in intercreek areas, "short Spartina."

Bozeman (1975) noted that the smooth cordgrass community of Cumberland Island, Georgia, was similar to that described in Teal (1958b), with short and tall smooth cordgrass ecotypes. Radford (1976) described the same community type from the regularly flooded marshes of Beaufort County, South Carolina, as did Eyles (1939) from several islands in the Savannah River estuary.

Adams (1963) described "die-out areas" in the medium Spartina zone. These areas,

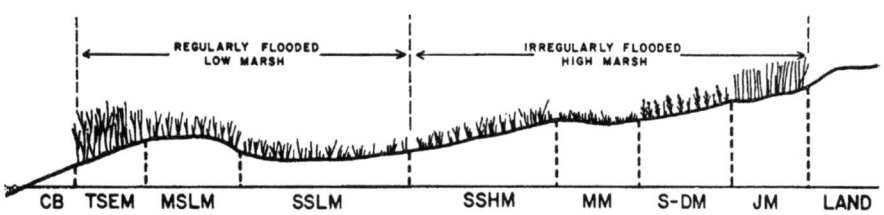

Figure 4-6. Diagram of saltmarsh vegetation zones (based on Teal 1958b).

CB = Creek bank
TSEM = Tall Spartina edge marsh
MSLM = Medium Spartina levee marsh
SSLM = Short Spartina low marsh

SSHM = Short Spartina high marsh
MM = Minax marsh
S-DM = Salicornia-Distichlis marsh
JM = Juncus marsh

213

which are probably caused by the accumulation of masses of dead cordgrass stems
("wrack"), occur as mud flats or non-
vegetated areas in the regularly flooded
zone. Smith (1970) listed other possible
causes of cordgrass "die-back" in
Louisiana. In time, the wrack either decays or is washed away and smooth cordgrass re-colonizes the barren area. (See
Intertidal Estuarine Subsystem - Flats for
further information on exposed mud areas
in the regularly flooded zone.)

Cooper (1974) pointed out that the
smooth cordgrass community is more extensive and well developed in the Sea Island
Coastal Region than elsewhere because of
tidal range [as much as 8 ft (2.5 m)].
Saltmeadow cordgrass becomes increasingly
dominant to the north.

(2) High Marsh. The irregularly flooded marsh ("high marsh") is,
unlike the monospecific regularly flooded
marsh, a mixture of several species of
grasses, forbs, and rushes. Teal (1958b)
illustrated four communities or marsh
types found in irregularly flooded marshes
of Sapelo Island, Georgia (Fig. 4-6):

1) Short Spartina High Marsh.
A transitional area between
high and low marsh, this type
is dominated by smooth
cordgrass that ranges from 10
to 50 cm (4 to 20 in) in
height. Here the sand content
of the substrate ranges from
40% to 70%.

2) Minax Marsh (termed
"minax" after the dominant
species of fiddler crab found
there, Uca minax). This type
occurs between types 1 and 3
and is dominated by extremely
short smooth cordgrass [10 -
20 cm (4 - 8 in)]. Rain
water accumulation and evaporation from this community
cause a salinity variance
from 10⁰/oo to 70⁰/oo).

3) Salicornia-Distichlis
Marsh. Glassworts and salt
grass dominate this community,
with scattered individuals of
smooth cordgrass present.
Parts of this community are
unvegetated sand flats where
high salinities and temperatures occur in summer. These
areas are sometimes referred
to as "salt pannes" or "salt
barrens." This community
(including the sand flats) is
only flooded during spring
and storm tides.

4) Juncus Marsh. Black
needlerush is the dominant
here. This community is

most common on the mainland
side of salt marshes and
in large estuaries.

Stalter (1968) divided the irregularly
flooded zone into middle marsh, low high
marsh, and high high marsh. The middle
marsh is a transitional community between
the regularly and irregularly flooded zones.
Salt grass, sea lavender, glassworts, and
short smooth cordgrass are present in this
zone. Table 4-17 shows species abundance
in Stalter's low high and high high marshes.
The low high marsh dominants were black
needlerush and glassworts; in the high high
marsh, dropseed, salt grass, saltmeadow
cordgrass, and black needlerush are the most
commonly found plants.

Radford (1976) described an irregularly flooded marsh in Beaufort County,
South Carolina, as having two communities,
1) a salt grass-glasswort community, and
2) a black needlerush community. Bozeman
(1975) termed high marsh vegetation on
Cumberland Island, Georgia, a "grass-forb-
rush" community. He listed glassworts,
saltwort, salt grass, salt-marsh bulrush,
black needlerush, sea lavender, and saltmeadow cordgrass as occurring in this community.

Tiner (1977) described four high marsh
communities in an inventory of coastal
marshes in South Carolina:

1) Short smooth cordgrass. At
the lowest level (near the mean
high water mark) of the high
marsh, smooth cordgrass less
than 1 m tall is found.

2) Sand barrens. Devoid of
vegetation, these areas are
slightly higher in elevation
than the short smooth cordgrass
community.

3) Vegetated sandflats. Topo-
graphically similar to the sand
barrens, this community may be
dominated by stands of glassworts or smooth cordgrass, salt
grass, and sea lavender mixed
with glassworts. Occasionally,
sea-blite and black needlerush
are present here.

4) Black needlerush. This
community is generally dominated by black needlerush, but
plants such as salt marsh
fimbristylis, marsh elder,
salt grass, sea ox-eye, and
saltmeadow cordgrass may be
present in openings between
stands of needlerush.

Other species occur in the irregularly
flooded zone, but salt marshes are not areas
of high plant species diversity. Plants
from salt marshes of South Carolina, and

Table 4-17. Species, density, frequency, and total number of plants found on m^2 quadrats located in different high marsh vegetation zones (adapted from Stalter 1968).

	Density	Range of Density	Frequency	Total Number of Plants
High Marsh				
a. High High Marsh (15 1-m² samples)				
Species				
(1) Salt grass	56.0	0– 644	20	840
(2) Salt marsh aster	1.1	0– 8	13	16
(3) Sea-blite	1.9	0– 28	7	28
(4) Saltmeadow cordgrass	44.8	0– 624	13	672
(5) Glasswort	4.3	0– 64	7	64
(6) Black needlerush	35.2	0– 524	13	528
(7) Sea ox-eye	16.5	0– 184	20	248
(8) Dropseed	327.2	0–2,020	27	4,908
(9) Sea lavender	2.4	0– 32	13	36
(10) Goldenrod	5.3	0– 80	7	80
(11) Smooth cordgrass	6.6	0– 84	13	100
			TOTAL	7,456
b. Low High Marsh (13 1-m² samples)				
Species				
(1) Salt grass	77.8	0– 320	62	1,012
(2) Sea ox-eye	61.2	0– 124	85	796
(3) Black needlerush	386.2	0–1,140	85	5,524
(4) Glasswort	326.0	0–1,856	64	3,836
(5) Smooth cordgrass	47.1	0– 212	100	612
(6) Sea lavender	16.9	0– 132	31	220
(7) Salt marsh aster	0.3	0– 4	8	4
			TOTAL	12,004

the zones in which they occur, are listed in Table 4-18. The "marsh-upland border" is a transitional zone that has been characterized by Shriner (1972), Boseman (1975), Radford (1976), Tiner (1977) and others. Woody species such as wax myrtle and sea myrtles are found here along with broom sedges, switchgrass, and orach. In Georgia, Bozeman (1975) noted Florida privet, milkvine, and common three-square in this zone. The marsh-upland transitional community usually grades into the maritime transitional shrub community.

b. Brackish Marsh. Brackish marshes, those occurring in salinities from 0.5°/oo to 10°/oo, occur between salt marshes and tidal freshwater wetlands (see Chapter Five). Although regularly and irregularly flooded zones exist in such marshes, plant zonation here seems to be more dependent on salinity than on duration of flooding. The most extensive areas of brackish marshes occur along the major river systems of the Sea Island

Coastal Region: the Altamaha, the Savannah, and the Santee.

Tiner (1977) listed the plants characteristic of brackish marshes in South Carolina (Table 4-19), and described the zonation of brackish marsh plants from the river mouths toward the freshwater line. The seaward limits of the brackish marsh are very similar to the high salt marsh and are dominated by black needlerush. Smooth cordgrass is found along river and creek banks where salinities are higher. Many typical high saltmarsh plants may be present, including salt-marsh bulrush, salt-meadow cordgrass, and salt marsh aster. Moving upstream, giant cordgrass replaces black needlerush as the dominant. Bulrushes and cat-tails become common, dominating the marsh in patches. In some areas, bulrushes exhibit vernal dominance on sites that are dominated by giant cordgrass during the late summer and fall. Finally, in the brackish-fresh marsh transition zone, giant cutgrass, wild rice, and other

215

Table 4-18. Species list of plants occurring in the salt marshes of South Carolina (adapted from Tiner 1977).

Scientific Name	Common Name	Location within Salt Marsh
Ampelopsis arborea	Pepper-vine	Marsh-upland border
Andropogon sp.	Broom-straw	Marsh-upland border
Andropogon scoparius	Little bluestem	Marsh-upland border
Andropogon virginicus	Broom sedge	Marsh-upland border
Aster subulatus	Annual salt marsh aster	High marsh
Aster tenuifolius	Salt marsh aster	High marsh
Atriplex patula	Orach	High marsh
Baccharis angustifolia	False willow	Marsh-upland border
Baccharis halimifolia	Sea myrtle	Marsh-upland border
Bacopa monnieri	Water-hyssop	High marsh
Batis maritima	Saltwort	High marsh
Borrichia frutescens	Sea ox-eye	High marsh
Chenopodium album	Lamb's quarters	Marsh-upland border
Chloris petraea	Finger grass	Marsh-upland border
Cladium jamaicense	Sawgrass	High marsh
Distichlis spicata	Salt grass	High marsh
Eleocharis sp.	Spikerush	High marsh
Fimbristylis spadicea	Salt marsh fimbristylis	High marsh
Hibiscus moscheutos	Rose mallow	High marsh
Iva frutescens	Marsh elder	High marsh
Juncus roemerianus	Black needlerush	High marsh
Kosteletskya virginica	Seashore mallow	High marsh
Liliaeopsis chinensis	Eastern lilaeopsis	High marsh
Limonium carolinianum	Sea lavender	High marsh
Limonium nashii	Sea lavender	High marsh
Lythrum lineare	Loosestrife	High marsh
Myrica cerifera	Wax myrtle	Marsh-upland border
Panicum virgatum	Panic grass	Marsh-upland border
Pluchea purpurascens	Marsh fleabane	High marsh
Rumex verticillatus	Swamp dock	High marsh
Ruppia maritima	Widgeon grass	Marsh ponds and potholes
Sabal palmetto	Cabbage palmetto	Marsh-upland border
Sabatia dodecandra	Sea pink	Marsh-upland border
Sabatia stellaris	Sabatia	High marsh
Salicornia bigelovii	Glasswort	High marsh
Salicornia europaea	Glasswort	High marsh
Salicornia virginica	Perennial glasswort	High marsh
Scirpus americanus	Common three-square	High marsh
Scirpus robustus	Salt-marsh bulrush	High marsh
Sesuvium maritimum	Sea purslane	High marsh
Sesuvium portulacastrum	Sea purslane	High marsh
Solidago sempervirens	Seaside goldenrod	High marsh
Spartina alterniflora	Smooth cordgrass	Low marsh and high marsh
Spartina cynosuroides	Giant cordgrass	High marsh
Spartina patens	Saltmeadow cordgrass	High marsh
Spergularia marina	Sand spurrey	High marsh
Sporobolus virginicus	Dropseed	High marsh
Suaeda linearis	Sea-blite	High marsh
Typha angustifolia	Narrow-leaved cat-tail	High marsh
Typha domingensis	Southern cat-tail	High marsh

Table 4-19. List of plants characteristic of brackish water marshes in South Carolina (adapted from Tiner 1977).

Alternanthera philoxeroides	Alligator-weed
Amaranthus cannabinus	Water hemp
Ammannia teres	- -
Andropogon sp.	Broom-straw
Apios americana	Groundnut/Potato bean
Aster tenuifolius	Salt marsh aster
Baccharis angustifolia	False willow
Baccharis halimifolia	Sea myrtle
Borrichia frutescens	Sea ox-eye
Carex sp.	Sedges
Cicuta maculata	Water hemlock
Cladium jamaicense	Saw grass
Cyperus spp.	Sedges
Dichromena colorata	Star-rush
Distichlis spicata	Salt grass
Eleocharis sp.	Spikerush
Fimbristylis spadicea	Salt marsh fimbristylis
Hibiscus militaris	Halberd-leaved marsh mallow
Hibiscus moscheutos	Rose mallow
Hymenocallis crassifolia	Spider-lily
Iris virginica	Blue flag
Iva frutescens	Marsh elder
Juncus effusus	Soft rush
Juncus roemerianus	Black needlerush
Juncus spp.	Rushes
Kosteletskya virginica	Seashore mallow
Lilaeopsis chinensis	Eastern lilaeopsis
Limonium carolinianum	Sea lavender
Lythrum lineare	Loosestrife
Panicum virgatum	Panic grass
Peltandra virginica	Arrow-arum
Pluchea purpurascens	Marsh fleabane
Polygonum spp.	Smartweeds
Pontederia cordata	Pickerelweed
Ptilimnium capillaceum	Mock-bishopweed
Rosa palustris	Swamp rose
Rumex verticillatus	Swamp dock
Ruppia maritima	Widgeon grass
Sagittaria spp.	Arrowheads
Scirpus americanus	Common three-square
Scirpus olneyi	Olney's three-square bulrush
Scirpus robustus	Salt-marsh bulrush
Scirpus validus	Soft-stem bulrush
Sesbania exaltata	Coffee-weed
Setaria geniculata	Foxtail grass
Setaria magna	Foxtail grass
Sium suave	Water parsnip
Solidago sempervirens	Seaside goldenrod
Spartina alterniflora	Smooth cordgrass
Spartina cynosuroides	Giant cordgrass
Spartina patens	Saltmeadow cordgrass

Table 4-19. Concluded

Scientific Name	Common Name
Typha angustifolia	Narrow-leaved cat-tail
Typha domingensis	Southern cat-tail
Zizania aquatica	Wild rice
Zizaniopsis miliacea	Giant cutgrass

freshwater species become increasingly
common. Havel (1976) listed giant cordgrass, salt-marsh bulrush, southern bulrush, soft-stem bulrush, and salt marsh
aster as dominants in the brackish marshes
of the lower Santee River flood plain.

Brackish marsh zonation in Georgia
seems to be very similar to that of South
Carolina. Johnson et al. (1974) pointed
out that giant cordgrass was the dominant
plant of Georgia brackish marshes, along
with salt-marsh bulrush to a lesser extent.
Gallagher and Reimold (1973) referred to
the brackish marshes in Georgia as "middle estuary" and pointed out that salinities may range from 0⁰/oo to 20⁰/oo.
Gallagher et al. (1975) mapped the fresh,
brackish, and salt marshes of the Altamaha
River Delta in Georgia (Fig. 4-7). They
found that giant cordgrass, black needlerush, cat-tails, and bulrushes dominate
the brackish marsh.

c. Ecology of Emergent Wetlands.
Plants of estuarine emergent wetlands vary
in the degree of seasonality they exhibit.
In South Carolina, smooth cordgrass of the
low marsh community usually dies back in
winter, although culms from the previous
year normally remain standing. Peak
height is reached in late summer, and
flowering occurs in September and October.
However, two crops of smooth cordgrass per
year may be produced in Georgia (Schelske
and Odum 1962). McIntire and Dunstan
(1975) studied seasonal growth variations
in three smooth cordgrass communities of
Georgia. Winter height and percentage of
dead material both show that the smooth
cordgrass is a seasonal community (Figs.
4-8 and 4-9). The high marsh community,
unlike that of the low marsh, exhibits
little or no visually perceptible seasonality. Equal amounts of dead (gray), dying (brown), and live (green) black needlerush stems seem to be present year around.
However, Foster (1968) has shown that this
is not actually the case. Primary production decreases in winter and leaf production varies throughout the year (Table
4-20). Black needlerush flowers from
April through May (Eleuterius 1975).

Unlike low marshes, high marshes are
subjected to occasional fires. Burned

areas of black needlerush quickly revege-
tate from rhizomes with no successional
sequence evident. Fire is much more common in brackish marshes. After these
areas are subjected to controlled burns in
late winter, various species of bulrushes
and spikerushes exert almost complete
dominance. Salt-marsh bulrush, common
three-square, soft-stem bulrush, and spikerushes are usually the dominants after
fire (Tiner 1977). In brackish marshes
not subjected to fire, bulrushes or cat-
tails may be the early season dominants,
but they are usually replaced by giant
cordgrass later in the growing season.

Estuarine emergent wetland communities exhibit little or no internal physical structure. The most significant
structural difference from community to
community is height. Black needlerush
marshes are generally taller than smooth
cordgrass communities, except for tall
smooth cordgrass stands, ranging from 1
to 2 m (3 to 6 ft) high. Brackish marshes
are usually taller than salt marshes, with
soft-stem bulrush and giant cordgrass
ranging as high as 5 m (15 ft), though
they and other species of this community
normally average from 1.7 to 2.0 m (5 to
6 ft) high (Odum and Fanning 1973).

Succession in salt marshes is the result of a combination of site salinity and
site elevation. On a given site, if the
salinity or the hydrology of the site is
not changed, existing vegetation will remain indefinitely. If the vegetation of
the site is destroyed and no changes in
salinity or hydrology have taken place, the
same vegetation normally returns. In some
cases, however, black needlerush will not
succeed itself. Stalter (1968) described
general successional trends along salinity
and elevation gradients.

An extensive literature exists concerning nutrient cycles and productivity in
estuarine emergent wetlands. Because all
of this information cannot be dealt with
here, major works from Georgia and South
Carolina are used as principal sources.

Pomeroy et al. (1969) and Pomeroy et
al. (1972) discussed the nutrient cycle in
salt marshes in Georgia. Pomeroy et al.

213

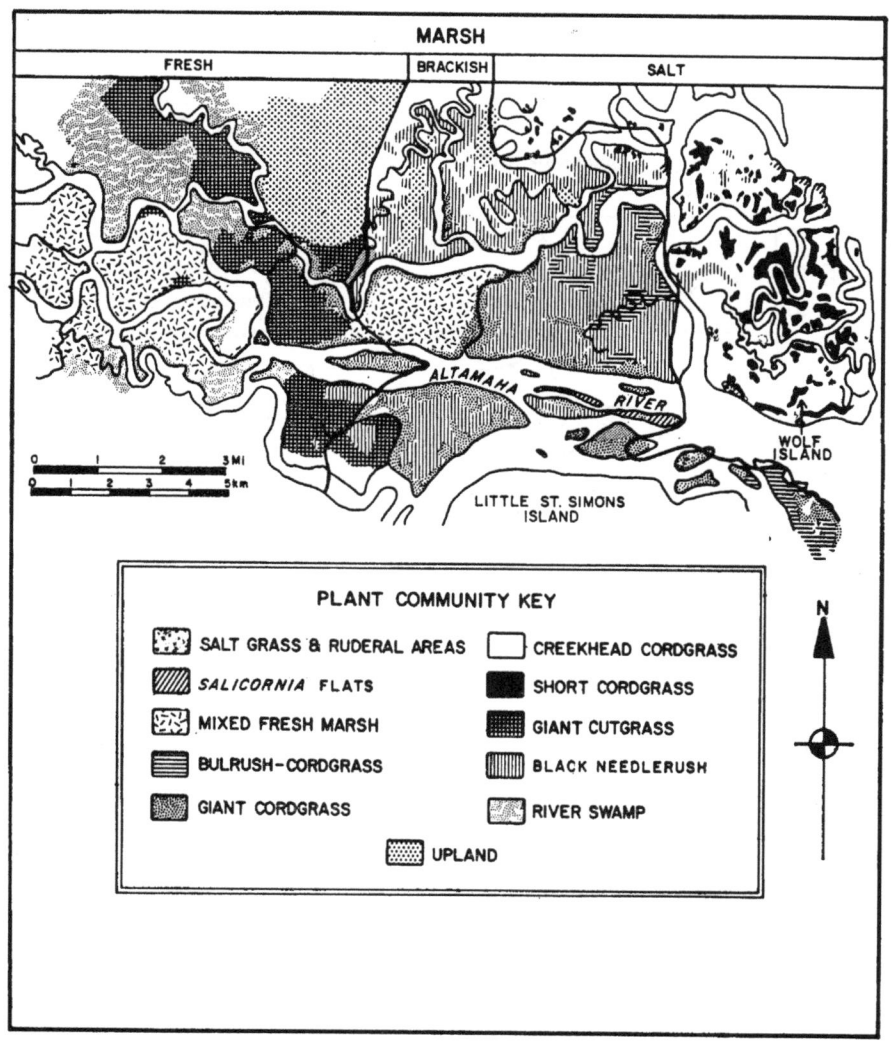

Figure 4-7. Fresh, brackish, and salt marshes of the Altamaha River Delta, Georgia (Gallagher et al. 1975).

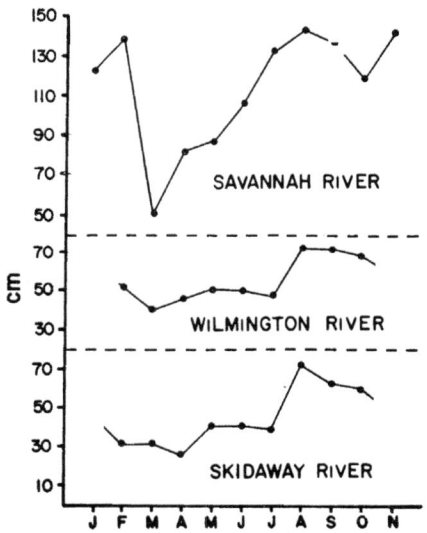

Figure 4-8. Seasonal variation of growth (cm) of three smooth cordgrass communities in Georgia (McIntire and Dunstan 1975).

(1969) pointed out that the uppermost meter of sediment in a smooth cordgrass marsh contains enough phosphorus to support 500 years of cordgrass production and enough zinc to support 5,000 years of production. The role of smooth cordgrass as a medium for the introduction of phosphorus and zinc into the food chain was also dealt with by Pomeroy et al. (1969). They pointed out that these minerals are passed on through bacteria and detritus feeders during smooth cordgrass decomposition. The extremely high concentration of these two elements in saltmarsh bacteria is explained here by the ability of smooth cordgrass to transfer zinc and phosphorus from sediment to organism.

The nutrient cycle and energy flow of the high marsh, black needlerush community, is relatively unknown. Williams and Murdoch (1966) noted no significant export from black needlerush marshes and pointed out that, unlike smooth cordgrass, black needlerush is not part of the principal food web of the salt marsh.

Teal (1962) analyzed energy flow in saltmarsh communities. Smooth cordgrass acts as a vital link in the nutrient cycle as well as in the flow of food energy through the salt marsh [see Gosselink and Kirby (1974) for further information on decomposition].

Schelske and Odum (1962) listed five tors that combine to maintain high ductivity in salt marshes:

1) Number of primary production units. Salt marshes have three primary production units, all maintaining high productivity. They are: a) mud algae, b) phytoplankton, and c) Spartina alterniflora and other emergent vegetation.

2) Tidal action. Tidal action increases the surface area in which phytoplankton photosynthesis may occur, transports decomposing Spartina to all areas of the salt marsh, and distributes nutrients throughout the marsh.

3) Abundant nutrients. The abundance of nutrients allows photosynthesis to continue at a high rate, with little or no nutrient limitations to productivity.

4) Conservation and rapid turnover of nutrients. The fact that nutrients (such as phosphorus) are rapidly turned over and remain in the estuary is more important than the actual concentrations of these nutrients.

5) Year-round production. In Georgia, Spartina produces two crops per year, while mud algae and phytoplankton are productive throughout the year.

These five factors make the salt marsh one of the most productive natural systems on earth. Primary production has probably been more intensively studied in salt marshes than in any other community. Turner (1976), Dawson (1977), and Reimold and Linthurst (1977) have reviewed salt-marsh productivity studies and methodology. Considerable variation in productivity values has occurred due to methodological and/or latitude differences.

Turner (1976) cited 29 references to productivity studies in smooth cordgrass (Spartina alterniflora) communities. Production varied from 3990 g dry wt/m^2 for tall Spartina in Georgia to 350 g dry wt/m^2 for short Spartina in Connecticut. Dawson (1977) summarized results of 10 studies of smooth cordgrass productivity from the Sea Island Coastal Region (Table 4-21). Standing crop biomass ranged from 70 to 3229 g/m^2.

The black needlerush community is generally less productive than the smooth cordgrass community. Production in black needlerush communities (Turner 1976) ranged from 370 g dry wt/m^2 in Connecticut to 2261 g dry wt/m^2 in Georgia. Dawson (1977)

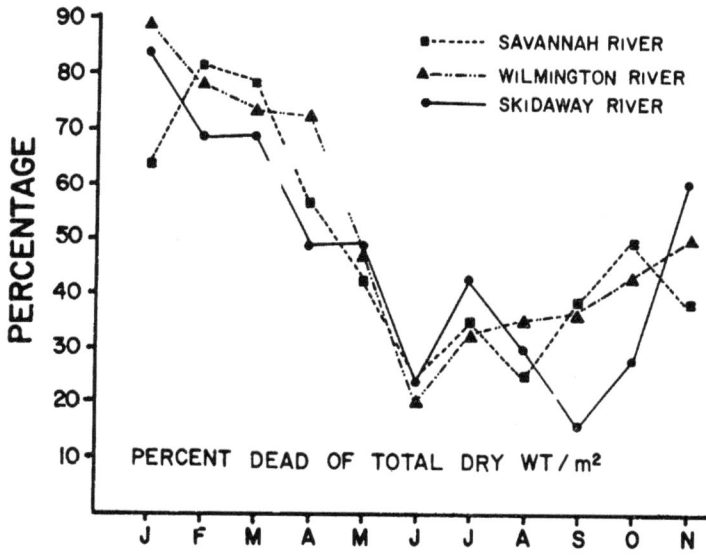

Figure 4-9. Monthly percentages of total dead dry wt/m² in three smooth cordgrass communities of Georgia (McIntire and Dunstan 1975).

Table 4-20. Production of new leaves by black needlerush at existing shoots in growth plots (leaves/100 shoots/28 days) (Foster 1968).

Period	New Leaves		
March 24 to May 14	18.8	October 23 to November 13	5.5
May 14 to June 4	12.5	November 13 to December 11	4.0
June 4 to July 2	21.2	December 11 to January 24	1.0
July 2 to July 31	10.9	January 24 to February 25	4.9
July 31 to August 28	8.9	February 25 to March 23	15.4
August 28 to September 25	15.1	March 23 to April 15	8.8
September 25 to October 23	9.6	April 15 to May 6	10.3

Table 4-21. Summary of information on marsh grass productivity from the Sea Island Coastal Region (adapted from Dawson 1977) (SSA - Short _Spartina alterniflora_, MSA - Medium _Spartina alterniflora_, TSA - Tall _Spartina alterniflora_, Avg. - Average).

SPECIES	STANDING CROP BIOMASS[a] Dry wt (g/m²)		NET ANNUAL PRIMARY PRODUCTIVITY[b] g/m/yr	EOST BIOMASS[c]	LOCATION	SOURCE
Smooth cordgrass	SSA avg.	325.6			S.C.	Stalter 1968
	TSA avg.	445.4				
Smooth cordgrass	1,297.0				S.C.	Duncan 1975
Smooth cordgrass	SSA avg.	401.3			S.C.	Coastal Consultants of Darien, Georgia 1975
	Marsh avg.	1,230.0				
Smooth cordgrass	High marsh		643.2	472	Ga.	Smalley 1958
	Streamside		1,098.0	1,230		
	Marsh avg.		870.6	762		
Smooth cordgrass	1,866.8				Ga.	Reimold et al. 1972
Smooth cordgrass	MSA	2,082.0		2,182	Ga.	Odum and Fanning 1973
	TSA	3,115.0		3,315		
	Marsh avg.	2,598.5		2,544		
Smooth cordgrass	Altamaha River				Ga.	Gallagher and Reimold 1973
	Lower estuary	303.0 - 496.0 (Apr)				
		356.0 - 2,398.0 (Aug)				
	Middle estuary	160.0 - 1,220.0 (Apr)				
		206.0 - 3,229.0 (Aug)				
	Upper estuary	70.0 - 350.0 (Apr)				
		494.0 - 1,507.0 (Aug)				
Smooth cordgrass	SSA	409.6			Ga.	Reimold et al. 1975
	TSA	1,411.0				
	avg.	910.3				
Smooth cordgrass	Savannah River	1,922.0	2,162.0		Ga.	McIntire and Dunstan 1975
	Wilmington River	328.0	369.0			
	Skidaway River	275.0	309.0			
	avg.	846.7	946.7			
Smooth cordgrass	High marsh		1,337.0		Ga.	Gallagher et al. 1976
	Streamside		3,773.0			
Black needlerush	2,430.0				S.C.	Duncan 1975
Black needlerush	1,410.0				S.C.	Coastal Consultants of Darien, Georgia 1975
Black needlerush	1,226.7				Ga.	Reimold et al. 1972
Black needlerush	2,131.5				Ga.	Reimold et al. 1975

222

Table 4-21. Continued

SPECIES		STANDING CROP BIOMASS[a] DRY wt (g/m²)	NET ANNUAL PRIMARY PRODUCTIVITY[b] g/m/yr	EOST BIOMASS[c]	LOCATION	SOURCE
Black needlerush			2,156		Ga.	Gallagher et al. 1976
Salt grass		362.2			Ga.	Reimold et al. 1972
Salt grass		741.0			S.C.	Coastal Consultants of Darien, Georgi
Sea ox-eye		461.0			S.C.	Coastal Consultants of Darien, Georgi
Glasswort		113.0			S.C.	Coastal Consultants of Darien, Georgi
Mixed salt marsh (Salt grass, Sea lavender, Glasswort, Smooth cordgrass)	avg.	200.0 – 398.8 / 297.4			S.C.	Stalter 1968
Mixed salt marsh (Salt grass, Sea ox-eye, Black needlerush, Glasswort, Smooth cordgrass, Sea lavender, Aster)	avg.	196.0 – 909.6 / 502.2			S.C.	Stalter 1968
Mixed salt marsh (Salt grass, Aster, Sea-blite, Saltmeadow cordgrass, Glasswort, Black needlerush, Dropseed, Sea lavender, Goldenrod, Smooth cordgrass)	avg.	181.0 – 1,095.0 / 550.6			S.C.	Stalter 1968
Giant cordgrass		872.0 – 2,092.0		1,028	Ga.	Odum and Fanning 1973
Giant cordgrass		1,924.0			Ga.	Gallagher and Reimold 1973
Common cat-tail		635.0			S.C.	Boyd 1970
Cat-tail		2,433.7			S.C.	Coastal Consultants of Darien, Georgi

223

Table 4-21. Concluded

SPECIES	STANDING CROP BIOMASS[a] Dry wt (g/m²)	NET ANNUAL PRIMARY PRODUCTIVITY[b] g/m/yr	EOST BIOMASS[c]	LOCATION	SOURCE
Bulrush	150.0			(Imp und-ment)	Boyd 1970
Wild rice	1,842.0			Ga.	Gallagher and Reimold 1973

a. Weight of plant material, living and dead, present in the system at any given instant (values are for aerial parts only)
b. Total amount of dry matter in above-ground plant parts per year.
c. End of season estimates of total biomass (both living and dead), taken from Turner (1976).

224

listed five studies of productivity in
black needlerush marshes in South
Carolina and Georgia (Table 4-21);
standing crop biomass ranged from 1,227
to 2,430 g/m².

Odum and Fanning (1973) compared
productivity in salt marshes with that of
brackish marshes. The smooth cordgrass
of salt marshes was more productive than
the giant cordgrass of brackish marshes,
although production in the giant cordgrass marsh was high by other standards,
with maximum production values of 3,990
g/m² for smooth cordgrass and 2,092 g/m²
for giant cordgrass. Dawson (1977) listed
several other studies of production from
brackish marshes in South Carolina and
Georgia (see Table 4-21). These investigations show that estuarine emergent wetlands are extremely productive environments.

Johnson et al. (1974) listed dredging,
filling, diking, and ditching as major
causes of the destruction of estuarine
emergent wetlands. Spinner (1969) pointed
out that approximately 0.7% of Georgia's
coastal marshes were destroyed (primarily
by dredge and fill) from 1954 to 1968.
Windom (1976) noted the impact of channel
construction and dredged material disposal
on estuarine wetlands. Dredged material
disposal areas may be open or confined by
dikes. Open disposal banks or "spill"
areas are recolonized by vegetation, but,
due to an increase in elevation, the dominant plants are usually high marsh (sea
ox-eye, sea myrtle, etc.) and upland
plants. Diked spoil areas eventually may
be recolonized by smooth cordgrass and
other estuarine plants, but large areas
of vegetation are often destroyed. However, revegetation of disposal areas may
be expedited by transplantation of estuarine plants, as shown by Dunstan et al.
(1975a) and Stalter and Batson (1969) in
Georgia and South Carolina, respectively.

Windom (1976) also noted the danger
of the deposition of polluted dredged
material in estuarine wetlands. Smooth
cordgrass often fails to recolonize areas
with high methylmercury concentrations.
In other areas, estuarine flora does re-
vegetate polluted spoil banks, but may
pass toxic materials into the estuarine
food chain by way of plant consumers and
decomposers (Dunstan et al. 1975b,
Windom 1975).

Baker (1970) pointed out the effects
of oils on the physiology of plants. The
potential disturbances in estuarine wetlands from oil spills is documented in the
final environmental impact statement for
South Atlantic OCS Oil and Gas Lease Sale
No. 43 (U.S. Department of Interior,
Bureau of Land Management 1977). Miller
and Egler (1950) described areas where
marsh grasses had been killed due to mosquito ditching. Johnson et al. (1974)

pointed out that estuarine wetlands in
Georgia have been destroyed during attempts to convert high marsh areas to
agricultural lands. Additionally, Chabreck
(1968) found that intensive use of marshes
by livestock (grazing) resulted in changes
in plant composition and reduction in
plant density. Smith (1970) suggested
that water pollution may play a significant role in what he termed Spartina "die-
back."

B. CONSUMERS

1. Benthic Meiofauna

Meiofauna (those animals passing
through a 500-μm sieve but retained by a
63-μm sieve) of intertidal estuarine areas
of the Sea Island Coastal Region have received relatively little attention. Most
of the available data are the result of
work in the high salinity marshes of the
North Inlet area, South Carolina, by B. C.
Coull and his associates at the University
of South Carolina (see also section on
meiofauna of estuarine subtidal environments).

Distribution patterns of meiofaunal
species in estuarine environments are not
well known. Coull (1973) noted that there
is commonly a reduction in the number of
meiofaunal species with decreasing salinity
up estuary. A similar phenomenon is seen
with most other estuarine faunas and is a
reflection of the fact that estuarine communities are primarily composed of marine
species with varying degrees of eury-
halinity. Coull (1973) also pointed out
that, in addition to salinity, a variety
of other factors, including sediment gran-
ulometric characteristics, redox potential,
and the distribution of bacteria (see
subtidal section), significantly influence
the distribution of meiofauna.

The distribution of meiobenthos across
saltmarsh gradients has been studied in
some detail in the Sea Island Coastal
Region. In Georgia, Teal and Weiser (1966)
examined the abundance and vertical distribution of nematodes at six stations
spread from the low tide level to the
landward edge of a salt marsh behind
Sapelo Island. They found nematodes to
be most abundant and distributed over a
greater sediment depth [12 - 14 cm (4.7 -
5.5 in)] at the highly productive edge of
the low marsh. Teal and Weiser (1966)
further found that nematodes accounted for
only 3% of the total respiration of Sapelo
marsh muds, compared to 25% - 33% for a
similar marsh area near Woods Hole,
Massachusetts. This suggested a much lesser role for the meiobenthic nematodes in
the energy flow of southern marshes. In
South Carolina, Coull et al. (1979) investigated the distribution of meiobenthic
copepods over a gradient from creek bottom to high marsh in North Inlet (Table
4-22). The copepods were second only to

Table 4-22. Densities of meiobenthic copepods at five sites on a saltmarsh gradient in North Inlet estuary, South Carolina (adapted from Coull et al, 1979).

| | Number copepods/10 cm^2 | | % copepods of total meiofauna |
	Range	Mean	
Creek Bottom (Low tide depth, 3 m)	69-97	83	12
Subtidal (Low tide depth, 0.5 - 1 m)	23-231	107	11
Mud flat (Exposed $\sim$ 5 h/ 12 h tidal cycle)	9-245	89	4
Low Spartina Marsh (Exposed 5 - 6 h/12 h tidal cycle; tall Spartina)	75-620	262	22
High Spartina Marsh (Exposed $\sim$ 7 h/12 h tidal cycle; short Spartina)	20-192	73	16

nematodes in importance in the meiobenthic assemblage. Coull et al. (1979) found densities of copepods to be highest in the low marsh and lowest in the high marsh (Table 4-22). They suggested that the peak numbers in the low marsh were the result of increased spatial heterogeneity (i.e., greater diversity of microhabitats) with only limited tidal exposure. In contrast, the much greater exposure, with consequent great fluctuations in physical factors, in the high marsh proved limiting to many copepods.

In a more detailed, micro-distributional study, Bell et al. (1978) examined variations in meiofaunal assemblages around and between Spartina culms and fiddler crab (Uca) burrows and between the surface [to 0.5 cm (to 0.2 in)] and deeper layers [0.5 - 3 cm (0.2 - 1.2 in)] of the sediment. They found more total meiofauna in the surface sediment regardless of location. During spring, surface meiofaunal densities were greater between rather than around Spartina culms. However, nematodes were more numerous in the surface layer around Uca burrows than between them, while the opposite case was true for copepods. In contrast to indications from Teal and Weiser's (1966) study of a low marsh area in Georgia, Bell et al. (1978) found no relationship or a negative association of meiobenthos with root biomass. They noted that the amount of root material in their high marsh area was 6 - 7 times that typically

seen in low marsh. Overall, Bell et al. (1978) showed that, at any given time, levels of variation in meiobenthic assemblages within their 5 m^2 study site were as great as reported seasonal variations in other studies. They concluded that small-scale spatial heterogeneity probably has a major influence on meiofaunal distribution, and thus on meiofaunal-macrofaunal relationships.

To date, only one long-term study of marsh meiobenthos has been conducted in the Sea Island Coastal Region. Bell (1979) looked at short- and long-term variation in the meiobenthos of a high Spartina marsh near North Inlet over a 22-month period. On the average, nematodes made up 73% of the meiofauna, with copepods second in abundance. Bell found that the high marsh meiobenthic assemblages she studied were characterized by marked variation, with little similarity of assemblages within seasons or between years. She reported peak densities of nematodes, copepods, and total meiofauna in fall, with a secondary peak in spring.

The trophic relationships of the meiobenthos are quite poorly known. Coull (1973) suggested that meiofauna may provide up to 5 times as much food as the macrofauna at any location. He further suggested three major trophic pathways for meiofauna: 1) consumption by benthic macrofauna such as shrimps and polychaetes; 2) consumption by nektonic forms; and

226

3) decomposition and nutrient regeneration through microbial action. Coull (1973) stated that the nutrient regeneration pathway is likely to be the most important, especially in view of the relative scarcity of meiofaunal predators. However, Bell and Coull (1978) have definitely shown that some macrofaunal predators, specifically the grass shrimp Palaemonetes pugio, can play a major role in regulating meiofaunal populations in marsh environments. Much more work is needed in the area of meiofaunal trophic dynamics and the trophic relationships between meiofauna and macrofauna.

2. Benthic Macroinvertebrates

The number of macroinvertebrate species represented in salt marshes is limited due to a number of environmental stresses, the more important of which include salinity, drainage, temperature, and exposure, together with rapid changes in environmental conditions (Teal 1962, Cooper 1974). Nevertheless, a variety of invertebrates are found in salt marshes (Table 4-23). While invertebrate communities are somewhat overshadowed by the productivity of plant assemblages in salt marshes, densities and biomass for some taxa such as decapods and mollusks are high (Tables 4-24, 4-25). Wolf et al. (1975) utilized a removal sampling technique to estimate densities of the fiddler crab Uca pugnax in marshes of the Duplin River estuary, Georgia. Numbers varied from location to location and showed extreme aggregation even within a given area, but total density was estimated at 205 individuals/m^2. Of these, an estimated 30/m^2 were larger individuals. Kraeuter (1976) provided estimated densities for several Georgia marsh invertebrate species, including Uca pugnax (30/ m^2), the gastropod Littorina irrotata (73/m^2), the fiddler crab Uca pugilator (6.6/m^2), and the mussel Geukensia demissa (6.4/m^2). Kuenzler (1961) estimated a population density of 7.8 specimens/m^2 of Geukensia demissa and a biomass of 11.5 g/m^2 in marshes near Sapelo Island. As with the fiddler crabs, Kuenzler found that ribbed mussels varied in density from one location to another and showed clumped distribution even in seemingly uniform areas. Vernberg and Sansbury (1972) studied the macroinvertebrates of several saltmarsh habitats in Port Royal Sound (Tables 4-23 and 4-24). The fiddler crab Uca pugnax was the most ubiquitous species in their samples. Another fiddler crab, Uca pugilator, was frequent near high tide in sandy substrates. Clumps or beds of oysters (Crassostrea virginica) were common and provided habitat for a variety of other invertebrates, including crabs and polychaetes. They found mussels (Geukensia demissa) in substrates of sandy mud and on oyster shells. The marsh periwinkle Littorina irrorata was frequent on Spartina in the high marsh,

while the mud snail Ilyanassa obsoleta occurred in the low marsh on wet and muddy substrates. Of the polychaetes, Nereis succinea was most widespread, being found in association with Spartina. In the marshes of Georgia studied by Dörjes (1977), dominant macroinvertebrates of the low marsh included the mollusks Geukensia demissa, Littorina irrotata, and Crassostrea virginica; the decapods Uca pugnax, Sesarma reticulatum, Panopeus herbstii, and Eurytium limosum; and the polychaete Nereis succinea. The high marsh was occupied primarily by Uca pugilator, although some local areas were inhabited by U. pugnax and U. minax.

Teal (1962) demonstrated that the fauna of marshes on Sapelo Island was made up of species from a number of different environments. Estuarine organisms were present from the Salicornia marsh to subtidal areas, but the number of estuarine species increased toward the low tide level. The number of terrestrial and freshwater species was small on the creek banks but became numerous elsewhere, particularly at the edge of the marsh and in levee areas. The number of species typically restricted largely to salt marshes was small; these organisms encompassed the region from the creek banks to the Salicornia marsh, but the largest number of species was present in the marsh proper.

Dörjes (1977) did not consider the creek banks to represent typical saltmarsh environments. The macroinvertebrates found there, including oysters (Crassostrea virginica), mud snails (Ilyanassa obsoleta), polychaetes (Diopatra cuprea and Heteromastus filiformis), and the decapod Upogebia affinis, occur elsewhere in intertidal or subtidal unvegetated areas. The extent of such habitat in Georgia and South Carolina is expansive, however, and many of the oyster beds in these States occur intertidally along the creek banks, particularly in higher salinities. Aspects of the energetics of intertidal oyster reefs on creek banks adjacent to salt marshes have been studied by Dame (1972a, b, 1974, 1976) and Bahr (1976).

Teal (1962) provided a detailed summary of energy flow in a Georgia marsh. Among the more important herbivorous invertebrates by Teal were the salt marsh grasshopper Orchelimum, the salt marsh plant hopper Prokelisia, the crabs Uca and Sesarma, the mollusks Littorina and Geukensia, various annelids including the polychaetes Streblospio and Capitella, and oligochaetes. Important invertebrate carnivores included spiders, dragonflies, and the xanthid crab Eurytium. Kraeuter (1976) emphasized the importance of invertebrate biodeposits in marsh biogeochemical cycles. He calculated that 53% of the yearly production in a marsh could be processed by invertebrates, although

227

Table 4-23. Species of macroinvertebrates typical of salt marshes and creek banks in estuaries of the Southeastern United States (Barnes 1953, Teal 1962, Davis and Gray 1966, Vernberg and Sansbury 1972, Roberts 1974b, Dörjes 1977).

Spiders	Insects	Miscellaneous Invertebrates
Lower Estuary,	**Spartina alterniflora association**	Mollusca
on Spartina alterniflora	Prokelisia marginata	Crassostrea virginica
	Sanctanus aestuarium	Geukensia demissa
Grammonota trivittata	Draeculacephala portola	Mercenaria mercenaria
Dictyna savanna	Conioscinella infesta	Polymesoda caroliniana
Eustala anastera	Chaetopsis fulvifrons	Petricola pholadiformis
Singa keyserlingi	Chaetopsis apicalis	Littorina irrorata
Hyctia pikei	Trigonotylus uhleri	Ilyanassa obsoleta
Poecilochroa unimaculata	Ischnodemus badius	Melampus bidentatus
	Tytthus vagus	
	Orchelimum fidicinium	
Upper Estuary	Conocephalus spp.	Crustacea
	Mordellistena spp.	
on Spartina alterniflora	Isohydnocera tabida	Uca pugnax
	Collops nigriceps	Uca pugilator
Eustala anastera	Chilcidoidea (undetermined)	Uca minax
Grammonota trivittata		Panopeus herbstii
Dictyna savanna		Eurypanopeus depressus
Argiope seminola		Callinectes sapidus
	Spartina - Salicornia -	Sesarma reticulatum
	Limonium association	Sesarma cinereum
		Clibanarius vittatus
Spartina - Distichlis -	Prokelisia marginata	Cyathura sp.
Salicornia beds	Sanctanus sanctus	Squilla empusa
	Conioscinella infesta	Chthamalus fragilis
Eustala anastera	Chaetopsis apicalis	Talorchestia longicornis
Dictyna savanna	Chaetopsis fulvifrons	Alpheus heterochaelis
Lycosa modesta	Dimecoenia austrina	Upogebia affinis
Pardosa floridana	Pelastoneurus lamellatus	Palaemonetes pugio
	Orphulella olivacea	Palaemonetes vulgaris
		Rhithropanopeus harrisii
Juncus beds		
	Juncus association	
Eustaka abastera		
Lycosa modesta	Keyflana hasta	Polychaeta
Pardosa floridana	Rhynchomitra microrhina	
	Chaetopsis fulvifrons	Heteromastus filiformis
	Cymus breviceps	Nereis succinea
Spartina patens beds	Conocephalus spp.	Laeonereis culveri
	Paroxya clavuliger	Phyllodoce fragilis
Tibellus duttori	Erythropliplax berenice	Haploscolopos fragilis
Hyctia pikei		Lumbrinereis tenuis
Eustala anastera		Glycera dibranchiata
Lycosa modesta	Distichlis association	Diopatra cuprea
Arcosta furtiva		Goniada maculata
	Amphicephalus littorális	Onuphis microcephala
	Delphacodes detecta	Capitella spp.
Mixed herbaceous	Spangbergiella vulnerata	
	Neomegamelanus dorsalis	
Mangora gibberosa	Tumidagena terminalis	
Hyctia pikei	Conioscinella infesta	
Phidippid (undetermined)	Oscinella ovalis	
Latrodectus mactans	Ceropsilopa costalis	
Pirata suwansus	Tomosvaryella coquilletti	
Lycosa sabida	Trigonotylus americanus	
	Rhytidolomia saucia	
	Cymus breviceps	
	Conocephalus spp.	
	Nemobius sparsalsus	
	Orphulella olivacea	
	Clinocephalus elegans	
	Naemia serriata	
	Scelionidea (undetermined)	
	Chalcidoidea (undetermined)	

Table 4-24. Maximum numbers/m^2 of some common macroinvertebrates in salt marshes adjacent to Port Royal Sound, South Carolina. D=decapod, B=bivalve, G=gastropod, P=polychaete (Vernberg and Sansbury 1972).

Species	Location				
	Chechessee Creek	Colleton River	Sawmill Creek	MacKay Creek	Hilton Head
Uca pugnax (D)	16	84	60	80	96
Uca pugilator (D)	28	12	16	20	48
Geukensia demissa (B)	4	8	12	4	8
Littorina irrorata (G)	28	120	84	4	108
Ilyanassa obsoleta (G)	868	1344	1000	476	256
Nereis succinea (P)	255	1146	764	1146	509
Laeonereis culveri (P)	382	129	509	129	0
Haploscoloplos fragilis (P)	1146	510	636	510	0
Maldanidae (undet.) (P)	129	2675	382	2675	1146

Table 4-25. Estimated numbers of adult crabs/m^2 in various saltmarsh zones of Sapelo Island, Georgia (Teal 1958b).

Species	Uca minax	Uca pugilator	Uca pugnax	Sesarma reticulatum	Sesarma cinereum	Eurytium limosum
Creek bank	--	52	13	·		6
Tall Spartina Edge Marsh	--	--		30	1	17
Medium Spartina Levee Marsh			61	21	--	4
Short Spartina Low Marsh	--	--	27	2	--	--
Short Spartina High Marsh	--	--	18	3		
Minax Marsh	32	--	4		2	--
Salicornia - Distichlis	--	43	12	--		
Juncus	·	·	·	--		--

he considered a substantially lower value to be more realistic. The insects were estimated by Smalley (1959) to consume 7% of the marsh grasses. The role of invertebrates in reworking the sediment was considered important by Kraeuter (1976) in view of their observed densities.

Cammen (1976) studied the colonization of invertebrates in _Spartina_ beds planted on dredge spoil in North Carolina. Factors controlling development of macroinvertebrate communities included the similarity of the spoil area in elevation and sediment type to natural marsh, rates of sedimentation, proximity to natural marsh areas, and the maturity of the community in adjacent natural marshes (see section on macrobenthos of estuarine subtidal environments for discussion of effects of perturbations on benthos).

3. Insects

To return to an aquatic existence, insects, primarily terrestrial or aerial organisms, have had to solve several ecological, physiological, and physical problems. For such a return to evolve, there has been a need for "bridging" habitats. Between land and sea, these environments are provided by such habitats as salt marshes and other similar intertidal zones (Cheng 1976). The majority of our so-called "marine" insects are still found in these habitats.

The insect fauna of Georgia and South Carolina salt marshes has not been studied comprehensively. Investigations on trophic relations of smooth cordgrass marshes show that insects are significant in energy flow within the marsh ecosystem (Smalley 1960, Teal 1962, Marples 1966, Foster and Treherne 1976). General surveys indicate that the saltmarsh insect fauna is varied and abundant (Davis and Gray 1966, Cameron 1972, Foster and Treherne 1976) (see Table 4-26). Davis (1978) listed insects collected from the following saltmarsh environments of South Carolina: a) the intertidal smooth cordgrass marsh, inundated at every high tide; b) the glasswort-smooth cordgrass marsh, in the vicinity of the high tide mark, where daily inundation is of brief duration; c) black needlerush and salt grass marshes, reached only by spring tides and storm tides; d) saltmeadow cordgrass, flooded only rarely (during storms or extraordinarily high tides); and e) panic grass, often appearing in dense stands on the landward side of black needlerush marshes, where freshwater drainage produces soil water of reduced salinity. Davis (1978) also listed the insects associated with plants occupying related "special" habitats or scattered among the dominant vegetation of salt marshes, including 1) sea ox-eye, often growing plentifully along the high tide mark; 2) common three-square and sawgrass, which

can form stands on the landward side of black needlerush marshes, 3) marsh elder, which can grow among saltmeadow cordgrass and black needlerush, and 4) sea myrtle, often appearing at the edge of the saltmeadow cordgrass and black needlerush marshes.

Species from all the major insect orders have been recorded from salt marshes, although Diptera (flies, mosquitoes, and midges), Coleoptera (beetles), and Hemiptera (true bugs) appear to predominate, comprising more than 75% of the species recorded. Vernberg and Sansbury (1972) collected representatives of several orders of insects, including the Homoptera, Hemiptera, Diptera, Odonata, and Lepidoptera, in high salinity marshes of the Port Royal Sound system, South Carolina. No density estimations were made, but on the basis of general observations, the overall insect community density appeared to be greatest during August.

The saltmarsh insect fauna can be divided into aquatic, subterranean, and surface-living groups (Foster and Treherne 1976). Water-oriented species in the marsh are invaders of freshwater origin. These include dipterous larvae (especially those of the families Culicidae, Chironomidae, and Ceratopogonidae), heteropterans, coleopterans, and certain trichopteran larvae (Nicol 1935, Balfour-Brown 1958, Sutcliffe 1961a, b, Green 1968b). The subterranean and surface-living forms are evidently of terrestrial origin and include representatives of most major insect orders.

Good drainage (dewatering) often occurs only at marsh edges and in the soil bordering estuarine creeks and salt-pannes (Chapman 1960, Foster and Treherne 1976). In edge regions, the soil has an enlarged pore structure, usually in the form of extensive cracks and fissures (Macleod 1967, Foster and Treherne 1975), which are suitable for colonization by insects and other marine and estuarine invertebrates. The physical factors associated with soil structure are probably important in restricting distribution of some insects within a marsh. A good example is the restriction of aphids in lower marshes to the fissured soil of regions bordering the edges of creeks and salt-pannes (Foster and Treherne 1975, 1976). A similar "edge effect" has also been described for other insect species, including a number of beetles (Evans et al. 1971) and collembolans (Green 1968b).

In intertidal marshes, insects of terrestrial origin may periodically be separated from their normal oxygen supply. Aquatic insects, however, face no unique respiratory problems in smooth saline waters (Foster and Treherne 1976). For surface-living species, the length of separation should roughly equate to the length of

Table 4-26. Seasonal distribution of dominant species of saltmarsh insects, listed by vegetative strata. Symbols: x = presence; xx = period of peak abundance (adapted from Davis and Gray 1966).

INSECT SPECIES AND TYPES OF VEGETATION	JAN	FEB	MAR	APR	MAY	JUN	JUL	AUG	SEPT	OCT	NOV	DEC
HOMOPTERA												
from: Smooth cordgrass												
Prokelisia marginata	x	x	x	x	x	x	xx	xx	xx	xx	xx	x
Sanctanus aestuarium					x	x	xx	xx	x	x	x	
Draeculacephala portola					x	x	x	xx	xx	x	x	
Graminella nigrifrons					x	x	x	x	x	x	x	xx
Delphacodes detecta	xx	x	xx		x	x	x	x	x	x	xx	xx
from: Cordgrass-Glasswort-Sea lavender												
Prokelisia marginata	x	x		x	x	x	x	x	x	x	xx	xx
Sanctanus sanctus					x	x	x	x	x	x	x	
Delphacodes detecta	x	x			x	x	x	x	x	x	xx	xx
from: Black needlerush												
Keyflana hasta	xx	x			x	xx	xx	x	x	x	x	x
Pissonotus albovenosus	x	x		xx	x	x	x	x	x	x	x	x
Rhynchomitra microrhina				x	x	x.	xx	x	x	x		
from: Salt grass												
Amphicephalus littoralis				x	x	x	xx	xx	x	x	x	x
Snangbergiella vulnerata	x		x	x	x	x	xx	xx	x	x	x	x
Delphacodes detecta				x	x	xx	xx	x	x	x		
Tumidagena terminalis				x	x	x	xx	x	x	x		
Neomegamelanus dorsalis				x	x	xx	xx	x	x	xx	x	x
Megamelus lobatus					x	x	x	x	x	x	x	
Graminella nigrifrons					x	x	x	x	x	x	x	x
Pissonotus albovenosus					x	x	x	x	x	x	x	
from: Saltmeadow cordgrass												
Delphacodes detecta	xx	x	xx	x	x	x	x	x	x	xx	x	xx
Tumidagena terminalis	x	x	x	x	x	x	x	x	x	x	x	x
Neomegamelanus dorsalis	x			x	x	x	x	x	x	x	x	x
Haplartus enotarus	xx	x		x		x	x	x	x	xx	xx	x
Aphelonema simplex	xx	x				x	x	x	x	x	x	x
Pentagramma vittatifrons	xx	x		x	x	x	x	x	x	x	x	xx
DIPTERA												
from: Smooth cordgrass												
Chaetopsis apicalis	x	x		x	x	xx	xx	xx	xx	x	x	x
Chaetopsis fulvifrons				x	x	xx	x	xx	xx	x	x	x
Conioscinella infesta	x			x	x	xx	xx	xx	xx	x	x	
Oscinella carbonaria				x	x	x	x	x	xx	xx		
Madiza trigramma?					x	x	x	x	x	x	x	
Oscinella nuda	x	x		x	x	x	x	x	x	x	x	x
Dictya oxybeles	x	x	x	x	x	x	x	x	x	x	x	x
from: Cordgrass-Glasswort-Sea lavender												
Dimeccenia austrina	x	x		x	xx	xx	xx	x	xx	xx	x	x
from: Salt grass												
Conioscinella infesta				x	x	x	xx	x	x	x	x	x
Oscinella ovalis												

231

Table 4-26. Continued

INSECT SPECIES AND TYPES OF VEGETATION	JAN	FEB	MAR	APR	MAY	JUN	JUL	AUG	SEPT	OCT	NOV	DEC
Ceropsilopa costalis	x			x	x	x	x	x	x	x		
Psilopa flavida		x		x	x	x	x	x	x	x	x x	x
Notiphila bispinosa	x	x x	x x	x x	x	x	x	x	x	x	x	
Neoscatella obscuriceps					x	x	x	x	x	x	x	x
Thinophilus ochrifacies					x x							
Tomosvaryella coquilletti												
from: Saltmeadow cordgrass				x								
Conioscinella infesta					x	x	x	x	x	x	x	
Oscinella ovalis					x	x	x	x	x			
Oscinella nuda						x	x	x x	x	x		
Hippelates particeps					x	x	x	x	x		x	
Hippelates dissidens					x	x	x	x	x	x		
Chrysotus discolor					x x	x	x	x				
Paraclius vicinus					x							x
from all types of vegetation:												
Aedes sollicitans	x	x x		x x	x	x	x	x	x x	x x	x	x
Hydrobaenus sp.			x x	x	x							
Pelastoneurus lamellatus[a]					x x	x	x	x	x	x	x	
HEMIPTERA												
from: Smooth cordgrass												
Trigonotylus uhleri					x x	x x	x x	x x	x x	x x	x x	x x
Ischnodemus badius			x	x	x	x	x	x	x x	x x	x x	
Tytthus vagus					x	x	x	x x	x x	x		
from: Black needlerush												
Cymus breviceps				x x	x x	x						
from: Salt grass												
Trigonotylus americanus				x x	x x	x x	x x	x x	x x	x		
Rhytidolomia saucia				x	x	x	x	x	x			
Cymus breviceps				x x	x x	x		x				
from: Saltmeadow cordgrass												
Trigonotylus uhleri					x	x x	x x	x x	x	x		
Cymus breviceps				x	x	x	x	x	x	x		
ORTHOPTERA												
from: Smooth cordgrass												
Orchelimum fidicinium				x	x x	x x	x	x	x	x		
from: Cordgrass-Glasswort-Sea lavender												
Orphulella olivacea					x	x	x	x	x			
from: Black needlerush												
Paroxya clavuliger						x	x	x	x x	x x		
Anakipha scia												
from: Salt grass												
Nemobius sparsaleus						x	x	x x	x x	x x		
Orphulella olivacea[b]					x x	x	x	x	x	x		
Clinocephalus elegans[b]												
from: Saltmeadow cordgrass												
Mermiria intertexta					x	x	x	x	x	x		

Table 4-26. Concluded

INSECT SPECIES AND TYPES OF VEGETATION	JAN	FEB	MAR	APR	MAY	JUN	JUL	AUG	SEPT	OCT	NOV	DEC
from all types of vegetation:												
Cycloptilum antillarum												
Oecanthus quadripunctatus						x	x	x	x	x		
Conocephalus spp.					x	xx	x	x	x	x		
COLEOPTERA												
from: Smooth cordgrass												
Collops nigriceps					x	x	x	x	x			
Isohydnocera tabida					x	x	x	x	x			
Mordellistena spp.				x	x	x	x	x	x	x		
from: Salt grass												
Naemia serriata (adults)				x	x	x	x	xx	x	x		
Naemia serriata (larvae)					x	x	x	xx	x	x		
from: Saltmeadow cordgrass												
Glyphonyx sp.						x	x	x				
Isohydnocera aegra				x	x	x	x	x				
Chaetocnema sp.						x	x	x	x			
Pachybrachys atomarius						xx	x	x				
Cryptocephalus venustus						xx	x	x				
HYMENOPTERA												
from: Saltmeadow cordgrass												
Tapinoma sessile					x	x	x	x	x			
Dorymyrmex pyramicus					x	x	x	x	x			
Iridomyrmex pruinosus				x	x	x	x	x	x			
Pseudomyrmex pallida					x	x	x	x	x	x	x	x
Monomorium minimum					x	x	x	x	x	x	x	
from all types of vegetation:												
Crematogaster clara	x			x	x	x	x	x	x	‡	x	x

a. Entry probably includes specimens of Paracitus claviculatus.
b. Young nymphs of Orphulella olivacea and Clinocephalus elegans are indistinguishable.

233

inundation by tidal waters. Some insects may avoid submergence by behavioral adaptations, while others do not. Avoidance was shown in laboratory and field responses of 12 insect species to a rising tide (Davis and Gray 1966). Insects exhibiting avoidance behavior included two Orthoptera, four Heteroptera, two Homoptera, two Diptera, and two Coleoptera. No insects were observed to be submerged by inundating tides. Submergence was avoided by flight, climbing, or movement on the surface film of the water. Other examples of insects avoiding tidal coverage are given by Arndt (1914), Teal (1962), Payne (1972), Ranwell (1972), and Polhemus (1976).

Some saltmarsh insects may be washed away during tidal coverage (Arndt 1914, Dexter 1943). Saltmarsh mosquitoes do not usually breed in marshes subject to tidal inundations, perhaps because they cannot withstand frequent flushing of their habitats by the tide (Connell 1940).

The generally limited number of plant species in salt marshes restricts the variety of potential food sources for herbivorous insects (Foster and Treherne 1976). Also, the large proportion of detritus swept away from the marsh surface by the tides (Teal 1962, Cameron 1972, Jeffries 1972) decreases the amount of food available to detritus-feeding insects. Algae provide a food source for insect invaders from terrestrial and freshwater environments (Foster and Treherne 1976). The detailed study by Davis and Gray (1966) of the insects of North Carolina marshes showed that these organisms were selective in their associations with particular plant zones (Table 4-27). Fifty percent of the plant tissue and sap feeders were associated with only one of the five major plant zones delineated.

Most of the insect species inhabiting brackish and salt marshes are herbivores and fall into three categories (see Table 4-27). The first group, those that feed on plant tissues, includes insects with chewing mouth parts. The principal herbivorous chewing insects in most, or all, types of saltmarsh vegetation are the grasshoppers, but several species of ants are also common in saltmeadow cordgrass. The second group, those that feed on plant sap, is composed of insects with piercing mouth parts, which enable them to penetrate plant tissues. These sap feeders are abundant in all types of marshes except black needlerush, and include many of the common coastal species of homopterans and hemipterans. The third group comprises those that principally subsist on fluids secreted by marsh plants. Included are several common species of flies (Diptera), particularly Chaetopsis fulvifrons, Chaetopsis apicalis, and Conioscinella infesta.

While the principal non-aquatic carnivorous arthropods in salt and brackish marshes are generally spiders (Barnes 1953, Marples and Odum 1964, Davis and Gray 1966), the most abundant carnivorous insects are odonates, beetles, asilids, mosquitoes, and reduviids. Predaceous insects can be divided into two groups, 1) those that feed primarily on solid tissues, and 2) those that extract the body fluids of their prey. Examples of insects belonging to the first group are dragonflies such as Erythrodiplax berenice, Pachydiplax longipennis, and Erythemis simplicicollis, individuals of which can often be seen fluttering about black needlerush marshes feeding on small flying insects. The second group of predators includes the asilid flies, which can sometimes be observed preying on marsh grasshoppers. Midges and culicids may be abundant in poorly drained marshes and especially in salt grass stands. Sciomyzid flies (e.g., Dictya oxybeles and Hoplodictya spinicornis) can be encountered in smooth cordgrass and cordgrass-glasswort-sea lavender communities. It has been suggested that all sciomyzid larvae attack snails (Berg et al. 1955). If so, it is likely that the larvae of Dictya oxybeles and Hoplodictya spinicornis feed upon the marsh periwinkle, Littorina. Predaceous bugs, such as the reduviids Doldina interjungens, Sinea diadema, Zelus cervicalis, and the nabid, Nabis capsiformis, can be widely distributed in the higher marsh zones.

Detritus, derived principally from plant tissues, is an abundant source of food in wetlands (Teal 1959b). Not surprisingly, then, the detritus-feeding ephyrid and dolichopodid flies are common in smooth cordgrass, cordgrass-glasswort-sea lavender, and salt grass stands.

Dipterous larvae parasitize saltmarsh plants, especially smooth cordgrass, where they live in the stems. The adults of parasitic Hymenoptera (chalcids, braconids, ichneumonids, tiphiids, and scelionids) occur in most or all types of saline marshes, and their larvae undoubtedly infect various types of saltmarsh insects (Davis and Gray 1966).

The major energy flow between autotrophic and heterotrophic levels in salt and brackish marshes is through the detritus rather than the grazing food chain (Odum and Smalley 1959, Smalley 1960, Teal 1962) (see Fig. 4-3). Since most saltmarsh insects are grazers (Marples 1966), their role in the consumption of primary producers should be small. However, Smalley (1960) estimated that a grasshopper (Orchelimum fidicinium) population of a Georgia salt marsh consumed 1% of the Spartina production, and Teal (1962) estimated that the herbivorous insects (Orchelimum, Prokelisia marginata) on a Georgia salt marsh assimilated approximately 4.6% of the Spartina production. A number of other insects have been shown to feed on Georgia saltmarsh plants (Davis

Table 4-27. Trophic relationships of characteristic insects and other associated invertebrates from the herbaceous strata of five types of North Carolina salt marshes (adapted from Davis and Gray 1966).

FEEDING HABITS FOOD	DOMINANT PLANTS				
	SMOOTH CORDGRASS	CORDGRASS - GLASSWORT - SEA LAVENDER	BLACK NEEDLERUSH	SALT GRASS	SALTMEADOW CORDGRASS
HERBIVOROUS					
Plant Tissues	Orchelimum fidicinium Conocephalus spp. Mordehistena spp.	Orphelella olivacea	Paroxya clavuliger Conocephalus spp.	Orphulella olivacea Conocephalus spp. Clinocephalus elegans Nemobius sparsalsus	Conocephalus spp. Mermiria intertexta Glyphonyx sp. Dorymyrmex pyramicus Pseudomyrmex pallida Iridomyrmex pruinosus
Plant Sap	Prokelisia marginata Sanctanus aestvarium Droeculacephala portola Ischnodemus badius Trigonotylus uhleri	Prokelisia marginata Sanctanus sanctus	Keyflena hasta Rhynchomitra micro- rhina	Amphicephalus littoralis Spangbergiella vulnerata Delphacodes detecta Tumidagena terminalis Neomegamelanus dorsalis Trigonotylus americanus Rhytidolomia saucia Cyrnus breviceps	Delphacodes detecta Tumidagena terminalis Neomegamelanus dorsalis Haplaxius enotatus Aphelonema simplex Trigonotylus uhleri Cyrnus breviceps
Plant Secretions	Chaetopsis apicalis Chaetopsis fulvifrons Conioscinella infesta	Chaetopsis apicalis Chaetopsis fulvifrons Conioscinella infesta		Conioscinella infesta Oscinella ovalis	Conioscinella infesta Oscinella ovalis Hippelates particeps
CARNIVOROUS					
Animal Tissues	Isohydnocera tabida Collops nigriceps	Spiders	Erythrodiplax berenice Spiders	Naemia serriata Spiders	Isohydnocera aegra Spiders
Animal Body Fluids	Dictya oxybeles Hoplodictya spinicornis Spiders	Dictya oxybeles Hoplodictya spinicornis Spiders	Reduviids Asilids Spiders	Tomosvaryella coquilletti Reduviids Culicids Asilids Spiders	Reduviids Asilids Spiders

Table 4-27. Concluded

DOMINANT PLANTS

FEEDING HABITS FOOD	SMOOTH CORDGRASS	CORDGRASS - GLASSWORT - SEA LAVENDER	BLACK NEEDLERUSH	SALT GRASS	
OMNIVOROUS					
Detritus	Ephydrids Dolichopodids Littorina irrorata	Ephydrids Dolichopodids		Ephydrids Dolichopodids	
PARASITIC					
Plant Tissues and Sap	Dipterous larvae	Dipterous larvae		Dipterous larvae	Dipterous larva
Animal Tissues and Body Fluids	Larvae of parasitic Hymenoptera	Larvae of parasitic Hymenoptera	Larvae of parasitic Hymenoptera	Larvae of parasitic Hymenoptera	Larvae of paras Hymenoptera

236

and Gray 1966, Marples 1966), and it may be reasonable to assume, following Kraeuter and Wolf (1974), that insects on these marshes at times may consume as much as 10% of the annual Spartina production.

Any treatment of salt and brackish marshes of South Carolina and Georgia should include the important saltmarsh grasshopper, Orchelimum fidicinium, an insect more frequently found in this habitat than in any other. Much of the following is summarized from the excellent paper by Smalley (1960).

Orchelimum fidicinium is the only species of grasshopper commonly found on open salt marshes of the Carolinas and Georgia, although other species occur near terrestrial vegetation (e.g., Orphulella olivacea, Conocephalus spp., Clinocephalus elegans, and Mermiria intertexta). The saltmarsh grasshopper is an abundant and conspicuous insect. Nymphs first appear in April and May, and by the end of summer only scattered individuals remain. One generation is produced each year, and hatching occurs within a short period. The saltmarsh grasshopper is most common in moderately tall, dense smooth cordgrass. Orchelimum is generally not found in extremes of cordgrass growth, in the very tall grass [>2 m (6.6 ft)] along larger creeks, or in very short grass[<1 m (3.3 ft)] growing in extensive areas of high and poorly drained marsh. During high tides, these grasshoppers cling to Spartina stems projecting above the water. They may dive under water at the approach of man or boat, but normally they are aerial. During very high tides, grasshoppers would appear to be vulnerable to predators, but unusually high mortality during spring tides has not been detected by quantitative sampling.

Species such as O. fidicinium, which hatch, grow, and die within 1 year, present an excellent opportunity for study of population energy flow in nature. Smalley (1960) studied the population density, growth rates, respiratory rates, caloric content, and rates of defecation for this grasshopper in Georgia salt marshes. Spartina ingested by the grasshoppers, assimilated with an efficiency of about 27%, amounts to approximately 2% of the net production of this marsh plant. Grass which is ingested but not assimilated does not enter the energy balance of the population. However, this process can be ecologically significant, since the grass is removed from the cordgrass stems and deposited on the marsh or in tidal waters in the form of feces, where it becomes available to a different group of organisms.

Insects serve as an important food source in the wetland community, and a wide variety of insect predators and parasites have been recorded from brackish and salt marshes (Foster and Treherne 1976). Birds, many species of which feed commonly on saltmarsh insects, are the most important vertebrate predators (Green 1968a). Birds prey particularly on herbivorous insects seeking refuge on tall vegetation during tidal submergence (Arndt 1914, Smalley 1960, Teal 1962, Davis and Gray 1966).

One of the birds which consumes large quantities of brackish and saltmarsh insects is the marsh wren. Kale (1965), in a study of the ecology and energetics of the marsh wren in Georgia salt marshes, provided a considerable list of insects consumed by this predator and their frequency of occurrence in the diet (Table 4-28). This information not only serves to reemphasize the important role of insects in marsh energetics, but also provides a useful listing of insects present in salt marshes of the study area. Kale (1965) also provided estimates of marsh insect and spider population densities, reporting a mean density of over 500 individuals/m^2. This density produced a mean dry biomass/m^2 of about 300 mg, and a mean energy content of 1,500 g cal/m^2. Kale concluded that the marsh wren is an important predator on 1) herbivorous insects of marsh vegetation; 2) hymenopterans, especially ants (Formicidae); 3) small parasitic wasps (Braconidae, Chalcidae, Ichneumonidae, Scelionidae); and 4) spiders that prey on insects. This wren may, therefore, be a major factor in the control of secondary consumers among the arthropods of the grazing food chain. It is reasonable to assume that a number of other coastal birds would similarly influence insect and other arthropod population dynamics in brackish and salt marshes.

Salt marshes provide breeding habitat for several medically and economically important insect species, particularly biting Diptera (Daiber 1974, Axtell 1976, Linley 1976, O'Meara 1976). The important biting insects along the coast of the eastern United States include several species of mosquitoes (Family Culicidae), gnats or sand flies (Family Ceratopogonidae), and tabanids such as greenhead flies and deer flies (Family Tabanidae). These species typically develop in different portions of the marsh and upland. In general (see Fig. 4-10), the saltmarsh (Aedes) mosquitoes breed in higher and only intermittently flooded portions of the marsh, often characterized overall by black needlerush vegetation. Mosquitoes do not breed in the low marsh, which is flooded by each high tide, but tabanid flies and gnats do. One species of gnat, Culicoides melleus, breeds in the very sandy margins of these marshes, as well as along the margins of creeks and sounds (Axtell 1974).

The socioeconomic impact of these insects is considerable. Control of

Table 4-28. Saltmarsh insects consumed by the marsh wren in Georgia salt marshes, and their average frequency of occurrence in pooled stomach samples (adapted from Kale 1965).

Insects	Frequency (percent)	Percent Total Volume
Total Hymenoptera	80.5	15.4
Formicidae	45.7	9.1
Ichneumonidae	5.1	0.2
Braconidae	21.6	2.7
Chalcididae	14.9	1.1
Others	48.7	2.5
Total Homoptera	68.2	23.7
Fulgoridae	63.6	23.0
Total Coleoptera	68.2	12.0
Curculionidae	35.4	5.4
Anthribidae	4.1	0.4
Cleridae	21.6	2.4
Melyridae	7.2	1.1
Mordellidae	14.4	1.4
Others	11.3	0.4
Larvae	9.2	1.1
Total Diptera	46.2	8.4
Ephydridae	13.8	3.6
Otitidae	13.3	1.2
Dolichopodidae	7.2	0.7
Others	19.5	2.8
Larvae	7.7	1.3
Total Hemiptera (Ischnodemus badius)	41.6	7.2
Total Lepidoptera	24.1	9.9
Larvae & Eggs	19.0	7.4
Total Orthoptera (Orchelimum fidicinium)	7.2	3.8
Total Insect Eggs	9.23	2.0

mosquitoes and biting flies in coastal areas has been a difficult, expensive, and continuous task over the past several decades. Such control has often been necessary to prevent disease transmission. It also provides relief from attack for citizens and helps various segments of the economy, especially the tourist and recreational industries.

With the advent of pesticides, particularly the chlorinated hydrocarbons such as DDT, control of nuisance insects has been aided, but not without major environmental complications. Ecological impacts of such pesticides have been well documented and will not be repeated here because the literature is massive. For further information, the reader is directed to Springer and Webster (1951) and Springer (1961). Also, the Office of

Biological Services (U.S. Department of Interior, Fish and Wildlife Service 1976) produced an extensive annotated bibliography on mosquito control procedures and practices and their effect on the environment. Reimold and Shealy (1976) monitored such pesticides reaching nontarget organisms from runoff into 11 estuaries representing all the major Atlantic drainage basins in Georgia and South Carolina. Low but detectable levels of DDT (10 - 33 ppb) and its metabolites DDE (10 - 23 ppb) and TDE (10 - 20 ppb) were found in these nontarget estuarine animals, despite the fact that use of DDT was discontinued in 1972.

Man has also at times aggravated the mosquito problem through coastal dredge and disposal activities. Extensive dredging to maintain channels in coastal rivers

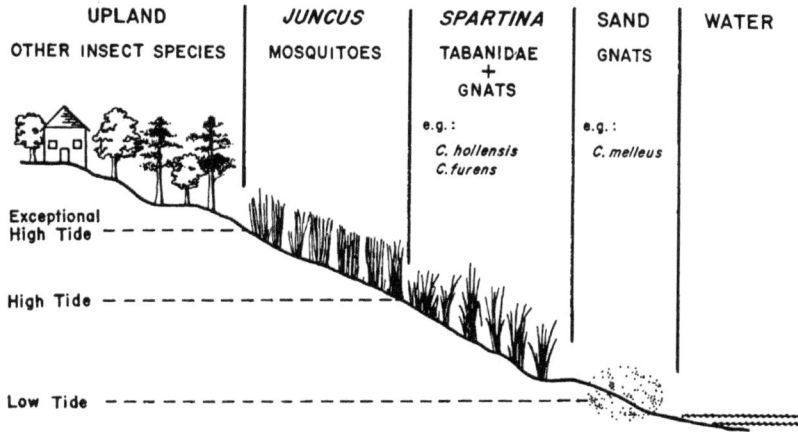

UPLAND	JUNCUS	SPARTINA	SAND	WATER
OTHER INSECT SPECIES	MOSQUITOES	TABANIDAE + GNATS	GNATS	

e.g.:
C. hollensis
C. furens

e.g.:
C. melleus

Exceptional
High Tide

High Tide

Low Tide

Figure 4-10. Diagram of saltmarsh habitats showing breeding sites of mosquitoes (Culicidae), deer flies and greenhead flies (Tabanidae), and biting gnats (Ceratopogonidae) in relation to dominant marsh vegetation and tide levels. This is generalized and the vertical scale is highly exaggerated (Axtell 1974).

and the Atlantic Intracoastal Waterway (AIWW) often brings about concurrent nearby deposition of dredged materials. These disposal sites often present ideal conditions for breeding mosquitoes. Dr. W. Bruce Ezell, Jr. and his associates in the Department of Biology, The Citadel, are studying insect succession in dredged material disposal sites. These data are being correlated with plant succession and soil conditions, including information on soil and water chemistry and plant productivity. These investigators are also researching coastal mosquitoes and are comparing habitats such as swales of barrier islands, dredge disposal sites, and duck ponds, including diked-off old rice fields (W. B. Ezell, 1979, The Citadel, Charleston, South Carolina, pers. comm.).

4. Fishes

A number of studies have been conducted on fishes inhabiting the emergent wetlands of the Sea Island Coastal Region. Dr. John M. Dean (ongoing research, University of South Carolina) is investigating the fishes of shallow tidal creeks in the Santee estuary using a variety of collecting methods, including small mesh stop nets, gill nets, bottom trawls, and nekton nets. In other areas of South Carolina, investigations of the fishes of intertidal creeks associated with emergent wetlands have been carried out. Turner and Johnson (1972, 1974) collected fishes

and invertebrates in small tidal creeks of Port Royal Sound and the Cooper River. In 1972, the Marine Resources Division conducted stop net surveys in small intertidal creeks at several locations along the South Carolina coast (South Carolina Marine Resources Division, 1972, Charleston, unpubl. data). Cain (1973) studied the annual occurrence, abundance, and diversity of fishes in an intertidal creek at North Inlet. Burns (1974) investigated the species and diversity of larval fishes in a high marsh tidal creek at North Inlet.

In Georgia, Dahlberg (1972) conducted investigations of fishes in tidal creeks and marshes of several estuaries. Miller and Jorgenson (1969) recorded 56 fish species from the high marsh habitat in Georgia. Mahood et al. (1974a, b, c, d) conducted seining in small tidal creeks throughout coastal Georgia over a 3-year period.

The National Marine Fisheries Service (NMFS) Beaufort Laboratory in North Carolina has conducted sampling for juvenile Atlantic menhaden in tidal creeks of Georgia and South Carolina since 1966. Very little of this information is in published form, although it has been used to calculate an annual crude index of year-class abundance. Tagging studies of juvenile Atlantic menhaden also have been conducted by NMFS along the South Atlantic (Kroger and Guthrie 1972).

239

Fishes which typically inhabit intertidal marshes of South Carolina and Georgia include those which move in and out with the tide and those which frequent potholes and standing pools of water remaining after the tide recedes. The most common resident species are the mummichog, sheepshead minnow, Atlantic silverside, and bay anchovy. Transient species which frequent intertidal wetlands during various stages of their life cycle include mullets and sciaenids (spotted seatrout, silver perch, and spot). Tidal pools in the high marsh are inhabited by cyprinodonts, sailfin molly, mosquitofish, and other stress-tolerant species. Other fishes at times found in tidal pools include ladyfish, juvenile tarpon, and snook.

In South Carolina, Cain and Dean (1976) collected 16,611 individuals representing 51 species of marine and estuarine fishes from a tidal creek in the North Inlet area over a 1-year period. They sampled with a small mesh stop seine. In early spring, the area was dominated by mummichog and Atlantic silverside, although bay anchovy, striped killifish, and larval or postlarval spot, Atlantic croaker, and mullet were numerous. The mummichog, striped killifish, and Atlantic silverside feed almost exclusively on larval fishes, whereas bay anchovy subsisted largely on planktonic crustaceans (copepods, etc.). In the summer, young spot were most abundant, followed by juvenile mullet, silver perch, and spotfin mojarra. Larger predatory species also were present, including Atlantic needlefish, bluefish, crevalle jack, and flounders, but these fishes were not numerous. In the fall, mummichog again became the most abundant species. Other abundant fishes were silversides, silver perch, anchovies, and mullets. Predators such as longnose gar, Atlantic needlefish, jacks, and flounders were numerous. In winter, Atlantic silversides were dominant, followed by mummichog, striped killifish, and pinfish; the first three species feed on spot and mullet larvae.

The diversity of fishes in the intertidal creek appears to be correlated with annual temperature cycles (Cain 1973). Only a limited number of common species (including mummichog, Atlantic silverside, striped killifish, and bay anchovy) would be considered year around, eurythermal residents.

During January 1977, Shenker and Dean (1979) intensively sampled a high salinity creek ($25^o/oo - 32^o/oo$) in the same North Inlet estuary. A total of 14,730 fishes representing at least 22 species were captured. Of this total, 11,051 were immature fish and represented at least 13 species.

Of the larval forms present, six species made up 99% of the total catch of immature fishes: larval spot, juvenile mullet, the leptocephalus of the speckled worm eel, larval pinfish, larval flounders, and larval Atlantic croaker. Larval spot was by far the most abundant but showed no apparent day-night pattern. No day-night pattern was evident for larval pinfish, juvenile mullet, and Atlantic croaker. Eel leptocephali were captured more frequently during daylight while larval flounders were taken more frequently at night.

The fishes utilizing the intertidal marsh-creek areas have a distinct advantage over other species in the estuary in that they have more available productive habitat. Species that can withstand the rigors of intertidal areas are extremely important ecologically.

Fishes which exhibit the highest annual biomass and abundance in intertidal estuarine wetlands are detritivores, omnivores, or primary consumers. None appear to be limited to an exclusive trophic level, since feeding behavior changes as they grow. Odum (1970a) stated that most food chains in shallow estuaries are based upon plant detritus and microalgae rather than on phytoplankton-zooplankton production. Fishes and other animals replace zooplankton as the critical herbivore link. For example, in marsh creeks, mullet feed largely on marsh detritus and living algae, and in turn are preyed upon by mid and top level carnivores.

Turner and Johnson (1973, 1974) conducted investigations of the standing crops of fishes and invertebrates inhabiting tidal creeks in the Port Royal and Cooper River estuaries, South Carolina, using small mesh stop nets. Although species diversity was greater in the high salinity Port Royal area (59 species) than in the brackish Cooper River area (45 species), total numbers of fishes and biomass were considerably higher in tidal creeks of the latter estuary (Table 4-29). Atlantic silverside, striped mullet, bay anchovy, mummichog, Atlantic menhaden, silver perch, and flounders were the most abundant species collected in the Port Royal creeks. In Cooper River tidal creeks, Atlantic croaker, mummichog, Atlantic menhaden, and spot predominated, with other abundant species including tidewater silverside, silver perch, striped mullet, freshwater goby, and southern flounder. Common invertebrate food organisms associated with the fishes in tidal creeks included grass shrimp (Palaemonetes spp.), penaeid shrimp (Penaeus spp.), juvenile blue crabs, and squid (Lolliguncula brevis). Salinity in Cooper River tidal creeks was very low during the study period ($0.0^o/oo - 3.0^o/oo$). This study is one of the few quantitative investigations conducted in low salinity tidal creeks in the upper reaches of an estuary in either

Table 4-29. Rough comparison of standing crop summaries (catch/surface area) of marine and estuarine fishes in three Cooper River, South Carolina tidal streams (Turner and Johnson 1974) and five tidal streams in the Port Royal Sound estuary, South Carolina (Turner and Johnson 1972). Numbers of organisms are subtended by weights (pounds) in parentheses.

FISHES	COOPER RIVER CREEKS			PORT ROYAL SOUND CREEKS		
	April	July	November	April	July	October
Carcharhinidae						
Lemon shark					1 (0.99)	
Dasyatidae						
Southern stingray				1 (0.95)	1 (1.05)	1 (0.35)
Lepisosteidae						
Longnose gar		2 6.4)	1 (0.1)			
Elopidae						
Ladyfish					1 (0.32)	
Anguillidae						
American eel	495 (10.4)	451 (37.5)	9 (0.9)	10 (0.23)	1 (0.05)	1 (0.09)
Clupeidae						
Atlantic menhaden	8,621 (2.7)	12,330 (35.1)	5 (0.1)	37 (1.56)	15 (1.25)	9 (1.32)
Atlantic thread herring				1 (0.01)	46 (0.35)	56 (1.14)
Blueback herring			384 (1.9)			
Gizzard shad	8 (0.3)	87 (0.5)	7 (1.4)	1 (0.05)		
Hickory shad	1 0.6)					
Threadfin shad	1 0.1)					

Table 4-29. Continued

FISHES	COOPER RIVER CREEKS			PORT ROYAL SOUND CREEKS		
	April	July	November	April	July	October
Engraulidae						
Bay anchovy	36 0.1)		4,083 (7.6)	13,715 (30.03)	1,573 (0.68)	3,997 (8.32)
Striped anchovy				1 (0.01)	32 (0.26)	27 (0.51)
Synodontidae						
Inshore lizardfish					1 (0.03)	
Ictaluridae						
White catfish	196 5.6)	407 (12.1)	154 (10.2)			
Ariidae						
Sea catfish				1 (0.50)	4 (1.59)	
Batrachoididae						
Oyster toadfish				1 (0.02)		
Gadidae						
White hake				1 (0.01)		
Hemiramphidae (see Exocoetidae)						
Halfbeak				1 (0.02)		1 (0.01)
Belonidae						
Atlantic needlefish	1 (0.3)	4 (0.1)	1 (0.1)			
Cyprinodontidae						
Mummichog	2,100 (7.0)	9,748 (18.3)	269 (0.3)	162 (0.86)	22 (0.11)	1 (0.01)
Striped killifish				45 (0.10)		1 (0.01)
Poeciliidae						
Mosquitofish		1 (0.1)				

242

Table 4-29. Continued

FISHES	COOPER RIVER CREEKS			PORT ROYAL SOUND CREEKS		
	April	July	November	April	July	October
Atherinidae						
Atlantic silverside	1 (0.1)	645 (0.6)		424 (4.44)	132 (0.21)	968 (6.44)
Tidewater silverside	414 (0.3)	541 (1.2)	696 (0.4)			
Rough silverside				17 (0.15)		
Syngnathidae						
Northern pipefish			1 (0.1)			
Chain pipefish					1 (0.01)	1 (0.01)
Centropomidae						
Snook			1 (0.1)			
Percichthyidae						
Striped bass			1 (0.4)			
Pomatomidae						
Bluefish		3 (0.1)		1 (0.01)	1 (0.06)	1 (0.03)
Carangidae						
Atlantic bumper					4 (0.1)	4 (0.15)
Crevalle jack			9 (0.2)		1 (0.01)	2 (0.01)
Leatherjacket					3 (0.01)	
Lookdown						1 (0.01)
Lutjanidae						
Gray snapper						1 (0.01)

Table 4-29. Continued

FISHES	COOPER RIVER CREEKS			PORT ROYAL SOUND CREEKS		
	April	July	November	April	July	October
Gerreidae						
Silver jenny			17 (0.1)			15 (0.25)
Pomadasyidae						
Pigfish					4 (0.10)	
Sparidae						
Pinfish						1 (0.05)
Sheepshead					1 (0.01)	1 (0.02)
Sciaenidae						
Atlantic croaker	32,473 (38.8)	3 (0.1)	9 (0.1)	48 (0.22)	<1 (0.11)	
Black drum					1 (0.07)	
Silver perch		1,701 (0.6)	1 (0.1)	21 (2.83)	622 (3.49)	470 (5.61)
Spot	12 (0.6)	1,375 (6.0)	1,987 (1.0)	25 (0.37)	52 (2.10)	12 (1.37)
Spotted seatrout			2 (0.1)	1 (0.19)	16 (0.95)	21 (1.89)
Weakfish					1 (0.01)	
Ephippidae						
Atlantic spadefish					1 (0.01)	
Mugilidae						
Striped mullet	1,488 (31.9)	429 (5.5)	9 (2.0)	302 (25.95)	948 (33.84)	140 (8.42)
Sphyraenidae						
Great barracuda						1 (0.01)

Table 4-29. Continued

FISHES	COOPER RIVER CREEKS			PORT ROYAL SOUND CREEKS		
	April	July	November	April	July	October
Blenniidae						
Feather blenny						1 (0.01)
Gobiidae						
Clown goby						1 (0.01)
Freshwater goby	1,386 (2.5)	1,403 (2.9)	404 (0.1)	1 (0.01)		
Green goby				1 (0.01)		
Lyre goby		1 (0.1)				
Naked goby	361 0.8)	267 (0.1)	139 (0.1)	3 (0.01)		
Sharptail goby		1 (0.1)	406 (0.2)			
Trichiuridae						
Atlantic cutlassfish				1 (0.01)		
Scombridae						
Spanish mackerel					6 (0.20)	1 (0.08)
Stromateidae						
Southern harvestfish					1 (0.01)	3 (0.09)
Butterfish				1 (0.05)		1 (0.05)
Triglidae						
Leopard searobin					1 (0.01)	
Bothidae						
Fringed flounder		4 (0.1)			2 (0.01)	1 (0.01)

245

Table 4-29. Concluded

FISHES	COOPER RIVER CREEKS			PORT ROYAL SOUND CREEKS		
	April	July	November	April	July	October
Bothidae (Continued)						
Southern flounder	1,383 (1.0)				1 (0.84)	1 (0.19)
Summer flounder			1 (0.2)	3 (1.58)	3 (1.65)	
Ocellated flounder				1 (0.01)		
Bay whiff					1 (0.01)	1 (0.01)
Gulf flounder				2 (0.61)	1 (0.06)	1 (0.12)
Unidentified young				230 (0.32)		
Soleidae						
Hogchoker	1 0.1)	1 (0.1)	1 (0.1)			
Cynoglossidae						
Blackcheek tonguefish					1 (0.01)	
Balistidae						
Orange filefish					1 (0.01)	
Pygmy filefish					1 (0.01)	
Diodontidae						
Striped burrfish					1 (0.01)	1 (0.01)
TOTAL FISHES	49,003 (102.0)	29,676 (137.7)	8,617 (30.9)	15,073 (71.45)	3,524 (51.05)	5,795 (39.03)
NUMBER OF SPECIES	24	30	32	32	40	36

South Carolina or Georgia, and indicates that the significance of these areas may be much greater than previously recognized.

In 1972, the South Carolina Marine Resources Division (unpubl. data) conducted tidal creek blockage sampling with a small mesh stop seine at four selected stations from Sewee Bay south to Toogoodoo Creek at the head of the North Edisto River. A total of 33 species of fish was collected. Atlantic silverside, mummichog, striped killifish, and striped mullet were by far the most abundant species collected (Table 4-30).

Mahood et al. (1974a, b, c, d) collected 46 species of fishes from seine collections made in small tidal creeks located in every major estuary of Georgia. The most common species encountered were mullet, spot, black drum, Atlantic croaker, and spotted seatrout. Total numbers and average sizes of juveniles of the 12 most important species taken are presented in Table 4-31. In the high marsh, Miller and Jorgenson (1969) recorded 56 species of fishes. Table 4-32 presents abundance figures for the 10 most common species collected during this investigation. In this habitat, the mummichog was the most abundant species. Dahlberg (1972) described the distribution of fishes in relation to environmental factors in a Georgia estuary and adjoining beach and in coastal plain creeks. The distributions of 168 fish species were related to nine recognizable habitats, including high marsh, low salinity tidal pool, and high salinity tidal pool. He recorded 22 species from the low salinity tidal pool habitat. Species characteristic of this habitat included ladyfish, tarpon, sheepshead minnow, mummichog, mosquitofish, sailfin molly, mullets, snook, and spot. Species restricted to low salinity tidal pools include marsh killifish, fat sleeper, and freshwater goby. Not surprisingly, high salinity tidal pools in Georgia supported a more diverse fish fauna (37 species). Large numbers of cyprinodontiformes (mummichog, sailfin molly, sheepshead minnow, striped killifish, mosquitofish) and spotfin mojarra were found in shallow pools of this type, while young silver perch, spot, and gobies were numerous in nearby deep pools. Ladyfish, mullets, and bay anchovy were found in large numbers in both types of pools.

During an investigation of saltmarsh fishes at Sapelo Island, Georgia (Linton and Rickards 1965), relatively large numbers of young-of-the-year snook were collected by small mesh seines in the upper reaches of tidal marsh creeks. Fishes usually taken in association with snook included juvenile tarpon, ladyfish, mojarras, mullets, spot, sailfin molly, mosquitofish, sheepshead minnow, and gobies.

Although much information is available on the species occurrence, diversity, and distribution of fishes in the intertidal wetlands of South Carolina and Georgia, little work has been done on their life histories, trophic relationships, or population dynamics. Likewise, the effects of man-induced changes on the fishes of this habitat are little known or understood. Fishes found in intertidal wetlands are highly susceptible to pesticides applied for insect control on adjacent coastal islands and agricultural lands, and kills due to the introduction of chemicals such as malathion, parathion and toxaphene have been frequent in recent years (South Carolina Marine Resources Division, Charleston, unpubl. data). Dredge and fill operations, drainage modifications on adjacent uplands, ditching, and other alterations can also have a significant but little known impact upon the marine and estuarine fishes inhabiting intertidal wetlands.

5. Amphibians and Reptiles

Few amphibians and reptiles are represented in the irregularly flooded intertidal areas of estuaries in Georgia and South Carolina, and none reported from this habitat are truly characteristic. Most records are of individuals reported from areas outside the Sea Island Coastal Region, but because of the similarity of habitats and conditions, these species would be expected to occasionally visit or utilize this habitat along the Georgia and South Carolina coasts, if only as transients. The paucity of herpetological research in saline habitats of the United States was noted by Neill (1958).

Southern toads were observed by Engels (1952) to invade irregularly flooded areas of Shackleford Banks, North Carolina, and Neill (1958) reported eastern narrowmouth toads on glasswort flats of Merritt Island, Florida. Eastern spadefoot toads are known to occur in the Altamaha Basin, Georgia (Georgia Department of Natural Resources, 1979, Atlanta, unpubl. data). Southern leopard frogs are often reported from saltmarsh habitats (Pearse 1936, Carr 1940, Neill 1958, Blaney 1971), and they are probably the most common amphibians in this habitat within the characterization area. Neill (1958) reported little grass frogs, green treefrogs, and chorus frogs from various halophytic flora such as "black rushes" (= Juncus?), cordgrass and "salt-marsh vegetation" on Merritt Island, Florida. Green treefrogs were noted in brackish habitats near Kiawah Island, South Carolina (Gibbons and Harrison 1975), and a similar species, the squirrel treefrog, was observed in saltmarsh vegetation in northwest Florida (Blaney 1971). The few reports of salamanders in estuarine intertidal situations involve circumstances which categorize the occurrence as accidental (see Neill 1958, and Amphiuma account under the subsection on flats of the intertidal estuarine subsystem).

Table 4-30. Total numbers and weights of marine and estuarine fishes collected by stop net at four selected tidal creek stations in South Carolina, May – September, 1972 (South Carolina Marine Resources Divison, unpubl. data).

FISHES	SEWEE CAMP No.	kg	BREACH INLET No.	kg	CHURCH CREEK No.	kg	TOOGOODOO CREEK No.	kg	TOTALS No.	kg
Atlantic stingray	2	2.7			2	3.3			4	6.0
Ladyfish					4	0.6			4	0.6
American eel	1	< 0.1							3	0.2
Speckled w rm eel	2	< 0.1							2	0.1
Atlantic menhaden	12	0.4							12	0.4
Atlantic thread herring	8	< 0.1	16	0.1					24	0.2
Bay anchovy			228	0.4	1	< 0.1	96	0.2	325	0.5
Inshore lizardfish			1	0.1					1	0.1
White catfish					3	0.8			3	0.8
Sea catfish	1	0.2			15	3.0			16	3.2
Oyster toadfish			3	0.1					3	0.1
Skilletfish			7						7	0.1
Atlantic needlefish	309	0.9	19	0.9	4,685	6.9	2	< 0.1	21	0.9
Mummichog	108	0.4	11,804	55.8			726	2.3	71,524	65.9
Striped killifish	3	< 0.1	15,500	143.3	1,501	2.9	8	< 0.1	17,117	146.7
Atlantic silverside	1	< 0.1	100,233	103.4			10	< 0.1	100,246	103.6
Northern pipefish			1	< 0.1	1	0.2			2	0.1
Striped bass									1	0.2
Bluefish	13	0.3	14	0.5	3	< 0.1			27	0.8
Horse-eye jack									8	0.1
Crevalle jack	8	0.1							3	0.1
Atlantic bumper							3	< 0.1	4	0.1
Gray snapper			4	0.1					5	0.1
Spotfin mojarra			5	0.1					17	0.5
Pigfish			17	0.5					2	0.4
Sheepshead			2	0.4					24	1.2
Pinfish			24	1.2					327	4.8
Silver perch	27	0.5	79	2.8	20	1.1	201	0.4	20	7.8
Spotted seatrout	9	2.0	6	0.9	3	0.2	2	0.2	399	1.1
Weakfish	179	0.2			220	0.9			35	0.2
Banded drum	35	0.2							357	7.8
Spot	94	6.8	92	7.8	169	5.8	2	0.1	75	0.7
Atlantic croaker	75	0.7	7	1.1					10	1.4
Black drum	1	0.2	5	2.4			2	< 0.1	6	2.5
Red drum							1	0.1	1	0.1
Star drum	1,547	44.2	366	40.4	229	9.5	4	0.9	2,146	95.1
Striped mullet	1,295	2.8	92	1.4	219	2.6			1,606	6.8
White mullet			1	0.1					1	0.1
Guaguanche									1	0.1

248

Table 4-30. Concluded

FISHES	SEWEE CAMP No.	kg	BREACH INLET No.	kg	CHURCH CREEK No.	kg	TOOGOODOO CREEK No.	kg	TOTALS No.	kg
Florida blenny			1	< 0.1					1	< 0.1
Blennies			3	< 0.1					3	< 0.1
Feather blenny			1	0.1					1	< 0.1
Naked goby							1	<0.1	1	< 0.1
Gobies			5	< 0.1					5	< 0.1
Darter g by							3	<0.1	3	< 0.1
Highfin g by							1	<0.1	1	< 0.1
Spanish mackerel			7	0.2					7	0.2
Bay whiff	7	0.3							7	0.3
Fringed flounder							2	<0.1	2	< 0.1
Summer flounder	1	< 0.1	15	1.9					16	0.1
Southern flounder	9	1.7	29	9.2	6	0.6	5	1.0	49	12.5
Blackcheek tonguefish					1	<0.1			1	< 0.1

Table 4-31. Relative abundance (i.e., total numbers of specimens collected) of the 12 most important species of juvenile fishes taken during monthly seining in tidal creeks throughout coastal Georgia (Mahood et al. 1974a, b, c, d). Second row of numbers for each species is average size in mm.

SPECIES	OCT	NOV	DEC	JAN	FEB	MAR	APR	MAY	JUN	JUL	AUG	SEPT	TOTAL/AVG. SIZE
Spotted seatrout	4 85	4 124	1 79					12 20	16 41	77 28	88 33	36 51	238 37
Spot	5 74			10 17	1,817 20	2,760 25	1,913 38	877 56	286 73	172 73	29 79	5 73	7,874 33
Striped mullet	13 96	13 40	29 74	1,616 28	1,205 32	670 46	927 46	580 60	297 67	272 59	35 101	93 104	5,750 43
Silver mullet	12 108	18 49	5 123	7 88				208 40	348 57	244 52	104 44		946 52
Weakfish								3 40	5 45	2 44	2 86		12 50
Atlantic croaker	40 22	115 19	181 26	135 31	7 40	10 42	115 43	36 58	10 88	7 87		1 21	657 32
Black drum		1 41					1 18	503 12	17 47	8 73	5 83	2 139	537 15
Southern kingfish										3 25			3 25
Summer flounder		1 42		2 145	1 20		6 51	14 54	4 72	6 102	2 84		36 69
Southern flounder						4 37	4 61	18 55	7 96	1 114			34 64
Red drum	3 34	18 57		5 67		3 92	1 105	1 164					31 65
Gray snapper	1 19	3 40	1 80							2 18	12 24		28 31

250

Table 4-32. Relative abundance (i.e., numbers of specimens collected) of the most common fish species encountered in Georgia salt marshes, March 1953 – May 1961 (Miller and Jorgenson 1969).

SPECIES	JAN	FEB	MAR	APR	MAY	JUN	JUL	AUG	SEPT	OCT	NOV	DEC	TOTAL BY SPECIES
Atherinidae													
Rough silverside	5	31	142	66		355	1,475	790	112	56	54	3	3,069
Atlantic silverside	672	434	330	226	4,040	1,444	720	1,614	1,753	562	349	605	12,749
Clupeidae													
Atlantic menhaden	1	17	767	5,788	3,804	701	216	153	21	29	71	1	11,647
Cyprinodontidae													
Mummichog	100	126	472		503	753	508	263	233	247	161	82	3,872
Striped killifish	5	16	60		223	295	103	109	78	33	21	37	1,103
Engraulidae													
Striped anchovy					9	175	29	102	35	4			359
Bay anchovy	15	40	339	1,784	3,036	7,680	2,848	1,070	2,153	635	212	296	20,108
Mugilidae													
Striped mullet	5,171	2,431	1,802	2,350	518	103	46	16	22	11	7	681	13,158
White mullet		1		1	1,713	678	249	195	43	14	12	1	2,907
Sciaenidae													
Spot	525	947	6,147	983	290	57	1	5	2	1		4	8,962

No snakes are common in irregularly flooded estuarine areas. Gibbons and Harrison (1975) occasionally found banded water snakes, cottonmouths, and yellow rat snakes in this habitat around Kiawah Island, South Carolina. Rough green snakes have been observed feeding in high marsh areas (i.e., short Spartina alterniflora) around James Island, Charleston County, South Carolina (C. M. Bearden, 1979, South Carolina Marine Resources Division, Charleston, pers. comm.), and Engels (1942) collected a rough green snake among black needlerush on Ocracoke Island, North Carolina. There exist several reports of brown water snakes in saline habitats of South Carolina. Along the lower Combahee River, this species was found in salt water where the associated fauna was mostly marine (Neill 1951). Jobson (1940) and Obrecht (1946) reported this species from tidal creeks or rivers in the Georgetown area, but tidal influence does not necessarily indicate saline conditions; these records may be marginal because the specimens were collected inland from the usual extent of the $1^o/oo$ isohaline (Estuarine Survey, 1978, South Carolina Marine Resources Division, Charleston, unpubl. data).

The eastern garter snake and the island glass lizard were found in association with fiddler crabs (Uca) near Bluffton, Beaufort County, South Carolina (Neill 1958). In northwest Florida, the eastern glass lizard was also noted to use Uca burrows (Blaney 1971). The typically freshwater eastern mud snake was reported by Neill (1958) to inhabit "inlets, estuaries, salt creeks, and salt marshes" along the Georgia and South Carolina coasts. In its normal habitats, this snake feeds almost exclusively on salamanders of the genera Amphiuma and Siren, but in these saline areas it is piscivorous (Neill 1958).

Carolina diamondback terrapins are perhaps the most characteristic reptile of irregularly flooded intertidal estuarine habitats. This area is utilized for sunning and egg laying (Coker 1906, Johnson et al. 1974), and for foraging during periods of high tide (Johnson et al. 1974). Other turtle species recorded from this habitat by Neill (1958) include the common snapping turtle, striped mud turtle [see Conant (1975)], eastern mud turtle, and chicken turtle.

American alligators are not infrequently encountered in estuarine emergent wetlands in the Sea Island Coastal Region (Obrecht 1946, Neill 1958, Gibbons and Harrison 1975). An individual was seen by Allen and Neill (1949) in a salt marsh (= Spartina?) in South Carolina, and Obrecht (1946) reported nesting in "tidal marshes" near Myrtle Beach, South Carolina, but he included no salinity data. Hillestad et al. (1975) reported

that alligators frequently move between the islands and mainland as well as between islands off the Georgia coast. Also, the larger alligators in island ponds move into tidal creeks in the salt marshes (Zweifel and Cole 1974).

6. Birds

The regularly flooded salt marshes of coastal South Carolina and Georgia provide a unique habitat for birds, and significant ecological relationships exist between large numbers of birds and the marsh vegetation. In such a pulse-stable ecosystem (Odum 1971), these relationships assume both direct and indirect dimensions (Shanholtzer 1974b).

Direct associations are those involving spatial and physical utilization of emergent wetlands by birds. The vegetation itself serves as a base for feeding, reproduction, and roosting activities for birds and other vertebrates (Johnson et al. 1974). Feeding habitat provided by emergent wetlands varies with seasons, tides, and plant species, height, and abundance. Shanholtzer (1974b) discussed seasonal dietary shifts in birds which result in different activity patterns. For example, the seaside sparrow and red-winged blackbird are carnivorous during the summer and granivorous during fall and winter. Thus, a vertical shift in space utilization is implied. Peak periods of feeding in smooth cordgrass by birds occurs during fall and spring migrations.

Nesting by marsh birds often takes place in conjunction with feeding activities. These rather discrete areas are many times defended against other members of the same species. Other marsh birds nest in colonies away from the feeding territories. The long-billed marsh wren is an example of the former (Kale 1965), and heron species assemblages exemplify the latter (Shanholtzer 1974a).

The use of Spartina marsh by nesting birds can remove a substantial amount of grass locally, as discussed by Shanholtzer (1974c). This is especially true of colonial avian nesting sites in the marsh. White ibis on South Carolina and Georgia coastal marsh islands are colonial nesters and remove substantial areas of marsh vegetation during nesting (Shanholtzer 1974c). In contrast, the long-billed marsh wren uses less grass in construction, does not nest in dense colonies, and removes relatively little marsh grass.

Salt marsh is also used as roosting or resting sites for breeding and nonbreeding marsh birds. Surface roosting species include the rails, while redwinged blackbirds, swallows, and wrens are plant roosting species. Also, various shorebirds use wracks of dead smooth cordgrass as resting sites. Most plant-based

252

roosting occurs in the medium to tall smooth cordgrass zone, where the grass presumably is better able to support the bird's weight (Shanholtzer 1974c).

Indirect relationships between birds and estuarine emergent wetlands involve nutrient and material cycling in the salt marsh and dispersal of halophyte seeds. Shanholtzer (1974c) investigated nutrient cycling and enrichment via fecal input from colonial nesting birds in a salt marsh near the Satilla River, Georgia. An estimated 3 - 4 metric tons (wet weight) of fecal material were excreted on a small marsh island from a heron colony of 3,000 - 4,000 active nests. Although the significance of this input has not been quantified, studies of artificially fertilized cordgrass marsh have shown a two-fold increase in live standing crop of stunted and short *Spartina*. Since avian fecal and urinary matter are concentrated with ammonium nitrate, the enrichment process from colonial nesting could be locally significant. Also, migratory species such as tree swallows and barn swallows are observed by the thousands feeding and migrating above the marsh surface. These birds, in addition to shorebirds and waterfowl, deposit fecal matter in the marsh and contribute to the nutrient cycle (Shanholtzer 1974b).

Approximately 70 species of birds use the estuarine emergent wetlands in the coastal zone of Georgia and South Carolina, but only about 27 should be considered as dominant in terms of relative abundance and trophic dynamics (Table 4-33). Most dominant species, such as herons, egrets, gulls, and raptors, are permanent residents of the estuarine emergent wetlands. Red-breasted merganser, lesser scaup, and tree swallow are winter residents. The only dominant summer resident is the least tern.

Ecologically, the dominant species can be grouped into eight trophic levels or classes - raptors, piscivores, scavengers, insectivores, aquatic herbivores, macrobenthivores, microbenthivores, and omnivores (Fig. 4-11). Of these trophic groups, the piscivores is probably the largest. Estuarine emergent wetlands are, in particular, the great feeding grounds of herons, egrets, and ibises. A characteristic sight on the South Carolina and Georgia coast is a linear or V-shaped formation of ibises flying out to the marshes in early morning and returning to roosting or nesting areas during evening.

The white ibis is apparently more abundant now than in historical times and certainly breeds more commonly in South Carolina and Georgia today than it formerly did (Sciple 1963, Burton 1970). Also, it has adapted as man's use of the environment has changed. Many earlier authors

indicated that freshwater crayfish constituted the principal food item of the white ibis (Audubon 1840, Wayne 1910, Bent 1963c, Sprunt and Chamberlain 1970). Most of these statements refer to a time when substantial areas of wetlands were freshwater or slightly brackish rice fields. Wayne (1910) even made the surprising statement that "the birds are very rarely to be met with on the salt marshes." Today, white ibis feed extensively on fiddler crabs and exhibit little dependence on crayfish.

The glossy ibis, although not a dominant species in this habitat, has spread northward dramatically during the present century. Wayne (1910) did not mention it, and Sprunt and Chamberlain (1949) considered it rare. However, Burton (1970), in the supplement to Sprunt and Chamberlain's work, changed its recorded status to permanent resident in South Carolina, while Shanholtzer (1974b) listed it as a fairly common summer resident in Georgia. At present, it appears to be almost as abundant as the white ibis, although precise counts are not available for comparison. Status of the glossy ibis has changed over the last 2 decades in coastal Georgia. Burleigh (1958) reported this species as extremely rare, and Sciple (1963) presented data in support of population increases. Hebard (1950) had published the only breeding record for the glossy ibis in Georgia prior to Shanholtzer's (1970) report on breeding records and distribution of this species. Shanholtzer found two Georgia nesting sites: the Darien rookery (McIntosh County) contained about 745 nesting pairs, and the Satilla River rookery (Camden County) contained 16 adults. This observation filled the hiatus in the breeding distribution for glossy ibis in Georgia, as noted by Teal (1959a). More recently, Odom (1976) reported eight active nests and estimated 40 adult birds present at the Satilla River rookery. The food habits of the glossy ibis have not been studied in this area, but Baynard (1912) investigated its food preference in Florida where it utilized freshwater habitats. It is now often seen in the marshes with the white ibis, and one could reasonably assume that its food habits in this environment do not differ appreciably from those of the white ibis.

The wood ibis is not an ibis in the strict sense, but a true stork. It is not abundant (Hamel 1977), but it is one of the most spectacular inhabitants of South Carolina and Georgia marshes in summer. When feeding in marshes, the wood ibis preys on fishes and crustaceans, including fiddler crabs. Trophic ecology of this species has been studied in recent years only in a predominantly freshwater environment of Florida (Kahl 1964). Audubon (1840) stated "besides the great quantity of fishes that these ibises destroy, they

Table 4-33. Birds of the estuarine emergent wetlands (Sprunt and Chamberlain 1949, 1970, Audubon Field Notes 1967 – 1970, Chamberlain 1968, American Birds 1971 – 1977, Forsythe 1978).

DOMINANT			MODERATE					
Great blue heron	C	PR	Green heron	C	PR	Yellow-crowned night heron	FC	PR
Little blue heron	C	PR	Black-crowned night heron	C	PR	Least bittern	FC	PR
Louisiana heron	C	PR	Glossy ibis	FC	PR	American bittern	U	PR
Great egret	C	PR	Black duck	C	WR Sept. – April	Wood stork	FC	PR
Snowy egret	C	PR	Bufflehead	FC	WR Nov. – April	Baldpate	C	WR Nov.
White ibis	C	PR	Hooded merganser	C	WR Nov. – April	Common goldeneye	FC	WR Nov.
Lesser scaup	C	WR Oct. – April	Virginia rail	FC	WR Aug. – Mar.	Ruddy duck	C	WR Oct.
Red-breasted merganser	C	WR Oct. – April	Sora	FC	WR Aug. – April	Sharp-shinned hawk	FC	PR
Marsh hawk	C	PR	American oystercatcher	C	PR	Osprey	U	PR
American kestrel	C	PR	Semipalmated plover	C	PR	Merlin	U	WR Aug.
Clapper rail	C	PR	Black-bellied plover	C	PR	Peregrine falcon	U	WR Aug.
Spotted sandpiper	C	PR	Ruddy turnstone	C	PR	American coot	C	PR
Herring gull	C	PR	Whimbrel	FC	PR	Lesser yellowlegs	FC	WR July
Ring-billed gull	C	PR	Willet	C	PR	Short-billed marsh wren	FC	WR Sept
Laughing gull	C	PR	Greater yellowlegs	C	PR	Bobolink	C	T April July
Forster's tern	C	PR	Least sandpiper	C	PR			
Least tern	C	SR Mar. – Oct.	Dunlin	C	PR			
Black skimmer	C	PR	Dowitcher	C	PR			
Belted kingfisher	C	PR	Semipalmated sandpiper	C	PR			
Tree swallow	C	WR July – June	Western sandpiper	C	PR			
Barn swallow	C	PR	Gull-billed tern	FC	SR April– Oct.			
Fish cr♦	C	PR	Common tern	C	PR			

254

Table 4-33. Concluded

DOMINANT			MODERATE			MINOR
Long-billed marsh wren	C	PR	Royal tern	C	PR	
Red-winged blackbird	C	PR	Sandwich tern	FC	SR April – Oct.	
Boat-tailed grackle	C	PR	Caspian tern	FC	PR	
Sharp-tailed sparr	C	WR Sept. – May	Black tern	C	SR May – Sept.	
Seaside sparrow	C	PR	Rough-winged swallow	C	SR Mar. – Sept.	
			Common grackle	C	PR	

Note: Dominance indicates relative importance of the species as a group in the community. This concept is not based necessarily on taxonomic relationships but rather on numbers, size, and trophic dynamics.

KEY: C – Common, seen in good numbers
FC – Fairly common, moderate numbers
U – Uncommon, small numbers irregularly
PR – Permanent resident, present year around
WR – Winter resident
SR – Summer resident
T – Transient resident

255

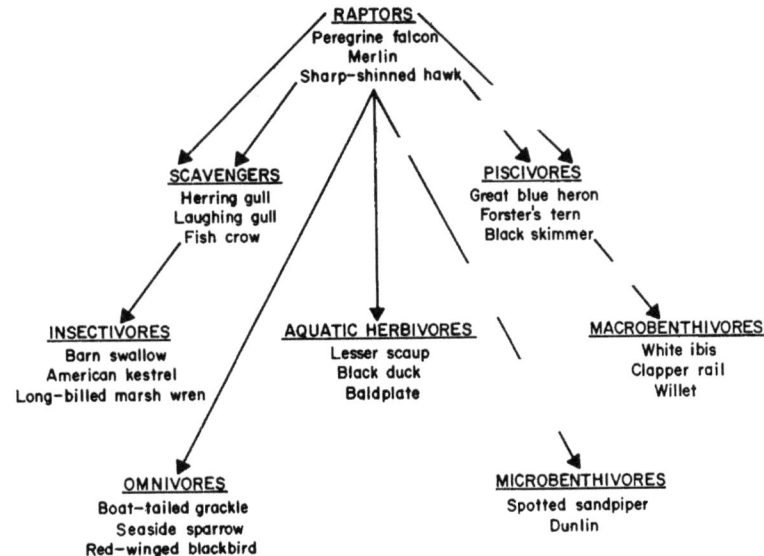

Figure 4-11. Generalized trophic relationships of representative birds of estuarine emergent wetlands.

also devour frogs, young alligators, wood rats, young rails and grackles, fiddlers and other crabs, as well as snakes and small turtles." However, Kahl (1964) was unable to induce captive wood ibis to feed on marsh rice rats even when other foods were withheld for a number of days, and Hansen (1975) illustrated the dependence of wood ibis on fish.

Egrets and herons (great egret, snowy egret, Louisiana heron, green heron, great blue heron, little blue heron, and night heron), by virtue of their size and abundance, are the dominant avian predators in estuarine emergent wetlands on a year around basis. Small fishes, shrimp, and fiddler crabs are the predominant food organisms of these wading birds. Principal fishes preyed upon in wetlands of South Carolina and Georgia are the mummichog, mullets, and juvenile Atlantic menhaden. White and brown shrimp are also taken extensively. Pfeiffer (1974) observed cattle egrets, which normally feed in cattle grazing fields, feeding in Sapelo Island salt marshes. According to his report, these birds were probably exploiting the cordgrass for saltmarsh grasshopper (Orchelimum fidicinium), wolf spiders

(Lycosidae), and other arthropods. Pfeiffer (1974) noted that the feeding of cattle egrets in this habitat might change the established food web in the marsh community on Sapelo Island. Since herons and egrets are not preyed upon by other marsh inhabitants to any appreciable extent, the energy they consume here is largely lost to the marsh and aquatic system. On the other hand, it could represent a significant transfer of energy to several terrestrial systems.

Wayne's clapper rail is one of the dominant components of the estuarine wetland avifauna of South Carolina and Georgia. It is a permanent resident of the wetlands, and feeds, roosts, nests, and raises its young within the Spartina marsh. Clapper rails feed extensively on small fiddler crabs, marsh crabs, and blue crabs and, unlike the herons and egrets, will probe in the mud for worms with their bills (Tomkins 1937, Burleigh 1958, Sprunt and Chamberlain 1970).

The clapper rail seems able to maintain high population levels despite a number of major mortality sources, including hunting. Not surprisingly, this bird has a high reproductive potential. Two broods per year (each with 9 - 12 eggs) are usual, although more than two can be produced when nests are destroyed by storms, tides, or predators (Sprunt and Chamberlain 1970). The breeding season is obviously a long one. Mink, otter, raccoon, and rats all prey on rail nests, and adult rails are occasionally caught by larger mammalian predators as well as by hawks. Domestic cats, which frequently enter marshes near human habitations, are able to capture adult rails with apparent ease.

Segrè et al. (1968) reported complex ecological interactions between clapper rails and laughing gulls in New Jersey coastal cordgrass marshes. When the two species nested in the same habitat, they exhibited ecological competition, predator-prey relationships, and simple propinquity which led to interspecific aggression. Segrè et al. (1968) found joint use of the same nests by gulls and rails and gave evidence that clapper rails prey upon the eggs of laughing gulls. In turn, gulls on occasion prey on the chicks of rails, and the two species compete at nest sites.

Several other rails occasionally occupy the coastal marshes, but because they are usually present in smaller numbers, they are not nearly as significant as the clapper rail.

A number of shorebirds, including the willet, greater yellowlegs, lesser yellowlegs, American oystercatchers, dowitchers, and several kinds of sandpipers, feed within the marsh on a seasonal basis. They consume mostly smaller crustaceans, mollusks, and annelids and do not offer much competition for those species already described.

A curious and poorly understood story surrounds the history of the curlews in coastal marshes of the study area. Prior to the latter part of the last century, the long-billed curlew was an abundant resident of this coast and, in fact, was considered more abundant than the Hudsonian curlew. Wayne (1910) indicated that after 1885, the long-billed curlew was virtually replaced by the Hudsonian. He attributed this to a shift in the migration pathway of the Hudsonian and speculated that the long-billed curlew could not compete with the new arrival for their primary food, the fiddler crab. While competition may have played a role, Wayne's explanation seems unlikely. At present, the long-billed curlew is only rarely seen in South Carolina or Georgia salt marshes. The Hudsonian curlew could not be considered common, but it is occasionally seen in small flocks.

The most detailed bioenergetics study of any bird population within the marsh community was that conducted by Kale (1965) on the long-billed marsh wren in Georgia. The same subspecies (Cistothorus palustris griseus) occupies marshes of the entire Sea Island Coastal Region, and most of Kale's findings can be reasonably expected to apply throughout this region. These wrens are highly territorial during the summer nesting period, but Kale found that the average territory size was only about 100 m^2 (120 yd^2). A number of false nests or resting nests, but only one breeding nest, are constructed within this territory. Nesting success varies widely from season to season; Kale's estimates over a 4-year period varied from 6.8% to 42.0%. The major source of mortality, aside from storm tide flooding, was nest predation by mammals. Principal predators in Georgia marshes were the raccoon, marsh rice rat, and mink. Since these mammals are present in all estuarine emergent wetlands of the study area, one would expect the same conditions to prevail throughout.

Long-billed marsh wrens are almost entirely insectivorous. They utilize a variety of insects, which Kale (1965) estimated to be 58% herbivores, 30% predators, and 12% detritus feeders by volume. Thus, the long-billed marsh wren is a secondary and tertiary consumer in the grazing food chain of the saltmarsh ecosystem.

Two sparrows are inhabitants of the marsh community. The sharp-tailed sparrow exists as several races which overwinter in the marshes of our region. These birds are primarily seed eaters, relying principally on the seeds of Spartina. However, Wayne (1910) reported that, in early spring, this sparrow feeds on a maritime moth common to the marshes. Macgillivray's seaside sparrow is a permanent resident of estuarine marshes, but it undergoes an interesting seasonal shift. During fall and winter, it is found almost entirely in high salinity marshes, especially those adjacent to the coastal islands. Then, in spring and early summer, it shifts inland to brackish or even freshwater marshes, where it nests. When the young are able to fly, they shift back to the salt marshes. This sparrow is reported (Sprunt and Chamberlain 1970) to be much more dependent on animal food than is the sharp-tailed sparrow. It utilizes seeds of Spartina but depends largely on marsh insects and even marine crustaceans and worms.

Several blackbirds are significant members of the marsh community by virtue of their numbers. The red-winged blackbird nests extensively in the coastal intertidal marshes, but its nesting is by no means limited to this habitat. According to feeding studies in environments other than salt marshes, the red-wing feeds on seeds as well as insects. The boat-tailed grackle also feeds extensively in wetlands, but does not nest there (Snyder and Snyder 1969). Its feeding is quite unlike other Icteridae, at least in the marsh environment, as it relies heavily on small fish and crustaceans. Thus, it functions as a secondary and tertiary consumer similar to the herons and egrets.

Several species of hawks and owls make use of salt marshes as a hunting area. Perhaps the best known in this area is the marsh hawk. This species is present in South Carolina and Georgia year around, but it is much more abundant in the fall and winter when northern migrants are added. Although the marsh hawk preys on clapper rails in coastal wetlands, it depends to a much greater extent on small mammals, particularly the marsh rice rat. The red-tailed hawk and red-shouldered hawk also feed in the marsh, and likewise depend predominantly on small mammals, including marsh rabbits and rodents. During darkness, great horned owls feed in the marsh. This owl occasionally captures ducks, primarily those already injured, during winter (Burleigh 1958, Sprunt and Chamberlain 1970).

There are few data available which specifically address man's impact on birds in the Sea Island Coastal Region. However, considering Odum's (1971) detritus-based ecosystem and the role of birds in a trophic sense, one can relate the general effects or impacts of man on birds and also apply studies from other areas. For example, the effects of dredging and filling in the estuarine wetlands have been well documented (Lunz 1938a, Ingle 1953, Mackin 1961, Odum 1963, Krenkel and Burdick 1976). The effects of pesticides on estuarine fauna have been reported and there are extensive data available on biological magnification of toxic compounds in estuarine organisms (Butler 1966, 1968a, b, 1969a, b, 1973; U.S. Department of Interior, Fish and Wildlife Service 1962, 1963, 1964, 1965).

During the past 2 decades, the effects of ditching tidal marshes for mosquito control have been a major concern in wetlands management. Bourn and Cottam (1939, 1950) found that ditching activities in coastal marshes adversely affected shorebirds and waterfowl by reducing food supply, primarily mollusks and crustaceans. Ditching also impacts birds such as the clapper rail when low wet marshes are altered. Other birds that need a constant water supply, such as the American and least bitterns, pied-billed grebe, and American coot, are also affected by drainage ditches. Urner (1935) reported similar deleterious effects on black ducks, willets, rails, and some of the herons. Bradbury (1938) reported drastic declines in shorebirds and waterfowl after ditching operations in coastal marshes.

In sharp contrast to the adverse impacts of ditching in wetlands are the beneficial effects on birds. Ditching can increase the "edge effect" through

the encouragement of smooth cordgrass in tidal areas. Thus, more shelter and nesting sites for species such as clapper rails and black ducks can be provided (Ferrigno 1961). Stewart (1951) actually recommended that ditches be sloped to produce the desired edge effect. He found nest densities to be greater with distinctive edges between tall and short forms of cordgrass.

The effects of impounding marshlands have been a controversial coastal zone management issue during recent years. While there are many beneficial and detrimental aspects to be considered, the overall impact on avifauna has probably been beneficial. Bradbury (1938), Griffith (1940), MacNamara (1949), Allen (1950), Chabreck (1960), Neely (1960), Morgan et al. (1975), and Landers et al. (1976) all discuss the significance of impoundments to waterfowl management. Other investigators have reported the increased usage of impoundments by birds (Catts 1957, Darsie and Springer 1957, Florschutz 1959, Tindall 1961, Mangold 1962, Shoemaker 1964, Lesser 1965, Provost 1968, Smith 1968). Darsie and Springer (1957) found some 86 species of birds in an impounded area, as compared to 55 prior to impoundment. Tindall (1961) identified a threefold increase in species diversity related to impoundment of marshes. Smith (1968) reported 62 species of birds on impoundments compared with 39 on natural marsh areas. Several of these investigators reported increases in annual broods following impoundment. Impoundments offer a variety of submergent and emergent foods for ducks; fish and invertebrates also serve as food for wading birds. In addition, impoundments provide open water for resting areas and flooded cover for nesting sites.

Conversely, some bird species have declined with the impounding of wetlands. The clapper rail, in response to the absence of fiddler crabs, is generally not found in impounded wetlands (Darsie and Springer 1957, Mangold 1962, Shoemaker 1964). Declines in small bird populations (song sparrows, seaside sparrows, sharp-tailed sparrows, and yellow-throated warblers) as a result of the loss of nesting sites and food have been reported by Mangold (1962) and Shoemaker (1964).

According to Smith (1968), marsh wrens and seaside sparrows were relatively abundant where high tide brush was present along ditches and pools. On Merritt Island, Florida, Provost (1969) found a decline in dusky seaside sparrows and fish-eating birds following impoundment. However, he concluded that there was an increase in numbers for 22 species of birds after impoundment. Thus, while most investigators have found enhanced bird usage following impoundment, it appears that more research is needed in this area.

There are also some beneficial impacts of dredging and filling in estuarine emergent wetlands. One of the major benefits to birds appears to be the disposal banks and dredge islands created by maintenance dredging and other water-related construction activities. When dredged material is deposited on the marshes, it generally forms a wide, cone-shaped mound. This mound often extends above the water level, forming an island. Such islands, usually called dredge-spoil islands, regularly occur adjacent to dredged channels in the South Carolina and Georgia study area. Generally, they have been considered as undesirable by-products of the dredging and filling process. However, studies on ecological succession of breeding birds in relation to plant succession on these islands have demonstrated their importance (Soots and Parnell 1975). Quay (1947, 1959) and Funderburg and Quay (1959) recorded dredge islands among the sites used by breeding birds in North Carolina and listed species of birds associated with various habitats found on the islands. Funderburg (1956) emphasized the importance of frequent deposition of dredge spoil on islands in maintaining early stages of vegetative succession necessary for ground-nesting birds. Soots and Parnell (1975) reported succession of breeding birds associated with vegetative changes on dredge islands. It appears that dredge islands provide nesting conditions comparable to those on natural sites. The year-round dense cover on these islands could be responsible for such diverse avian populations.

7. Mammals

Emergent estuarine wetlands are often treated as two separate habitats, the two being separated by the degree of regularity with which they are flooded. For the purpose of describing the role of mammals in estuarine marshes, regularly and irregularly flooded marshes will be treated together. At the same time, we recognize that some mammals are more at home in one type than the other. For most species, however, movement between high and low marsh is common, being controlled largely by the stage of the tide.

The herbivorous mammal most closely associated with the estuarine marsh, and the one of greatest ecological significance, is the marsh rabbit. This species occurs throughout the Sea Island Coastal Region of both Georgia and South Carolina and feeds on a variety of herbaceous materials, including cordgrass (Golley 1962). Nevertheless, only a minor component of marsh plant material is consumed and routed through the food web via this pathway, and it is doubtful that the marsh rabbit is ever sufficiently abundant to control marsh plant levels.

Young and adult marsh rabbits constitute an important link in the food

chain to birds of prey. Bent (1961), citing unpublished reports from Tomkins, reported that the marsh hawk in salt marshes of South Carolina and Georgia depends primarily on marsh rabbits during winter. In summer, other hawks no doubt exert considerable predation pressure. It is likely that predation rather than food supply is the principal population control on marsh rabbits. Specific population estimates have not been attempted, to our knowledge, but the marsh rabbit is known to be abundant in all coastal counties.

The marsh rabbit nests on the mainland adjacent to marsh, or on small brushy islands of dredged material scattered within the marsh, rather than within the intertidal portion of the marsh itself. The nest is usually concealed within dense brush or under fallen logs. Breeding may occur year around, but in Georgia, Tomkins (1935) found peak reproductive activity in late winter, with a depressed period in the fall. The gestation period in Louisiana is reported by Lowery (1974) to be 28 - 32 days. Lowery stated that up to six litters per year may be produced by a single female. Tomkins (1935) estimated an average litter size in Georgia of three young. Even with a litter size that is small for rabbits in general, the frequency of breeding is sufficient to insure a high reproductive capacity.

A marsh herbivore that one might expect in the coastal marshes of the Sea Island Coastal Region is the muskrat, but it is entirely absent. One reason for its absence may be the tidal range, because it occurs inland in both States (Golley 1962, 1966). This species is present in coastal areas both north and south of the characterization area and is sufficiently abundant in coastal Louisiana marshes to provide the basis of a valuable fur industry. The U.S. Fish and Wildlife Service attempted to establish a muskrat colony on Cape Island in the Cape Romain National Wildlife Refuge in 1950 but, according to Colley (1966), the population did not survive. An ecologically related exotic mammal, the nutria, was introduced about the same time on Blackbeard Island in Georgia. Neuhauser and Baker (1974) indicated that nutria survived into the late fifties, but were extirpated by 1960. Wilson (1968) suggested that a few nutria may exist in the marshes around Brunswick, Georgia.

The meadow vole is not generally associated with the Sea Island Coastal Region, but a population was found in Charleston County near the Santee River. Skulls of 59 individuals were found in barn owl pellets (Nelson 1934), and specimens were taken in low stands of saltmeadow cordgrass on Cape Island in 1939; the latter record is based on museum records and is cited by Sanders (1978).

White-tailed deer often graze in the high marsh, feeding on saltmeadow cordgrass

and on several species of glasswort.
This is most common where the marsh is
adjacent to dense cover. Unless pursued,
deer seldom venture into the lower marsh
due to its soft substrate. However, deer
are excellent swimmers and will cross
large marsh creeks. The distribution of
deer in the various coastal localities is
treated in connection with maritime forest
mammals (Chapter Three). In addition to
deer, several large domesticated herbivores such as horses, cattle, and goats,
may utilize the upper elevations of the
salt marsh. Florida manatees have also
been observed grazing in Georgia marshes
(M. Hardisky, 1978, Georgia Department of
Natural Resources, Brunswick, pers. comm.).

The principal omnivorous mammal of
the saltmarsh community is the marsh rice
rat, which is also among the most highly
aquatic of the coastal rodents. Unlike
most other marsh mammals in this area,
the marsh rice rat may remain permanently
in the marsh. Nests are often made of
cordgrass, but marsh rice rats also utilize abandoned nests of marsh wrens
(Golley 1962). The most detailed study
of the ecology of this species is that of
Negus et al. (1961) on Breton Island in
the Gulf of Mexico; no quantitative
studies have been conducted in this area.
Although regularly flooded salt marsh was
not a vital habitat on Breton Island,
much of the information provided by Negus
et al. (1961) would be applicable to the
Sea Island Coastal Region of Georgia and
South Carolina.

Although many rodents are predominantly herbivorous, such is not necessarily the case with the marsh rice rat.
Certainly, plant material is less important in its diet during some months than
others (Fig. 4-12). Golley (1962) reported that marsh rice rats feed on cordgrass in Georgia coastal marshes, but
they also utilize crabs (probably fiddler
and other marsh crabs) and insects. Kale
(1965) noted extensive predation by marsh
rice rats on the eggs and young of the
marsh wren. Sharp (1962) reported that
the major portion of the diet of the
marsh rice rat consisted of crabs and insect larvae. Such studies indicate that
the marsh rice rat is an opportunistic
omnivore.

Not only is the marsh rice rat an important predator within the estuarine wetlands, it is also an important prey species, especially for birds. In addition
to the recognized birds of prey (hawks and
owls) which seek rodents in less densely
vegetated sections of the marsh (Sprunt
and Chamberlain 1970), many of the larger
wading birds (e.g., great blue heron,
great egret, night herons, wood stork)
will also prey on rodents whenever they
have the opportunity to do so (Bent 1963c).

Marsh rice rats seldom live more
than 1 year, and they undergo dramatic

seasonal changes in abundance (Negus et
al. 1961). On Breton Island, decreases in
population density appeared to be related
to the duration and severity of the winter,
and Negus et al. speculated that harsh winters influenced rat populations by controlling the food supply. These authors
provided a simple model to show the relationship of environmental factors to population density (Fig. 4-13).

In most habitats, the raccoon is
properly considered an omnivore, but it
functions exclusively as a carnivore in
the salt marsh. Raccoons utilize practically all coastal plains habitats, but
their populations appear to be especially
high in marshes and in woodlands adjacent
to marshes. The raccoon is predominantly
nocturnal, generally spending the day in
its den in a large tree and leaving to
forage at night. This behavior pattern
may be somewhat modified in estuarine wetlands because of the tides. It is not unusual to see raccoons feeding in the marsh
in full daylight on isolated coastal islands, but such observations are much less
common near human habitations even though
raccoon populations may be quite large
there. The raccoon is without doubt the
most characteristic medium-sized mammal of
the coastal marshes. Within the marsh, it
depends rather heavily on crustaceans
(fiddler crabs, marsh crabs, juvenile blue
crabs), competing with the clapper rail
and white ibis for the same food resources.
In addition to crustaceans, marsh mollusks
are important food items, especially small
intertidal oysters and mussels. Kale
(1965) reported predation by raccoons on
the nests and young of marsh wrens. Raccoons also constitute a source of mortality
for clapper rail eggs and young. —

The raccoon has few or no predators
in high salinity wetlands, but alligators
may cause significant mortality in low
salinity marshes. Coastal areas and wetlands provide the raccoon a virtually unlimited food supply, so factors other than
food must control its population levels.
Hunting and trapping pressures on raccoons
in coastal areas are rather light. The
marsh or low country raccoon is considered
to have a low quality pelt compared to upland populations. Probably the greatest
single source of raccoon mortality, other
than disease, is the automobile. Yet, on
many coastal islands this ceases to be an
important element of mortality because
automobile traffic is either non-existent
or extremely limited. At present, disease
is probably the principal factor controlling raccoon populations. Raccoons are
quite susceptible to a distemper-like
respiratory disease, which is almost certainly density dependent.

In the characterization area, the
raccoon breeding season may range from
late February into August, but the raccoon
does not breed within the marsh because
den trees are generally required for

260

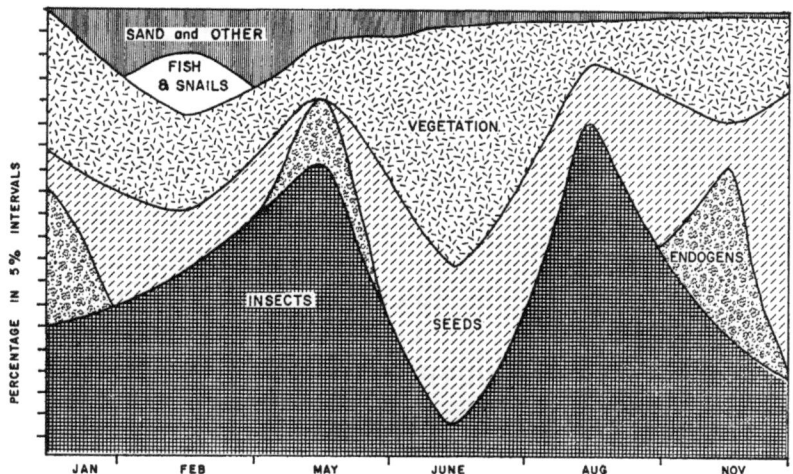

Figure 4-12. Food habits of the marsh rice rat on Breton Island, Gulf of Mexico, as determined by stomach analyses of 61 animals from various seasons, 1957-1959 (Negus et al. 1961).

nesting. The gestation period is 63 days, and usually a single litter is produced each year. About 50% of females will breed when 1 year of age, but full size and maturity may not be attained until the second year. The young remain with the mother for several months after they are weaned, and they are given close attention and training.

The river otter is relatively abundant in salt marshes of the Sea Island Coastal Region. Johnson et al. (1974) list this species as occurring on virtually all of the Georgia coastal islands. Sanders (1978) cited records from most South Carolina coastal counties, but it undoubtedly occurs in all. The river otter is entirely carnivorous in the marsh habitat and probably depends more on fishes and crabs than do the other marsh mammalian predators (Wilson 1954). Its numbers are too low to exert significant population controls on any of its prey organisms, and the otter itself is not subject to significant predation in this habitat. Factors controlling its population size are not known.

Like the river otter, the mink is a semi-aquatic mammal that frequently occupies coastal marshes, but it too is far from restricted to this environment. The diet of the mink is more varied and is likely to include marsh birds and marsh rodents along with fish and crustaceans (Colley 1966). Like the otter, this species occurs in relatively low population densities and is unlikely to control prey population levels.

The interface between the marsh and other habitats is extensively used as a feeding ground by mammals other than those already mentioned. These species will be treated in connection with other habitats.

The impacts of dredging, filling, ditching, and other human activities on marshes have been treated elsewhere. Habitat destruction of the marsh would generally have the same effects on mammals as on other faunal components, except that most mammals are not permanent residents of the marsh. A few exceptions, however, should be noted. Most mammals, except for the most aquatic, make more extensive use of the high marsh than they do of the low marsh. Thus, partial filling which converts low marsh to high marsh may be a favorable change for mammals, despite the resulting loss of productivity to the aquatic system. Likewise, the former practice of building small islands within the marsh with dredge spoil material may create more favorable conditions for mammals.

261

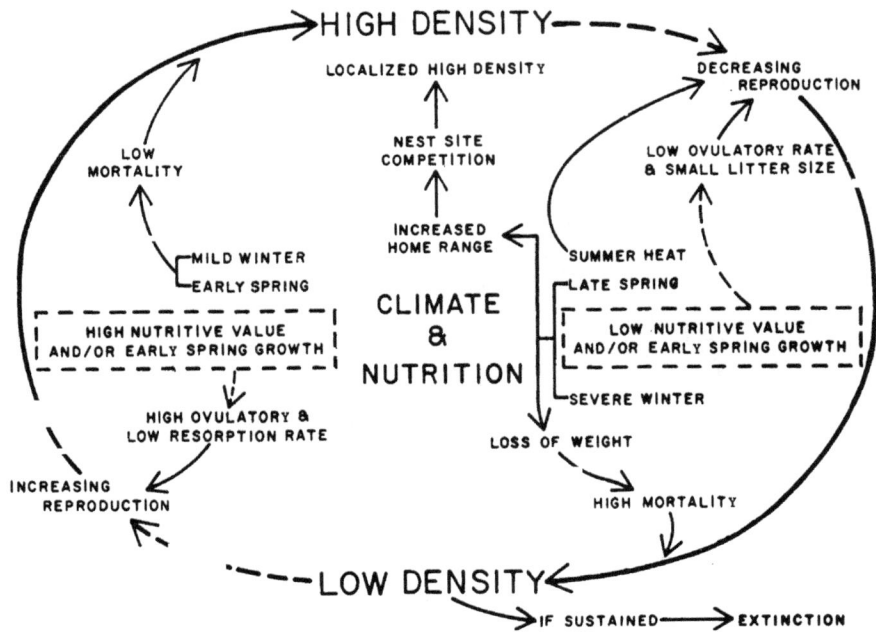

Figure 4-13. Schematic diagram of the Breton Island marsh rice rat population and its response to the environment. Dotted lines represent hypothetical aspects of the scheme (Negus et al. 1961).

With the notable exception of promoting or allowing domestic animals to graze in the marsh, most mammals on this coastline have little direct effect on the marsh habitat. Analyses of grazed and ungrazed transects suggest that grazing may not only crop down the vegetation but also may alter the zonation. Trampling by heavy mammals such as cattle and horses may have a direct unfavorable impact on some marsh plant species.

8. Decomposers

See sections on decomposers of estuarine subtidal waters and intertidal flats.

IV. INTERTIDAL ESTUARINE
SUBSYSTEM - FLATS

A. DESCRIPTION

According to Cowardin et al. (1977), flats are ". . . level landforms composed of unconsolidated sediments. Normally, Flats occur only in areas sheltered from strong currents and wave action. They may be irregularly shaped or elongate and continuous with the shore, whereas bars generally are elongate, parallel to the shoreline, and separated from the shore by water. Water regimes are restricted to irregularly exposed, regularly flooded, irregularly flooded, seasonally flooded, temporarily flooded and intermittently flooded."

Gray (1974) noted that flats have pronounced gradients of environmental conditions such as salinity, temperature, tidal influence, and bottom type. He observed that tides control water exchange and that the tidal range determines the extent and duration that flats are exposed or submerged. Tidal currents are important in determining the nature of the substrate, which in turn determines the types of organisms present. Although flats present a rigorous habitat for aquatic organisms, densities in some cases may be very high. Sanders et al. (1962) encountered estimated densities of 7,000 - 355,000 benthic invertebrates/m^2 on intertidal flats in Barnstable Harbor, Massachusetts.

Recovery from natural disasters is generally rapid on mud flats if the substrate has not been permanently altered (Gray 1974). According to Gray, permanent damage may result from filling, sewage contamination, disposal of industrial waste, and pesticide runoff. Intertidal areas are also particularly vulnerable to oil spills.

Most of the oysters in higher salinity regions of Georgia and South Carolina estuaries are restricted to the intertidal zone, where they form clusters or "rocks." Hard clams are also abundant in the lower intertidal zone of estuarine flats in the Sea Island Coastal Region.

B. PRODUCERS

1. Nonvascular Flora

Refer to the section on nonvascular flora of the emergent wetlands of the intertidal estuarine subsystem.

2. Vascular Flora

Sand and mud flats are common in the estuarine intertidal zone. Mud flats are generally below the level of the lowest stands of smooth cordgrass or within the smooth cordgrass community (Teal 1958b, Adams 1963). Creekbank mud flats are exposed during low tides, with additional areas of mud being exposed during neap tides. Although these flats are rich in productive mud algae, no vascular plants are found. Within smooth cordgrass communities, flats or "die-off" areas are occasionally present (Adams 1963, Smith 1970). One cause of these areas is the accumulation of dead smooth cordgrass culms ("wrack"). The wrack usually covers living rhizomes of smooth cordgrass in winter; the rhizomes fail to re-sprout in spring due to lack of light, and a barren area is formed. The wrack usually decomposes or is transported to another part of the marsh. Regeneration from rhizomes of the mud flat may begin immediately if the wrack has been in place for only a season. However, if the wrack remains for several years, a complete rhizome die-off usually occurs. Regeneration is then a longer process, with smooth cordgrass gradually re-invading the mud flat. Annual glassworts are common in mud flats until the smooth cordgrass re-colonizes.

Smith (1970) described mud flat areas within smooth cordgrass stands and along creeks. Termed Spartina "die-back" areas, these openings may be the result of several factors. Smith (1970) listed excess salinity, wrack accumulation (a minor factor, according to Smith), waterlogging, lack of available iron, hydrogen sulfide toxicity, oxygen deficiency (in roots), tidal regime changes, and water pollution as possible factors in die-back of smooth cordgrass.

Sand flats are commonly found in the transitional zone between low and high marsh (slightly above the mean high tide level), or between the low marsh and upland communities if no high marsh is present. In Figure 4-6, the zones in which sand flats may occur are the minax marsh and the Salicornia-Distichlis marsh (see Teal 1958b). Tiner (1977) pointed out that sand flats may be unvegetated ("salt pannes" or "sand barrens") or vegetated. In unvegetated sand flats, temperatures may reach as high as 45°C (113°F), causing rapid evaporation and producing a soil of high salinity (up to 70°/oo - thus the name "salt pannes") (Teal 1958b, Cooper 1974). Sand flats are components of the irregularly flooded marsh and are inundated only during spring and storm tides. For a discussion of sand flats in the context of saltmarsh zonation, see the subsection on emergent wetlands of the intertidal estuarine subsystem.

C. CONSUMERS

1. Benthic Meiofauna

Refer to section on benthic meiofauna of estuarine intertidal wetlands.

2. Benthic Macroinvertebrates

Rich infaunal communities inhabit many of the sheltered intertidal flats occurring in estuarine areas of the Sea Island Coastal Region. These communities are influenced by pronounced gradients and/or fluxes in a number of factors, including salinity, temperature, tidal influence, and substrate type (Gray 1974, Roberts 1974a). As a result of wide variations in temperature and salinity, the resident fauna is made up largely of eurythermal and euryhaline species. Water currents partly determine the nature of the substrate, and sediment type is known to be a major factor determining the assemblage of animals present at a given locale. The extent and time that a flat is exposed, as well as water exchange over the flat, are governed by tidal action.

Intertidal benthic communities recover rapidly following natural perturbations, such as those caused by hurricanes, if the substrate has not been seriously and permanently altered (Gray 1974, Roberts 1974a). Damage may be more permanent from disposal of sewage and industrial wastes, filling operations, or from pesticides carried in water runoff. Populations of some tidal flat species such as the polychaete Chaetopterus variopedatus and the commensal crab Pinnixa chaetopterana have been decimated by over-collecting in the vicinity of marine laboratories (H. Porter, 1978, Institute of Marine Science, University of North Carolina, Morehead City, pers. comm.).

A number of different "soft-bottom" estuarine intertidal habitats have been investigated in both Georgia and South Carolina (Table 4-34). On point bars studied by Dörjes (1977) in Georgia, different assemblages of invertebrates were found in different subenvironments. In

263

Table 4-34. Species of macroinvertebrates typical of several "soft-bottom" intertidal habitats in estuaries of Georgia and South Carolina (Land and Hoyt 1966, Holland 1974, Holland and Polgar 1976, Dörjes 1977, Holland and Dean 1977a, b).

Creek Banks

Nereis succinea
Diopatra cuprea
Heteromastus filiformis
Crassostrea virginica
Ilyanassa obsoleta
Uca pugilator
Panopeus herbstii
Alpheus heterochaelis
Upogebia affinis
Onuphis microcephala

Muddy Flats

Diopatra cuprea
Heteromastus filiformis
Onuphis microcephala
Glycera dibranchiata
Drilonereis longa
Nephtys picta
Spiophanes bombyx
Amphitrite ornata
Scoloplus sp.
Ilyanassa obsoleta
Mulinia lateralis
Nassarius vibex
Terebra dislocata
Busycon carica
Polinices duplicatus
Dosinia discus
Tagelus plebeius
Tagelus divisus
Callianassa sp.
Pagurus longicarpus
Clibanarius bittatus
Crassostrea virginica
Uca pugilator
Uca pugnax
Saccoglossus kowalevskii
Balanoglossus aurantiacus

Sandbars and Sandflats

Heteromastus filiformis
Haploscoloplos fragilis
Nephtys picta
Spiophanes bombyx
Nerinides sp.
Polydora sp.
Nereis succinea
Glycera dibranchiata
Donax variabilis
Solen viridis
Tellina sp.
Tagelus plebeius
Oliva sayana
Onuphis microcephala
Haustorius sp.
Acanthohaustorius milisi
Bathyporeia sp.
Parahaustorius longimerus
Pseudohaustorius carolinensis
Albunea paretii
Saccoglossus kowalevskii
Balanoglossus aurantiacus
Lepidactylus dytiscus

Low Salinity Point Bars

Nereis succinea
Diopatra cuprea
Heteromastus filiformis
Haploscoloplos fragilis
Polymesoda caroliniana
Ilyanassa obsoleta
Tagelus plebeius
Lepidactylus dytiscus

High Salinity Point Bars - Exposed

Bathyporeia sp.
Lepidactylus dytiscus
Haploscoloplos fragilis
Callianassa sp.
Acanthohaustorius sp.
Leaonereis culveri
Heteromastus filiformis
Onuphis microcephala
Tellina texana
Mercenaria mercenaria
Anadara ovalis
Mulinia lateralis

High Salinity Point Bars - Sheltered

Diopatra cuprea
Heteromastus filiformis
Onuphis microcephala
Sthenelais boa
Glycera dibranchiata
Ilyanassa obsoleta
Tagelus plebeius
Drilonereis longa
Upogebia affinis
Pagurus longicarpus
Clibanarius vittatus
Nassarius vibex
Terebra dislocata
Uca pugilator
Amphitrite ornata
Saccoglossus kowalevskii
Balanoglossus aurantiacus

clean sands, the haustoriids Acanthohaustorius, Lepidactylus, and Bathyporeia and the decapod Callianassa sp. predominated. Numbers of species increased, but the number of individuals declined with an increase in the percentage of clay and organic debris in the sediment. Typical inhabitants on the channel side of these point bars included such species as Laeonereis culveri, Heteromastus filiformis, and Onuphis microcephala (polychaetes), Tellina cf. texana (pelecypod), and several amphipods. Occasionally, a number of other species such as Nassarius vibex and Ilyanassa obsoleta (gastropods) and Uca spp. (decapods) were present. In muddier sediments toward the marsh, Diopatra cuprea, Sthenelais boa, Glycera dibranchiata, and Drilonereis longa (polychaetes), Tagelus plebeius (pelecypod), and Upogebia affinis (decapod) were prevalent.

Several studies on intertidal invertebrate communities have been conducted in the high salinity North Inlet estuary, South Carolina. Holland and Dean (1977a) recognized four subregions on intertidal sand bars of North Inlet, each of which was characterized by a discrete faunal assemblage. Along tidal creek fringes, the haustoriids Acanthohaustorius millsi and Pseudohaustorius carolinensis were dominant. In tidal channels and point areas, Acanthohaustorius millsi and Protohaustorius deichmannae were most important numerically. In the central region of the sandbars, Lepidactylus dytiscus was the most abundant species, with Haploscoloplos fragilis the only other important macroinvertebrate animal. Heteromastus filiformis was numerically dominant in lagoon areas, followed by Glycera dibranchiata and Pseudohaustorius carolinensis. Mean species diversity was low at each of the four subregions, ranging from a low of 0.23 in central regions to a high of 1.44 on tidal channels and point areas. Diversity decreased with elevation above MLW. While community structure differed in different substrates and along elevational gradients, seasonal changes at a sample site near the center of a large sandbar were small and were associated with the population dynamics of the numerically dominant species (Holland and Polgar 1976).

Dame (1976) conducted a seasonal abundance study of the macrobenthos of intertidal oyster reefs in North Inlet, South Carolina. Of 37 species identified, numerical dominants included the American oyster (Crassostrea virginica), the scorched mussel (Brachidontes exustus), a gastropod (Odostomia impressa), and two species of polychaetes (Heteromastus filiformis and Nereis succinea). Densities varied from 2,476 - 4,077 organisms/m^2, with maximum numbers in early summer.

Oyster clusters, formed by successive yearly sets of "spat" on older oysters,

are common in the intertidal zone of Georgia and South Carolina estuaries (Fig. 4-14). As new sets occur and oysters grow, the clusters develop, often becoming shaped like bushes. According to Keith and Gracy (1972), clusters attached to a solid matrix may attain a height of 45 cm (18 in) or more. As new growth occurs, the added weight pushes the lower oysters into the mud where they eventually suffocate. Only the outer and top-most oysters remain alive.

Intertidal oysters are also found in groups known as "oyster rocks" (Keith and Gracy 1972). These oysters evidently grow on a firm foundation and probably begin as a very thick mat of cluster oysters. Successive sets occur and eventually the clusters join to form a continuous group. Through a period of many years, tiers of oysters are laid one upon another. The lower-most oysters die of suffocation or starvation before they ever reach a harvestable size. Eventually the "oyster rocks" grow to be several feet thick (see Fig. 4-15). Only the uppermost oysters are alive and, due to their prolonged exposure to the elements, they usually remain of an unharvestable size. Locally, these oysters are sometimes referred to as "coon oysters." To be of economic value, such oysters must be broken into "seed" and distributed to better growing areas.

The ubiquitous intertidal oyster clusters and oyster rocks provide microhabitats and food for other benthic species. Species commonly occurring in crevices among the clustered oysters are mud crabs (Panopeus herbstii, Neopanope sayi, Eurypanopeus depressus), snapping shrimp (Alpheus spp.), amphipods, isopods, numerous polychaetes, and gobiid and blenniid fishes. Sessile and encrusting forms such as barnacles, bryozoans, and tubiculous worms are also common. Juvenile blue crabs (Callinectes sapidus) and oyster toadfish may occasionally be found among intertidal oysters, but are much more common where the oyster flats are submerged. Between clumps of oyster shells, burrows of the abundant and nearly ubiquitous fiddler crabs, Uca spp., may be present.

3. Insects

Refer to the section on insects of the emergent wetlands of the intertidal estuarine subsystem.

4. Fishes

Few previous investigations on fishes of estuarine intertidal flats in the Sea Island Coastal Region of South Carolina and Georgia have been conducted. This habitat is characterized by mud and/or sand flats, often interspersed with intertidal oyster reefs. Studies by Dahlberg (1972) and Mahood et al. (1974a, b, c, d) included collecting in Georgia intertidal areas by means of seines and gill nets. Hicks (1972) conducted seasonal gill net sampling in

Figure 4-14. Representative intertidal oyster beds in coastal South Carolina.

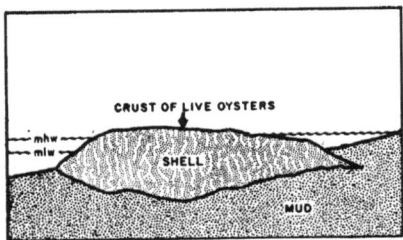

Figure 4-15. Generalized cross section of an intertidal oyster rock.

shore areas of Port Royal Sound, South Carolina. The South Carolina Marine Resources Division (Noe 1977) conducted a 1-year gill net sampling program on an intertidal flat in Charleston Harbor, South Carolina.

Resident fish species known to frequent the intertidal estuarine habitat in South Carolina and Georgia include silversides, bay anchovy, and mummichog (Dahlberg 1972, Turner and Johnson 1972, 1974). Transient fishes which frequent intertidal estuarine flats seasonally during various stages of their life cycles include sting rays, mullets, sea catfish, Atlantic menhaden, mojarras, pipefishes, pinfish, and sciaenids (spot, silver perch, and Atlantic croaker). Juveniles and adults of southern kingfish, carangids, pompanos, and Atlantic thread herring are also common during the warmer months over the intertidal flats in the lower reaches of estuaries.

Larger predatory fishes which invade the intertidal flats during higher stages of the tide for feeding purposes include small carcharinid and hammerhead sharks, longnose gar, striped bass, white perch, bluefish, spotted seatrout, red drum, black drum, and southern flounder. Table 4-35 presents numbers of individuals of various species taken monthly by gill net sampling on an intertidal flat in Charleston Harbor (Noe 1977). Major fish species taken during gill net sampling in shallow tidal areas of Georgia by Mahood et al. (1974a, b, c, d) were spotted seatrout, weakfish, spot, Atlantic croaker, red drum, southern kingfish, bluefish, sea catfish, Atlantic menhaden, Spanish mackerel, striped mullet, and crevalle jack.

Intertidal oyster reefs are commonly situated along the shorelines of estuaries or interspersed on mud flats, and form a unique habitat type. Fishes commonly associated with oyster reefs such as those found in the characterization area include gobies, blennies, skilletfish, killifishes, and oyster toadfish. Other fishes often associated with oyster reefs at high tide include mullets, spotted seatrout, red drum, and southern flounder. Adult spotted seatrout frequent oyster reefs for both food and cover. During colder weather, spotted seatrout and, to a certain extent small red drum, apparently utilize oyster reefs as protective cover from predators such as Atlantic bottlenose dolphin.

Dahlberg and Conyers (1973) conducted an ecological study of two species of gobies associated with shallow water oyster reefs near Sapelo Island, Georgia. This included both larger intertidal reefs and small subtidal reef patches. Sexual dimorphism,

Table 4-35. Number of individuals of each fish species caught per month and season by gill net from an intertidal area in Charleston Harbor, 1975-76 (Noe 1977).

Species	July	Aug.	Sept.	Summer Total	Oct.	Nov.	Dec.	Fall Total	Jan.	Feb.	Mar.	Winter Total	Apr.	May	June	Spring Total	Year Total
Atlantic menhaden	79	111	350	540	39	61	28	128	10	26	7	43	43	44	44	132	843
Spot	72		14	86	190	138	25	353	81	46	48	175	120	48	69	237	851
Striped mullet	100	78	87	265	15	57	38	110	22	57	67	146	97	79	111	287	808
Spotted seatrout	23	23	6	52	3	10	3	16	1	1	6	8	62	147	68	277	353
Bluefish	12	10	32	54	6	10	43	59		2		2	2	1	10	13	128
Pinfish	78	24	21	123	15	8		23					1	3	44	48	194
Atlantic croaker	13	18	19	50	3			3					8	19	39	66	120
Red drum	35	15	6	56		2	3	5	9		4	13	1		2	3	77
Black drum		1		1	6	6	8	20	2	1	7	10	23	11	11	45	76
Silver perch											5	5	51			51	56
Sea catfish	1	4	19	24									13	2	5	20	44
Longnose gar	1	5	1	7	6	2	1	9					12	3	4	19	35
Gizzard shad	3	3	5	11	2	8		10			2	2	2	7	1	10	33
Atlantic bumper	2	6	16	24	1	1		2						2		2	28
Southern flounder	3		1	13	1	1		2					3				15
Yellowfin menhaden			13	13													13
Crevalle jack	1	5	4	10	1	1		2									12
Harvestfish			12	12													12
Ladyfish			7	7	2	2		4									11
White mullet			8	8	2	2											11
Spanish mackerel	1	2	2	5		2		2									9
Striped anchovy												2					7
Blueback herring												2					2
Florida pompano		2		2													2
Striped bass								1			1	1					1
Tarp a																	1
Gafftopsail catfish				1	1			1									1
Sand perch						1		1									1
Gray snapper						1		1									1
Searobins					1			1									1
Lookdown					1			1									1
Summer flounder														1		1	1
Butterfish		1		2													1
TOTALS	425	317	623	1,365	292	312	149	753	125	135	150	410	438	365	409	1,212	3,740

267

nesting behavior, spawning, and fecundity of the naked goby and the seaboard goby were investigated. Seven species of resident fishes characteristic of the oyster reefs were collected. In addition to the two gobies, the following species were found: oyster toadfish, skilletfish, and blennies (feather blenny, striped blenny, and crested blenny).

The above fishes associated with oyster reefs were found to remain within the interstices of the reef throughout the tidal cycle during the warmer months, becoming scarce or absent in winter. Other fishes associated with the reefs were classified as temporary visitors. The oyster reef provides nesting sites, food, and protection from predators for the species which inhabit them. Food of gobies on the reefs included amphipods (Gammarus) and other crustaceans, small fish, eggs, annelid worms, and dying oysters.

On the intertidal estuarine mud and mud/sand flats, the movement of various species of fishes depends largely upon tidal stage. These flats are utilized by transient species for feeding to some extent, but many species move quickly across flats, once the tide has covered them, and into adjacent intertidal creeks and marshes. Generally, the first fishes to move onto flats at early stages of the flood tide are mullets, mummichogs, rays, pinfish, flounders such as southern flounder, and juvenile sciaenids (spot, silver perch). Wind and wave action on flats stir up the bottom sediments along with benthic food organisms. Motile food organisms (penaeid shrimp, grass shrimp, and other small crustaceans) are abundant during the lower stages of the tide, but animals such as these rapidly move into adjacent marshlands as the tide rises. Predators such as red drum, spotted seatrout, and bluefish move across those areas at high tidal stages. A similar pattern of movement occurs on the ebb tide, with mullet and other species tolerant of high temperature and turbidity remaining until the water becomes very shallow.

Intertidal flats could be considered as extensions of the subtidal estuarine habitat, and a number of fish species commonly found on flats are the same as those occurring in deeper waters and channels. However, some species common to the latter area are rarely, if ever, taken in the intertidal zone. For example, star drum, juvenile Atlantic croaker, weakfish, and black sea bass found in deeper subtidal waters are not usually found on intertidal flats. Conversely, mullet, mummichog, Atlantic silversides, and others appear to be abundant in the shallow waters of flats and smaller marsh creeks.

No published information could be found concerning the effects of environmental alterations on fishes of intertidal flats. It is assumed that these species would be susceptible to aerial pesticide application and pollutants entering estuaries from adjacent land drainage.

In the past, there has been a tendency to place more emphasis on the importance of vegetated marshlands in South Carolina and Georgia than on open water shoal habitat such as intertidal mud flats. In some instances, filling of flats to create new marshlands has been considered as an alternative for disposal of dredged materials. In view of the present high ratio of marshlands to open water shoal habitat in South Carolina and Georgia, and considering the ecological significance of flats and associated oyster reefs, this practice should be discouraged wherever other environmentally acceptable alternatives are available.

Further study of the fishes and other biological features of intertidal flats in the South Carolina and Georgia coastal areas is needed to determine the ecological significance of these areas. Investigations employing multiple gear sampling techniques during various tidal cycles will be required to obtain information about the occurrence, diversity, abundance, patterns of movements, and trophic relationships of the fishes utilizing these areas. Intertidal mud flats are a common habitat type in both States and present in an extremely difficult area for sampling due to the often soft nature of the bottom and the frequent presence of intertidal oyster mounds or reefs. Sampling of these areas might best be accomplished with variable mesh gill nets and large stop nets (seines).

5. Amphibians and Reptiles

Refer to the section on emergent wetlands of the intertidal estuarine subsystem.

6. Birds

Estuarine intertidal flats in the study area support diverse biotic communities, which in turn support a large and diverse avian population. At present, however, there exist no quantitative data on the numbers of birds utilizing intertidal flats in the South Carolina-Georgia coastal zone. On the other hand, the occurrence of various species in this habitat is well documented.

Approximately 23 species of birds are dominant residents that regularly utilize exposed intertidal flats (Table 4-36). Of this number, all but one, the least tern, are permanent residents. Perhaps the most conspicuous of these dominants, and certainly among the most important ecological elements of the intertidal flat community, are the wading birds. Five species of egrets and herons make extensive use of intertidal flats as feeding areas. The great egret, snowy egret, and Louisiana heron are the most abundant and appear to occur in nearly equal numbers in the study area. Somewhat less abundant are the

Table 4-36. Birds of Georgia and South Carolina estuarine intertidal flats (Sprunt and Chamberlain 1949, 1970, Burleigh 1958, Audubon Field Notes 1967 – 1970, Chamberlain 1968, American Birds 1971 – 1977, Shanholtzer 1974b, Forsythe 1978).

DOMINANT				MODERATE				MINOR			
Great blue heron	C	PR		Brown pelican	C	PR		Green heron	FC	PR	
Little blue heron	C	PR		Double-crested cormorant	C	PR		Black-crowned night heron	C	PR	
Louisiana heron	C	PR		Yellow-crowned night heron	FC	PR		Wood stork	FC	PR	
Great egret	C	PR		Glossy ibis	FC	PR		White ibis	C	PR	
Snowy egret	C	PR		Killdeer	C	PR		Spotted sandpiper	C	PR	
American oystercatcher	C	PR		Whimbrel	FC	PR		Lesser yellowlegs	FC	WR	July-May
Semipalmated plover	C	PR		Greater yellowlegs	C	PR		American avocet	U	WR	Oct.-June
Black-bellied plover	C	PR		Marbled godwit	FC	WR	Sept.–June				
Ruddy turnstone	C	PR		Bonaparte's gull	C	WR	Oct.-May				
Willet	C	PR		Gull-billed tern	FC	SR	April-Oct.				
Least sandpiper	C	PR		Common tern	FC	PR					
Dunlin	C	PR		Sandwich tern	FC	SR	April-Oct.				
Dowitcher	C	PR		Caspian tern	FC	PR					
Semipalmated sandpiper	C	PR		Fish crow	C	PR					
Western sandpiper	C	PR									
Herring gull	C	PR									
Ring-billed gull	C	PR									
Laughing gull	C	PR									
Forster's tern	C	PR									
Least tern	C	SR	Mar.-Oct.								
Royal tern	C	PR									
Black skimmer	C	PR									
Boat-tailed grackle	C	PR									

KEY:
C – Common, seen in good numbers
FC – Fairly common, moderate numbers
U – Uncommon, small numbers irregularly
PR – Permanent resident, present year around
WR – Winter resident
SR – Summer resident

Note: Dominance indicates relative importance of the species as a group in the community. This concept is not based necessarily on taxonomic relationships but rather on numbers, size, and trophic dynamics.

great blue heron and little blue heron. Black-crowned night herons and yellow-crowned night herons also utilize the flats, but because of their nocturnal feeding habits, it is difficult to estimate their relative abundance.

Altogether, estuarine intertidal flats are frequented by 37 permanent resident species, plus three summer residents and four winter residents (Table 4-36). Many of the permanent residents, although present year around, are least abundant during winter (e.g., the little blue heron, Louisiana heron, willet, laughing gull, royal tern, and black skimmer) (Sprunt and Chamberlain 1949, Burleigh 1958). Populations of other species increase during the winter season as northern birds migrate to the South Carolina and Georgia coasts to winter. Included in this group are the American oystercatcher, herring gull, ring-billed gull, and the double-crested cormorant (Sprunt and Chamberlain 1949, Burleigh 1958). Joining these birds are winter residents such as the marbled godwit, lesser yellowlegs, American avocet, and Bonaparte's gull.

During summer, the least tern, gull-billed tern, and sandwich tern are common residents that breed and feed in other habitats. Many of the shorebirds, although regarded as permanent residents, do not breed in this area and are generally less common in the summer season. Typical non-breeding permanent residents include the semipalmated plover, black-bellied plover, ruddy turnstone, least sandpiper, dunlin, dowitcher, semipalmated sandpiper, and the western sandpiper. Individuals of these species over-summering on the South Carolina-Georgia coastal flats are thought to be subadults, but more investigation of this matter is needed.

Many of the true shorebirds reach peak numbers in the Sea Island Coastal Region during spring and fall migration. An excellent example is the whimbrel or Hudsonian curlew. This species is a non-breeding permanent resident that is uncommon during mid-winter and mid-summer but abundant during migration. Large numbers are regularly observed near the intertidal flats from mid-April to mid-May and again from early July to mid-September (Sprunt and Chamberlain 1949, Burleigh 1958).

As pointed out above, most of the species that play a major role in the intertidal flat habitat are either wading birds or true shorebirds. These species spend much of their time feeding on the benthic and nektonic fauna of the flats. Juvenile penaeid shrimp, fiddler crabs, and small fish are major food items, particularly of the wading birds. The American oystercatcher feeds on oysters (Tomkins 1947, 1954). Other shorebirds, including the dowitcher, greater and lesser yellowlegs, marbled godwits, and willets,

probe in sediments of the flats, feeding on infaunal forms; willets also take small fishes that become trapped in tidal pools. Several other species (e.g., brown pelican, double-crested cormorant, and 12 species of gulls and terns) utilize flats for resting and feed primarily in adjacent habitats. Nevertheless, the gulls at least function to some degree as predators and scavengers in this habitat. Two species not generally thought of as aquatic, the boat-tailed grackle and the fish crow, also commonly feed on intertidal flats. The grackle is the more abundant, and both consume invertebrates (probably juvenile crabs, shrimp, and other crustaceans) and small fishes (Wayne 1910).

The trophic interactions of the avifauna, described in part in the preceding paragraph, are summarized in Figure 4-16. The small sandpipers utilize the base trophic level, where they eat small quantities of marine algae along with the amphipods, annelids, and other benthic organisms that compose the bulk of their diet. Larger crustaceans and mollusks are consumed by plovers, large sandpipers, and virtually all of the large wading birds and scavengers. Fishes are the main food source for the larger species. Herons and egrets are highly predatory and, although generally regarded as scavengers, gulls also can be predatory. The top predator is the peregrine falcon. This bird is not a typical species of this habitat, but in its role in adjacent habitats it will occasionally prey on birds of exposed intertidal flats.

One bird, the American oystercatcher, deserves special mention here. It is one of the largest and most spectacularly marked shorebirds. Although it nests on open ocean beaches, the oystercatcher feeds almost exclusively on estuarine intertidal banks and flats. This species is rare over most of its range, and its greatest concentrations occur in the Sea Island Coastal Region. Tomkins (1947) studied this species in South Carolina and Georgia coastal waters and reported on feeding habits and life history. The breeding oystercatchers of coastal South Carolina and Georgia are permanent residents, but their numbers are augmented in winter when breeding birds from Virginia and North Carolina arrive. The Cape Romain area appears to be the principal wintering grounds on the Atlantic coast. Even so, these birds are not abundant. Hunters reduced oystercatcher populations to near extinction levels around the turn of the century. Sprunt and Chamberlain (1970) estimated the winter concentration of American oystercatchers in the Cape Romain area to be around 1,500 birds in 1940; they considered this to be a record number. Population estimates for recent years are lacking, and efforts should be made to obtain such data. Single flocks exceeding 100 birds have been observed, and the total number of oystercatchers in the prime wintering ground is

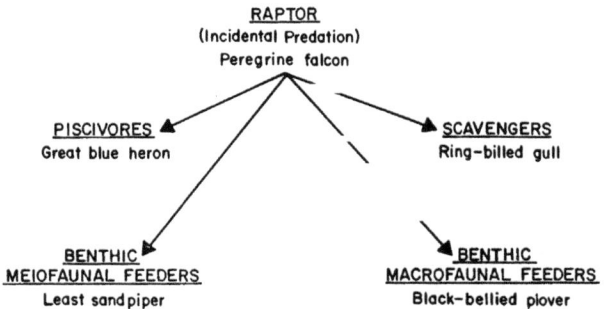

RAPTOR
(Incidental Predation)
Peregrine falcon

PISCIVORES
Great blue heron

SCAVENGERS
Ring-billed gull

BENTHIC
MEIOFAUNAL FEEDERS
Least sandpiper

BENTHIC
MACROFAUNAL FEEDERS
Black-bellied plover

Figure 4-16. Generalized trophic relationships of representative birds of estuarine intertidal flats.

believed to exceed several thousand individuals.

Man's impacts on birds of the estuarine intertidal flats are very similar to those discussed in the emergent wetlands section of this chapter. An additional impact, not prevalent in the intertidal marshes, has to do with the presence of people on the flats. Recreational use of flats has become more intensive with increases in numbers of boaters in the area. Flats are often used for picnicking, shellfish gathering, hunting, fishing, and other recreational activities. Also, commercial oyster and clam gathering is a major activity on the intertidal flats. Such activity is obviously disruptive to bird populations utilizing this habitat.

7. Mammals

Estuarine intertidal flats are predominantly aquatic environments, but one terrestrial mammal, the ubiquitous raccoon, makes extensive use of them as feeding grounds. Where intertidal flats are not widely separated from adjacent highlands, raccoons regularly visit the flats when exposed, especially at night, to forage for food. In isolated areas, raccoons may be seen on the flats in full daylight, although they are predominantly nocturnal wanderers.

Much of the success of the raccoon in coastal environments is due to its generalized habitat and food requirements. On intertidal flats, raccoons feed predominantly on crustaceans and mollusks. Blue crabs which become trapped in tide pool depressions by receding tides are especially sought. Many older accounts refer to raccoons feeding on intertidal

oysters and clams, although it is likely that crustaceans are of greater importance. On those intertidal flats over which fiddler crabs range, they are preyed upon heavily by raccoons.

The marsh rice rat occasionally ranges out onto the intertidal flats but only where there are adjacent marsh or high land areas. This rodent is largely carnivorous in certain habitats and will feed on fiddler crabs where they are available.

8. Decomposers

The roles of decomposers, principally bacteria and fungi, in estuarine intertidal flats and marshes have been reviewed recently by Peterson and Peterson (1979). Basically, these so-called decomposers perform three major functions in intertidal environments: 1) the decomposition of dead organic matter into inorganic nutrients and the cycling of these nutrients (see section on biogeochemical cycles in Chapter One); 2) the conversion of often indigestible plant materials (such as cellulose) into a form (i.e., microbial biomass) in which it can be assimilated readily by detritus- and deposit-feeding organisms; and 3) the conversion of dissolved organic and inorganic materials into consumable particulate matter. A number of investigators have noted that the colonizing macroflora appears to provide most of the nutritive value of detritus to consumers, since the bacteria and fungi appear to be assimilated during passage through the guts of detritivores while the detritus itself is defecated in enriched form. As Peterson and Peterson (1979) point out, the process of redevelopment of microbial populations on defecated detrital material may be "an important rate-limiting step which determines the abundance of various deposit-feeding species in marine benthic communities."

Little work has been done concerning the effects of perturbations on intertidal microbial populations in the Sea Island Coastal Region. Christian et al. (1978) subjected marsh plots near Sapelo Island, Georgia, to two kinds of perturbations: 1) clipping the shoots and pruning the below-ground parts of the smooth cordgrass cover for periods up to 18 months and 2) enrichment of defoliated and control plots with glucose, ammonium nitrate, or both. Microbial responses to these perturbation indicated that the soil microbial community was relatively unlinked to plant growth and quite resistant to change. Thus, perturbations which may drastically affect marsh vegetation may have relatively little impact on the intertidal microflora.

V. INTERTIDAL ESTUARINE SUBSYSTEM - IMPOUNDMENTS

A. DESCRIPTION

Numerous impoundments, ranging in size from a fraction of an acre to several thousand acres, occur principally within the South Carolina sector of the Sea Island Coastal Region. Most of these are former rice fields, although some may consist of newly diked brackish marsh (Johnson et al. 1974, Tiner 1977). Many have been maintained and managed as game preserves since the demise of commercial rice production in this area during the latter half of the last century. Most impoundments were constructed by diking off wetland areas intersected by tidal creeks. In some instances, entire marsh-creek areas were completely encircled by dikes, although the most common practice was to dike off the open end of a marsh slough bounded by high lands. These impoundments are usually equipped with flood gates or other structures for regulating water level and salinity. This is done in most cases to manage plant growth suitable for waterfowl utilization, but salinities in a few are controlled for aquacultural purposes (Morgan 1974, Tiner 1977). For a complete description of the history and development of rice field impoundments, see Volume II, Chapter Six.

Salinities in impoundments vary from completely fresh to as much as 25⁰/oo in those along the lower estuarine reaches. Impoundments with salinities averaging greater than 0.5⁰/oo are herein considered part of the estuarine subsystem. Because of their brackish nature, the flora and fauna of estuarine impoundments can best be compared and contrasted with biotic communities of the estuarine intertidal system, although technically these enclosures are artificially intertidal (i.e., tidal waters are regulated into and out of the impoundments).

New ecological systems replace old ones when portions of an estuary are impounded (Copeland 1974), and significant changes in hydrography accompany the impoundment of such an area. Water circulation is reduced and may become practically non-existent; increased sedimentation changes the nature of the substrate; smothering of aquatic vegetation may occur; and water salinity, temperature, oxygen, pH, and nutrient levels are altered (Copeland 1974, Dean 1975). Periodic draining and variations in hydrographic parameters limit the number of species occurring within impoundments, particularly in shallow rice field systems. The lack of adequate water circulation may be limiting to many filter feeding benthic organisms. Although such areas are characterized by low species diversity, overall productivity is high (Dean 1975).

In South Carolina, 14% - 16% of coastal marshes [approximately 70,000 acres (28,328.6 ha)] are functional impoundments. While being used primarily for waterfowl management, their capacity for other uses (e.g., aquaculture, waste treatment, and recreation), as well as their ecological importance as elements of marsh systems, has brought impoundments to the forefront of interest as ecological systems. The unique advantages of saltmarsh impoundments for aquaculture have been known for many years. The use of ponds for bivalve culture can be traced back to the Roman empire in the first century B.C., and may have originated even earlier with the Chinese (Yonge 1960). In the Southeastern United States, research in pond culture was stimulated by the observation of gigantism in blue crabs (Callinectes sapidus), and initial experimental success in the polyculture of fish, crabs, and oysters (Lunz 1951, 1968). This initial success was reiterated and quantified in more recent studies at several locations in South Carolina (Anderson 1976, Manzi et al. 1977b).

Despite their abundance and the increased pressure for reclamation, little research is presently underway to study the ecological processes of impounded wetlands. The general lack of knowledge concerning saltmarsh impoundments makes this area of marsh ecology a principal data gap.

B. PRODUCERS

1. Nonvascular Flora

The nonvascular microphytes and macrophytes which inhabit estuarine impoundments, and their role in impoundment processes, have been investigated only marginally. Dominant forms have been documented to some extent (Manzi and Zingmark 1978, Wiseman 1978), and in general seem to be correlated with estuarine/tidal creek population dynamics. Apparent deviations in microphyte population structure between impoundments and their adjacent tidal creeks include larger components of nannoplanktonic flagellata and benthonic blue-green algae in impounded areas (J. J. Manzi, 1978, South Carolina Marine Resources Division, Charleston, unpubl. data). In a recent study of the feasibility of bivalve culture in several South Carolina saltmarsh impoundments, phytoplankton concentrations were found to be generally higher in impounded areas than in adjacent tidal creeks (Manzi et al. 1977b). Figure 4-17 illustrates this dissimilarity and provides a comparison between the ponds and creeks encompassed by this study. As indicated by this information and other data (Anderson 1976), productivity is relatively high in low marsh impoundments and appears to reflect a classic nitrogen-limiting system.

Salt marshlands, particularly impounded areas with continuous or intermittent access to tidal creeks, normally act as sinks for both matter and energy (Odum and de la Cruz 1967, Odum 1970b, Pomeroy et al. 1972, de la Cruz 1973). The sinks are flushed or diluted regularly by spring tides and irregularly by storms, thus transporting nutrient-rich wastes and detritus to coastal waters (de la Cruz 1973). Periodic outwelling is indeed a primary factor in the high productivity of coastal waters. In a recent study, Manzi et al. (1977b) measured phytoplankton biomass, nutrient concentrations, and rates of primary production. Their data illustrate the concentrating properties of tidal creeks and impoundments (Fig. 4-17), and reflects this fertility in oyster growth and meat yields. Areas characterized by low marshlands and good tidal exchange (Blue Heron and Waring Creek, South Carolina) normally exhibited high concentrations of phytoplankton with resultant decreases in available nitrate. Hitchcocks Pond, while surrounded primarily by maritime forest, was fed by Adams Creek, a long narrow tidal inlet surrounded by extensive low marsh, and subsequently exhibited the same high phytoplankton, low nitrogen concentration characteristic. Orthophosphate was present in high concentrations at all locations and was probably not a limiting factor for phytoplankton populations. Estimates of primary productivity (Fig. 4-18) suggested strongly that increased fertility of saltmarsh impoundments led to increased oyster yields. Potential primary productivity was without exception higher in impoundments than in adjacent creeks or rivers, and was correlated directly with oyster growth among the impoundments.

2. Vascular Flora

Because of their value to waterfowl, the desired and usually dominant plant species found in managed brackish water impoundments are widgeon grass, salt-marsh bulrush, and dwarf spikerush (Baldwin 1956, Wilkinson 1970, Tiner 1977). Other desirable plants for waterfowl management are sago pondweed, dotted smartweed, muskgrasses (all nonvascular species), and vascular species such as soft-stem bulrush and common three-square (Baldwin 1956, Johnson et al. 1974). Tiner et al. (1976) interviewed impoundment managers in the Santee River estuary concerning management procedures and dominant plants; results are listed in Table 4-37. A list of plants common to brackish water impoundments in South Carolina is given in Table 4-38.

Baldwin (1968) discussed impoundments in regularly and irregularly flooded salt marshes. He concluded that diked but regularly flooded smooth cordgrass marsh can be managed for widgeon grass with the least effort. In irregularly flooded marshes, impoundments are generally shallower and tend to be vegetated with various bulrushes, dwarf spikerush, wild millets, panic grasses, and giant foxtail grass. Baldwin suggested managing these impoundments for salt-marsh bulrush for maximum waterfowl utilization.

Wilkinson (1970) carefully studied dominance in five newly diked brackish marshes on South Island, Georgetown County, South Carolina. Management procedures for each were as follows:

Impoundment I: "Drawn down in March to keep the marsh soil dry. Flooded in October just prior to the usual time of arrival of waterfowl."

Impoundment II: "Water levels maintained at ground level, which produced a saturated soil condition from March through September. This impoundment was also flooded in October just prior to the usual time of arrival of waterfowl."

Impoundment III: "Water level slowly raised from March through September to a depth of 24 inches, and drained during February of each year."

Impoundment IV: "Water level maintained at full pond depth (approximately 24 inches), except during each February, when it was drained."

Impoundment V: "The inflow and outflow gates were left open to allow the tide to flood and ebb in the impoundment from March through September of each year. During each October the impoundment was flooded to a depth of approximately 24 inches, and held at that depth until the following March."

Impoundment I was therefore basically dry, Impoundment II was saturated, Impoundment III was characterized by a slowly rising water level, Impoundment IV was fully

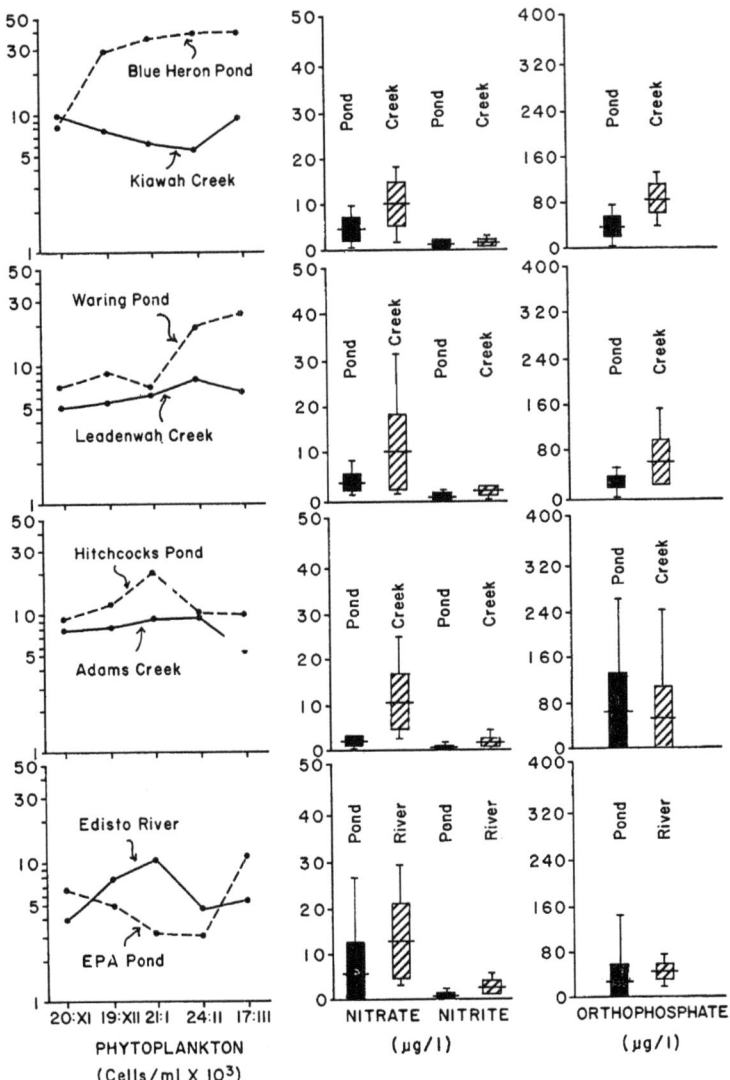

Figure 4-17. Estimates of monthly (ordinate scale) phytoplankton concentrations and means, ranges, and standard deviations of principal nutrients in four tidal impoundments and their adjacent feeder creeks in South Carolina (Manzi et al. 1977a). (Sampling dates given.)

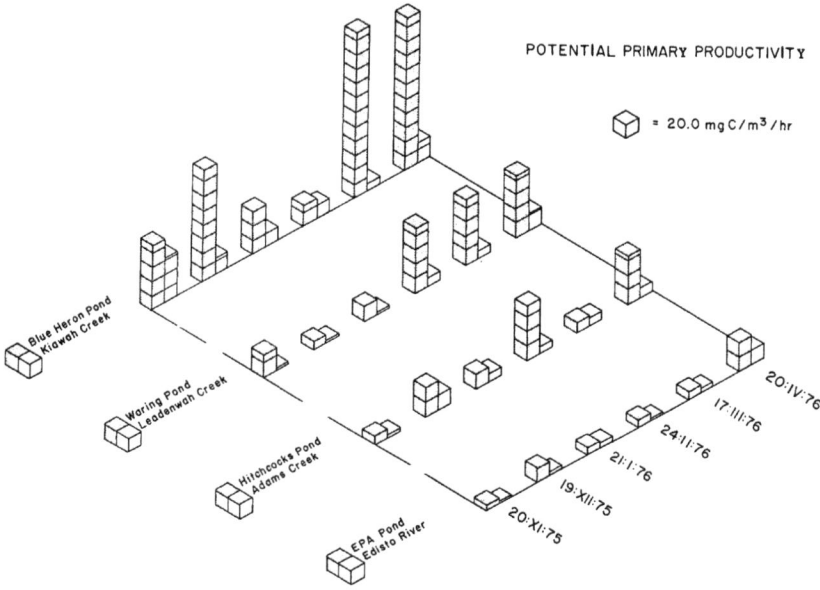

POTENTIAL PRIMARY PRODUCTIVITY

☐ = 20.0 mgC/m³/hr

Blue Heron Pond
Kiawah Creek

Waring Pond
Leadenwah Creek

Hitchcocks Pond
Adams Creek

EPA Pond
Edisto River

20:XI:75
19:XII:75
21:I:76
24:II:76
17:III:76
20:IV:76

Figure 4-18. Comparative monthly (ordinate scale) estimates of potential primary production in
four South Carolina tidal impoundments (shaded) and their adjacent feeder creeks
(clear) (Manzi et al. 1977b). (Sampling dates given.)

flooded, and Impoundment V was open to
normal tidal fluctuation. These procedures
are representative of the possible methods
of managing brackish impoundments in the
characterization area.

In Impoundment I, giant cordgrass increased its dominance dramatically, while
salt-marsh bulrush, saltmeadow cordgrass,
and other species increased in lesser proportions (Table 4-39). Changes in the
saturated impoundment are presented in
Table 4-40; here, dwarf spikerush very
rapidly increased in abundance, although
giant cordgrass maintained dominance. In
Impoundment III, widgeon grass and smooth
cordgrass both increased considerably in
abundance, while dwarf spikerush declined
(Table 4-41). In Impoundment IV, widgeon
grass and salt-marsh bulrush became dominant (Table 4-42). Narrow-leaved cat-tail
also increased in dominance, although most
other species declined. Impoundment V was
dominated by giant and smooth cordgrass
(Table 4-43). Wilkinson's conclusions
support the management strategies of most

local managers of brackish waterfowl impoundments in that Impoundments III and
IV were most successful in attracting
waterfowl.

Morgan (1974) described the three
basic methods by which the plant composition of brackish water impoundments in the
Ashepoo-Combahee-Edisto area of South
Carolina is managed. Cyclical fluctuations
in water level produce salt-marsh bulrush
dominance, while slowly rising water level
and permanent flooding favor widgeon grass.
Neely (1960) described in detail how careful fluctuations of impoundment water
levels at 4 - 6 in (10 - 15 cm) intervals
up to 12 in (30 cm) in depth produce saltmarsh bulrush dominance, with dwarf spikerush sometimes dominating former bare spots
in the impoundments.

If the flood gates of the brackish
water impoundment remain intact and the
water level is no longer managed, the impoundment will gradually change into a
fresher water environment with sago pondweed

275

Table 4-37. Results of a questionnaire survey of several coastal impoundment managers in the Santee River estuary, South Carolina, (Tiner et al. 1976).

Name Of Manager	Area Managed	Brackish-Water Impoundments (acres)	Dominant Plants	Freshwater Impoundments (acres)	Dominant Plants
Phillip M. Wilkinson	South Island Plantation	2,500	Widgeon grass, dwarf spikerush, salt-marsh bulrush, sago pondweed, and muskgrass	500 (dry)	Corn, wheat, barley, rye, Italian rye grass, clover, peas, milo, millet, brown-top millet, soybeans, and grain sorghum
Thomas H. Strange, Jr.	Santee Coastal Reserve	10,500	Widgeon grass, salt-marsh bulrush, dwarf spikerush, and giant cordgrass	1,500	
	Santee Delta Game Management Area				Swamp smartweed, asiatic day-flower, tear-thumbs and giant cordgrass
Kenneth Williams	Kinloch Plantation	4,000	Widgeon grass, dwarf spike-rush, salt-marsh bulrush, giant cordgrass, narrow-leaved cat-tail, and southern cat-tail	150	Not Reported
Graham Reeves	Annandale Plantation	1,500	Dwarf spikerush and widgeon grass		

Table 4-38. List of vascular plants common to brackish water impoundments of South Carolina (adapted from Tiner 1977).

Scientific Name	Common Name
Alternanthera philoxeroides	Alligator-weed
Aster subulatus	Annual salt marsh aster
Baccharis halimifolia	Sea myrtle
Bacopa monnieri	Water hyssop
Borrichia frutescens	Sea ox-eye
Ceratophyllum demersum	Coontail
Cladium jamaicense	Saw grass
Cyperus strigosus	Sedge
Distichlis spicata	Salt grass
Echinochloa walteri	Salt marsh millet
Eleocharis parvula	Dwarf spikerush
Eupatorium capillifolium	Dog fennel
Iva frutescens	Marsh elder
Juncus roemerianus	Black needlerush
Lemna spp.	Duckweeds
Leptochloa sp.	Sprangletop
Myrica cerifera	Wax myrtle
Najas guadalupensis	Bushy pondweed
Nymphaea mexicana	Banana water-lily
Nymphaea odorata	White water-lily
Panicum spp.	Panic grasses
Pluchea purpurascens	Marsh fleabane
Polygonum punctatum	Dotted smartweed
Polygonum sp.	Smartweed
Potamogeton berchtoldii	Narrow-leaved pondweed
Potamogeton pectinatus	Sago pondweed
Ruppia maritima	Widgeon grass
Salicornia europaea	Glasswort
Scirpus americanus	Common three-square
Scirpus olneyi	Olney's three-square bulrush
Scirpus robustus	Salt-marsh bulrush
Scirpus validus	Soft-stem bulrush
Sesbania exaltata	Coffee-weed
Setaria magna	Giant foxtail
Spartina alterniflora	Smooth cordgrass
Spartina cynosuroides	Giant cordgrass
Spartina patens	Saltmeadow cordgrass
Spirodela polyrrhiza	Big duckweed
Typha angustifolia	Narrow-leaved cat-tail
Typha domingensis	Southern cat-tail
Typha glauca	Blue cat-tail
Typha latifolia	Common cat-tail

Table 4-39. Vegetative analysis of Impoundment I (basically dry; see text for explanation of management procedures used in this impoundment) (adapted from Wilkinson 1970).

Smooth cordgrass	42.0	33.3	15.3
Bare	40.0	28.5	27.0
Giant cordgrass	8.0	13.8	36.3
Salt grass	4.0		0.9
Saltmeadow cordgrass	2.0	9.9	7.4
Glasswort	2.0		0.0
Salt-marsh bulrush	1.6	0.6	6.1
Black needlerush	0.4	1.1	0.0
Marsh fleabane	0.0	0.0	2.2
Sedge	0.0	0.0	3.1
Panic grasses	0.0	0.0	1.3
Marsh elder	0.0	0.0	0.4

Table 4-40. Vegetative analysis of Impoundment II (saturated soil; see text for explanation of management procedures used in this impoundment) (adapted from Wilkinson 1970).

Plant Species	Percent Occurrence		
	1967	1968	
Bare	65.3	35.6	24.3
Giant cordgrass	10.0	12.0	18.0
Soft-stem bulrush	5.8	0.0	0.0
Salt-marsh bulrush	5.0	26.5	4.6
Narrow-leaved cat-tail	5.0	0.0	0.0
Dwarf spikerush	3.8	10.7	16.7
Salt grass	3.2	7.4	14.9
Marsh fleabane	1.9	7.0	17.0
Smooth cordgrass	0.0	0.8	4.5

Table 4-41. Vegetative analysis of Impoundment III (slowly rising water level; see text for explanation of management procedures used in this impoundment) (adapted from Wilkinson 1970).

Plant Species	Percent Occurrence		
	1967	1968	1969
Salt grass	45.3	33.8	29.5
Narrow-leaved cat-tail	13.8	1.5	2.2
Giant cordgrass	13.3	16.6	13.4
Bare	5.3	0.5	11.2
Widgeon grass	14.6	33.3	15.2
Saltmeadow cordgrass	2.6		4.0
Smooth cordgrass	1.7	7.6	20.5
Salt-marsh bulrush	1.3	2.2	4.0
Marsh fleabane	1.3	0.0	0.0
Dwarf spikerush	0.4	4.5	0.0
Sedge	0.4	0.0	0.0

Table 4-42. Vegetative analysis of Impoundment IV (fully flooding; see text for explanation of management procedures used in this impoundment) (adapted from Wilkinson 1970).

Plant Species	Percent Occurrence		
	1967	1968	1969
Open water, nothing growing	25.8	2.4	1.1
Narrow-leaved cat-tail	25.8	13.3	20.5
Giant cordgrass	15.8	12.1	2.1
Dwarf spikerush	14.3	23.0	9.7
Widgeon grass	10.8	30.3	45.0
Salt grass	6.2	2.4	
Salt-marsh bulrush	5.0	14.5	21.6
Algae (Cladophora spp.)	4.3	2.0	0.0

Table 4-43. Vegetative analysis of Impoundment V (normal tidal fluctuation; see text for explanation of management procedures used in this impoundment) (adapted from Wilkinson 1970).

Plant Species	Percent Occurrence		
	1967	1968	1969
Narrow-leaved cat-tail	0.0	3.4	4.0
Giant cordgrass	38.0	49.1	48.3
Salt-marsh bulrush	0.0	10.1	9.2
Smooth cordgrass	0.0	1.1	22.7
Bare	62.0	36.3	15.8

and other pondweeds appearing first, and southern cat-tail and water-lily quickly following. In irregularly flooded, shallow impoundments, exposed soil species invade the unmanaged impoundment, with alligator-weed and various cat-tails asserting dominance (Baldwin 1968). If the water-control structures of the impoundments are no longer operable, or if the dikes are broken, the impoundment will gradually change into marsh environment corresponding with those of similar elevation and salinity. Baden et al. (1975) studied two abandoned rice fields in Georgetown County, South Carolina. In Thousand Acre Rice Field, they found a brackish marsh with plant zonation according to elevation; smooth cordgrass dominated the lower portions of the marsh, while giant cordgrass and black needlerush were dominants in the higher areas. Further inland, the much fresher upper marsh was found to be dominated by narrow-leaved cat-tail, giant cordgrass, common three-square, and soft-stem bulrush. These marsh types correspond closely to other marshes of similar salinities and elevations that have never been impounded.

Fritz (1975) studied aquatic primary productivity in an impoundment in Georgetown County, South Carolina, and measured the standing crop biomass of another impoundment dominated by black needlerush, widgeon grass, and salt grass. Fritz compared its total biomass to that of a nearby unimpounded smooth cordgrass marsh and concluded that the smooth cordgrass marsh was 1.3 to 1.8 times more productive than the impounded marsh (Fritz 1975). Further comparative studies are needed to gain a better understanding of nutrient cycles, total biomass, and primary productivity in estuarine impoundments.

C. CONSUMERS

1. Zooplankton

No in-depth studies of the zooplankton of estuarine impoundments of the Sea Island Coastal Region have been completed to date. However, certain parallels may be inferred from the work of Deevey (1948), who studied Great Pond, Massachusetts, an impoundment periodically opened and closed to the sea. When free exchange occurred, the pond fauna clearly resembled that present in adjacent open waters. Salinity alterations brought on by periodic closure to the sea restricted numbers of some zooplankters in the pond. In general, Deevey found that successful pond zooplankters were highly euryhaline and that temperature was most important in controlling seasonal succession.

Zooplankton is introduced when impoundments of the Sea Island Coastal Region are flooded with water from adjacent creeks, rivers, or inlets. Resistant eggs appear to play little if any role in establishing the zooplankton community. To be successful in these temporary aquatic habitats, zooplankters must be highly euryhaline because salinity is often drastically and deliberately altered. Meroplankton is probably less important in impoundments than in adjacent open water areas. In many cases, the larvae of benthic invertebrates are less tolerant of salinity fluctuations than the adults and, perhaps more importantly, the ponds do not necessarily receive estuarine water during peak spawning seasons for many species.

Overall, the degree of similarity between zooplankton of impoundments and adjacent open estuarine waters probably depends upon the following factors: 1) the time of year the impoundment is flooded, 2) the amount and frequency of water exchange permitted between impoundment and open estuary, 3) the mean salinity maintained in the impoundment and the variation about that mean, and 4) the amount and frequency of rainfall. Since the most successful estuarine holoplankters are strongly euryhaline, dominant zooplankters in impoundments may well be the same as those in adjacent estuarine waters. However, differences in abundance of individual species between the two habitats might be great because factors such as predation, or those mentioned above, could selectively favor one species over another.

Impoundments may enhance productivity of estuarine areas by providing protected nursery grounds and spawning sites for zooplankton, which are then periodically released into the open estuary when water is released. This would serve to concentrate food for planktivorous animals, and would permit zooplankters to reach larger size prior to dispersion by currents.

Recent studies indicate that phytoplankton production is higher in coastal brackish ponds than in their feeder creeks (Anderson 1976). Thus, impoundments may support large zooplankton populations. The zooplankton, in turn, may play a major role in recycling nutrients such as nitrogen, which is believed to limit primary production in estuarine impoundments during summer (Anderson 1976).

Predators such as small fishes, American eel, and juvenile crabs may enter impoundments when they are initially flooded and whenever additional water is taken in. These organisms probably control the abundance of zooplankton in impoundments. In summer, low levels of dissolved oxygen, which often result in fish kills, probably also reduce zooplankton numbers significantly.

In a study of a flooded former rice field adjacent to the North Santee estuary, Dean (1975) reported that a copepod Acartia (presumably Acartia tonsa), grass shrimp (Palaemonetes sp.), and several decapod crustacean larvae (Uca pugnax, Sesarma reticulatum, and Eurypanopeus depressus) were important zooplankters. He noted that density and diversity of zooplankton were low in this brackish impoundment, but gave no numerical data. Molluscan larvae were not reported from this habitat.

Knott (1980) compared zooplankton populations of two man-made ponds filled from the North Edisto River with those of the North Edisto estuary itself. The impoundments were completely isolated from the river; water input from the river was accomplished by pumping. Knott (1980) reported that the annual mean density of zooplankton was much greater in the river

(10,148 organisms/m^3) than in either of the ponds (3,417/m^3 and 5,450/m^3). Further, the river zooplankton community was more diverse and more stable over the year than those of the impoundments. The copepod Acartia tonsa was the dominant zooplankter in both environments, but was more important, and frequently more abundant, in the impoundments than in the river (Table 4-44). It also exhibited a marked seasonal variation in abundance in the impoundments, but such variation was much less pronounced in the river. Other important zooplankters in these environments are listed in Table 4-44.

The impact on zooplankton of environmental alterations in brackish-water ponds has not been investigated and can be predicted at present only from studies of open estuarine systems. The same degree of alteration would be expected to have a more pronounced effect in an impoundment than in an open estuary because: 1) animals have less chance to avoid a contaminant because of areal constraints, 2) dilution proceeds more slowly because of less circulation, and 3) temperatures often are higher, tending to accelerate response of organisms to toxic substances.

2. Benthic Meiofauna

No studies of the meiobenthos of impoundments have been conducted in the Sea Island Coastal Region. See the section on meiofauna of estuarine intertidal wetlands for information most likely to be pertinent to this environment.

3. Benthic Macroinvertebrates

Over the last three decades, impoundments have been studied with a view toward intensive cultivation of commercially important species of aquatic invertebrates and fishes in South Carolina. Studies on the rearing of shrimp, crabs, and oysters in ponds were undertaken at Bears Bluff Laboratories between 1946 and 1969. These investigations demonstrated that growth of shrimp was rapid in ponds and that high quality oysters could be grown in such areas (Lunz 1951, 1952a, 1955, 1956, 1957, 1958, Lunz and Bearden 1963). Ballard (1975a, b) studied growth and survival of oysters (Crassostrea virginica) and clams (Mercenaria mercenaria) in a 250 acre (101.2 ha) pond at Annandale Plantation, Georgetown County, South Carolina. Intensive studies also have been conducted by the South Carolina Marine Resources Research Institute (Charleston) on pond culture of the Malaysian prawn, Macrobrachium rosenbergii, at various locations in South Carolina, including Cayce (Richland County), Bonneau (Berkeley County), and Bears Bluff (Charleston County). Smith et al. (1976) observed low mortality and rapid growth of prawns. Duration of grow-out season varied from 5 to 6.5 months, depending upon the site.

Table 4-44. Numerically abundant zooplankters collected from the North Edisto River and two adjacent impoundments over a 1-year period (from Knott 1980).

Overall Rank		% Of Total Fauna	% Of Total Number		
			Pond 1	Pond 3	Edisto River
1.	Acartia tonsa	65.88	79.22	93.97	41.36
2.	Euterpina acutifrons	5.34	0.12	0.09	11.04
3.	Pseudodiaptomus coronatus	4.38	6.37	1.36	5.62
4.	Parvocalanus crassirostris	4.12	0.63	0.35	8.12
5.	Copepod nauplii	3.69	1.94	0.78	6.40
6.	Rotifera	3.67	0.02	< 0.01	7.66
7.	Cirripedia larvae	3.05	1.25	0.75	5.35
8.	Tortanus setacaudatus	1.77	1.09	1.36	2.33
9.	Gastropod veligers	1.21	0.88	0.12	2.08
10.	Oithona colcarva	1.14	1.62	0.14	1.62
11.	Decapod larvae	1.07	0.53	0.15	1.92
12.	"Saphirella tropica"	0.98	0.14	0.03	1.97
13.	Metis sp.	0.60	2.69	0.16	0.05
14.	Polychaete larvae	0.50	0.08	0.20	0.88
15.	Foramenifera	0.48	0.04	0.01	0.97
16.	Oikopleura sp.	0.36	0.00	0.00	0.77
17.	Nematoda	0.28	1.20	0.02	0.08
	TOTALS	98.52	97.82	99.49	98.22

An extensive data base exists on water quality and productivity of ponds used for fish culture, particularly for freshwater systems. However, little information is available on the benthos of estuarine impoundments, particularly for the coastal plains of Georgia and South Carolina. Ballard (1975a) observed high densities of Palaemonetes in a pond at Annandale Plantation. Also present were planktonic larvae of the decapods Uca pugnax, Sesarma reticulatum, and Eurypanopeus depressus. The absence of natural oyster beds in the pond was attributed to the lack of suitable substrate and to periodic draining of the impoundment. Blue crabs (Callinectes sapidus) and penaeid shrimp (Penaeus spp.) were shown to thrive in ponds at Bears Bluff Laboratories (Bears Bluff Laboratories, Inc. 1956).

Results of studies on culture of oysters in impounded environments along the Atlantic and Gulf coasts are described by Lunz (1951), Shaw (1965), and May (1969). Comprehensive reviews of early oyster culture and artificial propagation of larvae can be found in Dean (1892a, 1893), Baughman (1948), Loosanoff and Davis (1963), Galtsoff (1964), and Joyce (1972).

Oyster culture in enclosed tidal areas (Dean 1892, 1893) was first reported in South Carolina by Colson (1888) in his history of the mill pond oyster, a delicacy which proliferated in large sawmill ponds from 1830 to 1869. Successful production was attributed to tidal flushing and the presence of floating logs bearing oysters (Colson 1888). Battle (1892) proposed tidal pond cultivation in South Carolina during his comprehensive investigation for the U.S. Fishery Commission. However, cultivation of oysters and fishes in marsh impoundments using an experimental approach was not initiated in South Carolina until 1943 (Lunz 1968). Further experimentation illustrated congruent polyculture of fish, crabs, and oysters in the same pond. Ponds dug in the marsh appeared to be less productive than impounded marshlands. In one annual study, oyster yield was estimated to be 35.2 m^3 of shell stock/0.4 ha (Lunz 1968).

Not all attempts to culture oysters in saltwater ponds of South Carolina have been successful. Lunz (1955) reported a disastrous mortality resulting from what was later thought to be the pathogenic fungus, Perkinsia marina, and possibly other predators. Boring sponges (Cliona) and oyster drills (Urosalpinx cinerea and Eupleura caudata) are sometimes reported in impoundments where salinities are suitable. Other high salinity predators, such as whelks and starfishes, are less frequently observed due to the characteristically reduced impoundment salinities. Mussels (Brachidontes spp.) are often found growing on oysters in ponds, as are barnacles (Balanus eburneus, Balanus improvisus) and blisters of mud worms (Polydora websteri). Blue crabs (Callinectes sapidus) and occasionally stone crabs (Menippe mercenaria) inhabit the impoundments but are not usually found inside oyster trays (Anderson 1976). Lunz (1968) indicated that predators such as the boring sponge could be controlled by lowering the salinity or draining the pond and allowing the oysters to be exposed to air.

MacGregor (1970), using two groups of 2 - 3 year old Crassostrea virginica in a 0.27 ha (0.67 acre) pond at Sapelo Island, in a 4-week experiment reached no conclusions concerning the feasibility of commercial pond culturing of oysters in Georgia salt marshes. Ballard (1975a) speculated on techniques and the potential of impoundment oyster culture in South Carolina.

Recent studies in impoundments located on Wadmalaw and Kiawah islands, South Carolina (Anderson 1976, Manzi et al 1977b) have substantiated the accelerated growth rates and favorable survival observed by others (Lunz 1955, 1956, 1968, Badger 1968). These experiments, however, were not designed to establish the economic feasibility of oyster culture in impoundments.

Use of saltwater ponds for aquaculture has the following advantages over open estuarine areas: 1) protection from strong waves and adverse currents, 2) easier access to bottoms for planting and management, 3) predator control, 4) modification of tidal exchange, and 5) artificial fertilization (Bouchon-Brandely 1882, Gaarder and Spärck 1932, Turner 1951, Lunz 1955, Carriker 1956, 1959, Korringa and Postuma 1957, Binmore 1964, Loosanoff 1964, Shaw 1965).

4. Insects

The insect fauna of coastal impoundments in the characterization area has not been studied in detail but is expected to be similar to that of marshes (see section on insects of estuarine intertidal wetlands).

5. Fishes

Dean (1975) investigated the mariculture potential of several marine and estuarine fishes, including Atlantic croaker and ladyfish, in impoundments at Annandale Plantation, South Carolina. Theiling and Loyacano (1976) studied the age and growth of red drum from a saltwater marsh impoundment at South Island. In Georgia, investigations of the fisheries of natural brackish ponds occurring on coastal islands have been carried out by Hillestad et al. (1975).

The mariculture potential of saltwater impoundments for fishes was investigated at Bears Bluff Laboratories over a number of years (Lunz 1951, 1956; Bearden 1967; Bears Bluff Laboratories, Inc., Wadmalaw Island, South Carolina, unpubl. data). During the period of 1947 - 1967, 1-acre (0.4 ha) marsh impoundments at Bears Bluff Laboratories were stocked annually with marine fishes and invertebrates by tidal flooding through water control structures, and drained each fall. Biomass of fishes harvested from these ponds ranged from 61.5 to 382 lb/acre (68.9 to 428.1 kg/ha), averaging approximately 200 lb/acre (224.2 kg/ha). Mullet, spot, ladyfish, and mummichog were the most abundant species. Biomass of smaller fishes (mummichog, silverside) was not normally recorded. These data suggested that during certain times the mean biomass of fishes in impoundments may be greater than that of natural, unimpounded tidal marsh areas. Turner and Johnson (1974), for example, found a mean biomass of 92 lb/acre (103.1 kg/ha) for estuarine fishes in tidal creeks of the Cooper River estuary, with a range of from 7.3 to 257.1 lb/acre (8.2 to 288.2 kg/ha) during April through November. This does not imply that impoundments are necessarily more productive on an annual basis than are natural tidal marsh creeks, since the former are semi-enclosed systems from which little emigration may take place, whereas recruitment and emigration take place continually in the latter zone.

Sixty-one species of marine and estuarine fishes (Table 4-45) have been identified from saltwater impoundments in South Carolina (Bears Bluff Laboratories, Inc., Wadmalaw Island, South Carolina, unpubl. data). Such impoundments are typically

283

Table 4-45. Systematic listing of fish species known to occur in salt and brackish water impoundments in South Carolina (Bears Bluff Laboratories, Inc., Wadmalaw Island, South Carolina, unpubl. data).

Order Elopiformes	Order Atheriniformes (Cont.)	Order Perciformes (Cont.)
Family Elopidae	Family Cyprinodontidae	Family Sciaenidae
Ladyfish	Sheepshead minnow	Southern kingfish
Tarpon	Mummichog	Northern kingfish
	Striped killifish	Black drum
Order Anguilliformes	Family Poeciliidae	Atlantic croaker
Family Anguillidae	Sailfin molly	Red drum
American eel	Mosquitofish	Family Ephippidae
		Atlantic spadefish
Order Clupeiformes	Order Gasterosteiformes	Family Mugilidae
Family Clupeidae	Family Syngnathidae	Striped mullet
Atlantic menhaden	Chain pipefish	White mullet
Gizzard shad		Family Blenniidae
Family Engraulidae	Order Perciformes	Feather blenny
Bay anchovy	Family Percichthyidae	Striped blenny
	Striped bass	Family Gobiidae
Order Cypriniformes	Family Serranidae	Sharptail goby
Family Cyprinidae	Rock sea bass	Naked goby
Carp	Family Pomatomidae	Marked goby
	Bluefish	Highfin goby
Order Siluriformes	Family Carangidae	Family Eleotridae
Family Ariidae	Crevalle jack	Fat sleeper
Sea catfish	Atlantic bumper	Spinycheek sleeper
	Family Gerreidae	Family Triglidae
Order Batrachoidiformes	Irish pompano	Striped searobin
Family Batrachoididae	Mojarras	Leopard searobin
Oyster toadfish	Family Lutjanidae	
	Gray snapper	Order Pleuronectiformes
Order Gadiformes	Family Pomadasyidae	Family Bothidae
Family Gadidae	Pigfish	Ocellated flounder
Spotted hake	Family Sparidae	Bay whiff
Family Ophidiidae	Pinfish	Fringed flounder
Striped cusk-eel	Sheepshead	Summer flounder
	Family Sciaenidae	Southern flounder
Order Atheriniformes	Silver perch	Family Soleidae
Family Belonidae	Weakfish	Hogchoker
Atlantic needlefish	Spotted seatrout	Family Cynoglossidae
Family Atherinidae	Banded drum	Blackcheek tonguefish
Atlantic silverside	Spot	

inhabited both by year-round resident fish species and species which enter periodically from outside waters as larvae or postlarvae but are not capable of reproducing there. Resident fishes are usually numerically dominated by the mummichog, sheepshead minnow, mosquitofish, sailfin molly, and Atlantic silverside. The most abundant species introduced during the flooding of such impoundments are striped mullet, American eel, spot, Atlantic croaker, red drum, spotted seatrout, silver perch, Atlantic menhaden, bay anchovy, mojarras, pinfish, southern flounder, and ladyfish. Impoundments also provide prime habitat for the young of several species of fishes not commonly found in adjacent estuarine waters, including snook and tarpon. Large numbers of juvenile tarpon, ranging from 59 to 300 mm SL, have been collected from saltwater impoundments in South Carolina during late summer and fall.

One collection of 130 juvenile tarpon was made from an 8-acre (3.2 ha) impoundment near Adams Creek, South Carolina, in 1965 (C. M. Bearden, 1978, South Carolina Marine Resources Division, Charleston, unpubl. data). Juveniles of red drum and spotted seatrout are often more common in impoundments than in adjacent natural areas (Bears Bluff Laboratories, Inc., Wadmalaw Island, South Carolina, unpubl. data). Some of the low salinity impoundments in the Sea Island Coastal Region also contain populations of carp. These fish have been observed in impoundments on the Santee estuary and have apparently adapted to brackish water conditions.

On many of the South Carolina and Georgia coastal islands, naturally occurring ponds formed in shallow depressions and influenced by tidal action occasionally contain numbers of euryhaline fish species.

In Georgia, Hillestad et al. (1975) sampled the aquatic systems on Cumberland Island, including brackish and freshwater ponds and their drainage outflows. Several of these ponds are closely associated with the ocean and are subject to occasional tidal flooding. Eight species of eury-haline fishes were found in the brackish water ponds. Large numbers of the sheeps-head minnow, sailfin molly, mosquitofish, and striped mullet were present, feeding on the abundant organic detritus of pond bottoms. Mosquitofish occurred in the saline and freshwater ponds of the island as well as in the drainage outflow sys-tems. Sailfin mollies also occurred in the drainage systems and were found in eutrophic pools of water beneath oak trees along the drainage channels. The lower, tidally influenced portions of the pond drainage outflows contained mullets, mojarras, mummichogs, marsh killifish, and American eels.

Impoundments provide a rich habitat and an abundant food supply for many fish species. Growth rates of many species appear to be higher in impoundments than in surrounding waters (Bearden 1967, Dean 1975). Average growth rates for four fish species commonly found in impoundments are given in Table 4-46.

Food habits and trophic relationships of fishes occurring in salt or brackish water impoundments are not well understood, and additional research along these lines is needed. Predominately herbivorous species such as mullets, Atlantic menhaden, sheepshead minnows, and sailfin mollies would presumably feed primarily upon the large quantities of phytoplankton, benthic algae, and vascular plant material present. Odum (1975) estimated that striped mullet in a brackish pond fed largely on living algae and to a lesser extent upon plant detritus. Primary and mid-level carni-vores, such as mummichog, mosquitofish, spot, and Atlantic croaker, would be ex-pected to feed largely on smaller fishes, Palaemonetes shrimp, insects, and benthic invertebrates such as polychaete worms, as well as organic detritus and plant material. In brackish ponds on the Santee estuary, South Carolina, young Atlantic croaker were found to feed largely on in-sects, insect larvae, and crustaceans (Dean 1975). Top level carnivores, in-cluding ladyfish, tarpon, red drum, black drum, spotted seatrout, and flounders, are known to feed extensively on Palaemonetes and penaeid shrimp, mummichogs, mosquito-fish, sailfin mollies, mullets, silver-sides, and other small fishes (Bearden 1967; Dean 1975; Bears Bluff Laboratories, Inc., Wadmalaw Island, South Carolina, unpubl. data). Table 4-47 presents trophic levels of the most abundant fish species commonly found in coastal saltwater im-poundments.

Fluctuations of several environmental factors (e.g., temperature, salinity,

dissolved oxygen) in coastal impoundments are more extreme than in nearby estuarine waters. Mortalities due to low dissolved oxygen and low temperatures are common-place in impoundments. Fish kills associ-ated with pesticide applications on ad-jacent agricultural lands have frequently occurred in impoundments in past years in South Carolina (South Carolina Marine Re-sources Division, Charleston, unpubl. data), and the U.S. Environmental Pro-tection Agency (1971) conducted a study of the movement of the pesticide mirex in small impoundments near Charleston. The major limiting factor of shallow natural ponds on coastal islands is water level fluctuation. Alterations in drainage brought about by development could have disastrous effects on this habitat.

While salt and brackish water impound-ments provide valuable habitat for many marine and estuarine fish species, unless properly managed with respect to water ma-nipulation (flooding and lowering at stra-tegic times), many introduced species can-not survive the rigorous extremes of temperature and dissolved oxygen supply found within these areas. The drawdown or draining of coastal impoundments in the fall, provided such measures are compatible with waterfowl or mariculture activities, could result in the release of large num-bers of fishes and invertebrates to the natural estuarine system.

Much research is needed concerning the biology and ecological relationships of the fishes in coastal impoundments in South Carolina and Georgia. Comparison of pro-ductivity and significance as habitat of impoundments versus natural marsh creek systems should be conducted. The maricul-ture and sportfishing potential of impound-ments for marine and estuarine fishes also needs much further study.

6. Amphibians and Reptiles

Salinities in estuarine impoundments vary widely, depending on their location and how they are managed. Numbers and di-versity of amphibians and reptiles are much higher in low-salinity (<5°/oo) impound-ments. For purposes of this section, how-ever, discussion of amphibians and reptiles will be limited to impoundments exhibiting estuarine characteristics.

Amphibians are the only class of ver-tebrates which have not adapted to saline waters, and only a few reptiles have adapted to estuarine conditions. The only characteristically estuarine reptile along the Georgia and South Carolina coast is the Carolina diamondback terrapin, which in-habits the estuarine zone throughout its entire range (Conant 1975). This turtle is relatively common and feeds on mollusks and crustaceans (Coker 1906, 1920); the natural history of a Gulf coast subspecies was reviewed by Cagle (1952). North and south of the Sea Island Coastal Region,

Table 4-46. Average growth rates of four fish species in brackish ponds of South Carolina (Bearden 1967).

| Species | Age in Years | | | | | |
| | I | | II | | III | |
	Total Length (inches)[a]	Weight (pounds)[b]	Total Length (inches)	Weight (pounds)	Total Length (inches)	Weight (pounds)
Red drum	14.5	1.5	20.5	3.5	26.0	7.1
Spot	7.5	0.3	9.8	0.5	11.5	0.8
Black drum	9.3	0.6	15.0	1.8	18.5	3.8
Atlantic croaker	8.5	0.3	10.5	0.5	--	--

a. 1 in = 2.54 cm.
b. 1 lb = .45359 kg.

Table 4-47. Trophic levels of predominant fish species found in saltwater impoundments in the Sea Island Coastal Region (C. M. Bearden 1978, South Carolina Marine Resources Division, Charleston, unpubl. data).

I. HERBIVORES
   Sheepshead minnow
   Sailfin molly
   Atlantic menhaden
   Striped mullet

II. PRIMARY CARNIVORES
   Mosquitofish
   Silverside
   Bay anchovy

III. MID CARNIVORES
   Mummichog
   American eel

MID CARNIVORES (Cont.)
   Pinfish
   Spot
   Atlantic croaker
   Silver perch

IV. TOP CARNIVORES
   Ladyfish
   Tarpon
   Weakfish
   Spotted seatrout
   Red drum
   Black drum
   Southern flounder

two subspecies of water snakes have adapted to saline conditions and, within their range, are characteristic saltmarsh fauna. The Carolina salt marsh snake is found along the Outer Banks and adjacent mainland of North Carolina (Conant and Lazell 1973), and the Atlantic salt marsh snake is found along the north-central portion of Florida's east coast (Conant 1975).

The herpetofauna of estuarine impoundments in South Carolina and Georgia has not been investigated. Thus, much of this discussion is restricted to animals recorded from impoundment-like situations or their probable occurrence in such habitats. Anurans (frogs and toads) are the only group of amphibians found with some regularity in areas similar to estuarine impoundments. Pearse (1936) observed the southern leopard frog in salinities of greater than 21º/oo near Beaufort, North Carolina. Most records of this species, however, were in salinities of less than 5º/oo. Ruibal (1959) observed that salinities of greater than 5º/oo were lethal to the eggs of the closely related northern leopard frog. Other records of this species occurring in low-salinity waters along the Atlantic and Gulf coasts of the Southeastern United States have been published by Viosca (1923), Carr (1940), Hardy (1953), Liner (1954), Neill (1958), and Blaney (1971). On the west coast of Florida, a population of exceptionally large leopard frogs occurs, individuals of which commouly ingest fiddler crabs and are capable of swallowing week-old alligators (Springer 1938). Another distinct population of leopard frogs, but of diminutive size,

exists in the same general area and caused
Neill (1958) to stress the need for a study
of the herpetofauna of saltwater areas.
Other anurans reported from saline habitats of the South Atlantic and Gulf coasts
include the southern toad, oak toad, green
treefrog, squirrel treefrog, southern
cricket frog, and the eastern narrowmouth
toad (Viosca 1923, Allen 1932, Carr 1940,
Burger et al. 1949, Hardy 1953, Smith and
List 1955, Neill 1958).

The two-toed amphiuma has been recorded several times from the front beach
on Hatteras Island, North Carolina, but
in each case the occurrence appeared accidental, i.e., just after a hurricane or
heavy rains. The specimens were probably
washed from drainage canals along roads
or from typically freshwater ponds (J. R.
Bailey, 1978, Duke University, Durham,
North Carolina, pers. comm.).

The American alligator is the only
naturally occurring crocodilian in Georgia
and South Carolina, although the Florida
crocodile is found elsewhere in the Southeast. Alligators frequent saltmarsh impoundments (Obrecht 1946, Allen and Neill
1949, Engels 1952), but successful nesting
is probably limited to impoundments where
salinities are less than $10^o/oo$. A salinity of $17^o/oo$ was determined to be
lethal to newly hatched alligators (Joanen
et al. 1972). Dietary and physiological
changes resulting from increased salinities are not known, but Chabreck (1972)
found significantly less food in stomachs
of alligators taken in salinities of $3^o/oo$
to $16^o/oo$, compared to those of alligators
taken in fresh water. He suggested that
alligator growth may be reduced in saline
waters because of low food intake. Alligator survival and reproduction can be affected by management of impoundments for
waterfowl (e.g., through flooding of nests
and changes in salinity). Adults are also
subject to shifts in reproduction due to
thermal loading around nuclear power production reactors (T. Murphy, 1978, South
Carolina Wildlife and Marine Resources
Department, Green Pond, unpubl. data).

Where salinities are low, as on
Kinloch and South Island plantations,
Georgetown County, South Carolina, impoundments provide optimum habitat for
alligators. Bara (1971) consistently observed the highest concentrations of alligators in canals within marsh impoundments.
Abundant food supply, deep and shallow water, and creation of nesting sites on
dikes are several benefits of these impoundments (Chabreck 1960). In addition,
private lands and game management areas
provide protection from illegal hunting.
Of 17 nests studied by Bara (1976), 12
were associated with diked impoundments.
Most of the 12 nests were located on a
dike berm or directly on an old abandoned
dike. Principal nest materials of these
17 nests reflected the brackish nature of
the habitat.

Impoundments at the Savannah National
Wildlife Refuge (SNWR) are managed for waterfowl, and salinities in the feeder creeks
typically range from fresh water to about
$10^o/oo$ (R. H. Dunlap, Jr., 1978, South
Carolina Marine Resources Division,
Charleston, pers. comm.). For several successive years, Bara (1976) cruised a line
24 mi (38.6 km) in length within the SNWR,
and the mean number of alligators sighted
per mile was as follows: 1971-9.38; 1972-
8.33; 1973-7.54; 1974-10.21; 1975-8.02
[only 9.6 mi (15.4 km) of the transect were
surveyed in 1975]. The largest number of
individuals observed was 245 for the 24-mi
(38.6 km) transect (Bara 1976).

Newly hatched alligators weigh less
than 2 oz (62 g) and measure about 8 in
(10.3 cm) long. The young usually remain
in the natal area for 2 or 3 years, feeding
mainly on insects but also on crayfish and
snails when they are available (Valentine
et al. 1972). They are opportunistic
feeders, shifting to larger prey as they
mature. After attaining a length of 4 ft
(1.2 m), they usually disperse from the natal
area. Growth of alligators varies from 4 to
6 in/yr (10.2 to 15.2 cm/yr) in South
Carolina (Bara 1976), or about half the rate
observed in Louisiana and Florida (McIlhenny
1934). The slower growth rate in South
Carolina results in greater juvenile mortality because the young are exposed to
predators, such as herons, egrets, and predacious fish, for a longer period. Major
predators on young alligators are herons,
egrets, and predacious fish. Predators of
lesser importance include raccoons, bobcats,
and adult alligators (Neill 1971).

Alligators reach sexual maturity upon
attaining a length of about 6 ft (1.8 m).
Growth slows to about 2 in/yr (5.1 cm/yr)
thereafter, and becomes negligible on approaching maximum length. The normal maximum length is 9 ft (2.7 m) for females and
12 ft (3.7 m) for males. Weight gain is
rapid until sexual maturity is reached. Age
to sexual maturity in South Carolina alligators is delayed due to the slower growth
rate; such information is unavailable for
specimens from Georgia. Information on
growth rates and the time required to reach
sexual maturity is necessary to determine
the possibility of a regulated harvest
(Bara 1976). The reproductive cycle is seasonal and related to temperature; a cool
spring may delay reproduction, while a warm
one may initiate the process early. The
onset of spermatogenesis usually occurs during the last 2 weeks of May and the first
2 weeks of June. Open waters are necessary
for successful courtship and breeding
(Nichols et al. 1976).

Nest construction and egg laying
take place during the first 2 weeks of July.
Secluded areas are sought for nesting. The
nests, constructed from vegetation at the
site, are approximately 3 ft (0.92 m) in
diameter at the base and 2.5 ft (0.76 m)
tall. The nest interior provides a

microhabitat having a stable, high relative humidity as well as some heat generated through decomposition of plant material (Joanen 1969). Clutch size averages 40 eggs, and incubation takes about 60 days depending on the nest site, construction, and temperature. Eggs deposited in shaded or poorly constructed nests require a longer incubation time. Most eggs hatch during the first 2 weeks in September, and females may be aggressive around the nest site. Stable water level is an important factor affecting the hatching success; both drought and flooding are detrimental. The raccoon is the major predator on nests, and the number destroyed may exceed 50% where there is land access, as on the side of impoundment dikes and levees. Should misfortune befall a nest, reproduction by that female is lost for the year, for renesting is unknown.

Alligators generally become semidormant from the second week of October to the second week of March, although there may be limited basking on mild days in winter. Feeding activity begins again in spring only after water temperatures exceed ∿25°C (∿77°F).

According to Chabreck (1966), large adults constitute a small portion of the alligator population [e.g., only 20% exceed 6 ft (1.8 m) long]. Juveniles should comprise at least 80% of an expanding population, with a 60:40 sex ratio favoring males. Size class distribution and sex ratios are unknown for mature, stable populations.

Range requirements must be considered when habitat needs for the species are evaluated. In Louisiana, the home range of an adult male is 2,200 acres (890.3 ha); of nesting females, 21 acres (8.5 ha); and of 3 - 6 ft (0.92 - 1.83 m) animals, 500 acres (202.3 ha) (Joanen and McNease 1970, 1973). These figures may not necessarily apply to populations in South Carolina and Georgia, but such data are unavailable from these States.

The standing crop of alligators probably exceeds that of any other large carnivore because of the wide extremes in size (8 in to 12 ft) (20.3 cm to 3.7 m), and the different habitat utilization and niche requirements for different size classes.

Ecologically, alligators function as a top carnivore on many prey species. Chabreck (1972) found that vertebrates were important food items in freshwater areas, with birds comprising one-third of the diet by weight. Other prey organisms include fishes, turtles, snakes, and various mammals. In addition, alligators maintain open, deep water areas and open trails in the marsh that are utilized by other wildlife. Because of their longevity (some may live 40 years), alligators can be useful as an indicator species for monitoring natural systems.

Alligators are important to man aesthetically and economically. In recent years, alligator hides have brought as much as $17/linear foot, but fashion demands cause prices to fluctuate considerably. Commercial harvesting for hides reached a peak in the late 1800's (McIlhenny 1935), and by 1960 the alligator had been practically eliminated from its original range (Chabreck 1968). During the 1960's, protective legislation was enacted by all States within its range, and the alligator now receives full Federal protection under the 1973 Endangered Species Act. In recent years, numbers of alligators have increased in the Southeast (Powell 1971), including South Carolina (Bara 1971) and Georgia (Joanen 1974).

In February 1977, the status of alligator populations in South Carolina and Georgia south of Winyah Bay, east of highways 17A and Interstate 95, and north of the Florida State line was changed from "endangered" to "threatened." This change in status was based on population estimates. The status of other alligator populations in South Carolina and Georgia is still classified as endangered. (See Chapter One for additional information on endangered species.) However, census data alone do not provide all the information needed to manage the species. Areas of research that need to be addressed include: 1) ways of accurately aging individuals, 2) mortality rates and factors, 3) the importance of size distribution on reproduction, 4) habitat suitability, especially as it affects reproduction, and 5) the northern extent of its range.

As alligator and human populations continue to expand, there are certain to be interactions between the two species, some of which will be negative. Increased development, especially in coastal areas, is likely to be a limiting factor on alligator populations through direct habitat destruction. Research is presently underway to determine the type and amount of habitat needed to maintain healthy alligator populations and to ensure that this top carnivore does not suffer as it did in years past.

Population data for South Carolina coastal counties in 1973 (Table 4-48) show stable populations in Dorchester, Berkeley, and Georgetown counties, and increasing populations in Colleton, Beaufort, Jasper, and Charleston counties (Joanen 1974). Joanen's survey of Georgia the same year revealed increasing populations in all coastal counties (Table 4-48). The Nongame and Endangered Species Section of the South Carolina Wildlife and Marine Resources Department is conducting surveys of alligator populations on South, Murphy, and Cedar islands, and in the Bear Island Game Management Area. Results of these surveys

288

Table 4-48. Alligator population estimates by county for coastal South Carolina and Georgia (Joanen 1974).

SOUTH CAROLINA

COUNTY	NO. OF ALLIGATORS	NO. OF SQUARE MILES IN COUNTY	NO. OF SQUARE MILES OF ALLIGATOR HABITAT	NO. OF ALLIGATORS PER SQUARE MILE OF HABITAT	POPULATION TREND	SOURCE OF INFORMATION
Dorchester	1,500	569.00	150.0	10.0	Stable	Mark Bara
Colleton	10,000	1,048.00	500.0	20.0	Increasing	Mark Bara
Beaufort	4,000	637.00	200.0	20.0	Increasing	Mark Bara
Jasper	7,000	662.00	350.0	20.0	Increasing	Mark Bara
Charleston	3,000	945.00	200.0	15.0	Increasing	Mark Bara
Berkeley	2,500	1,100.00	250.0	10.0	Stable	Mark Bara
Georgetown	4,500	813.00	300.0	15.0	Stable	Mark Bara

GEORGIA

COUNTY	NO. OF ALLIGATORS	NO. OF SQUARE MILES IN COUNTY	NO. OF SQUARE MILES OF ALLIGATOR HABITAT	NO. OF ALLIGATORS PER SQUARE MILE OF HABITAT	POPULATION TREND	SOURCE OF INFORMATION
McIntosh	2,000	430.94	214.05	9.34	Increasing	Tip Hon
Effingham	300	480.00	No data	No data	Increasing	Tip Hon
Chatham	600	440.94	No data	No data	Increasing	Tip Hon
Bryan	500	439.06	128.87	3.88	Increasing	Tip Hon
Camden	600	655.94	240.60	2.49	Increasing	Tip Hon
Glynn	500	422.97	177.45	2.82	Increasing	Tip Hon
Liberty	500	510.00	162.50	3.08	Increasing	Tip Hon

will provide information on the size and structure of alligator populations in intertidal estuarine impoundments in South Carolina.

Other reptile species indigenous to the Georgia and/or South Carolina coast, and recorded in habitats similar to estuarine impoundments, include the common snapping turtle, eastern mud turtle, striped mud turtle (see Conant 1975), chicken turtle, Florida softshell turtle, striped crayfish snake, cottonmouth, yellow rat snake, banded water snake, eastern garter snake, and the eastern mud snake (Viosca 1923, Engels 1942, 1952, Neill 1958, Conant 1975, Gibbons and Harrison 1975, Gibbons 1978). These species are not characteristic of estuarine impoundments, and their occurrence in this habitat is considered marginal. Neill (1958) provided a detailed discussion and literature review of herpetofauna in saline areas.

7. Birds

The habitats formed by numerous impounded wetlands in South Carolina and Georgia are among the most dramatic and active ecological units for birds. Some 68 species commonly or occasionally occur in impoundments (Table 4-49). Trophic relationships of these birds are illustrated in Figure 4-19. Specific groups of birds use rather distinct areas within the impoundments. Waterfowl, for instance, use the open water areas for feeding, whereas shorebirds concentrate along the edges and adjacent shallow flats. Earthen dikes delimit the habitat and provide an excellent "edge effect" when fully stabilized with vegetation. Species such as the sparrows, long-billed marsh wren, and common snipe are found in border vegetation and ecotonal communities.

Impoundment border vegetation is a fundamental link among nearly all species and provides for feeding, roosting, nesting, and cover. Impoundments in coastal South Carolina and Georgia are generally managed for waterfowl and are characterized by the dominance of brackish or freshwater vegetation, especially desirable duck food plants. The management of coastal impoundments for attracting waterfowl has been documented by Chabreck (1960), Neely (1960, 1962), Baldwin (1968), and Morgan et al. (1975).

Waterfowl occurring in impoundments of the Sea Island Coastal Region can be divided into four major groups: 1) swans and true geese, 2).surface-feeding or puddle ducks, 3) diving ducks or pochards, and 4) sea ducks. Among these four groups, there are some 19 species which occur regularly in the impoundments each year.

Swans and true geese are represented by only one major species, the Canada goose. This species, a common visitor to coastal South Carolina and Georgia, has become more abundant in recent years due to intensive management. Geese forage in water and on land, and large crops of grain in agricultural fields have attracted them.

Puddle ducks, probably the most abundant waterfowl group in coastal impoundments, include mallards, blue-winged teal, green-winged teal, gadwall, baldpate, wood ducks, and shovelers. Among the favorite food plants of puddle ducks are wild rice, spikerush, pondweeds, smartweeds, bulrushes, widgeon grass, acorns, cyprus balls, and various fruits (Kerwin and Webb 1972). Animals such as mollusks, insects, small fish, crayfish, and small crabs are also utilized for food to a lesser extent by the puddle ducks (Sprunt and Chamberlain 1970).

Pochards, or diving ducks, occupy a different niche in coastal impoundments from that of the puddle ducks. They feed in deeper waters of open bays, sounds, and coastal waters and are gregarious, tending to raft up in large flocks. Commonly occurring pochards include the ring-necked duck, canvasback, lesser scaup, greater scaup, and redhead. The ring-necked duck, canvasback, and redhead are more herbivorous than carnivorous and consume seeds of the water-lily, water-shield, and spikerush. Scaup, however, feed on a wide variety of animal matter.

Sea ducks play a relatively minor role in estuarine impoundments. Like the pochards, they spend most of their time in open bays and sounds and the sea. Buffleheads, hooded mergansers, and ruddy ducks are common winter residents which utilize deeper waters of impoundments. They are largely carnivorous, feeding on fish, insects, mollusks, and crustaceans. Sprunt and Chamberlain (1970) reported two records of hooded mergansers nesting in South Carolina. The hooded merganser is also more common than the common merganser and the red-breasted merganser and seldom mixes with these other two species, since they generally feed in different areas. The red-breasted merganser and common merganser include more fish in their diets than does the hooded merganser.

Rails, gallinules, and coots are commonly found in estuarine impoundments. The king rail is rarely seen in areas other than those characterized by freshwater vegetation, such as cat-tails and rushes, and is therefore not considered to be a resident of estuarine impoundments. Conversely, the clapper rail is commonly found at the water margins of estuarine impoundments where smooth cordgrass grows. This species is a common, permanent salt-marsh resident. The common gallinule is another common permanent resident, occurring in both brackish and freshwater impoundments. Gallinules frequently intermingle with coots and ducks and feed on plant and

Table 4-49. Birds of estuarine intertidal impoundments (Sprunt and Chamberlain 1949, 1970, Burleigh 1958, Audubon Field Notes 1967 – 1970, Chamberlain 1968, American Birds 1971 – 1977, Shanholtzer 1974b, Forsythe 1978).

DOMINANT			MODERATE			MINOR		
Pied-billed grebe	C	PR	Horned grebe	C	WR	Brown pelican	C	PR
Great blue heron	C	PR	Green heron	FC	PR	Double-crested cormorant	C	PR
Louisiana heron	C	PR	Little blue heron	C	PR	Yellow-crowned night heron	FC	PR
Great egret	C	PR	Black-crowned night heron	C	PR	Wood stork	FC	PR
Snowy egret	C	PR	Least bittern	FC	SR Mar.-Sept.	Canada goose	C	WR
White ibis	C	PR	Glossy ibis	FC	PR	Fulvous whistling duck	U	WR Nov.-Jan.
Blue-winged teal	C	WR Aug.-May	Mallard	C	WR Sept.-April	Wood duck	C	PR
Baldpate	C	WR Nov.-April	Black duck	C	WR Sept.-April	Redhead	U	WR Nov.-Mar.
Scaup	C	WR Oct.-April	Gadwall	FC	WR Oct.-April	Canvasback	FC	WR Nov.-Mar.
Bufflehead	FC	WR Nov.-April	Pintail	C	WR Sept.-April	Osprey	U	PR
Hooded merganser	C	WR Nov.-April	Green-winged teal	C	WR Oct.-April	Black-bellied plover	C	PR
Red-breasted merganser	C	WR Oct.-April	Shoveler	C	WR Oct.-Mar.	Ruddy turnstone	C	PR
Clapper rail	C	PR	Ring-necked duck	C	WR Oct.-April	Dowitcher	C	PR
American coot	C	PR	Ruddy duck	C	WR Oct.-April	American avocet	U	WR Oct.-June
Spotted sandpiper	C	PR	Virginia rail	FC	WR Aug.-Mar.	Black-necked stilt	U	SR Mar.-Aug.
Willet	C	PR	Sora	FC	WR Aug.-April	Gull-billed tern	FC	SR Mar.-Sept.
Greater yellowlegs	C	PR	Common gallinule	C	PR	Common tern	FC	PR
Herring gull	C	PR	Semipalmated plover	C	PR	Bald eagle	U	PR
Ring-billed gull	C	PR	Lesser yellowlegs	FC	WR July-May			
Laughing gull	C	PR	Least sandpiper	C	PR			
Forster's tern	C	PR	Dunlin	C	PR			
Least tern	C	SR Mar.-Oct.	Semipalmated sandpiper	C	PR			

291

Table 4-49. Concluded

DOMINANT			MODERATE			MINOR	
Belted kingfisher	C	PR	Western sandpiper	C	PR		
			Bonaparte's gull	C	WR Oct.-May		
			Royal tern	C	PR		
			Black tern	FC	SR May-Oct.		
			Black skimmer	C	PR		

Note: Dominance indicates relative importance of the species as a group in the community. This concept is not based necessarily on taxonomic relationships but rather on numbers, size, and trophic dynamics.

KEY: C - Common, seen in good numbers
      FC - Fairly common, moderate numbers
      U - Uncommon, small numbers irregularly
      PR - Permanent resident, present year around
      WR - Winter resident
      SR - Summer resident

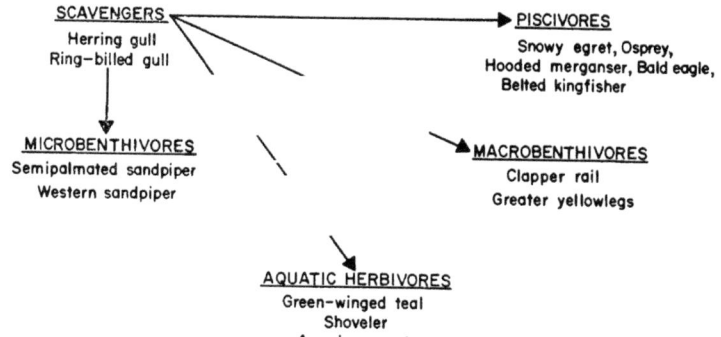

SCAVENGERS
Herring gull
Ring-billed gull

PISCIVORES
Snowy egret, Osprey,
Hooded merganser, Bald eagle,
Belted kingfisher

MICROBENTHIVORES
Semipalmated sandpiper
Western sandpiper

MACROBENTHIVORES
Clapper rail
Greater yellowlegs

AQUATIC HERBIVORES
Green-winged teal
Shoveler
American coot

Figure 4-19. Generalized trophic relationships of representative birds of estuarine impoundments.

animal matter. The American coot, also a permanent resident, is extremely abundant in estuarine impoundments. Its food consists primarily of seeds, roots, vegetative parts of aquatic plants, smartweed, small fish, snails, tadpoles, and insects. The Virginia rail and the sora are other common winter residents which frequent the marshes and marsh edges within impoundments, consuming mixtures of animal and vegetable matter.

The herons, storks, and ibises constitute another group of birds which are abundant throughout much of the coastal ecosystem but especially in estuarine impoundments. Among the dominant or characteristic species within this habitat are the great egret, the snowy egret, the Louisiana heron, and the white ibis. Also occurring, but playing a moderate-to-minor role in ecological interactions, are the great blue heron, little blue heron, glossy ibis, green heron, black-crowned night heron, yellow-crowned night heron, and the wood ibis or wood stork.

Both the great and snowy egrets are common permanent residents in coastal impoundments. These "plume birds" have made a dramatic comeback since the days of Wayne (1910), when they were slaughtered for millinery purposes. Both species nest in rookeries within coastal South Carolina and Georgia and feed in shallow water impoundments. The snowy egret appears to venture out into saltwater marshes and creeks more than does the great egret, which prefers freshwater ponds, marshes, and impoundments. These birds are commonly seen in communal roosts in trees adjacent to impoundments. The great egret

is a still hunter and can be observed in a motionless stance seeking its prey. Its diet consists of small fishes such as gizzard shad, minnows, and sunfishes. Sprunt and Chamberlain (1970) reported that frogs, mice, lizards, fiddler crabs, grasshoppers and other insects, and even small alligators are consumed. In contrast, the snowy egret is an active hunter, always moving and stabbing at fiddler crabs, shrimp, snails, small fish, insects, frogs, and lizards. No other egret or heron feeds in this manner.

The Louisiana heron is perhaps the most abundant heron in the study area. It is a common permanent resident which nests in rookeries with other herons and ibises, as well as in dissimilar locations such as washed oyster shell banks and cypress lagoons. Its diet consists of killifish, shrimp, crayfish, spiders, and insects.

The little blue heron, also a common permanent resident, exhibits nesting and feeding habits similar to those of the Louisiana heron and also eats frogs, turtles, and snakes. The green heron and black-crowned night heron also are common permanent residents of impoundments. The yellow-crowned night heron is a common summer resident, but it is not as numerous around impoundments as are the other herons. The latter three herons feed on small fishes, crustaceans, and insects in the impoundments and congregate with other herons in nesting.

One of the most distinctive shorebirds occurring within impoundments is the willet. This species is a permanent resident of the South Carolina and Georgia coast, occurring

293

in great abundance during summer. These birds can often be seen feeding on small mollusks, fiddler crabs, and insects along the banks, flats, and shorelines of estuarine impoundments. Willets frequently nest along sandy overgrown impoundment dikes, as well as on barrier islands or in open pastures. They generally prefer areas where vegetation is tall enough to conceal the nests. Bent (1962a) gave a detailed description of willet nesting habits near Bulls Bay, South Carolina.

The greater yellowlegs, a permanent resident, is also a typical shorebird of impoundments and waterways throughout the coastal region. Although this bird feeds in the shallows like other shorebirds, its long legs enable it to use deeper waters in catching minnows, insects, and snails.

A number of other shorebirds, including the lesser yellowlegs, semipalmated plover, black-bellied plover, ruddy turnstone, dowitcher, and sandpipers, occur commonly in estuarine impoundments. However, two relatively rare birds, the American avocet and black-necked stilt, are undoubtedly among the most spectacular of impoundment shorebirds. Wayne (1910) never observed an avocet in coastal South Carolina, but in recent years this bird has been observed on numerous occasions in impoundments. Sprunt and Chamberlain (1970) reported that one specimen was taken in the Santee River in 1923, with the greatest number (about 50) observed on South Island in 1946. Apparently, these birds overwintered there (Sprunt and Chamberlain 1970). American avocets are now observed annually in the South Island impoundments and elsewhere in the Sea Island Coastal Region of South Carolina and Georgia (Wilkinson 1970).

The black-necked stilt, a rare resident, is one of the most distinctive shorebirds in the coastal area. Wayne (1910) observed one pair of these birds during his many years in the field. Today, the stilt appears regularly in small numbers within the coastal area, and breeding records are now established (Sprunt and Chamberlain 1970).

The gulls and terns are represented in estuarine impoundments by the herring, ring-billed, and laughing gulls, and the common, least, royal, and Caspian terns. These birds feed to some extent in impoundments, particularly during summer fish kills caused by oxygen deficiencies. These birds also rest on open waters within impoundments.

The osprey and bald eagle, although uncommon in this habitat, have been observed to forage these impoundments in the Cape Romain-Santee Delta area of South Carolina (G. R. Garris, 1979, U.S. Fish and Wildlife Service, Awendaw, South Carolina, pers. comm.).

During the past 2 decades, there have been many ecological objections raised over the diking and impounding of wetlands. These objections are based on the following rationale: the blocking of tidal exchange results in a reduction of nutrient export; valuable marsh nursery grounds are lost for marine organisms; public interest factors are not considered. The objections could go further and in many cases the above rationale is reasonable. However, there are certain advantages to consider in evaluating coastal saltmarsh impoundments. Perhaps the greatest ecological advantage is the valuable habitat created for certain birds. Waterfowl, wading birds, shorebirds, and song birds find compatible niches in this ecosystem. As for adverse impacts, the use of pesticides in nearby agricultural areas (e.g., soybean fields, tomato crops, etc.) would appear to be the most damaging to avifauna. According to C. M. Bearden (South Carolina Marine Resources Division, 1978, Charleston, pers. comm.), there are fish kills annually in the coastal impoundments of lower South Carolina. Available evidence points to the use of pesticides in nearby agricultural lands as a leading cause. Many times, various birds are observed feeding on the dead fish, and occasionally a dead bird is found near the impoundments. This problem, although present in Georgia, is not prevalent because there are fewer coastal impoundments. The biological magnification of pesticides in avian populations is probably the greatest impact. These effects have been well documented over the years (U.S. Department of Interior, Fish and Wildlife Service 1962, 1963, 1964, 1965, Keith 1968). Borthwick et al. (1973) found mirex residues in 78% of birds collected from a study area near Charleston, South Carolina. The highest level of mirex was found in the belted kingfishers and demonstrated the fate of organochlorides in the estuarine environment.

Another important impact on birds in estuarine impoundments is hunting. Annually, there are thousands of waterfowl harvested from coastal impoundments in South Carolina and Georgia. For a detailed discussion of hunter participation and harvest records, the reader is referred to Volume II, Chapter Eight.

8. Mammals

Refer to the section on mammals of estuarine intertidal emergent wetlands.

CHAPTER FIVE

FRESHWATER ECOSYSTEMS

I. MAJOR CHARACTERISTICS

A. DEFINITIONS AND MODELS

Freshwater environments include all
wetland systems where average salinities
measure less than 0.5°/oo (Cowardin et al.
1977). Swamps, bays, savannahs, pocosins,
flood plains, freshwater marshes, lakes,
ponds, reservoirs, creeks, and rivers are
all considered freshwater environments.
Biologically, freshwater environments have
many similarities due to their common
aquatic nature. For this reason, this
chapter describes the three freshwater
ecosystems - palustrine, lacustrine, and
riverine systems - as a group. (See
Atlas Frontispiece for delineations of the
freshwater ecosystems.)

The palustrine system is defined
(Cowardin et al. 1977) as ". . . all
nontidal wetlands dominated by trees,
shrubs, persistent emergents, nonaquatic
mosses or lichens, and all such wetlands
that occur in tidal areas where salinity
due to ocean-derived salts is below 0.5°/oo.
It also includes wetlands lacking such
vegetation, but with all the following
characteristics: 1) size less than 8
hectares; 2) absence of an active wave-formed
or bedrock shoreline feature; 3) water
depth in the deepest part of basin less than
2 m at low water; and 4) salinity due to
ocean-derived salts less than 0.5°/oo."
In the study area, this palustrine definition would include most, if not all, of
the freshwater coastal impoundments, the
emergent and forested wetlands, and such
features as Carolina Bays, depressions,
pond margins, bogs, savannahs, and
ditches.

The diversity of the palustrine
ecosystem is reflected in the richness of
its species composition and the high rate
of production. In a stylized model of
energy flow in palustrine systems (Fig.
5-1), it is possible to reduce the complicated food webs and biogeochemical
cycles to a simple schematic indicating
major flows of energy and matter. Figure
5-1 indicates that major inputs into
palustrine systems include solar energy and
a variety of allochthonous material from
river water, groundwater, sediments, soil
drainage, precipitation, and tides (where
appropriate). The nonvascular and vascular
autotrophs incorporate this energy and
matter through photosynthesis in primary
production. This chemical bond energy is
then made available to various consumer
trophic levels. Final energy degradation
takes place through the decomposition of
organic materials by bacteria and fungi,
and the resultant release of various
nutrients for uptake by the primary

producers. It should be remembered that
palustrine systems are not closed but rely
heavily on allochthonous input and substantial export. Thus, the system must be
viewed in a broad context including both
autochthonous and allochthonous activity.

Cowardin et al. (1977) defined
lacustrine as those systems which ". . .
include wetlands and deep-water habitats
with all of the following characteristics:
1) situated in a topographic depression
or a dammed river channel; 2) lacking
trees, shrubs, persistent emergents,
nonaquatic mosses or lichens with greater
than 30 percent areal coverage; and
3) greater than 8 hectares (20 acres) in
size. Similar wetlands and deep-water
habitats smaller than 8 ha are also included in the Lacustrine System if an
active wave-formed or bedrock shoreline
feature forms all or part of the boundary
or if the water depth in the deepest
part of the basin is greater than 2 m at
low water." This definition includes
both tidal and nontidal waters but limits
ocean-derived salinity to less than 0.5°/oo.

Two subsystems are recognized components of the lacustrine system. The
limnetic subsystem includes all profundal
or deepwater habitats within the lake
and may be absent from many small lacustrine
systems. The littoral subsystem includes
all wetlands within the system and normally
extends from the shoreward boundary of the
lake to a depth of 2 m (6.5 ft) or to the
maximum extent of nonpersistent emergent
vegetation. The littoral lacustrine subsystem is difficult to separate from
palustrine systems in certain habitats of
the Sea Island Coastal Region. In many
areas palustrine wetlands may lie within
the boundaries of lacustrine systems.
Solution-formed ponds, oxbow lakes, old mill
ponds, rice field reserves, farm ponds,
and man-made lakes (such as Lake Moultrie)
are expressions of lacustrine systems in
the South Carolina - Georgia study area.

The general ecology of lacustrine systems
in the study area has not been well studied.
A simple model (Fig. 5-2 and Atlas plate
6), however, can indicate observed and
expected ecological relationships in the
lacustrine system. Figure 5-2 illustrates
the flow of energy and matter in a qualitative manner. Major influences on the
system include solar energy, water (through
rain as well as river and groundwater),
climatological influence (temperature,
wind, cloud cover, etc.), and man (through
economic activities, i.e., dams, sewage,
drainage, agriculture, lumber, etc.).
These and other influences impact upon the
entire system by regulating productivity
and stability. Net primary production,
through the principal autotrophs (nonvascular and vascular plants), drives the
system by supplying chemical bond energy
converted from solar energy and nutrients
by phytosynthesis. These organic molecules

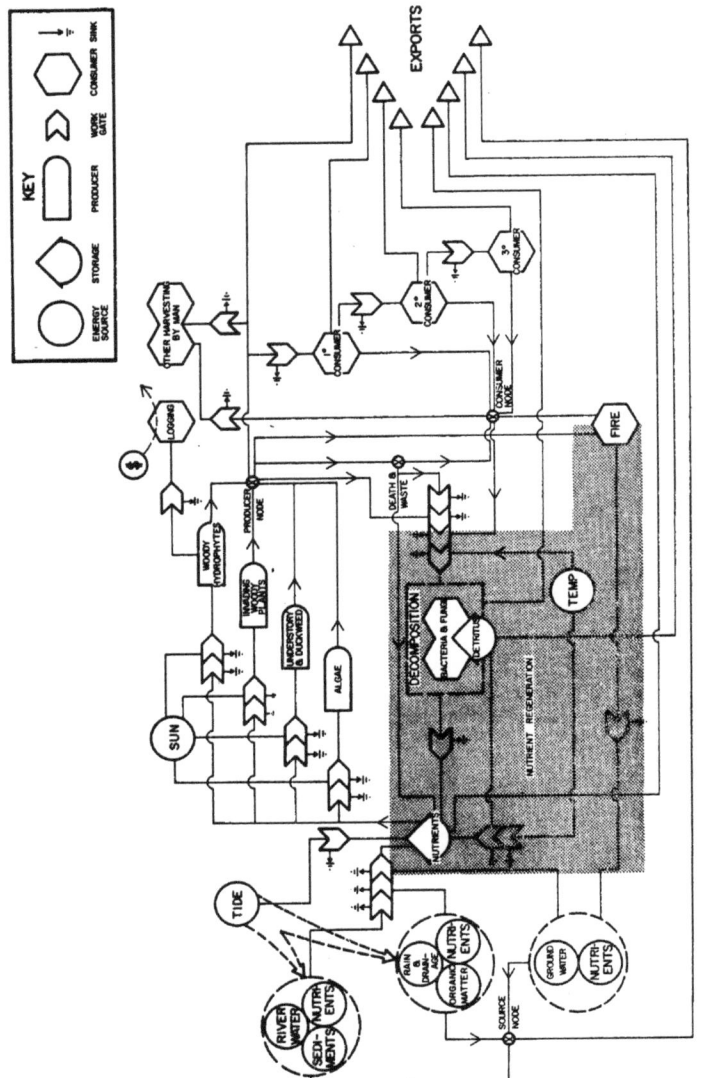

Figure 5-1. A generalized energese model of the palustrine ecosystem of the Sea Island Coastal Region.

296

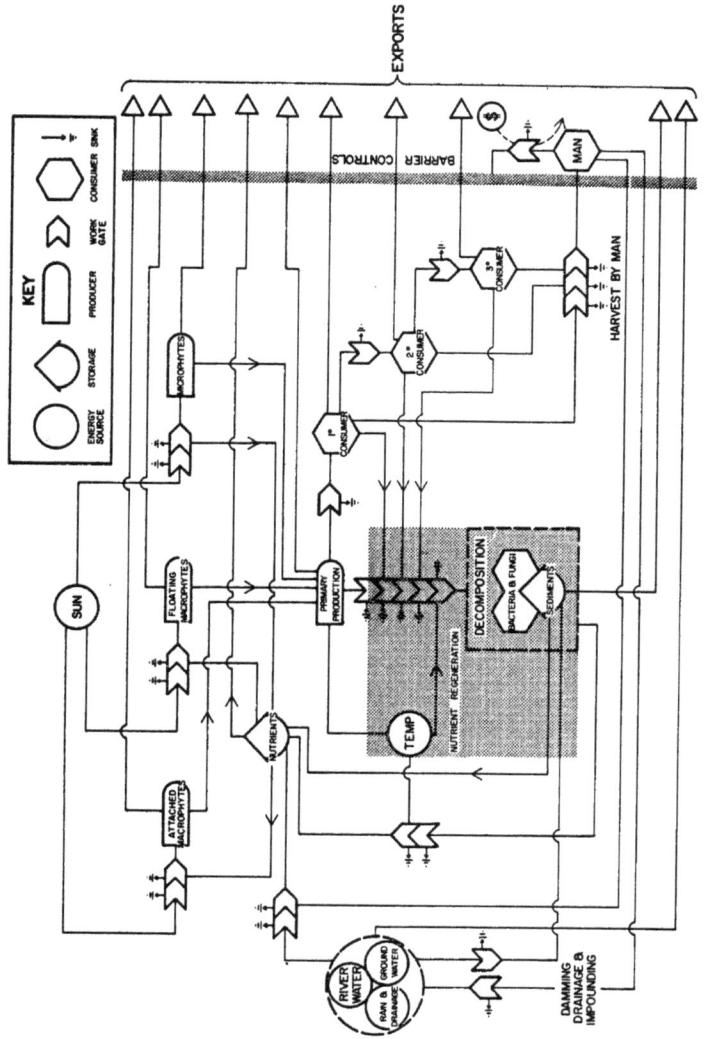

Figure 5-2. A generalized energese model of the lacustrine ecosystem of the Sea Island Coastal Region.

are utilized by primary consumers and passed up the trophic ladder. Final degradation of this energy is reached through decomposition of organics by bacteria and fungi. Inorganic nutrients are thus recycled, through decompostion, for uptake by autotrophs and the cycle continues.

All wetland and deepwater habitats contained within a channel are riverine ecosystems (see Atlas plate 7). Cowardin et al. (1977) qualify this rather broad definition by excluding wetlands dominated by trees, shrubs, persistent emergents, and nonaquatic mosses or lichens, and habitats with water containing ocean-derived salinity in excess of 0.5°/oo. On the shoreward side, the riverine system is bounded by upland, channel bank (both natural and man-made) or by nonaquatic plant-dominated wetlands or wetlands dominated by persistent emergents, trees, or shrubs. Downstream, the riverine system terminates at the oligohaline estuary (salinity > 0.5°/oo) at mean annual low flow or when the channel enters a·lake. Upstream, the system terminates at the stream origin or where the channel leaves a lake. The riverine system includes four subsystems: the tidal, lower perennial, upper perennial, and intermittent. In the Sea Island Coastal Region only the tidal subsystem is expressed. This subsystem is described as that section of the riverine system where water velocity is low and fluctuates under tidal influence.

A simple energese model is presented (Fig. 5-3) to illustrate, in a qualitative manner, the major flows of energy and matter through a riverine ecosystem. This system typifies the dependence on allochthonous material. Primary producers (nonpersistent emergents and phytoplankton), and most consumers are ephemeral components entering and leaving either passively (with flora) or actively (with most fauna). The driving forces or inputs include river and groundwater, rain and drainage, sediments, tide, and, of course, solar radiation. The nonvascular and vascular autotrophs synthesize organic substances with high chemical bond energy from sunlight and various inorganic (and some organic) compounds through photosynthesis. Energy is thus made available in usable form for the rest of the trophic structure. Nutrients are regenerated and energy degraded in the process of decomposition. In this process, bacteria and fungi catabolize organic wastes and dead tissue liberating inorganic nutrients which are then utilized by the autotrophs. It should be remembered that this cycling of energy and matter is not closed and that import and export are occurring constantly in all systems.

B. GENERALIZED FOOD WEB AND RELATIONSHIPS

The trophic dynamics of freshwater environments are extremely complex due to the almost infinite number of ecological niches existing in swamps, marshes, lakes, and rivers. Figure 5-4 is a simplied food web diagram for freshwater environments. The details of food webs for palustrine, lacustrine, and riverine environments, of course, vary considerably. This diagram, however, attempts to portray the similarities of these three ecosystems in a simple manner. The primary producers in freshwater environments are vascular and nonvascular macrophytes and phytoplankton. These groups, along with detritus, form the trophic foundation of the freshwater food web. The consumers of the freshwater system, on the other hand, occupy several trophic levels. Birds, for example, may be considered primary consumers because some birds are granivores or herbivores, secondary consumers because some birds are insectivores, or tertiary consumers because some birds are piscivores. Table 5-1 identifies common organisms occupying freshwater trophic levels and Figure 5-5 illustrates their relationships.

Bacteria and fungi decompose organic matter from the system, regenerating new nutrients. Rivers frequently bring large nutrient supplies into lacustrine and palustrine systems. Floodplain systems and lacustrine systems fed by rivers sometimes have extremely high productivity rates (Wharton 1970, Kitchens et al. 1975). However, diversity in freshwater environments is not directly related to productivity. Highly productive systems may produce natural monocultures of aquatic vegetation (such as water-lily ponds, water hyacinth beds, or Porazilia elodea) with low faunal species diversities. For further information on food webs in freshwater environments, see specific ecosystem sections and Atlas plates 5, 6, and 7.

The various interacting factors which result in the productivity cycle of a freshwater system are ultimately derived from the physiogeographic character of the system. Figure 5-6 is a modification of the Rawson Diagram (Rawson 1939). Although prepared many years ago, this diagram accurately portrays the physical characters of freshwater systems that directly or indirectly influence the biotic and abiotic cycles. Foremost among the various cycles associated with these systems are the organic cycle of productivity (the carbon cycle), nitrogen cycle, phosphorous cycle, and the trophic cycle of energy flow. All organisms occupying niches in a freshwater system are interdependent by means of nutrition or trophic relations. In a simple energy flow diagram (Fig. 5-7), we see that solar energy is converted to chemical bond energy by autotrophs (vascular and nonvascular plants). This energy is then used by autotrophs in their own cellular respiration processes or is passed on to consumers or decomposers. Thus, nutritional dependence of freshwater communities rests ultimately with solar energy. Community primary consumers

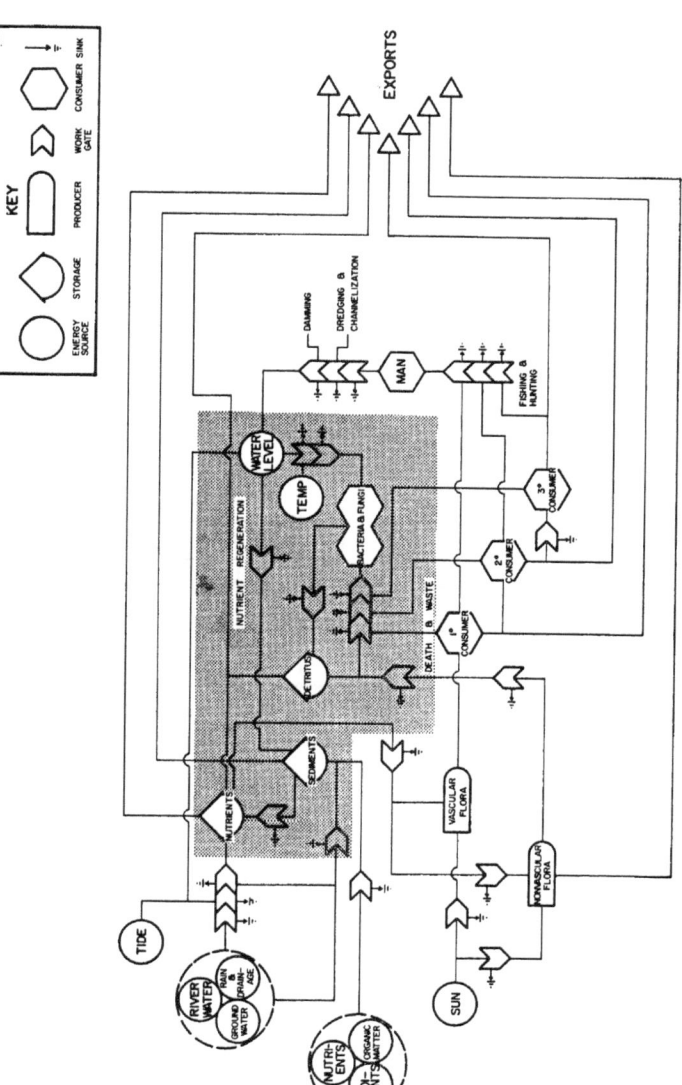

Figure 5-3. A generalized energese model of a riverine ecosystem of the Sea Island Coastal Region.

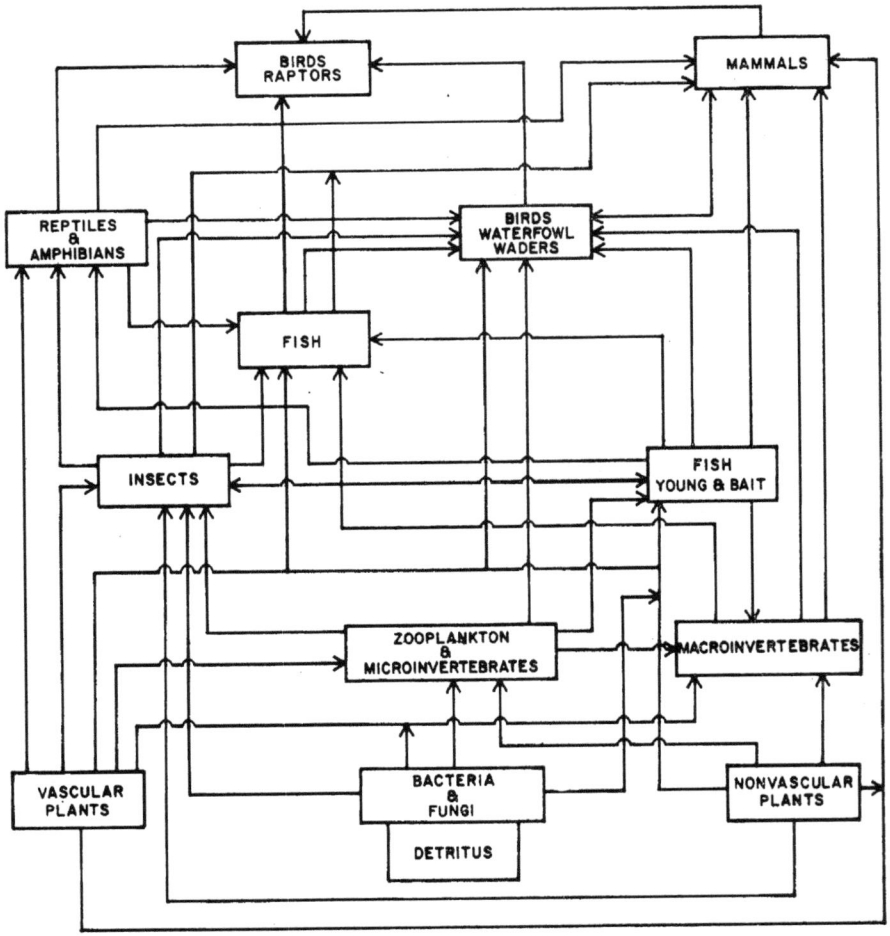

Figure 5-4. A generalized food web for freshwater ecosystems of the Sea Island Coastal Region.

Table 5-1. Trophic levels of freshwater consumers.

I. **PRIMARY CONSUMERS** (Feed directly on producers)	
Zooplankton	(cladocerans, rotifers, copepods)
Herbivorous invertebrates	(amphipods, mayfly larvae)
Granivorous and herbivorous birds	(Savannah sparrow, mallard)
Omnivorous vertebrates	(carp)
Granivorous and herbivorous mammals	(mice, shrews, deer)
II. **SECONDARY CONSUMERS** (Feed on primary consumers)	
Omnivorous invertebrates	(dragonfly larvae, isopods)
Omnivorous vertebrates	(salamanders, frogs)
Insectivorous birds	(short-billed marsh wren, northern parula)
Predacious fish	(crappie, bluegill)
Predacious reptiles and amphibians	(water snakes)
Mammals	(otter, raccoon, mink)
III. **TERTIARY CONSUMERS** (Feed on some secondary consumers)	
Omnivorous vertebrates	(salamanders, frogs)
Predacious reptiles and amphibians	(alligators, cottonmouths)
Predacious fish	(largemouth bass)
Piscivorous birds and birds of prey	(osprey, hawks, eagles)
Mammals	(bobcat, man)

(herbivores and omnivores) utilize auto-trophic producers for nutrition and, in turn, become food for secondary, tertiary, and quaternary consumers. Microorganisms obtain energy through decomposition of metabolic wastes and dead tissues of both consumers and producers, thus recycling minerals and nutrients back to producers and effecting final degradation of energy.

This interdependence of organisms in freshwater systems is exemplified by the cyclical pathways of specific nutrients. Biogeochemical cycles are important aspects of aquatic systems and serve as good models for illustrating organic interdependence. The reader is referred to Chapter One for a discussion of bio-geochemical cycles.

II. PALUSTRINE ECOSYSTEM

A. NONVASCULAR FLORA

The nonvascular plants of freshwater environments in South Carolina and Georgia have not been well studied. The earliest work in South Carolina and Georgia was performed by Ravenel, who did not publish until 1882. Bailey (1851) made collections from 60 sites in a trip through South Carolina, Georgia, and Florida, including many in Charleston and Jasper counties. He listed over 80 freshwater species from South Carolina and Georgia. Wood (1874) attempted the first comprehensive treatise on American freshwater algae, including many of his own collections from South Carolina as well as the records of Ravenel. Philson (1939) published a systematic survey of algae in South Carolina, listing 15 species of cyanophyta; later, Philson (1940) added six new species of Oédogonium. Brown (1930) included South Carolina and Georgia in his listing of desmids from the southeastern coastal plain, and Frohne (1942) published a report on the occurrence of Phymatodocis in Jasper County, South Carolina, and several counties in Georgia. Corbin (1951) identified 15 new species of algae in South Carolina, and Metcalf (1947) published a list of 54 algal genera collected from a freshwater pond on Wadmalaw Island, South Carolina. In a study of the algae of the Savannah River Plant area, Macfie and Swails (1957) discovered a new species of Micrasterias. Dillard (1967) listed 44 algal taxa in a summary of his records for South Carolina.

In more recent studies, Jacobs (1968, 1971) listed 585 taxa in her preliminary survey of the freshwater algae of the Baruch Plantation in Georgetown, South Carolina. Zingmark (1975) listed 114 taxa from a freshwater pond in an environmental inventory of Kiawah Island, and Grant (1974) reported a dominance of

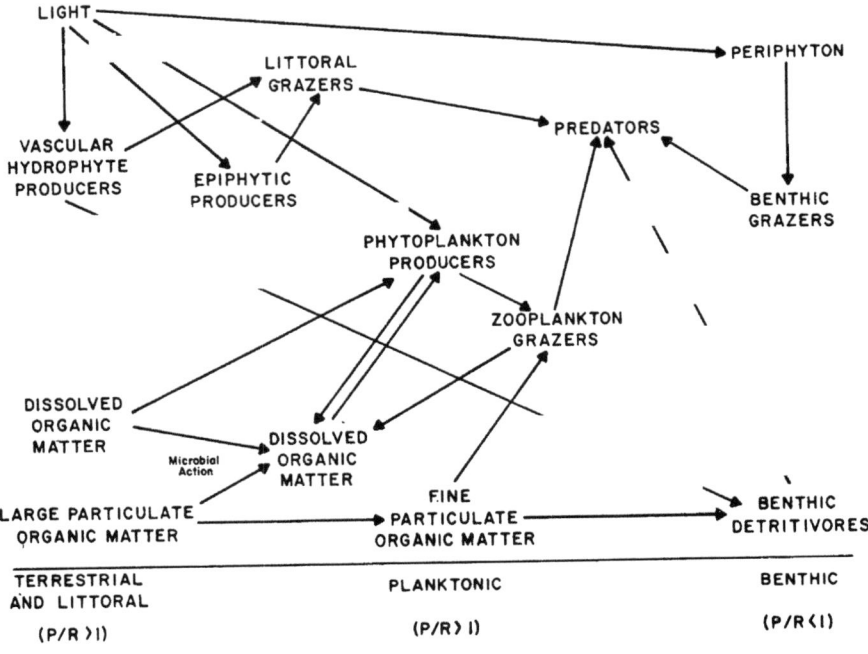

Figure 5-5. Simplified functional compartmentalization of freshwater ecosystems, with general production/respiration ratios (adapted from Cummins 1975).

diatoms (105 taxa) in the periphyton/ phytoplankton component of the upper Cooper River-Tailrace Canal system in South Carolina. In an environmental assessment report for the Amoco Chemicals Corporation (Dames and Moore Associates 1975), four divisions of algae were collected as periphyton in the Cooper River area of South Carolina. Included in the four divisions were 33 species of diatoms, 8 species of Chlorophytes, 2 species of Cyanophytes, and 1 euglenoid. In the same study, a total of 35 phytoplankton species were identified, with diatoms the most abundant form in both periphyton and phytoplankton samples. Goldstein and Manzi (1976) listed a total of 259 taxa identified from freshwater fish culture ponds in South Carolina. Identified taxa included 146 Chlorophyta, 11 Pyrrophyta, 46 Cyanophyta, 45 Chrysophyta, 9 Euglenophyta, and 1 Rhodophyta. Numerous Cryptophytes were noted, but none were identified to species. Among nonvascular flora, diatoms have received only limited detailed study in freshwater systems of the Southeast. In

a study of haptobenthic algal flora in two North Carolina streams, Dillard (1966) reported a total of 70 diatom taxa. In his review of algal research in South Carolina (Dillard 1967), he reported only 25 diatom taxa. The Savannah River, which serves as both a political and natural boundary between South Carolina and Georgia, has been the site of intensive diatom research by the Academy of Natural Sciences of Philadelphia (Reimer 1966, Patrick et al. 1967). In a recent study (Camburn et al. 1978), haptobenthic diatom flora were studied in Long Branch Creek, South Carolina, to provide a detailed floristic survey of the diatom flora in an area of North America where few such studies have been conducted. They reported 268 diatom taxa representing 31 genera, the most numerous of which included _Eunotia_, _Achnanthes_, _Navicula_, _Pinnularia_, _Gomphonema_, and _Nitzschia_. A complete listing of all freshwater species identified to date in the coastal counties of South Carolina has been published by Manzi and Zingmark (1978).

302

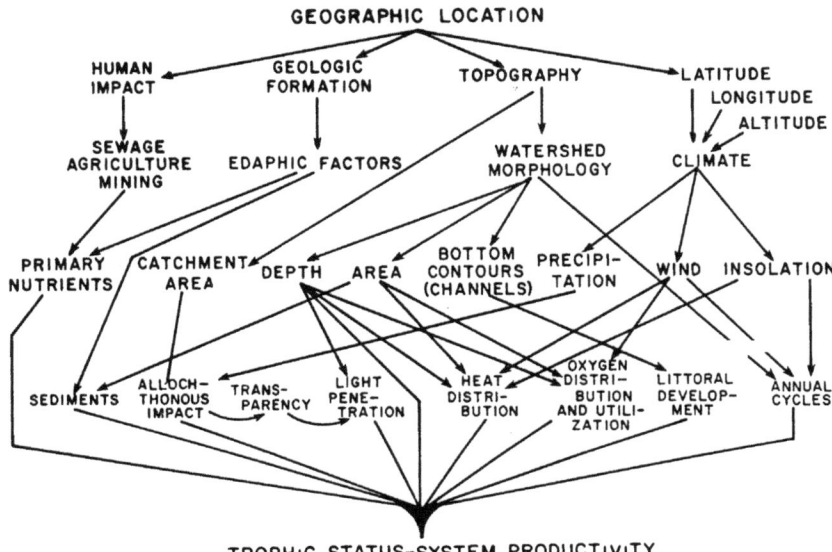

Figure 5-6. Productivity cycle of a freshwater system (adapted from Rawson 1939).

Algae that inhabit freshwater environments constitute a diverse assemblage with differing physiological requirements and variations in terms of tolerance to physical and chemical environmental parameters. The open water algae, phytoplankton, are regulated both spatially and temporally by several major classes of environmental factors. Light, temperature, and turbidity interact with a number of inorganic and organic nutrient factors in the succession of algal populations. Unlike marine systems, successional periodicity of phytoplankton biomass and productivity in undisturbed freshwater systems is fairly constant. Seasonal changes are muted in lower latitudes, although periodicity of phytoplankton biomass and productivity are often out of phase, e.g., growth rates of blue-greens are rapid and turnover times are shortened during summer months in South Carolina and Georgia.

Aside from the descriptive studies listed above, little is known about nonvascular plant associations and population dynamics in freshwater habitats of South Carolina and Georgia. Despite this paucity of data, certain generalizations can be made concerning algal associations in lakes with various nutrient loads and in various states of succession

from oligotrophy to eutrophy. Some of these associations are set out in Table 5-2. It must be remembered, however, that these are generalized data, and the table should be used only as a broad comparison for observed data. This is primarily because the categories do not present a satisfactory spectrum of all the intergradations between lake trophic levels and the shifts that occur seasonally at each level.

Seasonal and spatial population dynamics of algae result from a large and constantly changing array of environmental parameters interacting with physiological characteristics of the organisms. The succession of algal populations in freshwater systems must be analyzed in the context of several interacting factors, particularly: 1) nutrient limitations, 2) light, 3) temperature, 4) organic micronutrients, and 5) biological factors (competition, predation, etc.). Wetzel (1975) noted that generalizations concerning seasonal patterns and periodicity of nonvascular plants in fresh water are difficult, but suggested that several aspects were reasonably consistent. These include: 1) relatively constant seasonal periodicity of phytoplankton biomass - if the system is not subjected to outside perturbations,

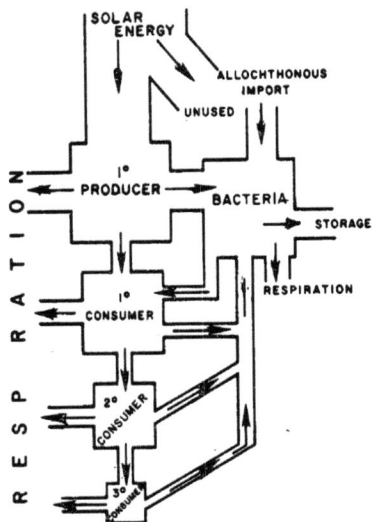

Figure 5-7. Energy flow through an
aquatic ecosystem
(adapted from Cole 1975).

such as the activities of man in watershed modifications, nutrient loading,
etc.; 2) the seasonal amplitude of phytoplankton biomass is usually large - up to
a thousandfold in polar and temperate
lakes and fivefold to a hundredfold in subtropical and tropical lakes; 3) the periodicity of rates of primary production is
quite often out of phase with maxima and
minima observed in biomass and numbers;
and 4) species composition fluctuates in
a regular way on an annual basis if the
system is not seriously perturbed.

B. VASCULAR FLORA

1. Tidal Impoundments

Most freshwater impoundments in
coastal South Carolina and Georgia occupy
former rice fields along major rivers
(impoundments with salinities less than
0.5⁰/oo are considered here). (See Chapter
Four for a discussion of impoundments
with salinities greater than 0.5⁰/oo.)
These impoundments generally occur upstream from the river estuary in the freshwater zone (consult Atlas plates 9 - 18).
Water control structures are present for
purposes of manipulating the impoundment
water level and keeping out brackish water
(if the impoundment is located near the
brackish-fresh transition zone of the

river). The principal use of these
impoundments is for waterfowl feeding;
however, some may be used for cattle
pasturage, snipe hunting, planting cypress,
or wildlife sanctuaries (Morgan 1974).

Morgan (1974) lists four possible
types of wetlands that may occur in
freshwater impoundments: 1) open water,
2) submerged plants, 3) pad plants, and
4) emergent plants. Baldwin (1956) lists
four slightly different types: 1) summer
drawdown edge; 2) shallowly flooded marsh;
3) pad plants, surface mats, and floating
plants; and 4) submerged aquatics.
Baldwin's type 1 and type 2 are generally
emergents and coincide with Morgan's type
4; and, although Baldwin does not include
an open water category, his type 3, which
includes floating plants, seems to be
broad enough to contain the open water
type.

Emergent communities in freshwater
impoundments are dominated by the smartweeds, spikerushes, red root, wild millet,
Asiatic dayflower, giant cutgrass, panic
grass, duck potato, cat-tails, alligator-
weed, wild rice, and soft-stem bulrush
(Baldwin 1956, Conrad 1966, Morgan 1974).

The submergent dominants are the
pondweeds, coontails, bladderworts, fan-
wort, and proliferating spikerush (Baldwin
1956, Morgan 1974). Floating communities
(pad, surface, and floating plants) are
dominated by duckweeds and water-shield
in open water areas, and by water-lily,
white water-lily, frog's bit, pennyworts,
and alligator-weed near shore (Baldwin
1956, Morgan 1974). (See Table 5-3 for
a list of common marsh plants associated
with freshwater impoundments in South
Carolina.) Percival (1968) studied the
ecology of six plant species commonly
found in freshwater impoundments: Asiatic
dayflower, water-shield, jointed spikerush,
square-stem spikerush, tearthumb, and
swamp smartweed. Here the influence of
water level and soil acidity on species
dominance can be seen. Table 5-4 presents summarized data from Percival (1968),
the only available work on nutrients in
freshwater impoundments in the Sea Island
Coastal Region. Quantitative nutrient
information is available in other tables
in that work.

Succession in managed freshwater
impoundments rarely proceeds in a natural
sequence because of water level mani-
pulations by impoundment managers.
However, impoundments with constantly
maintained water levels become dominated
by floating and submergent vegetation.
The dominants vary according to depth of
impoundment, but white water-lily, duckweeds, coontails, and bladderworts are
usually the most common species.
Succession here is comparable to succession
in shallow lakes; for a discussion of
succession in lakes, see the lacustrine
ecosystem section of this chapter.

Table 5-2. Characteristics of common major algal associations of phytoplankton in relation to increasing lake fertility (adapted from Wetzel 1975).

General Lake Trophy	Water Characteristics	Dominant Algae	Other Commonly Occurring Algae
Oligotrophic	Slightly acidic; very low salinity	Desmids Staurodesmus, Staurastrum	Sphaerocystis, Gloeocystis, Rhizosolenia, Tabellaria
Oligotrophic	Neutral to slightly alkaline; nutrient-poor lakes	Diatoms, especially Cyclotella and Tabellaria	Some Asterionella spp., some Melosira spp., Dinobryon
Oligotrophic	Neutral to slightly alkaline; nutrient-poor lakes or more productive lakes at seasons of nutrient reduction	Chrysophycean algae, especially Dinobryon, some Mallomonas	Other chrysophyceans, e.g., Synura, Uroglena; diatom Tabellaria
Oligotrophic	Neutral to slightly alkaline; nutrient-poor lakes	Chlorococcal Oocystis or chrysophycean Botryococcus	Oligotrophic diatoms
Oligotrophic	Neutral to slightly alkaline; generally nutrient-poor; common in shallow Arctic lakes	Dinoflagellates, especially some Peridinium and Ceratium spp.	Small chrysophytes, cryptophytes, and diatoms
Mesotrophic or Eutrophic	Neutral to slightly alkaline; annual dominants or in eutrophic lakes at certain seasons	Dinoflagellates, some Peridinium and Ceratium spp.	Glenodinium and many other algae
Eutrophic	Usually alkaline lakes with nutrient enrichment	Diatoms much of year, especially Asterionella spp., Fragilaria crotonensis, Synedra, Stephanodiscus, and Melosira granulata	Many other algae, especially greens and blue-greens during warmer periods of year; desmids if dissolved organic matter is fairly high
Eutrophic	Usually alkaline; nutrient enriched; common in warmer periods of temperate lakes or perennially in enriched tropical lakes	Blue-green algae, especially Anacystis (=Microcystis), Aphanizomenon, Anabaena	Other blue-green algae; euglenophytes if organically enriched or polluted

305

Table 5-3. List of vascular plants associated with freshwater
impoundments in South Carolina (adapted from Tiner
1977).

SCIENTIFIC NAME	COMMON NAME
Alternanthera philoxeroides	Alligator-weed
Aneilema keisak	Asiatic dayflower
Baccharis spp.	Sea myrtles
Brasenia schreberi	Water-shield
Cabomba caroliniana	Fanwort
Cephalanthus occidentalis	Button bush
Ceratophyllum spp.	Coontails
Cyperus erythrorhizos	Redrooted nutgrass
Cyperus odoratus	Sedge
Cyperus polystachos	Sedge
Cyperus spp.	Sedges
Echinochloa crusgalli	Wild millet
Echinochloa spp.	Millets
Egeria densa	Water-weed
Eichhornia crassipes	Water hyacinth
Eleocharis baldwinii	Proliferating spikerush
Eleocharis equisetoides	Jointed spikerush
Eleocharis quadrangulata	Square-stem spikerush
Elodea spp.	Water-weeds
Erianthus spp.	Plume grasses
Hydrochloa caroliniensis	Water grass
Hydrocotyle spp.	Pennyworts
Juncus effusus	Soft rush
Lachnanthes caroliniana	Redroot
Leersia hexandra	Rice cutgrass
Leersia oryzoides	Rice cutgrass
Lemna spp.	Duckweeds
Limnobium spongia	Frog's bit
Ludwigia peploides	Water-primrose
Melochia corchorifolia	Chocolate-weed
Myriophyllum heterophyllum	Water milfoil
Najas guadalupensis	Bushy pondweed
Nelumbo lutea	Lotus
Nelumbo pentapetela	Lotus
Nuphar advena	Spatter-dock
Nymphaea odorata	White water-lily
Panicum bisulcatum	Asiatic panic grass
Panicum dichotomiflorum	Fall panic grass
Panicum hemitomon	Maidencane
Paspalum boscianum	Bullgrass
Peltandra virginica	Arrow-arum
Pluchea spp.	Marsh fleabanes
Polygonum arifolium	Tearthumb
Polygonum densiflorum	Southern smartweed
Polygonum hydropiperoides	Swamp smartweed
Polygonum pensylvanicum	Large-seed smartweed
Polygonum portoricense	Southern smartweed
Polygonum sagittatum	Tearthumb
Polygonum setaceum	Swamp smartweed
Polygonum spp.	Smartweeds
Pontederia cordata	Pickerelweed
Potamogeton berchtoldii	Narrow-leaved pondweed
Potamogeton diversifolius	Variable-leaved pondweed
Potamogeton pectinatus	Sago pondweed

Table 5-3. Concluded

SCIENTIFIC NAME	COMMON NAME
Sagittaria graminea	Delta duck potato
Sagittaria latifolia	Duck potato
Sagittaria spp.	Arrowheads
Salix spp.	Willows
Scirpus validus	Soft-stem bulrush
Sesbania macrocarpa	Seban
Spartina cynosuroides	Giant cordgrass
Spirodela polyrrhiza	Big duckweed
Typha latifolia	Common cat-tail
Typha glauca	Blue cat-tail
Utricularia spp.	Bladderworts
Zizania aquatica	Wild rice
Zizaniopsis milliacea	Giant cutgrass

The most widespread utilization of successional trends by the impoundment manager is summer drawdown. Drawdown (the lowering or removal of water) insures germination of many annuals, allows for seasonal burning, and permits grazing or cultivation (Baldwin 1950). Cultivated crops, such as corn, brown-top millet, wheat, barley, rye, soybean, and grain sorghum may be planted after drawdown (Tiner 1977). If cultivation is not the desired use, the period of drawdown may be shortened, promoting growth of various smartweeds. Prolonged drawdowns sometimes allow for germination of cat-tails, willows, and button bush, all undesirable plants to waterfowl management. Fanwort-pondweed beds, also undesirable, may be controlled by prolonged drawdown, with the more desirable muskgrass, a nonvascular plant gaining dominance upon reflooding (Baldwin 1950).

In summary, successional trends are manipulated by impoundment managers to produce desired vegetation for the specific use the manager envisions, be it waterfowl management, grazing, or cultivation.

2. Tidal Emergent Wetlands

Commonly referred to as "tidal freshwater marsh" or "river marsh," palustrine tidal emergent wetlands are found in estuaries between palustrine tidal forested wetland and estuarine emergent wetland communities (i.e., between forested flood plain and brackish marshes) (Tiner 1977, Wharton 1978). The tidal freshwater marsh zone is relatively wide just upriver from a point where salinities are 0.5°/oo; but as one moves further upriver and tidal amplitude decreases, the tidal freshwater marsh becomes a narrow fringe along

the edge of the forested flood plain. The tidal freshwater marsh extends only to the edge of the river channel by definition; floating mats and nonpersistent emergents found in the actual channel are discussed under the section on vascular flora of the riverine system. Figure 5-8 illustrates the location of tidal freshwater marsh communities in an estuary.

Tiner (1977) described tidal freshwater marsh as a zone of coastal rivers where low or no salinity is present and where tidal amplitude fluctuates relatively little. Plant distribution here seems to be more contingent upon river floods (in the spring and after severe storms) than on brackish water (with spring tides), although both factors exert some influence on the vegetation (Tiner 1977). Although the boundary between brackish and tidal freshwater marsh is given in this text as the point at which salinities fall below 0.5°/oo [based on Cowardin et al. (1977)], this boundary, in reality, is a transitional zone and not a clear demarcation. Some plants of the upper (upriver) zones of brackish marsh (e.g., giant cutgrass, wild rice, water parsnip, cat-tails, and saw grass) become more common in tidal freshwater marsh. Giant cutgrass, wild rice, and cat-tails are the more obvious plants of this community, but plant diversity is greater here than in either salt or brackish marshes, as witnessed by the many co-dominants found in the tidal freshwater marsh (Tiner 1977). Table 5-5 lists the plants of South Carolina's tidal freshwater marshes.

Havel (1976) noted that smooth cordgrass (typically a saltmarsh species) giant cordgrass (typically a brackish marsh species), and

307

Table 5-4. Ecology of six common freshwater impoundment plants in the coastal plain of South Carolina (adapted from Percival 1968).

Taxa	Flood Regime	Ferrous Iron	Nitrate	Ammonium Nitrogen	Sulfate	pH	P	K	C	Mg	Dissolved Oxygen
Anellema keisak (Asiatic dayflower)	flooded occasionally during growing season	low[a]	high[a]	high	low	slightly acidic[a]	low	low	high	low	*
Brasenia schreberi (Water-shield)	standing water through-out growing season	-	high	high	-	acidic	low	low	low	low	*
Eleocharis equisetoides (Jointed spikerush)	shallow flooding	low	low	high	low	slightly acidic	low	low	low	low	*
						slightly basic	*	*	*	*	
Eleocharis quadrangulata (Square-stem spikerush)	variable	low	low	low	low	slightly acidic	*	*	*	*	low
Polygonum arifolium (Tearthumb)	dry site	*	*	*	*	acidic	low	low	low	low	*
Polygonum hydropiperoides (Swamp smartweed)	saturated to slightly above soil level	*	*	*	*	acidic	*	*	*	*	*

a. Terms low, high, and slightly are comparative terms taken from the text of the study. For quantitative information, see data in the original source (Percival 1968).
* Results were indeterminate.

308

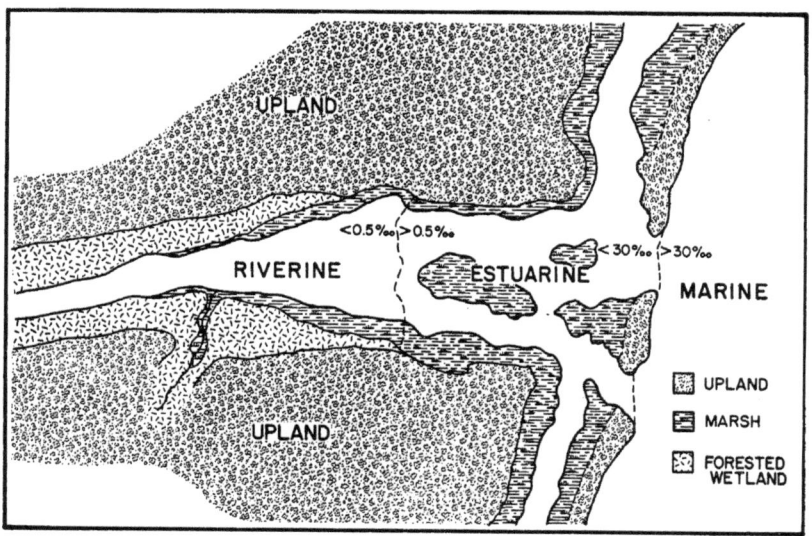

Figure 5-8. The location of the tidal freshwater marsh community in relation to marine, estuarine, riverine, upland, and forested wetland systems.

wild rice (typically a tidal freshwater marsh species) were all found in a transitional zone between the tidal freshwater marsh and the brackish marsh on the South Santee River. Havel pointed out that on the North Santee River freshwater plants were found growing nearer to the river mouth than were freshwater plants on the South Santee. He attributed this anomaly to the fact that most fresh water of the Santee River system flows out of the northern branch. The dominants of the Santee tidal freshwater marsh are wild rice, giant cutgrass, arrow-arum, arrow-heads, smartweeds, sedges, and pickerel-weed (Havel 1976).

In Georgia, tidal freshwater marshes of the Altamaha River Delta are dominated by giant cutgrass, wild rice, pickerelweed, soft-stem bulrush, arrow-arum, spikerush, and cat-tail (Gallagher and Reimold 1973, Gallagher et al. 1975). In other river systems, giant cutgrass also seems to be the tidal freshwater marsh dominant, while wild rice, pickerelweed, arrow-arum, plume grass, beak rush, water hemlock, and marsh daisy are often co-dominants (Wharton 1978).

Mellinger and Mellinger (1962) described tidal freshwater marsh communities of the Savannah National Wildlife Refuge along the Savannah River. Two types were

recognized, the giant cutgrass community and the arrow-arum community. The giant cutgrass community is a uniform, single-species community, whereas the arrow-arum community contains 60 other species (Mellinger and Mellinger 1962).

Many tidal freshwater marsh communities were rice fields in the eighteenth and nineteenth centuries. The dikes and canals that were necessary for rice cultivation are still obvious on aerial photographs of tidal freshwater and brackish marshes (Tiner 1977, Georgia Department of Natural Resources 1978).

Little or no data specific to productivity in tidal freshwater marshes are available on these subcategories. Jervis (1964) studied productivity of a nontidal freshwater marsh in New Jersey; Odum and Fanning (1973) compared productivity of salt and brackish marshes, but did not deal with productivity of tidal freshwater marshes; and Boyd (1970) and Boyd and Hess (1970) discussed the productivity of common cat-tail and common three-square in nontidal marshes. A general comparison between tidal freshwater and estuarine wetlands is presented in Table 5-6.

Table 5-5. Species list of characteristic plants in tidal
freshwater marshes of South Carolina (adapted
from Tiner 1977).

SCIENTIFIC NAME	COMMON NAME
Alnus serrulata	Tag alder
Alternanthera philoxeroides	Alligator-weed
Amaranthus cannabinus	Water hemp
Amorpha fruticosa	False indigo
Aneilema keisak	Asiatic dayflower
Arundo donax	Giant reed
Aster spp.	Asters
Azolla caroliniana	Mosquito fern
Baccharis halimifolia	Sea myrtle
Bidens spp.	Beggar ticks
Brasenia schreberi	Water-shield
Carex spp.	Sedges
Carpinus caroliniana	Ironwood
Cassia fasciculata	Partridge pea
Cephalanthus occidentalis	Button bush
Chenopodium album	Lamb's quarters
Cicuta maculata	Water hemlock
Cinna arundinacea	Wood reed
Cladium jamaicense	Saw grass
Clematis crispa	Leather-flower
Clethra alnifolia	Sweet pepperbush
Cuscuta sp.	Dodder
Cyperus spp.	Sedges
Dichromena colorata	Star-rush
Echinochloa crusagalli	Wild millet
Egeria densa	Water-weed
Eleocharis spp.	Spikerushes
Elodea spp.	Water-weeds
Elymus virginicus	Wild rye grass
Erianthus giganteus	Plume grass
Eryngium aquaticum	Marsh eryngo
Eupatorium capillifolium	Dog fennel
Gleditsia aquatica	Water locust
Hibiscus militaris	Halberd-leaved marsh mallow
Hibiscus moscheutos	Rose mallow
Hydrocotyle ranunculoides	Pennywort
Hydrocotyle spp.	Pennyworts
Hymenocallis crassifolia	Spider-lily
Impatiens capensis	Jewel-weed
Iris virginica	Blue flag
Lemna spp.	Duckweeds
Lilaeopsis chinensis	Eastern lilaeopsis
Limnobium spongia	Frog's bit
Liquidambar styraciflua	Sweet gum
Lobelia cardinalis	Cardinal flower
Ludwigia spp.	Water-primroses
Lythrum lineare	Loosestrife
Mikania scandens	Climbing hempweed
Myrica cerifera	Wax myrtle
Myriophyllum sp.	Water milfoil
Nuphar luteum	Yellow pond-lily
Nymphaea odorata	White water-lily
Nyssa aquatica	Water tupelo
Nyssa sylvatica	Black gum

Table 5-5. Concluded

SCIENTIFIC NAME	COMMON NAME
Orontium aquaticum	Golden club
Osmunda regalis	Royal fern
Panicum spp.	Panic grasses
Paspalum distichum	Bullgrass
Peltandra virginica	Arrow-arum
Phragmites communis	Reed
Pluchea spp.	Marsh fleabanes
Polygonum spp.	Smartweeds
Pontederia cordata	Pickerelweed
Potamogeton spp.	Pondweeds
Ptilimnium capillaceum	Mock-bishopweed
Rhynchospora sp.	Beak rush
Rosa palustris	Swamp rose
Rumex verticillatus	Swamp dock
Sacciolepis striata	Sacciolepis
Sagittaria spp.	Arrowheads
Salix caroliniana	Swamp willow
Sambucus canadensis	Elderberry
Saururus cernuus	Lizard's tail
Scirpus americanus	Common three-square
Scirpus cyperinus	Bulrush
Scirpus olneyi	Olney's three-square bulrush
Scirpus robustus	Salt-marsh bulrush
Scirpus validus	Soft-stem bulrush
Scutellaria sp.	Skullcap
Senecio sp.	Butterweed
Setaria magna	Giant foxtail grass
Sium suave	Water parsnip
Solidago sempervirens	Seaside goldenrod
Spartina cynosuroides	Giant cordgrass
Spirodela polyrrhiza	Big duckweed
Taxodium distichum	Bald cypress
Tripsacum dactyloides	Gamma grass
Typha angustifolia	Narrow-leaved cat-tail
Typha domingensis	Southern cat-tail
Typha glauca	Blue cat-tail
Typha latifolia	Common cat-tail
Uniola latifolia	Spike-grass
Uniola laxa	Spike-grass
Utricularia spp.	Bladderworts
Verbesina occidentalis	Crownbeard
Vernonia sp.	Ironweed
Viburnum dentatum	Arrowwood
Zizania aquatica	Wild rice
Zizaniopsis miliacea	Giant cutgrass

Table 5-6. A comparison of system characteristics between tidal freshwater and estuarine wetlands (Odum 1978).

System Characteristic	Tidal Freshwater Wetlands	Estuarine Wetlands
Location	Head of estuary	Mid and lower estuary
Sediments	Silt-clay, high organic content	More sand, lower organic
Dissolved Oxygen	Very low (summer)	Low (summer)
Vegetation	Freshwater species	Marine and estuarine species
Plant Diversity	High	Low
Plant Zonation	Present, but not always distinct	Pronounced
Seasonal Sequence of Plant Species	Pronounced	Absent or minor
Above Ground Primary Production	Very high (?)	High
Rate of Decomposition of Intertidal Plants	Extremely high	Moderate
Nutrient Cycles	High "leakage" in autumn and winter	Slow release
Primary Consumers	Larval insects, annelids, and amphipods	Mollusks, crustaceans, and polychaetes
Fish Community	Freshwater and Oligohaline species + anadromous larvae and juveniles	Estuarine and marine species
Waterfowl	High usage	Medium to low usage
Furbearers	High population densities	Medium to low population densities

3. Tidal Forested Wetlands.

All forested wetlands of the river systems of the Sea Island Coastal Region are included in this palustrine tidal subsection. The tidal zone of the palustrine system extends upriver to the point at which no measurable tidal fluctuation in water level is perceptible (salinity is not a factor in this definition) (Cowardin et al. 1977). Near the inland boundaries of the study area (especially in South Carolina), short sections of several rivers and their tributaries are not tidally influenced. Technically, these sections should be described under the nontidal palustrine subsection; but because these areas are not biologically distinct from the palustrine tidal forested wetlands, they are dealt with here.

The palustrine tidal forested wetlands are the "black" and "red" river flood plains of common parlance. Bozeman and Darrell (1975) described three classes of coastal plain "river swamps": Class I rivers originate in the mountains or Piedmont and flow through the coastal plain (these are the "red" or muddy rivers); Class II rivers (black rivers) originate in the upper coastal plain; and Class III rivers originate on lower coastal plain terraces and are dominated by tidal currents. Wharton (1978) separated coastal plain rivers into two general types (with several sub-types): 1) the blackwater river and swamp system and 2) the alluvial river and swamp system. Penfound (1952) divided southern flood plains into deep swamps (cypress-tupelo communities) and shallow swamps (bottomland hardwood communities). Palustrine tidal forested wetlands may therefore be broken down into two general floodplain or river swamp types, the alluvial and the blackwater (nonalluvial), both of which potentially harbor deep and shallow communities (Penfound 1952, Bozeman and Darrell 1975, Wharton 1978).

Alluvial rivers of the Sea Island Coastal Region include the Great Pee Dee, Santee, Savannah, Ogeechee, and Altamaha. The flood plains of these rivers may be from 3 to 12 miles (5 to 19 km) in width, with oxbow lakes, deep sloughs, natural levees, and high bluffs (along the edge of the flood plain) common. These flood plains are characterized by great fluctuations in water level and by heavy sediment load. The flood plain of alluvial rivers is far from being a flat plain. A high natural levee is often present along the river banks, with deep sloughs alternating with ridges throughout the flood plain. Frequently, back swamps occupy the broad sloughs adjacent to the floodplain bluffs. (See Bozeman and Darrell 1975 and Wharton et al. 1976 for additional information.)

Blackwater rivers are so-called because the high organic acid content of their waters gives them the appearance of being black in deep areas (the water is actually tea colored). The flood plains of these rivers are generally narrower than those of alluvial rivers and, unlike alluvial rivers, blackwater rivers carry a light sediment load (these rivers originate in the sands and clays of the coastal plain). Bozeman and Darrell (1975) pointed out that though natural levees are low and narrow in blackwater flood plains (his class II rivers), the same geomorphic processes that govern alluvial rivers (meander development, oxbow formation, cut-offs, etc.) are active in these smaller flood plains. All of the non-alluvial rivers and their flood plains (of the Sea Island Coastal Region) are included under the blackwater river category, along with most tributaries of alluvial rivers (those tributaries that originate in the coastal plain). These classifications are presented in Table 5-7. Because their terminology and subsystems are complex, no attempt will be made to integrate the classification systems. However, the three investigators agree that the two basic forest types found in the palustrine tidal forested wetlands are the bottomland hardwood and the cypress-tupelo forests. The following narrative describes each forest type.

Bozeman and Darrell (1975) noted that the cypress-tupelo forest is flooded by 1 ft (0.3 m) or more of water for 6 months or longer. This community usually has canopies with 100% cover, but cypress trees allow considerable light penetration. The understory is composed of only two to three species of small trees, and a shrub layer is absent. The cypress-tupelo swamp forest, also called the bald cypress-water tupelo community, occupies deep sloughs, margins of oxbows, and wet flats. The buttressed cypress bases and omnipresent cypress knees are characteristic of this community. The dominant plants of the cypress-tupelo forest type are listed in Table 5-8.

The bottomland hardwood forest type is drier than the cypress-tupelo forest with a considerably shorter hydroperiod. These communities are usually uneven-aged (except where cutting or fire has been present), with dense canopies (100% cover). The understory is less dense than the canopy, and shrub layers may be completely absent. Wharton (1978) noted that a slightly acid (5.1 pH) to alkaline soil was important to the survival of bottomland hardwoods. He also noted that extensive tracts of this community are absent from Georgia's blackwater flood plains. Figure 5-9 illustrates the physiography of Wharton's (1978) bottomland hardwoods subcommunities. Gaddy et al. (1975) listed 10 community types from an old growth area of the Congaree River flood plain (inner coastal plain, South Carolina) (Table 5-9). Table 5-10 summarizes the

313

Table 5-7. Classifications of coastal plain river swamps of the Sea Island Coastal Region.

Bozeman and Darrell 1975	Wharton 1978	Penfound 1952
Class I	Nonalluvial Clearwater River System	Deep Swamps
Rivers originating in mountains and Piedmont and flowing through the coastal plain	spring-fed stream	Freshwater woody communities with surface water throughout most or all of the growing season
	Nonalluvial (blackwater) River System	
Class II	blackwater river and swamp system	Shallow Swamps
Rivers originating in the coastal plain	blackwater branch or creek swamp	Freshwater woody communities, the soil of which is inundated for only short periods during the growing season
Class III	Alluvial River System	
Rivers originating on lower coastal terrace and dominated by tidal currents.	alluvial river and swamp system – Piedmont	
	alluvial river and swamp system – coastal plain	
	Coosa River and swamp system	
	Water Dam Systems	
	tidewater river and swamp system	
	backwater streams	
	river marsh and freshwater marsh	

Table 5-8. Dominant cypress-tupelo forest vegetation documented for palustrine tidal forested wetlands.

INVESTIGATOR	LOCATION	DOMINANTS
Bozeman and Darrell 1975	Altamaha River, Ga. Canooche River, Ga.	CANOPY Bald cypress; water tupelo, over-cup oak, green ash, water ash, black gum, pond cypress, swamp tupelo, Ogeechee plum, white ash, red maple, water oak, laurel oak, sweet bay UNDERSTORY Swamp privet, swamp dogwood, fetter-bush, sweet pepperbush, wax myrtle, titi, myrtle holly, possum haw, storax, Georgia fever bark
Wharton 1978	Altamaha River, Ga.	Bald cypress, sweet gum, water tupelo, black gum, swamp tupelo, red bay, dwarf palmetto
Woodwell 1958	North Carolina South Carolina	THREE DOMINANT COMMUNITIES 1) Cypress-bald cypress 2) Water tupelo - fetter-bush dominated by sweet bay, red maple, swamp tupelo 3) Water tupelo-cypress dominated by water tupelo, bald cypress, swamp tupelo
Klawitter 1962	Santee River, S.C.	Swamp tupelo, bald cypress, green ash, water tupelo, red maple, wax myrtle, titi
Penfound 1952	Southeastern U.S.	Bald cypress, pond cypress, water tupelo, swamp tupelo

the dominant vegetation reported by these and other investigators for the bottomland hardwood forests.

The obvious structural features of cypress-tupelo communities (especially bald cypress-dominated communities) are the buttressed bases of the cypresses and the presence of cypress "knees." Mattoon (1915) pointed out that buttress development differs with hydroperiod and that buttress shape is indicative of the swamp or floodplain type in which the tree is found. Low, very broad bases are typical of "non-alluvial wet swamps" (blackwater river swamps); trees with high, full basal swell are typical of areas subjected to deep inundations; and trees with small basal swell are typical of inland (palustrine nontidal forested wetlands)

and alluvial swamps. Kurz and Demaree (1934) asserted that buttress curve is a good indicator of relative period of inundation.

Penfound (1952) noted that dominants of cypress-tupelo communities (bald cypress, pond cypress, water tupelo, and swamp tupelo) all have pneumatophores or knee-like projections. Mattoon (1915) and Wells (1942) proposed that the purpose of cypress knees was for anchorage and gas exchange. When roots of the tree were submerged, the emergent knees could carry out respiration, they reported. Penfound (1952) rejected the gas exchange theory, pointing out that cypress does not produce knees on dry land or on soils continously inundated by water where knees would really be needed. Penfound analyzed the wood of

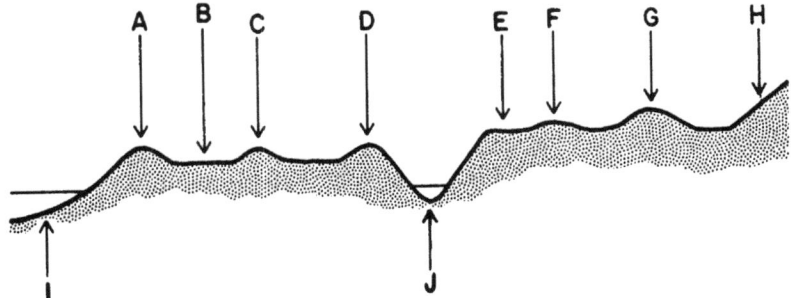

A– Natural levee or front
swamp cottonwood, silver maple, willows, river birch

B– Flats or back swamps
overcup oak, water hickory, sugarberry, green ash, red maple, American elm

C– Low ridge
sweet gum, willow oak, green ash

D– High ridge or old natural levee
white oak, black gum, winged elm, hickories, loblolly pine

E– Second bottom flats
sweet gum, laurel oak, willow oak, green ash, red maple

F– Second bottom low ridge
swamp chestnut oak, cherrybark oak, water oak, black gum, winged elm

G– Second bottom high ridge or old natural levee
white oak, black gum, winged elm, hickories, loblolly pine

H– High ground

I – River channel

J – Oxbow lake

Figure 5-9. The physiography of bottomland hardwoods of a Georgia flood plain (adapted from Wharton 1978).

Table 5-9. Plant communities of the Beidler Tract of the Congaree flood plain, South Carolina (adapted from Gaddy et al. 1975).

COMMUNITY	PHYSIOGRAPHY
1. Loblolly Pine-Mixed Hardwoods	High ridges ·
2. Riverbank Hardwoods	Levees and along creeks
3. Cherrybark Oak-Sweet Gum-Swamp Chestnut Oak	High inter-floodplain ridges
4. Loblolly Pine-Swamp Tupelo	Seepage areas at the edge of the flood plain
5. Sweet Gum-Mixed Hardwoods	Flats and low ridges
6. Laurel Oak-Sweet Gum	Low flats
7. Overcup Oak	Shallow sloughs
8. Ash-Red Maple	Low flats
9. Swamp Tupelo	Boggy seepage area near floodplain bluffs
10. Water Tupelo-Bald Cypress	Deep sloughs and creeks

Table 5-10. Dominant bottomland hardwood forest vegetation documented for palustrine tidal forested wetlands.

INVESTIGATOR	LOCATION	DOMINANTS
Bozeman and Darrell 1975	Altamaha River, Ga. Canoochee River, Ga.	CANOPY Overcup oak, water hickory, green ash, American elm, sweet gum, water oak, red maple, swamp chestnut oak, laurel oak, river birch, black gum, water ash, Ogeechee plum, water tupelo, sweet bay UNDERSTORY Water elm, ironwood, persimmon, swamp holly, buttonbush, Sebastian bush, Virginia willow, dwarf palmetto, greenbrier, blackberry, rattanvine
Wharton 1978	Georgia	See Figure 5-9
Klawitter 1962	Santee River, S.C.	Water oak, laurel oak, overcup oak, water hickory, sweet gum, green ash, loblolly pine, bald cypress
Gaddy et al. 1975	Congaree River, S.C.	Beech, laurel oak, hickories, cottonwoods, river birch, sweet gum, willows

cypress knees and found that no
aerenchymous tissue was present. The
role of cypress knees in stabilizing the
tree in soft sediments needs further investigation (see Mattoon 1915).

Water level is important in germination and regeneration of the cypress-tupelo
community. Mattoon (1915), Demaree (1932),
Shunk (1939), and Wells (1942) have pointed out that bald cypress and water tupelo
seeds will not germinate while submerged.
Soils of the cypress-tupelo community must
be exposed for a short period of time in
late summer or early fall if germination
is to occur. Demaree (1932) went on to
point out that young cypresses will not
survive if they are permanently inundated.
Cypress-tupelo communities that are subject to abnormal amounts of standing water
are characterized by an open understory
with little or no succession occurring
[succession may occur on stumps or on
floating logs, see Woodwell (1958) and
Dennis (1973)]; in areas where water has
been excluded due to damming or draining,
cypress-tupelo is invaded by sweet gum,
green ash, and other drier-site species.

Klawitter's (1962) conclusions concerning the ecological relationships
among sweet gum, swamp tupelo, and water
tupelo sites can be summarized as follows:

1. Sweet gum and swamp tupelo are
    found at higher elevations than
    water tupelo, but sweet gum sites
    tend to be better drained than
    swamp tupelo sites because the
    former are located primarily on
    first bottom slopes.

2. The water table is at about the
    same depth on swamp tupelo and
    water tupelo sites when the river
    is below flood stage, but the
    latter are more deeply inundated
    by the Santee River. Water table
    height on sweet gum sites is
    usually below that occurring on
    the other two types of sites
    when the river is not flooding.

3. Surface soil reaction is highest
    on sweet gum sites, intermediate
    on water tupelo sites, and lowest
    on swamp tupelo sites.

4. Organic matter content, as
    approximated by loss on ignition,
    is highest on swamp tupelo sites,
    lowest on sweet gum sites, and
    intermediate on water tupelo
    sites.

5. Silt plus clay content of the
    surface soil on sweet gum and
    water tupelo sites averages
    considerably higher than on
    swamp tupelo sites. Further-
    more, there is a decided
    decrease in the silt plus clay
    content of subsoils under each
    species.

An important structural feature of
palustrine tidal forested wetland
communities is the height and girth some
of the constituent trees attain. Alluvial
swamps harbor large, fast-growing trees.
In the Congaree Swamp near the Savannah
River Plant, Swails et al. (1957) found
trees 120 - 130 years old with breast
height diameters ranging from 25 to 40 in
(63.5 to 101.6 cm). In the Beidler
Tract of the Congaree River flood plain,
Gaddy et al. (1975) found that the canopy
averaged from 110 to 130 ft (33.5 to
39.6 m) in height, with some trees as tall
as 170 ft (51.8 m). Gaddy (1977) listed
14 State and national record trees from
this same old growth forest, many of which
were larger than 20 ft (6.1 m) in circum-
ference.

Bozeman and Darrell (1975) listed
willows, water elm, sycamore, water locust,
and river birch as early successional
species of bottomland hardwoods. Green
ash, silver maple, swamp cottonwood, oaks,
and water hickory were noted as plants of
later successional stages.

Gaddy et al. (1975) pointed out the
importance of windthrow as a regenerative
agent in old growth bottomland hardwoods.
Figure 5-10 is a generalized diagram of
succession on bottomland flats in the
Congaree River flood plain. Pines,
cottonwoods, willows, and river birches
are all pioneer species in the Congaree
flood plain; sweet gum occur in both early
and late successional forests (see Fig.
5-10). Beech, laurel oak, hickories,
and other shade-tolerant hardwoods are
found in the most undisturbed communities
(Gaddy et al. 1975). The role of fire
as a successional agent in bottomland
hardwoods needs more study.

Wharton (1970) pointed out that river
swamps (alluvial swamps) are among the
most productive of the world's environments. Alluvial swamps have an estimated
gross primary production potential of
between 20,000 and 40,000 kcal/m^2/yr.
Odum (1969) described the alluvial swamp
as a fluctuating water level environment
with pulses of high primary productivity.
A flood plain may produce as much as 30
metric tons/ha/yr of plant material
(Wharton 1970).

Conner and Day (1976) studied the productivity of Lac des Allemands swamp in
the Mississippi River flood plain (a
flood plain comparable in ecology, if not
size, to the alluvial swamps of the Sea
Island Coastal Region). Net primary productivity was 1,140 g/m^2/yr for cypress-
tupelo and 1,174 g/m^2/yr for bottomland
hardwoods. Herbaceous production was
much higher in bottomland hardwoods
than in cypress-tupelo swamps (high water
levels limit the growth of herbs in
cypress-tupelo communities). Table 5-11
compares the productivity of various
swamp communities in the Southeast.

318

1. After fire or clearcutting, sweet gum dominates regeneration.

2. As forest ages, laurel oaks and other shade-tolerant hardwoods appear in understory.

3. Laurel oaks and shade-tolerant hardwoods approach canopy size.

4. Sweet gum–mixed hardwood canopy eventually falls or is blown down (due to age). Laurel oaks and other shade-tolerant hardwoods assume dominance. In openings, sweet gums and other hardwoods renew the cycle.

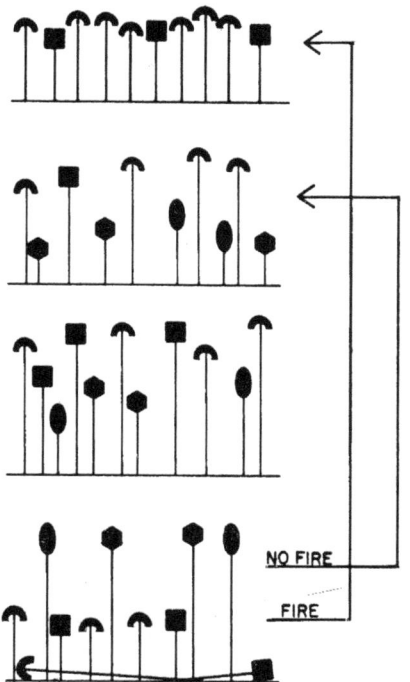

NO FIRE

FIRE

⬡ LAUREL OAK

⬣ MIXED HARDWOODS (SHADE TOLERANT)

◠ SWEET GUM

■ MIXED HARDWOODS (SHADE INTOLERANT)

Figure 5-10. Plant succession in bottomland hardwoods (adapted from Gaddy et al. 1975).

319

Table 5-11. Comparative swamp productivity for the Southeastern United States (Conner and Day 1976).

Area	References	Stem Growth g/m²/yr	Herbaceous Production g/m²/yr	Litterfall g/m²/yr	Net Primary Production g/m²/yr
des Allemands, La. (seasonal flooding) Cypress-tupelo Bottomland-hardwood	Conner and Day 1976	500	20	620	1,140[a]
Big Cypress Swamp, Fla. (riverine)	Carter et al. 1973				
Drained		120	-20	267	367[a]
Undrained-edge strand		485	312	373	1,170[a]
Undrained-central strand				756	
Withlacoochee State Forest, Fla. Combined riverine and cypress dome	Mitsch 1975				600[b, c]
Cypress Domes, Fla.	Mitsch 1975				
Drained					416[c]
Undrained (stagnant)					192[c]
Okefenokee Swamp, Ga. Very slowly flowing water	Schlesinger and Marks 1975	267	97	328	692[c]

a. Stem growth + herbaceous production + litterfall.
b. Average of 23 sites.
c. Stem growth + litterfall + root loss.

Monk (1968) analyzed mineral levels in wetland communities in north-central Florida. Mixed swamps (blackwater river swamps in his study) had the highest concentrations of calcium (8,131 ppm), magnesium (295 ppm), potassium (1,974 ppm), and phosphorus (75 ppm) of any of the wetland types.

Brinson (1975) studied nutrient cycling in an alluvial swamp in the coastal plain of North Carolina. Litterfall there was near the upper limits of the range for temperate forests (6,430 kg/ha/yr), and phosphorous and nitrogen cycling in the litterfall were higher than that reported for most ecosystems.

4. Nontidal Emergent Wetlands

Nontidal emergent wetlands are freshwater wetlands that are associated with Carolina Bays, depressions, pond margins, bogs, and ditches. They are, therefore, distinct from palustrine tidal emergent wetlands, although there are many plants that occur in both types of wetlands. The palustrine tidal emergent wetlands are almost always referred to as "freshwater marshes"; however, the nomenclature for palustrine nontidal emergent wetlands is much more complex. Penfound (1952) listed some of the possible names for palustrine nontidal emergent wetlands: 1) "rush marshes or wet prairies" (Davis 1946); 2) "wet meadows" (Hotchkiss and Stewart 1947); 3) "grass-sedge bogs" (Wells 1942, Garren 1943); 4) "pitcher plant lands" (Pessin 1933); and 5) "moist pine barrens or savannahs" (Wright and Wright 1932). Recently, Wharton (1978) referred to a similar environment as a "herb bog or pitcher plant bog" (in Georgia). Woodwell (1956) pointed out that, in the Southeast, the term "savannah" is often used to mean open woodland areas with a ground cover of herbaceous vegetation. In this text, the term "savannah" is differentiated from "forested savannah." Areas that fit Woodwell's (1956) description are referred to as forested savannahs; while the term savannah is used here to mean a treeless or almost treeless shallow wetland that is dominated by herbaceous vegetation.

Penfound (1952) divided the freshwater marshes of the South into two groups, deep marshes and shallow marshes. Deep marshes are found along rivers, in pond margins, in deep bays and natural depressions, and in ditches and man-made depressions. Shallow marshes, on the other hand, are usually found in or adjacent to upland communities. They are common in poorly drained flats, wet openings in pine woods, and shallow depressions in upland communities. In deep marshes, the water table is usually at or above the soil surface throughout the year; on the other hand, shallow marshes normally have standing water in

the winter and in the early part of the growing season, with drying out occurring in late summer and fall (Wells 1928, Penfound 1952).

Deepwater marshes may contain many species of emergents, but the most common dominants are cat-tails, woolgrass bulrush, giant plume grass, maidencane, black needlerush, beak rushes, and sedges. Wells (1928) described a "hydric fresh water marsh" association from the coastal plain of North Carolina that was dominated by cat-tails and bulrushes, though he pointed out that this association probably contained more than 450 species of plants. Penfound and Hathaway (1938) called the dominant deepwater marsh community of Louisiana a cat-tail - bulrush - maidencane community. In South Carolina, five deepwater marsh communities were recognized during the course of a survey of coastal wetlands (National Wetlands Inventory 1978). These communities are: 1) maidencane marsh, probably the shallowest of the deepwater marshes; 2) cat-tail marsh, common near pond edges and in disturbed wetlands; 3) woolgrass bulrush-giant plume grass marsh, found in ditches and near pond edges; 4) water loosestrife marsh, a low shrub-dominated deep marsh; and 5) button bush marsh, a high shrub deep marsh.

Occasionally, deep marshes are found in openings in swamps or bays. Porcher (1966) described several nonforested openings in a Carolina Bay that contained beak rush, yellow pitcher-plant, and blue flag, along with the woody St. John's-wort and a mat of peat moss. Big Openings, a nonforested marsh in Hellhole Swamp, South Carolina, is dominated by St. John's-wort, three-way sedge, beak rush, woolgrass bulrush, and various invading trees and shrubs (South Carolina Wildlife and Marine Resources Department 1974a).

The shallow marshes or wetlands (they are not commonly referred to as marshes) include the wet prairies, grass-sedge bogs, and wet meadows that were mentioned above. In the Sea Island Coastal Region, the shallow wetlands of the palustrine nontidal emergent grouping may be divided into: 1) the fern-sedge bogs and 2) the savannahs [the grass-sedge bogs of Wells (1942) and the herb or pitcher plant bogs of Wharton (1978)].

The fern-sedge bogs are dominated by either Virginia chain fern or Walter's sedge or both species. Sometimes blue flag or maidencane (usually found in deeper wetlands) may co-dominate. Woodwell (1956) noted the presence of a "Woodwardia union" in coastal wetlands of the Carolinas. He pointed out that Virginia chain fern was the dominant here, with a mat of peat moss almost always present. Bozeman (1965) described a "sedge-peat bog" community from McIntosh County, Georgia, dominated by Walter's sedge and peat moss. In the southeastern coastal plain of South Carolina, Virginia chain fern and Walter's sedge co-dominate in treeless areas that are underlain with a clay hardpan (National Wetlands Inventory 1978). In deeper areas of the same wetland type, blue flag dominates. The fern-sedge bogs are the wettest of the shallow marshes. The soil surface dries out for only a short duration each year; usually it is saturated or under water. The long hydroperiod and the fact that the clay hardpan is so close to the soil surface protects this community from invasion by woody species.

The savannahs are extremely shallow and have a shorter hydroperiod. Standing water is present in winter and during the early growing season, but later in the year, the soil may completely dry out. The water regime here would permit the invasion and dominance of woody plants if it were not for the occurrence of fire (Wells and Shunk 1928). Annual fires prevent the invasion of woody plants into this community and maintain a geophyte-dominated grass-herb population (geophytes are bulbous or rhizomatous perennials).

Savannahs are most commonly found in poorly drained interstream flats among pine flatwoods. Wells (1942) pointed out that savannahs are probably the result of the removal of pines from wet flatwoods that are annually burned. The annual fire prohibits the reestablishment of pines after they have been removed. The savannahs of the southeastern coastal plain of North Carolina and the northeastern coastal plain of South Carolina are good examples of this community. In Georgia, savannahs are limited to smaller areas interspersed in pine communities (Wharton 1978). Dominance in savannahs is usually seasonal or even monthly. Burning generally occurs in early spring (or late fall) and is followed by a progression of plants that flower for a short period of time and then disappear (they are still present, just no longer obvious).

In South Carolina, toothache grass, meadow beauty, gentians, white-bracted sedge, orchids, rosebud orchid, rose pogonia, pitcher-plants, milkworts, and rayless goldenrod are common in savannahs (South Carolina Wildlife and Marine Resources Department 1974b, 1975a, Radford 1976). In Georgia, Wharton (1978) listed meadow beauty, seed box, orchids, Barbara's buttons, clubmosses, pitcher-plants (including Parrot pitcher-plant, which does not occur in South Carolina), toothache grass, yellow-eyed grass, pipeworts, colic root, beak rushes, sticky tofieldia, and gentians as common species of the "herb bog."

Dominance in marsh communities is usually determined by hydroperiod. Wells and Shunk (1928) pointed out that the savannah community does not occur on peaty soils (even after fire) but is found on mineral soils. Wells (1928) noted that a deep marsh soil type could also play an important role. Sandy and clayey soils may be dominated by different communities, even though both soil types occur in the same marsh with the same hydroperiod.

Nontidal emergent wetlands are dormant in winter and productive in summer months. No year-round producers are present. However, the savannah communities exhibit considerable seasonal and intraseasonal changes in dominance. Savannah communities are burned in late winter or early spring (by man), or in late summer or early fall (naturally). In February or March, when the first spring dominants appear, sun-bonnets usually appear and bloom first. As soon as these early dominants begin to fruit, another group of plants becomes dominant. This flowering sequence makes the savannah one of the most picturesque of the coastal plain communities (Wells 1932, Wharton 1978). Table 5-12 is a list of the plant dominants in a coastal plain savannah for each month of the growing season. If fire does not occur, the flowering sequence is altered with woody plants invading during the latter months of the growing season.

Most communities of the nontidal emergent wetlands are single-tiered, but, occasionally, two-tiered communities with a scattered shrub overstory and emergents beneath may be found (tall shrub communities are discussed in the Upland sections). Monocots are the dominant floristic form in the nontidal emergent wetlands. In savannahs, the presence of annual fire has prompted the predominance of plants with special modifications. Wells and Shunk (1928) pointed out that 95% of the species in a North Carolina grass-sedge bog were perennials (the seeds of most annuals are destroyed by fire). Wells and Shunk also noted the abundance of geophytes (bulbous or rhizomatous plants) in the savannah community (see Radford 1976).

Savannahs are known to have a low soil nitrogen content (Plummer 1963). Wells and Shunk (1928) found no legumes in their study. Many of the savannah plants here, therefore, are adapted to a nitrogen-deficient environment. Plants such as sundews, pitcher-plants, butterworts, and Venus' fly traps have taken on structural leaf modifications and have become insectivorous as a means of overcoming nutrient deficiencies. The insectivorous pitcher-plants, for example, use the nutrient gained from their prey to compensate for nutrient deficiency (Hepburn et al. 1920, Plummer and Kethley 1964). Christensen (1976), however, studied yellow pitcher-plants and found that the "availability of insects resulted in significantly higher

Table 5-12. Seasonal plant dominance in a coastal plain savannah
(adapted from Wells and Shunk 1928, Wells 1932).

MONTH	DOMINANTS
March	Chaptalia tomentosa (sun-bonnets) Sarracenia flava (trumpet-plant) Pinguicula spp. (butterworts) Utricularia subulata (bladderwort) Drosera spp. (sundews)
April	Sarracenia minor (hooded pitcher-plant) Sarracenia purpurea (flytrap pitcher-plant) Sisyrinchium mucronatum (blue-eyed grass) Amianthium muscaetoxicum (fly-poison) Aletris farinosa (colic root)
May	Pogonia ophioglossoides (rose pogonia) Cleistes divaricata (spreading pogonia) Spiranthes praecox (grass-leaved ladies' tresses) Calopogon spp. (grass-pinks) Polygala lutea (red-hot poker) Helenium vernale (sneeze-weed)
June	Eriocaulon spp. (pipeworts) Erigeron vernus (fleabane) Rhexia spp. (meadow beauties) Sabatia brachiata (sabatia) Lycopodium alopecuroides (foxtail clubmoss) Ctenium aromaticum (toothache grass) Dichromena latifolia (white-bracted sedge) Coreopsis falcata (coreopsis)
July	Ludwigia spp. (water-primroses) Cynoctonum sessilifolium (miterwort) Habenaria spp. (orchids) Xyris spp. (yellow-eyed grasses)
August	Zigadenus densus (crow-poison) Oxypolis rigidior (dropwort) Eryngium integrifolium (eryngo) Coreopsis spp. (coreopsis) Habenaria spp. (orchids)
September	Liatris spp. (blazing stars) Tofieldia glabra (false asphodel) Chondrophora nudata (rayless goldenrod)
October	Aster spp. (asters) Solidago spp. (goldenrods)

tissue concentrations of nitrogen and phosphorus, but had no effect on calcium, magnesium, and potassium [concentrations]."

Palustrine nontidal emergent wetlands may be pioneer communities, fire-maintained communities, or communities maintained by a long hydroperiod. Most deepwater communities are maintained by long hydroperiod; they are successionally static communities due to the presence of water year around. However, long-term aggradation of sediments from decaying vegetation may eventually shorten the hydroperiod and change the nature of the communities (Lindeman 1942,

Vogl 1969). Emergent communities may be pioneer communities where cutting of swamps has taken place. Woolgrass bulrush is a common dominant of communities of this type. After a few years, however, woody vegetation shades out emergents and becomes dominant if the hydroperiod is suitable. Savannah communities are fire-maintained and gradually succeed into evergreen shrub bog communities (holly - myrtle - titi) if fire is absent (Wells and Shunk 1928). The evergreen shrub bog may be maintained by recurrent fire or succeed to a bay forest community (see Fig. 5-11).

323

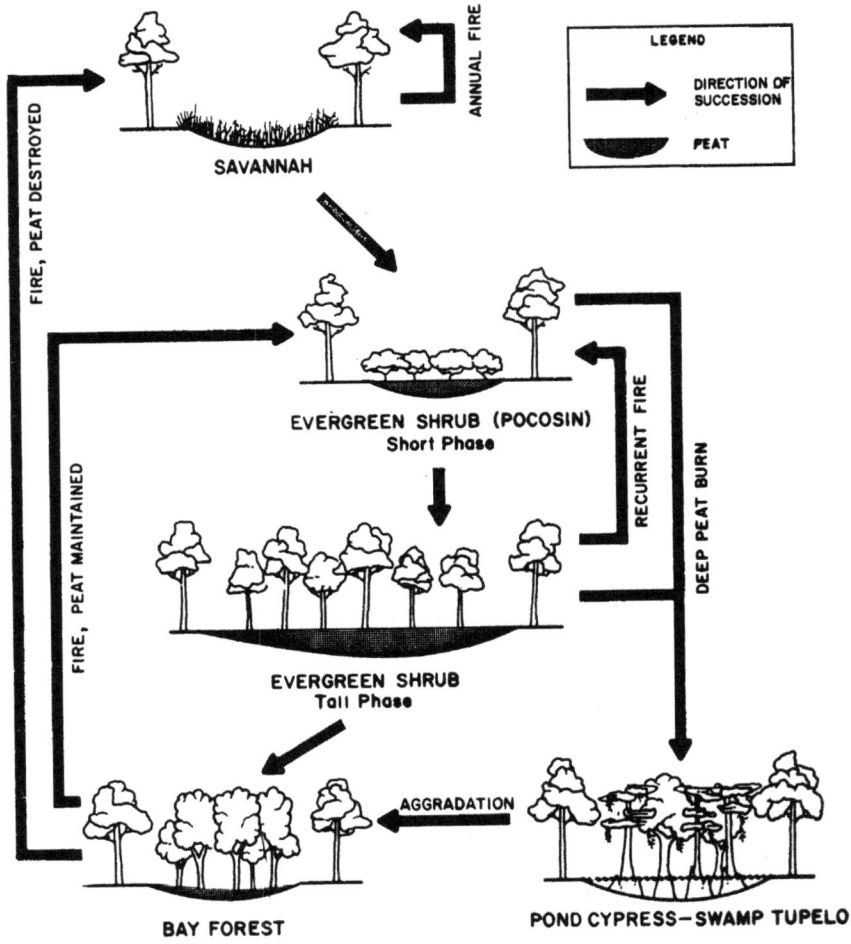

LEGEND

→ DIRECTION OF SUCCESSION

▬ PEAT

SAVANNAH

ANNUAL FIRE

FIRE, PEAT DESTROYED

EVERGREEN SHRUB (POCOSIN)
Short Phase

RECURRENT FIRE

DEEP PEAT BURN

FIRE, PEAT MAINTAINED

EVERGREEN SHRUB
Tall Phase

AGGRADATION

BAY FOREST

POND CYPRESS—SWAMP TUPELO

Figure 5-11. Successional relationships in poorly drained interstream flats (Penfound 1952).

324

Jervis (1964) studied production in a New Jersey deep marsh. Cat-tail communities averaged 10.88 g/m^2/day, as opposed to 9.72 g/m^2/day for shrub communities, and 8.52 g/m^2/day for sedge swale communities. Table 5-13 compares net primary production and seasonal production within four community types. Figure 5-12 illustrates graphically the difference in seasonal production rates among these four communities; Jervis (1964) noted that marshes may become nutrient sinks when they are located in depressions surrounded by uplands.

Cat-tails marshes seem to be the most productive nontidal emergent wetlands according to Jervis (1964), Boyd and Hess (1970), and Boyd (1970). Savannahs, on the other hand, seem to be the most nutrient-deficient of these wetlands. (See nonvascular plant section of this chapter for additional freshwater marsh productivity data.)

5. Nontidal Forested Wetlands

Palustrine nontidal forested wetlands are widespread in the Sea Island Coastal Region. These wetlands, commonly called swamps or bogs, may be found in a variety of physiographic situations. Poorly drained interstream flats, depressions, sinks, Carolina Bays, the "bays" of ridge and bay topography, and flood plains all harbor palustrine nontidal forested communities. Table 5-14 lists eight basic vegetation types and the physiography with which they are generally associated. Each vegetation type listed in the table includes one or more community types. If a particular vegetation type includes only one community type, the terms community type and vegetation type may be used synonymously. As may be seen in the table, some vegetation types occur in more than one physiographic type. The particular forest type present on any site is dependent upon several factors: 1) fire regime, 2) hydroperiod, 3) presence and depth of peat or organic muck, and 4) degree of disturbance (cutting, etc.). Below, the vegetation types listed in Table 5-13 are discussed in the sequence given in the table.

a. Pine Savannah Communities. These community types are forested ("forested" is defined here as 30% or greater canopy coverage) phases of the savannah community described earlier (see nontidal emergent wetlands section). The canopy is sparse and is dominated by longleaf pine with pond pine often present. Frequent fires have eliminated the sub-canopy and shrub layers, leaving a savannah-like understory. Two herbaceous dominants are recognized in the pine savannah: toothache grass and three-awn grass. Pine savannahs are, therefore, referred to as winter pine - toothache grass or pine - three-awn grass

savannahs (Kohlsatt 1974). The pine - toothache grass savannahs are slightly wetter than pine - three-awn grass savannahs. Orchids, gentians, and meadow beauties are common in both communities (Wells and Shunk 1928, Kohlsatt 1974, Radford 1976) (see palustrine nontidal emergent wetlands for a complete discussion of savannah plants).

Pine savannahs are found on poorly drained interstream flats and are often transitional between treeless savannahs and pine flatwoods communities. Recurrent fires have eliminated the organic layer of the soil, exposing a mineral soil (Radford 1976).

b. Pond Cypress Communities. Communities dominated by pond cypress may occur in shallow or deep depressions, sinks, or in Carolina Bays. Pond cypress savannahs are found in shallow depressions in pine flatwoods where fire is able to invade the community (Wharton 1978). Due to the presence of fire, sub-canopy and shrub layers are absent, as in the pine savannah above. The herbaceous layer is dominated by grasses, sedges, pitcher-plants, orchids, and other common savannah plants. (See the palustrine nontidal emergent wetlands section and Wharton 1978.) Pond cypress savannahs have also been reported from Carolina Bays in South Carolina (Dennis 1975a). The community here is considerably wetter than that reported by Wharton (1978) in Georgia. Walter's sedge, yellow pitcher-plant, blue flag, milkwort, and various grasses are some dominants of the Carolina Bays of the Santee Coastal Reserve in Charleston County, South Carolina (Dennis 1975a). Porcher (1966) described two pond cypress savannah communities from Wambaw Creek Bay (Berkeley County, South Carolina). The pond cypress-St. John's-wort community is dominated in the understory by a sparse low shrub layer of shrubby St. John's-wort with milkwort, beak rush, and yellow pitcher-plant also common. The pond cypress-beak rush community is present in shallower water than is the pond cypress-St. John's-wort community. Both communities are typical of pond cypress savannahs with a sparse canopy, no sub-canopy, little or no shrub layer, and a dense layer of herbaceous vegetation. Fire undoubtedly is responsible for maintaining this community (Fig. 5-11).

Radford (1976) described a lime sink in Berkely County, South Carolina, and listed the dominant tree as pond cypress (98% dominance). Only three other woody species were found in the sink itself: dahoon, swamp tupelo (2% dominance), and pond spice. Along the edge of the sink, red maple, sweet gum, loblolly pine, water oak, dahoon, pond spice, and wax myrtle were common. Standing water was present in the sink all year, but no well-developed aquatic community was present.

Table 5-13. Summary of seasonal and intercommunity differences in net primary production in a New Jersey deep marsh (aerial parts only). Seasonal rates are given in g/m^2/day. Annual net production is in g/m^2/yr (Jervis 1964).

	COMMUNITIES				
	OPEN AQUATIC	CAT-TAIL	SEDGE SWALE	SEDGE-SHRUB	MARSH AVERAGE
Spring	3.16	17.31	5.48	7.03	8.25
Early Summer	18.35	20.86	22.78	21.77	20.94
Midsummer	14.91	6.79	-0.48	0.11	5.33
Late Summer	3.37	-7.64	-10.27	-11.06	-6.40
Fall	-17.27	-5.86	1.00	-6.49	-7.16
Growing Season	8.85	10.88	8.52	9.72	9.50
Annual Net Production	1,547.14	1,904.76	1,491.65	1,698.99	1,665.63

As one moves southward in the Sea Island Coastal Region, the pond cypress-dominated depression becomes more common. In Beaufort and Jasper counties (South Carolina), small circular depressions (domes?) are frequently found to be dominated by pond cypress, with dahoon and fetter-bush common in the shrub layer. These communities are only found where fire has been present. Without fire, the depression is invaded by swamp tupelo (National Wetlands Inventory 1978).

In Georgia, Wharton (1978) listed two general types of "cypress ponds" (cypress domes and irregularly shaped cypress ponds). The cypress dome or "head" is a circular depression with the tallest trees in the center of the depression, forming a dome. The cypress dome is extremely common in northcentral Florida, where extensive work has been carried out in this community. Brown (1963), Monk and Brown (1965), Monk (1968), Ewel (1976), Ewel and Mitsch (1975), and Sroka (1975) are all excellent studies of cypress domes in Florida. From a cross-sectional survey of the Austin Cary Forest cypress dome in Florida, Sroka (1975) listed 22 plant species according to abundance, water depth, and location in the dome. Table 5-15 lists his results. The dominants were pond cypress (note that only three tree species are present), fetter-bush, beak rush, and Virginia chain fern.

McEwan (1976) studied the floating log communities of the Austin Cary cypress dome and found 18 species of vascular plants. The dominants (highest frequency) were St. John's-wort, bugleweed, beggar ticks, and peat moss. Ewel and Mitsch (1975) observed the effects of fire on two domes in Florida. They pointed out that fire destroys pines and hardwoods that have invaded the domes, acting as an agent in the perpetuation of pond cypress domination.

Cypress ponds, Wharton (1978) pointed out, differ from cypress domes in that they are irregular in outline. According to Faircloth (1971), an almost pure pond cypress community occurs in the deepest part of cypress ponds (see the pond cypress-swamp tupelo section of this chapter for a discussion of the shallower sections). Many ponds have permanent standing water with aquatic communities dominated by water-shield, water-lily, yellow pond-lily, floating hearts, lotus, pondweeds, and fanworts (Faircloth 1971, Wharton 1978). (See earlier discussion of lime sinks in South Carolina for comparison of communities.)

Bozeman (1965) described several cypress ponds occurring on a fluvial sand ridge adjacent to the Altamaha River. They varied from ponds completely dominated by pond cypress (dahoon was the most common shrub here) to pond cypress stands mixed with bay species (a "cypress-bay forest," according to Bozeman). (See pond cypress-swamp tupelo and bay forest sections for additional discussions of the cypress-bay community.) Porcher (1966) noted the presence of a cypress pond adjacent to a Carolina Bay in Berkeley County, South Carolina.

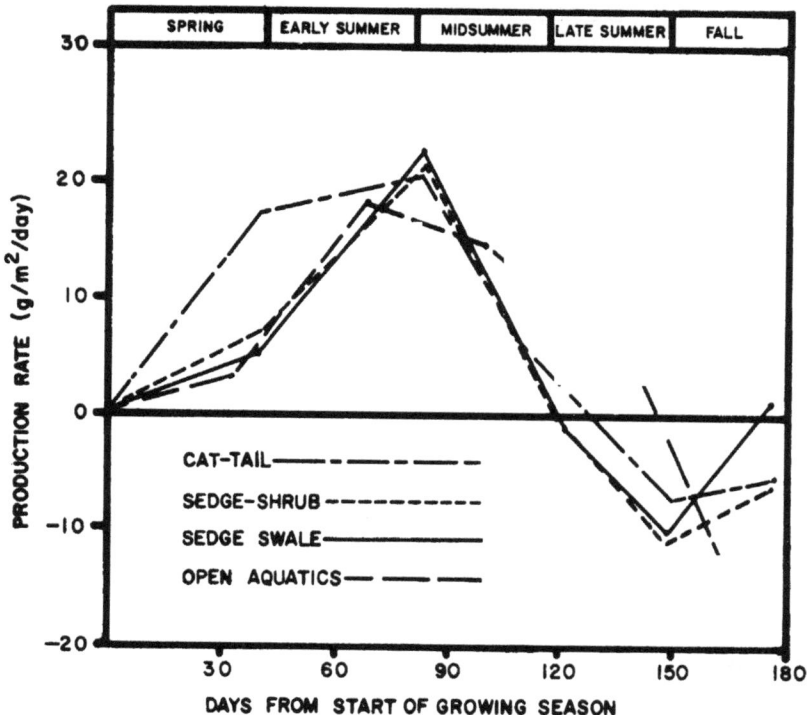

Figure 5-12. Seasonal production rates for four freshwater marsh communities (adapted from (Jervis 1964).

Table 5-14. Vegetative types and associated physiography of nontidal forested wetlands.

VEGETATION TYPES	OCCURRENCE
1. Pine savannahs	A.
2. Pond cypress	A, B, C.
3. Swamp tupelo	A, B, C, D, E.
4. Pond cypress-swamp tupelo	A, B, C, D, E.
5. Evergreen shrub (pocosin)	A, B, C, D.
6. Bay forest	A, B, C, D.
*7. Bottomland hardwoods	E.
*8. Bald cypress-water tupelo	E.
9. Miscellaneous	

PHYSIOGRAPHIC TYPES

A. Poorly drained interstream flat

B. Depressions, sinks, and ponds

C. Carolina Bays

D. Ridge and bay topography

E. Flood plains

*These community types are discussed in the previous section on tidal forested wetlands.

c. **Swamp Tupelo Communities.** Swamp tupelo-dominated forests occur on poorly drained interstream flats, in depressions and ponds, and in the flood plains of small creeks and rivers. Wharton (1978) described the "gum ponds" of the Georgia coastal plain as being dominated by swamp tupelo (also called swamp black gum). The understory composition of three gum ponds in Jasper County, South Carolina, varied from a dense understory of button bush to an open savannah-like understory (probably maintained by water level fluctuations, not fire) to an herb layer dominated by cinnamon fern (Wharton 1978). Gum ponds are common in South Carolina, frequently occurring where fire has not been present in recent years (L. L. Gaddy and D. A. Rayner, 1978, South Carolina Wildlife and Marine Resources Department, Columbia, unpubl. data). Little Wambaw Swamp (Berkeley County, South Carolina), located on a poorly drained interstream flat area, was described as a swamp tupelo-dominated swamp (South Carolina

Wildlife and Marine Resources Department 1974c). Red maple, sweet bay, and red bay were also common there.

Woodwell (1956) pointed out that swamp tupelo dominates swamps of relatively short hydroperiod, such as poorly drained interstream flats and small flood plains. Based on observations from 32 stands in North and South Carolina, Woodwell described a swamp tupelo union. Red maple and cypress (pond?) occurred as co-dominants in some stands. The abundance of the red maple was inversely related to increases in hydroperiod, while the abundance of cypress was directly related. Associated species of the understory were sweet bay, fetterbush, and Virginia chain fern (Woodwell 1956). (For a more complete discussion of swamp communities associated with flood plains, see the earlier section on tidal emergent wetlands.)

328

Table 5-15. Distribution of 22 species of plants in the Austin Cary Forest cypress dome in Florida (adapted from Sroka 1975).

Species	Number of Individuals	Depth of Water Where Found (m) X̄	S.E.	Rank (1 = Deepest)	Distance from Center Where Found (m) X̄	S.E.	Rank (1=Closest to Center)
Lizard's tail	19	0.290 ±	0.028	1	47.79 ±	5.41	3
Button bush[a]	4	0.175 ±	0.109	2	35.20 ±	11.16	1
Pickerelweed	19	0.147 ±	0.009	3	82.58 ±	12.26	9
Beak rush[c]	29	0.120 ±	0.006	4	83.13 ±	3.36	10
Twig-rush	9	0.100 ±	0.012	5	96.14 ±	7.48	15
Pond cypress[b]	24	0.090 ±	0.020	6	80.77 ±	6.51	8
Unidentified sedge	3	0.080 ±	0.060	7	100.16 ±	27.83	17
Swamp tupelo	11	0.075 ±	0.022	8	75.13 ±	4.27	6
Bamboo[c]	4	0.072 ±	0.043	9	79.52 ±	8.86	7
Virginia chain fern	27	0.065 ±	0.012	10	95.86 ±	4.67	14
Swamp willow[b]	1	0.050		11.5	104.80	-----	18
Stagger bush	2	0.050 ±	0.050	11.5	109.00 ±	12.00	20
Three-way sedge	1	0.020		13	125.8	-----	22
Wax myrtle	14	0.010 ±	0.011	14	94.62 ±	8.73	13
Fetter-bush[a]	90	0.008 ±	0.003	15	85.69 ±	3.50	11
Virginia willow[a]	13	0.004 ±	0.004	16	68.23 ±	12.06	5
Cinnamon fern	1	0		19.5	45.90	-----	2
Netted chain fern	5	0		19.5	52.64 ±	6.09	4
Poison ivy[c]	2	0		19.5	91.25 ±	0.85	12
Bitter gallberry[a]	2	0		19.5	97.20 ±	3.80	16
Dahoon[a]	1	0		19.5	107.40	-----	19
St. John's-wort	1	0		19.5	120.30	-----	21

a. shrub
b. tree
c. vine

329

d. **Pond Cypress-Swamp Tupelo Communities**. One must be careful to distinguish the pond cypress-swamp tupelo of depressions, ponds, and Carolina Bays from the bald cypress-water tupelo communities of large floodplain systems (Penfound 1952). (See the section on palustrine tidal forested wetlands for a discussion of the latter community.) Bald cypress and pond cypress are rarely found growing together. Bald cypress is found where alkaline (or circumneutral) soil conditions prevail, while pond cypress is found in acidic soils (Laessle 1942, Monk and Brown 1965, Monk 1966).

Monk (1968) listed swamp tupelo as present in every one of 15 cypress domes in a north-central Florida study area, although he gathered no density or basal area data. Brown (1963) also studied cypress domes in north-central Florida and listed pond cypress as having an importance value of 182 compared to a value of 47 for swamp tupelo (an importance value of 300 would mean the community was a pure stand). From these two studies, it may be concluded that pond cypress-swamp tupelo communities occur in cypress domes, and that pond cypress is the most abundant tree of the association. It may further be noted that both the depth of the depression and the presence of fire are limiting factors for swamp tupelo (see Woodwell 1956, Mitsch and Ewel 1976).

The most common physiographic situation in which the pond cypress-swamp tupelo community is found is the Carolina Bay. Porcher (1966) studied nine Carolina Bays in Berkeley County, South Carolina; five of these bays had well-developed pond cypress-swamp tupelo communities. Buell (1939) pointed out that pond cypress is a pioneer tree, colonizing the rim of bay lakes (in deep bays) and gradually encroaching on the lake. The bases of the cypresses are islands for zenobia, titi, sweet pepper-bush, red bay, and red maple. Evidently, the Carolina Bays in the Georgia coastal plain are deeper and younger geologically (based on Buell's theory) than those of South Carolina. Wharton (1978) described three Carolina Bays from Georgia that contained pond cypress and aquatic communities; none contained the pond cypress-swamp tupelo community.

Bozeman and Darrell (1975) listed the pond cypress-swamp tupelo community as one of the dominant associations of rivers that originate in the coastal plain (strongly acidic flood plains flooded 6 months or more). Woodwell (1956) also described this community from the flood plains of the coastal plain of North and South Carolina. Because most of the length of coastal river systems in the Sea Island Coastal Region are under the influence of tidal pressure, the pond cypress-swamp floodplain community is discussed in the section on palustrine tidal forested wetlands.

e. **Evergreen Shrub Communities** (pocosins). Commonly called "pocosins," evergreen shrub communities may be found on poorly drained interstream flats, in depressions, in Carolina Bays, and in the "bays" of ridge and bay topography. Two phases of the evergreen shrub community have been described: the tall (high) shrub zone and the short (low) shrub zone (Penfound 1952, Woodwell 1956, Porcher 1966, Radford 1976). Some authors have called this community the pond pine-evergreen shrub community, due to the presence of an open canopy of pond pines (Woodwell 1956). However, most authors use the terminology "evergreen shrub bog" or "pocosin," both of which will be used here.

Penfound (1952) called the pocosin a "peaty fresh water swamp" and listed the dominants of these "evergreen shrub swamps" as sweet gallberry, titi, and zenobia. Woodwell (1956) noted that pocosins occurred on certain upland flats in North and South Carolina. Woodwell listed pond pine as the dominant tree (of 56 study sites), with three geographically distinct shrub unions occurring in the understory: 1) the titi union, dominant in North Carolina pocosins; 2) the fetter-bush union, dominant in South Carolina pocosins; and 3) the zenobia union, common in an overlapping area of the two other unions. Sweet gallberry and fetter-bush are common in the titi union; sweet gallberry may be codominant with fetter-bush in the fetter-bush union; and fetter-bush, sweet gallberry, sweet bay, red bay, and titi occur in the zenobia union.

In Georgia, Wharton (1978) pointed out that a "low diversity" shrub bog may form a ring community around cypress ponds (see the section on pond cypress communities in this chapter). This bog is generally composed of a single species-- either titi or black titi (black titi does not occur north of Georgia).

The best examples of the pocosin (evergreen shrub) community in South Carolina and Georgia are found in Carolina Bays. Buell (1946) documented the presence of the evergreen shrub community in a peaty Carolina Bay in North Carolina. Buell noted a "low shrub" community dominated by zenobia (with leather-leaf and sweet pepperbush also common), and a "tall shrub" zone ("simply a later stage. . .of the former") dominated by pond pine, fetter-bush, bamboo brier, zenobia, smooth winterberry, lambkill, male-berry, and Virginia willow.

Porcher (1966) studied nine Carolina Bays in Berkeley County (South Carolina) and listed the following species as common in pocosin communities: 1) low shrub

zone--sweet gallberry, titi, zenobia,
fetter-bush and wax myrtle; and 2) high
shrub zone--pine, loblolly bay, red bay,
sweet bay, red maple, and swamp tupelo,
with bamboo brier growing over the top
of the high shrub layer. Porcher found
peat moss as a ground cover in openings.
Wharton (1978) listed 15 shrubs as co-
dominants of Georgia pocosin communities
of Carolina Bays (pond pine, sweet bay,
red bay, red maple, and loblolly bay
are common trees, with black titi and
titi as the dominant shrubs). Co-
dominants of Wharton's "high diversity
bog" not common in North and South
Carolina pocosins are: rusty lyonia,
odorless wax myrtle, possum haw, and
dwarf laurel. According to Wharton et
al. (1976), the length of the interval
between fires is responsible for low
or high diversity in shrub communities.

The bays or depressions of ridge and
bay topography also contain pocosin or
evergreen shrub vegetation. Woodwell
(1956) included this physiographic type
among his 56 pocosin study areas. Site
specific data are unavailable for these
bays, but the bays of the ridge and bay
areas of the upper Sea Island Coastal
Region (Georgetown County, South
Carolina) probably are dominated by
Woodwell's (1956) zenobia union and his
fetter-bush union, with a scattered
overstory of pond pine (Woodwell 1956,
National Wetlands Inventory 1978).

Wells (1942) pointed out that the
pocosin community always occurs on peat.
In most forested wetlands, the soil surface is considerably lower than the
surrounding uplands; however, in the
pocosin community, the soil surface is
nearly level with that of surrounding
uplands. The depression has become filled
by peat (see Fig. 5-13; note that the
peat is deeper in Carolina Bays than in
poorly drained interstream areas). Wells
(1942, 1946) explained the role of the
peat level in controlling hydroperiod.
During wet seasons (usually winter and
spring), the water table is near the
surface, but during dry seasons (summer
and fall), the surface is very dry and
susceptible to burning. Therefore, fire
is common in pocosins. Recurrent fires
sweep through the evergreen communities,
often burning the shrubs back to ground
level. However, if a deep peat burn does
not occur, the evergreen shrubs return.
(Wells 1946). Pond pine is fire-adapted
in that its pine cones open only after
hot fires; consequently, if pines are
destroyed in pocosin fires, new seeds
are made available to regenerate the
community. Thus, alternate exposure
of plant roots to standing water and
drought, along with the effects of
fire, determine the ecology of the pocosin
community (Wells 1942).

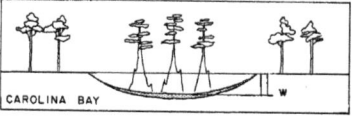

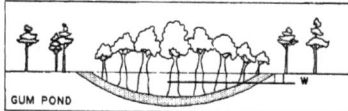

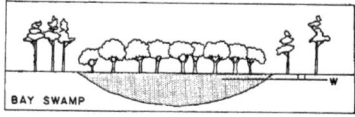

Figure 5-13. A comparison of peat depth
in various forested wetland
communities (stippled portion = peat layer; W = depth
of water table.

f. Bay Forest Communities. Bay
forests or "bayheads" are successionally
related to pocosins (evergreen shrub) and
pond cypress-swamp tupelo communities.
Most researchers think that the bay forest
is a later stage of the pond cypress-swamp
tupelo community (following aggradation
of organic matter), and that bay forests
follow pocosins where fire has been excluded (Wells 1928, Penfound 1952, Monk
1968, Radford 1976). Consequently, the
bay forest is common in the same physiographic types as are the pond cypress-
swamp tupelo and pocosin communities--
poorly drained flats, shallow depressions,
Carolina Bays, and ridge and bay topography. Buell (1946) pointed out the
relationship between the high or tall
shrub stage of the pocosin and the bay
forest (see the preceeding section on
evergreen shrub communities). He listed

331

sweet bay, red maple, highbush blueberry, fetter-bush, Virginia willow, red bay, and bamboo brier as the components of the bay forest. Red bay and sweet bay comprised the dominant bay association given by Penfound (1952). This association was one of the climax phases of his "peaty swamp complex."

Although Porcher (1966) did not use the term "bay forest," some of the high shrub communities he described from Carolina Bays seem to fit the description of the bay forest community. He noted that red bay, loblolly bay, and sweet bay were common in the high shrub zone. Radford (1976) said that the bay forest is dominated by red bay, sweet bay, loblolly bay, swamp tupelo, pond cypress, red maple, and pond pine in the canopy and by titi, hollies, sweet pepperbush, zenobia, fetter-bushes, myrtles, and red chokeberry in the shrub layer. He also pointed out that bay forests are often found along the margins of tall shrub pocosins (Buell 1946, Radford 1976).

In Georgia and farther south, bay forest communities are called "bayheads." Bozeman (1965) described two communities from an Altamaha sand ridge that appear to be related to the bay forest. Another bay forest-like community called a "cypress-bay forest" occurred in ponds on the sand ridge and appeared to be similar to previously described bay forest communities except for the dominance of pond cypress. Monk (1968) pointed out that bayheads in north-central Florida were dominated by sweet, red, and loblolly bays.

g. Bottomland Hardwood Communities. The bottomland hardwood community occurs in the flood plains of major river systems and their tributaries. Although some creeks and small rivers of the study area may be nontidal, most of the bottomland hardwood communities of the Sea Island Coastal Region are associated with tidal rivers and are discussed in the section on palustrine tidal forested wetlands.

h. Bald Cypress-Water Tupelo Communities. The bald cypress and water tupelo community is associated with the deep sloughs of river flood plains, and is discussed in the section on palustrine tidal forested wetlands.

i. Succession. Pine savannahs and pond cypress savannahs are fire-maintained communities that exhibit considerable seasonal or monthly change (see the section on palustrine nontidal emergent wetlands). Pond cypress, swamp tupelo, pond cypress-swamp tupelo, bottomland hardwood, bald cypress-water communities are characterized by normal winter - spring - summer - fall seasonality common in temperate zones. Evergreen shrub (pocosin) and bay forest

communities, on the other hand, are dominated by broad-leaved evergreens and show little or no seasonal fluctuations in appearance (Wells 1932).

The pine savannah is thought to have arisen from the cutting of wet pine savannahs (Wells 1942). If annual fire is present, the community remains a treeless savannah (pine seeds are destroyed by annual fire); however, if fire is excluded, the evergreen shrub community (short shrub phase) becomes dominant (Wells 1942, 1946). If fire is absent for a longer period of time, the tall shrub phase of the evergreen shrub (pocosin) community dominates. Ultimately, the bay forest community will dominate if fire is not present. If the tall shrub phase of the evergreen shrub community is burned, the area reverts to the short shrub phase (after sprouting takes place). A deep peat burn in the short shrub phase often results in pond cypress-swamp tupelo community, which may gradually aggrade into a bay forest (without fire) or become a pond cypress savannah (with fire) (Penfound 1952) (see Fig. 5-14).

In Carolina Bays, the bay lake is gradually filled in by encroachment of a pond cypress-island community (Fig. 5-14). Pond cypress colonizes bare sand along the lake edge during late summer and early fall; gradually, communities of shrubs and small trees (e.g., titi, zenobia, sweet pepperbush, and red maple) establish themselves on the buttressed bases of the cypresses (Buell 1939). As the cypress shrub fringe encroaches upon the lake, a mat of roots is extended into the water (a marsh zone is not present as in northern bog lakes). Material is washed from the fringe and the mat into the lake bottom, gradually filling it in. After the lake becomes peat-filled, the pond cypress is eliminated by the shortening of the hydroperiod, resulting in the dominance of the evergreen shrub (pocosin) community (Fig. 5-14). Succession in the Carolina Bay evergreen shrub community proceeds similarly to succession in a poorly drained interstream flat (see Fig. 5-11).

Monk (1968) viewed the southern mixed hardwood (hydric) community as the climax in north-central Florida wetland communities. The bayhead community was preceded successionally by the pond pine phase of what Monk termed the "flatwood complex," or by cypress swamps. Figure 5-15 illustrates Monk's successional relationships. Succession in bottomland hardwood and bald cypress-water tupelo communities has been discussed under palustrine tidal forested wetlands.

Mitsch and Ewel (1976) pointed out that in pond cypress (cypress domes) communities, tree growth is slowest in still standing water. Flowing water in pond cypress communities brings in

332

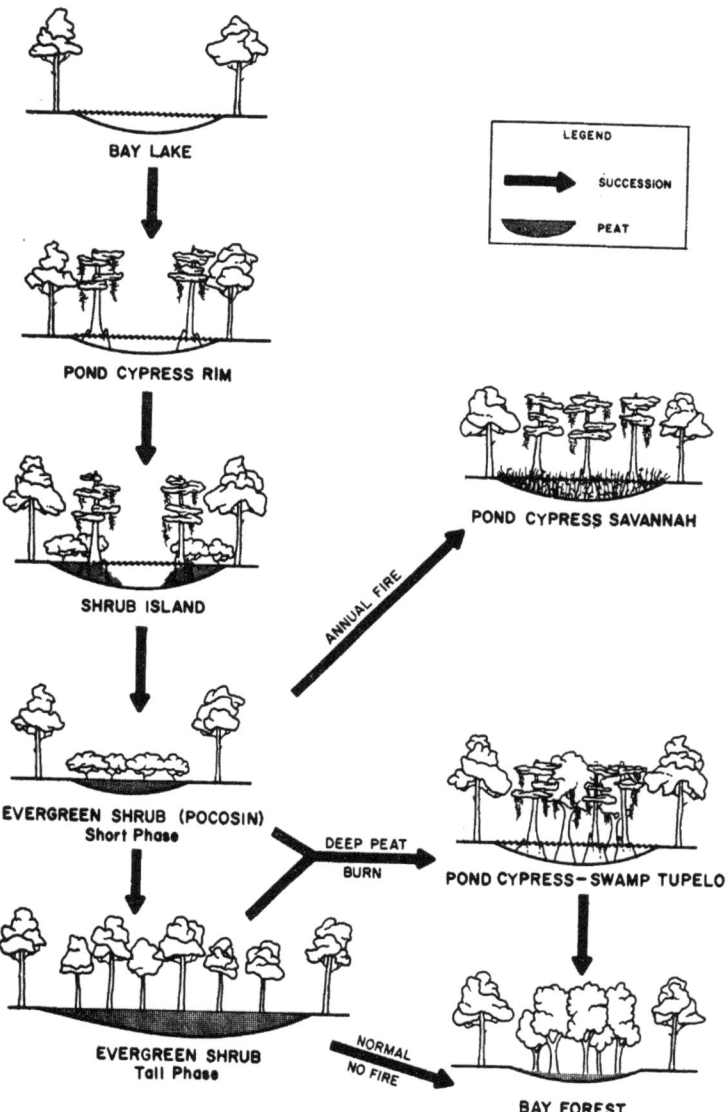

BAY LAKE

LEGEND
→ SUCCESSION
⏝ PEAT

POND CYPRESS RIM

SHRUB ISLAND

POND CYPRESS SAVANNAH

EVERGREEN SHRUB (POCOSIN)
Short Phase

ANNUAL FIRE

DEEP PEAT
BURN

POND CYPRESS—SWAMP TUPELO

EVERGREEN SHRUB
Tall Phase

NORMAL
NO FIRE

BAY FOREST

Figure 5-14. Successional relationships in Carolina Bays.

333

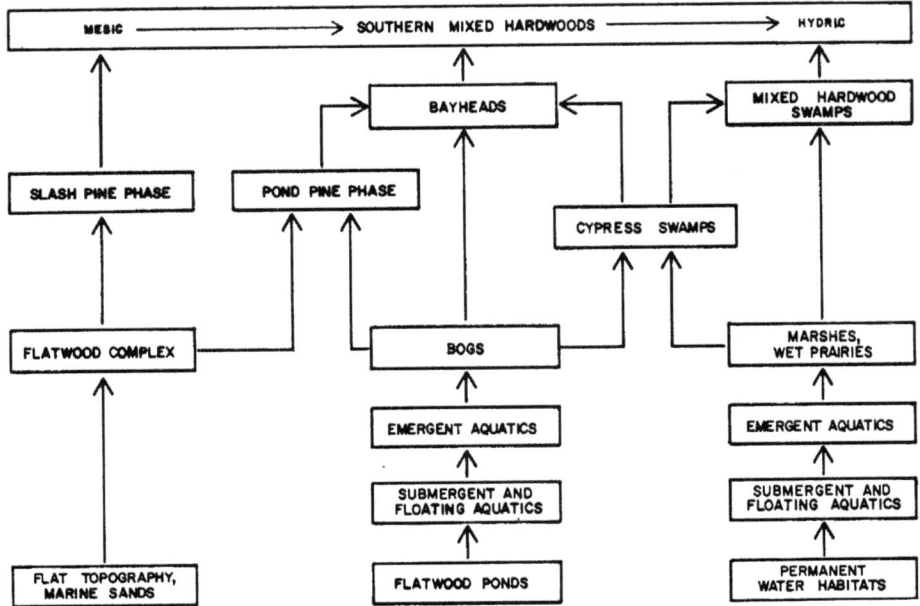

Figure 5-15. Successional relationships in the southern mixed hardwoods complex (adapted from Monk 1968).

nutrients and reduces competition from pines and hardwoods. They further noted that cypress growth is faster in drained cypress domes, but that pines and hardwoods eventually invade the dome and eliminate the cypresses. Finally, Mitsch and Ewel (1976) assert that the floodplain or riverine cypress (bald cypress?) communities are the most productive of all cypress communities.

Woodwell (1958) explained the slow growth of pond pines and evergreen shrubs in pocosins and Carolina Bays by pointing out the presence of gliotoxins in peat. These toxins inhibit mycorhizal fungi, thus slowing tree growth. Furthermore, the high acidity in peat excludes most bacteria that carry out nitrification. Most nitrogen in pocosins is probably atmospherically fixed (Woodwell 1958).

The effects of cutting, fire, and other disturbances have been more carefully studied in the cypress domes of north-central Florida than in other wetland environments. Timber harvesting and fire often result in a community reverting to an earlier successional stage. This is especially true in the evergreen

shrub (pocosin)-bay forest sequence; if a bay forest is burned, an evergreen shrub community follows. The drainage of palustrine nontidal forested wetlands has become a common practice in poorly drained interstream flats, Carolina Bays, and pine savannahs. Acres of these wetland types have been ditched, drained, and planted in slash and loblolly pines. Large Carolina Bays have been ditched and diked and are now being farmed (National Wetlands Inventory 1978).

C. INVERTEBRATES

1. Zooplankton

Studies of the zooplankton of palustrine habitats in the Sea Island Coastal Region are limited to the work of Turner (1910). Zooplankton of palustrine tidal habitats are expected to closely parallel that of the riverine tidal habitats. (See the section on invertebrates of the riverine ecosystem.) Riverine tidal waters flood into emergent and forested wetland communities of palustrine tidal wetlands, distributing zooplankton from one ecosystem to another. Zooplankton of nontidal palustrine habitats are more closely related to that

334

of lacustrine habitats. (See the section on invertebrates of the lacustrine ecosystem.)

Water pH greatly influences zooplankton communities. Separate faunas are found in acidic (as in lacustrine and palustrine nontidal habitats) and in alkaline (or basic) waters (as in riverine and palustrine tidal habitats). Zooplankton assemblages of temporary ponds are characterized by groups which have very short life cycles and often exhibit desiccation-resistant stages. Such species are generally successful in temporary water bodies until predators become established in them. For example, the fairy shrimp Streptocephalus seali is a very common inhabitant of drainage ditches and other temporary waters but is rarely found in the stable, predator-rich waters of other coastal habitats (Moore 1955). N. A. Chamberlain (1978, College of Charleston, Charleston, South Carolina, pers. comm.) has found this species in two ponds in the Francis Marion National Forest (Berkeley County, South Carolina), and Coker (1938) reported it from a ditch near Society Hill, South Carolina. The freshwater decapod shrimp Palaemonetes paludosus is common in coastal South Carolina (P. A. Sandifer, 1978, South Carolina Marine Resources Division, Charleston, pers. comm.) and Georgia (Wharton 1978). Although larval development in this shrimp is abbreviated, it still exhibits a meroplanktonic larval phase. Certain zooplanktonic copepods are associated with bogs, swamps, and temporarily flooded areas. Robertson (1972), in studies of Oklahoma calanoid copepods, found that the preferred habitat of Osphanticum laboronectum was swampy areas. He also reported Diaptomus salticulinus present in temporary ponds, and D. saskatchewanensis present in swampy areas. Coker (1938) found Cyclops crassicaudi regularly occurred in wagon ruts near Chapel Hill, North Carolina. Insect larvae are also present in palustrine zooplankton but will be treated below.

Turner (1910) listed 4 species of calanoid copepods, 10 cyclopoids, 1 harpacticoid, and 24 species of Cladocera from temporary and permanent ponds, ditches, and "holes fed by creeks" in the vicinity of Augusta, Georgia. While most Entomostraca were collected in submerged vegetation, many of these species would likely be planktonic on occasion. The copepod Cyclops serulatus and the cladoceran Simocephalus serrulatus were the most widely distributed species in Turner's samples. No cladocera bearing "winter eggs" were found in these samples, even though temperatures were on occasion just above zero.

2. Insects

In the Sea Island Coastal Region, various insects of freshwater environments have a number of common life patterns and requirements, regardless of whether they are associated with palustrine, riverine, or lacustrine habitats. Much of the following introductory material, which holds true for all of these environs, is summarized from Pennak (1953), Borror and Delong (1964), and Gosner (1971). Table 5-16 summarizes orders of hydrophilic insects, their basic life history patterns, and whether any species of each order are associated with salt (marine or estuarine) or fresh (riverine, lacustrine, or palustrine) waters, or both.

Eleven of the 30 to 35 orders of insects (depending upon the classification system followed) contain species that are partially or totally limnetic. Among the primitive insects that have no metamorphosis, Collembola (springtails) is the only order in which freshwater species occur. A few of these occur on the surface of ponds and pools.

The Hemiptera (true bugs) is the only paurometabolous (gradual metamorphosis) order containing freshwater forms. In this group, metamorphosis is gradual and the series of immature forms (nymphs) resemble adults except in size, body proportions, and wing development. Many species of Hemiptera occur below the surface in both nymphal and adult stages, while others move about the surface film.

The mayflies (Ephemeroptera), dragonflies and damselflies (Odonata), and stoneflies (Plecoptera) are hemimetabolous insects. The adults are terrestrial, but a series of aquatic nymphs (naiads) occurs, usually possessing accessory gills.

The remaining six aquatic orders are holometabolous, with developmental stages consisting of the egg, active larva, acquiescent pupa, and adult. A few lacewings (Neuroptera), moths and butterflies (Lepidoptera), numerous beetles (Coleoptera), flies, mosquitoes, and midges (Diptera), and all alderflies, dobsonflies, and fishflies (Megaloptera), and caddisflies (Trichoptera) have aquatic larvae. All caddisflies and many dipterans have aquatic pupae. A very few hymenopterans are egg parasites, entering water only long enough to find and oviposit on eggs of their aquatic hosts.

With few exceptions, aquatic insects are found near shorelines, in shallow waters, and where an adequate supply of oxygen exists. Only a few dipterans are

Table 5-16. Summary of hydrophilic insect life history patterns and habitat distribution (adapted from Gosner 1971).

CLASS INSECTA	COMMON NAME	FRESH[a]	SALINE[b]	LIFE HISTORY[c]
Subclass Apterygota				
Order Collembola	Springtails	+		am
Order Thysanura	Bristle tails	shorelines		am
Subclass Pterygota				
Order Plecoptera	Stoneflies	+	-	he
Order Odonata	Dragonflies	+	+	he
Order Ephemeroptera	Mayflies	+	+	he
Order Hemiptera	True bugs	+	+	pa
Order Trichoptera	Caddisflies	+	+	ho
Order Megaloptera	Alderflies	+		ho
Order Neuroptera	Lacewings	+	-	ho
Order Lepidoptera	Butterflies, moths	+	+	ho
Order Diptera	Flies, mosquitoes	+	+	ho
Order Coleoptera	Beetles	+	+	ho
Order Hymenoptera	Wasps	Parasitic		ho

[a]Fresh = riverine, lacustrine, or palustrine waters and associated environs.
[b]Saline = marine or estuarine waters and associated environs.
[c]am = ametabolous; pa = paurometabolous; he = hemimetabolous; ho = holometabolous.

consistently found in deeper lakes and in waters possessing reduced oxygen supplies. The free-moving plankton has evolved only in some of the midge (Diptera) larvae. The majority of insects aquatic in the adult stage are nektonic, as exemplified by the water beetles (Coleoptera) or neustonic and pleustonic as exemplified by the water striders (Hemiptera).

Hillestad et al. (1975) provided a listing of orders and families of insects from freshwater shrub marsh on Cumberland Island, Georgia (Table 5-17). Included in their collections were representatives of the insect orders Orthoptera, Hemiptera, Homoptera, Coleoptera, Diptera, and Hymenoptera. The Orthoptera were dominated by crickets (Gryllidae), followed by short-horned grasshoppers (Acrididae), roaches (Blattidae), and long-horned grasshoppers (Tettigoniidae). These were undoubtedly found in and around aerial portions of emergent vegetation, as these insects are not aquatic in any life stage (Pennak 1953, Borror and Delong 1964). This would also be true of the homopteran leafhoppers (Cicadellidae), froghoppers and spittlebugs (Cercopidae), and the hemipteran assassin bugs (Reduviidae). Toad bugs (Hemiptera: Gelastocoridae) were also present and are generally associated with moist shores of ponds, marshes, and streams (Pennak 1953). The beetles (Coleoptera) were the most

diverse in numbers of families (9), with the ground beetles (Carabidae) being predominant. Dipterans (true flies) and hymenopterans (ants and sawflies) were also common.

3. Benthic Invertebrates

Commonly occurring benthic invertebrates of palustrine ecosystems include sponges, hydrozoans, turbellarians, nematodes, rotifers, bryozoans, oligochaetes, leeches, pelecypods, gastropods, ostracods, harpacticoid and cyclopoid copepods, isopods, amphipods, crayfish, numerous aquatic insects, several species of adult bugs and beetles, and arachnids such as water mites and fisher spiders.

Invertebrates of temperate ponds undergo pronounced seasonal patterns of activity and faunal density. Activity is minimal during the winter, and the fauna is variously adapted for survival of cold water temperatures. Several types of "resting stages" are known, particularly in the lower invertebrate phyla. Gemmules, consisting of a mass of cells protected by a hard inner membrane and an outer layer of columnar spicules, are formed by sponges during autumn. Certain hydroids and entoprocts lose their hydranths and calyces, respectively, during late autumn or winter and enter a dormant stage. Species in a number of taxa, including the Hydrozoa, Turbellaria, and Rotifera, produce "winter eggs" which survive the cold. Freshwater

Table 5-17. Numbers of orders and families of insects collected from freshwater shrub marsh on Cumberland Island, Georgia (adapted from Hillestad et al. 1975).

COLLECTION DATA (1973)

INSECTS	June 14	June 26	July 3	July 10	July 17	July 24	July 31	Aug. 7	Aug. 14	Aug. 22	Aug. 27	TOTAL
ORTHOPTERA												
Acrididae	-	1	-	2	-	-	5	-	7	-	-	15
Gryllidae	7	2	25	18	30	17	15	18	-	18	13	163
Tettigoniidae	1	1	-	-	-	-	-	-	-	-	-	1
Blattidae	1	-	-	-	-	2	-	-	-	-	1	3
HEMIPTERA AND HOMOPTERA												
Cicadellidae	1	-	-	-	-	-	-	-	-	-	-	1
Reduviidae	-	-	-	-	-	-	-	1	-	-	-	1
Cercopidae	-	1	-	1	1	2	-	1	-	2	-	7
Gelastocoridae	-	1	-	-	2	1	-	1	-	-	-	5
COLEOPTERA												
Histeridae	4	-	-	-	-	-	-	1	1	-	-	6
Carabidae	18	4	9	13	12	11	13	4	8	1	3	96
Curculionidae	4	-	-	-	1	-	1	2	2	-	-	11
Elateridae	2	-	1	2	2	3	2	1	-	-	-	12
Scarabaeidae	-	-	1	1	1	2	3	1	2	2	-	13
Staphylinidae	-	1	1	-	-	-	-	1	1	1	-	5
Hydrophilidae	-	1	-	1	-	-	-	-	1	-	-	5
Tenebrionidae	-	-	-	-	-	-	-	2	-	-	-	2
Cicindelidae	-	-	-	-	-	-	-	1	-	-	-	1
DIPTERA	1	-	8	5	-	4	3	2	6	-	5	34
HYMENOPTERA												
Formicidae	1	-	5	7	7	15	12	6	12	14	8	87
Pompilidae	-	1	-	-	-	1	-	1	-	-	-	3
TOTAL	39	10	49	51	56	58	54	43	39	38	30	467

bryozoans form dormant statoblasts, consisting of a cell mass covered by chitinous valves. Produced in enormous numbers during summer and autumn, these statoblasts provide an important means of dispersal as well as being tolerant of both low temperatures and desiccation. With the return of favorable environmental conditions, each of these "resting stages" opens and begins development. Other species, including some annelids, mollusks, and arthropods, burrow into the sediments and hibernate. A few, such as certain oligochaetes, mollusks, insect larvae, isopods, and amphipods, may remain active all winter. As temperatures rise in late winter and early spring, faunal activity increases markedly and reproductive cycles commence for many species. Insects such as mayflies, caddisflies, and dragonflies, which constitute an important part of the palustrine benthos as larvae, metamorphose into adults and leave the water. In contrast, several other insects, including water bugs (Notonectidae, Corixidae, Belostomatidae, Naucoridae) and water beetles (Haliplidae, Dytiscidae, Noteridae, Gyrinidae, Hydrophilidae), are common pond inhabitants as adults.

Water levels frequently drop in palustrine environments during summer, and some periodically dry up. Despite this, such ponds support a community of benthic invertebrates made up of both temporary and permanent residents. Many temporary residents are insect larvae whose development is sufficiently rapid to ensure metamorphosis prior to the time when such ponds normally go dry. Permanent residents either burrow in sediments and aestivate, or survive as cysts or other stages adapted to withstand drying. Species numbers are usually lower in these habitats than in "permanent" ponds, and are also reduced in ponds having low levels of dissolved oxygen. Nevertheless, species adapted to such conditions are often present in large numbers.

Among palustrine areas of the Sea Island Coastal Region are lakes of uncertain origin known as "Carolina Bays." Most of these have become filled with deposits or have been drained (Hutchinson 1957), but a few still constitute lakes. They are typically shallow, usually acidic (pH of 4 - 5), and have bottoms of peat and sand (Yount 1966). Although productivity is low, the fish fauna is often quite extensive (Frey 1951). The benthic invertebrates of several Carolina Bays have been examined by Frey (1948, 1949). In Lake Waccamaw, North Carolina, an average density of 579 organisms/m^2 was found. Of these, the most abundant taxa included mollusks (208/m^2), annelids (179/m^2), and insects (160/m^2). Fewer invertebrates were found in the more acidic lakes, and mollusks in particular were more scarce.

Dragonflies (Odonata) and other invertebrates of a 1-ha (2.5 acre) farm pond at the Savannah River Plant near Aiken, South Carolina, were studied by Cross (1955), Benke (1972, 1976), and Benke and Benke (1975). Benke (1976) observed a density of 10,000 larval midges (Chironomidae)/m^2 in Ekman grab samples taken from May to September. Biomass of larval midges and mayflies (largely Caenis sp.) amounted to 0.6 g/m^2 (dry weight), or about two-thirds of the total macrobenthos other than dragonflies. The remaining third consisted mostly of beetles (Coleoptera) and horseflies and deer flies (Tabanidae), although caddisflies (Trichoptera), biting midges (Ceratopogonidae), damselflies (Zygoptera), predacious bugs (Hemiptera), and various microcrustaceans were also present. Biomass of the dominant larval dragonflies (Ladona deplanata, Eiptheca spp., and Celithemis fasciata) during May-September was about 6 g/m^2, while total odonate biomass was estimated at 8 g/m^2. Such high predator-to-prey ratios were possible because of high turnover rates in prey populations. Sufficient refuges were also believed by Benke (1976) to be responsible for preventing the annihilation of prey populations. Since high populations of larval dragonflies have been observed elsewhere on the Savannah River Plant and in a lagoon near Athens, Georgia, Benke and Benke (1975) suggested that they may be a predominant feature of pond ecosystems in the Southeastern United States.

The food chain in a managed fish pond in Georgia was included in Odum (1971) from Welch (1967). The number of food chains in such ponds was intentionally reduced to increase production of desired species of fish. Growth of vascular plants is discouraged and producers are restricted to phytoplankton, which are preyed upon by zooplanktonic crustaceans. Detritus from plankton is fed on by benthic invertebrates, including large numbers of larval midges (chironomids), which are food for sunfish. Sunfish, in turn, are preyed upon by bass, a desirable game fish.

The ecology and community structure of benthic invertebrates in palustrine ecosystems of the Sea Island Coastal Region have generally been ignored and constitute a major data gap.

D. FISHES

Few in-depth studies concerning freshwater fishes have been conducted in the coastal counties of Georgia and South Carolina. Early faunal reports (Jordan 1877, 1878, 1885, Welsh 1916, Fowler 1935, 1945) have proved useful, but must be used with some caution due to possible misidentifications and the uncertain

taxonomic status of several species at
that time. Loyacano (1975) and Dahlberg
and Scott (1971) provided much needed
information on species distribution in
their reports on freshwater fishes of
South Carolina and Georgia, respectively.
These works were primarily based on a review of earlier collections such as those
at the University of Georgia, Cornell
University, and Tulane University. Inas-
much as most of the early collections were
inland collections, a significant data
gap exists for the coastal zone of Georgia
and South Carolina.

Recent freshwater fisheries research
in the Sea Island Coastal Region has
dealt primarily with the commercially or
recreationally important anadromous species
(Vaughn 1967, Rees 1968, Street 1969,
1970, White 1969, 1970, 1972, Wade 1971).

Numerous other unpublished, unsummarized
project reports containing freshwater fish
data are available from each State. Fund-
ing for these projects came from the
Anadromous Commercial Fisheries Act 89-304
and D-J funds (Dingell-Johnson) from
the Federal Aid in Fish and Wildlife
Restoration Acts. The Georgia reports are
available from the Georgia Department of
Natural Resources, Game and Fish Division,
and the South Carolina reports are available from the South Carolina Wildlife
and Marine Resources Department, Division
of Wildlife and Freshwater Fisheries.
(See Directory of Information Sources for
complete addresses.)

Anderson (1964a, b) and Seehorn (1975)
provide more recent information on species
abundance and distribution for the
Combahee and Ashepoo rivers and the Santee
and Cooper rivers, respectively. Anderson
based his reports solely on his own
collections, while Seehorn's report was
divided evenly between actual collections
and review of reference collections.
Another source of distributional information
is the unpublished collection data on incidental catches made in conjunction with
various programs conducted by the Game
and Freshwater Fish Division of the Georgia
Department of Natural Resources and the
South Carolina Wildlife and Marine Resources
Department (C. W. Hall, 1978, Georgia
Department of Natural Resources, Richmond
Hill, pers. comm.; M. G. White, 1978,
South Carolina Wildlife and Marine Resources
Department, Bonneau, pers. comm.).

Wharton (1978) provides an attempt at
classifying, defining, and describing the
natural environments of Georgia. While
this work provides a comprehensive review
of the composition of various ecosystems,
it contains significant data gaps as far
as distribution and life history of the
ichthyofauna. Trophic and reproduction
information is provided on major species,
but the basic species distribution information is taken from Dahlberg and Scott
(1971), which does not effectively delineate

the range of the species within a given
watershed. This is a basic data gap
in the research of these areas.

Poole (1978) provides a limited
literature and collection review of the
fishes of the coastal zone of South
Carolina, and gives one of the first
attempts at generating a checklist for
coastal zone freshwater fishes in the
State. Limited collection data are provided on most species. This list should
be viewed as an initial start in achieving
a comprehensive checklist of coastal
freshwater fishes.

Morrow's (1972) attempt to quantify
Georgia's wildlife and available habitat
generated data on aquatic acreage of
impoundments and streams within the coastal
plain, along with the carrying capacity,
dominant species information, and harvest
statistics for the major species. This
information should be considered a general
guideline and should be used with caution.

Eleven river drainages are considered
in this characterization (Table 5-18).
Many smaller drainages occur within the
study area, but all possess a salinity
>0.5°/oo during times of low flow; con-
sequently, they are discussed in the
estuarine system. The Georgia drainages
discussed in this section are the St.
Marys, Satilla, Altamaha, Ogeechee, and
Savannah. The South Carolina river
systems considered here are the Combahee,
Ashepoo, Edisto, Cooper (including Lake
Moultrie and the lower portion of Lake
Marion), Santee, and Pee Dee, which
includes the Black River. Ponds, impoundments, and other bodies of water
within the study area which drain into
each system are included as part of that
system.

The ichthyofaunal distribution among
these drainage systems is presented in
Table 5-18. The table includes both
documented species and undocumented
species, i.e., species which are believed
to be present, though no record is
currently known. A total of 104 species
of freshwater fishes, representing 18
families, has been reported for the 11
major watersheds of the Sea Island Coastal
Region (Table 5-18). Anadromous,
catadromous, and predominantly marine
species which range into fresh water, such
as striped bass, blueback herring, American
eel, hogchoker, and Atlantic needlefish,
were not included here as these species
are discussed in the sections dealing
with their primary habitats.

The observed species diversity among
these systems may be more a reflection of
the concentration of research towards some
of the larger rivers - e.g., Altamaha,
Ogeechee, Savannah, Edisto, and Cooper -
than a true indication of actual diversity.
The Savannah River drainage shows the
highest diversity, with 75 documented species

339

Table 5-18. Distribution for 104 species of freshwater fishes of the Sea Island Coastal Region by river drainage. (x = Documented, O = Probable Occurrence)

Species	St. Marys	Satilla	Altamaha	Ogeechee	Savannah	Combahee	Ashepoo	Edisto	Cooper	Santee	Pee Dee
Lepisosteidae											
Longnose gar	x	x	x	x	x	x	x	x	x	x	x
Florida gar	o	o	x	x	x						
Amiidae											
Bowfin	x	x	x	x	x	x	x	x	x	x	x
Clupeidae											
Gizzard shad			x	x	x	x	x	x	x	x	x
Threadfin shad			x	o	x	o	o	x	x	x	x
Umbridae											
Eastern mudminnow	x	x	o	x	x	x	x	x	x	x	x
Esocidae											
Redfin pickerel	x	x	x	x	x	x	x	x	x	x	x
Chain pickerel	x	x	x	x	x	x	x	x	x	x	x
Cyprinidae											
Goldfish	o	o	x	o	x	x	o	x	x	x	x
Carp	o	o	x	x	x	x	x	x	x	x	x
Cypress minn ♥	x		x		x			x			
Silvery minnow	x	x	x	x	x						
Bluehead chub					x						
Highfin shiner					x	x		x			
Rosefin shiner						x					
Ocmulgee shiner	x	x	x	x	x	x	x	x	x	x	x
Ironcolor shiner	x	x	x	x	x	x		x	x		
Greenhead shiner	o	o	x	x	x	x	o	x	x	o	
Dusky shiner	o	x	x	x	x	x	x	x	x	x	x
Pugnose minn ♥	x	x	x	x	x	x	x	x	x	x	x
Spottail shiner	x	x	x	x	x		x	x	x	x	x
Sailfin shiner	o	x	x	x	o			x			
？e shiner	o	o	o	x	x	x	x	x	x	x	x
Taillight shiner	x	x	x	x	x	x	x	x	x	x	x
Whitefin shiner	x	x	x							x	x
Coastal shiner				x	x	x	x	x	x	x	x
Swallowtail shiner				o	x	o	o	x	o	x	o
Yellowfin shiner		x	x	x	x	o	o	o	o	o	o
Golden shiner	x		x	x	o	x	x	x	o	o	o
Fathead minnow				o	o	x	x	x	x	x	x
Catostomidae											
River carpsucker			x		x						

340

Table 5-18. Continued

Species	St. Marys	Satilla	Altamaha	Ogeechee	Savannah	Combahee	Ashepoo	Edisto	Cooper	Santee	Pee Dee
Catostomidae (Continued)											
Highfin carpsucker			x		o						
Creek chubsucker	x		x	x	x	x	x	x	x	x	x
Lake chubsucker	x	x	x	x	x	x	x	x	x	x	x
Spotted sucker		x	x	x	x				x	x	
Silver redhorse		x	x					x			
Black jumprock								x			
Shorthead redhorse					x						
Suckermouth redhorse	o				o	x	o	x	o	o	o
Ictaluridae											
Snail bullhead	o	o	x	x	x	x	x	x	x	x	x
White catfish	x	x	x	x	x	x			x	x	x
B uel catfish							x	x	x	x	x
Yell w bullhead	x	x	x	x	x	x	x	x	x	x	x
Br wn bullhead	x	x	x	x	x	x	o	x	x	x	x
Flat bullhead			x	x	x	x	o	x	x	x	x
Channel catfish	x	x	x	x	x	x	x	x	x	x	x
Black madtom							x				
Tadpole madtom	o	o	x	x	x	x	x	x	x	x	x
Margined madtom							x	x	x	x	x
Speckled madtom	x	x	x	x	x	x		x			
Flathead catfish			x	x	x				x	x	
Amblyopsidae											
Swampfish	o	o	x	x	x	x	x	x	x	x	x
Aphredoderidae											
Pirate perch	x	x	x	x	x	x	x	x	x	x	x
Cyprinodontidae											
Golden topminnow	x	o	x	x	x	x	x	x	x	x	x
Banded topminnow	x	x		x							
Marsh killifish				x							
Lined topminnow											
Spotfin killifish				x							
Starhead topminnow	x	x	x	x	x	x	x	x	x	x	
Pygmy killifish	x	x	x	x					x	x	
Bluefin killifish	o	o	x	x	o				x		
Rainwater killifish						x		x	x	x	x
Poeciliidae											
Mosquitofish	x	x	x	x	x	x	x	x	x	x	x
Least killifish	x	x	x	x	x	x	x	x	x	x	x
Sailfin molly	x	x	x	x	x	x	x	x			

341

Table 5-18. Continued

Species	St. Marys	Satilla	Altamaha	Ogeechee	Savannah	Combahee	Ashepoo	Edisto	Cooper	Santee	Pee Dee
Atherinidae											
Brook silverside			x		x				x	x	x
Percichthyidae											
White perch									x	x	x
White bass			x		x				x	x	x
Centrarchidae											
Mud sunfish	o	x	x	x	x	x	x	x	x	x	x
Flier	x	x	x	x	x	x	x	x	x	x	x
Everglades pygmy sunfish	x	x	x	x	x	x	x	x			o
Okefenokee pygmy sunfish	x	x	x	o	o	o	x	x		x	x
Banded pygmy sunfish	x	x	x	x	x	x	x	x	x	x	x
Blackbanded sunfish	o	o	o	o	x	x	x	x	x	x	x
Bluespotted sunfish	x	x	x	x	x	x	x	x	x	x	x
Banded sunfish	x	x	x	x	x	x	x	x	x	x	x
Redbreast sunfish		x	x	o	x	x	x	x	x	x	x
Green sunfish		o	o	o	x	x	x	x	x	x	o
Pumpkinseed		o	x	o	x	x	x	x	x	x	x
Warmouth	x	x	x	x	x	x	x	x	x	x	x
Bluegill	x	x	x	x	x	x	x	x	x	x	x
Dollar sunfish		x	x	x	x	x	x	x	x	x	x
Longear sunfish		o	o	o	x	x	x	x	x	x	x
Redear sunfish	x	x	x	x	x	x	x	x	x	x	x
Spotted sunfish	x	x	x	x	x	x	o	x	x	x	x
Largemouth bass	x	x	x	x	x	x	x	x	x	x	x
White crappie							o		x	x	x
Black crappie	x	x	x	x	x	x	x	x	x	x	x
Percidae											
Savannah darter	x				x	x	x	x	x	x	x
Swamp darter		x	x	x	x	x	x	x			
Christmas darter			x	x	x	x		x			
Niangua darter					x	x	x	x			
Tessellated darter			x	x	x	x	x	x	x	x	x
Sawcheek darter			o		x	x	x	x	o	o	o
Glassy darter								x			
Banded darter					x			x			
Yellow perch			x	x	x	x	o	x	x	x	x
Logperch											
Piedmont darter			o	o	o	x		x	x	x	x
Blackbanded darter			o	x	o	x		x	o		o
Shield darter			o	o	o	o	x				x

Table 5-18. Concluded

Species	St. Marys	Satilla	Altamaha	Ogeechee	Savannah	Combahee	Ashepoo	Edisto	Cooper	Santee	Pee Dee
Eleotridae											
Fat sleeper							o	o	x	x	
Spinycheek sleeper							o	o	x	x	
Gobiidae											
Freshwater goby				x							

343

and nine other species of probable occurrence.

Thirty-one species of fishes representing 11 families are common to all of the drainages in the coastal zone of Georgia and South Carolina. Nine other species representing two additional families are listed as probably occurring in each of these systems. Thus, approximately 38% of the species noted for the study area are found throughout it.

Habitat utilization for 82 fish species belonging to 16 families found within the study area is given in Table 5-19. Each species is referred to as being common, frequent, occasional, or rare. When considering habitat distribution, it should be remembered that fish are extremely motile animals and as such may occur in any habitat provided no natural or artificial barriers are present to prevent movement. Thirteen species are listed as frequent, occasional, or rare, indicating they are not abundant in any freshwater habitat of the Sea Island Coastal Region. Sixteen species occur throughout the coastal zone as common or frequent components of the ichthyofauna in all major freshwater habitats.

Food habits of fishes usually vary not only with season but also with changing habitat. Holder (1973), in studies in the Suwannee River, Georgia, found that diet of the bowfin would change from piscivorous during periods of low water when they were confined within the main river and streams to a diet consisting almost entirely of crayfish during periods of high water when they foraged on the flood plain and in the swamps. Wilbur (1969) studied the redear sunfish in two similar lakes in Florida. He found that while the overall types of food (gastropods, crustaceans, and aquatic insect larvae) were similar, species composition varied significantly. He stated that seasonal changes in the redear sunfish's diet were directly related to the food organisms available. Bass and Hitt (1974) arrived at the similar conclusion that redbreast sunfish tend to feed on any type food item of correct size that is present.

Considering the variability of the diets of fishes, it would be erroneous to apply food habit studies conducted in one area to other areas except in general terms. Food habits of every fish species in the various ecological communities are potentially different, and specific studies would be required for each.

Food habits of 37 species and/or genera of fishes which occur in the Sea Island Coastal Region are presented in Table 5-20. This information is primarily drawn from studies outside the study area and from unpublished observations. Trophic analysis represents a gap in the ecology of coastal freshwater fishes. The importance of zooplankton as a primary food during the early development of fishes is shown in Table 5-20. Zooplankton is consumed by every species and is a major forage item of 22 species and/or genera. Benthic invertebrates and insects are utilized by 35 species and/or genera, and serve as a major item in the diet of 27 and 34 species and/or genera, respectively. Fishes serve as food to 22 species and/or genera of fishes and act as a significant forage item to 14 of these. Organic detritus, vascular plants, and/or algae are found in the diet of 19 species and/or genera and serve as a major forage to seven of these. Phytoplankton is utilized by the lowest number of fishes, with only six of the species and/or genera regularly feeding on it.

As with all life, there exists an intricate web of interactions between fishes and other life around them. Other than man, piscivorous birds are probably the dominant predators on freshwater fishes in the coastal zone. Such forms as grebes, anhingas, loons, egrets, herons, gulls, terns, eagles, ospreys, and vultures are common fish predators. Reptiles and amphibians such as alligators, turtles, water snakes, frogs, and salamanders also feed regularly on freshwater fishes. Among the mammals which consume fish (foxes, bobcats, raccoons, opossums, minks, and otters), the otter utilizes fish in its diet to a higher degree than the rest. Fish also figure into the diet of many aquatic invertebrates such as the giant waterbug, leeches, and crayfish. Comprehensive studies of these interactions within the study area do not exist and must be drawn from species accounts.

1. Impoundments

Most palustrine impoundments in the Sea Island Coastal Region are former rice fields with dikes in varying stages of erosion, allowing free exchange of water and ichthyofauna with the various other subsystems such as palustrine emergent wetlands. The small portion of former rice fields that have maintained dikes are managed primarily for waterfowl. Shallow water impoundments managed for waterfowl average 30 - 45 cm (12 - 18 in) in depth and are dry at varying intervals, some annually, others no more than once every 10 years (R. J. Rhodes, 1978, South Carolina Marine Resources Division, Charleston, pers. comm.). Centrarchids are by far the dominant fish family occurring here. Redfin and chain pickerel, bowfin, largemouth bass, carp, longnose gar, mosquitofish, golden shiner, bullheads, gizzard shad, and threadfin shad are the most prominent species of this impoundment type (Table 5-19).

The vast majority of fish species inhabiting palustrine impoundments are nest building spawners. Carp, redfin pickerel,

Table 5-19. Habitat utilization by 82 species of freshwater fishes inhabiting the coastal counties of Georgia and South Carolina. (C-Common, F-Frequent, O-Occasional, R-Rare)

| Species | Lacustrine | | Forested Wetlands | Palustrine | | Riverine |
	Littoral	Limnetic		Emergent Wetlands	Impoundments	Open Water
Lepisosteidae						
Longnose gar	F	C	R	O	C	C
Florida gar	O	F	R	O	O	F
Amiidae						
Bowfin	C	O	C	C	C	F
Clupeidae						
Gizzard shad	F	C	F	O	C	C
Threadfin shad	O	C	O	O	C	C
Umbridae						
Eastern mudminnow	C	R	C	C	C	R
Esocidae						
Redfin pickerel	C	R	C	O	R	R
Chain pickerel	C	O	C	C	C	O
Cyprinidae						
Goldfish	C	O	C	C	C	F
Carp	C	O	C	C	C	C
Cypress minnow	O	C	C	C	O	O
Silvery minnow	C	C	O	O	O	O
Rosyface chub	R	R	F	F	R	F
Bluehead chub	O	O	C	C	O	C
Golden shiner	C	C	C	C	C	C
Ironcolor shiner	O	O	C	C	F	O
Dusky shiner	C	O	C	O	O	O
Pugnose minnow	F	F	F	F	F	F
Ohoopee shiner	R	R	R	R	R	F
Taillight shiner	F	R	R	R	R	F
Whitefin shiner	C	F	O	O	O	O
Coastal shiner	C	F	F	F	O	C
Fathead minnow	C	C	F	F	F	C
Catostomidae						
Creek chubsucker	C	C	O	C	C	C
Lake chubsucker	C	C	O	C	C	C
Spotted sucker	F	F	C	C	F	C
Black jumprock	O	O	F	F	O	F
Suckermouth redhorse	O	O	F	F	O	F
Ictaluridae						
Snail bullhead	F	O	O	O	F	C

Table 5-19. Continued

Species	Lacustrine		Palustrine			Riverine
	Littoral	Limnetic	Forested Wetlands	Emergent Wetlands	Impoundments	Open Water
Ictaluridae (Continued)						
White catfish	C	C	C	C	C	C
Blue catfish	C	C	O	F	F	C
Yellow bullhead	C	C	F	C	C	C
Brown bullhead	C	C	F	C	C	F
Flat bullhead	C	C	F	F	C	C
Channel catfish	C	C	F	F	C	C
Black madtom	O	R	C	C	C	O
Tadpole madtom	O	R	C	C	O	O
Margined madtom	O	R	C	C	O	O
Speckled madtom	O	R	C	C	O	O
Flathead catfish	O	O	F	F	R.	C
Amblyopsidae						
Swampfish	F	R	F	F	F	R
Aphredoderidae						
Pirate perch	O	R	C	C	C	R
Cyprinodontidae						
Golden topminnow	F	O	C	C	F	R
Banded topminnow	O	R	F	F	O	F
Marsh killifish	O	O	O	O	F	R
Starhead topminnow	C	O	O	O	C	F
Spotfin killifish	O	O	C	C	F	R
Pygmy killifish	F	R	O	O	F	R
Bluefin killifish	F	R	R	F	F	R
Rainwater killifish	C	R	F	C	C	R
Poeciliidae						
Mosquitofish	C	C	C	C	C	F
Least killifish	C	O	C	C	F	O
Sailfin molly	F	O	C	C	C	O
Atherinidae						
Brook silverside	C	F	F	C	C	F
Percichthyidae						
White perch	C	C	O	O	C	C
White bass	C	C	O	O	C	C
Centrarchidae						
Mud sunfish	C	O	C	C	C	R
Flier	C	F	C	C	C	O
Everglades pygmy sunfish	C	O	F	C	C	R

346

Table 5-19. Concluded

| Species | Lacustrine | | Forested Wetlands | Palustrine | | Riverine |
	Littoral	Limnetic		Emergent Wetlands	Impoundments	Open Water
Centrarchidae (Continued)						
Okefenokee pygmy sunfish	C	O	F	C	C	R
Banded pygmy sunfish	C	O	F	C	C	R
Blackbanded sunfish	F	R	F	C	F	R
Bluespotted sunfish	F	O	C	C	F	R
Banded sunfish	C	R	F	C	F	R
Redbreast sunfish	F	O	C	C	F	C
Green sunfish	C	F	F	C	C	O
Pumpkinseed	C	O	F	C	C	O
Warmouth	C	F	C	C	C	F
Bluegill	C	F	C	C	C	F
Dollar sunfish	C	F	C	C	C	F
Longear sunfish	C	F	C	C	C	F
Redear sunfish	C	C	F	C	C	F
Spotted sunfish	C	F	C	C	C	C
Largemouth bass	C	C	C	C	C	F
White crappie	C	C	C	C	C	C
Black crappie	C	F	C	C	F	C
Percidae						
Swamp darter	O	O	C	C	O	O
Tessellated darter	O	O	C	F	F	O
Sawcheek darter	F	O	C	F	F	O
Yellow perch	O	F	C	C	F	C
Logperch	O	R	C	O	O	O
Blackbanded darter	F	R	C	O	O	R

Table 5-20. Major food items of some freshwater fishes common to the coastal counties of Georgia and South Carolina (compiled from Lagler 1956, Buntz 1966, Carlander 1969, Olmsted and Kilambi 1970, Holder 1971, 1973, McSwain 1971, Tucker 1972, Johnson and McSwain 1973, 1974, Pasch 1973, Sandow 1973, 1974, Ager 1975, M. G. White, 1978, South Carolina Wildlife and Marine Resources Department, Bonneau, pers. comm.). (o = Primary, x = Secondary Source)

Species	Organic Detritus	Vascular Plants	Algae	Phyto-plankton	Zoo-plankton	Benthic Invertebrates	Insects	Fishes
Lepisosteidae								
Lepisosteus spp.					x	x	x	o
Amiidae								
Bowfin					x	o	o	o
Clupeidae								
Dorosoma spp.	x				o	x		
Umbridae								
Eastern mudminnow					o	o	o	
Esocidae								
Esox spp.					x	o	o	
Cyprinidae								
Carp	o		x		o	x	o	
Golden shiner			o		o	o	o	
Notropis spp.	o		o	o	o	o	o	
Fathead minnow			o	o	o			
Catostomidae								
Creek chubsucker	x				x	o	o	
Lake chubsucker	x				x	o	o	
Spotted sucker	x				x	o	o	
Ictaluridae								
White catfish	x	x	x	x	x	o	o	
Yell w bullhead		o	o	x	x	o	o	
Brown bullhead	x	o	o		x	o	o	
Channel catfish		x			x	x	o	
Noturus spp.					x	o	o	
Amblyopsidae								
Swampfish					o	o	o	
Aphredoderidae								
Pirate perch					o	o	o	
Cyprinodontidae								
Fundulus spp.					o	o	o	

348

Table 5-20. Concluded

Species	Organic Detritus	Vascular Plants	Algae	Phyto-plankton	Zoo-plankton	Benthic Invertebrates	Insects	Fishes
Poeciliidae								
Mosquitofish	x				x	x	o	
Sailfin molly		o	x		x		o	
Atherinidae								
Brook silverside					o	o		
Percichthyidae								
White perch					x	x	o	o
White bass					x	x	o	o
Centrarchidae								
Mud sunfish		x	x		x	o	o	x
Okefenokee pygmy sunfish		x	x		x	o	o	
Blackbanded sunfish		x	x		x	o	o	
Redbreast sunfish			x		x	o	o	x
Warmouth					x	o	o	o
Bluegill		x	x		x	o	o	x
Redear sunfish		x	x		x	o	o	x
Largemouth bass					x	o	o	o
Pomoxis spp.					x	x	o	o
Percidae								
Etheostoma spp.					o	o	o	
Yellow perch					o	o	o	
Percina spp.					o	o		x

349

and longnose gar are among the few exceptions. The sunfish family is especially successful in reproducing in this habitat and is susceptible to over-population (Swingle 1950, Lagler 1956). Most species prefer to nest near or among submerged vegetation or obstructions, though spawning will occur throughout the habitat.

During periods of drawdown in former rice fields, or during low water in small impoundments, piscivorous birds, mammals, reptiles, and amphibians concentrate in large numbers around these canals and pools. Vultures, foxes, bobcats, and opossums regularly feed on dead fish washed ashore during periods when die-off occurs.

2. Emergent Wetlands

The shallow water edges along river banks where various aquatic plants abound provide a wide array of niches and food organisms for the ichthyofauna utilizing this area. Because of the diversity of niches and food present in this community, coupled with the immediate access to the deep water riverine habitat, this area has the highest diversity of freshwater fishes common to the coastal zone (see Table 5-19). A total of 52 of 82 genera and/or species listed in Table 5-19 are commonly found in this community, with another 14 genera and/or species frequently occupying this area.

Former ricefields with dikes which allow free exchange with contiguous riverine habitat are present on most of the river systems in the Sea Island Coastal Region. Some of the best examples of these can be found on the Cooper and Altamaha rivers. This habitat has ichthyofauna very similar to that of the lacustrine littoral environment, though it tends to be somewhat richer in species diversity. This diversity is linked in part to its intermediate position between the riverine habitat and the palustrine forested wetlands environment. Members of the sunfish, catfish, and minnow families, along with gars, bowfins, suckers, and pickerels dominate this highly productive area.

Most of the available fish information of freshwater environments is not clearly differentiated by habitat. This is especially true for emergent wetlands. Thus, while Anderson (1964a) and Bayless (1968) present physical decriptions of their collection sites, most life history and related studies [e.g., Buntz (1966), McSwain (1971), Holder (1973), Bass and Hitt (1974), Sandow et al. (1974), and Germann et al. (1975)] provide little information as to specific habitat descriptions of collection sites. For example, Sandow et al. (1974) define the collection area for redbreast as ". . . mainstream of the Satilla River from

river mile 206 near Pearson, Georgia, downstream to river mile 52. . ." Thus, there is no way to know if all of the specimens occurred within the riverine system or if most specimens are taken from palustrine emergent wetlands. This problem results from research being conducted on a river system basis, rather than on an ecological community basis.

Some of the most frequently utilized food organisms of fishes in this habitat are insects, insect larvae, crustaceans, and small fishes. In Florida, Wilbur (1969) found insects and crustaceans predominant in the diet of redear sunfish, a prominent species of the emergent wetlands. Food habits of Satilla River redbreast sunfish are presented in Table 5-21, and are an example of the general diet of several species of the genus Lepomis, which dominates in this habitat. Chain pickerel; redfin pickerel; bowfin; largemouth bass; longnose gar; black crappie; white catfish; yellow, brown, and flat bullheads; and channel catfish represent the dominant predators which regularly forage on smaller fishes such as shiners, killifish, chubsuckers, small sunfishes, darters, and the brook silverside. These latter forage species, in turn, feed heavily on insects and insect larvae, zooplankton, and crustaceans. The mosquitofish, which abounds in this habitat, feeds almost entirely on mosquito larvae when they are available. The carp, which is common to this area, is the most prominent herbivore. Many fish species such as white bass, white perch, and flathead catfish, along with large individuals of largemouth bass, channel catfish, blue catfish, and carp, though not residing in this shallow water community, will regularly utilize this area as a feeding ground.

As Holder (1970) points out for a similar ecological community on the Suwannee River in Georgia, this is a significant spawning and nursery ground providing stock for the mainstream fauna. As is the case with fish in the lacustrine littoral environment, fishes spawning in this habitat fall into two primary categories, the nest builders and the broadcast spawners. (See the discussion in the lacustrine littoral subsystem on spawning practices for the description of each spawning type.) Bowfin, largemouth bass, redear sunfish, warmouth, bluegill, fliers, and the various bullheads represent some of the dominant nesting species recognized as important to man. Some of the more prominent broadcast spawners utilizing this area are the chain pickerel, redfin pickerel, longnose gar, Florida gar, white bass, and the gizzard and threadfin shads.

The ichthyofauna present in this community interacts with a wide array of predatory reptiles such as American

Table 5-21. Relative occurrence of food items in stomachs of redbreast sunfish collected from the Satilla and Little Satilla rivers, Georgia, 15 July 1971 - 5 January 1972 (adapted from Sandow et al. 1974).

Food Item	Frequency of Occurrence[a]	
	Number of Stomachs	Percentage of Stomachs Containing Food
Decapoda	14	8.00
freshwater shrimp	4	2.29
crayfish	10	5.71
Insecta	119	68.00
Isoptera	1	0.57
Orthoptera	1	0.57
Odonata	16	9.14
Hemiptera	4	2.29
Coleoptera	8	4.57
Tricoptera	7	4.00
Diptera	21	12.00
Hymenoptera	1	0.57
Unidentifiable	75	42.86
Osteichthyes	4	2.29
Notropis sp.	1	0.57
fish scales	3	1.71
Annelida	2	0.57
Vegetationb	46	26.29
Miscellaneous	22	12.57
organic matter & detritus[b]	12	6.86
sandb	12	6.86
Stomachs containing food	175	45.45
Empty stomachs	210	54.55
Total number fish examined	385	100.00

a. Note that data reflects co-occurrences, i.e., stomachs often contained more than one food item.
b. These are considered to be consumed incidental to normal feeding.

alligators, common snapping turtles, and cottonmouths; amphibians such as the bull-frog and northern and southern leopard frogs; and mammals such as fox, bobcat, opossum, raccoon, mink, and river otter. The reptiles and amphibians, along with mink and otter, actively seek out fish as prey, while the opossum, fox, bobcat, and to a lesser extent raccoons, feed primarily on dead fish that wash ashore. Piscivorous birds (gulls, terns, egrets, herons, anhingas, kingfishers, ospreys, and bald eagles), however, are the primary predators of this community.

3. Forested Wetlands

Forested wetland communities of the palustrine system account for a significant proportion of the freshwater habitat within the coastal zone of South Carolina and

Georgia. Two of the more prominent and best known features of this community are the cypress-tupelo swamps (which occur along all 11 drainage systems discussed herein), and the Carolina Bays which are scattered throughout the coastal zone of both States. Wharton (1978) provides the best available information on these areas, though the ecosystem breakdown varies considerably from that used in this study.

This community is commonly utilized by over half of the 82 genera and/or species of fishes listed in Table 5-19, with an additional 24 genera and/or species frequently utilizing the area, indicating a wide diversity of fishes.

This community, although inadequately studied, has received more attention than most of the freshwater communities in the coastal zone. This has probably resulted from the intrigue held for one area of this community, the cypress swamp. The primary work conducted in this area has been inventory or survey type work by Anderson (1964a), Bayless (1968), and Seehorn (1975). Wharton (1978) gives information on some of the major species inhabiting this area. Segments of studies conducted in other areas which are similar to this area can be applied. Such studies conducted on the lower Suwannee River, are especially useful. Holder (1973), Sandow et al. (1974), Germann et al. (1975), and Coomer et al. (1978) provide studies dealing with various life history facets of redbreast sunfish, warmouth, bowfin, and spotted sucker.

Bayless (1968), in his survey of the Edisto River, collected 30 species of fishes in one rotenone sample of a 0.4 acre area (0.2 ha) in Timothy Swamp (Berkeley County, South Carolina) (Table 5-22). A total of 585 specimens, weighing in excess of 41.5 lb (19 kg), was collected, indicating a wide diversity in the ichthyofauna and the preference of smaller sunfish species for this habitat. The largemouth bass, redbreast sunfish, warmouth, bowfin, and redfin pickerel represent the more abundant predators occurring in this community (Table 5-19). This habitat in the Edisto River has virtually become synonymous with recreational angling for redbreast sunfish, spotted sunfish, dollar sunfish, and the flier. Forage fish commonly found in this habitat include the dusky shiner, ironcolor shiner, golden shiner, and creek chubsucker, along with the smaller species of sunfish (Table 5-19).

Fishes residing in forested wetland habitats feed primarily on insects, insect larvae, crustaceans, or other fishes. Holder (1971) found that bowfin in the Suwannee River switched from a heavily piscivorous riverine diet to a crustacean (principally crayfish) diet during times of floodplain inundation. Warmouth collected in the Okefenokee Swamp were found to feed principally on crustacea (primarily crayfish) and insects (odonata being the most abundant order) (Holder 1971). Sandow et al. (1974) found that insects comprised 60% of the diet of redbreast sunfish in the Satilla River. Crustacea and insects play an equally important role in the diets of smaller fishes such as shiners, creek chubsucker, brook silverside, and the pygmy sunfishes. During times of flooding, food organisms of terrestrial origin were found to play an increasingly important role in the diets of these fishes (Woodall et al. 1975). Although fishes do not play a significant role in the diets of warmouth and redbreast, they are a major item in the diets of largemouth bass and redfin pickerel.

Forested swamplands serve as a major spawning and nursery ground for a significant proportion of the fishes inhabiting each river. Holder (1970) and Wyatt and Holder (1969) identified such an area on the Suwannee River as a prime spawning and nursery area. The majority of spawning activity that occurs here is carried out in nests (see lacustrine littoral subsystem) by such species as the bowfin, members of the sunfish family, swampfish, and the pirate perch. Most nesting activity begins in early spring and continues through the end of summer, with decreasing amplitude from late June. Many nesting species such as largemouth bass, redbreast sunfish, and bowfin construct their nests near or among submerged obstructions, while bullheads and madtoms prefer submerged objects like logs and stumps.

The ichthyofauna of this area probably has a higher interaction level with mammals, reptiles, and amphibians than most of the other communities. Such animals as otters, raccoons, water snakes, alligators, snapping turtles, sirens, and bullfrogs actively feed, to varying degrees, on fishes in the area. Herons, egrets, ibises, bitterns, anhingas, and kingfishers are the dominant piscivorous birds utilizing the area.

E. AMPHIBIANS AND REPTILES

1. Tidal Impoundments

Because of their juxtaposition to riverine systems, herptiles inhabiting freshwater tidal impoundments do not differ substantially from those of the river itself. Combinations of abundant food, diverse aquatic vegetation, restricted water flow, and proximity of dry land in impoundments, however, tend to create an ecotonal effect. Thus, certain species such as greater sirens, dwarf waterdogs, and two-toed amphiumas find this habitat more favorable than the adjacent river. These aquatic amphibians are generally found among subtidal vegetation, bottom debris, or roots of floating aquatics (Conant 1975, Harrison 1978). Food items of amphiumas and sirens include crustaceans, mollusks, worms, insects, small fish, etc.(Conant 1975). Freshwater impoundments also provide good habitat for the aquatic rainbow snake and eastern mud snake, which feed primarily on the American eel and eel-like salamanders, respectively (Conant 1975, Wharton 1978).

Several species of turtles exhibit relatively generalized requirements for aquatic habitats and are found regularly in impoundment communities. Such species include the common snapping turtle, eastern mud turtle, and stinkpots (Conant 1975). These species are nocturnal and seldom bask (J. R. Harrison, 1978, College

352

Table 5-22. Fishes taken in a rotenone collection from a 0.4 acre area in Timothy Swamp, Berkeley County, South Carolina (Bayless 1968).

| Species | Length Class | | | Total Wt. (pounds)[b] |
	< 4 inches[a]	4-6 inches	> 6 inches	
Tadpole madtom	40	3	0	0.1
Largemouth bass	1	1	12	10.5
Chain pickerel	6	0	17	3.5
Redfin pickerel	4	0	12	1.5
Warmouth	4	6	24	8.5
Spotted sunfish	1	13	3	2.5
Redbreast sunfish	9	0	10	3.1
Dollar sunfish	3	0	0	Trace
Bluespotted sunfish	18	0	0	Trace
Banded sunfish	4	0	0	Trace
Banded pygmy sunfish	71	0	0	Trace
Mud sunfish	1	0	2	0.2
Bluegill	0	1	0	0.1
Pumpkinseed	0	1	0	Trace
Creek chubsucker	3	0	20	8.3
Longnose gar	0	0	2	0.1
Brook silverside	3	0	0	Trace
Bowfin	0	0	1	0.4
Brown bullhead	1	0	6	1.2
Golden shiner	1	1	2	0.1
Mosquitofish	4	0	0	Trace
Pirate perch	102	0	0	0.3
Swampfish	19	0	0	Trace
Ironcolor shiner	30	0	0	Trace
Dusky shiner	104	0	0	0.1
Eastern mudminnow	2	0	0	Trace
Swallowtail shiner	8	0	0	Trace
American eel	Present			
Pinewoods shiner	6	0	0	Trace
Piedmont darter	1	0	0	Trace

a. 1 in = 2.54 cm
b. 1 lb = 453.59 g

of Charleston, Charleston, South Carolina, pers. comm.). The striped mud turtle is found in extreme southern Georgia and would probably inhabit impoundments found within its range. Other turtles which occupy impoundments include Florida cooters, yellowbelly sliders, eastern chicken turtles, Florida softshells, and Gulf Coast spiny softshells (Conant 1975, Mount 1975, Gibbons 1978). Young Florida cooters are omnivorous, whereas the young of yellowbelly sliders are primarily carnivorous; adults of both species are herbivorous. Eastern chicken turtles, Florida softshells, and Gulf Coast spiny softshells are carnivorous (Mount 1975). Turtles lay eggs above the normal high-water line, with the emydids and trionychids preferring sandy, friable soil, while kinosternids select soil of high organic content for nesting sites (Mount 1975). Levees provide nesting substrates.

Aquatic vegetation in impoundments is relatively abundant with many emergent (pickerelweed, arrowhead, cat-tail, cutgrass), floating (alligator-weed), and floating-leaved plants (water-lily, floating heart). (See the section on palustrine tidal impoundments for additional vegetation information.) These plants often occur in dense stands or mats and, if not too expansive, improve the habitat for many frogs and water snakes. Hylids occurring among emergent or floating vegetation in palustrine impoundments throughout the Sea Island Coastal Region include southern cricket frogs, Cope's gray treefrogs, green treefrogs, and squirrel treefrogs (Conant 1975, Harrison 1978). The Florida cricket frog is found throughout peninsular Florida and along the immediate Georgia coast north to the Savannah River, while the southern cricket frog is found throughout most of the remaining coastal plain of Georgia and all of that of South Carolina (see Conant 1975 for distribution). In the Savannah National Wildlife Refuge, the northern cricket frog has invaded old rice fields along the Savannah River (J. R. Harrison, 1978, College of Charleston, Charleston, South Carolina, pers. comm.).

This species is considered rare in the coastal zone (Harrison 1978).

Floating vegetation and levees provide temporary habitat for ranid frogs. Common inhabitants include bullfrogs, pig frogs, bronze frogs, and southern leopard frogs. Presently within the Sea Island Coastal Region, palustrine impoundments are perhaps the most ideal habitat for pig frogs. It is in this habitat that they are most abundant, and males can be heard calling among floating and herbaceous emergent vegetation day and night during spring and summer (Wright and Wright 1949).

The close proximity of dry land, water, abundant vegetation, and food contribute to make palustrine impoundments prime habitat for aquatic serpents. Redbelly water snakes, banded water snakes, brown water snakes, and cottonmouths are common or even locally abundant (Conant 1975, Gibbons 1978). Floating vegetation near levees also provides good habitat for rough green snakes and eastern ribbon snakes (Conant 1975). The green anole is abundant among vegetation near aquatic environments (Gibbons 1978) and would not be unexpected among emergent aquatics in impoundments. Species uncommon or rare in impoundments are generally uncommon or rare throughout the coastal zone. These species include the Florida green water snake, glossy crayfish snake, and the Carolina and north Florida swamp snakes (Martof 1956, Conant 1975, Gibbons 1978). In extreme southeastern Georgia, adjacent to Florida, the striped crayfish snake is found in impoundment-type habitats, albeit uncommonly (Martof 1956, Conant 1975).

The area immediately north and south of the Savannah River is a transition zone for several subspecies of herptiles common to palustrine impoundments. In addition to the southern cricket frogs already mentioned, subspecies of ribbon snakes, swamp snakes, and cottonmouths differ between most of coastal Georgia and most of coastal South Carolina. The peninsula ribbon snake, north Florida swamp snake, and Florida cottonmouth inhabit most of coastal Georgia, whereas the eastern ribbon snake, Carolina swamp snake, and eastern cottonmouth inhabit most of coastal South Carolina (Conant 1975).

2. Tidal Emergent Wetlands

During times of high tide, this habitat may be occupied by any species of herpetofauna in the adjacent river. The emergent portions of cat-tails, pickerel-weeds, arrowheads, and similar aquatic vegetation are inhabited by several hylid frogs, including southern cricket frogs, green treefrogs, and squirrel treefrogs. Although these frogs are seldom seen unless purposely sought, their presence and numbers are revealed by their mating

calls during the breeding season. These calls are often the major component of night sounds heard along rivers in spring and summer.

More conspicuous, visually, are the true frogs (Ranidae) which are generally larger and less secretive than the hylids. Several species attain a size to merit commerical importance for human food and research. The bullfrog and pig frog are the most common large ranids. Both species are nocturnal and feed on most motile animals of swallowable size. Diets of bullfrogs are known to include other frogs, birds, snakes, turtles, and mice (Mount 1975). Other ranid representatives which are locally common in this habitat include bronze frogs and southern leopard frogs (Conant 1975, Mount 1975, Harrison 1978). In coastal South Carolina, the pickerel frog inhabits grassy areas along river or stream banks, but it is considered rare (Harrison 1978). It has not been reported from coastal Georgia (Martof 1956).

Water snakes and several aquatic turtles (primarily of the genus Chrysemys) utilize emergent palustrine wetlands as feeding sites. Many snakes seek prey species including frogs, amphiumas, or small fish among herbaceous vegetation. The ribbon snake and rough green snake are semiaquatic species frequenting vegetated edges of rivers and streams (Mount 1975).

3. Tidal Forested Wetlands

A considerable number of herpetofaunal species endemic to the coastal plain are found in flood plain forests; however, few are restricted to this habitat. Aquatic salamanders inhabiting backwaters, sloughs, or shallow ponds throughout the Sea Island Coastal Region include the eel-like greater siren, two-toed amphiuma, broken-striped newt, and central newt (Martof 1956, Harrison 1978, Wharton 1978). These species range in abundance from uncommon to locally common throughout most of the Sea Island Coastal Region. A number of ambystomid and plethodontid salamanders are common or even abundant in floodplain forests. Spotted, marbled, Mabee's, mole, southern dusky, southern two-lined, dwarf, and slimy salamanders are commonly found under decaying logs, among damp forest debris, or near margins of aquatic areas (Martof 1956, Conant 1975). The mud salamander is found uncommonly in coastal Georgia and South Carolina and typically inhabits seepage slopes or springs in hardwoods (Martof 1956, Harrison 1978). Another species considered uncommon in coastal South Carolina is the three-lined salamander. This species is an associate of the more common two-lined salamander and is found in river swamps, stream margins, seepage slopes, and springs, particularly in areas near limestone outcrops

(Harrison 1978). Martof (1956) did not report this species as occurring in the lower coastal plain of Georgia. Terrestrial eft stages of broken-striped, central, and striped newts are also found near aquatic areas in floodplain forests (Conant 1975).

A number of anuran species are common in this habitat. Southern toads are ubiquitous throughout the coastal plain of both States (Martof 1956, Harrison 1978), and the eastern narrowmouth toad is associated with a wide variety of aquatic situations (Martof 1956, Conant 1975, Harrison 1978). Among swamps or floodplain forests near water, the following hylid species are commonly found in the Sea Island Coastal Region: southern cricket frog, Cope's gray treefrog, green treefrog, barking treefrog, and squirrel treefrog (Martof 1956, Conant 1975, Harrison 1978). Other inhabitants whose distribution is limited within the study area include the northern cricket frog, upland chorus frog, Brimley's chorus frog, and the eastern bird-voiced treefrog. Northern cricket frogs are rare in coastal South Carolina and are restricted to cool, moist ravines along major rivers (Harrison 1978) (see Chapter Six, Upland Ecosystem). Coastal populations of northern cricket frogs may be a relict from the last glacial period (Harrison 1970). Upland chorus frogs are typical of the Piedmont province, but disjunct populations occur in coastal South Carolina. Schwartz (1957) recorded this species in the counties of Berkeley, Charleston, and Dorchester. Neither the northern cricket frog nor the upland chorus frog has been reported from the coastal region of Georgia.

The eastern bird-voiced treefrog is common in its range, but it is restricted to river swamps along the Savannah River drainage in Georgia and South Carolina (Conant 1975, Harrison 1978). Brimley's chorus frog inhabits low, wet hardwood forests and river swamps throughout coastal South Carolina and a small area of coastal Georgia adjacent to the Savannah River (Martof 1956, Conant 1975, Harrison 1978). Because of their ambulant nature and often changing environmental conditions (rain, drought, day, night, etc.), hylid frogs occur in a number of microhabitats, but they are usually found near water or humid situations. During day or in times of drought, they seek moist places such as under boards, logs, bark of standing dead trees, shallow edges of ponds, etc.

Bullfrogs, bronze frogs, river frogs, and southern leopard frogs are the most common ranids in forested flood plains in coastal Georgia and South Carolina (Martof 1956, Conant 1975, Harrison 1978). These species are more aquatic than most hylids, and frequently are found along banks or vegetated areas of swamps, wet areas, or oxbow lakes. The southern leopard frog is the most terrestrial of the group and, during summer, can be found considerable distances from water (Conant 1975). Swamp stream banks are favorite environs of the river frog (Wharton 1978).

Several turtles common to floodplain forests exhibit relatively generalized habitat preferences and occur in a variety of aquatic environments such as small sloughs, streams, back-water areas, and swamps. The common snapping turtle, eastern mud turtle, stinkpot, Florida cooter, yellowbelly slider, and eastern chicken turtle are considered common or locally common throughout the Sea Island Coastal Region of Georgia and South Carolina (Martof 1956, Schwartz 1956, Conant 1975, Gibbons 1978). In larger and deeper bodies of water (oxbows and back waters, for example), Florida softshells and Gulf coast spiny softshells may also be found. The rare spotted turtle is more frequently associated with small streams and branch swamps than large river flood plains (Wharton 1978). The distribution of softshells and the spotted turtle in the study area is restricted; Conant (1975) provides a range map for each.

Almost any freshwater turtle may be occasionally encountered in terrestrial situations in floodplain forests, particularly near standing water, oxbow lakes, or rivers. Some species are more frequently encountered than others (Gibbons 1978), but all are considered transients. The only species encountered commonly and considered terrestrial is the eastern box turtle, which is partial to more mesic hardwood forests (J. R. Harrison, 1978, College of Charleston, Charleston, South Carolina, pers. comm.). (See also Chapter Six, Upland Ecosystem.)

Snake species generally restricted to dry, sandy areas are the only species which would be unexpected in floodplain forest environments. Most other species indigenous to coastal Georgia or South Carolina occur in this habitat at least occasionally. Characteristic of, but not confined to, non-flowing waters in river flood plains and swamps are the glossy crayfish snake, north Florida swamp snake, Carolina swamp snake, and striped crayfish snake (Wharton 1978). These relatively small snakes are reported to favor floating mats of aquatic vegetation such as those formed by water hyacinth (Conant 1975) or alligator-weed. Perhaps most characteristic of river swamps is the rainbow snake (Wharton 1978). Other species inhabiting non-flowing aquatic habitats include the banded water snake, the red-belly water snake, the uncommon Florida green water snake, and the cottonmouth (Martof 1956, Gibbons 1978); the brown water snake may occur here, but it is more abundant near flowing waters (Gibbons 1978).

A number of small, secretive snakes are found in and among forest-floor debris of forested flood plains. Although these species are not considered aquatic, their habitats are moist and, because of their location, are subject to periodic inundation. Some of these cryptic species are rarely found and none are considered more than locally common (Gibbons 1978). The following species inhabit forested flood plains, but none are restricted solely to this environment: eastern worm snake, southern ringneck snake, redbelly snake, rough earth snake, and eastern earth snake. The scarlet kingsnake and the pine woods snake are also secretive species which occur chiefly under bark or in the interior of decaying pine trees, but may be found in floodplain situations (Martof 1956, Conant 1975).

Active terrestrial snake species such as the southern black racer, corn snake, yellow rat snake, and eastern kingsnake are common in forested flood plains and, together with the water snakes, comprise the most conspicuous components of the herpetofauna. Eastern ribbon snakes, eastern garter snakes, and rough green snakes are also well represented in this habitat.

All crotalids inhabiting coastal Georgia and South Carolina occur on high ground of forested palustrine environments. Most abundant, however, are the canebrake rattlesnake and the southern copperhead. Pigmy rattlesnakes and eastern diamondback rattlesnakes occur less frequently in this habitat type. These species are generally secretive and not as active as the larger colubrids and, therefore, do not form a conspicuous component of the herpetofauna.

The American alligator is prevalent throughout the floodplain forests of the Sea Island Coastal Region. All coastal counties of South Carolina and Georgia have stable, prosperous populations. For a detailed discussion of the alligator, refer to Chapter One (endangered species section) and also to Chapter Four (section on impoundments of the intertidal estuarine subsystem).

In South Carolina, palustrine forested wetlands seem to support smaller alligator populations than palustrine impoundments. In Georgia, however, the higher populations of alligators are not found in the coastal section of the study area but in more inland locations. Population estimates range from 300 animals in Effingham County to 2,000 animals in McIntosh County. Where habitat changes from the estuarine to palustrine habitats, as in Ware, Wayne, Bulloch, and Charleston counties,

the population estimates in 1973 were in the thousands, to a high of 5,000 for Ware County (this county is included in the Okefenokee Swamp) (Joanen 1974).

4. **Nontidal Forested Wetlands**

Nontidal forested wetlands encompass a variety of wet and seasonally wet communities such as cypress ponds, gum ponds, Carolina Bays, bay swamps, shrub bogs, and pocosins. Typically, herpetofauna inhabiting these communities are similar to those of riverine forested flood plains, but during dry seasons, amphibians and reptiles from surrounding pine flatwoods may intrude (Wharton 1978). The edges of these wetlands are ecotones and many typically terrestrial forms may be encountered.

In and near cypress ponds in coastal Georgia, Wharton (1978) considered the following herptiles as dominant: southern dusky salamander, dwarf salamander, southern cricket frog, little grass frog, eastern narrowmouth toad, oak toad, southern black racer, banded water snake, cottonmouth, and southern ringneck snake. Also reported to occur are the many-lined salamander, slimy salamander, carpenter frog, eastern slender grass lizard, eastern glass lizard, and the eastern indigo snake (Wharton 1978).

The edges of cypress ponds appear to be important habitats for several species considered rare or uncommon; earth snakes, scarlet kingsnakes, northern scarlet snakes, pine woods snakes, and the southeastern crowned snake are fossorial or semi-fossorial species frequently found in this transition zone (Wharton 1978).

The presence of wetlands in pine forests is crucial to the maintenance and perpetuation of many amphibian and reptile species. Recent qualitative and quantitative investigations have shown that there exists a dynamic relationship among herpetofauna populations in surrounding pinelands and wetlands. Species composition varies from pond to pond, and the number inhabiting a single pond can be quite impressive. Drift fences, funnel traps, and can traps revealed the presence of nine species of salamanders, nine species of snakes, and three species of turtles in one small gum pond (Wharton 1978).

Surface water is necessary for the reproduction of all amphibians and many reptiles and wetlands in pine flatwoods serve as reproductive sites for many herpetofaunal populations. Utilization of wetlands for reproduction is cyclic and begins with the arrival of breeding adults, followed in sequence by egg deposition, adult departure, egg hatching, larval growth, and finally immature amphibian (or reptile) emigration (Wharton 1978).

Thus, these wetlands serve not only as
nursery areas for amphibians and many
reptiles but also as net exporters to the
surrounding landscape (Jetter and Harris
1976). The role of these wetlands,
however, is reversed by eutrophication.
When treated sewage effluent was released
into a cypress dome in Florida, the
habitat switched from producer to a
"consumer" of amphibians. This was
attributed to the fact that few larvae
survive in anaerobic waters and that increased vegetation resulting from eutrophication enhanced insect populations
which, in turn, attracted anurans. The
overall result was that standing biomass
of anurans was greater, but production
lower, in the cypress dome receiving
treated effluent than in the dome receiving groundwater. Also, anurans were more
concentrated in the eutrophicated cypress
dome and, as a consequence, predation
increased (Jetter and Harris 1976).

Because of their isolated nature,
these wetlands are particularly vulnerable
to a number of perturbations. Fire is a
major agent in modifying and maintaining
herb bogs, shrub bogs, cypress ponds,
gum ponds, bog swamps, and bay swamps.
Burning intervals may be yearly for herb
bogs or as long as 150 years for some
cypress ponds (Wharton 1978). These
burns are regulated by droughts which
also affect herpetofaunal populations.
Wharton (1978) reported observing several
cottonmouths around a drying cypress pond
and suggested that the larger frogs
(Rana spp.) may bury themselves in
moist litter during such times. Bennett
et al. (1970) found that mud turtles
emigrated as far as 600 m (656 yd) from a
Carolina Bay and burrowed into sand or
litter; time on land ranged up to 142
days. Droughts subject the fauna to
stress and increase their vulnerability.
Charred carapaces of striped mud turtles
were found in Florida cypress domes after
severe burns (Jetter and Harris 1976).

F. BIRDS

1. Colonial Wading Bird Rookeries

Supported by an abundance of estuarine
and freshwater swamp habitat, the Sea
Island Coastal Region of South Carolina
and Georgia maintains a high population
of colonial wading birds. Reinforced by
large food supplies, stable water regimes,
and freedom from disturbance, colonial
wading birds continue to thrive and reproduce in nesting colonies scattered
throughout the area.

Although colonial wading bird
rookeries have been known from this area
for well over 100 years, little documentation existed prior to 1975 for comparisons
of present and past populations. However,
recent studies by Odum (1976), Custer and
Osborn (1977), Osborn and Custer (1978),
and Sprunt et al. (1978) have located
and censused approximately 291 colonies
of egrets, herons, and their allies
along the Atlantic coast of the United
States. (See the Atlas for rookery
locations.)

Twelve avian species are commonly
associated with wading bird rookeries,
as indicated in Table 5-23. Of these,
four species are considered dominant,
the white ibis, cattle egret, Louisiana
heron, and snowy egret. All are common
residents that occur most frequently in
rookeries.

There are eight types of colonies
based on the selection of habitat (see
Table 5-24). These can be broadly
classed as upland sites, inland swamps,
estuarine islands, and small ponds.
Upland sites are the least common of
wading bird rookeries, as their use is
largely confined to the great blue
heron and the great egret. This type of
colony is typically small, with less than
150 pairs of birds, and has no standing
water. Nests are usually constructed in
tall pine trees (loblolly pine and slash
pine) with a very dense understory often
composed of myrtle (wax myrtle), cabbage
palmetto, or saw palmetto. (See Chapter
Six for additional information on birds
of upland systems.)

Inland swamp sites can be divided
into two distinct types, natural swamps
and man-made swamps or reserves. Natural
swamp locations are commonly sloughs
where standing water has accumulated.
Nests are often constructed in bald
cypress, black gum, sweet bay, water
tupelo, or willows. Nesting success and
site tenacity are highly variable in
this type and are highly dependent on the
availability of standing water.

Man-made or artificial swamps are
primarily the remnants of early attempts
to cultivate rice in the eighteenth
century. Old rice fields gradually
undergoing succession are sometimes
utilized, if standing water is present
and adequate nesting platforms are provided by water-tolerant trees. Most
commonly, old reserves or water storage
areas provide the rookery sites. These
areas are often large, up to several
hundred acres, and constantly maintain
standing water several feet deep through
a system of dikes and ditches. Long
abandoned, old rice field reserves frequently contain dense old growth stands
of bald cypress, water tupelo, swamp
tupelo, and red maple. The understory is
sparse, limited chiefly to button bush,
sweet bay, and fetter-bush. The
stability of the reserve rookery made it
the most favored site for many wading
birds prior to the advent of spoil islands
in the 1940's. Populations in reserve
colonies were typically large, with

Table 5-23. Colonial wading birds commonly associated with rookeries in the Sea Island Coastal Region (Sprunt and Chamberlain 1949, 1970, Burleigh 1958, Audubon Field Notes 1967 - 1970, Chamberlain 1968, American Birds 1971 - 1977, Shanholtzer 1974b, Forsythe 1978).

DOMINANT		MODERATE		MINOR	
White ibis	C PR	Great egret	C PR	Great blue heron	FC PR
Cattle egret	C PR	Little blue heron	C PR	Yellow-crowned night heron	FC PR
Louisiana heron	C PR	Glossy ibis	FC PR	Green heron	C PR
Snowy egret	C PR	Black-crowned night heron	C PR	Anhinga	C PR

Note: Dominance indicates relative importance of the species as a group in the community. This concept is not based necessarily on taxonomic relationships but rather on numbers, size, and trophic dynamics.

KEY: C - Common, seen in good numbers.
FC - Fairly common, moderate numbers.
PR - Permanent resident, present year around.

Table 5-24. Number of colonial wading bird rookeries by habitat types in the Sea Island Coastal Region (compiled from Odum 1976; T. A. Beckett, 1978, Charleston, South Carolina, pers. comm.; P. W. Wilkinson, 1978, South Carolina Wildlife and Marine Resources Department, Charleston, pers. comm.).

	Upland	Natural Swamps	Reserves	Alligator Holes	Island Sloughs	Artificial Ponds	Dredge Islands	Estuarine Islands
SOUTH CAROLINA	2	6	5	1	1	2	2	5
GEORGIA	8	2	-	3	2	8	-	2

thousands of pairs of herons, egrets, and ibises a common sight. All locally breeding species of colonial wading birds were represented, with the exception of the great blue heron, which is locally distributed.

Small ponds are also utilized by colonial wading species, although in reduced numbers. As a general type, small ponds can be divided into alligator holes, island sloughs, and artificial ponds. Alligator holes are commonly found throughout the lower coastal plain and can be characterized by their small size and presence of a hole 5 - 10 ft (1.5 - 3.0 m) deep created by the alligator in times of drought. Vegetation around these holes is typically composed of willows, wax myrtle, and cabbage palmetto. Alligator holes are often used by such species as the green heron, anhinga, black-crowned night heron, and yellow-crowned night heron, which often form small isolated colonies. For this reason, there are doubtlessly many more colonies of this type than are currently known.

Island sloughs are limited to barrier islands where they are formed between dune ridges. Also known as cat-eye ponds (Hayes et al. 1975), these sloughs are frequently without standing water and are subject to rapid succession. Vegetation may be composed of cat-tails, willows, wax myrtles, popcorn trees, or cabbage palmettos. Island sloughs often support a variety of wading birds in moderate numbers (Chamberlain and Chamberlain 1975). Artificial or man-made ponds include farm ponds, waterfowl impoundments, and diggings of the remnant phosphate industry of the latter portion of the last century. These ponds vary widely in size and shape and in the numbers of birds using them. Vegetation is primarily wax myrtle, willow, cabbage palmetto, and button bush.

Estuarine islands also play an important role in rookery site selection, and these may be classed as natural islands or dredge spoil islands. Natural islands afford isolation and reduced predation, but they are also subject to storm overwash and erosion. Vegetation is often sparse, dominated by smooth cordgrass, black needlerush, saltmeadow cordgrass, seabeach panic grass, and occasionally wax myrtle. Man-made spoil islands are a recent addition to nesting sites selected in estuarine areas. Beginning in the late 1940's, these areas received periodic spoil disposal until they were significantly higher than surrounding marsh islands. As vegetation became established, spoil islands became attractive to colonial wading species. Such islands are utilized highly by wading birds in the early stages of vegetative succession when sea myrtle and wax myrtle are dominant. Utilization continues as

species such as sugarberry and white mulberry dominate, but bird populations decline as these species develop a closed canopy (T. A. Beckett, 1978, Charleston, South Carolina, pers. comm.; E. Cutts, 1978, Charleston, South Carolina, pers. comm.).

A review of rookeries by type is given in Table 5-24. At present, rookeries in Georgia outnumber those known in South Carolina with emphasis on upland sites and artificial ponds. This reflects a larger number of great blue heron rookeries as well as a greater number of small, mixed-species colonies associated with man-made impoundments. In South Carolina, however, the greatest numbers of wading birds are concentrated on spoil islands, natural swamps, and old reserves. Individual colony locations with estimates of species composition and population levels are given in Table 5-25.

The distribution of wading bird colonies in the coastal zone of South Carolina and Georgia is subject to yearly fluctuation. Small colonies are more vulnerable to such factors as disturbance and predation than are large, well-established rookeries that have been active for several years (Buckley and Buckley 1976). In support of this observation, one small colony studied in coastal Georgia suffered a minimum nestling mortality of 50% in four of its five species. Predation resulted in nest destruction, and when no attempt to re-nest was made, the colony was abandoned (Teal 1958a). Weather conditions are also a major cause of colonial fluctuation, particularly when drought reduces the available food supply. The white ibis is extremely sensitive to drought and often responds with massive population shifts (Dusi and Dusi 1968). Tidal overwash can also force population shifts on estuarine islands, particularly when eggs are washed during the critical incubation period (P. J. DeCoursey, 1978, University of South Carolina, Columbia, pers. comm.). White ibis are also known to make dramatic shifts in nesting locations without apparent cause. In 1950, about 1,000 pairs of white ibis deserted a well-known South Carolina rookery that had been occupied continuously for 28 years (Sprunt 1922, Denton et al. 1950). Yearly fluctuations between rookeries are also common, as noted between Drum Island and Pumpkinseed Island in South Carolina (T. A. Beckett, 1978, Charleston, South Carolina, pers. comm; P. J. DeCoursey, 1978, University of South Carolina, Columbia, pers. comm.). A more unusual shift in populations was noted in 1975 when the white ibis from the two above-mentioned rookeries relocated in the Okefenokee Swamp, Georgia, over 200 miles (322 km) in distance (Ogden 1978).

Table 5-25. Wading bird colonies in South Carolina and Georgia (compiled from Chamberlain and Chamberlain 1975; Odom 1976; Custer and Osborn 1977; T. A. Beckett, 1978, Charleston, pers. comm.; L. Blue, 1978, U.S. Fish and Wildlife Service, Patuxent Wildlife Research Center, pers. comm.; L. L. Gaddy, 1978, South Carolina Marine Resources Division, Charleston, pers. Comm.; J. Shuler, 1978, South Carolina Wildlife and Marine Resources Department, Columbia, pers. comm.; P. M. Wilkinson, 1978, South Carolina Wildlife and Marine Resources Department, Charleston, pers. comm.).

LOCATION	COORDINATES	SPECIES FEATURED	ESTIMATED NESTS (ADULTS)	COMMENT
South Carolina 1 Pumpkinseed Island Georgetown County	33°17'0 " 79°12'30"	Little blue heron Cattle egret Great egret Snowy egret Louisiana heron Black-crowned night heron Glossy ibis White ibis	200 (400) 250 (500) 500 (1,000) 450 (900) 1,000 (2,000) 40 (80) 125 (250) 19,500 (39,000)	1978 down due to overwash
South Carolina 2 Cat Island Plantation Georgetown County	33°14'0 " 79°15'30"	Great blue heron	25 (50)	
South Carolina 3 Esterville Plantation Georgetown County	33°15'30" 79°17'00"	Great egret Anhinga	UNKNOWN	
South Carolina 4 Kinloch Plantation Georgetown County	33°12'42" 79°19'32"	Great blue heron Great egret Little blue heron Yellow-crowned night heron Anhinga	200 total nests. Specifics unknown	
South Carolina 5 Blake's Reserve Charleston County	33°08'30" 79°23'0 "	Great blue her a Great egret Snowy egret White ibis Black-crowned night heron Yellow-crowned night heron Anhinga Cattle egret Glossy ibis	200 pairs of GBH and GE. No recent population figures available	
South Carolina 6 Marsh Island Charleston County	32°58'00" 79°37'00"	Great egret Louisiana her a Snowy egret Glossy ibis	26 (52) 398 (796) 2)	Louisiana heron and snowy egret figures combined

360

Table 5-25. Continued

LOCATION	COORDINATES	SPECIES FEATURED	ESTIMATED NESTS (ADULTS)	COMMENT
South Carolina 7 White Banks Charleston County	33°01'00" 79°30'00"	Cattle egret Glossy ibis Little blue heron Louisiana heron Snowy egret	2 (4) 9 (18) 3 (6) 730 (1,460)	Figures for snowy egret and Louisiana heron combined
South Carolina 8 Avendaw Charleston County	33°03'00" 79°33'15"	Snowy egret Great egret Little blue heron Louisiana heron	200 300 UNKNOWN UNKNOWN	Estimates now out of date
South Carolina 9 Penny Dam Charleston County	32°58'50" 79°44'15"	Great egret Snowy egret Little blue heron Anhinga White ibis	UNKNOWN	No recent surveys
South Carolina 10 Richmond Plantation Berkeley County	33°03'0 " 79°58'0 "	Great blue heron	UNKNOWN	No recent surveys
South Carolina 11 Daniel's Island Berkeley County	32°50'15" 79°55'30"	Louisiana heron Little blue heron Black-crowned night heron Yellow-crowned night heron White ibis	UNKNOW	No recent surveys
South Carolina 12 Drum Island Charleston County	32°48'45" 79°55'30"	Great egret Snowy egret Louisiana heron Little blue heron Cattle egret Black-crowned night heron Yellow-crowned night heron Green heron White ibis Glossy ibis	200 (400) 2,000 (4,000) UNKNOWN UNKNOWN 200 (400) 25 (50) UNKNOWN 20,000 (40,000) 100 (200)	Largest rookery in S.C. No recent accurate survey-population figures given are 1974. Cattle egrets were well into the hundreds in 1978.

Table 5-25. Continued

LOCATION	COORDINATES	SPECIES FEATURED	ESTIMATED NESTS (ADULTS)	COMMENT
South Carolina 13 Castle Pinckney Charleston County	33°48'40" 79°55'30"	White ibis	25 (50)	
South Carolina 14 Magnolia Gardens Charleston County	32°52'17" 80°04'17"	Snowy egret Louisiana heron Little blue heron White ibis Black-crowned night heron Cattle egret Anhinga	350 (700) 12 (24) 10 (20) 6 (12) 7 (14) 350 (700) 50 (100)	
South Carolina 15 Kiawah Island Charleston County	32°27'0 " 80°02'30"	Great egret Snowy egret Little blue heron Louisiana heron Green heron Cattle egret Anhinga	7 (14) 50 (100) 5 (10) 50 (100)+ 5 (10)	An additional 28 nests were probably cattle egrets
South Carolina 16 Deveaux Bank Charleston County	32°32'30" 80°10'0 "	Snowy egret Louisiana heron Cattle egret Glossy ibis	350 (700) 400 (800) 12 (24) 2 (4)	
South Carolina 17 St. Helena Beaufort County	32°18'05" 80°38'20"	Great egret Little blue heron Cattle egret Black-crowned night heron Anhinga	15 (30) 4 (8) 25 (50) 4 (8) 3 (6)	
South Carolina 18 Buckfield Plantation Beaufort County	32°42'55" 80°50'40"	White ibis Snowy egret Great egret Cattle egret Little blue heron Black-crowned night heron Anhinga	UNKNOWN	No recent survey

362

Table 5-25. Continued

LOCATION	COORDINATES	SPECIES FEATURED	ESTIMATED NESTS (ADULTS)	COMMENT
South Carolina 19 Pinckney Island Plantation Beaufort County	32°17'0 " 80°47'30"	Great egret Cattle egret Black-crowned night heron Yellow-crowned night heron Anhinga	2 (4) 15 (30) 2 (4) 12 (24) 2 (4)	
South Carolina 20 Daufuskie Island	32°7 '0 "	UNKNOWN	UNKNOWN	Small rookery known to have herons and egrets but no recent survey available
South Carolina 21 Whooping Crane Pond Hilton Head Island Beaufort County	32°14'07" 80°43'11"	White ibis Great egret Cattle egret Snowy egret Louisiana heron Green heron Anhinga	1,000 (2,000) 100 (200) UNKNOWN 500 (1,000) UNKNOWN UNKNOWN 12 (12)	White ibis estimates range to 2,000 pairs; however all nesting reduced during 1978 season
South Carolina 22 Cypress Pond Hilton Head Island Beaufort County	32°14'07" 80°44'17"	White ibis Cattle egret	UNKNOWN	No survey available but activity reduced in 1978 season
South Carolina 23 Taylor Property Hilton Head Island Beaufort County	32°12'30" 80°42'24"	White ibis Cattle egret Great blue heron Yellow-crowned night heron Green heron Anhinga	UNKNOWN	No recent survey
South Carolina 24 Arcadia Plantation Georgetown County	33°23'30" 79°18'55"	UNKNOWN	(+ 100)	No recent survey but large numbers of herons are known to breed
Georgia 1 Crooked River Camden County	32°51'10" 81°31'10"	Great blue heron	15 (30)	Not active 1976
Georgia 2 Cumberland Island West Camden County	30°52'15" 81°27'30"	Great blue heron	16 (32)	

363

Table 5-25. Continued

LOCATION	COORDINATES	SPECIES FEATURED	ESTIMATED NESTS (ADULTS)	COMMENT
Georgia 3 Cumberland Island East Camden County	30°53'01" 81°26'0 "	Little blue heron Cattle egret Snowy egret Great egret Black-crowned night heron White ibis Anhinga	15 (30) 40 (80) 15 (30) 45 (90) 12 (24) 300 (60) 3 (6)	
Georgia 4 Satilla River Camden County	30°58'0 " 81°29'30"	Cattle egret Snowy egret Great egret Louisiana heron Black-crowned night heron Glossy ibis White ibis	(40) 250 (500) 1,000 (2,000) 800 (1,600) 5 (10) 8 (16) 1,500 (3,000)	Largest rookery in Georgia
Georgia 5 Darien McIntosh County	31°20'05" 81°22'15"	Snowy egret Great egret Louisiana heron	25 (50) 25 (50) 5 (10)	Anhinga probably also nests
Georgia 6 North River McIntosh County	31°24'0 " 81°23'30"	Great blue heron Great egret	18 (36) 40 (80)	
Georgia 7 Sapelo Island, South Tanner Trail McIntosh County	31°29'35" 81°15'07"	Great blue heron Great egret Black-crowned night heron Anhinga Green heron	1 (2) 10 (20) 1 (2) 3 (6) 1 (2)	
Georgia 8 Sapelo Island, North Tanner Trail McIntosh County	31°29'45" 81°15'03"	Little blue heron Snowy egret Louisiana heron Black-crowned night heron Anhinga Green heron	30 (60) 3 (6) 5 (10) 20 (40) 1 (2) 1 (2)	
Georgia 9 Sapelo Island, North Duckpond McIntosh County	31°31'0 " 81°14'0 "	Great blue heron Anhinga	16 (32) 4 (8)	Federal survey shows only great egret in 1975 with 32 adults

Table 5-25. Continued

LOCATION	COORDINATES	SPECIES FEATURED	ESTIMATED NESTS (ADULTS)		COMMENT
Georgia 10 Blackbeard Island	31°31'10" 81°11'45"	Little blue heron Cattle egret Snowy egret Great egret Louisiana heron Black-crowned night heron White ibis Anhinga	50 175 125 300 600 40 900 10	(100) (350) (250) (600) (1,200) (80) (1,800) (20)	
Georgia 11 Oldnor Island McIntosh County	31°34'30" 81°12'30"	Great blue heron Great egret	1 4	(2) (8)	Abandoned 1975, not listed 1976
Georgia 12 Wahoo Island McIntosh County	31°36'21" 81°13'17"	Great blue heron	15	(30)	Not listed 1976
Georgia 13 Moss Island, St. Catherines Island Liberty County	31°37'30" 81°12'55"	Great blue heron	6	(12)	
Georgia 14 Harris Neck McIntosh	31°37'20" 81°16'45"	Little blue heron Cattle egret Snowy egret Great egret Louisiana heron Black-crowned night heron Anhinga	5 65 30 200 4 10 6	(10) (130) (60) (400) (8) (20) (12)	White ibis and green heron were present 1975, nesting not verified. Harris Neck goose pond active 1976, no figures
Georgia 15 St. Catherines Island-A Liberty County	31°37'35" 81°10'05"	Little blue heron Cattle egret Snowy egret Great egret Louisiana heron	25 50 40 60 35	(50) (100) (80) (120) (70)	
Georgia 16 St. Catherines Island-B	31°40'05" 81°08'43"	Snowy egret Great egret	12 48	(24) (96)	Not active 1976
Georgia 17 St. Catherines Island-C	31°40'35" 81°08'35"	Great egret	8	(16)	Not active 1976

Table 5-25. Continued

LOCATION	COORDINATES	SPECIES FEATURED	ESTIMATED NESTS (ADULTS)		COMMENT
Georgia 18 Egret Pond-Ossabaw Island Chatham County	31°47'34" 81°07'0 "	Little blue heron Cattle egret Snowy egret Great egret Louisiana heron Black-crowned night heron White ibis Anhinga	6 (25 (25 (175 (4 (20 (30 (20 (	12) [32] 50) [00] 50) [350] 350) [700] 8) [208] 40) [90] 60) [160] 40)	Federal survey (numbers in brackets) greatly different in estimates
Georgia 19 Middle Place-Ossabaw Island Chatham County	31°47'35" 81°07'0 "	Little blue heron Cattle egret Snowy egret Great egret Louisiana heron Black-crowned night heron White ibis	10 (25 (150 (175 (100 (25 (50 (	20) 150) 300) 350) 200) 50) 100)	
Georgia 20 Wassaw Island Chatham County	31°53'10" 80°57'50"	Little blue heron Cattle egret Snowy egret Great egret Louisiana heron Black-crowned night heron	30 (50 (150 (150 (240 (30 (	60) 100) 300) 300) 480) 60)	
Georgia 21 Flora Hammock Chatham County	31°53'45" 81°01'00"	Great blue heron Great egret	35 (100 (	70) 200)	
Georgia 22 Petit Gauke Hammock Chatham County	31°55'00" 81°07'00"	Great blue heron Great egret	25 (25 (	50) 50)	
Georgia 23 Skidaway Island-Lewis Property Chatham County	81°03'00" 31°65'00"	Little blue heron Snowy egret Great egret Black-crowned night heron	3 (50 (75 (2 (	6) 100) 150) 4)	Not listed 1976 survey

366

Table 5-25. Concluded

LOCATION	COORDINATES	SPECIES FEATURED	ESTIMATED NESTS (ADULTS)		COMMENT
Georgia 24					
Skidaway Island–Union Camp	81°03'00"	Snowy egret	7	(14)	Not listed 1976 survey
Chatham County	31°75'00"	Great egret	40	(80)	
		Black-crowned night heron	15	(30)	
Georgia 25					
Little Tybee Island	31°58'25"	Great blue heron	12	(24)	
Chatham County	80°54'45"	Great egret	10	(20)	

In spite of yearly population shifts, several large rookeries have been in continuous use for over a half century. Blake's Reserve, known to be an active rookery since 1823, had a population of five species with 1,125 pairs in 1922, not radically different in size from its present population (Sprunt and Chamberlain 1949; Charleston Museum, 1922, Charleston, South Carolina, unpubl. data).

Variation in rookery populations is closely tied to both species composition and history. At the turn of the century, wading birds were under extreme pressure from plume hunters and egg collectors. Breeding populations were reduced to the point that formerly abundant species such as the great egret and snowy egret were almost extinct (Wayne 1910). By the 1930's, these birds had made a strong comeback, with the little blue heron the most abundant breeder (Sprunt and Chamberlain 1949, Burleigh 1958, Ogden 1978). Beginning in the 1920's, the gradual northern range extension of the white ibis began. This massive population movement was an important influence on the character of present-day rookeries in both South Carolina and Georgia. While Florida was the recognized center for breeding white ibis, this species was known to breed as far north as the Altamaha Swamp region of Georgia in the 1860's (Burleigh 1958, Bent 1962a). By 1922, however, the white ibis was breeding in South Carolina and was increasing in numbers from the original discovery of a few hundred to nearly 3,000 birds in 1947 (Sprunt 1922, Sprunt and Chamberlain 1949). Today, the white ibis is the dominant breeding colonial wading bird in South Carolina and Georgia, with a total population estimated at 65,000 birds (Ogden 1978).

The number of wading birds breeding in the Southeast was also supplemented by the natural introduction of the cattle egret from Africa via South America. Although this species arrived in North America about 1942, the first breeding record for Florida was 1953 (Sprunt 1954). The cattle egret extended its range rapidly, reaching Georgia in 1954 and South Carolina in 1953 (Burleigh 1958, Burton 1970). The first record of the cattle egret breeding in South Carolina was in 1956, when 2 pairs were found on Drum Island in Charleston Harbor (Burton 1970). The present breeding population is estimated at 25,000 individuals for both Georgia and South Carolina, while in Florida the cattle egret population exceeds all other wading birds in the Eastern United States by 70,000 individuals (Custer and Osborn 1977, Ogden 1978).

The cattle egret has not only extended its range north along the Atlantic coast, but has also moved inland to become a dominant breeding wader in the upper coastal plain and Piedmont regions of South Carolina and Georgia (Davis 1960, Post 1970). In inland areas, the cattle egret does not compete with other waders to any significant level since the other species are less common. However, it has been accused of competing successfully against the little blue heron in coastal areas, although there is currently little evidence to support such a belief (Ogden 1978).

One other species, the glossy ibis, is also a recent addition to the wading birds breeding in the coastal area. Arriving in South Carolina in 1947 and Georgia in 1949, the glossy ibis populations have remained well below that of the white ibis and cattle egret (Sprunt and Chamberlain 1949, Burleigh 1958).

The breeding season for wading species in the Sea Island Coastal Region begins in late February and early March when the larger species begin concentrating as a prelude to the actual nesting process. Depending on the severity of winter, great blue herons begin nesting from mid-to-late March. Great egrets follow in late April to early May, as do the other species with the exception of the little blue heron. Nesting of the little blue heron normally occurs in late May or June, resulting in increased competition with the final arrival of the cattle egret in June or July (Sprunt and Chamberlain 1949, Burleigh 1958). Competition in the rookery commonly involes nest site selection and stealing of nesting material.

Although there are several types of rookeries based on habitat, there are only three general types based on composition: upland colonies, mixed species colonies, and night heron colonies. Upland colonies are dominated by the great blue heron and occasionally include nesting great egrets. Mixed species colonies usually contain great blue herons or great egrets which occupy taller nest sites throughout the colony (Burger 1978, Wiese 1978). Smaller species fill out the balance of the colony, with green herons and yellow-crowned night herons occupying the outer edge, if they are present (Sprunt and Chamberlain 1949). Due to the solitary nature of the yellow-crowned and black-crowned night herons, they often nest in small, remote colonies. This is particularly common of the yellow-crowned night heron, which is much less social than the black-crowned night heron (Wayne 1910, Sprunt and Chamberlain 1949, Bent 1963c).

Several species are commonly found in small numbers in association with colonial waders. The most common is the anhinga or snake bird. The anhinga was found in 28% of Georgia's colonies and 46% of South Carolina's colonies, as listed in Table 5-25. Common gallinules are also found in freshwater rookeries, as are clapper rails in estuarine colonies (T. A. Beckett, 1960 - 1977, Charleston, South Carolina, unpubl. data). Common grackles are also commonly associated with wading bird colonies where they nest at the fringe of the colony and occasionally prey on the eggs of un-guarded nests (Sprunt and Chamberlain 1949).

Two of the more unusual species associated with wading bird colonies are the osprey and the great horned owl. The osprey has colonized Blake's Reserve for many years and now boasts a population of approximately 39 pairs (P. M. Wilkinson, 1978, South Carolina Wildlife and Marine Resources Department, Charleston, pers. comm.; T. A. Beckett, 1978, Charleston, South Carolina, pers. comm.). On rare occasions, the great horned owl has also been known to inhabit wading bird colonies, rebuilding abandoned great blue heron or osprey nests (Bent 1963c; T. A. Beckett, 1969, Charleston, South Carolina, unpubl. data).

One of the principal causes of low productivity in wading birds is the loss of eggs and young to predators. Figure 5-16 gives a simplistic view of the trophic relationships commonly associated with wading bird colonies. Avian pre-dators include such raptors as the red-tailed hawk and the barred owl, but the fish crow is commonly the most destructive (Sprunt and Chamberlain 1949; Dusi and Dusi 1968; T. A. Beckett, 1978, Charleston, South Carolina, pers. comm.). Roving in large flocks, fish crows can virtually eliminate a rookery by de-stroying unguarded eggs. Such behavior was responsible for the loss of one South Carolina rookery in the 1950's (Cutts 1955). At the intermediate level, two predators are also members of the wading bird colony. Both the black-crowned night heron and yellow-crowned night heron are known to prey on young of other herons, egrets, and ibis. On Drum Island in South Carolina, the ground beneath the nests of night herons is often strewn with partially digested nestling white ibis and cattle egrets which the young night herons are unable to swallow (T. A. Beckett, 1978, Charleston, South Carolina, pers. comm.). External predators include such familiar animals as raccoons and American alligators, but snakes and even man also play important roles (Teal 1958a, Bent 1963c, Dusi and Dusi 1968). In the re-cent past, local crabbers had to be pre-vented from using young herons and egrets for bait in South Carolina (T. A. Beckett, 1978, Charleston, South Carolina, pers. comm.).

Although predation takes a heavy toll on young wading birds, other factors reduce breeding productivity.

Poor nest site selection and poor nest construction result in the loss of some eggs and young, as does cannibalism in some species such as the cattle egret (Dusi and Dusi 1970). On a much greater scale, site disturbance during the early, critical portion of the breeding season can cause nest desertion and wholesale loss of young (Buckley and Buckley 1976). Introduction of certain environmental pollutants has also caused infertility and eggshell thinning, further contributing to low reproductive success (Ohlendorf et al. 1978).

The future of wading birds in Georgia and South Carolina is generally projected to be good. If the application of chlorinated hydrocarbon pesticides remains under strict control, there should be no reduction of nesting productivity as experienced in the 1950's and 1960's. While coastal (particularly estuarine) colonies are expanding in numbers, in-land sites are undergoing population reductions as freshwater swamp habitat is coming under increased developmental pressure (Ogden 1978). Although the range extension of the cattle egret has masked this problem to a degree, the cattle egret has also expanded at such a rate that it may seriously threaten native species such as the little blue heron and snowy egret through nesting competition (Dusi and Dusi 1970, Ogden 1978).

2. Nonforested Wetlands

The following discussion of birds in palustrine nonforested wetlands includes treatment of emergent wetlands and im-poundments together as one ecological unit. Separation of avifauna occurring in emergent wetlands and impoundments would be purely artificial, justifiable only as a convenience. In reality, they are inseparable and ecological distur-bances which might effect birds in one habitat inevitably affect birds in the other habitat.

The palustrine nonforested wetlands of the upper Santee, Edisto, Combahee, Savannah, Ogeechee, Altamaha, Satilla, and St. Marys rivers of the coastal plain province are ideally suited to the needs of a variety of birds. The subtle transition from brackish water to fresh water produces an abundant natural food supply through a diversity of vegetation (Tomkins 1958, Wharton 1978). The emergent wetland plants, together with those in adjacent natural upland and man-made levees, create habitat and structural foundations for feeding, roosting, and breeding activities of many birds. Peak periods of utiliza-tion of palustrine nonforested wetlands by birds occur during spring and fall migrations.

369

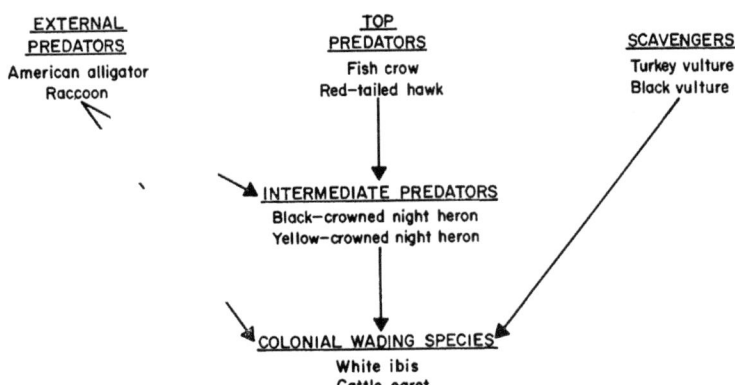

EXTERNAL
PREDATORS

American alligator
Raccoon

TOP
PREDATORS

Fish crow
Red-tailed hawk

SCAVENGERS

Turkey vulture
Black vulture

INTERMEDIATE PREDATORS

Black-crowned night heron
Yellow-crowned night heron

COLONIAL WADING SPECIES

White ibis
Cattle egret

Figure 5-16. Trophic levels associated with colonial wading birds.

Feeding habitats may be quite seasonal, coinciding with shifts in diet. For example, diets of the red-winged blackbird and seaside sparrow shift from a carnivorous diet in spring and summer to a granivorous diet in fall and winter when wild rice seed is readily available. Meanley (1972) studied the importance of wild rice and other freshwater marsh plants to the red-winged blackbird and found that seeds of smartweed, wild rice, millet, and corn formed the bulk of its diet during late summer and fall.

Often, nesting occurs in wetland areas where feeding also occurs; the long-billed marsh wren is a prime example. Other birds, such as herons, assemble in nesting colonies but feed primarily in a variety of locations some distance away (see Lacustrine ecosystem). Both breeding and non-breeding species use the palustrine nonforested wetlands as roosting and/or resting sites. Swallows, marsh wrens, and red-winged blackbirds are examples of plant roosting species, whereas the king rail is a ground roosting species. Racks of dead marsh grass also act as resting sites for shorebirds.

There is an obvious overlap of habitat requirements for many of the birds found in salt, brackish, and freshwater wetlands. Birds of prey such as the marsh hawk, osprey, and bald eagle are frequently observed soaring over estuarine and palustrine emergent wetlands and impoundments. Perching birds such as the red-winged blackbirds, long-billed marsh

wren, sparrows, and grackles, are also found in both kinds of wetlands. On the other hand, some species are more habitat selective. For instance, the boat-tailed grackle is a familiar bird in the estuarine area, but it rarely overlaps with the common grackle, a permanent resident of the coastal plain which nests in colonies near freshwater marshes. Macgillivray's seaside sparrow provides another interesting example of habitat selectivity. This species is a permanent resident of the coastal plain and, because of its prevalence in the salt marshes during fall and winter, Wayne (1910) looked in these areas for nesting birds. However, Wayne's efforts were fruitless. Later, Sprunt (1924) accidentally found this species nesting in a brackish/freshwater marsh area. Since then, the nesting habitats of Macgillivary's seaside sparrow have been well documented in palustrine nonforested wetlands rather than in salt marshes.

Approximately 78 species of birds occur in this habitat (Table 5-26). Of these, only 22 species should be considered as dominant, based on relative abundance and their ecological roles in this habitat. Dominant permanent residents include the belted kingfisher, barn swallow, long-billed marsh wren, great blue heron, great egret, white ibis, yellowthroat, eastern meadowlark, common grackle, and red-winged blackbird. Dominant winter residents include the marsh hawk, American kestrel, eastern phoebe, tree swallow, short-billed marsh wren, Savannah sparrow, and the swamp sparrow.

Table 5-26. Dominant, moderate, and minor bird species of palustrine nonforested wetlands in the Sea Island Coastal Region (Sprunt and Chamberlain 1949, 1970, Audubon Field Notes 1967 - 1970, Chamberlain 1968, American Birds 1971 - 1977, Forsythe 1978).

DOMINANT		MODERATE		MINOR	
Marsh hawk	C WR	Black vulture	C PR	Swallow-tailed kite	U SR
American kestrel	C WR	Turkey vulture	C PR	Mississippi kite	FC SR
Chimney swift	C SR	Sharp-shinned hawk	FC WR	Bald eagle	R PR
Belted kingfisher	C PR	Red-tailed hawk	C PR	Osprey	FC PR
Eastern phoebe	C WR	Red-shouldered hawk	C PR	Merlin	U WR
Tree swallow	C WR	King rail	FC PR	Yellow rail	R WR
Barn swallow	C PR	Field sparrow	C PR	Black rail	R SR
Long-billed marsh wren	C PR	Sora	FC WR	Short-eared owl	FC WR
Short-billed marsh wren	FC WR	Rough-winged swallow	C SR	Bank swallow	U T
Great blue heron	C PR	Purple martin	C SR	Cliff swallow	U T
Great egret	C PR	White-throated sparrow	C WR	House wren	C WR
Green heron	C PR	Song sparrow	C WR	Virginia rail	FC WR
Snowy egret	C PR	Black-crowned night heron	C PR	Purple gallinule	FC SR
White ibis	C PR	Least bittern	FC SR	Common gallinule	FC PR
Louisiana heron	C PR	Glossy ibis	C SR	Yellow-crowned night heron	FC SR
Little blue heron	C PR	Mallard	C WR	American bittern	U SR
Yellowthroat	C PR	Blue-winged teal	C WR	Wood stork	FC PR
Eastern meadowlark	C PR	Green-winged teal	C WR	Black duck	C WR
Red-winged blackbird	C PR	Palm warbler	C WR	Gadwall	U WR
Common grackle	C PR	Bobolink	C T	Pintail	C WR

Table 5-26. Concluded

DOMINANT		MODERATE		MINOR	
Savannah sparrow	C WR	Rusty blackbird	C WR	Shoveler	C WR
Swamp sparrow	C WR	Painted bunting	C SR	Ring-necked duck	C WR
				Greater scaup	C WR
				Lesser scaup	C WR
				Bufflehead	C WR
				Ruddy duck	C WR
				Hooded merganser	C WR
				Red-breasted merganser	C WR
				American coot	C WR
				Grasshopper sparrow	FC WR
				Henslow's sparrow	FC WR
				Le Conte's sparrow	U WR

Note: Dominance indicates relative importance of the species as a group in the community. This concept is not based necessarily on taxonomic relationships but rather on numbers, size, and trophic dynamics.

KEY: C – Common, seen in good numbers
      FC – Fairly common, moderate numbers
      U – Uncommon, small numbers irregularly
      PR – Permanent resident, present year around
      WR – Winter resident
      SR – Summer resident
      T. – Transient

Ecologically, avifauna of the palustrine nonforested wetlands can be divided into seven trophic levels. These are the predators, omnivores, granivores, insectivores, herbivores, piscivores, and scavengers (Fig. 5-17). The marsh hawk and sparrow hawk are the more common birds of prey in this habitat. However, the osprey and bald eagle occupy the highest avian trophic levels in the coastal plain.

While the osprey is fairly common in both South Carolina and Georgia, the bald eagle is reportedly observed more in South Carolina than in Georgia. Recently, there have been few reports of bald eagles nesting in Georgia. One of the last reported active bald eagle nests in Georgia was on St. Catherines Island in 1970 (W. D. Chamberlain, 1978, South Carolina Marine Resources Division, Charleston, unpubl. data). However, Burleigh (1958) reported bald eagles nesting previously on the Georgia coast at St. Marys, Cumberland Island, Blackbeard Island, Darien, Savannah, and Little Tybee Island. Hebard (1941) cites a number of records for the Okefenokee Swamp in Georgia. The large number of impoundments in South Carolina has been suggested as a major factor in the number of nesting eagles in that State.

The bald eagle is a piscivore as well as a raptor, preferring fish as a stable diet item when available, although carrion is also readily taken. The bald eagle also catches some birds, especially waterfowl and American coots, and some mammals. Bald eagles will often force ospreys to drop fish, which then are caught in mid-air by the eagle (Burleigh 1958). In the coastal plain, nests are usually constructed in living pines, mainly slash or loblolly. Large, old trees with big crowns are usually selected. Such trees are seldom less than 70 years old (Chamberlain 1974). Perch trees are apparently a necessary component of the nesting habitat. They may be located as far as one-quarter mile from the nest and, generally, define the nesting territory. The territory size varies from 28 to 112 acres (12 to 47 ha), with an average of 57 acres (24 ha). There are approximately 18 nesting territories in the South Carolina Sea Island Coastal Region (Table 5-27).

A number of factors have contributed to the decline in bald eagle populations (e.g., shooting, electrocution, loss of suitable nesting areas, and severe weather). The greatest single factor at this time, however, seems to be the lowering of reproduction caused by pesticide build-up in the food chain. The effect of such accumulation in bald eagles has caused an almsot complete lack of reproduction in many nests. Key habitat requirements for the bald eagle include suitable nest trees and roost sites, and water areas which can provide adequate supplies of suitable food, mainly fish. During migration, the bald eagle will travel considerable distances from water and is then sometimes seen in the mountains, but at all other times the eagle shows a strong preference for coastal areas or for large inland bodies of water. It does not tolerate intense human activity, hence requiring relatively large areas with little disturbance (Chamberlain 1974).

As shown in Figure 5-17, there is a common ecological bond between the marsh hawk and American kestrel and the typical omnivores, granivores, and insectivores in the palustrine nonforested habitat. The bobolink, or ricebird, a rather abundant granivore in the spring and fall, is a target species for the birds of prey.

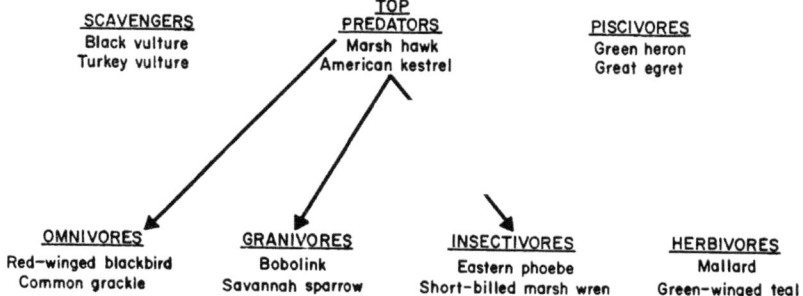

SCAVENGERS
Black vulture
Turkey vulture

TOP PREDATORS
Marsh hawk
American kestrel

PISCIVORES
Green heron
Great egret

OMNIVORES
Red-winged blackbird
Common grackle

GRANIVORES
Bobolink
Savannah sparrow

INSECTIVORES
Eastern phoebe
Short-billed marsh wren

HERBIVORES
Mallard
Green-winged teal

Figure 5-17.   Generalized trophic relationships of representative birds of palustrine nonforested wetlands of the Sea Island Coastal Region.

Table 5-27. Active Southern bald eagle nesting territories in South Carolina during 1978 season[a] (T. M. Murphy, 1979, South Carolina Wildlife and Marine Resources Department, Green Pond, unpubl. data).

Savannah River east of Highway I-95

Hunting Island

Two on Combahee River between Highway 17 and Highway 17-A

Combahee River east of Highway 17

Chehaw River

Two on Ashepoo River east of Highway 17

Dawhoo Creek

Four on Cooper River north of the Tee

Two in Santee Coastal Reserve

South Island

Cat Island

Winyah Bay, east of Highway 17

a. Lake Marion and Lake Wateree may be considered as probable nesting territories outside the Sea Island Coastal Region.

The bobolink has been characterized as a destructive bird in the lower coastal plain, due to its depredations on the rice crops in the mid-1800's. These birds were directly responsible for the loss of millions of dollars. The Eastern phoebe, an insectivore, is also linked to the birds of prey. This species is a flycatcher and consumes mostly insects in its diet. The red-winged blackbird and common grackle (omnivores) are also common components of the palustrine nonforested trophic stuctures.

Waterfowl are well represented in this habitat by dabbling ducks such as the mallard, gadwall, blue-winged teal, green-winged teal, pintail, and wood duck. For these ducks, the freshwater vegetation present in palustrine nonforested wetlands is more important for feeding than that of salt marsh areas (Kerwin and Webb 1972). Of the pochards, the ring-necked duck is more commonly found in freshwater areas. This is probably due to its food preferences, as it feeds on seeds of the water-lily, water-shield, etc. The canvasback is also commonly found in this habitat, where it feeds on vegetable matter.

Closely associated with the waterfowl are the American coot, king rail, yellow rail, Virginia rail, and sora. All of the above species have similarities but also major differences in their diets. According to Horak (1970), the sora, having a short heavy beak, consumes about 73% seeds in its diet. The Virginia rail, with its long, slender decurved beak, eats approximately 62% insects. These differences in food habits demonstrate that avian fauna in this habitat can live together without serious food competition. The king rail occupies a unique niche in this habitat and is considered to be nonspecific with the clapper rail (a saltwater resident), as they both freely interbreed in coastal areas where fresh and salt water mix.

The gallinules, close relatives of the rails and coots, are also well represented in the palustrine nonforested habitat. Both the purple gallinule and common gallinule nest in freshwater marsh of this habitat.

The wading birds, particularly the herons, are quite euryphagous and frequently feed in the palustrine nonforested wetlands on frogs, fish, snakes, field mice, and insects. All three of the common permanent residents (Louisiana heron, great blue heron, and little blue heron) occupy large rookeries in the Sea Island Coastal Region. Although herons and egrets are more commonly found in salt marshes, they do feed along the shorelines and tidally exposed banks of this freshwater habitat. The white ibis is a common summer resident of palustrine wetlands and feeds on crayfish and insects. However, in late fall, the white ibis feeds more in salt marshes (on fiddler crabs) than in palustrine areas. Many of these rookeries, which may also include ibises and night herons, are located near the rice field - marsh - swamp-land complex. (See the lacustrine ecosystem section and the section on colonial wading birds of the palustrine ecosystem.) Here, one needs to consider the indirect relationships between the avifauna and the wetlands. For example, the herons must cycle large amounts of organic matter and nutrients from impounded waters to the marshes, swamps, and land (Shanholtzer 1974b). This enrichment process may be locally significant and may partially account for large standing crops of Southern wild rice, cattails, etc. A similar situation probably exists in the estuarine impoundments and emergent wetlands.

3. Forested Wetlands

The palustrine forested wetlands are perhaps the most ideally suited habitat for birds in the Sea Island Coastal Region. The length of time the forest floor is covered with water controls or determines the species diversity and numbers of individuals found in this habitat. Odum (1969) identified palustrine forested wetlands (river swamps) as fluctuating water level ecosystems. Such an environment, according to Odum, has "pulses" of primary productivity. The river swamp is a hydrologic community, where dramatic fluctuations occur within the framework of the natural sequence of high and low water levels.

Avian productivity in this environment is based on the various substrates for life forms found in this hydrologic stratum. The variability of wet and dry sites, of mesic and hydric forest tree species, and of grasses and closed canopy sites all contributes to the diversity of avifauna in this environment. Approximately 122 species of birds occur in the palustrine forested wetlands of the Sea Island Coastal Region (Table 5-28). This represents the greatest diversity found in any of the natural environments of the study area. About 42 of these species are dominants in this habitat.

Dennis (1966) observed birds in the old growth Congaree flood plain of South Carolina and found that the most abundant breeding birds were, in order of abundance: red-eyed vireo, northern parula, cardinal, and Carolina wren. Other permanent or breeding

residents were the yellow-crowned night heron, wood duck, red-shouldered hawk, chimney swift, American woodcock, barred owl, pileated woodpecker, hairy woodpecker, blue jay, Carolina chickadee, tufted titmouse, Swainson's warbler, common grackle, and rufous-sided towhee. Additional summer or spring residents were green heron, swallow-tailed kite, Mississippi kite, Acadian flycatcher, veery, white-eyed vireo, prothonotary warbler, hooded warbler, and Swainson's warbler. A summer survey of unaltered cypress-gum swamps of the Savannah River (Aiken County, South Carolina) by Briese and Smith (1974) revealed a similar avifauna. Beckett (1975) studied the Santee-Cooper Basin and found 235 species of birds, many of which were similar to those on the Dennis (1966) list. The above listing is generally true for flood plains of the outer coastal plain of the Sea Island Coastal Region, with the addition of herons and egrets as dominant species (Table 5-28).

In considering avian trophic dynamics in this environment, eight major trophic levels can be clearly defined (Fig. 5-18). The predatory birds are best represented by the red-shouldered hawk, barred owl, sharp-shinned hawk, and red-tailed hawk. These birds prey on smaller species of birds as well as on reptiles and amphibians for the major portion of their diet. Waterfowl such as the mallard, black duck, and wood duck overwinter in the river swamp habitat where they can feed on acorns, hickory nuts, and cypress balls, as well as fruits of gum, water elm, holly, dogwood, and some species of insects (Martin et al. 1951, Sprunt and Chamberlain 1970). The larger wading birds such as herons, egrets, and bitterns feed on fish and other small vertebrates, while smaller birds feed primarily on insects. There is an abundance of food for avian species in this environment because the bulk of the life stages of many insects occurs in the river swamps. Also, large aggregations of salamanders, frogs, and other reptiles and amphibians occur during avian breeding seasons.

Endangered and threatened species are intimately associated with coastal swamps and flood plains. (For additional information on endangered species, see Chapter One.) For example, Bachman's warbler, a rare summer resident, appears to be dependent upon swamps for survival (Shuler 1977). Swainson's warbler, a more common species, is closely associated with patches of river cane in coastal flood plains. Formerly, the Carolina parakeet and the ivory-billed woodpecker were inhabitants of this environment. The last of the Carolina parakeets are believed by some ornithologists to have existed in the Santee River Swamp during 1936 - 1938. This species was closely

Table 5-28. Dominant, moderate, and minor bird species of palustrine forested wetlands in the Sea Island Coastal Region (Sprunt and Chamberlain 1949, 1970, Audubon Field Notes 1967 - 1970, Chamberlain 1968, American Birds 1971 - 1977, Forsythe 1978).

DOMINANT		MODERATE		MINOR	
Belted kingfisher	C PR	Chuck-will's-widow	C SR	Red-headed woodpecker	FC PR
Pileated woodpecker	FC PR	Chimney swift	C SR	Fish crow	C PR
Red-bellied woodpecker	C PR	Ruby-throated hummingbird	C SR	White-breasted nuthatch	FC PR
Downy woodpecker	C PR	Yellow-shafted flicker	C PR	Louisiana waterthrush	U T
Great crested flycatcher	C SR	Yellow-bellied sapsucker	FC WR	Fox sparrow	FC WR
Acadian flycatcher	C SR	Hairy woodpecker	FC PR	Gray-cheeked thrush	FC T
Common crow	C PR	Brown creeper	FC WR	Golden-crowned kinglet	FC WR
Carolina chickadee	C PR	Catbird	FC PR	Swainson's warbler	U T
Tufted titmouse	C PR	House wren	C WR	Worm-eating warbler	R T
Carolina wren	C PR	Winter wren	FC WR	Golden-winged warbler	R T
Robin	C WR	Eastern phoebe	C WR	Blue-winged warbler	R T
Yellowthroat	C PR	Ovenbird	C T	Bachman's warbler	R T
Hooded warbler	C SR	Northern waterthrush	C T	Tennessee warbler	R T
Red-winged blackbird	C PR	Kentucky warbler	FC SR	Blackpoll warbler	FC T
Common grackle	C PR	American redstart	C T	Yellow-crowned night heron	FC SR
Cardinal	C PR	Rusty blackbird	C WR	American bittern	U SR
American goldfinch	C WR	Painted bunting	C SR	Black duck	C WR
Rufous-sided towhee	C PR	Purple finch	FC WR	Gadwall	U WR

376

Table 5-28. Continued

DOMINANT		MODERATE		MINOR	
Dark-eyed junco	C WR	Song sparrow	C WR	Pintail	C WR
White-throated sparrow	C WR	Hermit thrush	FC T	Green-winged teal	C WR
Swamp sparrow	C WR	Swainson's thrush	FC T	Shoveler	C WR
Wood thrush	C SR	Veery	FC T	Ring-necked duck	C WR
Blue-gray gnatcatcher	C PR	Cedar waxwing	C WR	Bufflehead	C WR
Ruby-crowned kinglet	C WR	Blue-headed vireo	FC WR	Ruddy duck	C WR
White-eyed vireo	C SR	Black and white warbler	FC T	Red-breasted merganser	C WR
Red-eyed vireo	SR	Yellow warbler	C T	Swallow-tailed kite	FC SR
Prothonotary warbler	FC SR	Black-throated blue warbler	C T	Mississippi kite	FC SR
Northern parula	C SR	Black-throated green warbler	FC SR	Cooper's hawk	U PR
Yellow-rumped warbler	C WR	Pied-billed grebe	C PR	Broad-winged hawk	U WR
Yellow-throated warbler	C PR	Great blue heron	C PR	Bald eagle	R PR
Anhinga	C PR	Little blue heron	C PR	Turkey	U PR
Great egret	C PR	Louisiana heron	C PR	Limpkin	R SR[a]
Green heron	C PR	Wood stork	FC PR	Black-billed cuckoo	U T
Snowy egret	C PR	Glossy ibis	C SR		
Black-crowned night heron	C PR	Mallard	C WR		
White ibis	C PR	Blue-winged teal	C WR		
Wood duck	C PR	Greater scaup	C WR		
Black vulture	C PR	Lesser scaup	C WR		
Red-shouldered hawk	C PR	Hooded merganser	C WR		

Table 5-28. Concluded

DOMINANT		MODERATE		MINOR
Common gallinule	C PR	Turkey vulture	C PR	
Yellow-billed cuckoo	C SR	Sharp-shinned hawk	FC WR	
Barred owl	C PR	Red-tailed hawk	C PR	
		Osprey		
		American woodcock	FC PR	
		Spotted sandpiper	C PR	
		Mourning dove	C PR	
		Screech owl	C PR	

Note: Dominance indicates relative importance of the species as a group in the community. This concept is not based necessarily on taxonomic relationships but rather on numbers, size, and trophic dynamics.

a. Lower Georgia.

KEY:
C – Common, seen in good numbers
FC – Fairly common, moderate numbers
U – Uncommon, small numbers irregularly
R – Rare
PR – Permanent resident, present year around
WR – Winter resident
SR – Summer resident
T – Transient

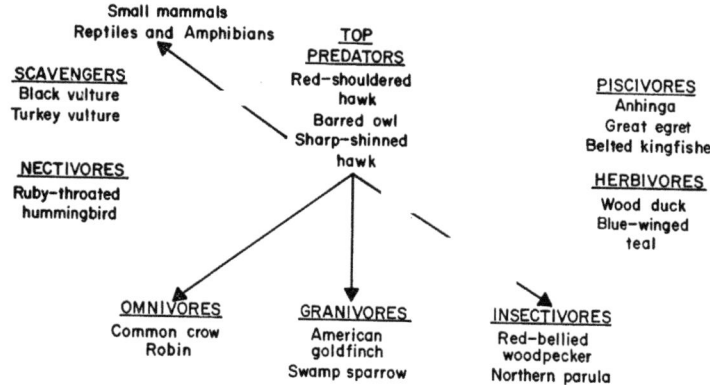

Figure 5-18.   Trophic structure of avifauna in palustrine forested wetlands.

associated with cypress trees (Sprunt and Chamberlain 1970). The ivory-billed woodpecker, on the other hand, required vast acreages of bottomland hardwoods such as found in the coastal flood plains. These birds are thought to have disappeared from the Santee River flood plain with construction of the Santee-Cooper Diversion Project in the late 1930's. This project resulted in a drastic alteration of the flood plain and, combined with development and logging operations, the species vanished. (See Volume I, Chapter Six for a discussion of the Santee-Cooper Diversion and Rediversion projects.) The food habits of the ivory-billed woodpecker were, in part, responsible for its disappearance. It was a specialized feeder, feeding on wood-boring insects such as larval Coleoptera which inhabit dying trees. According to Tanner (1942), the real threat to this species came from the cutting of big trees in virgin forests. This essentially destroyed the bird's food supply.

Feeding habits of woodpeckers occupying the river swamp habitat are quite different. The pileated woodpecker feeds on boring grubs much like the ivory-billed woodpecker. The most important difference between the two birds is that the ivory-billed woodpecker feeds mostly on borers occupying the area just beneath the bark of decaying trees. The pileated woodpecker feeds mostly on borers deep within the sap and heart of dead trees (Tanner 1942). The red-bellied woodpecker occupies a similar niche in the river swamp, but, unlike the pileated woodpecker, it feeds on larval beetles in live trees.

Many secretive species of birds also breed in palustrine forested wetlands to hide their nests in dense foliage or in tree hollows. Colonial nesters, such as the herons, egrets, etc., are far too conspicuous to hide their nests and, therefore, construct their nests over or near water for protection from invaders. Blake's Reserve serves as a good example of such nesting habitat. This area deserves special attention since it has been of great significance in the history of bird survival in this country and continues today as a haven for several endangered and threatened species.

At the turn of the century, several of our most attractive wading birds, including the snowy egret and the great egret, were near extinction because of the heavy mortality due to plume hunters. Wayne (1910) stated that he had not seen a snowy egret in 10 years. This is difficult to imagine considering their abundance today and considering that Wayne was almost continuously in the field and concentrated his efforts on the coast. He also indicated that the great egret was generally absent except in those areas where it was afforded protection. Blake's Reserve, or Washoe as it was formerly known, has a continuous record of being used as a rookery by these two species since 1823. Since it was included in the property of a private hunting club prior to the turn of the century, it was afforded protection at a most critical time.

In addition to egrets, the white ibis, glossy ibis, and great blue heron also nest in the reserve. The history of these two ibises is difficult to determine. The first report of white ibis using this rookery was not provided until 1943, and the first serious suggestion that glossy ibis nested here was in 1947. It is generally agreed that the glossy ibis is a recent addition to the avifauna of this coastal area, but the white ibis is a long-recognized common resident (Wayne 1910). Burton (1970) reported that although the number of white ibis nesting in Blake's Reserve had been estimated in recent years to be as high as 5,000 pairs, they had more recently abandoned that area in favor of Drum Island in the Cooper River. He also indicated that as many as 200 pairs of glossy ibis had recently been recorded as using the rookery of Blake's Reserve. More recent observations on nesting of ibises in the reserve, although unpublished, indicate that Blake's Reserve is also a significant nesting site for the osprey, which at the present time seems to be succeeding quite well in this area. Here in the reserve, the osprey follows its usual pattern of nesting in dead trees, generally dead cypress in this case. The most recent published counts of nests in the reserve were 30 provided by Henry and Noltemeier (1975) and 39 by T. M. Murphy (1978, South Carolina Wildlife and Marine Resources Department, Green Pond, pers. comm.). The osprey does not participate significantly in the trophic relationships of this community except it does, of course, feed its young here. Its fishing tactics are much better suited to open waters which are close by.

One of the more spectacular wading birds in this habitat is the wood stork. This species, the only true stork of regular occurrence in the nation, is a permanent resident in the South Carolina and Georgia Sea Island Coastal Region. Sprunt and Chamberlain (1970) and Burleigh (1958) discuss the occurrence of this species in the freshwater forested wetlands such as the Santee River and Okefenokee swamps, and Hamel (1977) reviewed the occurrence of the wood stork in South Carolina and emphasized the highly volatile and uncertain status of the species. Kahl (1964) stressed the wood stork's dependence upon highly specific conditions of water level for feeding and nesting.

There is a rather distinct movement of wood storks from freshwater forested wetlands to salt marshes in July, where they feed throughout the summer. Feeding habits are interesting in that a number of birds will seek out a pool and, by stirring up mud and water with their feet, bring food to the surface where it is available. The wood stork's diet consists of fish (mostly minnows), frogs, small turtles, snakes, fiddler crabs, and wood rats. Young rails and grackles have also been found in their stomachs (Sprunt and Chamberlain 1970).

Although a variety of ducks utilize swamps and flood plains, the wood duck is without question the most characteristic. This is our only major duck that nests in the Sea Island Coastal Region, and it is a permanent resident. Palustrine forested wetlands are not only major feeding areas, but also important nesting areas. The wood duck is one of the few ducks that does not nest on the ground, and nest boxes have been very successful in helping this species survive the lack of available nesting trees (hollow standing dead trees). Hester and Dermid (1973) found that over 60% of the females nesting in boxes one year nested there again the following year. The food preferences of the wood duck are well known. It feeds almost exclusively on aquatic plant material; duckweed and scales of cypress balls are extensively utilized. A number of the other plants and plant parts are also favored, including arrow-arum, pondweeds, water-lily seeds, wild grapes, water elm seeds, and buttonbush. The animal material consumed by wood ducks seldom exceeds 10% of their diet and in this environment would consist principally of aquatic insects (Sprunt and Chamberlain 1970).

Odum et al. (1976) found 63 species of birds in a cypress groundwater dome during a 2.5 year sampling period in Florida. Aquatic feeders and a large number of uncommon birds (e.g., brown creeper, prothonotary warbler, Cape May warbler, and blackpoll warbler) were frequently sighted in this cypress dome.

Overall, the avifauna in nontidal palustrine areas is not as diverse as in floodplain (tidal palustrine) areas. However, the nontidal palustrine habitat is extremely important to birds of surrounding pine flatwoods.

G. MAMMALS

1. Impoundments

The mammals associated with palustrine impoundments can be considered in two groups. First, and most numerous, are those which utilize the dikes and the emergent or shrub areas of the land-water interface; the second group consists of a few species which enter the water and feed on aquatic prey.

The principal herbivores of this habitat include the marsh rabbit and a variety of small rodents. Marsh rabbits are good swimmers and will not hesitate to enter water; however, their feeding activities are largely confined to grasses and herbaceous plants of the moist edges. Pelton (1975) trapped a number of rodents along the edge of an impoundment on Kiawah Island. These included eastern wood rats, cotton rats, cotton mice, and marsh rice rats. The presence of these small mammals attracts a number of predators, including reptiles and raptorial birds.

The principal omnivores of this habitat are the raccoon and opossum. The raccoon, of course, enters the water to feed on aquatic forms. If the water is fresh, crayfish and frogs constitute significant portions of the raccoon's diet. In estuarine impoundments, fiddler crabs, marsh crabs, and blue crabs assume great importance in the diet. In southeastern Georgia, the nine-banded armadillo should be included along with the opossum and raccoon as a user of the land-water interface. The nine-banded armadillo is largely insectivorous.

Several small carnivorous mammals are common to the emergent impoundment edge environment. These include the eastern mole, the star-nosed mole, and all three native species of shrews: the short-tailed, least, and southeastern shrew.

The principal predatory mammals, and the only ones to feed extensively within impoundment waters, are the mink and the river otter. The predatory habits of both species were studied extensively by Wilson (1954) in eastern North Carolina, and his findings would almost certainly be appropriate to the Sea Island Coastal Region. The mink is a more generalized predator than the river otter, utilizing a wide selection of small mammals, birds, reptiles, amphibians, arthropods, and fishes. The river otter, on the other hand, feeds almost exclusively on fishes and crustaceans in the same general environment.

2. Forested Wetlands

Palustrine forested wetlands include a wide variety of forest types, differing with respect to frequency and consistency with which the forest floor is flooded. Consequently, most terrestrial mammals of the Sea Island Coastal Region will be found at one time or another in forests fitting this broad classification. The principal exceptions would be those ground-dwelling mammals which are ecologically linked to sandy soils or dry grassy areas.

Although small herbivorous mammals are not well represented in palustrine forested wetlands, the eastern wood rat shows an interesting adaptation to this environment. This species is well known for the large nests or dens which are constructed of sticks and twigs. In woodlands that are seldom flooded, these dens are usually built on the forest floor near the base of a large tree. However, the eastern wood rat also constructs arboreal dens, and in regularly or permanently flooded forests, the tree den is the typical form (Lowery 1974). In palustrine forested woodlands where the eastern wood

rat is abundant, it is an important link in the food web. It is entirely herbivorous and constitutes an important source of prey for reptiles and predacious mammals. Because the eastern wood rat is largely nocturnal, it is preyed upon extensively by owls instead of hawks.

Another herbivore that at one time was prevalent throughout both States is the beaver. Although the beaver's population has been decimated, this species is making a comeback in some areas because of protection and reintroduction (Golley 1962, 1966). The beaver occurs primarily in the Georgia portion of the Sea Island Coastal Region. According to Parrish (1960), Golley (1962, 1966), and Hicks (1977), the beaver is found along rivers, streams, and lakes primarily along the Savannah and Altamaha river drainages. In South Carolina, the closest reported population to the coast is also along the Savannah River at the Savannah River Plant (Golley 1966; D. Shipes, 1979, South Carolina Wildlife and Marine Resources Department, Columbia, pers. comm.). However, reports have recently been confirmed of range extensions into the Georgetown County area (P. M. Wilkinson, 1979, South Carolina Wildlife and Marine Resources Department, Charleston, pers. comm.).

Food items for beaver consist of woody and aquatic plants, especially sweet gum, willow, alder, blue beach, dogwood, yellow poplar, maple, water-lily, and corn (Hicks 1977). Some control over localized concentrations is deemed necessary; therefore both States permit the harvesting of beaver. For additional information concerning harvest, see Chapter Eight of Volume II.

The mouse most characteristic of floodplain forests is the cotton mouse. This mouse is not fully herbivorous, as is the case of the eastern wood rat. During summer, insects and small invertebrates constitute a significant portion of its diet. The marsh rabbit is also an important component of the mammalian fauna of palustrine forested wetlands. According to Tomkins (1955), the marsh rabbit is less abundant in floodplain forests than in the brackish-water marshes further downstream. As in other habitats, this rabbit constitutes an important source of prey for larger hawks and owls and may be the principal food source for bobcats in forested wetlands. The gray squirrel is the most abundant arboreal mammal of palustrine forested wetlands and can exist where the forest floor is permanently flooded only if tree cover is dense enough to allow transit from tree to tree. As in other habitats, the gray squirrel is almost entirely herbivorous, existing on a wide variety of nuts, buds, and seeds.

The large herbivore of the swamps is the white-tailed deer. Deer occupy virtually all types of palustrine forested wetlands, as long as islands of high ground are available. They are probably most abundant in floodplain forests where shrubs are available for browse. Deer have no serious predators other than man; it is in the infrequented floodplain forests that they are least susceptible to hunting pressure. Older and larger white-tailed deer would be expected to be found in these environments.

The opossum and raccoon are the two principal omnivorous mammals of forested wetlands of the Sea Island Coastal Region. Both species are abundant in every coastal county. Caldwell (1963) found the raccoon to be more abundant in palustrine forested wetlands than in other habitats in north-central Florida. In the Sea Island Coastal Region, however, raccoon populations are probably most dense in the estuarine environment (see Chapter Four, Estuarine Ecosystem).

The most detailed analysis of raccoon food habits in forested areas in the Southeast are those conducted by Johnson (1970) in Alabama. One of Johnson's study areas (Fred T. Stimson Game Sanctuary) was a coastal plain area that included forested swamps. As expected, a variety of plant and animal materials was consumed. Fruits, berries, and acorns were the principal plant materials. Insects, especially beetles and grasshoppers, were the most commonly encountered animal remains in stomach content analyses. During fall, the raccoons that Johnson studied utilized plant materials almost exclusively. Johnson's food analyses are likely to be appropriate to Georgia and South Carolina forested wetlands, with crayfish, amphibians, and fishes playing a greater role in the raccoon diet in wetter forested wetlands. The opossum is more likely to be found in drier habitats. It seems to rely on insects to a greater extent than does the raccoon.

Palustrine forested wetlands were historically the home of another omnivore, the black bear. These large omnivores were formerly widespread in forested wetlands, but today are restricted to a few large floodplain forests. According to records cited by Golley (1962), as many as 500 bears existed as recently as 1953 in the Okefenokee Swamp. It is unlikely that bears are sufficiently abundant in any of the palustrine forested wetlands of the Sea Island Coastal Region today to have appreciable ecological significance.

The top mammalian carnivore of forested wetlands today is the bobcat. This species still occurs within the larger flood plains in all coastal counties of South Carolina, but its distribution is less clear in Georgia. Although we find no recent records, it is likely that floodplain forests bordering the major rivers of the Georgia coast still harbor

populations of this cat. Certainly, it still exists in the Okefenokee Swamp just west of the Sea Island Coastal Region.

The bobcat is more likely to be found in the larger flood plains and spends most of its time in drier environments. However, the bobcat is not averse to entering the water. Within the floodplain environment, the marsh rabbit can be expected to be the principal item of diet, though rats, mice, and birds may also be consumed. Bobcats will occasionally prey upon young fawns, but this is a relatively infrequent occurrence.

The river otter and mink are smaller but more numerous predators within the palustrine forested wetland habitat and are quite at home in swamps with permanently standing water. Although quantitative estimates are lacking, it is generally believed that mink and river otter are more abundant in estuarine environments than in swamps. Food studies have not been conducted on mink or otter in South Carolina or Georgia swamps, but a North Carolina study by Wilson (1954) did include this habitat. Small fish (principally cyprinodonts) and a variety of small mammals, birds, reptiles, amphibians, and aquatic invertebrates (mostly crayfish) constituted the principal dietary items for mink. In general, the food of mink reflected the faunal makeup of the swamp environment. In contrast, mink depended relatively more on rodents when feeding in the marsh environment. River otter depend to a much higher degree on fishes, whether in swamp or marsh environments.

The gray fox is another common predator but is generally restricted to those portions of the swamp which are flooded irregularly. While in this environment, small mammals, birds, and insects constitute the principal foods. In the lower coastal counties of Georgia, the nine-banded armadillo is an additional predator in irregularly flooded forests. The nine-banded armadillo is largely insectivorous.

Several small fossorial mammal predators, including the southeastern shrew and the eastern mole, are common in swamp margins and in irregularly flooded forests. These animals obviously cannot exist where the ground is completely saturated or flooded.

III. LACUSTRINE ECOSYSTEM

A. VASCULAR FLORA (for nonvascular plants, see the palustrine section)

1. Littoral Subsystem

The lacustrine littoral subsystem is defined as the zone that extends from the shoreward boundary of the lacustrine body to the point at which depths are 2 m

(6.5 ft) or greater, or to the maximum extent of nonpersistent emergents, i.e., if they extend out beyond the 2 m (6.5 ft) depth zone. Some lacustrine water bodies may be completely littoral, not being greater than 2 m deep at any point. Some depressions and sinks in the Sea Island Coastal Region fit this description. These areas must, however, be greater than 20 acres (8 ha) to be called lacustrine.

Most lacustrine bodies have a littoral zone; however, the littoral zone is generally better defined in mill ponds, solution ponds, and farm ponds than in oxbow lakes and rice field reserves. The general ecology of the littoral zone in Southeastern lacustrine systems is complex and has not been dealt with adequately in the literature. The few sources available are concerned primarily with species composition. Ecological succession in the lacustrine system will be treated in the discussion of the limnetic subsystem.

Eyles (1941) described the flora of a series of solution-formed "boggy ponds" located in the Georgia coastal plain just outside of the Sea Island Coastal Region. These ponds are shallow (less than 2 m deep) with peaty bottoms, and range up to 30 acres (12.5 ha) in size. The dominant plants of these ponds were white waterlily and water milfoil in every case. Common plants of the Nymphaea-Myriophyllum community of these ponds are listed in Table 5-29 in order of abundance. Eyles pointed out that big floating heart, yellow-eyed grass, water milfoil, and spikerush are most characteristic of the solution pond, while many other species found in the solution ponds are also found in other types of lacustrine environments.

The zonation of plants in the solution ponds was not commented upon by Eyles, but it was noted that the entire surface of each pond was covered with white waterlily. Based on structural characteristics of the flora described, the ponds probably exhibit some shore-to-center zonation. Of the plants listed in Table 5-29, 13 are nonpersistent emergents, four are floating-leaved (rooted), one is a submergent, and one is a shrub. The nonpersistent emergents probably are more common near the shore, while the floating-leaved (rooted) and submergent plants are probably found in deeper water.

Radford (1976) and South Carolina Wildlife and Marine Resources Department (1974d) described a series of sinks in Berkeley County, South Carolina. Radford (1976) called these depressions "lime" sinks, pointing out that they resulted from limestone solution. South Carolina Wildlife and Marine Resources Department (1974d) implied that some of the larger ponds may be deep pocosins. Radford reported burmannia, hedge hyssop, St. John's-wort, bladderwort, Muhlenberg's

amphicarpum, and sedge from sinks; no submergents or floating-leaved plants were listed. South Carolina Wildlife and Marine Resources Department (1974d) listed big floating heart, water-hoarhounds, and water spider orchid among lacustrine herbs of the sinks (because some of these sinks have a canopy coverage of more than 30%, they are also dealt with in the section on palustrine forested wetlands).

Dennis (1975b) surveyed the flora of Blake's Reserve, a former rice field impoundment in Charleston County, South Carolina. This lacustrine feature is surrounded by palustrine wetlands in the form of bald cypress-water tupelo communities. In areas where canopy density is less than 30% areal coverage, the depth of the reserve is such that it limits the growth of emergent vegetation. The littoral vegetation in Blake's Reserve, therefore, is significantly different from that of other lacustrine bodies. Due to the depth of the littoral zone, most of the lacustrine flora at the Reserve is consequently submergent, floating-leaved (rooted), or floating macrophytic vegetation. Table 5-30 lists the common plants of Blake's Reserve. In total numbers of plants, the floating-leaved (rooted) and the floating macrophytes dominate the littoral (and limnetic) zone of Blake's Reserve. Surface vegetation is frequently so thick that a boat cannot be paddled through it. Floating log and mat communities are present in the littoral subsystem of the Reserve, but because they are more common in the deeper water, they will be dealt with in the limnetic subsection.

The littoral zone of oxbow lakes is very limited or completely absent. Lizard's tail, dayflower, pennyworts, arrowheads, and pickerelweed are often present in the nonpersistent emergent zone. Literature concerning the floating and submergent flora of oxbow lakes in the Sea Island Coastal Region is nonexistent. In North Carolina, lacustrine features (bay lakes) are present in large Carolina Bays. Many South Carolina and Georgia pond cypress-dominated deep bays have many lacustrine characteristics (see Wharton 1978) but do not fit the definition of the lacustrine system used here (Cowardin et al. 1977). For a discussion of deepwater Carolina Bays and cypress ponds, see the section on vascular flora of the palustrine system.

The structure of littoral lacustrine subsystems is generalized in Figure 5-19. This structure is determined by water depth. In most littoral zones, nonpersistent emergents are found in the shallowest water, submergents dominate slightly in deeper water, and floating-leaved (rooted) plants and floating macrophytes are dominant in the deepest littoral and shallowest limnetic zones. The number and density of growth forms found in littoral lacustrine zones is usually indicative of the presence

383

Table 5-29. Flora of the <u>Nymphaea-Myriophyllum</u> community, in order of abundance, in a series of solution-formed "boggy ponds" located in the Georgia coastal plain (Eyles 1941).

Scientific Name	Common Name
<u>Myriophyllum pinnatum</u>	water milfoil
<u>Nymphaea odorata</u>	white water-lily
<u>Xyris smalliana</u>	yellow-eyed grass
<u>Utricularia</u> sp.	bladderwort
<u>Nymphoides aquatica</u>	big floating heart
<u>Eleocharis</u> sp.	spikerush
<u>Eleocharis elongata</u>	spikerush
<u>Eleocharis robbinsii</u>	spikerush
<u>Bacopa caroliniana</u>	lemon bacopa
<u>Nymphaea bombycina</u>	water-lily
<u>Eriocaulon compressum</u>	pipewort
<u>Eleocharis equisetoides</u>	jointed spikerush
<u>Sphagnum</u> sp.	peat moss
<u>Hydrocotyle umbellata</u>	marsh pennywort
<u>Psilocarya scirpoides</u>	bald rush
<u>Brasenia schreberi</u>	water-shield
<u>Cephalanthus occidentalis</u>	button bush
<u>Panicum hemitomon</u>	maidencane
<u>Polygonum hirsutum</u>	smartweed
<u>Scirpus etuberculatus</u>	bulrush

or absence of clearly defined zonation. In Eyles'(1941) study, 13 emergent species, 1 submergent species, and 4 floating-leaved (rooted) species were found in a series of shallow solution ponds. The high number of emergent species was probably due to the very shallow nature of the sinks. In Blake's Reserve, on the other hand, the number and density of floating-leaved (rooted) and floating macrophytic plants far exceeded that of emergent and submergent plants. This was because the water depth was 1 - 3 ft (0.3 - 0.9 m) at the beginning of the littoral zone.

2. Limnetic Subsystem

The limnetic subsystem includes those areas of lacustrine systems where the water is deeper than 2 m (6.5 ft) at low water, except when nonpersistent emergents grow beyond the 2 m depth zone; then, the limnetic zone begins at the maximum extent of these plants. Vegetation here consists entirely of submergents, floating-leaved (rooted) plants, floating macrophytes, floating mats, and vegetation of floating logs.

Table 5-30 lists common submergent, floating-leaved (rooted), floating macrophytes, floating log, and floating mat plants of Blake's Reserve. Floating-leaved (rooted) plants and floating macrophytes are the most common growth-forms encountered in the limnetic zone, with floating hearts, frog's-bit, mosquito fern, and duckweeds the most commonly seen species (Dennis 1975b). Water spider orchid

and false nettle are very common on floating logs in Blake's Reserve, and swamp smartweed is the most abundant plant of the floating mats. Figure 5-19 illustrates plant zonation in Blake's Reserve.

Hunt (1943) described four zones of vegetation on floating mats in a reservoir near Charleston, South Carolina. The pioneer zone, the advancing portion of the mat, was dominated by water-primrose, alligator-weed, and buttercup-leaved pennywort. This zone averaged about 50 ft (15 m) in width. These plants formed the mat base on which other floristic zones rest. The second zone, the cat-tail zone, was dominated by common cat-tail and varied from 25 to several hundred feet in width. A narrow shrub zone of wax myrtle and black willow was found between the cat-tail zone and the next zone, which Hunt referred to as the "main body." The main body of the mat had reached the forest stage and was dominated by woody plants. Red maple was common with red bay and bald cypress occasionally being found. The surface of the main body of the mat was made up of a thick layer of peat moss (Hunt 1943).

Dennis (1975b) is the only available reference concerning floating log communities in Southeastern lacustrine bodies, although Dennis (1973), Dennis and Batson (1974), and McEwan (1976) discussed floating log communities in palustrine environments. Several species are common to floating logs in both palustrine and lacustrine systems (notably water spider orchid and false nettle). Table 5-30 lists

Table 5-30. Common plants of the lacustrine system of Blake's Reserve, Charleston
County, South Carolina (Dennis 1975b, National Wetlands Inventory
1978).

EMERGENT	SUBMERGENT	FLOATING-LEAVED (ROOTED)
Decondon verticillatus (water loosestrife)	Cabomba caroliniana (fanwort)	Hydrocotyle ranunculoides (pennywort)
Polygonum hydropiperoides (swamp smartweed)	Ceratophyllum echinatum (coontail)	Alternanthera philoxeroides (alligator-weed)
Bacopa caroliniana (lemon bacopa)	Utricularia inflata (bladderwort)	Nymphoides aquatica (big floating heart)
Taxodium distichum (bald cypress)	Potamogeton berch.. .. li (narrow-leaved po . .)	Limnobium spongia (frog's-bit)
		Nymphaea odorata (white water-lily)
		Nymphoides cordata (little floating heart)

FLOATING MACROPHYTES	FLOATING LOGS	FLOATING MATS
Azolla caroliniana (mosquito fern)	Habenaria repens (water spider orchid)	Alternanthera philoxeroides (alligator-weed)
Spirodela polyrrhiza (big duckweed)	Centella asiatica (Chinaman's shield)	Polygonum hydropiperoides (swamp smartweed)
Lemna spp. (duckweeds)	Dulichium arundinac a (three-way sedge)	Décodon verticillatus (water loosestrife)
Wolffiella floridana (eastern wolffiella)	Itea virginica (virginia willow)	Sacciolepis striata (sacciolepsis)
	Boehmeria cylindrica (false nettle)	
	Carex decomposita (sedge)	

385

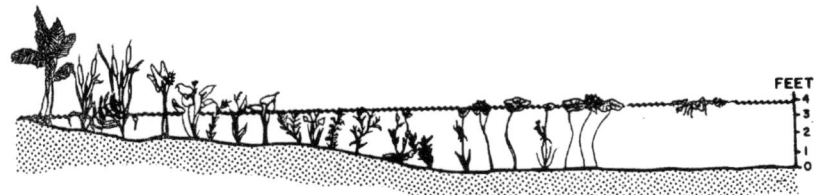

PALUSTRINE		LACUSTRINE		
EMERGENT ZONE		SUBMERGENT ZONE	FLOATING-LEAVED ZONE (ROOTED)	FLOATING MACROPHYTES
Decodon verticillatus (water loosestrife)	*Alternanthera philoxeroides* (alligator--weed)	*Potamogeton spp.* (pondweeds)	*Nymphaea ordorata* (water lily)	*Spirodela spp.* (duckweeds)
Salix spp. (willows)	*Sagittaria spp.* (arrowheads)	*Ceratophyllum spp.* (coontails)	*Nymphoides spp.* (floating hearts)	*Lemna spp.* (duckweeds)
Cephalanthus occidentalis (button bush)	*Pontederia cordata* (pickerelweed)	*Myriophyllum spp.* (water milfoils)	*Limnobium spongia* (frog's bit)	*Wolffiella floridana* (Eastern wolffiella)
Typha spp. (cat-tails)	*Polygonum spp.* (knotweeds)	*Najas spp.* (bushy pondweeds)		

Figure 5-19. Littoral lacustrine plant zonation of Blake's Reserve (Charleston County, South Carolina).

common plants found on floating logs in Blake's Reserve.

Succession in Southeastern lacustrine environments is poorly documented. Little or no literature exists concerning succession in sinks, rice field reserves, and other static water-level lacustrine features. Hunt (1943) discussed succession in Goose Creek Reservoir (South Carolina), an impounded flood plain. Succession there proceeded from marsh to forest on floating mats. Hunt (1943) compared succession in Southeastern reservoirs with that of Northern bog lakes. Reservoirs (such as Goose Creek) do not go through the "bog phase," he noted, because their waters are actively circulating. Water circulation keeps the pH of the reservoirs at a low acid level, in contrast to the highly acidic water of bog lakes (Oosting 1956). Oxbow lakes undergo a gradual successional trend from river to lake to a bald cypress–water tupelo community. Repeated floods bring in silt, filling the lakes in a slow process.

Lacustrine systems may be classified as oligotrophic or eutrophic, based on their productivity (Table 5-2). Oligotrophic lakes are low in productivity due to phosphorous limitations (the carbon content in these lakes is characteristically high); eutrophic lakes are high in phosphorus and are highly productive (Wetzel 1975). Most of the lacustrine bodies in the Sea Island Coastal Region generally exhibit eutrophic conditions. Studies of nutrients and productivity in lacustrine environments are unavailable for the sinks, oxbow lakes, and rice field reserves of the Sea Island Coastal Region; however, limited information does exist concerning the reservoirs of the area.

Boyd (1971), referring to Parr Pond in Aiken County, South Carolina, pointed out that oligotrophic lakes in the Southeast may have a well-developed vascular flora. Rooted vascular plants do not have to compete with phytoplankton and nonrooted macrophytes for nutrients and, therefore, may flourish in bodies of water with a low supply of dissolved nutrients (rooted vascular plants absorb ions through both root and leaf tissues; phytoplankton and nonrooted macrophytes absorb only dissolved nutrients).

Penfound and Earle (1948) studied the productivity of water hyacinth in Louisiana. Biomass ranged from 630 - 1,472 g dry wt/m^2 and annual productivity ranged from 1,500 - 4,400 g/m^2/yr. Floating-leaved and submergent vascular plants normally have standing crop biomasses of less than 500 g dry wt/m^2. Emergent plants have biomass values that range from 500 - 1,500 g/m^2 (Boyd 1971).

B. INVERTEBRATES

1. Zooplankton

Pennak (1953) described freshwater limnetic zooplankton communities as remarkably simple from a species diversity standpoint. In studies of 79 lakes around the world, including 37 in Colorado, he found that copepods were usually represented by one or two species, cladocerans by one to three species, and rotifers by three to seven species in any given lake. The individual species present may change seasonally, but in a "typical" lake situation 80% of all copepods present at a given time are of one species, as are 78% of the cladocerans and 64% of the rotifers. Pennak also found that larger lakes tended to have more species present at a given time than did smaller lakes. He further reported that in any given lacustrine environment, it is unusual for more than one species of a particular genus to be represented in the zooplankton at a time. In rare instances when congeneric species did occur, one species almost always outnumbered the other by at least 20 to one.

The zooplankton of lacustrine habitats is generally composed of the same groups found in riverine situations; those of greatest importance are the free-living nonphotosynthetic protista, rotatoria, and crustaceans (see the section on invertebrates of the riverine eocsystem for a detailed discussion of these groups). However, species composition of these groups in lacustrine habitats on occasion may be quite different from that observed in riverine situations. Further, lake zooplankton populations are generally denser than those of riverine environments.

The chemistry of lacustrine waters may be a limiting factor for some species of limnetic plankters. Some cladocerans such as Holopedium gibberum are usually found in soft saltwater lakes and almost never in hard water. Some, particularly pond species, are tolerant of low oxygen concentrations (Hutchinson 1967). Ahlstrom (1938) found typically alkaline or acid water rotifer species as well as

a third group having a wide pH tolerance. Acid water, which is common in coastal areas, has greater zooplankton diversity than hard alkaline water, which is typical of the Piedmont region.

Zooplankton of lakes may be subject to predation from many sources. Most prominent zooplankton feeders are fish, larvae of insects such as Chaoborus, and other plankters. Fedorenko (1975) found that Chaoborus larvae preyed heavily on copepods, consuming 2% of the nauplii, 3% of Diaptomus tyrelli, 9% of Diaptomus kenai, and 4% of Diaphanosoma in a Canadian lake. The cyclopoid copepods include many predacious species (Fryer 1957), and Anderson (1970) reports predation by larger diaptomid copepods. The cladocerans Leptodora kindtii and Polyphemus sp. are also highly predatory plankters (Davis 1955). While no studies of the lacustrine zooplankton of the study area have been made, Hudson (1975) described the zooplankton of Keowee Reservoir, a South Carolina impoundment of the Piedmont region supplying cooling water for the Oconee Nuclear Station. Of the 53 species of copepods and cladocerans identified from the reservoir, only about 15 species are common in the plankton while the remainder are littoral or benthic species. Diaptomus mississippiensis, Mesocyclops edax, and Tropocyclops prasinus were the most abundant copepods, while Diaphanosoma branchyurum, Holopedium amazonicum, Daphnia ambigua, and two species of Bosmina were the most abundant planktonic cladocerans in 1975 (Hudson 1975).

Ahlstrom (1938) described planktonic rotifers found in North Carolina, and Coker (1938) reviewed the distribution of planktonic crustacea in both North and South Carolina. Coker found many copepods and cladoceran species of the region to be worldwide in distribution. Coker (1926) found no crustacean plankton in a swift water creek feeding Lake James, North Carolina, while the lake itself had a comparatively rich planktonic Entomostraca. Turner (1910) described copepods and cladocerans in mostly palustrine habitats near Augusta, Georgia; however, several permanent ponds were sampled and he listed species that are found in lacustrine systems elsewhere (e.g., Diaptomus mississippiensis, Daphnia hyalina, Bosmina longirostris).

2. Benthic Invertebrates

Most published accounts of lacustrine benthic invertebrate ecology in North America have been based on studies conducted in the Northern States and Canada. Scattered investigations have been undertaken in the Southeast, especially in the lakes of the Tennessee Valley Authority (see various TVA reports), but no publications or documents dealing with the benthic ecology of such systems in the Sea Island Coastal Region were found in our literature survey. It seems improbable that the species composition and community structure of lakes in the Sea Island Coastal Region could accurately be inferred from investigations made in lacustrine systems far removed from the study area. Invertebrate assemblages are known to vary from one lake to another because each lacustrine habitat is unique in its physical and chemical characteristics. Faunal composition appears to depend to some extent upon lake size, morphology, fertility, sediment types, and abundance of plant life.

Although community composition cannot be inferred precisely from studies conducted in other areas, major taxa known to occur in such ecosystems include sponges, hydrozoans, flatworms, nematodes, rotifers, bryozoans, entoprocts, oligochaetes, leeches, ostracods, cladocerans, mysids, isopods, amphipods, decapods, bivalves, gastropods, and larval and adult insects (orders Odonata, Ephemeroptera, Plecoptera, Hemiptera, Diptera, Trichoptera, Neuroptera, Megaloptera, Coleoptera) (Hart and Fuller 1974, Cole 1975, Ursin 1975, Wetzel 1975). Soft bottom sediments typically support populations of oligochaete worms, pelecypods, crustaceans, and various insect larvae. Another assemblage of organisms, including sponges, hydrozoans, flatworms, bryozoans, leeches, gastropods, and a variety of crustaceans and larval insects, occurs in hard substrates such as rocks, logs, and aquatic vegetation. Other species, including water striders, collembolans, beetles, bugs, and spiders, live on or near the water surface.

Seasonal cycles of activity and abundance, similar to those described earlier (see the section on invertebrates of the palustrine ecosystem), occur in the benthic invertebrate fauna of lakes. After the winter lull, species numbers, numbers of individuals, feeding activity and growth rates all increase in spring as eggs and dormant stages hatch and hibernating species return to activity. As water temperatures rise, reproductive cycles commence for many species as well. Seasonal changes in faunal density occur, especially within populations of larval insects. The invertebrate fauna is important in the food web of lakes, and the significance of insect larvae in particular as prey for fishes has long been recognized. Encrusting species such as sponges and bryozoans, which overwinter as gemmules and statoblasts, respectively, increase greatly in colony size throughout the summer and early autumn. In addition to seasonal changes in faunal density, invertebrate populations are known to undergo fluctuations in abundance from one year to another.

With increasing depth, the benthos of lakes typically decreases in terms of species numbers, species diversity, and biomass. This has been attributed to a combination of factors including greater diversity of habitat and prey selection in shallow areas, the presence of numerous air-breathing insects and arachnids near the water surface, and the colder temperature and periodically low oxygen concentration of the waters in the profundal zone. Aquatic vegetation is especially important to the benthos as food, substrate, and shelter, and macroscopic plants are restricted to shallow water areas. Organisms typical of the profundal zone include oligochaetes, chironomids, chaoborids, nematodes, and bivalves of the family Sphaeriidae. The fauna of the littoral zone is much more diverse, especially in sheltered areas having beds of aquatic vegetation. Insect larvae, crustaceans, oligochaetes, and mollusks are typically well represented in the upper regions of a lake. A relatively depauperate fauna may occur in sandy, wave-exposed areas, and fresh waters severely impacted by domestic pollution may have a benthic fauna dominated by stress-tolerant tubificid oligochaetes and chironomids.

The macrobenthic communities of Lake Murray, located in the South Carolina coastal plain near the study area, were characterized by Environmental Research Center, Inc. (1976). Thirty-five taxa were distinguished in samples from the lake, and overall diversity appeared to be quite low. The fauna was dominated numerically by larval dipterans (44.6%), oligochaetes (34.4%), and mollusks (12.6%), while mollusks were dominant in terms of biomass. Oligochaetes, largely represented by Limnodrilus sp. and Peloscolex sp., predominated in deeper regions of the lake. Nematodes and the sphaeriid clam Pisidium sp. were also frequent at deepwater sites. Sediments at these depths were silty, and dissolved oxygen levels approached zero during summer stratification. Under such conditions, species diversity was low. In shallow waters, oligochaetes were of minor importance except at one sampling site, and a more diverse fauna comprised largely of larval insects, mollusks, and nematodes was observed. The most abundant larval insects were tube dwelling chironomids Chironomus sp., a largely herbivorous organism of importance in the food web as prey for fishes, and the phantom midge Chaoborus sp., whose habits are as much planktonic as benthic. Mayflies, principally belonging to the genus Hexagenia, were prevalent in areas having substrates of gravel and organic debris. Although the fingernail clam Pisidium sp. was well represented in deeper waters, other mollusks were most abundant along the shoreline. Mussels were plentiful at many locations in shallow waters, and the introduced asiatic clam Corbicula manilensis was frequent.

The paucity of adequate studies on benthic invertebrates of lakes within the Sea Island Coastal Region is not unusual. According to Brinkhurst et al. (1974), ". . . there has never been a study of the benthos of a lake which, in the senior author's opinion, the sampling methodology and schedule have been properly evaluated, most of the major species identified, and which extended over all seasons of the year for a consecutive number of years."

3. Insects

The insect fauna of the littoral zone of shallow lakes and ponds in South Carolina and Georgia is generally rich in abundance and diveristy. The emergent and submergent vegetation in littoral lacustrine waters provide suitable subsurface strata for physical support of immatures and adults of several aquatic insect orders, while still other adult insects capable of submerging themselves can often be seen in sunlit shallows diving or swimming below the surface. Lake and pond bottoms also provide habitat for many insects and other invertebrates whose immature stages require or prefer a benthic mode. Still other insects are associated with the surface of these relatively quiet, lentic waters. Aerial adults of many of these insects having aquatic stages as immatures can be seen flying about, hovering and darting above the water surface or shoreline or resting on emergent vegetation. Much of the following treatment of insect orders common to lacustrine habitat is taken from Pennak (1953).

With the exception of the mosquitoes and flies, the dragonflies (Odonata: Anisoptera) and damselflies (Odonata: Zygoptera) are perhaps the most recognizable insects of the lacustrine environment. The aerial adults are often called "darning needles" or "mosquito hawks" because of their size, shape, and habits. They are medium to large insects, with slender abdomens, and often are handsomely colored. From early morning until late evening, dragonflies may be found flying back and forth or darting about erratically along the shores and over the open waters of lacustrine environments. (See also the section on insects of the riverine ecosystem.)

Only the series of nymphal stages, or naiads, of dragonflies and damselflies are aquatic. These are grotesque creatures, robust or elongated, and gray, greenish, or somber-colored. The naiads are commonly found on submerged vegetation and the bottoms of ponds, marshes, and the littoral zone of the lakes. Rarely found in polluted waters, naiads are good indicators of clean lakes.

Adult odonates are often seen mating, with male and female in tandem, either while at rest or in flight. Among some odonates, more than one generation per year can occur, while more than 4 years may be required to complete the life cycle in other species. Odonate nymphs may be roughly classified as aquatic climbers, sprawlers, or burrowers. The climbers move about slowly in dense vegetation or debris in still waters. To this group belong the families Coenagrionidae (narrow-winged damselflies) and the Aeschnidae (darners). Most of the sprawlers are long-legged, sluggish, dull-colored creatures occurring on many types of bottoms. To this group belong the Agrionidae (broad-winged damselflies) and Libellulidae (common skimmers). Although not true burrowers, the Cordulegasteridae (biddies) lie almost hidden in sand or silt bottoms. The Gomphidae (clubtails) and Petaluridae (graybacks) burrow into silt, mud, and sand so that usually only the tip of the abdomen and the eyes are above the substrate. Just before transformation to the imago (new adult), the nymph crawls out of the water, usually on emergent vegetation, where the old exoskeleton splits and releases the adult stage. The new adult can often be seen clinging to the nymph exoskeleton for an hour or so while the wings and body become stiff and dry.

Dragonflies (Odonata) and other invertebrates of a 1 ha (2.5 acre) farm pond at the Savannah River Plant near Aiken, South Carolina, were studied by Cross (1955), Benke (1972, 1976), and Benke and Benke (1975). Benke (1976) observed a density of 10,000 larval midges (Chironomidae)/m^2 (8,300 larval midges/yd^2) in Ekman grab samples taken from May to September. Biomass of larval midges and mayflies (largely Caenis sp.) amounted to 0.6 g/m^2 (dry weight), or about two-thirds of the total macrobenthos other than dragonflies. The remaining one-third consisted mostly of beetles (Coleoptera) and horseflies and deer flies (Tabanidae), although caddisflies (Trichoptera), biting midges (Ceratopogonidae), damselflies (Zygoptera), predacious bugs (Hemiptera), and various microcrustaceans were also present. Biomass of the dominant larval dragonflies (Ladena deplanata, Epitheca spp., and Celithemis fasciata) during May – September was about 6 g/m^2, while total odonate biomass was estimated at 8 g/m^2. Such high predator to prey ratios were possible because of high turnover rates in prey populations. Sufficient refuges were also believed by Benke to be responsible for preventing the annihilation of prey populations. Because high populations of larval dragonflies have been observed elsewhere on the Savannah River Plant and in a lagoon near Athens, Georgia, Benke and Benke (1975) suggested that they may be a predominant feature

of pond ecosystems in the Southeastern
United States.

Among the true bugs (order Hemiptera),
there exists a group of semiaquatic and
aquatic families which show a gradual
transition in their freshwater-associated
habitats from the damp shores of ponds
and lakes to the subsurface waters. The
shore bugs (Saldiciae), velvety shore
bugs (Ochteridae), and toad bugs
(Gelastocoridae), for example, live primarily on the shore, running and jumping
in varying degrees, making short flights,
but generally lighting on the water only
by accident. The velvet water bugs
(Hebridae) run about at the water's edge
on floating vegetation, surface of the
water, and the adjacent damp ground.
The marsh treaders and water measurers
(Hydrometridae), water treaders
(Mesoveliidae), and some broad-shouldered
water striders (Veliidae) venture farther
out and are almost always found on floating algae and plant rafts. Other veliids
and the water striders, pond skaters, and
wherrymen (Gerridae) skate rapidly over
the surface of the water. In the velvet
water bugs (Hebridae), the body is covered
with a velvety pile which effectively
sheds water, but the other surface-
living families usually have only the
legs (especially the tarsi) covered with
pile.

Six families of true bugs are truly
aquatic and normally found below the
surface. Of these, the water scorpions
(Nepidae) and giant water bugs (Belostomatidae) usually cling to their substrate
but remain more or less in contact with
the surface film. The water boatmen
(Corixidae) and the back swimmers
(Notonectidae) are the true bugs that ex-
cel at swimming and are among the best
known of all water bugs. Nearly all
aquatic Hemiptera are strict predators,
the particular prey (depending on the
specific habitats of the various families)
being chiefly small terrestrial and
aquatic insects and entomostraca. Although some of the semiaquatic Hemiptera
oviposit on sphagnum and shore grasses,
the eggs of the aquatic families are
normally laid on the surface or within
the tissues of submerged plants. A few
water boatmen oviposit on crayfish,
snails, and dragonfly nymphs. All molt-
ings of the aquatic forms occur under
water. The great majority of species
overwinter as adults hidden in mud,
debris, or vegetation.

Another insect order typical in
freshwater settings of the Sea Island
Coastal Region is the beetles (Coleoptera).
While many of the more than 150 families
of beetles are terrestrial, some families
are totally or partially aquatic in the
adult or larvae stages. The more common
families include the crawling water
beetles (Haliplidae), with all adults

and larvae aquatic; the predacious diving
beetles (Dytiscidae), with all adults
and larvae aquatic; the whirligig beetles
(Gyrinidae), with all adults and larvae
aquatic; the water scavenger beetles
(Hydrophilidae), with the majority of
species aquatic both as larvae and
adults; the elmid beetles (Elmidae),
with all larvae and most adults
aquatic; the helodid beetles (Helodidae=
Cyphonidae), with larvae aquatic, but
adults terrestrial; the leaf beetles
(Chrysomelidae), with larvae of several
genera found underwater or on emergent
vegetation; and the weevils (Curculionidae),
with adults and immature stages of a few
genera feeding on emergent aquatic
vegetation, and adults occasionally found
swimming under water.

Many adult aquatic coleopterans are
known to fly, and in this way they migrate
from one body of water to another. Most
adult aquatic beetles are fundamentally
dependent upon atmospheric oxygen and,
when submerged, carry their supply of
oxygen with them. Adult beetles are
very common in ponds and protected
bays of lakes, as well as in streams and
rivers. They occur in the shallows near
shore, particularly where there exist
quantities of debris and aquatic vegetation, but are generally absent from
wave-swept shores and the deeper waters
of lakes and rivers. With few exceptions,
the pupal stage is terrestrial. So far
as is known, all adult aquatic beetles
have aquatic larvae, the eggs usually
being deposited below the water surface.
In general, the larvae occur in the
same habitats as the adults.

The two-wing flies (Diptera) are
particularly common in and around freshwater habitats and include such familiar
insect groups as mosquitoes, crane flies,
midges, and horseflies. Although the
adults of the order Diptera are never
aquatic, many families have members which
have aquatic immature stages. Such
larvae and pupae occur in every type
of freshwater habitat, often in enormous
numbers. Immatures of only a few
families, however, such as the mosquitoes
(Culicidae) and black flies (Simuliidae),
are exclusively aquatic. The egg-laying
habits of aquatic Diptera are diverse.
The females of some species scatter their
eggs just below the surface of the water
on vegetation or debris. Others oviposit
in regular or irregular gelatinous masses
or strings below the surface, at the
surface, or on objects just above it.

Aquatic dipteran larvae show greater
variability in structure and habitat than
any other order of aquatic insects.
Usually, however, they are distinguished
by their elongated, wormlike body, and
the absence of eyes and jointed thoracic
legs. The body is usually soft and
flexible, and the most common colors are

white, gray, yellow, reddish, brown, and black. Many larvae swim by rapid wriggling movements of the body. In some families, such as the black flies and mosquitoes, the food consists of minute organisms and particles which are strained from the water. In other families, the mouth parts are modified for scraping debris from rocks, for feeding on aquatic plants, for predatory food habits, or simply for consuming plant and animal debris.

The larval stage of aquatic dipterans may last for only several weeks or it may persist for at least 2 years, depending on the species, temperature, and food conditions. Although mosquito and midge pupae are capable of rapid wriggling movements, most other pupae are inactive. Some of the black fly pupae occur in loosely built cocoons. The pupal stage of aquatic dipterans usually lasts less than 2 weeks, although a few types may overwinter.

Some of the more common families of dipterans having aquatic immature stages, or aerial or terrestrial adults near fresh water, include the true crane flies (Tipulidae), phantom crane flies (Ptychopteridae), moth flies (Psychodidae), marsh flies (Tetanoceridae), shore flies (Ephydridae), phantom midges (Culicidae), midges (Tendipedidae), and black flies (Simuliidae), although the latter are most often found around flowing waters. Also present are the biting midges, commonly called "no-see-ums" or "punkies" (Ceratopogonidae); mosquitoes (Culicidae); and horseflies (Tabanidae). These latter three families all contain species which inflict irritating bites and, in some lacustrine environments, they are often present in such enormous numbers as to be serious pests.

Mayflies (Ephemeroptera) represent another order of insects common to this and other freshwater habitats. Adult mayflies are small to medium-sized terrestrial insects with delicate, many-veined, transparent wings which are held together vertically when at rest. They are found only in the vicinity of bodies of fresh water in which the immature stages are passed. Much of the time, particularly in strong winds, the adults remain clinging to vegetation; but on calm, sunny, spring and summer days (or in the evening), they take to the air in great hovering swarms. Such swarms are composed almost entirely of males. Females enter the swarm singly and emerge almost immediately accompanied by a male. Mating then occurs during flight.

Mayfly nymphs are almost entirely herbivorous, although a few have been observed feeding on exuviae (shed exoskeletons), the bodies of dead nymphs, and small invertebrates. Generally, they browse on the substrate, feeding upon algae or tissues of higher aquatic plants. Unlike the adult, nymphal stages constitute a relatively long portion of the mayfly life history. Most species have an annual life cycle, although a few live 2 and possibly 3 years.

Another order of insects typical of freshwater habitats is Megaloptera, containing the alderflies, dobsonflies, and fishflies. Megalopterans are dull-colored and medium to large in size, the adult length ranging from about 10 - 70 mm (0.4 - 2.8 in). The two pairs of wings are similar and are held flat or rooflike over the body when at rest. Flight is rather weak. The head bears long, slender antennae and biting mouth parts.

The alderflies are small and diurnal. The dobsonflies and fishflies, on the other hand, are long and nocturnal. Megalopterans are widely distributed but rarely found in large numbers. Larval stages are often found in lacustrine environments. Pupation occurs on shore with adults appearing in spring and early summer. The minute, elongated eggs are laid in rows forming masses of several thousand on vegetation and objects overhanging the water; upon hatching, the larvae fall into the water. In most cases the larval stages last for 2 or 3 years. Considering this long period of time, surprisingly few larval instars, generally 10, occur. "Mature" megalopteran larvae are among the most striking of aquatic insects. The body is stout, elongated, from 25 - 90 mm (1.0 - 3.6 in) in length, dull yellowish, brown, or dusky, and often mottled. The head bears four-segmented antennae, rather small compound eyes, and strong mouth parts. The legs are well developed.

C. FISHES

1. Littoral Subsystem

The shallow areas of lakes where light penetrates to the bottom are rich in vegetation. This rich bottom area provides food and shelter for a diverse array of fishes. Species within this area are very similar to the fauna found in the rivers that drain these lakes (Table 5-18). This habitat is the second highest in diversity of freshwater fish species with 43 (52%) of 82 genera and/or species listed for the coastal area common to this habitat (Table 5-19).

The productivity of the littoral subsystem is indicated by the sampling of two coves on Lake Moultrie, where an average of 2,813 fish/acre (1,173 fish/ha) of littoral zone was found (Table 5-31). Density levels over a 14-year period ranged from a low of 343 fish/acre (101 fish/ha) in 1974 to a high of 6,490 fish/acre (2,704 fish/ha) in 1967 (Table 5-32).

Table 5-31. Results of two rotenone samples taken from Lake Moultrie in 1971 (White 1972).

	Cove #1 S.C. Highway 6 September 29, 1971 Avg. Depth: 4 feet[a] Area: 1 acre[b]		Cove #2 Lion's Beach September 30, 1971 Avg. Depth: 5 feet Area: 1 acre	
Number of fish/acre	2,986.0		2,639.0	
Pounds of fish/acre[c]	179.8		101.0	
Young-of-the-year largemouth bass/acre	147.0		36.0	
Pounds of shad/acre	90.7		50.4	
Species	% Total Number	% Total Weight	% Total Number	% Total Weight
Largemouth bass	5.6	15.8	1.4	2.2
Suckers	1.5	12.2	0.6	7.4
Pickerel	0.3	1.2	0.1	0.2
Catfishes	1.1	1.6	1.5	3.1
Bream	24.2	11.6	66.0	20.5
Shad	56.0	50.5	13.9	49.9
Yellow perch	11.0	6.5	15.9	12.9
Crappie	0.3	0.6	0.5	0.7
Bowfin	--	--	0.1	3.1
TOTAL	100.0	100.0	100.0	100.0

a. 1 ft = 0.3048 m
b. 1 acre = 0.4047 ha
c. 1 lb = 453.59 g

Studies conducted in the littoral subsystem of the coastal area have dealt primarily with monitoring game fish and herring stocks, tagging anadromous species, and conducting creel censuses of sport fishes (White 1971, 1972, 1973, 1974, 1975, 1976). The lack of specific life history information on the fishes of this area makes it necessary to extrapolate general or applicable information from studies conducted in other areas. Segments of works such as Wilbur (1969) on redear sunfish; Sandow et al. (1974) on redbreast sunfish; Jester (1974) on carp; and Mathur (1970) on channel catfish, can be applied to the fishes of the Sea Island Coastal Region.

The sunfish family (Centrarchidae) is probably the most important fish family of this habitat. All centrarchid species known to occur in the coastal area utilize this habitat. Excluding anadromous species, White (1971, 1972, 1973, 1974, 1975, 1976) found centrarchids to consistently rank the highest in number of individuals collected in the littoral areas of Lake Moultrie and Lake Marion, South Carolina. Pickerels (Esocidae), minnows (Cyprinidae), suckers (Catostomidae), catfishes (Ictaluridae),

bowfins (Amiidae), and yellow perch (Percidae) were also regular components of these collections.

Centrarchid family members are primarily carnivorous; however, a few species may consume plant material when other food supplies diminish. Their diet, as with other fishes, is heavily influenced by seasonal fluctuation in abundance of the various forage items. The first food consumed by the young of this family is zooplankton. As they grow, they gradually shift to a diet favoring small crustaceans, insects, small fishes, and insect larvae.

Members of the genera Acantharchus, Centrarchus, Elassoma, Enneacanthus, and Lepomis heavily utilize crustaceans, insects, and insect larvae throughout their life when these are available, though a wide array of items from vegetation and annelids to fish eggs and small fishes may be consumed. As very young fishes, crappies and largemouth bass utilize insects and insect larvae as a substantial part of their diet; however, at a relatively early age, they switch to a diet composed predominantly of fishes (Pasch 1974). The diet of the adult

Table 5-32. Comparison of rotenone samples on Lake Moultrie, 1960 - 1974 (White 1975).

Year	Avg. No. Fish/Acre	Avg. Wt. Fish/Acre (lb)	Avg. No. Y.O.Y. Lm. Bass/Acre[b]	Avg. Wt. Shad/Acre (lb)	Percent Shad by Number	Percent Shad by Weight
1960	2,114	104	40	44	60	42
1961	2,371	202	55	88	15	44
1962	5,808	236	45	70	17	30
1963	3,212	93	44	29	24	31
1964	1,420	120	39	60	42	42
1965	2,993	120	39	60	50	38
1966	1,744	54	45	23	42	42
1967	6,490	83	209	12	1	15
1969	2,436	50	45	19	41	38
1970	1,209	25	36	8	6	16
1971	2,812	140	91	70	53	50
1972	1,818	24	14	16	70	67
1973	1,712	47	15	24	35	36
1974	343	8	0	5	55	60

a. 1 lb = 453.59 g
b. Young-of-the-year largemouth bass

largemouth bass is probably one of the most diversified of this family. Given the opportunity, they are known to eat reptiles, amphibians, small birds, and even small mammals on occasion.

Individuals of the genera Acantharchus, Centrarchus, Elassoma, Enneacanthus, and Lepomis serve extensively as forage for larger carnivores such as striped bass, pickerels, catfishes, and bowfins. Members of the centrarchid family will also readily prey on smaller family members (e.g., largemouth bass prey on small Lepomis).

Members of the minnow family (Cyprinidae) are among the more common forage fishes found in this habitat and are represented in the study area by at least 25 species. Many of these are prominent members of the littoral community such as carp, golden shiner, and silvery minnow. These fish are omnivorous in their feeding habits, consuming terrestrial and aquatic insects and insect larvae along with small crustaceans and members of the zooplankton community. Algae and aquatic plants serve as major items in their diets, as does organic detritus to a lesser degree. The relative importance of plant and animal matter in the diet varies among minnow species, and even with the seasons. With the exception of carp, which reaches lengths in excess of 90 cm (36 in), these fishes seldom exceed 25 cm (10 in) in length.

This habitat (i.e., the littoral subsystem) is the primary spawning ground for the majority of species inhabiting the lacustrine system. The southwest side of Lake Moultrie is a primary example of a littoral area utilized for spawning. Fishes using this habitat for spawning generally fall into two categories: those which deposit their eggs in a nest, and those which broadcast their eggs among the vegetation. These fishes are primarily spring and summer spawners.

Fish nests are generally constructed in or near vegetation or submerged obstructions in 0.3 - 2.0 m (1.0 - 6.5 ft) of water, though they may occur in deeper water. The male of most lacustrine species constructs the nest by fanning out a depression in the substrate with its fins. The male may guard the nest until hatching occurs in many species such as the bluegill, redbreast sunfish, green sunfish, blackbanded sunfish, some darters, and various minnows. Some species, such as the bowfin, will continue to guard the young for a varying period following hatching. Some species, such as longear sunfish and redear sunfish will form distinct colonies while others, like largemouth bass and spotted sunfish, form loose-knit nesting colonies. The bullhead catfishes and madtoms vary from this form of spawning by nesting under logs and other obstructions.

Non-nesting species such as carp, redfin pickerel, chain pickerel, threadfin shad, killifishes, and longnose gar spawn by scattering their eggs among shallow-water vegetation. The eggs of most broadcast spawners are adhesive and attach to

393

vegetation or submerged obstructions, where they remain until hatching. Species utilizing this spawning method provide no parental care to the eggs or offspring. Upon hatching, the fry seek shelter among the food-rich vegetation for protection from predators.

Many species spend their entire lives in this community, moving into limnetic habitats only when forced to do so by unfavorable conditions such as excessively high or low temperature. Even then the departure is only for a short period of time. The pygmy sunfishes, largemouth bass, redfin pickerel, chain pickerel, mosquitofish, and least killifish are some of the fishes which may spend their entire lives in this habitat. Others, such as adults of white bass and large catfish, are transients utilizing this area primarily for feeding purposes.

Water level drawdowns can cause severe damage to a balanced population by producing an increase in predation. Small fishes may even be stranded in some cases (Bennett 1962). Drawdown has been one method employed to control carp populations by exposing the shallow vegetated spawning area immediately following the spawn (Sprague 1961).

2. Limnetic Subsystem

The limnetic subsystem exhibits over a 50% reduction in ichthyofaunal diversity compared to the littoral community (Table 5-19). This is directly related to the drop in diversity of habitat and food types available in the subsystem. The limnetic zone of lakes offers a more stable, though less productive, habitat than does the littoral zone. Consequently, many predominantly littoral species such as bluegill, dollar sunfish, eastern mudminnow, and starhead topminnow will retreat to these more stable waters during periods of unfavorable conditions (e.g., extreme temperatures) in the littoral habitat. The use of this habitat as a retreat is reflected by the high number of species which are listed as frequently or occasionally inhabiting this area (Table 5-19).

Dominant predatory fishes commonly inhabiting these waters include white bass, white crappie, largemouth bass, and long-nose gar, along with many of the larger catfishes such as the white catfish, blue catfish, yellow bullhead, brown bullhead, and channel catfish. Anadromous herring species are by far the most abundant forage species inhabiting the area. Gizzard shad and threadfin shad are also very abundant in this habitat and serve as major forage species. Other small forage species which are commonly found in this area include the silvery minnow, golden shiner, coastal shiner, creek chubsucker, and lake chubsucker.

Catfishes are the most abundant bottom-dwelling predator inhabiting this community. Stevens (1959), using data from gill net and trotline collections in lakes Marion and Moultrie, showed that white catfish outnumbered channel catfish four to one. Channel catfish were rare in Lake Marion with only nine collected compared to 173 in Lake Moultrie for the same period. White (1975) found white and channel catfish to be even in relative abundance in these two lakes during 1974-75. Collections by Stevens (1959) and by White (1975) indicate that a major expansion in the channel catfish population occurred in these reservoirs during this 16-year period. However, White (1975), while stating that his collections were combined for both lakes, makes no reference to variation in the relative abundance of various species between the two reservoirs. Consequently, the channel catfish is assumed to be more common in the current ichthyofauna of Lake Marion. Morrow (1972) found white catfish to be of little importance in Georgia impoundments, while the channel catfish is considered very important.

The majority of fishes which normally feed in this community utilize two major food groups, plankton (both plant and animal) and small fishes. The various species of anadromous herrings, the gizzard shad, and the threadfin shad are the most abundant consumers of plankton in this community. Young of these primary consumers provide a major source of food for predators like white bass and larger catfishes, which regularly reside and forage in this community. Stevens (1959) showed that members of the herring family (Clupeidae) comprised roughly 28% of the diet of white and channel catfishes. Trotline collections of channel catfish were 59% higher using cut herring than octagon soap, and they were 210% higher using cut herring than dough bait, showing a definite preference for cut herring by channel catfish (White 1975). Stevens (1959) found that fish constituted 64% and 76% of the diet in white and channel catfishes, respectively. Mollusks, insects, crustaceans, and some plant material were also found to be a part of their diet. Young channel catfish tend to feed mainly on aquatic insects (Bailey and Harrison 1948) or on bottom arthropods (Darnell 1958); but fish over 100 mm (3.9 in) in length are usually omnivorous (Bailey and Harrison 1948, Darnell 1958) or piscivorous (Stevens 1959).

Many species of fish inhabiting this zone spawn in the upstream riverine systems such as those that drain into Lake Marion and Lake Moultrie (e.g., Congaree and Wateree rivers). Considerably fewer species of fish utilize this area as a spawning ground than they do the littoral subsystem. Many commonly littoral nesting species, such as the redear sunfish, the

crayfish, small fish) are found, and that
they may seek protection from a primary
predator, the eastern mud snake. In
lacustrine environments supporting sizeable
populations of American eels, rainbow
snakes may also be common.

Emergent vegetation tend to increase
the area of the "edge effect," where
numbers and species of animals are usually
greater (Odum 1971). This is particularly
true for air-breathing amphibians and
reptiles. Redbelly water snakes, banded
water snakes, and cottonmouths are common
in this habitat (Conant 1975, Gibbons
1978). Although they do not obtain their
greatest density here, cottonmouths appear
to be more abundant along vegetated lake
shallows and adjoining sloughs than
along rivers. Florida green water snakes,
glossy crayfish snakes, and black swamp
snakes are uncommon throughout coastal
Georgia and South Carolina (Martof 1956,
Gibbons 1978). In the extreme southeastern portion of Georgia adjacent to
Florida, the striped crayfish snake
inhabits vegetated lake edges, albeit
uncommonly. The semi-aquatic eastern
ribbon snake and rough green snake are
frequently found among emergent vegetation
(Conant 1975, Mount 1975). Emergent
and floating-leaved vegetation (cat-
tails, pickerelweed, arrowhead, alligator-
weed, water hyacinth, etc.) provides
aquatic snakes protection from predators,
easy access to water, and an environment
in which food is abundant. Green anoles
are also common among cat-tails and other
emergents near shore.

Turtles occurring in open waters
generally frequent areas of emergent
vegetation for feeding or as transients.
Species more characteristic of vegetated
lake shallows are eastern mud turtles,
striped mud turtles, stinkpots, yellowbelly sliders, eastern chicken turtles, and
common snapping turtles (Conant 1975,
Gibbons 1978).

The American alligator is common in
the littoral zone of lacustrine environments. Where well-developed emergent
zones are associated with floating vegetation, large populations of alligators
are found. Bara (1976) recorded the
following observations (alligators per
mile) on Goose Creek Reservoir in
Berkeley County, South Carolina: 1972-
6.44; 1973-2.79; 1974-3.53; and 1975-
4.56. (See vegetation of the lacustrine limnetic subsystem for description of the floating vegetation of Goose
Creek Reservoir.) Murphy (1977) observed,
in Parr Pond in Barnwell County, South
Carolina, that alligators were nearly always seen within 100 yd (91.4 m) of the
shoreline and were seldom visible in open
water.

2. Limnetic Subsystem

The two-toed amphiuma occurs here uncommonly, whereas the greater siren is found relatively frequently (Mount 1975). Numbers of both increase toward shore where submergent vegetation and bottom debris are more abundant. Nearly all freshwater turtles occurring in the coastal zone can be found in open lake waters. Species most common to this habitat are the eastern chicken turtle, common snapping turtle, Florida cooter, yellowbelly slider, and the Florida softshell (Gibbons 1978). Eastern mud turtles, stinkpots, river cooters, and gulf coast spiny softshells (see Conant 1975) occur infrequently (Gibbons 1978). Water snakes of the genera _Nerodia_, _Seminatrix_, and _Farancia_ utilize the periphery of large lakes and are probably only transients in open waters. Where habitat is undisturbed and sizable populations exist, American alligators frequently spend time in open waters; however, they are generally more common in littoral areas.

E. BIRDS

Approximately 71 species of birds occur in lacustrine systems (Table 5-33). Because there is such an extensive overlap between birds in littoral and limnetic habitats within this system, they will be treated as one ecological unit. About 23 species dominate the avifauna of the lacustrine ecosystem and include the following permanent residents: pied-billed grebe, anhinga, great blue heron, green heron, little blue heron, great egret, snowy egret, white ibis, wood duck, common gallinule, American coot, belted kingfisher, and tree swallow. Dominant winter residents include the mallard, green-winged teal, blue-winged teal, baldpate, ring-necked duck, greater scaup, lesser scaup, ruddy duck, and hooded merganser.

The birds of prey occurring in lacustrine habitats include the peregrine falcon and merlin, both of which are of incidental significance in the trophic structure (Fig. 5-20). The peregrine falcon has experienced dramatic declines within the continental United States. Within South Carolina and Georgia, this species is regularly reported only during migration and in the winter (Gauthreaux et al. 1979).

The merlin, formerly known as the pigeon hawk, is a fairly common winter resident occurring in this habitat. The merlin's diet consists chiefly of birds, usually warblers, sparrows, and vireos. It also takes insects, dragonflies, grasshoppers, beetles, and small mammals.

Because of the great similarity between lacustrine and palustrine habitats and because of the high degree of mobility possessed by birds, the avifauna of these two ecosystems is quite similar. For additional information on the birds common to both systems, the reader is referred to the sections on birds of the palustrine ecosystem.

F. MAMMALS

See the section on mammals of the palustrine ecosystem for information on mammals in freshwater ecosystems.

IV. RIVERINE ECOSYSTEM

A. VASCULAR FLORA (for nonvascular plants, see the palustrine section)

Vascular flora of the riverine ecosystem includes only the nonpersistent emergent, submergent, and floating vegetation found within river or tributary stream channels. Wetlands dominated by trees, shrubs, or persistent emergents are included under the palustrine system where salinities are less than 0.5°/oo and under the estuarine system where salinities are greater than 0.5°/oo. If the river or stream channel is not easily defined, the beginning of persistent emergents (e.g., _Typha_ spp.) marks the palustrine boundary of the riverine system (Cowardin et al. 1977).

Very little information is available on the ecology of river channel plant communities. Several species lists, however, exist for this habitat in South Carolina. A transect survey of the aquatic vegetation of the Santee and Cooper rivers was carried out in 1971 and 1972 (Curtis 1972). Water-weed and alligator-weed, both introduced aquatics, dominated the flora of the Cooper River channel, while alligator-weed dominated in the Santee. Table 5-34 lists the species and their occurrence from the above study. Nelson (1974) listed pickerelweed, arrowheads, lizard's tail, swamp dock, water hemlock, leather-flower, spider-lily, water-primroses, eastern lilaeopsis, and arrow-arum as nonpersistent emergents from the Cooper River; alligator-weed was listed as the only floating-leaved plant.

Havel (1976) added water hemp, marsh fleabane, two smartweeds, knotweed, and tearthumb to the list of nonpersistent emergents from the Santee River channel. Tiner (1977) reported approximately 33 nonpersistent emergents, 3 submergents, and 13 floating plants that potentially occur in the river channels of South Carolina (see Tables 5-3 and 5-5). Common floating aquatics listed were frog's-bit, yellow pond-lily, white water-lily, pennyworts,

396

Table 5-33. Dominant, moderate, and minor bird species of the lacustrine ecosystem (Sprunt and Chamberlain,1949, 1970, Audubon Field Notes 1967 - 1970, Chamberlain 1968, American Birds 1971 - 1977, Forsythe 1978).

DOMINANT		MODERATE		MINOR	
Pied-billed grebe	C PR	Black-crowned night heron	C PR	Horned grebe	C WR
Anhinga	C PR	Glossy ibis	C SR	Yellow-crowned night heron	FC SR
Great b uel her α	C PR	Black duck	C WR	Wood stork	FC PR
Green heron	C PR	Pintail	C WR	Whistling swan	R WR
Little blue her α	C PR	Shoveler	C WR	Canada goose	FC WR
Great egret	C PR	Bufflehead	FC WR	White-fronted goose	R WR
Snowy egret	C PR	Red-breasted merganser	C WR	Snow goose	R WR
White ibis	C PR	Greater yellowlegs	C PR	Fulvous whistling duck	U WR
Mallard	C WR	Spotted sandpiper	C PR	Gadwall	U WR
Green-winged teal	C WR	Ring-billed gull	C PR	Cinnamon teal	R WR
Blue-winged teal	C WR	Black tern	C SR	European wigeon	R WR
Baldpate	C WR	Rough-winged swallow	C SR	Redhead	U WR
Wood duck	C PR	Barn swallow	C PR	Canvasback	FC WR
Ring-necked duck	C WR			Common goldeneye	U WR
Greater scaup	C WR			Oldsquaw	U WR
Lesser Scaup	C WR			Common merganser	U WR
Ruddy duck	C WR			Black vulture	C PR
Hooded merganser	C WR			Turkey vulture	C PR
Common gallinule	C PR			Osprey	U PR
American C o	C PR			Peregrine falcon	R WR

397

Table 5-33. Concluded

DOMINANT		MODERATE	MINOR	
Chimney swift	C SR		Merlin	R WR
Belted kingfisher	C PR		Limpkin	R BR[a]
Tree swallow	C PR		Purple gallinule	FC T
			Lesser yellowlegs	FC T
			Solitary sandpiper	FC T
			American avocet	R WR
			Herring gull	C PR
			Laughing gull	C PR
			Bonaparte's gull	C WR
			Forster's tern	C PR
			Bank swallow	U T
			Cliff swallow	U T
			Purple martin	C SR
			Common crow	C PR
			Fish crow	C PR

Note: Dominance indicates relative importance of the species as a group in the community. This concept is not based necessarily on taxonomic relationships but rather on numbers, size, and trophic dynamics.

a. Lower Georgia.

KEY: PR – Permanent resident, present year around   C – Common, seen in good numbers
     WR – Winter resident   FC – Fairly common, moderate numbers
     SR – Summer resident   U – Uncommon, small numbers irregularly
     BR – Breeding resident   R – Rare
     T – Transient

398

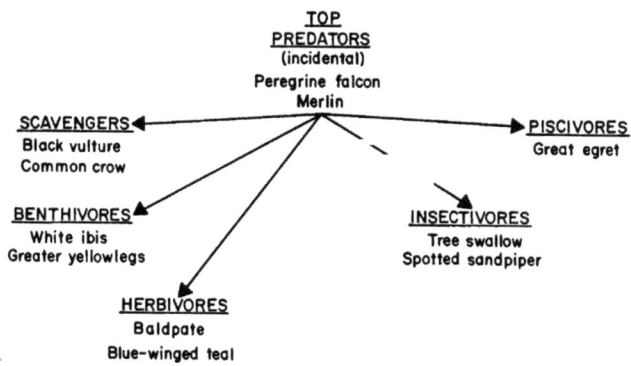

Figure 5-20. Trophic structure of the avifauna of lacustrine environments.

duckweeds, big duckweed, water-weeds, golden club, and mosquito fern. Common submergents included parrot-feather and pondweeds. Nonpersistent emergents listed by Tiner (1977) as common were lizard's tail, mock-bishopweed, and sacciolepis. In protected river channels, floating mats of vegetation often occur. Butter-leaved pennywort, smartweeds, and alligator-weed are common in these mats which are most common along black-water rivers (National Wetlands Inventory 1978).

The ecology of southeastern riverine or channel vegetation has not been well documented. Most vascular vegetation in the river channels of the study area clings close to the channel edge, rarely hindering navigation. Submergent vegetation is common in still, deep water, but the flow rate of most rivers prevents colonization of the deeper parts of the channel. Nonpersistent emergent, submergent, and floating vegetation is common within the channels of slow-flowing tributary streams of the Sea Island Coastal Region, but its distribution and ecology are poorly documented.

B. INVERTEBRATES

1. Zooplankton

While much of the zooplankton of riverine environments is washed in from drainage basin lakes, ponds, and back waters, there is a resident holoplanktonic component. Zooplankton development is most pronounced in the slower moving portions of a river system where reduced current velocity and silt deposition tend to make these habitats nearly indistinguishable from typical lentic habitats. These

same situations, however, may be dominated by dense floating or rooted vegetation, which may effectively filter zooplankton out.

Some tidal riverine zooplankton is chiefly of marine ancestry and is characterized by the ability to maintain internal fluids hypertonic to the surrounding water (Pennak 1953). These organisms generally are considered stenohaline, but very euryhaline estuarine species may occur regularly in riverine situations. These euryhaline forms differ from marine zooplankters in that they tend to have fewer eggs and more often are carried by the parent until they hatch. Floating eggs almost never occur in freshwater species (except in cladocerans and insects), whereas they are common among marine zooplankton species. No large zooplankton taxonomic groups are exclusively riverine (fresh water), but the Rotatoria, Nematomorpha, Cladocera, and Hydroacorina are generally considered freshwater organisms because there are so few marine and estuarine species in these groups.

The problem of maintaining zooplankton populations in riverine systems when currents are always strong enough to carry the zooplankton seaward may be solved in three ways: 1) resting eggs may be transported upstream by fishes, birds, or other animals; 2) repopulation may be from populations present in lacustrine or palustrine elements of a drainage basin; and 3) the organisms may exhibit high reproductive rates coupled with short life cycles. In Sea Island Coastal Region rivers, probably each of these mechanisms is involved, with the extensive contiguous palustrine forested wetlands most important as reservoirs of seed stock.

Table 5-34. Checklist of aquatic vegetation collected in 1971 and 1972 from the Santee and Cooper rivers (adapted from Curtis 1972).[a]

Scientific Name	Common Name	Cooper River	Santee River
Alisma plantago-aquatica	Broadleaf waterplantain		X
Alternanthera philoxeroides	Alligator-weed	X	X
Elodea canadensis	Water-weed	X	X
Egeria densa	Water-weed	X	X
Cabomba caroliniana	Fanwort	X	
Cabomba sp.	Fanwort	X	
Ceratophyllum demersum	Coontail	X	
Chara sp. (non-vascular)	Muskgrass	X	
Cladium jamaicense	Saw grass		X
Eleocharis parvula	Dwarf spikerush	X	
Eleocharis sp.	Spikerush	X	X
Eryngium aquaticum	Marsh eryngo		X
Bacopa caroliniana	Lemon bacopa		X
Juncus biflorus	Rush	X	
Leersia hexandra	Cutgrass		X
Lemna minor	Duckweed	X	
Lycopus sessilifolius	Bugleweed		X
Myriophyllum heterophyllum	Water milfoil		X
Nelumbo lutea	Lotus	X	
Nitella sp. (non-vascular)	Nitella	X	
Nymphaea odorata	White water-lily	X	
Polygonum sp.	Smartweed	X	X
Pontederia cordata	Pickerelweed	X	X
Potamogeton diversifolius	Variable-leaved pondweed	X	
Potamogeton sp.	Pondweed	X	
Ruppia maritima	Widgeon grass	X	
Rhynchospora careyana	Beak rush		X
Rhynchospora macrostachya	Beak rush		X
Sagittaria latifolia	Duck potato	X	X
Samolus parviflorus	Water pimpernel	X	X
Scirpus etuberculatus	Bulrush		X
Scirpus validus	Bulrush	X	X
Spartina sp.	Cordgrass	X	X
Typha angustifolia	Narrow-leaved cat-tail	X	
Typha latifolia	Common cat-tail	X	
Utricularia sp.	Bladderwort	X	
Xyris caroliniana	Yellow-eyed grass	X	
Zannichellia palustris	Horned-pondweed	X	
Zostera marina	Eel grass	X	

a. Includes brackish water plants.

Protozoa may be found in all aquatic habitats, and many are planktonic. Class Sporozoa has only parasitic members, and although these may parasitize plankton, they are not true plankters themselves. A number of soft-bodied forms belonging to the Sarcodina and Infusoria are found in the plankton. However, most freshwater zooplanktonic protista belong to the class Mastigophora; in many instances, they are second only to green algae in importance. Free-living protozoa exhibit varied types of nutrition, from holophytial (plant-like) photosynthesis to saprophytic (absorb salts and organic compounds from surroundings) to holozoic (herbivores, carnivores, and omnivores). The protozoa, in turn, are important as food for many species of rotifers, cladocerans, and copepods. Kathorobic protozoa are found in streams and rivers rich in oxygen and low in organic matter and would likely be prominent in riverine situations. Patrick et al. (1967) listed a large number of protozoa from the Savannah River, some of which would likely be planktonic. Asexual reproductive bodies of sponges may be tychoplanktonic ("accidental" plankton) at times, with reproduction occurring in the fall.

Ploimate rotatoria are the most characteristic taxonomic group of zooplankton in a stable river system. This is one of the few groups to have originated in fresh water, and of 1,700 known species, more than 1,600 species occur there (Winner 1975). Rotifers feed on periphyton, detritus, and small planktonic organisms such as algae, small cladocera, and larvae of other rotifers. Parthenogenetic reproduction takes place most of the year, with the population chiefly female. Males, usually appearing only at certain times of the year, are all planktonic, and live a very short time. The males are specialized for reproduction and often lack a digestive tract. Fertilized or resting eggs may be produced in response to environmental change. Some species may have one abundance peak per year, while others exhibit two or more. Most riverine plankton communities average betwen 40 - 500 rotifers/liter (150 - 2,000 rotifers/gallon).

Cladocera are primarily freshwater crustaceans and are found in all fresh water except rapid streams, brooks, and heavily polluted waters. They rank just behind rotifers in importance as riverine zooplankters. They feed on algae, protozoa, organic detritus, and bacteria. Some are predacious (e.g., Leptodora and Polyphemus), feeding on rotifers and small crustaceans. Parthenogenetic reproduction occurs during most of the year, with only females produced, except during spring and fall when males appear and fertilized or resting eggs are formed. One or more population maxima may occur during the year. Patrick et al. (1967) reported two species of cladocera from the Savannah River, Georgia.

Copepods are the third important component of riverine zooplankton. While of marine origin, five of the seven orders have freshwater representatives. Only freshwater calanoids, cyclopoids, and euryhaline harpacticoids such as Scottolana canadensis are present as holoplankters. Most calanoids are filter feeders on algae; however, cyclopoids have mouth parts modified for seizing other zooplankters. Eggs are held in ovisacs attached to the female until they hatch, and only fertilized eggs are viable. Larvae hatch out as nauplii and, after five molts, metamorphose into copepodids. There are five copepodid instars prior to the adult stage. Special thick-walled eggs are produced to withstand adverse environmental conditions, although ordinary eggs are thought to overwinter in an extended incubation period. Cyclops encysts as a copepodid; its cysts are resistant to desiccation. Seasonal copepod maxima in temperate climates probably vary with temperature, as was the case in the Pamunkey River, Virginia (Burrell 1972).

Neomysis americanus, a euryhaline estuarine mysid shrimp, may occur in tidal riverine zooplankton , particularly as juveniles. This species was found regularly at a freshwater station in the Pamunkey River by Burrell (1968), and is known to be abundant in South Carolina estuaries (Kelley 1978). Banner (1953) described a freshwater mysid shrimp species, Taphromysis louisianae, from southern waters of the United States.

Gammarid amphipods may occur in riverine plankton collections, but most are true plankters and may be carried off the bottom by currents. Patrick et al. (1967) listed Gammarus fasciatus and Crangonyx gracilis from the Savannah River. Fox (1978) reported Crangonyx pseudo gracilis and species of the Gammarus fasciatus tigiurus complex to be present in the Southeast.

Larvae of shrimp and xanthid crabs are the most common riverine decapod plankters. Species reported or thought to be present in the Sea Island Coastal Region are Palaemonetes paludosus, Macrobrachium ohione, Macrobrachium acanthurus, Uca minax, and Rhithropanopeus harrisii. Wharton (1978) listed Macrobrachium ohione and Macrobrachium acanthurus as present in the Ochlockonee River and Palaemonetes paludosus as present in this system and in the Altamaha River, both in Georgia. He also listed Palaemonetes kadiakensis as widely distributed in Georgia streams, but this species is not known from South Carolina and is perhaps a misidentification (P. A. Sandifer, 1978, South Carolina Marine Resources Division, Charleston, pers. comm.). Patrick et al. (1967) also found Macrobrachium ohione, Macrobrachium acanthurus, and Palaemonetes paludosus present in the Savannah River, Georgia.

Riverine zooplankton of coastal South Carolina and Georgia has not been studied intensively as a group in any system. The most indepth study was of the Cooper River and adjacent streams from a March 1975 sampling program (Dames and Moore Associates 1975). Twelve taxa of Rotatoria, four taxa of copepods, and two taxa of Cladocera were identified from samples taken near high tide and near low tide at six midchannel sites. Total density of zooplankton ranged from 14.4 organisms/liter to 38.2/liter. Rotifers and copepods were dominant at each station, together comprising 88% - 100% of the zooplankton community. The numbers of rotifers/liter ranged from 6.6 to 20.4, with Polyarthra sp. and Keratella cochlearis most abundant. Copepods varied from 6.6 to 20.7/liter (25 to 79/gallon), with the most abundant taxa identified only as nauplii. The only copepod genus differentiated was Diaptomus, which was present at five stations. Cladocerans varied from 4% to 12% of the zooplankton at each station and consisted of two species, Bosmina longirostris (most numerous) and Alonella sp. Six of the 12 genera of rotifers found usually

401

dominate the zooplankton of running waters; in this study, however, copepods were equally abundant. Heard and Heard (1971) found the zooplankton of Riceboro Creek, Georgia, to be chiefly unidentified species, freshwater copepods, and cladocerans during wet seasons, and larvae of estuarine species during dry seasons.

In studies of the zooplankton of Keowee Reservoir, South Carolina, Hudson (1975) found several of the same species reported from riverine reaches of the Pamunkey River, Virginia, by Burrell (1968, 1972). Those species reported from both areas are probably also present in coastal riverine environments in South Carolina. They include the cyclopoid copepods Eucyclops agilis, Cyclops vernalis, and Mesocyclops edax, and the cladocerans Leptodora kindti, Sida crystallina, Diaphanosoma branchyurum, Ilyocryptus sordidus, and Simocephalus exspinosus. Williams (1966) listed Keratella sp., Polyarthra sp., and Trichocera sp. as the most abundant rotifers present in the Savannah River at North Augusta, South Carolina.

2. Benthic Invertebrates

Literature dealing with the benthic invertebrates of riverine environments in the United States is largely taxonomic in approach, and papers dealing with individual taxa are numerous. Keys, descriptions, and illustrations of species, and lists of important references to the invertebrate fauna of such habitats are provided in publications such as those of Pennak (1953) and Edmondson (1959). Invertebrate phyla having free-living representatives in freshwater environments include the Porifera, Cnidaria, Platyhelminthes, Rhynchocoela, Rotifera, Nematoda, Nematomorpha, Bryozoa, Gastrotricha, Annelida, Mollusca, Tardigrada, and Arthropoda.

Hynes (1970) noted that invertebrates characteristic of soft riverine substrates include tubificids (annelids), chironomids (insects), burrowing mayflies, prosobranch gastropods, and pelecypods. The fauna on hard substrates commonly includes the sessile sponges, hydrozoans, and bryozoans as well as more motile forms such as flatworms, insect larvae, gastropods, amphipods, and isopods. Nematodes are ubiquitous but easily overlooked because of their size. Other small and easily overlooked species include rotifers (common on detritus and aquatic plants during the warmer months), gastrotrichs (frequent on detritus, submerged vegetation, and in sandy substrates), and tardigrades (occasional on mosses, algae, aquatic tracheophytes, detritus, and substrates of sand and mud). Various species of invertebrates are known to

be especially abundant in tributaries immediately below lakes because of an abundant food supply in the form of plankton (Hynes 1970).

Crayfish, among the more conspicuous freshwater invertebrates because of their size, are represented by 24 species in South Carolina (Wishart and Loyacano 1974). Although Wishart and Loyacano found crayfish within the Sea Island Coastal Region, they did not observe concentrations large enough to support a commercial fishery. The spread of the introduced Asiatic clam Corbicula manilensis into rivers of the Georgia and South Carolina coastal plain is of particular concern. This species, a nuisance in many areas of the country, is now known to occur in many of the riverine systems of the study area. One of the more curious insects known from the study area is a mayfly of the genus Tortopus. Larvae of this species construct U-shaped burrows in clay banks along the Savannah River (Scott 1959, Hynes 1970).

Factors limiting the occurrence and distribution of benthic invertebrates in lotic habitats include current velocity, water temperature (including seasonal and altitudinal differences), substrate type (including vegetation and degree of siltation), dissolved substances, drought and floods, food, interspecific competition, shade, and zoogeography (Hynes 1970), as well as pollution. Benthic community composition typically changes along the reach from mouth to head of a river, reflecting changes in sediments, water currents, vegetation, turbidity, food, and water chemistry. Significant changes also occur across a river from the shore to the channel. Beds of aquatic plants, detritus accumulations, and the availability of diverse substrates combine to provide food and shelter for a diverse benthic fauna in the shallows. Fewer species normally occur in the predominantly sand and clay bottoms of current-scoured channels.

Most freshwater animals are stenohaline, and few penetrate very far into estuarine areas (Remane and Schlieper 1971). Remane (1934) demonstrated that species numbers decline progressively as salinity declines, to a minimum near the juncture of the oligohaline and mesohaline zones. The number of species then increases rapidly into fresh waters. Freshwater and marine faunas are essentially distinct, despite the occurrence of a few holeuryhaline species. In a study of the hydroids, only one species (Cordylophora caspia) was found in both limnetic and estuarine zones of various rivers in coastal South Carolina (Calder 1976). As a group, the hydroids are largely marine, and the fewest number of species encountered along the halocline was found in freshwater areas.

Seasonal changes in benthic community structure are well known in temperate riverine environments (Hynes 1970). Such changes are closely tied to life histories of the species comprising the invertebrate assemblages, and to predation. Major seasonal fluctuations in numbers are especially evident in the larval insect component of the fauna. Invertebrate densities are often lowest during the summer, a time when predation is high and many of the insect larvae have metamorphosed into adults.

A number of year-round studies on the benthic invertebrates of lotic environments in the mountains and Piedmont areas of the Carolinas and Georgia have been published (e.g., Tebo and Hassler 1961, Nelson and Scott 1962). A few field studies on invertebrate community ecology have been undertaken in freshwater environments of the Sea Island Coastal Region, mostly as part of impact assessments, but little has been published in journals or in other readily available sources. Patrick et al. (1967) listed the flora and fauna, including species of benthic invertebrates, from the Savannah River near the Savannah River Plant. Dorjes (1977) observed that the dominant macrobenthic invertebrates from freshwater areas of the Ogeechee estuary were the amphipod _Lepidactylus dytiscus_ and the polychaete _Scolecolepides viridis_. Several studies have been undertaken in the Charleston area for various industrial concerns (Westvaco Corporation 1972, Academy of Natural Sciences of Philadelphia 1974, Enwright Associates, Inc. 1977). Through the cooperation of DuPont, Westvaco, and South Carolina Electric and Gas corporations, we have been granted permission to include information from these reports (Tables 5-35 and 5-36). The most comprehensive of these studies is that by Enwright Associates, Inc., who sampled the fauna quantitatively over several seasons near the Arthur M. Williams station of South Carolina Electric and Gas Company. A total of 176 taxa were distinguished in their samples, including 78 insects, 28 oligochaetes, 16 gastropods, and 10 amphipods. In addition, 88 species of macroinvertebrates collected in ichthyoplankton samples were also identified and counted.

Some data are also available from benthic studies conducted during the mid-1970's at stations in fresh waters by the Estuarine Survey Program of the South Carolina Marine Resources Research Institute. The amphipod _Lepidactylus dytiscus_ and the polychaete _Scolecolepides viridis_ accounted for nearly 60% by number of the macroinvertebrates from a limnetic (occasionally oligohaline) station on the South Edisto River (Table 5-37). Sampling was conducted on a sandy bottom, and _Lepidactylus_ densities averaged 139/m for the year (Calder et al. 1977b).

Community structure was considerably different at the Tee on the Cooper River (Table 5-38) and reflected the predominantly hard clay bottom at this location (Calder and Boothe 1977). At a sandy bottom location in the Waccamaw River, oligochaetes, the bivalve _Corbicula manilensis_, the polychaete _Scolecolepides viridis_, chironomid larvae, and the amphipod _Lepidactylus dytiscus_ accounted for over 85% of the numbers of macroinvertebrates (Table 5-39). All the areas sampled by the Estuarine Survey Program near the limnetic-oligohaline border (i.e., within the riverine tidal reach) have been characterized by the presence of relatively few species of macroinvertebrates and rather low species diversity, particularly in comparison with shelly bottom areas in high and stable salinities along the coast. (See also the following section on insects of the riverine ecosystem.)

3. Insects

Several investigators have included riverine insects among their field surveys of the study area. Curtis (1971) conducted a survey of macrobenthos of both the Santee and Cooper rivers, South Carolina, to determine the relative abundance and species composition of bottom fauna in these rivers. Results of the insect portion of that survey are shown in Table 5-40. All benthic riverine insects obtained during that sampling study were immature stages, either nymphs, larvae, or pupae. Midges (Diptera) outnumbered all other insects, and only oligochaete worms (not tabled) were as numerous among all the other benthic macrofauna collected.

A number of other macroinvertebrate studies have resulted in a useful compilation of South Carolina riverine insect information (Westvaco Corporation 1972, Richardson 1974, Enwright Associates, Inc. 1977). Richardson (1974) studied insects of the Santee-Cooper River System and found that the riverine odonate (dragonfly and damselfly) fauna was well distributed. Damselflies were particularly abundant among beds of Brazilian elodea, but were also found on trailing roots, submerged and floating wood, and entrapped debris. Representatives of each type (climbers, crawlers, and burrowers) of dragonfly naiad were found. Among the most common species of climbers were _Nasiaeschna pentacantha_ and _Coryphaeschna ingens_, which clambered among the erect stems of the coarse marsh vegetation. The sprawlers and burrowers were much more abundant than the climbing species, and the soft flocculent mud near riverine creek inlets and heavy deposits of decaying vegetation elsewhere were well suited for a number of species (e.g. _Epicordulia princeps_ and _Tetragoneuria cynosura_).

403

Table 5-35.	Taxa of macroinvertebrates (exclusive of Insecta) from stations in or near the limnetic zone of the Cooper River, Charleston County, South Carolina (Fuller 1974).

TAXON	NUMBER OF SPECIES
Porifera	1
Platyhelminthes	1
Nematoda	1
Bryozoa (Ectoprocta)	2
Annelida	
Oligochaeta	3
Hirudinea	6
Mollusca	
Pelecypoda	6
Gastropoda	7
Arthropoda	
Isopoda	3
Amphipoda	3
Decapoda	4
Acarina	4

Table 5-36.	Taxa of macroinvertebrates identified from the limnetic zone of the Cooper River, Charleston County, South Carolina (Enwright Associates, Inc. 1977).

TAXON	NUMBER OF SPECIES
Porifera	1
Cnidaria	2
Platyhelminthes	4
Rhynchocoela	1
Nematoda	?
Entoprocta	1
Bryozoa (Ectoprocta)	2
Annelida	
Oligochaeta	28
Polychaeta	1
Hirudinea	9
Arthropoda	
Branchiura	1
Isopoda	6
Amphipoda	10
Decapoda	8
Arachnida	1
Insecta	78
Mollusca	
Gastropoda	16
Pelecypoda	6

Three families of mayflies (Ephemeroptera) were found by Richardson. These families are particularly well suited to slow-flow riverine conditions. The Baetidae, particularly Callibaetis, were very numerous in Brazilian elodea beds along the river margins. The Caenidae were represented by two prolific species, Caenis sp. and Tricorythodes sp., both of which are most often associated with slow (but still lotic), silted, backwater habitats. Members of the mayfly family Heptageniidae, represented by Stenonema, were restricted mainly to floating wood, but, when present, their populations were large.

The Diptera, or true flies, were dominated by the midges (Chironomidae), although the true crane flies (Tipulidae), biting midges (Ceratopogonidae), horseflies (Tabanidae) and marsh flies (Sciomyzidae) were also represented. Richardson (1974) found that species diversities of true bugs (Hemiptera), beetles (Coleoptera), and caddisflies (Trichoptera) in permanently flowing waters were most abundant inland adjacent to palustrine areas and decreased in the direction of the coast.

Among the true bugs, large populations of giant water beetles (Belastomatidae) and creeping water bugs (Naucoridae) were harbored where vast beds of aquatic macrophytes were present. Along river margins, water measurers (Hydrometridae) and marsh treaders (Mesoveliidae) were exceedingly common. At upper riverine locations having low flow, the most numerous true bugs were the striders (Gerridae).

Representative of the beetles were the whirligig beetles (Gyrinidae) Gyrinus analis and Dineutus assimilis, the elmid beetle Dubiraphia bivittata, and the crawling water beetles (Haliplidae) Haliplus triopsis and Peltodytes sexmaculatus. Among the caddisflies, the dominant group was the leptocerids, with Triaenoides injusta in great abundance.

Enwright Associates, Inc. (1977) also surveyed riverine insects and included bottom (ponar dredge) and water column (plankton net) collections, as well as qualitative sampling of shoreline and submergent vegetation. Results of that insect survey are provided in Table 5-41 for comparison of findings with those of Curtis (1971) and Richardson (1974), and to augment the summary of known riverine insect taxa for the Sea Island Coastal Region.

Table 5-37. Species of macroinvertebrates collected during four seasons at limnetic stations on the South Edisto River, and their estimated densities in numbers/m². Percent of total fauna, cumulative percent, and rank by number are given for each species. A = Amphipoda, P = Polychaete, I = Isopod, In = Insect larva, B = Bivalve, An = Anthozoan (Calder et al. 1977b).

SPECIES	JUNE	OCT.	JAN.	APRIL	% OF FAUNA	CUMUL. %	RANK BY NUMBER
Lepidactylus dytiscus (A)	59	236	172	87	43.55	43.55	1
Scolecolepides viridis (P)	102			182	14.31	57.86	2
Polychaeta A (undet.)	38	8	5	41	12.26	70.12	3
Chiridotea sp. (I)		38		5	6.37	76.49	4
Polychaeta B (undet.)				72	5.66	82.15	5
Ceratopogonidae (undet.) (In)	5	46	20	10	5.58	87.73	6
Gammaridae (undet.) (A)		5	18	23	2.59	90.32	7
Parapleustes aestuarius (A)				13	1.81	92.13	8
Corbicula manilensis (B)		5			1.42	93.55	9
Gammarus fasciatus (A)	8	8			1.26	94.81	10
Cumacea (undet.)		5		10	1.18	95.99	11
Paraprionospio pinnata (P)		3	8		0.86	96.85	12
Actiniaria (undet.) (An)		10			0.79	97.64	13
Polydora ligni (P)			8		0.63	98.27	14
Cyathura polita (I)	3	5			0.63	98.90	14
Nereis succinea (P)			5		0.39	99.29	15
Oligochaeta (undet.)	3				0.24	99.53	16
Monoculodes edwardsi (A)		3			0.24	99.77	16
Diptera (undet.) (In)		3			0.24	100.01	16
No. Individuals	218	375	236	443			
No. Species	7	13	7	9			
Species Diversity (H')	1.93	2.00	1.48	2.42			
Species Richness	1.11	2.02	1.10	1.31			
Evenness (J')	0.69	0.54	0.53	0.76			

Table 5-38. Species of macroinvertebrates collected at a freshwater riverine station on the Cooper River in July 1973 and January 1974, and their estimated densities in numbers/ m^2. Percent of fauna, cumulative percent, and rank by number are given for each species. A = Amphipod, I = Isopod, B = Bivalve, In = Insect larvae and pupae (Calder and Boothe 1977).

Species	July	January	% of fauna	Cumul. %	Rank by Number
Gammarus sp. (A)	210	369	67.09	67.09	1
Cyathura polita (I)	77	44	14.02	81.11	2
Corbicula manilensis (B)	41	8	5.68	86.79	3
Diptera larva (undet.) (In)	49		5.68	92.47	3
Diptera pupae (undet.) (In)	26		3.01	95.48	4
Unidentified Taxon	15		1.74	97.22	5
Chironomidae (undet.) (In)		15	1.74	98.96	5
Corophium lacustre (A)		3	0.35	99.31	6
Mytilopsis leucophaeata (B)			0.35	99.66	6
Gastropoda (undet.)	-		0.35	100.01	6
No. Individuals	424	439			
No. Species	8	5			
Species Richness	1.16	0.66			
Species Diversity (H')	2.15	0.86			
Evenness (J')	0.72	0.37			

In a 5-year riverine study of macro-scopic bottom-dwelling invertebrates conducted for the Westvaco Corporation by the Institute of Paper Chemistry (Westvaco Corporation 1972), dipteran (true fly) insects were included. This study is particularly useful in that it provides density estimates for these riverine insects. The midge family (Chironomidae) dominated the dipteran assemblage, with densities ranging from 0 to 324 individuals/m^2 (0 to 270 individuals/yd^2) for Calopsectra, 0 to 54 individuals/m^2 (0 to 45 individuals/yd^2) for Procladius and Cryptochironomus, and 0 to 81 individuals/m^2 (0 to 68 individuals/yd^2) for Polypedilum. The Georgia Water Quality Control Board (1971, 1972) surveyed a number of sites along the Chattahoochee and Flint rivers in Georgia, and included riverine insects in their sampling. Insects present at various Chattanoochee River stations were sub-divided into groups generally considered to be intolerant, partially tolerant,

or tolerant of polluted river waters. A summary of these results is provided in Hammer, Siler and George Associates (1975).

A common order of insects found in and around aquatic habitats (particularly lotic environments) of the study area is the caddisflies (Trichoptera). Adult caddisflies are small-to-medium-sized moth-like insects found near streams (and also ponds and lakes), particularly from spring to early fall. Flight in most species is rapid, with well-developed dodging movements. Mouth parts are feeble and specialized for the ingestion of liquid foods (Pennak 1953).

Caddisfly larvae are chiefly omnivorous, although the Limnephilidae, Hydropsychidae, and many Hydroptilidae feed primarily on diatoms, other algae, and higher plants. Some other trichopterans are thought to be generally carnivorous. Animal food of these latter caddisfly larvae consists of

Table 5-39. Taxa of macroinvertebrates collected during January and April 1977 at a station in the Waccamaw River 6 miles above Georgetown, South Carolina. Estimated densities are given in numbers/m². Percent of fauna, cumulative percent, and rank by number are given for each taxon. B = bivalve; P = polychaete; A = amphipod; I = isopod (D. R. Calder and B. B. Boothe, 1978, South Carolina Marine Resources Division, Charleston, unpubl. data).

Taxon	January	April	% of Fauna	Cumul. %	Rank by Number
Oligochaeta (undet.)	265	445	40.97	40.97	1
Corbicula manilensis (B)	128	185	18.06	59.03	2
Scolecolepides viridis (P)	5	183	10.85	69.88	3
Chironomidae (undet.)	123	23	8.42	78.30	4
Lepidactylus dytiscus (A)	110	20	7.50	85.80	5
Acetes americanus carolinae	110		6.35	92.15	6
Ceratopogonidae (undet.)	35	43	4.50	96.65	7
Gammarus sp. (A)		25	1.44	98.09	8
Chiridotea sp. (I)	3	8	0.63	98.72	9
Mysidacea (undet.)		5	0.29	99.01	10
Batea catharinensis (A)		5	0.29	99.30	10
Hirudinea (undet.)		3	0.17	99.47	11
Cumacea (undet.)			0.17	99.64	11
Isopoda (undet.)		3	0.17	99.81	11
Amphipoda (undet.)	3		0.17	99.98	11
No. Individuals	785	948			
No. Species	10	12			
Species Richness	1.35	1.60			
Species Diversity (H')	2.51	2.21			
Evenness (J')	0.76	0.62			

small crustaceans, annelids, and insect larvae. Most caddisfly larvae move about actively in search of food, but those stream forms that build catch-nets simply eat the plant and animal material that collects on the inner surface of the nets.

Another insect order characteristic of riverine habitats in the Sea Island Coastal Region is the stoneflies (Plecoptera). Pennak (1953) pointed out that adult stoneflies are actually terrestrial, but that they are seldom found very far from running water, the habitat of the immature stages. The adults are somber-colored, elongated, somewhat flattened, medium to large, and decidedly primitive in structure. The legs are well developed. The two pairs of long wings are folded over the back when at rest. Although mouth parts are of the biting type, they are rather weak and, in some cases, the mandibles are reduced. Stoneflies are poor fliers and are usually found resting on objects along the shores of streams (or lakes). In temperate climates, such

Table 5-40. Number of riverine insects obtained from macrobenthic sampling of the Santee and Cooper rivers (adapted from Curtis 1971).

Taxa	Santee	Cooper
Collembola		
Isotomurus		2
Ephemeroptera		
Caenis (nymph)		4
Paraleptophlebia (nymph)		1
Cloeon (nymph)		7
Odonata		
Helocordulia (nymph)		1
Lestes (nymph)		4
Agrion (nymph)		1
Gomphus (nymph)	1	
Trichoptera	6	
Hydropsyche (larvae)	3	1
Triaenodes (larvae)		2
Neureclipsis (larvae)	5	2
Odontoceridae (larvae)		1
Coleoptera		
Dystiscidae (larvae)	3	
Cybister (larvae)	1	
Diptera		
Tendipedidae		
Pentaneura (larvae)	44	36
Pentaneura (pupae)	10	8
Hydrobaneus (larvae)	9	5
Hydrobaneus (pupae)		5
Metriocnemus (larvae)	66	27
Tendipes (larvae)	8	38
Tendipes (pupae)	6	7
telmatogen (larvae)		4
tetnans (larvae)		15
Calospectra (larvae)	53	
Coryneura (larvae)	1	
Pelopiinae (pupae)	1	
Culicidae		
Chaoborus (larvae)	2	4
Ceratopogonidae		
Culicoides (larvae)		3
Palpomyia (larvae)	18	1

as that of the study area, most adults are found between mid-fall and late summer, although the specific time of appearance of the adult stage varies from species to species. As adults, stoneflies do not live more than several weeks (Pennak 1953).

For the most part, stonefly nymphs are sluggish. They occur in debris, masses of leaves and algae, and under stones in practically every kind of lotic environment. In general, they are only found where dissolved oxygen is abundant. Many species are quite specific in their ecological preferences.

Some occur only in small streams; others, especially the larger forms, occur in larger rivers. Some are found only where the current is swiftest; others, only in pools.

Additional taxonomic, life history, ecological, behavioral and economic information may also be found in the section on insects of the lacustrine ecosystem for the orders Ephemeroptera (mayflies), Odonata (dragonflies and damselflies), Hemiptera (true bugs), Megaloptera (alderflies, dobsonflies, and fishflies), Coleoptera (beetles), and Diptera (flies, mosquitoes, and

Table 5-41. Summary of riverine insects collected from the bottom, shoreline and submergent vegetation, and the water column of the Santee-Cooper River system (Cooper portion) (adapted from Enwright Associates, Inc. 1977).

Insects	Bottom	Riverine Habitat Shoreline & Submergent Vegetation	Water Column
Collembola			
Isotomidae			
Isotomus palustris			X
Ephemeroptera			
Caenidae			
Caenis sp.		X	X
Tricorythodes			X
Heptageniidae			
Stenonema sp.			X
Baetidae			
Baetis sp.		X	
Callibaetis fluctuans		X	
C. sp.		X	
Unidentified Baetidae			X
Odonata			
Aeschnidae			
Anax junius		X	
Libellulidae			
Macrodiplax sp.		X	
Pseudoleon sp.	X		
Unidentified Libellulidae			X
Coenagrionidae			
Enallagma signatum		X	
E. sp.	X	X	X
Ischnura ramburii		X	
I. sp.	X	X	
Unidentified Coenagrionidae	X	X	X
Hemiptera			
Corixidae			
Trichocorixa sp.		X	X
Unidentified Corixidae		X	X
Gerridae			
Rheumatobates sp.		X	
Trepobates sp.		X	
Mesoveliidae			
Mesovelia mulsanti		X	
Belostomatidae			
Belostoma flumineum		X	
Neuroptera			
Sisyridae			
Climacia areolaris			X
Trichoptera			
Psychomyiidae			
Cyrnellus fraternus		X	X
Polycentropus cinereus			X
P. sp.	X		X
Hydropsychidae			
Hydropsyche orris		X	X
Hydroptilidae			
Agraylea sp.			X
Hydroptila sp.		X	X
Orthotrichia sp.			X
Oxyethira sp.			X
Leptoceridae			
Arthripsodes transversus		X	
Leptocella candida			X
Leptocerus americanus		X	X
Cecetis cinerascens		X	X
C. sp.			X
Triaenodes injusta		X	

Table 5-41. Concluded

	Riverine Habitat	
	Shoreline & Submergent Vegetation	Water Column
Lepidoptera		
Pyralidae		
Nymphula maculalis	X	
Synclita sp.	X	
Coleoptera		
Hydrophilidae		
Berosus aculaetus	X	
Elmidae		
Stenelmis sp.		X
Chrysomelidae		
Galerucella nymphaeae	X	
Diptera		
Chaoboridae		
Chaoborus punctipennis	X	X
C. sp.	X	X
Chironomidae		
Tanypodinae		
Ablabesmyia sp.	X	
Clinotanypus sp.	X	X
Coelotanypus concinnus	X	X
C. scapularis	X	X
C. sp.	X	X
Labrundinia sp.		X
Larsia sp.		X
Procladius (Psilotanypus) bellus	X	X
P. (s.s.) sp.		X
Corynoneurinae		
Thienemanniella sp.	X	X
Orthocladiinae		
Cricotopus sp.	X	X	X
Heterotrissocladius sp.			X
Nanocladius sp.	X	X	X
Parakiefferiella	X	X
Chironominae		
Chironomus near attenuatus	X	
C. sp.		X
Cryptochironomus sp.	X	
Dicrotendipes modestus		X	X
D. nervosus·		X	X
D. sp.		X	X
Endochironomus sp.			X
Paratanytarsus sp.	X	X
Polypedilum (s.s.) "convictum" group		X	X
P. (s.s.) "simulans" group	X	
Pseudochironomus sp.			X
Robackia sp.			X
Tanytarsus sp.		X
Xenochironomus (Anceus) sp.	X	
Ceratopogonidae		
Culicoides sp.		X
Dasyhelea sp.			X
Bezzia/Palpomyia complex	X	
Unidentified Ceratopogonidae	X	
Tetanoceridae		
Sepedon fuscipennis		X

midges), all of which are typical of
both riverine and lacustrine habitats.

C. FISHES

1. Freshwater Species

Approximately 24 fish species are
common in the riverine system, with an
additional 22 species utilizing this
habitat less frequently (Table 5-19).
This system is thus comparable to the
limnetic lacustrine subsystem in the
total number of common species. Morrow
(1972) gives an average carrying capacity
range of 113 - 125 lb (51 - 57 kg) of
fishes per acre for streams in the
coastal plain of Georgia. It is assumed
that this figure probably includes the
contiguous palustrine emergent wetlands
and, subsequently, is not the carrying
capacity of the riverine habitat or
the palustrine emergent wetlands,
individually.

Fish studies conducted in this zone
of the Sea Island Coastal Region,
especially those conducted by State
government agencies, have dealt primarily with anadromous herrings and
striped bass (Cadieu and Bayless
1968; Rees 1968; White 1969; Curtis
1970 a, b; Wade 1971; Crochet 1977;
Dudley et al. 1977). Life histories
of fishes occupying this zone in both
States have received little attention.
Although limited census work has been
conducted on the Cooper, Santee, and
Combahee rivers (Curtis 1970b, 1973,
1975), not enough data are available to
discern the ecological communities to
which these fishes belong. Curtis
(1970b) presents limited age-growth
information on redbreast sunfish, bluegill, and largemouth bass in the
Combahee River.

Catfishes (e.g., white, channel,
blue, and flathead catfishes, plus the
various bullheads), largemouth bass,
black crappie, white bass, and yellow
perch are among the most common fish
species considered important to man in
the riverine system. Other species such
as gizzard and threadfin shad, creek
and lake chubsucker, spotted sucker,
golden shiner, green sunfish, carp,
whitefin shiner, coastal shiner, and
longnose gar are also common constituents
of the riverine ichthyofauna.

Most species residing in this
community regularly utilize the shallow,
food-rich, palustrine emergent wetland
communities as foraging grounds. The
diversity of food items available in
this community is rather restricted compared to the emergent wetlands area.
Plankton, benthic invertebrates,
crustaceans, limited insect drift, and
fishes are the primary available foods.
In a comparison of the diets of the
redbreast sunfish and the spotted sucker
in the Satilla River, Coomer et al.
(1978) found a portion of the diet of
these two species to overlap, especially
in the fall. But due to their
different feeding habits, these two
species did not enter into intense
competition for these food items (Table
5-42). While the spotted sucker feeds
on benthic organisms, the redbreast will
feed at the surface or bottom and is believed to utilize food organisms
colonizing snags (Coomer et al. 1978).
It was also found that when the diet of
these species was compared to the total
available food supply (Tables 5-43 and
5-44), little selection could be shown
for either species, indicating that they
are opportunistic feeders.

Stevens (1959), in a study of the
white and channel catfish of the upper
Cooper River and lakes Marion and
Moultrie, found fishes to be the principal food item consumed by both (Table
5-45). Gizzard and threadfin shad were
the most prominent species in the diet
of white catfish, while other catfishes
and herrings were most prominent in
the diet of channel catfish. One interesting item was that pondweed occurred
in 23% of the stomachs of white catfish
and in less than 2% of the channel catfish stomachs examined.

Accounts of spawning activities of
the various species within the riverine
system for species other than the
anadromous stocks are virtually non-
existant. J. Bayless (1978, South
Carolina Wildlife and Marine Resources
Department, Bonneau, pers. comm.)
suggests that the riverine system is the
primary spawning ground for blue, channel,
and flathead catfishes. This area is
also utilized to a lesser degree for
spawning by largemouth bass, black
crappie, white crappie, redbreast sunfish,
and redear sunfish.

Because of its relative inaccessibi-
lity from high ground, the interaction
of riverine ichthyofauna with other predatory animals is more limited than that
occurring in the palustrine system.
Prominent among the predators of fishes
are otters, alligators, turtles, amphiuma,
and sirens. Piscivorous birds, primarily gulls, terns, kingfishers, and
ospreys, are the most significant predators in this area.

2. Anadromous Species

The riverine systems of the Sea
Island Coastal Region are important to
six species of anadromous fishes and one
catadromous species. The anadromous
species are American shad, hickory shad,
blueback herring, striped bass, Atlantic
sturgeon, and shortnose sturgeon; the
American eel is the sole catadromous

411

Table 5-42. Percent composition by weight of food items found in the stomachs of redbreast sunfish (RS) and spotted suckers (SS) captured from the Satilla River, Georgia, during the different seasons (Coomer et al. 1978).

Food Items	Winter RS	Winter SS	Spring RS	Spring SS	Summer RS	Summer SS	Fall RS	Fall SS
Annelida	T[a]	48.7	0.2	42.5	5.8	53.0		
Arachnoidea			T		T		T	
Crustacea								
Cladocera	T	0.2	T	0.3	T	0.7	0.1	3.1
Copepoda	T	2.7	T	9.4	T	15.3	3.0	24.6
Other	8.6	0.1	2.9	T	0.2	0.4		0.1
Insecta								
Coleoptera	0.7	0.1	0.9	2.0	3.1	5.7	5.4	1.2
Diptera								
Ceratopogonidae	0.7	1.0	1.9	1.2	.5	3.4	0.2	3.4
Chironomidae	13.5	38.4	45.7	33.7	8.2	14.0	30.0	52.6
Culicidae						3.7		
Simulidae	0.1	T			0.2	0.1		
Ephemeroptera	12.8	2.0	0.7	0.3	2.3	0.5	2.3	8.8
Hemiptera	0.3		1.9		2.6		0.3	
Hymenoptera	0.1		0.2		0.1			
Odonata	39.5	3.2	32.4	7.7	43.4		2.5	
Plecoptera	2.3		1.5		T		0.4	1.8
Trichoptera	17.8	1.3	11.0	1.4	30.5	3.0	15.2	4.4
Insect pupa	0.1	0.1	0.7	1.5	0.1	0.2	T	T
Osteichthyes	3.5	2.2			3.0		40.6	
Total	100.0	100.0	100.0	100.0	100.0	100.0	100.0	100.0

a. T = Trace

species. Anadromous species annually utilize the rivers as spawning grounds as well as nursery grounds for developing larvae and juveniles, while the catadromous American eel spends most of its adult life in this and adjoining freshwater ecosystems.

a. American shad. Adult American shad spend most of their lives in the ocean, but migrate up coastal rivers to spawn. The spawning migration occurs in the spring in Georgia and South Carolina, beginning in early January when water temperatures are 10° - $15^\circ C$ and ending by late April. Most American shad mature at 3 to 6 years of age, with the majority of males entering the rivers for the first time at 4 years of age and the majority of the females at 5 years. It is thought that most American shad return to their natal rivers to spawn (Hollis 1948).

Spawning generally occurs from the tidal portion of rivers to the headwaters, if natural or man-made obstructions do not restrict upstream movement; however, most American shad seem to spawn in tidally influenced fresh water (Pacheco 1968). Walburg and Nichols (1967) reported major spawning grounds in the Waccamaw River to be near Conway. Crochet et al. (1976),

working in the Waccamaw and Pee Dee rivers, determined that the Pee Dee was probably the major spawning stream of the Winyah Bay system, with possible important spawning sites near Hunt's Bluff and the Hasty Point-Thoroughfare Creek area. In the Lynches River, Walburg and Nichols (1967) determined the major spawning ground to be near the U.S. Hwy. 378 bridge near Lake City and Hannah, South Carolina; they also reported that ripe females indicated spawning grounds ranging from near Andrews to Kingstree, South Carolina, in the Black River. In the Santee River, Walburg and Nichols (1967) delineated major spawning grounds as being between the U.S. Hwy. 52 bridge and Wilson Dam. The major spawning area in the Cooper River was near Stony Landing, just below the tailrace canal of Pinopolis Dam (Walburg and Nichols 1967). Curtis (1970b) reported that American shad spawn above Sland's Bridge (Hwy. 17-A, river mile 34) in the Ashley River. Several studies have been made on the Edisto River to locate spawning grounds: Hildebrand and Cable (1938) found spawning to take place between Givhans Ferry State Park and West Bank; Walburg (1956) produced similar results, although his samples were small; and Wade (1971) reported that 92% of spawning activity in the Edisto occurred between

Table 5-43. Estimated number of organisms/10 ,m^3 by season calculated from drift samples taken in the Satilla River, Georgia, from 20 September 1973 to 26 September 1974 (Coomer et al. 1978).

Classification	Winter	Spring	Summer	Fall	Average
Annelida	2.0	8.0	34.7	17.6	15.6
Arachnoidea				1.1	
Insecta					
Coleoptera	2.4	6.8	5.6	0.4	3.8
Dytiscidae (larvae)			1.5		0.4
Elmidae (adult)		0.8	0.8	0.4	0.5
Elmidae (larvae)	2.4	6.0	3.3		2.9
Diptera	61.0	58.5	101.7	197.7	104.7
Ceratopogonidae	37.5	39.0	14.0	81.0	42.9
Chironomidae	23.5	19.1	86.2	103.3	58.0
Culicidae			0.5	9.9	2.6
Tipulidae		0.4	1.0	3.5	1.2
Ephemeroptera			0.8	75.0	19.0
Heptagenidae			0.8		0.2
Othera				75.0	18.8
Odonata	1.6	0.4	0.5	1.9	1.1
Coenagrionidae	1.2	0.4		1.9	0.9
Gomphidae			0.5		0.1
Other	0.4				0.1
Plecoptera	2.0			17.6	4.9
Trichoptera	0.8		11.1	45.7	14.4
Hydropsychidae			4.3		1.1
Psychomiidae			3.0		0.7
Philopotamidae			3.8		1.0
Other	0.8			45.7	11.6
Insect pupae	0.4		1.0	4.1	1.4
Osteichthyes	0.4			0.6	0.3

a. Individuals placed in this category could only be identified as to order. They may or may not be members of the families listed.

West Bank Landing and Jellico's Landing, and that 8% occurred between Jellico's and Givhans Ferry State Park. Walburg and Nichols (1967) reported spawning grounds in the Ashepoo River from 20 mi (32.2 km) below Walterboro, South Carolina, to the headwaters near Walterboro, and in the Combahee River from 40 to 60 mi (64.4 to 96.6 km) upstream near Miley, South Carolina. Curtis (1970b) could not delineate spawning grounds due to tidal influence at his lower sampling station (Hwy. 17 bridge) in the Combahee River, but he did report that there were indications of spawning within 10 mi (16.1 km) of the bridge; sampling at the U.S. Hwy. 17-A bridge on the Combahee indicated that the bridge was near the lower end of the major spawning grounds (Curtis 1970b).

In the Savannah River, Walburg and Nichols (1967) reported that American shad spawned from the U.S. Hwy. 301 bridge upstream to the Savannah Lock with some spawning in Brier Creek, a tributary entering the river between the mouth of the river and the lower lock. White (1970) corroborated Walburg

and Nichols' findings. Spawning grounds in the Ogeechee River are located between Kings Ferry and Midville, Georgia (Walburg and Nichols 1967). In the Altamaha River, Walburg and Nichols (1967) recorded spawning from Hwy. 144 bridge upstream to both tributaries (the Oconee and Ocmulgee rivers). Godwin and McBay (1967) and McBay (1967) delineated spawning grounds in the Altamaha River from the vicinity of Doctortown, Georgia, upstream to both tributaries. In the Satilla and St. Marys rivers, Walburg and Nichols (1967) reported major spawning grounds near Owens Ferry and between Traders Hill and Folkston, Georgia, respectively.

American shad are free spawners, broadcasting their eggs and milt in open water when water temperatures are 12° – 20°C (54 – 68°F), with median temperatures for spawning between 16° – 17°C (61° – 63°F) (Pacheco 1968). Fecundity ranges from 100,000 to 600,000 eggs, depending on body size and origin of the fish (Davis 1957, Cheek 1968). Davis (1957) determined a fecundity of 360,000 – 480,000 for shad from the Edisto River,

Table 5-44. Estimated number of organisms/m^2 by season calculated from benthic samples taken in the Satilla River, Georgia, from 20 September 1973 to 26 September 1974 (Coomer et al. 1978).

Classification	Winter	Spring	Summer	Fall	Average
Annelida	0.1	0.1	0.1	0.2	0.1
Arachnoidea	0.1	0.6	0.2	0.3	0.3
Crustacea	2.2	8.6	2.4	9.3	5.6
Cladocera	1.6	1.9	0.2	7.4	2.8
Copepoda	0.5	6.6	2.2	1.8	2.8
Decapoda	0.1	0.1		0.1	0.1
Isopoda		T^a			T
Insecta	34.7	42.9	78.0	39.1	48.7
Coleoptera	0.9	4.4	9.5	2.2	4.3
Dytiscidae (adult)	0.2	0.5	0.8	1.1	0.7
Dytiscidae (larvae)	0.1	0.2	1.8	T	0.5
Elmidae (adult)	0.4	2.8	5.0	0.7	2.2
Elmidae (larvae)	0.2	0.7	2.6	0.4	1.0
Gyrinidae (larvae)		0.2	0.1		0.1
Staphylinidae		T			T
Diptera	13.7	9.2	13.7	7.1	10.9
Ceratopogonidae	0.1		0.1		0.1
Chironomidae	2.1	1.2	3.1	6.1	3.1
Culicidae	0.2	0.2	0.4	0.1	0.2
Empididae		T	0.5		0.1
Simulidae	11.3	7.8	9.6	0.9	7.4
Tipulidae				T	T
Ephemeroptera	0.7	2.4	3.1	5.7	3.0
Ephemeridae	0.1	0.6	0.4	0.5	0.4
Heptagenidae		0.1	0.5		0.2
Baetidae	0.2	0.5	2.2	1.4	1.1
Other	0.4	1.2		3.8	1.4
Megaloptera (Corydalidae)	T	T	0.1		0.1
Neuroptera (Sisyridae)	T	T	0.1		0.1
Odonata	0.2	0.1	0.9	0.1	0.3
Coenagrionidae	0.1	0.1	0.2	T	0.1
Gomphidae	0.1		0.1	T	0.1
Libellulidae			0.5	0.1	0.2
Others[b]			0.1	T	0.1
Plecoptera	2.0	0.4	0.1	0.2	0.7
Trichoptera	0.5	4.6	10.8	4.1	5.0
Hydropsychidae	0.2	1.7	8.8	1.3	3.0
Leptoceridae	0.1	0.1	0.4	T	0.2
Psycomiidae	T	0.2	0.9	0.2	0.3
Philopotamidae	T	0.1	0.7	0.1	0.2
Other	0.2	2.5		0.1	0.7
Insect pupae	0.7	1.1	1.9	0.5	1.1
Non-aquatic	0.3	0.4	0.5		0.3
Osteichthyes	0.1	0.5	0.4	T	0.3

a. T = Trace
b. Individuals placed in this category could only be identified as to order. They may or may not be members of the families listed.

South Carolina, and 359,000 - 501,000 for Ogeechee River, Georgia, shad. Vaughn (1967) reported fecundities ranging from 273,000 to 486,700, with a mean of 364,700 for shad from the Altamaha River, Georgia.

When deposited, the eggs are pink or amber transparent spheres and average about 1.3 mm in diameter. After fertilization, they "water-harden" to a diameter of

approximately 2.5 - 3.8 mm. The eggs are slightly heavier than water and are non-adhesive, sinking to the bottom where they are carried by currents. Hatching occurs within a temperature range of 12° - 29°C (54° - 84°F), with optimum hatching success at 17°C (63°F) in 3 - 8 days (Leim 1924, Leach 1925).

The larvae, approximately 9 - 10 mm long when hatched, grow rapidly and reach

Table 5-45. A list of food items found in 178 full white catfish stomachs and 111 full channel catfish stomachs taken in Lake Moultrie, Lake Marion, and the Tailrace Sanctuary between 1 January 1958 and 30 June 1959 (Stevens 1959).

Food Item	White Catfish Frequency		Channel Catfish Frequency	
	Number	Percent	Number	Percent
Shada	17	9.6		
Gizzard shad	8	4.5	1	0.9
Threadfin shad	8	4.5	5	4.5
Herring	5	2.8		
Hickory shad			12	10.8
Unidentified clupeoids	7	3.9	1	0.9
Bream (bluegill)	4	2.2	3	2.7
Crappie			10	9.0
Yellow perch			2	1.8
Catfish	5	2.8	1	0.9
Atlantic needlefish	1	0.6	17	15.3
Mullet			2	1.8
American eel			7	6.3
Unidentified fish	51	28.7	21	18.9
Fish scales	7	3.9	2	1.8
Fish eggs	1	0.6		
Mussel	2	1.1		
Crayfish			2	1.8
Freshwater shrimp	3	1.7		
Mayfly larvae	28	15.7	16	14.4
Dragonfly larvae	1	0.6	1	0.9
Adult beetles	1	0.6	2	1.8
Diptera	3	1.7	1	0.9
Hemiptera	1	0.6		
Hymenoptera	1	0.6		
Unidentified insects	5	2.8	2	1.8
Annelid worm	1	0.6		
Filamentous algae	3	1.7		
Potamogeton	41	23.0	1	1.8
Seeds	2	1.7		
Debris	5	2.8		

a. Includes undifferentiated threadfin and gizzard shad.

the juvenile stage, approximately 25 mm long, in 4 - 5 weeks. Juveniles spend their first summer in the river of their birth, dispersing throughout the nursery area. Crochet et al. (1976) found primary nursery areas for American shad were located between river mile 40 and 80 on the Pee Dee, and between river mile 40 to a point approximately 15 mi (24.1 km) above Conway, South Carolina, on the Waccamaw.

As fall approaches, the juveniles congregate in the lower portions of the rivers and estuaries. When water temperatures drop below 15.5°C (60°F), they move to sea (Walburg and Nichols 1967). Once out at sea, they probably overwinter off the Middle Atlantic region and migrate to the Gulf of Maine with the adults the following summer.

Most spawning shad seem to prefer tidal fresh water with extensive flats of sandy or pebbly shallows near creek mouths (Pacheco 1968). However, in the Altamaha River, Adams (1970) found the main river channel to be the most productive spawning area for shad. This channel is characterized by moderate flow [2 - 4 mi/h (3.2 - 6.4 km/h)] near sand bars, with an average depth of 4 - 6 ft (1.2 - 1.8 m). These areas were usually over a clean sandy bottom.

b. Hickory Shad. The hickory shad, ranging along the Atlantic coast from the Bay of Fundy to Florida, spends most of its life in the ocean, ascending coastal rivers in the spring to spawn. In Georgia and South Carolina, the runs normally begin in early January and continue through early May, with the majority of spawning occurring in late March and April (Cadieu and Bayless 1968, Street 1970, White 1970, Curtis 1973, 1974, 1975, 1976, Crochet et al. 1976).

Hickory shad in South Carolina and Georgia usually mature at age 2 and 3. White and Curtis (1969), working in the Pee Dee and Black rivers, found that male hickory shad matured at age 2, and females at age 2 and 3. Wade (1971) found that females in the Edisto spawned first at age 3. In the Savannah, White (1970) reported that males spawned first at age 3, and females at age 2 and 3. Street and Adams (1969), working in the Altamaha River, concluded that most females mature at age 2 and that males made their first spawning runs at age 2 and 3.

Spawning usually occurs in back water areas off the main channel of the river. Street (1970) reported that hickory shad seem to spawn in larger tributaries and lakes of the Altamaha River system; but they do spawn in lakes of the upper region and probably in lakes and tributaries of the whole river. Adams (1970) also reported similar findings in the Altamaha.

Female hickory shad produce an average of 500,519 eggs, with a range of 252,693 - 730,213 eggs (Street 1969). Fecundity is mainly dependent on size and weight, but age is also important. Hickory shad are free spawners, releasing eggs and milt into the water where fertilization takes place. Unfertilized eggs are asymmetrical and amber-colored, averaging 1.1 mm in diameter. After fertilization, they become transparent spheres with a diameter of 1.3 mm. The eggs are slightly adhesive but are easily dislodged by currents, becoming semi-demersal in slow-moving waters and buoyant under turbulent conditions. Hatching occurs in about 70 hours at 18°C (65°F). The larvae at hatch average 6.0 mm in length and reach the juvenile stage at approximately 35 mm.

Juvenile hickory shad move out of the riverine system much earlier than other species of herring. The bulk of the young-of-the-year leave the rivers in early summer, utilizing adjacent estuaries as nursery grounds. (See the section on fishes of the subtidal estuarine system.)

c. Blueback Herring. Adult blueback herring usually inhabit a narrow band of coastal water, but enter fresh or brackish water during spring to spawn. In South Carolina, these spawning runs occur from late March to late April, peaking in mid-April (Curtis 1972).

Most blueback herring enter the fishery at 3 to 6 years of age, with the majority of males maturing at 3 and 4

years and the majority of females at 4 years. Curtis (1973, 1974, 1975, 1976), working in the Santee and Cooper rivers, found spawning populations to be dominated by 3- and 4-year-old fish. Bulak and Curtis (1978) found that 83.3% of blueback herring in the Santee River were 4 years old. White (1970) found that 10% of the males from the Savannah River spawned first at age 2, 30% at age 3, and 60% at age 4; 47% of the females spawned first at age 3 and 53% at age 4. Street and Adams (1969) reported that the majority of males from the Altamaha, Ogeechee, and Savannah rivers spawned for the first time at 3 years (67%), and the majority of females at 3 (44%) and 4 (47%) years.

Blueback herring are repeat spawners in South Carolina and Georgia. Bulak and Curtis (1978) found 8.3% of the herring in the Santee River to be repeaters. White (1970), working the Savannah, reported that 10% of the males had spawned once previously, and 10% had spawned twice previously. Of the females, he found that 10% were spawning for the second time and 5% for the third time. Street and Adams (1969) reported that only a small proportion of female blueback herring in Georgia survive after their initial spawning migration, while males return one or two more times.

Spawning grounds of blueback herring in South Carolina and Georgia seem to be located in flooded swamps and backwaters off the main channel of the rivers (see palustrine system). White and Curtis (1969) reported that spawning probably occurred immediately upstream from the U.S. Hwy. 301 bridge on the Pee Dee River. Bulak and Curtis (1978) found blueback herring utilizing the entire Santee River from Jamestown to Wilson Dam for some degree of spawning, and that spawning in the Cooper River occurs mainly upstream of river mile 44 in the west branch. Godwin and Adams (1969) found no herring eggs in the main channel of the Altamaha River, Georgia, concluding that spawning probably occurred in flooded river swamps. Street (1970) corroborated these results when he collected spawning blueback herring in oxbow lakes and flooded woods well above the normal level of the river.

Fecundity of blueback herring varies mainly with weight and age. In work done on the Altamaha, Ogeechee, and Savannah rivers, Street (1970) determined that fecundity ranged from 121,126 to 399,735 eggs, with an average of 244,152. After fertilization, the eggs are demersal and adhesive, sticking to rocks and bottom debris. They are semi-transparent and yellowish, with a diameter of approximately 1.2 mm. Incubation time is 2 to 6 days depending on temperature, with hatch occurring in 50 hours at 22°C. The

newly hatched larvae, approximately
3.5 mm long, reach the larval stage
in about 4 days at 5.2 mm; the pre-
juvenile stage at a length of 20.5 -
25.0 mm; and the juvenile stage at
30.0 mm.

Juveniles remain in the rivers
during their first summer. Godwin and
Adams (1969) concluded that the main
nursery area for blueback herring in
the Altamaha River was in an area
between river mile 10 and 30. Street
(1970) came to the same conclusion in
his studies on the Altamaha River.

d. Striped bass. Striped bass are
native to the Atlantic coast from the
St. Lawrence River in Canada to the
St. Johns River, Florida. They are
usually found in coastal habitats not
far from shore, generally in coastal
bays, rivers, and estuaries (Raney 1952).
Larger striped bass in areas north of
North Carolina undertake extensive
coastal migration, while smaller bass
tend to remain in the vicinity of their
natal streams. Southern populations
are thought to contribute little to the
migratory stocks, remaining in sounds
near their native streams. In the
Savannah River, tagging studies indicate that striped bass populations
are riverine, migrating upstream after
spawning, remaining in all parts of
the river during winter, and returning
downstream to spawning areas in spring
(Dudley et al. 1977).

Spawning migrations of striped bass
occur in late winter and early spring.
Temperature seems to be the controlling
factor in spawning migrations, with the
majority of activity occurring between
15.5° - 19.5°C (60 - 67°F) (Raney 1952).

Most striped bass mature at 2 to 6
years of age. Most males are mature
at 2 years, and almost 100% are mature
at 3 years of age. Females usually
mature later than males, with 25%
reaching maturity at 4 years of age,
75% by 5, and 95% by 6 years of age
(Merrimen 1941).

Spawning usually occurs in tidal
fresh water proximal to estuarine zones.
Crochet et al. (1976), although unable
to pinpoint actual grounds, found eggs
only in the lower Waccamaw River.
White and Curtis (1969) reported that
major spawning grounds for striped bass
in the Pee Dee River were located upstream from the U.S. Hwy. 301 bridge.
In the Black River, they found spawning
immediately upstream from the U.S. Hwy.
701 bridge, and in the Lynches River,
spawning grounds were located above
the Hwy. 41 bridge. Cadieu and Bayless
(1968) observed that principal spawning
areas for striped bass in the Cooper
River were located in the vicinity of the

lower end of Tail Race Canal. In the
Ashley River, Curtis (1970a) found striped
bass spawning near the U.S. Hwy. 17-A
bridge (Slands Bridge: river mile 34).
Curtis (1970b) maintained that spawning
occurred between the U.S. Hwy. 17 and 17-A
bridges in the Combahee River, although
he was unable to pinpoint the grounds.
In the Savannah River, Smith (1970) reported major spawning grounds at the
mouth of Back River and at the U.S. Hwy.
17 bridge, 23 mi (37 km) upstream from
the sound. In the Altamaha River,
Smith (1970) found indications of spawning
at the U.S. Hwy. 17 bridge, 10 mi
(16.1 km) upstream from the sound.

Striped bass are free spawners;
usually 20 to 50 males gather around one
large female, which then broadcasts her
eggs into the water where they are fertilized. Female striped bass produce
180,000 - 700,000 eggs, the number varying with age and size (Raney 1952).
The eggs, approximately 1.28 - 1.36
mm in diameter after fertilization, are
spherical, non-adhesive, and slightly
heavier than water, requiring a slight
current to keep them off the bottom.
After 12 hours, when water absorption is
usually complete, the eggs range from
3.2 - 3.8 mm in size. Hatching is
dependent on water temperature, with
hatch occurring in 48 hours at 12.8°C
(55°F) and in 70 - 74 hours at 11.6° -
12.0°C (53° - 54°F). Larvae are approximately 2.5 mm long and reach the postlarval stage in about 10 days. After
4 to 5 weeks, the juvenile stage is
reached at a length of approximately
36 mm.

Juveniles are found dispersed throughout the tidal zones of their native
streams, with the majority spending the
summer in estuarine systems (see the
section on fishes of the subtidal estuarine
system). Most adult striped bass seem to
prefer upper tidal reaches of freshwater
rivers, where the bottom is usually sand
or mud and current velocities are 2.5 -
3.0 mi/h (4 - 5 km/h). Juveniles seemingly
show a preference for gravelly beaches or
mud-sand bottom with little gravel and
a few scattered rocks (Merrimen 1941).

e. Sturgeon. For discussion of the
sturgeons, see Chapter Four.

3. Catadromous American Eel

The freshwater distribution of the
catadromous American eel, Anguilla rostrata
(LeSueur), ranges from Greenland to Trinidad
(Jensen 1937). Recently, American eels
have recieved much attention as Oriental
and European stocks have been reduced
significantly due to overexploitation.

The life cycle of the American eel
is reported as somewhat complex and inconsistent. The following life cycle

synopsis is reported by Bayless and Loyacano (1979). In general, females live in fresh water and males in brackish to salt water (Vladykov 1966). When sexually mature, the females migrate downstream in late summer to early winter and meet mature males at the mouth of rivers.

According to Schmidt (1922), spawning occurs in the southeast North Atlantic, east of Florida and the Bahamas and south of Bermuda, in the Sargasso Sea.

Sheldon (1974) stated that females produce up to 15 - 20 million eggs, each about 1 mm in diameter. After fertilization, they develop into surface dwelling larvae or leptocephali which are transparent and shaped like willow leaves. The adults are believed to die at sea. The leptocephalic larvae drift with the currents and, about 1 year after hatching, reach the continental shelf when they are from 6 to 8 cm long. At this time, they metamorphose to the glass eel stage which is shaped like the adult but lacks all pigment. As they enter the river mouths, glass eels begin to obtain pigment and are called elvers. Several investigators, including Smith and Saunders (1955), have shown that all elvers do not leave saltwater habitats, and it is generally believed (Bigelow and Welsh 1925) that mainly female elvers move up rivers above the influence of tides, and most male elvers remain in brackish to saltwater habitats. However, Smith and Saunders (1955) pointed out that both sexes may develop in salt water and that there may not be any clear-cut segregation of sexes during elver migrations into fresh water.

Vladykov (1966) and Wenner and Musick (1974) found a relationship between size of American eel elvers and latitude. Elvers from northern parts of North America were larger than those collected at southern latitudes. Vladykov (1966) further postulated that larger elvers become females and smaller individuals become males, thereby explaining the increase in relative abundance of female eels over males as one moves from Southern United States to Canada. Wenner and Musick (1974) explained this south-to-north size gradient of elvers by assuming that since leptocephali reach southern shores earlier in their development than in more northern region, they metamorphose at smaller sizes thereby producing smaller elvers.

Most elvers begin to enter the mouths of rivers in large masses in late winter and spring. According to Godfrey (1951), they usually enter with the high tide and generally remain near the water surface close to the river bank. Godfrey (1951) contends that elvers may run predominantly at night or day depending on the location.

Godfrey (1951) also stated that elvers almost completely disregard disturbances during migration and are capable of surmounting or by-passing most obstacles. In some areas, elver migrations may be dispersed and less conspicuous than mass migrations that often occur.

Limited data exist on the biology of the American eel for South Carolina and Georgia. Few published references are available for South Carolina and our literature review found no references for Georgia. Some unpublished data exist from both States' natural resource agencies. (See Directory of Information Sources.)

P. Christian (1980, Georgia Marine Extension Service, Brunswick, pers. comm.) reports that Georgia is just beginning more intensive eel research. The State is in the process of conducting an experimental eel fishery study, which began during February 1980. This is the first major study conducted in Georgia. Additionally, a 2-year study is planned to investigate the biological aspects of the American eel to include population estimates and migratory behavior. In South Carolina, the majority of research has been conducted on the Cooper River. The following discussion summarizes this research.

The food habits of American eels in the Cooper River were found to vary with size of eels, season, and prey availability. Fish were the most abundant food item, followed by crustaceans, mollusks, and insects. Blueback herring was an important food item in winter and spring. Elvers were used as food year around. Crustaceans were taken in the spring, and insects and mollusks were eaten in the spring, summer, and fall (McCord 1977). Food habits for larger elvers included chironomid larvae and adults, small benthic crustaceans, cladocerans, amphipods, and some fish (McCord 1977).

American eels taken from the Cooper River averaged 5.1 years and ranged up to 15 years old. Total lengths ranged from 98 to 834 mm and weights ranged from 1 to 1,224 gm. Males constituted only 1.5% of the population in the freshwater regions of the Cooper River (Harrell 1977).

Several commercial fishing operations exist in North and South Carolina for both adult eels and elver stages (Harrell 1977, McCord 1977, Hornberger 1979). These fisheries are primarily located on the Edisto River, Cooper River, and in the Winyah Bay system (J. V. Miglarese, 1977, South Carolina Marine Resources Division, Charleston, unpubl. data; Hornberger 1979; D. E. Marchette, 1979, South Carolina Marine Resources Division, Charleston, unpubl. data).

Much controversy has centered around the recently developing American eel industry in South Carolina. Fishery biologists for the State are concerned with the impact of this fishery on sportfishes. However, Hornberger (1979) found that eels were not an essential diet item to either largemouth or striped bass and that preferred fishing gear (fyke nets and other traps) of certain mesh and throat sizes were not detrimental to non-commercial species of fishes.

Although no seasonal laws exist for the taking of eels (those laws that did exist were created for other species such as catfish), the most productive adult eel fishing takes place from March through August with March - May being by far the most productive period. The fishery for the elvers takes place from December through May with January through April being the more productive months (D. E. Marchette, 1979, South Carolina Marine Resources Division, Charleston, unpubl. data). No information is available concerning the economics or commercial value of either the adult or elver American eel fishery in South Carolina. Abbas (1977) reports a tremendous economic advantage when fishing part-time in the North Carolina fishery.

D. AMPHIBIANS AND REPTILES

Amphibians inhabiting riverine habitats in the study area are represented primarily by three species, each belonging to a different family. The two-toed amphiuma and greater siren are aquatic eel-like amphibians found among organic debris and aquatic vegetation. Amphiumas are relatively common throughout the coastal area (Martof 1956, Harrison 1978), but in Alabama (and presumably elsewhere), they are rarely found in main river channels with substantial current (Mount 1975). Greater sirens, unlike amphiumas, are found in considerable numbers in open bodies of water (Mount 1975). The dwarf waterdog is an uncommon inhabitant of coastal rivers and resides in and among sunken logs and bottom debris (Harrison 1978). All three of these species are active nocturnally and feed on crustaceans, worms, insects, and small fish. They are occasionally caught on hook and line by fishermen.

Reptile species inhabiting the riverine ecosystem are more numerous and far more conspicuous than amphibians. The river cooter, yellowbelly slider, gulf coast spiny softshell (see Conant 1975 for distribution), brown water snake, and banded water snake can be considered characteristic. Gulf coast spiny softshells and river cooters seldom leave the water except for basking and egg laying (Mount 1975, Gibbons 1978).

River cooters are extremely wary and difficult to approach while basking, and the ecology and distribution of this species is poorly known in the coastal plain of South Carolina (Gibbons 1978; J. R. Harrison, 1978, College of Charleston, Charleston, South Carolina, pers. comm.). River cooters are herbivorous, whereas gulf coast spiny softshells feed on aquatic invertebrates and possibly fishes (Mount 1975). Brown water snakes and banded water snakes are strictly carnivorous, feeding on fishes, frogs, tadpoles, and occasionally salamanders (Mount 1975). In Georgetown County, South Carolina, large brown water snakes have been captured in herring nets set in tidal freshwater creeks (Jobson 1940). Glossy crayfish snakes, north Florida swamp snakes, striped crayfish snakes, southern cricket frogs, eastern ribbon snakes, and rough green snakes may occur in or among floating vegetation in the main river channel. Floating vegetation near the river edge or in backwaters and sloughs would be more favorable habitat for these species, however. Glossy crayfish snakes and north Florida swamp snakes are not common in coastal Georgia and South Carolina.

Other turtle species commonly found in riverine habitats include common snapping turtles, stinkpots, Florida cooters, and the Florida softshell. These turtles generally inhabit the shallow edges of slower flowing streams and are not necessarily characteristic of open river waters. Redbelly water snakes, Florida green water snakes, and occasionally cottonmouths occur in this habitat (Mount 1975, Gibbons 1978).

Additional reptile fauna of riverine habitats include two unusual species of snakes. The rainbow snake and eastern mud snake are among the most beautifully colored and inconspicuous aquatic serpents in the United States. They feed primarily on amphiumas, sirens, and American eels. They are unusual in having a particular anatomical adaptation to facilitate capturing their prey. The terminal scale is spine-like and is used to aid retaining or maneuvering food items into swallowing position. It is this spine-like tail and the snake's penchant for "pricking" its captor which is responsible for the stinging-snake legend. Rainbow snakes and mud snakes are uncommon to common throughout the Sea Island Coastal Region of Georgia and South Carolina (Martof 1956, Conant 1975, Mount 1975, Gibbons 1978).

American alligators are not infrequently observed in riverine habitats of coastal rivers and streams. Both sexes utilize open waters in the spring during breeding season (Joanen and McNease 1970, 1973). After breeding, the males remain in deep, open waters, while the females return to their dens (Chabreck

419

1966, Joanen and McNease 1970, 1973).
Alligators have responded well to rigid
protection provided them in recent
years and are now relatively common
components of coastal river fauna.

Herpetofauna inhabiting coastal
rivers experience relatively few adverse
conditons not induced by man. Predation by many common carnivores is a
major population check. Juvenile
turtles, alligators, snakes, frogs, and
salamanders are preyed upon by skunks,
river otters, raccoons, mink, adult
alligators, hawks, snapping turtles,
adult bullfrogs, large predatory fishes,
great horned owls, herons, and snakes
(Arthur 1928; Allen and Swindell 1948;
Barbour 1956; Wharton 1969; Neill 1971;
Lowery 1974; Mount 1975; Garrick and
Lang 1977; R. E. Mancke, 1977, South
Carolina State Museum Commission,
Columbia, pers. comm.). Periodic floods
result in displaced individuals and
possibly local populations depending
on the flood's severity. Increased
salinities from drought conditions can
stress animals physiologically and affect
their food supply, but these situations
are generally temporary and restricted
geographically.

E. BIRDS

Discussion of avifauna in this
habitat is limited to open waters of
the river system and does not include
adjacent freshwater wetlands (which are
discussed in the palustrine section of
this chapter). There are several important factors which strongly influence avifauna in this particular
ecosystem. The rate of river flow is
perhaps one of the most important
limiting factors, along with tidal stage.
The relatively strong tidal and upstream run-off flows, which are
characteristic in rivers such as the
Santee, Edisto, Savannah, Altamaha,
Ogeechee, and Satilla, probably account
for the generally low diversity of
birdlife occurring in the river proper.
There are approximately 47 species of
birds which frequent this habitat in
the study area, and only 11 can be considered as dominant species (Table 5-46).
There are relatively low numbers of
birds in this habitat as compared to
other habitats discussed previously.
Literature, published and unpublished,
is scarce on birds in this habitat.
Overall, the highest numbers of birds
in the riverine system would coincide
with spring and fall migrations, while
the highest diversity would correspond
with the presence of winter waterfowl
populations.

Trophic relationships are relatively
simple here (Fig. 5-21) compared to
some of the other habitats. There is
apprently no active predation in this
habitat, with scavengers occupying the
highest trophic level.

Most birds occurring in the
riverine areas also occur in other
habitats and use open water areas
for resting and feeding. Dominant birds,
which are permanent residents, include
the pied-billed grebe, great blue heron,
green heron, wood duck, spotted sandpiper,
and belted kingfisher. Common winter
residents which are dominant species
include blue-winged teal, ring-necked
duck, hooded merganser, and tree swallow.

The osprey deserves special attention
in this habitat because it frequently nests
in bald cypress trees in adjacent palustrine
habitats, and on dead snags, channel markers,
and power line poles in the riverine system.
Henry and Noltemeier (1975) surveyed osprey
nesting populations in the coastal Carolinas
and found 111 active nests in South Carolina.
T. D. Murphy (1978, South Carolina Wildlife
and Marine Resources Department, Green
Pond, pers. comm.) estimated that there
are approximately 200 osprey nests in
coastal South Carolina. Most of these
nests are concentrated north of Charleston
in three areas: the Santee Coastal Reserve
near the mouth of the South Santee River,
the Waccamaw River near Georgetown, and in
the Charleston vicinity. According to
Murphy, there were 29 active nests along
the Waccamaw River in 1978. Henry and
Noltemeier (1975) reported that only 12
active nests were believed to exist in
South Carolina south of Charleston, and
that verified nesting records for Georgia
were limited, except in the Savannah
region. Denton (1977), however, reported
50 to 55 active osprey nests in Georgia
during 1976. These nests were generally
distributed throughout the coastal area
of Georgia and also in three inland
areas, the Okefenokee National Wildlife
Refuge, Lake Seminole, and the Ocmulgee
River. Ospreys were absent from some
of the highly developed barrier islands
such as Tybee and Jekyll, while the
Wassaw Island National Wildlife Refuge
had the highest number of active nests
(14) in the State (Denton 1977).

F. MAMMALS

See the section on mammals of the
palustrine ecosystem for information on
mammals in freshwater ecosystems.

V. PERTURBATIONS

A. INTRODUCTION

Freshwater environments are probably
subjected to more man-induced disturbances
than any other ecosystem or ecosystem
complex. Rivers are impounded, channelized,
polluted, or diverted; lakes and ponds may
be drained, filled, polluted, or deepened;
nontidal swamps are drained, filled,

Table 5-46. Dominant, moderate, and minor bird species of the riverine ecosystem (Sprunt and Chamberlain 1949, 1970; Burleigh 1958; Audubon Field Notes 1967 – 1970; Chamberlain 1968; American Birds 1971 – 1977; Forsythe 1978).

DOMINANT		MODERATE		MINOR	
Pied-billed grebe	C PR	Common loon	C WR	Little blue her a	C PR
Great blue heron	C PR	Double-crested cormorant	C PR	Great egret	C PR
Green her a	C PR	Anhinga	C PR	Black-crowned night heron	C PR
Blue-winged teal	C WR	Snowy egret	C PR	Mallard	C WR
W d duck	C PR	Louisiana heron	C PR	Black duck	C WR
Ring-necked duck	C WR	White ibis	C PR	Pintail	C WR
H ded merganser	C WR	Green-winged teal	C WR	Baldpate	C WR
Spotted sandpiper	C PR	Shoveler	C WR	Bufflehead	C WR
Black tern	FC SR	Greater scaup	C WR	Ruddy duck	C WR
Belted kingfisher	C PR	Lesser scaup	C WR	American merganser	U WR
Tree swallow	C WR	American c b	C PR	Osprey	U WR
		Semipalmated plover	C PR	Common gallinule	C PR
		Greater yellowlegs	C PR	Lesser yellowlegs	FC WR
		Least sandpiper	C PR	Ring-billed gull	C PR
		Herring gull	C PR	Bonaparte's gull	C WR
		Forster's tern	C PR	Bank swallow	U T
		Chimney swift	C SR	Cliff swallow	U T
		Rough-winged swallow	C SR		
		Barn swallow	C PR		

421

Table 5-46. Concluded

Note: Dominance indicates relative importance of the species as a group in the community. This concept is not
based necessarily on taxonomic relationships but rather on numbers, size, and trophic dynamics.

KEY: C – Common, seen in good numbers
      FC – Fairly common, moderate numbers
      U – Uncommon, small numbers irregularly
      PR – Permanent resident, present year around
      WR – Winter resident
      SR – Summer resident
      T – Transient

422

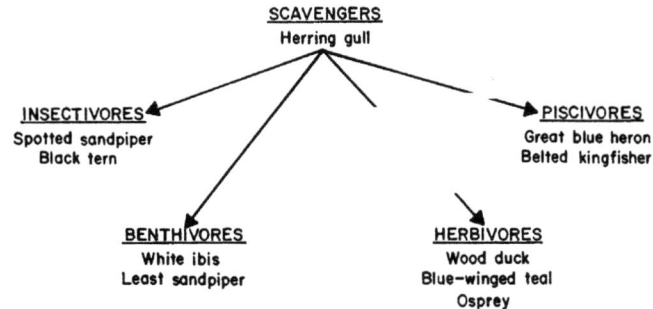

SCAVENGERS
Herring gull

INSECTIVORES
Spotted sandpiper
Black tern

PISCIVORES
Great blue heron
Belted kingfisher

BENTHIVORES
White ibis
Least sandpiper

HERBIVORES
Wood duck
Blue-winged teal
Osprey

Figure 5-21. Generalized trophic relationships of representative birds of the
riverine ecosystem.

clearcut, or impounded; and tidal flood plains may be channelized, polluted, clearcut, or disturbed through the alteration of their flooding regimes. Most perturbations (excluding pollution and clearcutting) are permanent; consequently, few unaltered freshwater environments still exist in the Sea Island Coastal Region.

Of the major river systems in the Sea Island Coastal Region, the Altamaha stands as the only relatively undisturbed system. Recently, a study team from the Georgia Department of Natural Resources proposed a 76 mi (122.3 km) stretch of the Altamaha as a "Scenic and Recreational" River (Georgia Department of Natural Resources 1978). The integrity of the only natural lakes of the study area, oxbows and barrier island cat-eye ponds, is threatened by water level alterations. Carolina Bays were once easily located in the coastal plain; but today, many have been drained and converted into pine plantations or farmland. At the source of these perturbations is the fact that no legal protection exists for most wetlands (today, navigable freshwater waterways and contiguous wetlands are under the permitting jurisdiction of the U.S. Army Corps of Engineers). Unlike estuarine wetlands, freshwater environments are often considered expendable ecosystems (see Georgia Department of Natural Resources 1976).

B. PALUSTRINE ECOSYSTEM

The greatest threats to palustrine wetlands are channelization, draining, and logging. Other potential disturbances to palustrine wetlands include thermal pollution, permanent flooding, and the introduction of solid pollutants to these wetlands.

Wharton (1970, 1978) discussed the problems caused by channelization of southeastern streams and rivers. An unnaturally high levee is usually formed by dredged material from the river or stream bottom, and this often prevents flood waters from reentering the river after periods of high water. Consequently, trees of sloughs and backswamps often die because of this extended hydroperiod. Tarplee et al. (1971), comparing natural and channelized streams in North Carolina, found that natural streams not only supported a greater diveristy of fish species, with over three times the pounds of fish per acre [155.37 lb (70 kg) for natural areas and 49.41 lb (22 kg) for the channelized areas], but that the average size of fishes in natural streams was almost 200% larger than in channelized streams. One reason for the severe drop in the standing crop of fish was that a 78.8% reduction of macrobenthic invertebrates occurred as a result of the alteration of bottom type and stream flow associated with channelization. In any ecosystem where components of lower trophic levels are so reduced, consumer biomass at higher trophic levels will likewise be reduced.

Tarplee et al. (1971) indicate that without channel or bank maintenance, a channelized stream could revert reasonably close to its natural state in a 15-year period. This is contrary to the findings of Arner et al. (1976), in a study of three areas of the Luxapalila River, Mississippi, which showed that, even after a period of 52 years, a channelized stream segment had not achieved its original natural state.

The cutting and/or draining of palustrine wetlands may have drastic effects on floral and faunal populations.

Allen (1958) studied succession
following clearcutting in a "tidewater
swamp forest" [similar to Bozeman and
Darrell's (1975) Class III, tidally
influenced blackwater swamp type - see
palustrine section, nontidal emergent
wetlands]. Black willow and cat-tails
were pioneer plant dominants, with bald
cypress, swamp and water tupelo, red
maple, and water ash succeeding in the
shade of willow communities. Clear-
cutting may affect drainage of flood-
plains, as creeks and sloughs are often
diked or dammed to facilitate road
access into wet forest communities.

Cutting and habitat alteration have
detrimental effects on the habitats of
many bird species. Meanley (1972)
stated that great canebrakes of southern
forested wetlands are disappearing,
directly affecting Swainson's warbler,
which is dependent upon these canebreaks
for nesting. Meanley also states that
the disappearance of Bachman's warbler
may have been related not only to the
cutting of virgin swamps in the early
1900's but also to the clearing of
canebreaks which provided understory in
primitive bottomlands.

Mahan et al. (1975) attempted to
evaluate the effects on wildlife
resources of timber harvesting on
Santee Swamp, located at the headwaters
of Lake Marion. They concluded that
most mammals would increase in numbers
with a selective timber harvest.
Opossums, raccoons, and other ground-
dwelling mammals would benefit from the
harvest; however, the number of mast-
producing trees (oaks and hickories) would
be reduced, lowering the potential gray
squirrel and white-tailed deer populations.
Drainage of palustrine wetlands would have
obviously adverse effects on mammals such
as otters and minks

Sharitz et al. (1974) analyzed the
effects of reactor effluents from the
Atomic Energy Commission's Savannah
River Plant (Aiken County, South Carolina)
on a section of the Savannah River flood
plain. A natural area, and a post-
thermal area were studied. Figure 5-22
illustrates the results of this study, and
reveals that most trees are basically
intolerant of thermal pollution. Thermal
pollution, both hyperthermal and hypo-
thermal, in mild forms usually tends to
shock a system, killing only the most
sensitive organisms and sending resident
species of fish and other aquatic organisms
into the riverine system or some other unaffected community. McFarlane (1976a)
studied streams receiving hyperthermal
effluents from the Savannah River plant.
He found fish diversity and abundance
reduced not only in the affected area
but also in first and second order tributaries of the stream as well.

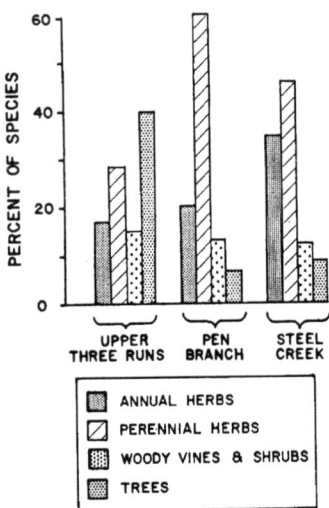

Figure 5-22. The influence of thermal
effluents on vegetative
structure of a natural area
(Upper Three Runs), a thermal
area (Pen Branch), and a post-
thermal area (Steel Creek)
in a flood plain at the
Savannah River Plant
(adapted from Sharitz et
al. 1974).

The effects of permanent flooding in
flood plains has been documented by Yeager
(1949) (in Illinois). He found that only
the trees that are most tolerant of prolonged flooding survive in areas where
higher than normal water levels persist.
Teskey and Hinckley (1977) listed button
bush, swamp privet, green ash, possum
haw, water tupelo, swamp tupelo, cottonwoods, black willow, and bald cypress as
"very tolerant" of inundation. Table 5-
47 lists 36 common tree species of the
flood plains of the Sea Island Coastal
Region and their tolerance to water level
changes.

Dennis (1973, 1975a) described the
flora of Santee Swamp. Here, impoundment
of the Santee River to form Lake Marion
has altered the water level in this
former flood plain. The successional
trends now tend toward a wetter type of
forest (cypress-tupelo). Aquatic plant
and floating log communities are more
common here than they are in flood plains

424

Table 5-47. Vegetative species responses to water level changes (Teskey and Hinckley 1977). (A – Total Submersion, B – Partial Submersion, C – Soil Saturation, 1 – During growing season, 2 – During dormant season, 3 – Year-round)

Species	Size	Root Development During Inundation	Survival Under Constant Inundation	Comments
Acer negundo (box elder)	Mature tree		B.1. 99% after 73 days (13).	
	Seedling		A.1. 100% for 2 weeks, 70% for 3 weeks, 36% for 4 weeks (26). A.1. 0% after 32 days (18).	Chlorotic leaves after 4 days, recovered slowly (18).
Acer rubrum (red maple)	Mature tree		B.1. Less than 41% of growing season (11).	Remained healthy if flooded less than 37% of growing season (11). Impoundment drained July 1 increased radial growth 85% (3).
	Seedling	Good adventitious root development (21).	B.1. 100% for 5 days, 90% for 10 days, 0% for 20 days (20). C.1. 100% after 32 days (29).	Adventitious roots developed after 15 days (21). Seeds did not germinate under water (17). Height growth decreased in saturated soil (21).
Acer saccharinum (silver maple)	Mature tree		B.3. 3 years, most died 2nd year (10). B.1. 98% after 73 days (13).	Trees above waterline improved (10).
	Seedling	Good adventitious root development (21).	A.1. 100% after 3 weeks (26). 100% after 60 days (21).	Wilted lower leaves after 2 days. Recovery slow (18). Flooding prevents seed germination. Good germination after flooding (17). Height growth better under saturated conditions than field capacity (21).
Asimina triloba (pawpaw)	Mature tree		B.1. Less than 14% of growing season (11).	Trees quickly died after tolerance period (11).
Betula nigra (river birch)	Mature tree		B.3. All trees died 2nd year, good survival first year (10).	Trees remain healthy if flooded less than 24% of growing season (10).

425

Species	Size	Root Development During Inundation	Survival Under Constant Inundation	Comments
Betula nigra (river birch)	Seedling		C.1. 100% for up to 32 days (29).	Severely stunted growth (29).
Carpinus caroliniana (ironwood)	Mature tree		B.1. Alive if flooded less than 26% of growing season (11).	Remains healthy if flooded less than 24% of growing season (11).
Carya aquatica (water hickory)	Mature tree			Shallow water impoundment drained July 1 increased radial growth 45% (3).
Celtis laevigata (sugarberry)	Mature tree		B.1. All trees died after 2 years (5).	Raising water table to 12" below soil surface increased radial growth 44% over drought years (4).
	Seedling		C.1. No mortality in 60 days (21).	No adventitious root formation, and only primary root lived after 60 days (21). Survived because species resists desiccation (21).
Cephalanthus occidentalis (button bush)	Mature tree		B.1. Hardy after 4 years (10). Trees died after flooding 53% of growing season (11).	
	Seedling		A.1. 100% for 30 days (20).	Better germination when seeds were submerged (9).
Crataegus spp. (hawthorn)	Mature tree		B.1. 69% after 73 days (13).	Healthy if flooded less than 33% of growing season (13).
Diospyros virginiana (persimmon)	Mature tree		B.3. All died after 2 years (5).	Healthy if flooded less than 31% of growing season (11).
			B.1. 93% after 73 days (13).	Impoundment drained July 1 increased radial growth 51% (3).
Fagus grandifolia (American beech)	Mature tree		B.1. All died if flooded more than 16% of growing season (11).	Healthy if flooded less than 4% of growing season (11).

Table 5-47. Continued

Species	Size	Root Development During Inundation	Survival Under Constant Inundation	Comments
Forestiera acuminata (swamp privet)	Mature tree		B.3. Hardy more than 4 years (10).	
Fraxinus americana (White ash)	Mature tree			Growth began with standing water on site. Growth was steady when water table was 2' - 5' deep (28).
Fraxinus caroliniana (water ash)				46% germination when inundated but only 5% under saturated conditions (9).
Fraxinus pennsylvanica (green ash)	Mature tree	Adventitious and secondary root development (16).	B.3. All died after 3-4 years (5). B.1. 99% after 73 days (13).	Raised water table to 46" improved growth 90% - 96% (3,4). Observed accelerated anaerobic respiration (16).
	Seedling	Adventitious and secondary root development (16).	A.1. 100% after 5 days, 90% after 10 days, 73% after 20 days, 20% after 30 days (20). B.1. 100% after 14 days (18). C.1. 100% after 60 days (21).	Chlorotic lower leaves after 8 days (18). Better growth in saturated soil than soil at field capacity (21). Good germination in swampy areas (4, 9).
Fraxinus tomentosa (pumpkin ash)	Seedling		C.1. 100% after 60 days (21).	Height growth improved in saturated soil over soil in field capacity (21).
Ilex decidua (deciduous holly)	Mature tree		B.1. Greater than 4 years (10).	Healthy if flooded less than 35% of growing season (11).
Liquidambar styraciflua (sweet gum)	Mature tree		B.3. All died 3rd year (5). B.1. Died if flooded 44% of growing season (11).	Impoundment drained July 1 increased radial growth 77% - 86% (3). Raising water table to 16" improved radial growth 60% (4). No adventitious or secondary root development when flooded (16).

Table 5-47. Continued

Species	Size	Root Development During Inundation	Survival Under Constant Inundation	Comments
Liquidambar styraciflua (sweet gum)	Seedling	No adventitious root formation (21).	A.1. All died in 32 days (18). B.1. All died in 3 months (20).	After 8 days, recovery slow (18). Poor germination when inundated (9).
Nyssa aquatica (water tupelo)	Mature tree		B.1. Very tolerant (15).	Best diameter growth when flooded (14). Metabolic adaptations (14). Survives moderate siltation 0"–3" (5,23).
	Seedling	Secondary root formation (16). Adventitious root formation (14).	B.1. 90% – 100% survival over growing season (15). B.1. Very deep flooding reduced survival to 32% (23).	Best growth when water table fluctuates (15). No seed germination under water (9).
Nyssa sylvatica var. biflora (swamp tupelo)	Mature tree		B.1. Tolerant (13).	Metabolic adaptations (14). Survives moderate siltation—0"–3" (5).
	Seedling	Secondary root development	B.1. 90% – 100% for growing season (15). C.1. 90% – 100% for growing season (15).	Poor root growth in stagnant water (12,15). Best growth in saturated soil (15). Poor germination under water (7,9).
Pinus taeda (loblolly pine)	Mature tree		B.1. Died if flooded more than 17% of growing season (11).	
	Seedling	No adventitious root development; root system died (22).		Reduced stem and root growth during flooding (22). Dormant season flooding increased height and diameter growth (6).
Planera aquatica (water elm)	Seedling		B.1. Died if flooded more than 54% of growing season (11).	Remained healthy if flooded less than 43% of growing season (11).

428

Table 5-47. Continued

Species	Size	Root Development During Inundation	Survival Under Constant Inundation	Comments
Platanus occidentalis (sycamore)	Mature tree		B.1. 94% after 73 days (13).	Healthy if flooded less than 24% of growing season (11). Raised water table to 48" increased radial growth 75% (4). Metabolic adaptations. Can withstand considerable siltation (4).
	Seedling	Secondary root development (16). Adventitious root development (8,16,21).	A.1. 100% after 10 days, 0% after 30 days (20). 95% after 32 days (21).	Germination retarded by inundation (9). Growth decreased by saturated soil (29).
Populus deltoides (eastern cottonwood)	Mature tree		B.3. Died after 2 years (10). B.1. 100% after 73 days (13).	Raising water table to 40" increased radial growth 54% (4). Impoundment drained July 1 increased radial growth 90% (3). Survives moderate siltation— 0" - 3" (5).
	Seedling	Adventitious root development (19,21).	A.1. 0% after 16 days (18). 90% after 10 days, 70% after 20 days, 47% after 30 days, (20, 21).	Seeds germinate quickly in water (17). Best height growth when water table is 2' below surface (4). Heavy mortality when deeply flooded (18).
Quercus falcata var. pagodaefolia (cherrybark oak)	Mature tree		B.1. All died first year (4).	Raising water table to 46" increased growth 20% (5).
	Seedling	Root system died (21).	A.1. 87% after 5 days, 6% after 10 days, 0% after 20 days (20). C.1. 89% after 15 days, 47% after 30 days, 13% after 60 days (21).	Height growth decreased by saturated soil (21). Submergence significantly lowers germination % (2).

429

Table 5-47. Continued

Species	Size	Root Development During Inundation	Survival Under Constant Inundation	Comments
Quercus lyrata (overcup oak)	Mature tree		B.3. Most die after 3 years (5).	Remains healthy if flooded less than 41% of growing season (11). Impoundment drained July 1 increased radial growth 20% (3). Submergence lowers seed germination (25).
Quercus phellos (willow oak)	Mature tree		B.1.2. All died after 3 years (5).	Remained healthy if flooded less than 31% of growing season (11). Increasing water table to 46" increased radial growth 40% (3). Shallow water impoundment drained July 1 improved growth 10% (3).
	Seedling	Few adventitious roots developed (21).	C.1. 100% after 60 days (21).	Much poorer growth in saturated soil than soil at field capacity.
Quercus shumardii (shumard oak)	Mature tree		B.1. 100% after 2 months (24).	Not damaged by 2 mo. flood (6 - 10 ft. deep) in spring (24).
	Seedling		A.1. 100% after 50 days, 90% after 10 days, 6% after 20 days (20). C.1. 100% after 30 days, 66% after 60 days (21).	Height growth poorer in saturated soil than soil at field capacity (21).
Quercus michauxii (swamp chestnut oak)	Mature tree		B.1. Dies if flooded more than 34% of growing season (11).	Remains healthy if flooded less than 28% of growing season (11).
Quercus nigra (water oak)	Mature tree		B.3. All died within 4 years (5).	Raising water table to 46" increased radial growth 48% (4). Remains healthy if flooded less than 17% of growing seasons (11).

Table 5-47. Concluded

Species	Size	Root Development During Inundation	Survival Under Constant Inundation	Comments
Salix nigra (black willow)	Mature tree		B.1. 100% after 73 days (13). B.3. All died after 3 years (10).	Can survive moderate siltation (5).
	Seedling	Good development of adventitious roots (21).	A.1. 100% after 30 days (18, 24). C.1. 100% after 60 days (21).	Height growth much better in saturated soil than at field capacity (21). Seed germinates in water (17).
Taxodium distichum (bald cypress)	Mature tree			Improved growth under flooded condition (8). Survives moderate siltation—0" - 3" (5).
	Seedling		A.1. 100% after 4 weeks (25).	No germination while seed is submerged (9).
Ulmus alata (winged elm)	Mature tree		B.3. All died after 2 years (5).	Remained healthy if flooded less than 24% of growing season (11).
	Seedling		C.1. 100% after 32 days (29).	Saturated soil decreased growth after 16 days (29).
Ulmus americana (American elm)	Mature tree		B.3. Died after 2 years (10). B.1. 95% after 73 days (13).	Impoundment drained July 1 increased radial growth 35% (3).
	Seedling		A.1. 100% after 10 days, 27% after 20 days, 0% after 30 days (20). C.1. 100% after 15 days, 94% after 60 days (21, 29).	Decreased height growth in saturated soil (21).

REFERENCES:

1. Allen and Scarbrough 1960
2. Briscoe 1961
3. Broadfoot 1967
4. Broadfoot 1973
5. Broadfoot and Williston 1973
6. Burton 1972
7. Debell and Auld 1971
8. Dickson et al. 1965
9. Dubarry 1963
10. Green 1947
11. Hall and Smith 1955
12. Harms 1973
13. Harris 1975
14. Hook et al. 1973
15. Hook et al. 1970
16. Hook and Brown 1973
17. Hosner 1957
18. Hosner 1958
19. Hosner 1959
20. Hosner 1960
21. Hosner and Boyce 1962
22. Hunt 1951
23. Kennedy 1970
24. Kennedy and Kinard 1974
25. Larsen 1963
26. Loucks and Keen 1973
27. McAlpine 1961
28. McClurkin 1965
29. McDermott 1954
30. McMinn and McNab 1971

further upstream from Santee Swamp [see Dennis (1973) and Gaddy et al. (1975)].

Research to date indicates that nutrient addition in the form of sewage to palustrine wetlands does not have uniformly detrimental effects on these wetlands. Wharton (1970) pointed out the value of the floodplain forest as a natural absorption system for sewage and organic chemicals (see also Kitchens et al. 1975). Odum et al. (1975, 1976) have analyzed the feasibility of cypress domes as wastewater recycling systems. Selected domes have been used for this purpose; but as ecological processes are altered here (the domes become anaerobic in late summer with a significant reduction in faunal populations), this approach appears to be destined only for limited application.

C. LACUSTRINE ECOSYSTEM

In natural lacustrine environments, the term "eutrophication" is used to describe the natural aging process of a lake as it accumulates nutrients. When this process is accelerated by man, "cultural eutrophication" occurs (Hasler 1947, Likens 1972). Nutrient loads from residential (sewage) and agricultural (organic compounds, especially phosphates) run-off speed up the eutrophication process, causing algal blooms and anaerobic conditions. Culturally eutrophicated lacustrine bodies may be significantly different in species composition than undisturbed lacustrine environments. Vascular macrophytes may become more in culturally eutrophicated lakes, though species diversity decreases drastically (Lind and Cottam 1969). Phytoplankton production may increase threefold in culturally eutrophicated lakes, altering the ecological stability of the entire lacustrine body (Likens 1972). As the dense growths of vegetation bloom and subsequently decay, periods of oxygen imbalance at varying levels of the lake occur. Local anaerobic conditions result in fish kills and drastically alter the species composition of bottom-dwelling invertebrates (Swingle 1966, Mackenthun and Ingram 1967).

Thermal changes in lacustrine environments also produce significant disturbances. Vigerstad and Tilly (1980) found that nuclear hyperthermal effluent from the Savannah River Plant (Aiken, South Carolina) resulted in higher density of two cladocerans, Cereodaphnia spp. and Diaphanosoma brachyum, while lowering numbers of another species, Bosmina longirostris. The same authors, from additional unpublished data, found that heated effluents appeared to restrict vertical movement of Bosmina. This cladoceran resided in the water column where heated water entered the pond, but at a depth where the temperature was the same as unheated water.

Pollution of lacustrine environments by the introduction of toxic chemicals consitutes a major disturbance to this system. Sherberger and Buikema (1976) analyzed the effects of chromium on the cladoceran, Daphnia pulex. Gannon and Stemberger (1978) discussed zooplankton as indicators of water quality in various lakes, concluding that quantitative data offer more promise than do qualitative data; the ratio of one group to another often is more meaningful than presence or absence of a group. They noted that knowledge of zooplankton as indicators of trophic conditions of lakes and ponds is derived entirely from studies of cold temperate lakes, and little is known regarding warm temperate and tropical lacustrine habitats. Herricks and Buikema (1977) have summarized and evaluated present-day knowledge of the effects of pollutants on freshwater invertebrates.

D. RIVERINE ECOSYSTEM

The most obvious disturbance in riverine environments is the destruction of the environment by impoundment of the river. The Santee-Cooper project in South Carolina resulting in the construction of lakes Marion and Moultrie, is an example of riverine alteration on a large scale. Acres of bottomland hardwood habitat were inundated, possibly affecting the last remaining nesting areas of the ivory-billed woodpecker (Sprunt and Chamberlain 1970). D. A. Wood (1945, South Carolina Wildlife and Marine Resources Department, Columbia, unpubl. data) reported adverse effects of this project on waterfowl. (For discussion of alterations caused by Santee-Cooper Diversion and Rediversion projects, see Volume I, Chapter Six.)

Since 1896, available spawning grounds for anadromous fishes in the Pee Dee and Santee rivers have been reduced by 199 and 200 miles, respectively, by dams (Walburg and Nichols 1967). In Georgia, dams on the Savannah River reduced available area by 204 miles (Walburg and Nichols 1967). Fish passage facilities are the only way of maintaining runs above the dams in these rivers, but to date none have been constructed. On the Cooper River, however, navigation locks at Pinopolis Dam are operated during the blueback herring run to lift these fish into the Santee-Cooper lakes, where they are important food fish for resident landlocked striped bass populations.

Churchill (1958) discussed many of the problems that may result from the discharge of reservoir water into river systems. Water temperature and oxygen content in the released water may be substantially different from that of the river. During certain periods of the year, this fact may significantly alter species composition in the stretches of river just below the dam (Dendy and Stroud 1949, Pfitzer 1954).

Channelization of river tributaries is discussed by Wharton (1970, 1978). Channelization wreaks massive destruction on vegetation, while reducing organisms that comprise the lower trophic levels. The fish population is completely restructured from one where high quality food and game fish dominate to one where numerous rough or undesirable fish dominate (Tarplee et al. 1971). Dredging of riverine environments, while not as drastic as channelization, has a similar affect in the removal of macrobenthic invertebrates and other low trophic level organisms. The practice of snagging and clearing of river channels, such that proposed for the Altamaha River (U.S. Army Corps of Engineers 1977), not only removes habitat utilized by fish and benthic invertebrates, but effectively removes spawning areas for such species as largemouth bass, redbreast sunfish, crappies, and catfishes.

Eddy (1932), in a classic study of the Sangamon River of Illinois, described effects of impoundments, municipal sewage, and sewage treatment plants on riverine zooplankton. He found that cities on the river would impound the water above the municipality and dump their raw wastes below. This practice results in long stretches of river almost devoid of zooplankton. Sewage treatment plants were installed and, within a few years, plankton typical of clean water dominated waters below municipalities. Williams (1966) studied zooplankton in major United States rivers and the Great Lakes. He found that in winter and early spring turbulent silt and other edaphic factors influenced abundance of dominant species more than industrial or domestic wastes, whereas in summer and fall, abundance of dominants was closely related to water quality.

Coastal plain rivers of the Sea Island Coastal Region receive herbicides, insecticides, fertilizers, and other chemicals associated with agriculture and silvculture, as well as wastes from industrial activities located on or near their banks. Zooplankton abundance would be expected to be lower in the vicinity of outfalls or tributary mouths where flushing rates may be high or toxicants or algal mortalities (increased due to eutrophication) may be concentrated (Gannon and Stemberger 1978). Herricks and Buikema (1977) have recently reviewed studies concerning the effects of pollution on invertebrates. This work gives a good update on studies of zooplankton reaction to substances that are presently being introduced into rivers and streams of the study area.

No studies have been made that quantify the effects of pollution on anadromous fishes, but this is considered to be a factor in limiting runs. One form of pollution that has been studied recently is thermal pollution, particularly from nuclear power plants. (See previous discussion on effects of thermal pollution in the section on perturbations of the palustrine ecosystem, this chapter.) Products of industrial and domestic pollution reduce water quality directly and indirectly through reduction in dissolved oxygen. Tagatz (1961) found that young American shad could tolerate limited exposure to reductions in dissolved oxygen to 2 - 4 ppm in natural waters. Concentrations of 91 mg/l gasoline, 167 mg/l diesel fuel oil, and 2.417 mg/l bunker oil were found to be toxic to young shad at the 48-hour median tolerance limit (Tagatz 1961).

Schubel (1974) found that most American shad, blueback herring, and striped bass eggs can survive the typical time-temperature exposures experienced during passage through cooling systems or entrainment in thermal plumes of power plants. Juvenile American shad were found to actively avoid rapid temperature gradients of 4°C (39°F), and to avoid effluent temperatures of 30°C (86°F) or more (Moss 1970, Marcy et al. 1972). If a temperature barrier does not extend entirely across the river, young shad are capable of traversing heated effluents.

Poor forestry practices (such as clearcutting without leaving a riverbank buffer zone) result in increased siltation and higher river turbidity. Anadromous fish populations may be reduced where siltation occurs because 1) eggs may be trapped on silted bottoms; 2) turbidity may delay hatching; and 3) the biochemical oxygen demand may be increased (by organic compounds), resulting in lower dissolved oxygen levels. Ellis (1937) noted that high sediment loads may destroy insect and mollusk populations.

Pollution from pesticides and herbicides has a tremendous overall impact on the ecology of riverine systems. Residual toxic chemicals such as DDT and DDE appear in populations of the highest trophic levels through biomagnification. Populations of piscivores such as the osprey, the bald eagle, and, in estuarine waters, the brown pelican, may be decimated by eggshell thinning, which results from high concentrations of DDE (Ohlendorf et al. 1978).

CHAPTER SIX

UPLAND ECOSYSTEM

I. GENERAL DESCRIPTION

A. DEFINITION

In general, uplands include all
lands that are not part of previously defined wetland or aquatic systems (i.e.,
marine, maritime, estuarine, lacustrine,
palustrine, and riverine ecosystems).
For this ecological characterization,
"uplands" have been divided into two distinct ecosystems: 1) an upland ecosystem,
affected by fresh water, and 2) a maritime ecosystem, comprising all upland
areas located on barrier islands. The
latter ecosystem has been discussed in a
previous chapter.

Soils of upland areas are predominantly non-hydric, and the vegetation is
predominantly mesophytic or xerophytic,
rather than hydrophytic. Uplands are
further characterized as lands that are
never flooded during years of normal precipitation. The upland ecosystem is divided into four major subsystems: 1) old
field, 2) pine forest, 3) pine-mixed
hardwood forest, and 4) mixed hardwood
forest.

B. FOOD WEB AND MODEL

Grazing food chains and detritus
food chains are both important in upland
habitats. According to Odum (1971), 90%
or more of the net primary production in
forest areas is typically used in the
detritus food chain. On the other hand,
he observed that 50% or more of the primary production of a heavily grazed pasture may be utilized in the grazing food
chain. In the latter case, the amount of
plant material actually assimilated is an
important consideration because undi-
gested material enters the detritus food
chain. The principal type of food chain
may therefore vary from one upland habitat to another, although the detritus
food chain probably predominates in most
non-agricultural situations.

Much of the primary production from
plants in forests and old fields eventually falls to the ground, where it is
acted upon by bacterial and fungal decomposers (Fig. 6-1). Soil invertebrates
are also believed to be important in the
breakdown and utilization of detritus.
While the soil fauna serves as a link between detritus and higher trophic levels
in the food chain, these animals are also
believed to play an important role in
mineral recycling, a process of importance to plant growth (see biogeochemical
cycles, Chapter One).

The types and number of links in upland food chains vary considerably, and

many species function at more than one
trophic level (Fig. 6-2). Energy transfer and loss at each trophic level becomes of direct significance to consumers
and their food supply, including humans.
Odum (1971) observed that the number of
humans that can be supported by a given
output of primary production depends upon
the number of trophic levels in the food
chain because of losses in available
energy at each level. Thus, more people
can be supported on a diet consisting primarily of grain and vegetables than on a
diet consisting primarily of meat and
other animal products.

The activities of man and his domesticated animals have drastically altered
the upland landscape in much of the Sea
Island Coastal Region, thereby altering
the biotic community structure and the
food webs of these habitats.

II. PRODUCERS

A. VASCULAR FLORA

Because the four major divisions of
this ecosystem are based on vegetative
characteristics, Table 6-1 presents a summary of the community structure of the
upland vascular flora. The discussion of
the vascular flora will follow this organization.

1. Old Field Community

Most research on old fields in this
geographic area has been conducted on
sandy and sandy loam soils within the
Energy Research and Development Administration's Savannah River Plant near Aiken,
South Carolina. Odum's (1960) classic
old field study formed the basis for other
studies such as those of Colley (1965),
Gabrielson (1968), and Pinder (1975).
Studies from old fields at the Savannah
River Plant are not applicable to all old
field soil types in the Sea Island
Coastal Region, but the basic successional
processes can be understood and extrap-
olated.

Odum (1960) studied two old field
types, a well-drained upland with sandy
soil and a poorly drained lowland of
sandy loam. His work, begun immediately
after agricultural abandonment, is still
ongoing. During spring of the first growing season, toad-flax was an early dominant in both old field types. Other
shared dominants included horseweed, ever-
lastings, and crab grass. Only after the
first growing season did vegetational
dominance begin to reflect a difference
in soil types. Odum (1960) used net production of dry matter rather than frequency or density of individual plants as
the measure of vegetational dominance.
He noted that net productivity for old
fields was approximately 500 g dry matter/
m^2/year during the first year of

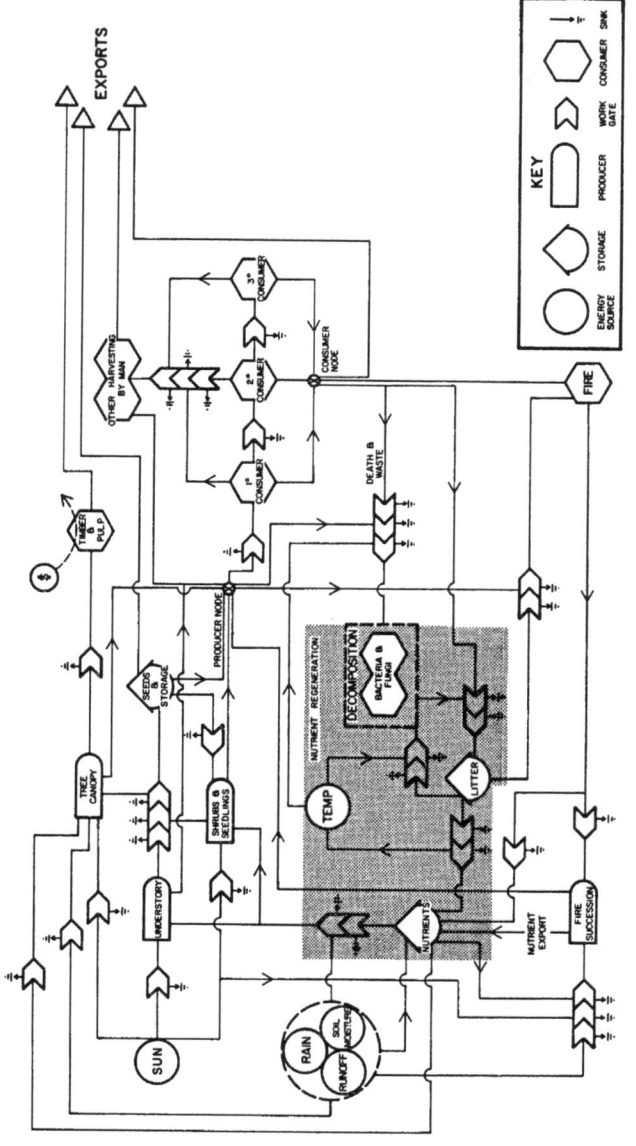

Figure 6-1. A generalized energese model of the upland forest ecosystem of the Sea Island Coastal Region.

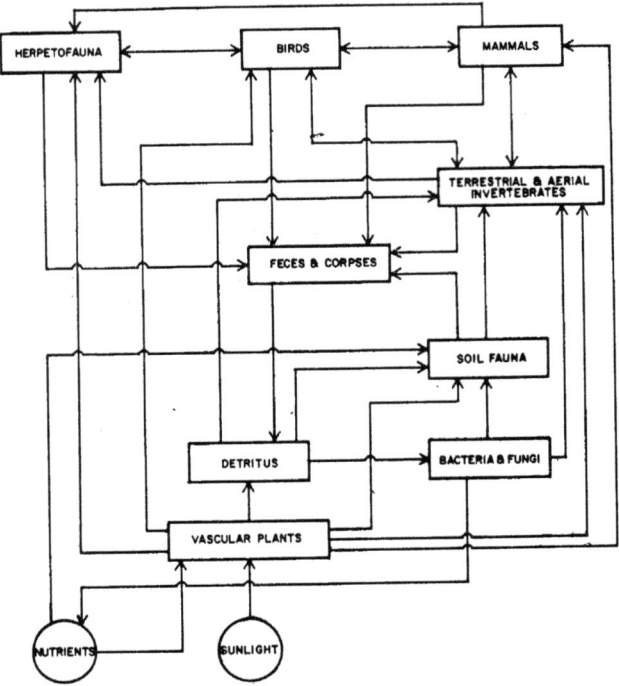

Figure 6-2. A generalized food web for the upland ecosystem of the Sea Island Coastal Region.

abandonment, with production levelling off to 300 g dry matter/m^2/year in subsequent years. Nutrient analyses revealed that high net production rates during the initial year probably resulted from residual nutrients, especially phosphorus, from fertilizers.

Productivity varied directly with the percentage of silt-clay in the subsoil of the field (Fig. 6-3). Productivity levels during the first year of abandonment were comparable to that of an average corn or wheat crop, but were less from year 2 through 7 (Odum 1960). Utilizing data from Odum (1960), Golley (1965) summarized production trends in old field communities for 1952 - 1962 (Fig. 6-4). Pinder (1975) analyzed the effects of species removal in an old field that had been abandoned for 5 years. He removed the dominants, broomstraw and three awn grass, and noted changes in production. Removal increased the net production of subordinate

species, but the transfer of production was not enough to equal the production of the unaltered community.

Dominants of well-drained and poorly drained sites during the first years of succession are presented in Table 6-2. Succession in Odum's old field sites leads to the formation of a broom-straw community. Broom-straw became important during years 5 and 6, but it did not form closed stands until the 8th year after abandonment on heavier soil types. Working with the broom-sedge community, Golley (1965) found three distinct layers present: 1) the algal-moss layer; 2) the rosette plants, including sour-grass, pineweed, sleepy catchfly, fescue, sedge, and three awn grass; 3) the tall strata plants, including sour-grass, everlastings, poor-mans pepper, daisy fleabane, partridge pea, Bermuda grass, Johnson grass, and horseweed. These strata exhibited seasonal dominance cycles. The algal-moss

Table 6-1. Community structure of the vascular flora of the upland ecosystem.

I. OLDFIELD COMMUNITY	II. PINE FOREST COMMUNITIES
Upland (well-drained) Lowland (poorly drained)	Longleaf Slash Loblolly Mixed pine
III. PINE-MIXED HARDWOOD COMMUNITIES	IV. MIXED HARDWOOD COMMUNITIES

III. PINE-MIXED HARDWOOD COMMUNITIES

   A. Mesic - loblolly dominant

      1. loblolly-mixed hardwoods community
        a. white oak phase
        b. ravine slope phase
        c. sweet gum-tupelo phase

      2. mixed pine-mixed hardwoods community
        a. shortleaf
        b. loblolly

   B. Xeric - longleaf dominant

      1. longleaf pine-turkey oak community
        a. turkey, blackjack, and shrubby
           post oak, plus longleaf dominate
        b. Chapman, live, turkey, shrubby
           post oak, longleaf dominate

      2. longleaf-shortleaf-turkey oak
        communities
        a. above, plus slash pine, Chapman
           and laurel oak, mockernut
           hickory
        b. blackjack and turkey oak,
           shrubby post oak, longleaf pine

IV. MIXED HARDWOOD COMMUNITIES

   A. Mesic slope hardwoods

      1. ravine slope hardwoods
      2. beech ravine
      3. mixed mesophytic hardwoods
      4. bluff and slope forest
      5. beech-bull bay community

   B. Upland mesic hardwoods (or oak-
      hickory forest, southern mixed
      hardwood forest, or mesophytic
      broadleafed forest)

   C. Hammock community

      1. lowland broadleaf evergreen
        forest
      2. upland broadleaf evergreen
        forest

   D. Scrub forest community

      1. evergreen scrub forest
      2. evergreen scrub-lichen forest

   E. Dwarfed oak-mixed hardwood
      community

      1. turkey oak-longleaf pine
        association
      2. turkey oak association
      3. blackjack oak association
      4. mockernut hickory association

   F. Live oak-mixed hardwood community

layer and the rosette plants were winter-early spring dominants, while species of the tall strata reached peak dominance from mid-summer to late autumn. Changing dominance patterns from 1959 to 1962 are presented in Table 6-3. After the first year of abandonment, horseweed and ever-lastings dominated the herb layer, while crab grass dominated the lower tier of plants. From the 2nd to the 5th years of abandonment, Haplopappus sp., golden aster, and other forbs formed an "over-story," while St. John's-wort, buttonweed, and Bermuda grass dominated the "under-story." Odum (1960) pointed out that well-drained sandy fields ultimately give rise to longleaf pine-turkey oak communities, while the poorly drained loamier fields produce loblolly pine and, eventually, mixed hardwood communities.

2. Pine Forest Communities

Pine communities dominate the upland ecosystem in the study area. The casual visitor to the Sea Island Coastal Region typically thinks of the forest as either live oak or pine. The original upland forests of the study area were probably mixed hardwood and pyric climax pine forests (Garren 1943, Quarterman and Keever 1962). Subsequent logging and clearing for agriculture resulted in the creation and maintenance of a pine-dominated upland. Loblolly pine and mixed pine communities probably occupy more land than they did in colonial days. Forestry management practices today maintain thousands of acres of former upland hardwood areas, flood plains, and swamp lands in pine communities. Many of the original forests

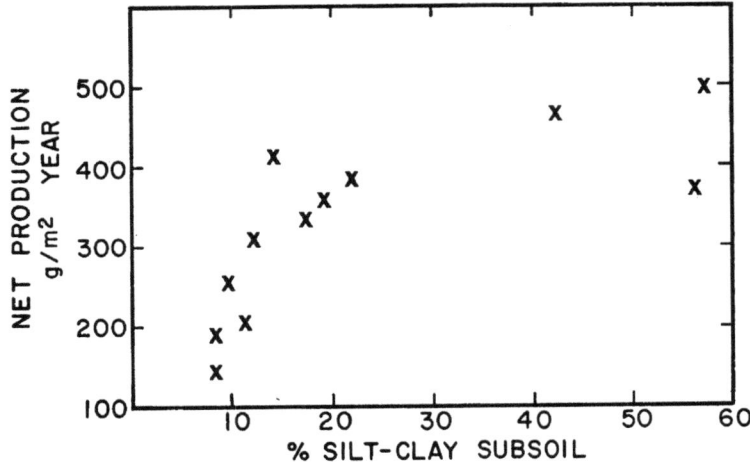

Figure 6-3. Relationship between net production and the silt-clay fraction of the subsoil (A_2 or B_1 horizon) at the Savannah River Plant. Each point on the graph represents annual production of a different field during one of the three years (1953, 1955, and 1956) when production was stabilized and rainfall was normal (Odum 1960).

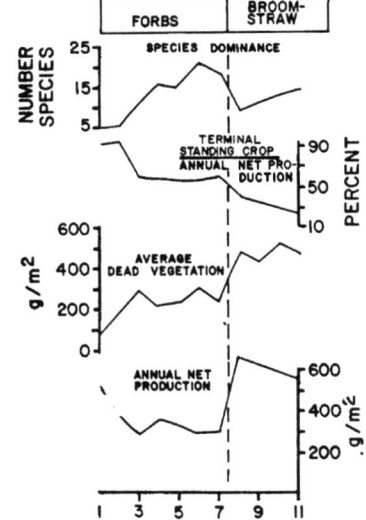

Figure 6-4. Production trends in old field communities at the Savannah River Plant, 1952 - 1962 (Golley 1965).

of longleaf pine have either been selectively logged or have been replaced by increasingly popular slash pine plantations (Wharton 1978).

Six native species of pines occur in the Sea Island Coastal Region, slash, spruce, loblolly, pond, longleaf, and shortleaf. Slash and spruce pines both reach their northern limit in South Carolina, where they are less common than in Georgia. The Sea Island Coastal Region has long been known as a major part of the "yellow" or longleaf pine belt and the "flatwoods" complex. Based on economic considerations, foresters have divided the pines into the turpentine group (slash and longleaf) and the non-turpentine group (pond, spruce, loblolly, and shortleaf) (Lehrbas and Eldredge 1941).

Basic pine communities of the upland Sea Island Coastal Region are: 1) the longleaf community, 2) the slash community, 3) the loblolly community, and 4) the mixed pine community. Spruce and pond pine communities are not discussed in this section because of their affinity for lowland sites. (For a description of these communities, see Chapter Five, Freshwater Ecosystems.) Loblolly, slash, shortleaf, and longleaf pines may be found in wetland communities, although they are more common in upland situations.

Table 6-2. Comparison of species net production (g dry wt/m^2/yr) during first 2 years of secondary succession on two soil types at the Savannah River Plant (adapted from Odum 1960).

SPECIES	WELL-DRAINED UPLAND FIELD (SAND)				POORLY DRAINED LOWLAND FIELD (SANDY-LOAM)	
	FIRST YEAR	SECOND YEAR	THIRD YEAR	FIFTH YEAR	FIRST YEAR	SECOND YEAR
Toad-flax	18	--	--	--	9	--
Horseweed	353	40	1	--	83	9
Cudweed	2	2	8	11	5	3
Rabbit tobacco	4	--	6	46	5	25
Crab grass	75	--	--	6	150	--
Sour-grass	--	4	4	30	5	10
Pin-weed	--	111	95	18	--	--
St. John's-wort	--	20	60	33	7	36
Evening primrose	--	1	8	2	--	1
Sedges		1	--	1	53	50
Haplopappus sp.		149	24	72	80	71
Bermuda grass	--	--	--	--	9	66
Diodia		--	--	14	0	5
Polypremum		--	--	--	--	21
Camphorweed		--	--	14	--	75
Venus' looking glass	--	--	1	--	--	--
Poor-mans pepper	--	--	3	--	--	--
Broom-straw			1	17	--	--
Fescue	--	16	--	--		
Love grass	--	45	--	--		
Sand grass	--	--	--	21		
Witch grass		--	--	3	--	--

a. Longleaf Pine Community. This community may be found on sandy ridge soils or in dry flatwoods. Most researchers believe that longleaf communities owe their origin and perpetuation to the maintenance, either by man or lightning, of an annual or biennial fire regime (Chapman 1932b, Garren 1943). Chapman (1926, 1932a, b, 1936) demonstrated that fire exposes mineral soils in which germination of longleaf pine seedlings occurs much more frequently than in organic litter. Heyward (1939) explained the relationship between longleaf pines and fire, noting that longleaf pine seedlings are susceptible to destruction by fire at two stages in their growth. Susceptible stages occur during the first year after germination, and again during height elongation when the tree is approximately 45 cm (18 in) tall. As a young tree, the pine is protected from fire by a dense growth of needles surrounding the terminal growth bud (frequently called the "grass" stage of development due to its resemblance to a tuft of grass). As the tree matures, fire protection is afforded by the bark.

The longleaf community is present in two phases, a fire phase and a successional phase. The fire phase of longleaf communities on upland sandy soils comprises only a canopy and a herbaceous layer. The herb layer rarely reaches over 1 m (39 in) in height. In Georgetown County, South Carolina, Almeida (1969) reported nut rush, camphorweed, beggar ticks, Amorpha herbacea, panic grass, broom-straw, and thoroughwort as the most common herbs of burned longleaf pine communities. In Charleston County, South Carolina, three awn grass was common on well-drained sandy soils, but on soils underlain with hardpans, bracken fern, goat's rue, Trilisa paniculata, beggar ticks, aster, and thoroughworts were the dominants. The shrub layer present in burned stands was composed of scattered clumps of bitter gallberry and huckleberries (L. L. Gaddy and D. A. Rayner, 1978, South Carolina Wildlife and Marine Resources Department, Columbia, unpubl. data).

In Georgia, Wharton (1978) noted that the longleaf pine-three awn grass community is more common inland than in coastal counties. However, longleaf pine communities do occur in coastal pine flatwoods on drier sites. Shrub layers of longleaf flatwoods are more common in Georgia than in South Carolina. This shrub layer is well developed and has a herbaceous layer below the shrub level. While appearing two-tiered, with canopy and shrub layers, the system is therefore actually

Table 6-3. Changing dominance patterns in an old field at the Savannah River Plant, 1959 – 1962 (adapted from Golley 1965).

SPECIES	PEAK ABOVEGROUND GREEN BIOMASS, g/m^2		
	1959	1960	1962
GRASSES			
Muhly grass			1.3
Three awn grass			14.4
Bermuda grass	10.8	33.2	20.0
Beard grass	4.8	4.0	
Switchgrass/panic grass			0.6
Sandspurs			0.4
Broom-sedge and broom-straw	250.0	191.0	95.0
Johnson grass			2.3
FORBS			
Sedge	2.0	0.4	4.8
Sour-grass	2.8	22.8	16.7
Poor-mans pepper			1.0
Partridge pea		27.2	
Japanese clover	3.6	3.2	8.4
Sericea			116.0
Pineweed		0.4	0.1
English plantain		0.4	0.3
Venus' looking-glass	0.8		0.1
Haplopappus divaricatus		1.2	
Heterotheca	1.2	5.6	17.0
Frost aster			1.3
Daisy fleabane			1.5
Horseweed			2.3
Pussy-toes	1.6		
Rabbit tobacco	0.8	6.4	0.9
Dog fennel		18.8	
TOTAL IDENTIFIED SPECIES	11	14	21
TOTAL DOMINANT SPECIES	8	11	14

three-tiered. The shrub layer is dominated by saw palmetto, which is absent from most of coastal South Carolina, as well as bitter gallberry, running oak, stagger bush, blueberries, and huckleberries. Where the shrub is more open or absent, grasses dominate with Curtiss' dropseed, three awn grass, toothache grass, Florida dropseed, muhly grass, and vanilla-plant most common (Biswell et al. 1943, Lemon 1949, Halls et al. 1952). Three awn grass and Curtiss' dropseed are fire tolerant, having buried leaf meristems and protective leaf sheaths. Other plants common in open longleaf communities are adapted to the fire-maintained community in having extensive rhizome systems for vegetative reproduction, and/or in producing many small seeds, allowing for "high mobility and rapid colonization" (Lemon 1949).

The successional phase of the longleaf community, occurring in the absence of fire, differs considerably in structure from fire phase forests. Here, a tall shrub-transgressive layer dominated by shrubs and young trees ranging in height from 1 to 7 m (3 to 23 ft) is prevalent. Almeida (1969) noted a change in herbaceous dominance in longleaf communities not subjected to the influence of fire. Broomstraw and panic grass increased in numbers, while camphorweed, thoroughworts, and beggar ticks decreased significantly. Almeida also illustrated the eventual invasion of shrubs and young trees that occur in unburned longleaf forests.

Heyward (1939) studied burned and unburned longleaf communities in Florida, Georgia, and South Carolina. Communities protected from fire become pine-mixed hardwood stands with sweet gum, black gum, loblolly pine, and wax myrtle the typical principal invaders. In Georgia, loblolly may be replaced by slash pine. A comparison between burned and unburned stands in the coastal plain of South Carolina is given in Table 6-4.

b. Slash Pine Community. This community is more common in Georgia than in South Carolina. Occupying as much as 90% of the Pleistocene flatwoods, it is the dominant pine community of Georgia coastal pine flatwoods (Wharton 1978). The soil of pine flatwoods is underlain with an organic hardpan 45 - 62 cm (18 to 24 in) beneath the surface. This soil alternately dries out and becomes waterlogged (Wharton 1978). Dominant understory plants of the slash pine community are dwarf laurel and saw palmetto (both absent in most of South Carolina), wax myrtle, three awn grass, and dwarf blueberry. Although natural fire usually occurs every 3 - 5 years, slash pine communities cannot tolerate fires as intense and frequent as those occurring in longleaf communities; therefore, they are found more frequently on the wettest part of the flatwoods.

Ineson and Eldredge (1938) and Wharton (1978) pointed out that the slash pine fire climax is found on wetter flatwoods than that of longleaf pine. The moisture excludes annual fire and allows slash pine to maintain itself; slash pine is more tolerant of fire than hardwoods, but less tolerant than longleaf pine. Chapman (1932b) noted that fire-free intervals of at least 10 years are necessary for slash pine perpetuation.

Slash pine communities of coastal Georgia flatwoods bear greater resemblance to the pine communities of Florida than those of South Carolina. Gano (1917) defined the flatwoods complex of northern Florida in terms of shrub layer density. She noted that longleaf and slash pine occur in communities having well-developed shrub layers dominated by saw palmetto and bitter gallberry. Working in slash pine communities in north central Florida, Edmisten (1963) found that the following trees, shrubs, and herbs occurred in 50% or more of his sampling plots: red maple, wax myrtle, slash pine, water oak, milk pea, dangleberry, bitter gallberry, cinnamon fern, polygala, running oak, saw palmetto, and vanilla plant.

In the absence of fire, slash pine communities succeed to a southern mixed hardwoods community on drier soils, or to a bayhead type community on wetter soils (Monk 1968). Many of the slash pine communities present in Georgia and Florida flatwoods have not been burned in 5 years or more, and are gradually succeeding into a slash pine-scrubby understory phase. This "scrub" is almost impenetrable due to the mix of saw palmetto and invading hardwood species.

c. Loblolly Pine Community. These communities generally are most common in old fields that have been left to succeed into forest. Dense thickets of loblollies produce sufficient shade for hardwood species to invade, if fire is absent. By the time pines reach maturity, a well-developed hardwood understory is present. However, managed loblolly communities are commonly controlled with fire. The stands are periodically thinned out and burned, preventing the invasion of hardwoods. These managed loblolly communities are more widespread in counties inland of the study area. Porcher (1974) noted and mapped the occurrence of loblolly stands in Hampton County, South Carolina, just inland of the study area.

d. Mixed Pine Community. The mixed pine communities are similar to loblolly communities in that they generally occur in areas of disturbance. Porcher (1974) described five different types of upland mixed pine: 1) slash-loblolly, 2) loblolly-shortleaf, 3) slash-longleaf, 4) loblolly-longleaf, and 5) mixed pine. Wharton (1978) noted that loblolly-shortleaf communities may follow longleaf after the site has been cultivated. Mixed pine communities are common on the Santee Coastal Reserve in Charleston County, South Carolina, where loblolly and longleaf are dominant (L. L. Gaddy and D. A. Rayner, 1978, South Carolina Wildlife and Marine Resources Department, Columbia, unpubl. data).

Excluding the deciduous understory, the various successional phases of most pine communities exhibit little seasonal change. However, communities in the fire phase are noticeably seasonal.

Prescribed burning in longleaf communities takes place in February or March, destroying all of the leaf litter from the previous season. During spring, the herbaceous flora is dominated by bracken fern, which may persist into July. Legumes (Fabaceae) begin to dominate in summer (goat's rue is a seasonal dominant), and grasses appear. Many legumes are dominants from summer until frost, but the most conspicuous dominants of late summer and autumn are the composites (Asteraceae). Sunflowers, thoroughworts, asters, and goldenrods are commonly encountered (L. L. Gaddy and D. A. Rayner, 1978, South Carolina Wildlife and Marine Resources Department, Columbia, unpubl. data). Lemon (1949) and Almeida (1969) pointed out changes in the herbaceous layer a year after fire had been excluded from longleaf pine communities. Heyward (1939) found that loblolly pine and black gum were dominant understory invaders in a longleaf community that remained unburned for 25 years (Table 6-4). Garren (1943) summarized succession in longleaf communities where fire has been excluded (Table 6-5).

In slash pine communities protected from fire, results are not as immediately obvious. Slash pine, unlike longleaf, may continue to survive in the understory,

Table 6-4. Comparison of the number of trees per acre for second-growth longleaf pine unburned for 25 years and for an adjoining stand burned annually on a moist site in the upper coastal plain, Summerville, South Carolina (adapted from Heyward 1939).

UNBURNED FOR 25 YEARS

d.b.h.[a] (inches)	Longleaf pine	Loblolly pine	Black gum	Sweet gum	Red maple	Wax myrtle	Laurel oak	Black jack oak	Southern red oak	Water oak	Sweet bay	American holly	Sumac
1	15	42	22	15	5	8	12				10		
2	38	40	18	3				2					
3	52	25	5					2					
4	62	10	10										
6	120	12	5										
8	107												
10	67	2											
12	18												
14	2												
TOTAL	481	131	50	18	5	8	12		2		10	–	
REPROD.	12	70	18	12	8	10	2				18		

BURNED ANNUALLY

d.b.h.[a] (inches)	Longleaf pine	Loblolly pine	Black gum	Sweet gum	Red maple	Wax myrtle	Laurel oak	Black jack oak	Southern red oak	Water oak	Sweet bay	American holly	Sumac
1	7												
2	58												
3	38												
4	65							2					
6	113												
8	67												
10	39							2					
12	2												
14	2												
TOTAL	381							4			–		
REPROD.	12												

a. Diameter at breast height.

442

even among hardwood invaders (Edmisten 1963).

Edmisten (1963), who studied the ecology of Florida pine flatwoods, concentrated on indicator species while studying the effects of fire exclusion. Saw palmetto and bitter gallberry reached a peak similar to that of slash pine, doing well up to the 12th to 15th year after fire before declining. Stagger bush populations increased until approximately the 24th year, but wax myrtle prospered more after each additional year of fire exclusion. Frequency of the three common herbs of the fire phase slash forest (vanilla-plant, black-root, and three awn grass) dropped drastically when fire was absent.

Laurel oak and sweet gum both eventually dominate stands where fire has been excluded for 25 years or more. The bays (loblolly, sweet, and red), however, follow a trend similar to saw palmetto and slash pine. They peak in population and dominance after about 18 years, and then decline in numbers. The canopy becomes progressively denser with time following fire exclusion. Diversity of slash pine woods excluded from fire increases until approximately the 15th year, and then declines rapidly. Garren (1943) summarized successional trends in slash pine communities following fire exclusion (Table 6-6).

Shortleaf-loblolly communities may persist, like longleaf and slash communities, as pyric climaxes after they are established. However, loblolly-dominated communities are communities of secondary succession. They follow longleaf pine, hardwood, or old field communities (Wharton 1978). The successional dynamics of old field to oak-hickory by way of a loblolly pine system are discussed in the subsections "Old Field Community" and "Pine-Mixed Hardwood Communities."

The relationship between soil nutrients and successional trends in Florida pine flatwoods communities was discussed by Edmisten (1963). He pointed out that upon fire exclusion soil microflora increased, subsequently increasing phosphorous content of the soil. Hardwood species use this increased phosphorus to their advantage, as do species such as wax myrtle. A fungus is a symbiont of the roots of the wax myrtle, and Edmisten (1963) noted that the wax myrtle may be able to fix nitrogen at a rate higher than that of legumes.

Several other studies of pine communities in the Sea Island Coastal Region have been conducted. Andrews (1917) explained the role of fire in the propagation of longleaf pine; Wahlenberg (1965) discussed loblolly and slash management in the Southeast; Barry (1968) listed the vascular flora of the Baruch Plantation in Georgetown County, South Carolina; and

Barry and Batson (1969) related vegetation of the Baruch Plantation to soil types.

3. Pine-Mixed Hardwood Communities

Pine-mixed hardwood forests are very common in the upland ecosystem of the Sea Island Coastal Region. They may be found from extremely xeric sites (longleaf pine-mixed hardwood community) to mesic sites (loblolly pine-mixed hardwood community). Near hydric pine-mixed hardwood communities, such as slash pine-mixed swamp hardwood and spruce pine-mixed swamp hardwood, are discussed in Chapter Five, Freshwater Ecosystems. Upland pine-mixed hardwood forests are dominated by two major community groupings, mesic pine-mixed hardwoods and xeric pine-mixed hardwoods.

a. Mesic Pine-Mixed Hardwood Community. These communities are generally made up of a supercanopy of older, taller pines dominated by loblolly. This community contains a subcanopy of pines (slash, spruce, or shortleaf) and hardwoods. The understory is tall and rather dense, with a shrub layer present. Because of its composition, this community is frequently referred to as a "mixed pine-mixed hardwood community."

Xeric pine-mixed hardwood communities are more complex than mesic pine-mixed hardwoods. The predominant pine is longleaf, although shortleaf and loblolly occasionally occur in the canopy. The history of fire and logging in the community determines canopy height, understory height, and density of both. The shrub and herb layers of the xeric pine-mixed hardwood communities are usually very open, regardless of fire and logging history, and areas of exposed soil are common (Wells and Shunk 1931, Kohlsaat 1974).

Braun (1950) reported that the "loblolly pine-hardwoods" community is more widespread in the Atlantic Coastal Plain region than it formerly was. As a second-growth forest, the loblolly-hardwoods community has replaced many acres of flatwoods that were previously dominated by longleaf communities. Data on a longleaf community invaded by loblollies and hardwoods after fire was excluded are presented in Table 6-4. Braun (1950) listed white oak, willow oak, southern red oak, water oak, hickories, sweet gum, black gum, and red maple as common hardwood codominants of the loblolly-hardwoods community.

The loblolly-mixed hardwood community is common within the Sea Island Coastal Region of South Carolina, although it has not been as well-documented here as it has in the North Carolina coastal plain. In a study of the southern mixed hardwood forest, Quarterman and Keever (1962) analyzed four loblolly pine-mixed hardwood communities in the coastal plain of South Carolina (three of which are in the Sea

Table 6-5. Succession in longleaf pine forests after fire exclusion (Garren 1943).

Type of longleaf forest	Next stage(s) in succession. (With years, if noted.)
1. Lowland, abandoned fields	Wax myrtle Laurel oak (12 - 15 years)
2. Rough, poorly drained lowlands	Water oak Sweet gum
3. Virgin forest, lowlands	Laurel oak, live oak (and varieties) with undergrowth of wax myrtle, red bay and holly (75 years)
4. Areas near lakes, in lowlands	Slash pine ——► Laurel oak, water oak, sweet gum
5. Well drained areas in upper coastal plain	Southern red oak Post oak Loblolly pine Shortleaf pine
6. Dry, loamy flatwoods	Southern red oak ——► oak-hickory, sweet gum, post oak ——► Beech-maple
7. Moist, sandy flatwoods	Cypress-tupelo swamps
8. Loamy flatwoods, moister slopes	Black gum, loblolly pine, sweet gum in thickets ↓ Oak-hickory

Table 6-6. Succession in slash pine forests after fire exclusion (Garren 1943).

Condition	Next Stage or Stages in Succession
Slash pine, scrub live oak, scrub oak, with fire protection	Live oak
Slash pine, saw palmetto, bitter gallberry. Gray on moist sandy flatwoods. Cutting of all pines, then frequent fires	Toothache grass Yellow-eyed grass } in bogs Meadow beauty Pitcher-plant
Slash pine, with fire protection	Laurel oak Water oak Gums
Slash pine. Cutting of pine, then fire more often than 10-year intervals.	Sprouting scrub-hardwoods

444

Island Coastal Region). In Jasper County, South Carolina, canopy dominants were water oak, black gum, and loblolly pine with laurel oak, sweet gum, bull bay, hickory, and dogwood in the understory.

Hartshorn (1972) and Radford (1976) described loblolly-mixed hardwood communities from the coastal plain of North Carolina. Hartshorn (1972) listed three possible phases of loblolly-mixed hardwood communities, 1) white oak phase, 2) ravine slope phase, and 3) sweet gum-swamp tupelo phase (discussed in Chapter Five on Freshwater Ecosystems). The white oak phase is dominated by white oak, southern red oak, post oak, beech, and loblolly. Hartshorn's ravine slope phase, which has a very low percentage of loblolly pine in the canopy, is discussed under "Mixed Hardwood Communities." A North Carolina coastal plain community dominated by sweet gum, tulip tree, swamp chestnut oak, and loblolly pine was described by Radford (1976). Subcanopy trees were flowering dogwood, red maple, sourwood, American holly, and black cherry. Blueberry, southern lady fern, rattlesnake fern, partridge berry, and crane-fly orchid were the dominants of shrub and herb layers.

Barry (1968) correlated three variations of the loblolly pine-mixed hardwood community with soil types in Georgetown County, South Carolina. On coarse, sandy Daufuskie soils, loblolly pine dominates with live oak, while persimmon and wax myrtle dominate the understory. Loblolly pine and sweet gum dominate on St. Johns soils, which consist of fine sand with a clay hardpan 20 - 50 cm (8 - 20 in) below the surface; red bay and bitter gallberry are common in the understory. Finally, loblolly pine, sweet gum, and laurel oak are the canopy dominants on Onslow loamy sand, with persimmon and yaupon holly common in the understory. In Table 6-7, tree, shrub, and herbaceous dominants are listed by soil types (see also Barry and Batson 1969).

Mesic pine-mixed hardwood communities are common in Charleston County, South Carolina, where two pine-mixed hardwood communities, a mixed pine-mixed hardwood community, and a live oak-water oak-mixed pine community were observed (L. L. Gaddy and D. A. Rayner, 1978, South Carolina Wildlife and Marine Resources Department, Columbia, unpubl. data). Loblolly and slash were the dominant pines in these communities, while water oak, laurel oak, sweet gum, and live oak were the most common hardwood species. Porcher (1974) also described a mixed pine-mixed hardwood community in the coastal plain of South Carolina.

In Georgia, Wharton (1978) described coastal plain pine-mixed hardwood communities as a "broadleaf deciduous-needleleaf evergreen upland forest." He noted that this forest type was more common in the inland coastal plain counties of Georgia (outside the Sea Island Coastal Region). Wharton (1978) listed shortleaf pine as the common pine of this forest type. He also listed mockernut hickory, southern red oak, sweet gum, laurel oak, and white oak as possible co-dominants with pine in this forest type.

b. Xeric Pine-Mixed Hardwood Communities. This designation is used here to describe only the driest of the pine-mixed hardwood communities. The dominants here are usually longleaf pine and turkey oak, although other oaks may co-dominate with turkey oak. The xeric pine-mixed hardwood communities of the Sea Island Coastal Region are generally found on ridges of coarse sand. These ridges are usually associated with rivers, but are similar in appearance to the coarse grain sandhills of the fall line. These areas are known by various names, including shrubby oak lands (Catesby 1771), and pine barrens, scrub oak lands, and turkey oak barrens (Kohlsaat 1974). All of these refer to barren lands vegetated with pine and/or small oaks.

Recently, Wharton (1978) termed the longleaf pine-turkey oak community type a "dwarf oak forest" in Georgia. He listed turkey oak, bluejack oak, shrubby post oak, and longleaf pine as possible dominant trees. Kohlsaat (1974) called the same community a "turkey oak barren" in South Carolina. The dominants here were longleaf pine, turkey oak, and three awn grass. Kohlsaat noted the openness of the community by describing bare sandy areas that were being invaded by spike-moss and reindeer lichen.

Moving southward in the Sea Island Coastal Region, the character of the longleaf pine-turkey oak community changes slightly. Dwarfed Chapman oak, gopher apples, saw palmetto, and Satureja calamintha begin to appear in the community. These changes begin approximately at the Savannah River, although several fluvial ridges on the South Carolina side of the river contain some of these plants. South Carolina Wildlife and Marine Resources Department (1974c) found live oak, turkey oak, shrubby post oak, and longleaf pine to be dominants of a fluvial sand ridge on the South Carolina side of the Savannah River (Jasper County). Persimmon, sparkleberry, and saw palmetto were also common.

Bozeman (1965) studied the flora of a turkey oak-longleaf pine community on a fluvial sand ridge along the Altamaha River in Long and McIntosh counties, Georgia. Turkey oak, longleaf pine, dwarfed live oak, bluejack oak, and shrubby post oak were the dominants. In the shrub layer, Bozeman found Satureja calamintha, Georgia plume, and gopher apples. Common trees, shrubs, herbs, and grasses from

Table 6-7. List of dominant (>60% occurrence) vegetation for major soil types of Baruch Plantation, Georgetown County, South Carolina (Barry and Batson 1969).

| | SOIL TYPES | | |
	Daufuskie	Onslow Loamy Sand	St. Johns
TREE SPECIES			
Loblolly pine	X	X	X
Live oak	X		
Persimmon	X	X	
Sweet gum		X	X
Red bay			X
Laurel oak		X	
SHRUBS			
Wax myrtle	X		
Bitter gallberry			X
Yaupon holly		X	
HERBACEOUS SPECIES			
Camphorweed	X		
Panic grass	X	X	X
Spanish moss	X		
Greenbriars	X		
Amorpha herbacea	X		
Beggar ticks	X		
Elephant's-foot	X		
Blackberries		X	
Black root			X
Bracken fern			X

Bozeman's (1965) study are listed in Table 6-8.

Xeric pine-mixed hardwoods may be dominated by trees other than longleaf pine and turkey oak. Radford (1976) described a longleaf pine-shortleaf pine-turkey oak community from Jasper County, South Carolina. The South Carolina Wildlife and Marine Resources Department (1974e) listed laurel oak, Chapman oak, and mockernut hickory as common associates of slash and longleaf pine on a sand ridge in Jasper County, South Carolina. In a study of Georgia sand ridge vegetation, Bozeman (1971) described an alliance of black jack oak and turkey oak on "upland sandy loams, clay loams, and sand ridges." The characteristic plants of this alliance are longleaf pine, shrubby post oak, bluejack oak, three awn grass, _Bonamia patens_, dog-tongue, _Tephrosia florida_, persimmon, _Tragia urens_, sparkleberry, squaw-huckleberry, and saw palmetto, along with turkey oak and black jack oak. These plants form associations that are found in different zones of the coastal plain. Sand ridge communities, in which pines are not dominant, are discussed under the following section on "Mixed Hardwood Communities."

Quarterman and Keever (1962) carried out a comprehensive study on the successional relationships of the "southern mixed hardwood forest" on moist to wet uplands. In this work, they described several stands, three of which were in South Carolina, having 30% or more dominance by loblolly pine. By analyzing understory species, Quarterman and Keever documented that pines were being replaced by hardwoods in all of these stands (loblolly pine is intolerant of shade, see Bormann 1956). According to Quarterman and Keever, a well-developed understory of hardwoods is present by the time mesic pine stands are 30 - 50 years old. Between the approximate ages of 50 - 75 years, the stand becomes a pine-mixed hardwood forest.

Gano (1917) described old field succession in northern Florida, and noted that communities dominated by loblolly, slash, and shortleaf pines and mixed hardwoods occurred at a mid-successional stage. Wells (1928) discussed the role of loblolly pine in the "Pinus Consocies" of the "Meso-Xeric Pine Forest." He pointed out that loblolly pine forests have taken over many former longleaf pine communities (i.e., those that have been cleared)

446

Table 6-8. Common plants of turkey oak-longleaf pine communities on a coastal plain fluvial ridge in Long and McIntosh counties, Georgia (adapted from Bozeman 1965).

TREES	HERBS AND GRASSES
Turkey oak	Spikemoss
Longleaf pine	Broom-straw
Shrubby post oak	Three awn grass
Bluejack oak	Indian grass
Live oak (usually dwarfed)	Beak rush
Mockernut hickory	Dog tongue
	Cottonweed
SHRUBS	Cleome aldenella
	Butterfly pea
Saw palmetto	Lupine
Gopher apples	Summer-farewell
Dangleberry	Stinging needle
Dwarf huckleberry	Queen's delight
Squaw-huckleberry	Blue star
Satureja calamintha	Balduina angustifolia
Georgia plume	Golden aster
	Blazing star

in the coastal plain. Chapman (1932a) and Heyward (1939) documented this fact (see Table 6-4). In the loblolly pine-mixed hardwood forest, referred to in this work as mesic pine-mixed hardwood communities, Wells (1928) noted that the loblolly occupies an "ecologically evanescent" role, being an intermediate stage between the old field or disturbed flatwoods and the oak-hickory forest.

In xeric pine-mixed hardwood communities, fire is significant in the succession of the community. Wells and Shunk (1931) pointed out that once the wire grass sod of old growth longleaf pine communities was disturbed, it never grew back. Where fire was absent, turkey oak quickly invaded, with black jack oak following after a leaf litter layer had formed. Black jack oak appeared sooner in communities where three awn grass (wire grass) had not been destroyed. The most common types of disturbance bringing about these changes in longleaf fire climax communities were agriculture (which destroyed the wire grass) and logging (which usually did not destroy the wire grass). The longleaf canopy was often not completely removed, and a community of small longleaf pines and invading scrub oaks persisted.

Wells and Shunk (1931) quantitatively confirmed the accepted observation that xeric pine-mixed hardwood communities result from lack of soil moisture and nutrients. The sterility of longleaf pine-turkey oak sands have led to the term "edaphic" climax for this community, although most observers, Wells and Shunk included, think the climax here (without fire) is oak-hickory. They described the coarse sand soils as a type low in

bacterial, fungal, and moisture content. The carbon to nitrogen ratio ranged from 25:1 to 40:1, compared with an average of 10:1 for agricultural soils. Wells and Shunk (1931) also observed various adaptations of the longleaf pine-turkey oak community to lack of water and to frequent fire. Such adaptations included root elongation, leaf tissue modification, wind-day leaf plane orientation, and seasonal dominance.

Garren (1943) noted that shrub oak forests predominated where all longleaf pine seed trees had been removed. However, Garren noted that after selective cutting, infrequent fires produce a longleaf pine subclimax normally having a low shrub layer of oak "sprouts."

4. Mixed Hardwood Communities

The southern mixed hardwood forest is an extremely complex entity, with dominance varying from stand to stand. Wells (1928) described the climax forest of the North Carolina coastal plain as a beech-maple association, with oaks and hickories as sub-dominants. Braun (1950) included the Sea Island Coastal Region within her "Southeastern Evergreen Forest Region," which is dominated by pines and evergreen hardwoods. As many as 20 different community variations have been described for mixed hardwood forests in the Sea Island Coastal Region, but six basic community types form the framework of the mixed hardwood forest in this area. These basic community types are the mesic slope hardwoods community, the upland mesic hardwoods community, the hammock community, the scrub forest community, the live oak-mixed hardwood community, and the dwarfed oak-mixed hardwoods community.

The mesic slope hardwoods, upland mesic hardwoods, hammock, and live oak-mixed hardwood communities are multi-tiered forests with relatively high species diversity in the canopy, sub-canopy, shrub, and herbaceous layers (Quarterman and Keever 1962, Monk 1968). The scrub forest and dwarfed oak-mixed hardwood communities, on the other hand, exhibit little structural or species diversity. The scrub forest is usually a single or double-tiered community with tree and shrub species blending into one canopy less than 6.1 m (20 ft) in height; the herbaceous layer, less than 15 cm (6 in) tall, is dominated by lichens and spikemoss. The dwarf oak-mixed hardwood community, seldom over 4.6 m (15 ft) tall, is also single- or double-tiered in structure, and extensive areas of bare sand are common (Wharton 1978). The hammock community originates from scrub forest communities, and is therefore usually found on soils low in mineral content; Monk (1966) suggested that the evolution towards evergreens in this community may have been a response to the sterility of the soil.

Coarse sand communities (shrub forest and dwarfed oak-mixed hardwood communities) occur on soils extremely low in organic content, in microflora and micro-fauna, and in soil nutrient content (Wells and Shunk 1931). The live oak-mixed hardwood community occurs on soils that are richer than scrub forest, hammock, and dwarfed oak-mixed hardwood communities, but lower in mineral content than mesic slope hardwoods and upland mesic hardwoods communities. According to Art (1976), meteorological nutrients may be important to this community type (see Maritime Ecosystems, Chapter 3).

a. **Mesic Slope Hardwoods Community.** This community is known by a variety of names, such as ravine slope hardwoods (Hartshorn 1972), beech ravine (Kohlsaat 1974), mixed mesophytic hardwoods (Braun 1950, South Carolina Wildlife and Marine Resources Department 1974f), and bluff and slope forest (Wharton 1978). In North Carolina, Hartshorn (1972) found white oak, beech, mockernut hickory, tulip tree, pignut hickory, black oak, and scarlet oak all occurring in the canopy of several "ravine slope" communities. In South Carolina, some of the same trees were found by South Carolina Wildlife and Marine Resources Department (1974f) to be dominants in this community, along with water oak and post oak. Braun (1950) noted that beech, red maple, and sweet gum were among the dominants of what she termed "mixed mesophytic hardwoods." Braun also described a beech-bull bay (magnolia) community from the Southeast. Commonly found on dissected riverbluffs and ravines, dominants of such communities are beech, bull bay, white oak, water oak, laurel oak, American holly, southern sugar maple, and redbud.

Batson et al. (1957) found a beech-bull bay community within the Savannah River drainage system in Allendale County, South Carolina. They pointed out that this assemblage was not a bluff community, but was found on high bottomland. Beech, bull bay, laurel oak, red maple, black gum, tulip tree, sweet gum, and loblolly pine were dominants.

Kohlsaat (1974) called the north-facing bluff forests of the coastal plain of South Carolina "beech ravines." Beech ravine dominants are similar to those described by Hartshorn (1972), for ravine slope communities, as well as the "bluff forest with northern affinities" of Wharton (1978). This community is not common in the Sea Island Coastal Region of Georgia.

Wharton's (1978) "bluff and slope forests" may be compared with the mesic slope and beech-bull bay communities discussed above. Wharton listed "dry" and "seepage" variations of the bluff community. A dry bluff on the Altamaha River was dominated by spruce pine, beech, bull bay, white oak, and laurel oak. On a seepage bluff at Magnolia Bluff, bull bay dominated along with swamp chestnut oak, beech, water oak, overcup oak, and other bottomland hardwoods colonizing the moist slope.

b. **Upland Mesic Hardwoods Community.** This community has also been called the "oak-hickory forest" (Braun 1950), the "southern mixed hardwood forest" (Quarterman and Keever 1962), and the "mesophytic broadleaved forest" (Wells 1928). Wells (1928) listed white oak and black oak as the dominants of this community, with southern red oak, mockernut hickory, and pignut hickory as sub-dominants. Quarterman and Keever (1962) pointed out that 14 tree species were "structurally important" in this community. The 10 most important, in order of importance, were beech, laurel oak, bull bay, white oak, sweet gum, mockernut hickory, water oak, southern red oak, pig-nut hickory, and black gum.

Hartshorn (1972) described an upland mesic hardwoods community dominated by white oak, beech, sweet gum, southern red oak, and post oak from the coastal plain of North Carolina. L. L. Gaddy and D. A. Rayner (1978, South Carolina Wildlife and Marine Resources Department, Columbia, unpubl. data) listed a water oak-mockernut hickory community from Charleston County, South Carolina. Quarterman and Keever (1962) and Monk (1968) have pointed out that the southern mixed hardwood community varies from site to site, where soil types and/or soil moisture differs.

c. **Hammock Community.** Monk (1968) noted that the southern mixed hardwood community is dominated by broadleaf deciduous trees on fertile, mesic sites, and by broadleaf evergreens on sterile xeric

sites. The mixed hardwood/broadleaf evergreen community is generally referred to
as the "hammock" community.

Wharton (1978) divided the hammock
community into two sub-types, a lowland
broadleaf evergreen forest and an upland
broadleaf evergreen forest. He noted
that the lowland hammock is very common
in Chatham County, Georgia. Dominants include water oak, live oak, laurel oak,
bull bay, American holly, and spruce pine,
with saw palmetto, stagger bush, and blueberries in the understory. The upland
hammock is completely dominated by evergreen species (live oak, laurel oak, bull
bay, American holly, and spruce pine),
with pignut hickory being the only deciduous tree reported in the canopy. Wild
olive, saw palmetto, red bay, sparkleberry, and drier-site shrubs are found in
the understory.

Bozeman (1965) described a hammock
community from a sand ridge in McIntosh
County parallel to the Altamaha River in
which laurel oak, water oak, live oak,
white oak, swamp chestnut oak, pignut
hickory, sweet pignut hickory, bull bay,
spruce pine, and saw palmetto were the
most important species. He noted that
this community corresponds to Laessle's
(1942) "mesic hammock," and to what is
termed "lowland hammock" here. Among the
shrubs and vines found by Bozeman on the
hammock were dwarf palmetto, chinquapin,
witch-hazel, red buckeye, Hercules' club,
dogwood, fetter-bush, wild azalea, sparkleberry, horse sugar, storax, wild olive,
French mulberry, summer grape, and coral
honeysuckle. A low number of herbs was
found by Bozeman in undisturbed mesic
hammocks. An even lower hammock, corresponding to Laessle's (1942) "hydric
hammock," was described from the same
ridge. The dominant there was cabbage
palmetto, with mixed hardwoods.

Bozeman (1971) described a laurel
oak-wild olive association that seems to
be comparable to Wharton's (1978) upland
broadleaf evergreen forest. The characteristic plants of the laurel oak-wild
olive association were laurel oak, wild
olive, bull bay, and dwarf pawpaw. Also
listed for this community were live oak,
spruce pine, partridge berry, mockernut
hickory, horse sugar, and squaw-root.
Monk (1960) described a mesic hammock
from northcentral Florida. The dominants
were laurel oak, bull bay, mockernut
hickory, southern red oak, hop hornbeam,
American hornbeam, flowering dogwood, and
American holly. This community seems to
fit neither of Wharton's (1978) hammock
types. Laessle and Monk (1961) noted the
possible variations of the live oak hammock in Florida, and pointed out that it
occurs with accompanying variations on
flatwoods, sandhills, and dune ridges.

Although common in Georgia, the hammock community does not range far into

South Carolina. Hamilton Ridge along the
Savannah River in Hampton County has typical hammock species such as laurel oak,
spruce pine, and bull bay in the canopy,
but they are not dominants (South
Carolina Wildlife and Marine Resources
Department 1974f). In Jasper County,
several lowland or hydric hammocks are
found near the Savannah River (Wharton
1978).

d. Scrub Forest Community. This
community of sand ridges is common in
Florida, uncommon in Georgia, and extremely
rare in South Carolina. Bozeman (1971)
described a myrtle oak-Chapman oak association consisting of myrtle oak,
Chapman oak, stagger bush, and giant
seeded beak rush as the community dominants, though other species including
Georgia plume may be present. In Georgia,
Bozeman pointed out that the scrub forest
community often occurs as a transitional
area between turkey oak and hammock communities. In Florida, the scrub forest
occurs on the highest, most sterile sands
with dwarfed oak-mixed hardwood communities found on lower more fertile sands
(Laessle 1958). Wharton (1978) described
two types of scrub in Georgia: 1) the
evergreen scrub forest, and 2) the evergreen scrub-lichen forest. Wharton noted
that evergreen scrub forests resemble the
dune-oak scrub communities of maritime
dunes (See Maritime Ecosystem, Chapter
Three) in being a dwarfed oak-dominated
community. Although scrubby post oak,
live oak, bull bay, and mockernut hickory
are present, the dominants are shrubs such
as myrtle oak, squaw huckleberry, Georgia
plume, and sparkleberry. Saw palmetto,
wild olive, and various blueberries may
also be present. The evergreen scrub-
lichen forest is found on the driest,
coarsest sand ridges. Here, dominants are
dwarfed live oak, dwarfed laurel oak, and
dwarfed red bay, along with sparkleberry,
three species of _Cladonia_ lichens, and
spikemoss.

Laessle (1958) documented the relationship between the longleaf pine-turkey
oak and scrub communities in Florida. His
"hilltop scrubs" resemble the evergreen
scrub forest communities described by
Bozeman (1971) and Wharton (1978). Myrtle
oak, Chapman oak, and saw palmetto dominate along with sand pine, which is absent
from the Georgia communities. The extent
of this community in Florida has led to
the use of the name "Florida scrub" to
signify a community with a dense layer of
dwarfed tree species mixed with saw palmetto and shrubs. This name has also been
applied to flatwoods communities that have
been invaded by shrubs and small trees
upon the exclusion of fire (Laessle 1958).

e. Dwarfed Oak-Mixed Hardwood
Community. On most sand ridges the turkey
oak-longleaf pine community is dominant.
Because pine is the dominant tree in natural, fire-maintained longleaf pine-turkey

449

oak communities, this community is discussed in the section on pine-mixed hardwood communities elsewhere in this chapter. After selective cutting of the pines, this community may become a mixed hardwood community. Historically, it seems that longleaf pine may have been a dominant.

Wells and Shunk (1931) listed three (dwarfed) mixed hardwood communities from the coarse sands of North Carolina, a turkey oak association, a black jack oak association, and a mockernut hickory association. Wells and Shunk explained these non-pine dominated communities in terms of successional stages. They pointed out that turkey oak dominates upon the exclusion of fire, black jack oak becomes more numerous as a layer of leaf litter develops, and a more mesic forest dominated by mockernut hickory will finally prevail.

Black jack oak-mixed hardwood communities occur in the inland portion of the Sea Island Coastal Region of South Carolina on Miocene clayey ridges, but they do not occur in the coastal counties of Georgia (Bozeman 1971, Wharton 1978).

f. Live Oak-Mixed Hardwood Community. This community is most commonly found on barrier islands (see Maritime Ecosystem, Chapter Three). However, live oak forests are also frequent along the outermost edge of the mainland, especially adjacent to salt marshes. In Charleston County, South Carolina, live oak, water oak, hickories, and loblolly pines dominate mature forests near the edge of the marsh (L. L. Gaddy and D. A. Rayner, 1978, South Carolina Wildlife and Marine Resources Department, Columbia, unpubl. data). Wharton (1978) called this community the "upland maritime forest." He found that live oak and laurel oak were dominants, with American holly, red bay, bull bay, water oak, and pignut hickory also present. Bozeman (1975) termed the same assemblage a "mixed oak-hardwood forest" community, and considered it a climax (see Maritime Ecosystem, Chapter Three). Slash pines may be in with the canopy in Georgia (Bozeman 1971), while loblolly pines may be present in South Carolina (L. L. Gaddy and D. A. Rayner, 1978, South Carolina Wildlife and Marine Resources Department, Columbia, unpubl. data).

Quarterman and Keever (1962) and Monk (1965) regard the "southern mixed hardwood forest" as a climax forest for the southeastern coastal plain. Beech-bull bay (magnolia) (Gano 1917) and beech-maple (Wells 1928) have also been listed as climax communities for the region. Quarterman and Keever (1962) noted that the "conspicuous appearance" of beech and bull bay (magnolia) may have led field workers to overestimate their importance. The role of bull bay in the climax

southern mixed hardwood forest has been further placed in doubt by the assertion that it does not reproduce itself in shade.

Quarterman and Keever's (1962) field work applies primarily to the upland mesic hardwoods community and, to a lesser degree, to the mesic slope hardwoods community. They noted that the climax for this community is a mixed hardwood community in which shade-intolerant pines have been completely eliminated. The 10 most important tree species here in order of importance are: beech, laurel oak, bull bay (magnolia), white oak, sweet gum, mockernut hickory, water oak, southern red oak, pignut hickory, and black gum.

The hammock, scrub forest, and dwarfed oak-mixed hardwood communities are closely related successionally. Bozeman (1971) noted that hammock communities follow sandhill communities (both dwarfed oak-mixed hardwood and scrub forest) after fire exclusion. Bozeman (1965) described hammock communities as being common in transition fire-protected areas (between sandhill communities and swamp or bay forests). Working in north-central Florida, Monk (1968) suggested that sandhill communities consisting of dwarfed oak-mixed hardwoods gradually become scrub forests, and eventually hammock communities, when fire is excluded from the community. Laessle (1958), on the other hand, stated that sandhill communities and scrub forest communities are successionally distinct and are differentiated as a result of edaphic conditions. Laessle and Monk (1961) pointed out that in live oak hammocks of Florida, a "scrubby flatwoods" stage was generally maintained until fire was excluded. Thereafter, a subclimax live oak hammock became dominant and eventually led to the establishment of a mesic (climax) hammock. In the terminology used here, this translates into a scrub forest to upland (xeric) hammock to lowland (mesic) hammock successional sequence.

The live oak-mixed hardwood community is considered a climax for upland areas adjacent to salt marshes on the mainland of the Sea Island Coastal Region (Bozeman 1975, Wharton 1978).

Upland mixed hardwood communities have been selectively eliminated in many areas of the Sea Island Coastal Region by cutting and subsequent pine management practices. Because of their accessibility, the upland hardwood forests were the first ones to be cut in colonial times. These forests were extensively used for building materials and firewood. In modern times, pine management programs, including the use of fire and "weed" tree destruction campaigns, have prevented hardwoods from becoming re-established in upland forests. Steep bluffs and hammocks surrounded by swamps are among the few locations harboring good hardwood stands today. (See

Volume II, Chapter Six for additional forestry information.)

III. CONSUMERS

A. INVERTEBRATES

1. Soil Fauna

According to Kevan (1968), the soil fauna ". . .includes those animals that occur in the soil during their whole life, or in one or more of their developmental stages. If a species only spends part of its life in the soil, then only this stage is considered to belong to the soil fauna." For a discussion of the physical characteristics of soils of the Sea Island Coastal Region, see Volume I, Chapter Three.

A diverse and abundant invertebrate fauna exists in the soil of terrestrial habitats such as old fields and forests (Kevan 1955, 1968, Murphy 1962, Burges and Rawe 1967, Wallwork 1970, Richards 1974). Representatives of several invertebrate phyla occur in the soil fauna, including Platyhelminthes, Rhynchocoela, Rotifera, Gastrotricha, Nematoda, Nematomorpha, Annelida, Tardigrada, Mollusca, and Arthropoda (Kühnelt 1955). The nematodes are often present in immense numbers, reaching estimated densities of 20 million individuals/m^2 (Peters 1955). The most important soil arthropods are the mites (Acarina), springtails (Collembola) and certain other insects, and myriapods (diplopods, pauropods, chilopods, and symphylids), although copepods, amphipods, isopods, pseudoscorpions, and phalangids are also present (Kevan 1968). An idea of the density of soil arthropods may be obtained from a study of pasture soil by Salt et al. (1948), in which an estimated total of 263.6 x 10^3 individuals/m^2 was reported. Populations of soil animals are aggregated, however, and Kevan (1968) observed that densities estimated on an acreage basis have little real meaning.

Soil invertebrates are believed to be important in the breakdown of detritus in forest ecosystems, and are therefore important in mineral recycling (Murphy 1955, Kowal and Crossley 1971, Edwards et al. 1973). According to Edwards et al. (1973), soil invertebrates play a role in the decomposition of organic matter in that:

"1. they disintegrate plant and animal tissues and make them more easily invaded by microorganisms;

2. they selectively decompose and chemically change parts of organic residues;

3. they transform plant residues into humic substances;

4. they increase the surface area available for bacterial and fungal action;

5. they form complex aggregates of organic matter with the mineral part of the soil; and

6. they mix the organic matter thoroughly into the upper layers of soil."

Earthworms have long been recognized as being beneficial to agriculture. They partially break down organic matter; mix clay, humus, and lime; eject ingested material as castings; enhance drainage, aeration, and root penetration by burrowing; and produce a soil structure favorable to small soil animals. According to Kevan (1968), "the diversity of fauna in a good mull soil may be one of its greatest assets because the humification process flows through a wide variety of varying metabolisms and offers better opportunities for degradation of intractible compounds." From the studies that have been made on soil animals has come a realization of the influence of faunal activity on soil structure and their role in the mechanical and biochemical breakdown of organic material, making it available for re-use by plants (Murphy 1955). The discovery of their role in humus formation (Kubiena 1955), and the damage that some can do to agricultural crops (Brown 1955), have further served to substantiate their importance. The literature on soil animals is rather extensive; Kühnelt (1963) noted that more than 500 papers were available on soil arthropods alone.

Major factors influencing the soil fauna include topography; soil structure, texture, and color; concentration of organic matter; electrolyte content; pH; light; temperature; moisture content; relative humidity; air composition in pore spaces; and possibly atmospheric pressure (Kevan 1968). Interrelationships with other organisms, including both plants and animals, are also significant. Kevan noted that alterations to the soil through tilling, building, excavating, draining, flooding, or dumping influence the fauna, as do fertilization and the rotation of crops. Increased plant growth stimulates the soil fauna indirectly; manure is also stimulating to these animals and introduces the dung fauna (e.g., Collembola, dipteran larvae, oribatid mites, enchytraeid worms, staphylinid beetles, etc.) to the soil. In "mor" soils, where organic matter occurs as a discrete layer at the surface and is little mixed with the mineral horizon below, large numbers of small species occur (Murphy 1955). The lack of mixing between these layers is largely attributable to a lack of earthworms. In "mull" soils, where organic and mineral matter are more thoroughly mixed in vertical profile, the fauna is richer with greater biomass, although the number of individuals may be smaller. The effects of insecticides on soil invertebrates vary, depending upon the type and concentration of the pesticide (Sheals 1955, Tinkham 1955). In an experimental apple orchard that had been subjected to longterm pesticide use in the State of New York, Menhinick (1962) found fewer taxa; fewer numbers and species of larger organisms; an increase in the numbers of small, tolerant animals; and a shift in trophic structure compared with control areas free of pesticides.

Microarthropods of old fields are reasonably well-known (see Engelmann 1961), and the fauna of litter was studied at the Savannah River Plant near Aiken, South Carolina, by Wiegert (1974). Densities of invertebrates in three fields were found by Wiegert to be highest in a 1-year abandoned cornfield, intermediate in a 12-year lespedeza field, and lowest in a 12-year broom sedge field. Densities differed significantly from field to field, as well as with season and litter type. Density was high during periods when moisture content was high in the litter, and low during dry periods. Animals identified from old fields in litter bag collections are listed in Table 6-9. The ants of old fields at the Savannah River Plant were studied by Van Pelt (1966). A total of 52 species or subspecies of formicids were identified, the most abundant of which was Dorymyrmex pyramicus.

Little information is available on the invertebrate fauna of coastal pine forests in the Sea Island Coastal Region, although studies have been conducted by several researchers in North Carolina forests. In the Duke Forest, Pearse (1943, 1946, 1953) reported that the Acarina and Collembola together accounted for 93% of the fauna in loblolly pine habitats with clay soil, 95% of the fauna in shortleaf pine habitats with sand soil, 90% of the fauna in white oak habitats with clay soil, and 92% of the fauna in white oak-post oak habitats with sandy soil (Pearse 1946). Orbatid mites (Acarina) and several other arthropod groups in pine litter and the underlying mineral soil were studied at Oak Ridge, Tennessee, by Crossley and Bohnsack (1960). Of the total number of animals collected, 82.9% were mites (Acarina), 12.2% were springtails (Collembola), and 3.6% were insects other than Collembola. Trophic studies of soil microarthropods on pine forest lands were undertaken at Oak Ridge by Kowal and Crossley (1971). Dominant among the saprophytes were Collembola, orbatid mites, Thysanoptera, and pselaphognathid millipedes (Diplopoda), while predators included centipedes (Chilopoda), predatory mites, beetles, spiders, and pseudoscorpions. Ants were important as both saprophytes and predators. Kowal (1969) observed

that the food web is relatively simple in pine mor compared with other forest floor types, consisting largely of decomposing pine litter, fungi, and microarthropods (primarily mites and springtails).

Finally, collections of insects and gastropods have been made on Cumberland Island, Georgia (Hillestad et al. 1975), including areas occupied by loblolly pine or in pine-oak scrub forest as well as hardwood areas. Collections of spiders were made, but no data were given on their habitats. Gastropods included Polygyra septemvolva and Succinea campestris; insects are listed in Tables 6-10 and 6-11.

2. Aerial and Crawling Insects

In addition to the soil invertebrates, a number of aerial and/or crawling insects can be expected in upland forest habitats, including dragonflies and damselflies (Odonata) and mayflies (Ephemeroptera), particularly near fresh water; true bugs (Hemiptera); lacewings (Neuroptera); butterflies and moths (Lepidoptera); beetles (Coleoptera); flies, mosquitoes, and midges (Diptera); wasps, bees, sawflies, ichneumons, chalcids, and ants (Hymenoptera); grasshoppers, katydids, crickets, cockroaches, mantids, and walking sticks (Orthoptera); termites (Isoptera); earwigs (Dermaptera); web-spinners (Embioptera); psocids (Psocoptera); zorapterans (Zoraptera); sucking lice (Anoplura) and fleas (Siphonaptera), particularly in the vicinity of farm stock or game species; thrips (Thysanoptera), especially on flower heads and fruit trees; cicadas, hoppers, whiteflies, aphids, and scale insects (Homoptera); and scorpionflies (Mecoptera).

Among destructive insects causing greatest damage to trees in the study area are the southern pine beetle, Ips beetles, and the black turpentine beetle. Aerial surveys are sometimes necessary for detection and evaluation of these forest pests (U.S. Department of Agriculture, Forest Service 1972).

Destructive insect infestations, particularly by Ips beetles, are likely to occur on and around pines damaged by lightning. If control is desired, lightning-struck trees may be either felled and removed from the area, or felled and thoroughly sprayed before beetle broods emerge. Quick action may prevent beetle brood development in the trees initially attacked. Construction or other mechanical activity by man, particularly in areas with extensive pine stands, may damage roots and trunks and may predispose trees to attack by insects. Care should be exercised in this regard.

Since insect damage is a constantly occurring natural phenomenon, control of infestations should be considered only in extreme cases. Insect invasion of a tree

Table 6-9. Microarthropods in litter bag samples from an old field in the coastal plain of South Carolina (Wiegert 1974).

Acarina	Insecta	Chilopoda
	Thysanura	
Asca piloja	Japygidae (undet.)	
Asca garmani	Lepismatidae (undet.)	Lithobiomorpha (undet.)
Rhodacrus spp.	Collembola	
Amblyseus spp.	Hypogastrura armata	
Typhlodromus sp.	Brachystomela sp.	
Hypoaspis spp.	Isotoma viridis	
Laelaps sp.	Entomobrya nivalis	
Tarsonemidae (undet.)	Entomobrya purpurascens	
Pyemotidae (undet.)	Lepidocyrtus languinosus	
Eupodes sp.	Lepidocyrtus cyaneus	
Lorryia sp.	Folsomia sp.	
Tydeus sp.	Orchesella zebra	
Cunaxa sp.	Sminthurus sp.	
Cunaxoides sp.	Bourletiella sp.	
Bdellidae (undet.)	Psocoptera	
Raphignathus sp.	Psyllipsocidae (undet.)	
Neophyllobius sp.	Hemiptera	
Ledermeullaria sp.	Lygaeidae (undet.)	
Eryngiopus sp.	Coleoptera	
Petrobia latens	Carabidae (undet.)	
Aplonobia sp.	Tachysellus spp.	
Oligonychus sp.	Harpalus sp.	
Schizotetranychus sp.	Anisodactylus sp.	
Bryobia praetiosa	Calisoma sp.	
Adamystis sarae	Pterostichus sp.	
Anystidae 1 (undet.)	Suprinus sp.	
Anystidae 2 (undet.)	Plegaderus sp.	
Cheyletidae (undet.)	Conoderus lividus	
Bimichaelia sp.	Elateridae (undet.)	
Tyrophagus spp.	Notovis sp.	
Passalozetes sp.	Anthicus sp.	
Tectocepheus sp.	Helops spp.	
Scapheremaeus spp.	Blapstinus sp.	
Oppia sp.	Capnochra sp.	
Galumna sp.	Xylobiops sp.	
Pergalumna sp.	Oedionychus sp.	
Peloribates sp.	Philonthus sp.	
Trhypochthonius sp.	Lepidoptera	
Cultroribula sp.	Epizeus aenula	
Podoribates sp.	Peridromus margaritosa	
Mochlozetidae (undet.)	Bucculatrix spp.	
	Aluctidae (undet.)	
	Diptera	
	Dasyneura leguminicola	

stand is usually the result of naturally induced habitat changes within the stand (e.g., lightning strikes, wind damage). In most instances, the resulting insect invasion changes the stand composition only on a minor scale. However, the change may provide new feeding or reproduction areas for various other wildlife. Woodpeckers, for example, are particularly attracted to bark beetle-infested timber stands, and at times feed actively in such areas following an attack. The lightning-induced removal of part of the overstory may permit light penetration into the stand, resulting in greater production of groundstory plants. These sites provide additional browsing habitat for deer (Hillestad et al. 1975).

Any treatment of insects of the study area would be incomplete without a discussion of the imported fire ant (Solenopsis saevissima). This ant is native to South America, but has been in this country since at least 1918. It is believed to have been accidentally imported as a stowaway in ship cargo or ballast at the port of Mobile, Alabama. By the 1950's, it had spread to nine Southeastern States, including South

Table 6-10. Insects collected in pitfall traps at two sites on Cumberland Island, Georgia (Hillestad et al. 1975).

Taxon	Loblolly Pine Forest	Pine-Oak Scrub Forest
Orthoptera		
Tettigoniidae	5	6
Gryllidae	64	48
Blattidae	68	17
Homoptera		
Fulgoridae	1	0
Cixidae	1	0
Coleoptera		
Carabidae	11	14
Histeridae	112	3
Silphidae	6	1
Staphylinidae	92	9
Elateridae	4	0
Pedilidae	2	0
Tenebrionidae	2	0
Alleculidae	3	3
Scarabaeidae	1,413	961
Chrysomelidae	0	1
Curculionidae	8	2
Diptera	61	48
Hymenoptera		
Tiphiidae	1	1
Formicidae	167	151
Vespidae	0	1
Sphecidae	4	0
Pompillidae	12	0

Carolina and Georgia. Fire ants are now common in the study area.

The economic impact of fire ants on agriculture is difficult to assess (Butler 1968c). Large mounds built in fields by these ants can interfere with the operation of combines, mowers, and other farm machinery. Primary losses appear to occur through reduced efficiency of labor, due to the nuisance factor, and damage to machinery. The ants sting field workers harvesting crops, and mower blades may be clogged on fire ant mounds.

Mounds are found in most types of coastal terrestrial habitats. However, fire ants prefer to build in open areas exposed to the sun, such as farm lands and pastures. In heavily infested areas, as many as 50 mounds per acre (0.4 ha) may occur (U.S. Department of Agriculture 1973). Colonies of this insect are common along impoundment dikes and other exposed grounds of the Sea Island Coastal Region. In more populated areas, this pest builds its mounds in lawns, parks, playgrounds, cemeteries, and other similar areas.

Looking much like other ants, fire ants are 2 - 7 mm (0.13 - 0.25 in) long and may be either reddish-brown or dark brown to black in color. The fully mature colony, or mound, consists of a queen ant, winged males and females (potential queens), and worker ants. An average mound may contain several thousand winged forms and 50,000 to 100,000 workers.

The method of reproduction involves a nuptial flight in which winged males and females leave the mound and mate in the air. Mated females are carried by the wind, and new colonies have been found as far as 12 miles from the nearest mature mound. After descending to the ground, the new queen breaks off her wings and digs a shallow burrow in which she lays her eggs (Markin et al. 1972). The new queen cares for the eggs and resulting larvae until the first brood of worker ants develops. Worker ants then assume the duties of the new mound, including foraging for food and building, maintaining, and protecting the colony. After the first year, winged sexual forms may be produced. At the proper time of development, they take off for the mating flight

Table 6-11. Insects collected in pitfall traps at three sites having predominantly hardwood cover on Cumberland Island, Georgia (Hillestad et al. 1975).

Taxon	Oak-Palmetto	Mixed Oak-Hardwood	Oak-Pine
Orthoptera			
Acrididae	3	0	0
Tettigoniidae	11	2	2
Gryllicrididae	4	0	0
Gryllidae	35	70	111
Blattidae	140	137	46
Hemiptera			
Aradidae	0	0	1
Coleoptera			
Carabidae	5	40	28
Histeridae	1	43	147
Silphidae	20	54	3
Staphylinidae	626	240	25
Lycidae	2	0	0
Elateridae	5	1	4
Tenebrionidae	1	1	0
Alleculidae	1	2	0
Scarabaeidae	812	5,265	11,789
Curculionidae	8	6	0
Diptera	353	74	50
Hymenoptera			
Braconidae	1	0	0
Mutillidae	1	0	0
Formicidae	260	113	252
Pompillidae	1	1	0

and repeat the reproductive cycle. Mating flights occur most frequently in late spring and early summer (Green 1967).

Over the past 20 years, a continuing controversy has raged over the fire ant, the magnitude of the problem it presents, the question of whether to make concerted attempts to eradicate the insect, and, if so, how to go about it. The problem is complex and remains unresolved.

In 1957, the first "war" against the fire ant, begun as a response to congressional and public demand (primarily farming interests), was initiated by the Pest Control Division of the U.S. Department of Agriculture and the chemical industry (Davis 1974). Large acreages in the Southeast were treated aerially with the insecticides dieldrin and/or heptachlor. The program was expensive, the pesticides were relatively ineffective, fire ant populations were not reduced, and some nontarget species, including livestock poisoned from farm ponds treated by the blanket aerial spraying, were inadvertently killed.

Reports from environmental scientists and concerned public, as well as the

banning of any residues of heptachlor in food by the Food and Drug Administration, enforced demands for more refined methods. By 1962, mirex, a chlorinated hydrocarbon insecticide, was developed specifically for use against the invading ant. Again, major aerial spraying programs were planned and initiated. Mirex is mixed with soybean oil and corncob grits to form a bait to be carried down into the mounds by the ants for consumption. While mirex has been shown to be partially effective in some localized situations, fire ant populations continue to spread, and the use of mirex has brought with it a number of environmental complications.

Following its widespread introduction to coastal environments, mirex was found to be toxic to decapod crustaceans, including juveniles of commercially important blue crabs and penaeid shrimp (McKenzie 1970, Lowe et al. 1971, Bookhout et al. 1972). In conjunction with a fire ant eradication program using aerially applied mirex in selected coastal areas of South Carolina, field studies were conducted to monitor the movement and accumulation of the pesticide in estuaries (Borthwick et al. 1973). Collections of background and periodic post-treatment samples of water, bottom sediments, shrimps, crabs, fishes, and estuarine-dependent birds and mammals were analyzed for mirex. These data showed that 1) mirex was translocated from treated lands and high marsh to estuarine biota (all estuarine animal groups sampled contained mirex), and 2) biological concentration of mirex occurred, especially in predators such as raccoons and birds. Mirex residues for respective sample categories were water ≤10.01 ppb, sediment ≤0.07 ppm, crabs ≤0.60 ppm, fishes ≤0.82 ppm, shrimps ≤1.3 ppm, mammals ≤4.4 ppm, and birds ≤17.0 ppm. Other investigators have since added to the considerable documentation now available showing ecological effects of mirex (Cooley et al. 1972, Collins et al. 1973, Redmann 1973).

A further environmental complication is that mirex, in the presence of light, can decompose to several degradation products including kepone, a major health hazard in aquatic and terrestrial ecosystems (Ivie et al. 1974). Because of these and other problems, including concern that application of mirex to marsh areas and/or adjacent fields could adversely affect fishery resources, aerial application of mirex in coastal counties was suspended. Regardless of whether the final decision is to expand mirex use or restrict it further, some adverse economic and ecologic effects are likely to occur until a nonharmful means of controlling fire ants is discovered. In the meantime, the fire ant-mirex debate continues.

B. VERTEBRATES

1. Amphibians and Reptiles

Although many amphibians and reptiles are restricted to wooded uplands, they require moisture to exist. Many of these "terrestrial" forms live in micro-habitats where humidity approaches 100%. Such habitats include accumulated forest litter, spring seeps, bogs, margins of tupelo-cypress ponds, pocosins, and areas under decaying logs. Also, many salamanders are fossorial and occupy old root channels, crayfish burrows, shrew or mole burrows, and the like (J. R. Harrison, 1978, The College of Charleston, Charleston, South Carolina, pers. comm.).

Temporary pools and semi-permanent ponds are necessary for the perpetuation of most aquatic amphibian species inhabiting upland woodlands. Many of the terrestrial salamanders, hylid frogs, ranid frogs, and toads utilize these palustrine habitats for rearing of larvae because they are free of fish predators. According to Martof (1956) and Harrison (1978), adult species associated with temporary ponds include newts, southern cricket frogs, treefrogs, ornate chorus frogs, southern chorus frogs, ranid frogs, and eastern narrowmouth toads. Newts indigenous to the characterization area include the broken-striped newt found

northeast of the Santee River in coastal South Carolina, and the central newt ranging from just south of the Santee River to northern Florida (Conant 1975, Harrison 1978). Also, the striped newt occurs in extreme southeast Georgia at isolated locations on the Georgia coastal plain (Conant 1975). (For additional information on these and other herpetofaunal species in palustrine habitats, see Chapter Five.)

Predators on herpetofaunal species are numerous. Skunks, raccoons, mink, shrews, and snakes naturally prey on amphibians (Lowery 1974, Mount 1975, Nowak and Brodie 1977). Snakes are preyed upon by certain other species of snakes, plus skunks, hawks (broad-winged hawk, Cooper's hawk, sharp-shinned hawk), great horned owl, swallow-tailed kite, feral pigs, and raccoons (R. E. Mancke, 1978, South Carolina Museum Commission, Columbia, pers. comm.).

a. Salamanders. Salamander species characteristic and commonly found in pine flatwoods include Mabee's salamander, mole salamander, and dwarf salamander. Species characteristic but uncommon in pine flatwoods include the flatwoods salamander, eastern tiger salamander, and many-lined salamander (Harrison 1978). The many-lined salamander is most often found in pocosins, tupelo-cypress ponds, and similar habitats where sphagnum moss is abundant (J. R. Harrison, 1978, The College of Charleston, Charleston, South Carolina, pers. comm.). The marbled salamander, slimy salamander, and mud salamander are species not typically characteristic of pine flatwoods, but occasionally are found in this habitat (Harrison 1978). Most salamander species typical of pine flatwoods breed in the fall or winter in temporary ponds, and salamander populations within a certain geographical area utilize particular ponds annually (Weller 1977). Salamanders are nocturnally active and feed on earthworms, insects, and other invertebrates (Conant 1975).

Salamander species having adult terrestrial stages in hardwood forests live under decaying logs, among accumulated forest litter, or near spring seeps and bogs where relative humidity always approaches 100%. Species commonly found in this habitat throughout most of the study area include the spotted salamander, marbled salamander, mole salamander, southern dusky salamander, slimy salamander, and dwarf salamander; infrequently observed species include the mud salamander (see Conant 1975 for distribution of subspecies in coastal Georgia), and three-lined salamander (Conant 1975, Harrison 1978).

b. Frogs and Toads. Hylid frogs are common in upland communities, particularly near wet areas. These frogs occupy humid micro-habitats and are somewhat secretive during daylight. Green treefrogs, squirrel treefrogs, Cope's gray treefrogs, and barking treefrogs are most frequently encountered. Brimley's chorus frog is indigenous from just south of the Savannah River in Georgia throughout the coastal plain to southeast Virginia; in South Carolina it inhabits low, wet hardwood forests (Conant 1975, Harrison 1978).

The pine woods treefrog is reported to live in the tops of pine trees during the warm summer months (Mount 1975), where its call is heard throughout the day and night. Harper (1932) provided an interesting review of his experiences with this species in Georgia's Okefenokee Swamp. The pine woods treefrog, barking treefrog, and the ubiquitous squirrel treefrog are found under bark in rotting stumps and pine logs, as well as among foliage of shrubs or bushes located near water (Mount 1975). Two other hylids common in pine flatwoods are the ornate chorus frog and the southern chorus frog (Harrison 1978). Both species inhabit ditches and shallow ponds, and breed in fall and winter (Conant 1975, Harrison 1978). The carpenter frog and the crawfish frogs are characteristic pine forest species associated with tupelo-cypress ponds, pocosins, and emergent vegetation (Conant 1975; J. R. Harrison, 1978, The College of Charleston, Charleston, South Carolina, pers. comm.). The crawfish frogs, two subspecies of which occur in the characterization area, exhibit relatively unique behavior among ranids, inhabiting burrows, root channels, and stump holes during daylight. The Carolina crawfish frog ranges from mid-coastal North Carolina to mid-coastal Georgia, and the Florida gopher frog from mid-coastal Georgia throughout most of northern Florida (see Conant 1975 for map). Florida crawfish frogs are frequently commensal in burrows of gopher tortoises.

In old field environments, the absence of permanent water limits the distribution of many species, particularly the ranids. Wandering subadult southern leopard frogs may be found in small, temporarily wet depressions (J. R. Harrison, 1978, The College of Charleston, Charleston, South Carolina, pers. comm.). During rainy seasons, these areas are utilized by the upland chorus frog and the ornate chorus frog. The most abundant treefrog in old fields is the squirrel treefrog; in late winter, northern spring peepers and ornate chorus frogs use wet, abandoned fields as breeding sites.

Toads are relatively common in hardwood forests, but only two species, southern toads and eastern narrowmouth toads, occur with regularity, and neither is habitat specific (Harrison 1978). Eastern narrowmouth toads are not as resistant to desiccation as southern toads, and are closely associated with hydric

micro-habitats (Conant 1975). The oak toad is abundant and generally restricted to pine flatwoods (Harper 1932, Harrison 1978), whereas southern toads are ubiquitous throughout the Sea Island Coastal Region of Georgia and South Carolina (Martof 1956, Harrison 1978). In old fields, toads easily dominate the herpetofauna. The southern toad is probably the most common and abundant species. Eastern spadefoot toads and eastern narrowmouth toads are other common inhabitants of old fields, while the oak toad may be found along field margins near pine woods.

c. Turtles. Few species of terrestrial turtles occur in coastal Georgia and South Carolina. Eastern box turtles may be encountered occasionally, and are typically hardwood forest inhabitants (J. R. Harrison, 1978, The College of Charleston, Charleston, South Carolina, pers. comm.). Gopher tortoises occur in coastal pinelands on sandy soil in Georgia (Martof 1956) and in the extreme southern portion of coastal South Carolina (Gibbons 1978). They tend to aggregate in loose colonies, and construct burrows which several species of herpetofauna utilize commensally as well as for overwintering. The distribution of these commensals is not dependent on that of the gopher tortoise, but where tortoises occur, these species commonly utilize the burrows as temporary retreats. Commensals in the Sea Island Coastal Region include the previously discussed crawfish frogs, Florida pine snake, eastern indigo snake, eastern coachwhip, and eastern diamondback rattlesnake. Mount (1975) and Speake and Mount (1974) discussed the relationships of these species in Alabama. Although the eastern indigo snake is said to occur in coastal South Carolina (Allen and Neill 1952), there are no documented records. However, this species is present in Effingham County, Georgia, which is just across the Savannah River from Jasper County, South Carolina (Williamson and Moulis 1979). Concern has been expressed for the welfare of these commensals because of decreasing tortoise numbers and the practice of pouring gasoline down tortoise burrows during rattlesnake roundups, usually held in late winter (Speake and Mount 1974). Turtles are relatively uncommon in fields, although an occasional eastern box turtle may be encountered along field margins near pine woods.

Almost any of the freshwater turtles may be occasionally encountered in upland habitats, particularly near lakes, rivers, or swamps. Some species are more frequently encountered than others (Gibbons 1978), but all are considered transients.

d. Lizards. The most commonly encountered lizards include the green anole, ground skink, and the six-lined racerunner (Conant 1975, Gibbons 1978). Near field edges and abandoned home sites, the southern fence lizard may be relatively common. Two glass lizards commonly utilize abandoned fields. The eastern glass lizard and eastern slender glass lizard are fossorial and somewhat secretive in their habits. Their diets consist primarily of insects and spiders (Mount 1975).

Accumulated forest litter provides habitat for several other lizards and serpents. Lizards common among litter and organic debris include the southeastern five-lined skink, ground skink, eastern slender glass lizard, and the eastern glass lizard (Gibbons 1978). The five-lined skink and broadhead skink are also found in pine woodlands, but are more common in hardwood forests (J. R. Harrison, 1978, The College of Charleston, Charleston, South Carolina, pers. comm.). Throughout the coastal zone of Georgia, the northern mole skink inhabits well-drained, sandy areas typically supporting scrub vegetation (Mount 1965, Conant 1975). Broadhead skinks are more arboreal than other skinks of the family Scincidae, and readily climb trees when pursued; they usually inhabit hollow trees or cavities of rotting logs or stumps (Mount 1975). Foliage of the forest subcanopy, particularly near water, is an ideal habitat for the green anole, as well as the rough green snake and hylid frogs (Mount 1975, Gibbons 1978).

One of the most characteristic reptiles of open pine woodlands is the southern fence lizard. This species is primarily insectivorous (Hamilton and Pollack 1961) and often basks in open areas (Conant 1975, Mount 1975). Its life history in Georgia was reviewed by Crenshaw (1955). The substory of pine forests also harbors green anoles and an occasional rough green snake (Mount 1975; J. R. Harrison, 1978, The College of Charleston, Charleston, South Carolina, pers. comm.).

The prey of lizards from the area of Ft. Benning, Georgia, was found to consist of several faunal groups. Stomach contents of the five-lined skink, southeastern five-lined skink, and broadhead skink consisted of insects, spiders, lizards, mollusks, and myriapods. The ground skink frequently ingested insects, mollusks, myriapods, spiders, isopods, and mites. Of the various lizard diets studied, that of the eastern glass lizard was the most diverse. Commonly consumed fauna included not only insects, lizards, snails, myriapods, and spiders, but also crustaceans, annelids, snakes, and a mammal (Hamilton and Pollack 1961).

e. Snakes. Snakes inhabiting ground litter in pine forests include the northern scarlet snake, mole kingsnake, brown snake (see Conant 1975, for distribution of subspecies in Sea Island Coastal Region), northern redbelly snake, southeastern crown snake, rough earth snake, and eastern coral snake (Gibbons 1978).

The pine woods snake is also a denizen of the litter environment, and is found most often at the margins of pocosins, ponds, and other aquatic environments in pine flatwoods. The scarlet kingsnake lives beneath bark of standing dead trees or within rotten stumps (J. R. Harrison, 1978, The College of Charleston, Charleston, South Carolina, pers. comm.). These snakes are generally small (<1 m) (<3 ft) and infrequently observed. Their diets include earthworms, frogs, salamanders, arthropods, slugs, lizards, and mice (Conant 1975, Mount 1975).

Larger snakes are generally less secretive and consequently are among the more conspicuous animals of open pine woods. Species such as the southern black racer, corn snake, yellow rat snake, eastern hognose snake, southern hognose snake, eastern kingsnake, eastern coachwhip, and eastern garter snake are commonly encountered (Conant 1975, Mount 1975, Gibbons 1978). Among the less common species in pine woodlands of the Sea Island Coastal Region are the eastern indigo snake in Georgia, and pine snakes and eastern diamondback rattlesnakes throughout the coastal plain of both States (Martof 1956, Gibbons 1978). Commonly ingested food items of the larger snakes include small mammals (particularly rodents and lagomorphs), birds, bird eggs, frogs, lizards, and occasionally turtles. Black racers, kingsnakes, pine woods snakes, coachwhips, eastern coral snakes, and indigo snakes feed on snakes (Hamilton and Pollack 1956, Mount 1975).

Snakes characteristic of open dry woodlands also frequent old fields, which provide habitat for their prey. The corn snake, yellow rat snake, southern black racer, eastern kingsnake, mole kingsnake, eastern coachwhip, pine snake, and eastern diamondback rattlesnake commonly feed on lizards, mice, and/or small birds (Mount 1975). The eastern hognose snake and the southern hognose snake feed primarily on toads and some frogs. Both species are semi-fossorial (Mount 1975). Eastern garter snakes have very generalized habitat preferences and are found in almost any terrestrial habitat (Mount 1975). Old fields with loose friable soil are utilized by the fossorial smooth earth snake and eastern coral snake. Coral snakes, seldom seen, eat lizards, frogs, and other snakes (Conant 1975, Mount 1975).

Snakes inhabiting forest litter are all relatively small and exhibit semi-fossorial behavior. They are nocturnally active, but may leave their protective daytime habitats during spring or summer rains (Conant 1975, Mount 1975). Those considered to be typical inhabitants of hardwood forests include the eastern worm snake, northern scarlet snake, scarlet kingsnake, southern ringneck snake, brown snake, redbelly snake, southeastern crowned snake, eastern earth snake, and eastern coral snake.

Larger snakes are generally active diurnally and are far more noticeable components of the herpetofauna than those inhabiting forest-floor litter. Depending on species, food items include rodents, young birds, lizards, frogs, toads, and other snakes. The southern black racer is common in a wide variety of habitats. Corn snakes, yellow rat snakes, and eastern kingsnakes also occur in many habitat types, but are more common in hardwoods, particularly those near aquatic areas (Gibbons 1978). Corn snakes and rat snakes are excellent climbers (Conant 1975).

A variety of venomous snakes inhabit hardwood forests. In addition to an occasional eastern coral snake, southern copperheads, cottonmouths, pigmy rattlesnakes, and canebrake rattlesnakes occur with greater frequency in this habitat than in other areas (Gibbons 1978). While cottonmouths are aquatic, concentrations are often found near drying, sluggish flowing streams in bottomland habitats.

Rattlesnakes at Ft. Benning, Georgia, were found to feed on a variety of animals (Hamilton and Pollack 1955). Frequency of occurrence of mammals (rodents and lagomorphs), reptiles (lizards), and birds in 26 canebrake rattlesnakes containing food was 73.1%, 23.1%, and 7.7%, respectively. Of 13 southern copperheads that had fed, 46.1% contained mammals (rodents), 30.8% contained reptiles (lizards, a snake, and a turtle), 23.1% contained insects, and 7.7% contained birds. Of nine eastern cottonmouths examined having stomach contents, reptiles (snakes and a turtle) were present in 44.4%, amphibians (frogs) were present in 44.4%, and mammals (rodents) were present in 11.1%; surprisingly, none contained fish. Twelve Carolina pigmy rattlesnakes having food in their stomachs were examined; of these, 50.0% contained reptiles (lizards and a snake), 33.0% contained chilopods, and 17.0% contained mammals (rodents) (Hamilton and Pollack 1955).

f. Man's Impacts. Perhaps the greatest impingements on amphibian and reptile populations are those resulting from man's activities. Lumbering and agricultural practices such as clearcutting, land clearing, site preparation prior to replanting, timber stand improvement, and elimination of hardwoods affect all reptiles directly and indirectly (Mount 1976). Several populations of the gopher tortoise have suffered drastically from recent land clearing for agriculture, and from timber harvesting in South Carolina (R. Montanucci, 1978, Clemson University, Clemson, South Carolina, pers. comm.). Pine lands are frequently burned and, although controlled burns reportedly have little direct effect on herpetofauna (Komarek 1969), indirect effects such as elimination of food items or desiccation of the soil surface have not been

assessed. Wild fire in which pine stands are decimated may cause heavy mortality of many herpetofaunal species, but fires in the Okefenokee Swamp were not as detrimental to amphibians or reptiles as the extreme drought preceding them (Cypert 1961).

During spring and fall when the herpetofauna are very active, traffic on highways traversing upland habitats may account for a large number of deaths. Salamanders are particularly vulnerable during fall and winter spawning migrations when highways separate breeding ponds from areas inhabited by adults. Frogs and toads are attracted to warm pavement after heavy rains, and large numbers are killed by highway traffic. Urban sprawl from increasing human population results in environmental degradation and deliberate persecution of snakes (Mount 1976). Urbanization has resulted in the extirpation of several populations of the mud salamander in coastal South Carolina (Harrison et al. 1979); populations of other herpetofaunal species have been affected similarly (J. R. Harrison, 1978, The College of Charleston, Charleston, South Carolina, pers. comm.). Selective killing of the eastern diamondback rattlesnake has resulted in reduced populations throughout its range, and intensive collecting for the pet trade is thought to be one of the reasons for the present rarity of indigo snakes (Mount 1976). Corn snakes, kingsnakes, scarlet snakes, rat snakes, and pine snakes are other highly sought species in the pet trade. Although not a problem in South Carolina, the practice of pouring gasoline down gopher-tortoise burrows during rattlesnake roundups may result in pine snake mortalities (Speake and Mount 1974).

Large-scale farming is not widely practiced in the study area, and a number of typically terrestrial amphibians and reptiles can be found along field margins bordering woods. These species include the corn snake, yellow rat snake, southern black racer, eastern coachwhip, eastern kingsnake, eastern garter snake, eastern diamondback rattlesnake, and various lizards and toads discussed earlier.

Occasionally some unexpected species are found in fields. Richmond (1945) and Knepton (1954) reported the rainbow snake and the two-toed amphiuma, respectively, being plowed up in fields. Twelve eastern glass lizards were collected during the plowing of a small garden in Phenix City, Alabama (Hamilton and Pollack 1961), and Jobson (1940) reported that eastern glass lizards were commonly plowed up during early spring in Georgetown County, South Carolina. Viosca (1924) found a greater siren under a board in a wet field about 1 mile from a marsh, and Hamilton and Pollack (1955) reported the killing of an eastern cottonmouth by a tractor mower at least a mile from the nearest water.

Studies comparing herpetofauna of the various stages of forest succession are few. The Tall Timbers Research Station, located in northwest Florida just south of Thomasville, Georgia, has for several years been studying the biology and life history of amphibians, reptiles, birds, and mammals on areas burned at selected schedules. These study areas include seral stages of pine and hardwoods (D. B. Means, 1978, Tall Timbers Research Station, Thomasville, Georgia, pers. comm.).

2. Birds

Upland succession in the Sea Island Coastal Region generally includes four broad stages, each dominated by four distinct plant life forms which gradually succeed one another (Odum 1959). These stages are 1) grasslands (including old field communities), 2) grassland-shrub edges, 3) pines, and 4) hardwoods. Each stage has distinctive breeding populations and characteristic densities of birds.

a. Grassland and Edge Communities.
Some of the highest concentrations of birds in the Sea Island Coastal Region are found in the ecotones of grasslands and adjacent edge communities. The old field or grassland community is characterized by broom sedge in the study area.

Old field communities include old croplands, fields, and pastures where secondary succession is in its early stages. Obviously, there are many environmental variables (i.e., vegetation present, stand vigor, soil moisture, size, etc.) which influence bird population density and composition in old fields. Vegetation of old field habitats in the study area has been discussed earlier in this chapter.

Johnson and Odum (1956) described trends in population density and diversity of breeding birds correlated with the advancement from abandoned agricultural fields to climax forest. Population numbers increased with succession from bare ground to the sub-climax, whereas a low density was encountered in young pine forests. Johnson and Odum (1956) found grasshopper sparrows and meadowlarks in early field succession (1-3 years old) characterized by a forb-grass stage. Five other species were found in old fields 15 years of age (grass-shrub stage), and additional species were found as the community progressed. Maximum species diversity was found in the pine-mixed hardwoods stage of succession.

The phenomenon of the edge community has been well-documented by Odum (1971). The "edge effect" occurs at the point of contact between two communities such as field and forest. Edges enable easy access to feeding areas, quick escape cover, and a prime nesting area for some species due to the interspersion of vegetation types. There is considerable overlap in species use of this habitat. Johnson and Odum (1956) found that from 40% to 50% of breeding birds in the Piedmont of Georgia were forest edge birds in their habitat requirements.

Dominant, moderate, and minor species of the upland old field community are listed in Table 6-12. Of the 54 species, the red-tailed hawk and sparrow hawk occupy the highest trophic level (Fig. 6-5). Red-tailed hawks are commonly seen soaring over the open fields in search of prey. The sparrow hawk or kestrel is the smallest of the hawks occurring in South Carolina and Georgia. According to Sprunt and Chamberlain (1970), it should be named "grasshopper hawk," since grasshoppers, rather than sparrows, are its principal food. This species usually perches where it can observe the open field, and it often returns to the perch with its prey. Brush fires attract large numbers of sparrow hawks, which feed on the hordes of insects trying to escape the fires. Apparently they are not bothered by the smoke, as they dart through it in search of insects. Although the sparrow hawk subsists largely on insects, it does feed on vertebrates such as mice and other mammals, reptiles, and other birds (Sprunt and Chamberlain 1970).

The Carolina wren is an important insectivore in old field communities. This species is of considerable agricultural value, since approximately 95% of its diet consists of insects such as caterpillars, moths, beetles, grasshoppers, boll weevils, crickets, ants, and flies.

The mockingbird is an abundant permanent resident of the coastal area. This species prefers the old field habitat, where it takes advantage of the "edge effect." Mockingbirds defend their established territories with great vigor against other birds, mammals, and reptiles. This behavior is quite apparent during the nesting season (April-August). Mockingbirds feed primarily on vegetable matter; berries (palmetto, wax myrtle, bitter gallberry, and butterfly pea), figs, and wild fruit form the bulk of its diet. During late fall and winter, coastal mockingbirds subsist largely on the chinaberry. Only a few insects are eaten by this species (Sprunt and Chamberlain 1970).

The mourning dove is not a resident of the old field community per se, but nests along the edges of fields, pastures, and other clearings. These birds feed on weed seeds throughout the year, and about 99% of their food consists of plant matter. Waste grain left in the fields after harvesting is an important source of food for these birds in the study area. Indispensable items in the dove's diet are gravel

Table 6-12. Dominant, moderate, and minor bird species of old field communities of the upland ecosystem (from Sprunt and Chamberlain 1949, 1970, Burleigh 1958, Audubon Field Notes 1967 – 1970, Chamberlain 1968, American Birds 1971 – 1977, Shanholtzer 1974b, Forsythe 1978).

DOMINANT				MODERATE				MINOR			
Red-tailed hawk	C	PR		Sharp-shinned hawk	C	PR		Cooper's hawk	U	PR	
American kestrel	C	PR		Red-shouldered hawk	C	PR		Merlin	U	WR	Sept.-April
Eastern bobwhite	C	PR		Marsh hawk	C	PR		Short-eared owl	FC	WR	Oct.-April
Eastern kingbird	C	SR	Mar.-Oct.	Killdeer	C	PR		B bolink	C	T	April-May July-Oct.
Tree swallow	C	WR	July-June	Wilson's snipe	C	WR	Sept.-April	Grasshopper sparrow	FC	WR	Oct.-April
Barn swallow	C	PR		Common flicker	C	PR		Henslow's sparrow	FC	WR	Oct.-April
Common crow	C	PR		Rough-winged swallow	C	SR	Mar.-Sept.	Fox sparrow	FC	WR	Nov.-Mar.
House wren	C	WR	Sept.-April	Fish crow	C	Pk					
Carolina wren	C	PR		Winter wren	FC	WR	Sept.-April				
Mockingbird	C	PR		Short-billed marsh wren	FC	WR	Sept.-May				
Loggerhead shrike	C	PR		Catbird	C	PR					
Yellow-rumped warbler	C	WR	Oct.-April	Brown thrasher	C	PR					
Prairie warbler	C	PR		R bin	C	WR	Oct.-April				
Yellowthroat	C	PR		Ruby-crowned kinglet	C	WR	Oct.-April				
Yellow-breasted chat	C	SR	April-Sept.	Orange crowned warbler	FC	WR	Oct.-A				
Eastern meadowlark	C	PR		Palm warbler	C	WR	Aug.-April				
Red-winged blackbird	C	PR		Boat-tailed grackle	C	PR					
Cardinal	C	PR		Common grackle	C	PR					
Painted bunting	C	SR	April-Oct.	Brown-headed cowbird	C	PR					
Rufous-sided towhee	C	PR		Blue grosbeak	FC	SR	April-Oct.				
Savannah sparrow	C	WR	Oct.-April	Indigo bunting	C	SR	April-Oct.				
Dark-eyed junco	C	WR	Oct.-May	Vesper sparrow	C	WR	Sept.-May				
Chipping sparrow	C	PR									

Table 6-12. Concluded

	DOMINANT			MODERATE
White-throated sparrow	C	WR	Oct.-May	
Song sparrow	C	WR	Oct.-May	

Note: Dominance indicates relative importance of the species as a group in the community. This concept is not based necessarily on taxonomic relationships but rather on numbers, size, and trophic dynamics.

KEY: C – Common, seen in good numbers
      FC – Fairly common, moderate numbers
      U – Uncommon, small numbers irregularly
      PR – Permanent resident, present year ar und
      WR – Winter resident
      SR – Summer resident
      T – Transient resident

462

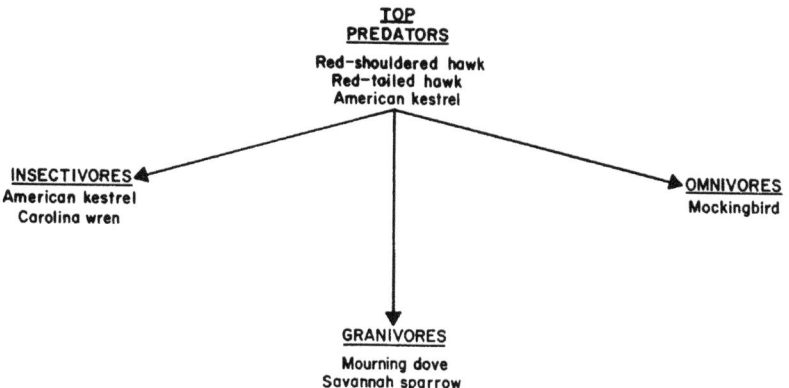

Figure 6-5. Generalized trophic relationships of representative birds of old field communities
in the upland ecosystem of the Sea Island Coastal Region.

(for grinding food in the gizzard) and water. The mourning dove is an important game bird, as are the eastern bobwhite, Wilson's snipe, and American woodcock, which also occur in the old field habitat.

A number of sparrows occur in the old field community, including the chipping sparrow, field sparrow, white-throated sparrow, swamp sparrow, song sparrow, and Savannah sparrow as dominant species.

The impacts of man on bird populations in this habitat vary from adverse to beneficial. The use of uncontrolled fire has a decidedly negative impact on certain birds. Burning off old fields destroys nesting populations of species such as meadowlarks, wrens, and sparrows.

In contrast to the destructive forces of man, clearing of land for agriculture and creation of hedgerows are of great benefit to birds. Once agricultural lands are abandoned, they continue to support the avifauna through natural succession as in the case of "old fields." The old field community presents a transitional habitat where birds may feed, nest, and find protection.

In recent years, farmers have realized that the most conspicuous benefits which birds contribute to agriculture result from their direct consumption of insects. The smaller species of birds associated with agriculture have the highest metabolic rates of all animals

(McFarlane 1976b). To support such large expenditures of energy, birds must ingest large quantities of food. Insects and other invertebrates provide a major source of food for birds in the agroecosystem. For example, Stewart (1975) reported that six avian species effectively removed tobacco hornworms and tobacco budworms from tobacco plants. McFarlane (1976b) summarized published work on the effects of woodpeckers on corn borers where bird predation was detectable and quantifiable.

b. Pine Forest Communities. The upland pine forest or pine flatwoods of the Sea Island Coastal Region is characterized by a relatively low density of breeding birds, compared with various wetland and terrestrial habitats discussed previously. Johnson and Odum (1956) found low avian density in 25 to 60-year-old Georgia pine forests, and Odum (1947) made similar observations for young loblolly-shortleaf pine communities. According to their information, few species of breeding birds appear to be specifically adapted to the pine forest community in the Southeast. Perhaps the most typical breeding birds of this particular habitat are the abundant pine warbler, the less abundant brown-headed nuthatch, and the lesser abundant red-cockaded woodpecker. The above species are somewhat restricted to pine forests, while other birds occurring in this habitat are more associated with the understory. For example, Bachman's sparrow and the eastern bobwhite are characteristic of the grass-shrub forest floor, rather than the upper pine trees.

463

There are 13 dominant species of birds in the pine forest community (Table 6-13). Of these, the screech owl is the representative predator in the trophic pyramid (Fig. 6-6). Insectivores, omnivores, and granivores are represented by the pine warbler, brown-headed nuthatch, and eastern bobwhite, respectively. There are some 16 other commonly occurring species which play moderate ecological roles in the pine forest community.

There appear to be fewer available habitats in southern pine forests compared with the previous habitats discussed. This is reflected in the low number of species of breeding birds in the pine overstory. Bird populations in pine forests are probably determined to a large extent by the nature of the understory. Coastal plain pine forests seldom develop heavy understories due to disturbances superimposed on the natural community. For example, fires and poor soil conditions directly influence understory development and interfere with ecological succession (Johnson and Odum 1956). Actual studies on the interrelationships between fire and non-game birds are rare. Marshall (1963) compared differences in the avifauna of a mixed pine-hardwood forest subjected to frequent fires with a similar forest where fire was excluded. Park-like forests were produced where repeated understory fires had occurred, as opposed to stunted and tangled stands where fire had been excluded. Species adjusted to open understory and scattered trees were found in the burned area, whereas those preferring heavy cover were more common in the unburned forest. Each bird is best suited to particular stages or combinations of stages in forest succession. Therefore, natural succession provides a constantly changing series of niches and habitats. Such variables as food availability, manner of feeding, nesting requirements, and physical characteristics are determining factors in habitat selection. Salt (1953), who worked in three coniferous forest types, noted that the nearer vegetational systems moved toward climax, the greater the biomass of birds supported. In Georgia, Johnson and Odum (1956) reported that a peak in bird density might be expected in the late subclimaxes on moist sites.

The role of bird populations in the functional framework of the upland pine forest is reflected by the foliage-gleaners (e.g., warblers) and timber drillers (e.g., woodpeckers). The pine warbler is a common permanent resident throughout the coastal plain, closely associated with pine trees. This species forages around the trunks of trees like a brown creeper or nuthatch. Pine warblers always nest in pine trees at varying heights; Wayne (1910) found nests from 4.6 to 41.1 m (15 - 135 ft) above ground. The nests are constructed of bark strippings, twigs, and grasses. Lined with pine needles and feathers, the nests are held to the limbs by cobwebs. The pine warbler is insectivorous and, as such, plays an obvious system-related function.

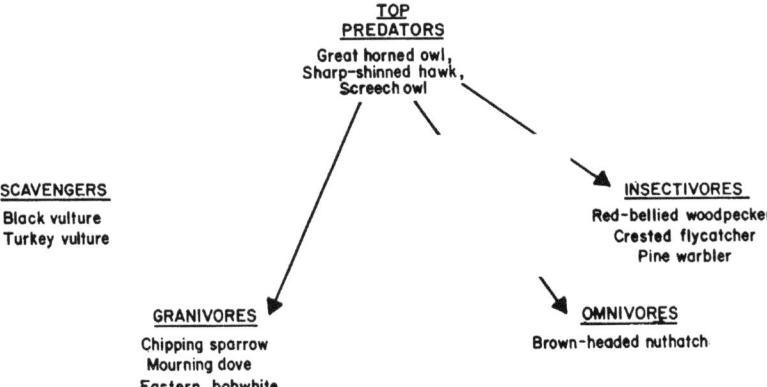

Figure 6-6. Generalized trophic relationships of representative birds of upland pine forest communities of the Sea Island Coastal Region.

Table 6-13. Dominant, moderate, and minor bird species of upland pine forest communities in the Sea Island Coastal Region (from Sprunt and Chamberlain 1949, 1970, Burleigh 1958, Audubon Field Notes 1967 – 1970, Chamberlain 1968, American Birds 1971 – 1977, Shamholtzer 1974b, Forsythe 1978).

DOMINANT			
Eastern bobwhite	C	PR	
Screech owl	C	PR	
Red-bellied woodpecker	C	PR	
Eastern wood pewee	C	SR	April-Oct.
Southern crested flycatcher	C	SR	April-Aug.
Common crow	C	PR	
Carolina chickadee	C	PR	
Brown-headed nuthatch	C	PR	
Eastern bluebird	C	PR	
Yellow-throated warbler	C	PR	
Pine warbler	C	PR	
Summer tanager	C	SR	Mar.-Oct.
Bachman's sparrow	FC	PR	

MODERATE			
Turkey vulture	C	PR	
Black vulture	C	PR	
Sharp-shinned hawk	C	PR	
Red-tailed hawk	C	PR	
Red-shouldered hawk	C	PR	
Mourning dove	C	PR	
Chuck-will's-widow	C	SR	Mar.-Sept.
Common flicker	C	PR	
Blue jay	C	PR	
Tufted titmouse	C	PR	
White-breasted nuthatch	FC	PR	
Brown creeper	FC	WR	Oct.-April
Carolina wren	C	PR	
Ruby-crowned kinglet	C	WR	Sept.-May
Prairie warbler	C	PR	
Yellowthroat	C	PR	
Common grackle	C	PR	
Chipping sparrow	C	PR	

MINOR			
Cooper's hawk	U	PR	
Turkey	U	PR	
Great horned owl	FC	PR	
Red-headed woodpecker	FC	PR	
Red-cockaded woodpecker	FC	PR	
Red-breasted nuthatch	U	WR	Sept.-April
Golden-crowned kinglet	FC	WR	Sept.-April
Pine siskin	FC	WR	Oct.-April

Note: Dominance indicates relative importance of the species as a group in the community. This concept is not based necessarily on taxonomic relationships but rather on numbers, size, and trophic dynamics.

KEY: C – Common, seen in good numbers
      FC – Fairly Common, moderate numbers
      U – Uncommon, small numbers irregularly
      PR – Permanent resident, present year around
      WR – Winter resident
      SR – Summer resident

465

According to Martin et al. (1951), its principal diet consists of ants and other Hymenoptera, bugs, beetles, caterpillars, spiders, grasshoppers, and flies. Considering the estimated annual loss of millions of dollars due to forest insect activity (Craighead 1950), the role of insectivorous species such as the warblers, nuthatches, chickadees, creepers, and woodpeckers becomes even more important. The pine warbler also consumes considerable vegetable matter in the winter, including the seeds of various pine trees and sumac, berries of the dogwood, and grapes.

The brown-headed nuthatch, smallest of the nuthatches found in the coastal plain, is a permanent resident commonly associated with the pine warbler and other birds characteristic of pine forest. Unlike the pine warbler, the brown-headed nuthatch builds its nests in a hole which both the male and female dig. They usually select decaying trees and stubs for nesting cavities, which are usually built at elevations of 4.6 - 7.6 m (15 - 25 ft) off the ground, although Wayne (in Sprunt and Chamberlain 1970) found a nest 15 cm (6 in) off the ground. Its favorite nest lining is the thin, transparent sheath of pine seeds. This species feeds mostly on pine seeds in the coastal plain, but it commonly forages around pine tree trunks for ants and other Hymenoptera, moth eggs, caterpillars and cocoons, and scale insects (Martin et al. 1951, Sprunt and Chamberlain 1970).

The red-cockaded woodpecker is also an important breeding species adapted for existence in pinelands. Although it is not a dominant bird, it is a permanent resident of the Sea Island Coastal Region and is the subject of special forest management techniques in much of the study area. Because of its habitat requirements, site fidelity, and declining numbers, this species is endangered. Within South Carolina, and especially in the Francis Marion National Forest, populations of the species are in very good condition. It is known that diseased pine trees often provide nesting cavities for this species, and much has been written concerning the apparent woodpecker-red heart rot fungus association found throughout southern pine forests (Ligon 1970). Pine plantations, especially in South Carolina and Georgia, and mature pine stands in the National Forest System, could play an important part in the preservation of this species. Trees more than 75 years old are used mostly for nesting, because they are more likely to have portions softened by heart rot disease (Lay and Russell 1970). The birds usually nest in loblolly pine and return to the same nest hole year after year (Sprunt and Chamberlain 1970).

The red-cockaded woodpecker is also an economically important species. It feeds largely on larvae of wood-boring insects, grubs, and beetles. Dingle (1926) also noted this species feeding in cornfields on boring worms. Martin et al. (1951) also listed grasshoppers, crickets, egg cases of roaches, caterpillars, termites, and spiders in the diet of this species.

The Bachman's sparrow is a dominant and characteristic bird of the pine forest understory. The grass-shrub undergrowths are essential to its needs, as it is a ground nesting bird. This species is a fairly common permanent resident in coastal plains pine woods. It is more insectivorous than many of the other sparrows, preferring grasshoppers, crickets, spiders, beetles, moths, caterpillars, and leaf hoppers. Grass seeds and those of sedges and pines are also eaten by this species (Sprunt and Chamberlain 1970).

Another characteristic species of the pine forest understory is the eastern bobwhite. This is a much studied and popular game bird, and is very valuable to farmers. Its diet consists of weed seeds, grains, berries, and fruit, as well as grasshoppers, boll weevils, army worms, tobacco worms, cutworms, potato beetles, cucumber beetles, and squash beetles. The dietary needs of the bobwhite have probably received more study than any other wild American bird (Stoddard 1931, Martin 1935, Martin et al. 1951).

Of the tyrant flycatchers occurring in upland pine forests of the Sea Island Coastal Region, the eastern wood pewee and the southern crested flycatcher are the most characteristic, and both species are common summer residents here. The eastern wood pewee is the most exclusively insectivorous bird of the flycatchers, with about 99% of its diet comprised of insects such as wasps, ants, sandflies, flies, caterpillars, moths, beetles, and grasshoppers (Sprunt and Chamberlain 1970).

The southern crested flycatcher, although a bird of the open woodlands, takes readily to man-made structures (e.g., mail boxes, bird houses, etc.). Grass, pine needles, and fur are common nesting materials, as well as one unusual item which is always present in the nest, a discarded snakeskin. Like most flycatchers, this bird consumes many and varied insect species during its life time.

Other species which might be considered dominant or characteristic of the upland pine forest in the Sea Island Coastal Region include the red-bellied woodpecker, common crow, eastern bluebird, Carolina chickadee, yellow-throated warbler, and summer tanager (Table 6-13).

Species having moderate to minor roles in this habitat (Table 6-13) deserve mention since there is much overlap between birds of the upland pine forest, mixed hardwoods, and open fields. Although the raptors prefer broken forests for hunting, species such as the sharp-shinned hawk, Cooper's hawk, red-tailed hawk, and Florida red-shouldered hawk nest in the upland pine forest. The red-tailed hawk is a resident commonly observed soaring above open country, and is the most frequently seen hawk in the air. The red-shouldered hawk, on the other hand, frequents areas with deciduous trees, such as wooded swamps, as well as pine forests. Hawks are highly valuable in controlling populations of rodents, mainly rats and mice.

Owls are among the most beneficial of all avian predators to humans. The screech owl and great horned owl are common permanent residents in the Sea Island Coastal Region. The southern screech owl nests in cavities of trees, usually old woodpecker holes. Although this species preys upon small birds (mostly sparrows and warblers), it is economically beneficial because of the mice and insects it consumes. The great horned owl is a carnivore of the upland pine forest. Sprunt and Chamberlain (1970) called this species a "tiger among birds" because of its fearless predatory habits. These birds consistently prey upon skunks and other members of the weasel family. The great horned owl is probably the only avian predator of crows. Rather than building its own nest, this species usually occupies the deserted nest of a hawk, crow, or osprey. This owl reportedly will occupy parts of a bald eagle's nest, and Wayne (in Sprunt and Chamberlain 1970) found eggs of the great horned owl on the bare wood of a deep crotch of a forest tree. In addition to rodents, skunks, and crows, great horned owls prey upon game birds, rabbits, squirrels, and poultry (chickens, turkeys, ducks, etc.). Therefore, its economic value is questioned by many farmers.

Vultures play moderate to minor roles in the upland pine forest. Two species, the turkey vulture and black vulture, occur in the study area. The black vulture is the more abundant of the two in this area, and feeds primarily on carrion. It is often found around coastal rookeries, where it feeds on the fish dropped by such birds as herons and pelicans. It also preys on eggs and young birds from the rookeries (Sprunt and Chamberlain 1970).

Avian communities in upland pine forests of the coastal plains area are fairly limited to specific breeding species, with low to moderate densities (Burleigh 1958, Sprunt and Chamberlain 1970, Forsythe 1978). The dominant group appears to be the small foliage insectivores. These species exhibit important predator-prey relationships, of which we know very little. In fact, we know very little of the linkages between avian consumers and the dynamics of their prey populations, and how these relationships are in turn linked to other components of the upland pine forest ecosystem.

The status of nongame bird species has recently received national attention in forest and range habitat management (Noble 1974, U.S. Department of Agriculture, Forest Service 1975, Noble and Hamilton 1976). Studies have shown that intensively managed, pure, even-aged pine plantations support significant bird populations (Noble and Hamilton 1976). It is generally accepted that both the kinds and density of birds can be increased on any tract of land by permitting more vegetative strata to develop. Maximum numbers of species and individuals of nongame birds can be attained through a diversified habitat, with all vegetative strata from ground level to tree-top height represented throughout the forest stand (Roberts 1963). Dead trees and stumps are important to many species such as woodpeckers (except for the red-cockaded), eastern bluebird, brown-headed nuthatch, tufted titmouse, barred owl, flycatchers, prothonotary warbler, eastern wood pewee, Carolina chickadee, and sparrow hawk (Noble and Hamilton 1976).

Large blocks of conifers such as slash pine, managed in short rotations to supply paper mills, are unnatural and may be considered poor wildlife habitat (Stoddard 1963). One of the most harmful things modern man has done to birds has been the exclusion of fire as a management practice in such pine forests. Stoddard (1963) reported the benefits derived from the use of fire in "fire type" pine forests of the Southeast. Pine forests often choke up with brush and lose their prairie-like vegetation in a few years without burning. Thus, birds which are dependent on burning for controlling cover can no longer find food and shelter. Availability of food appears to be the most significant attraction to recently burned areas. Small birds find suitable food such as fruits, seeds, and green matter in burned-over forests. Birds especially attracted to burned-over pine woods include robins, eastern bluebirds, mourning doves, sparrows, flickers and other woodpeckers, pine warblers, and many others. Such birds flock on burns during their northward migration in March and April, when most controlled burning is done (Stoddard 1963). According to Wharton (1978), the longleaf pine forest, if burned regularly, can support up to one quail per acre, plus other dominant birds such as the red-cockaded woodpecker, Bachman's sparrow, and brown-headed nuthatch. The red-cockaded woodpecker, which nests only in live pines with heart disease, usually leaves an area if fire is excluded for 5 years. Recent special management techniques for this species in southern forests have been very beneficial for sustained longevity of an endangered species (Ligon 1970, Hopkins and Lynn 1971).

Conversely, the use of "clear-cutting" techniques in forest management has in many cases been detrimental to birds of the pine forest. Even though the clear-cutting stage of even-aged timber management is now an accepted practice among foresters, its use is controversial. Hundreds of acres are often cleared in one large block, resulting in the destruction of important wildlife habitat. Wharton (1978) discussed man's impact on this habitat and observed that old growth pine can be removed with little effect on the other vegetation. Considering this factor, the effects of scientific lumbering on birds in this habitat are not permanent.

c. Pine-Mixed Hardwood Communities. The pine-mixed hardwood forest, characteristic of uplands in much of the Sea Island Coastal Region, represents an important habitat for avifauna. Many birds that breed in the northern half of North America leave the colder temperatures to winter in the Southeast. Pine-mixed hardwood forests of the coastal plain are used as stopping points for feeding and resting during these migrations. This forest community is approximately midway between pioneer and climax community development, and a well-developed and actively growing subcanopy is present. The presence of such an understory has a profound impact on avian population density and species composition. Johnson and Odum (1956) presented data from Georgia showing an upward trend in bird populations with the progression of plant succession, except in the young pine forest. General increases in avian diversity and density through progressive successional stages toward climax vegetation have been further documented by Odum (1950) in North Carolina. Increases generally occur in the number of individuals and species of birds in the Sea Island Coastal Region as plant succession progresses. This, of course, would be dependent upon sufficient moisture and other conditions for complete understory development. Available habitats probably increase in number with increases in vegetative height, volume, and diversity.

Approximately 32 dominant species of birds commonly occur in this habitat, while another 20 commonly occurring species play moderate ecological roles (Table 6-14). Although the number of dominant species in this habitat generally exceeds that for young pine flatwoods, there is noticeable overlap between some species. For example, the pine warbler, yellow-throated warbler, southern crow, and Carolina chickadee occur in the pine-mixed hardwood and mixed hardwood forests.

In the pine-mixed hardwood forest community, there is an obvious increase in the number of warblers that parallels the addition of a subcanopy of sweet gum, oaks, dogwood, and trees with Spanish moss to the community structure.

From the standpoint of trophic dynamics (Fig. 6-7), the role of birds as insectivorous predators in this habitat should be recognized. It is generally accepted that birds consume tremendous numbers of insects and similar invertebrates. Considered as facultative feeders, insectivorous birds exploit a variety of prey species as opportunities arise. Recent studies have shown that birds may act as agents in the actual control of insects (McFarlane 1976b).

Woodpeckers have been reported as insectivores in southern forests (Knight 1958, Solomon 1969). Dominant woodpeckers in pine-mixed hardwood forest communities of the study area include the pileated woodpecker, red-bellied woodpecker, and downy woodpecker. The pileated woodpecker has specific habitat requirements for this type of forest. It requires a mature forest with dead trees scattered throughout. In search for food, this species removes large chunks of bark and decaying wood. Considered a valuable asset to forests and timber, the pileated woodpecker eats wood-boring beetles, ants, and grubs. It also eats wild fruits, berries, and seeds of bull bay (magnolia) (Sprunt and Chamberlain 1970). Solomon (1969) investigated woodpecker predation on boring insects of hardwoods and found that these birds removed 39% of white oak borers, 39% of living beech borers, and 13% of poplar borers.

The tufted titmouse is a common companion of the warblers in this forest habitat. It nests in old woodpecker holes or other cavities. The tufted titmouse is insectivorous, taking caterpillars, wasps, bees, sawfly larvae, and boll weevils; it also eats acorns, mulberries, and wax myrtle berries (Sprunt and Chamberlain 1970). The tufted titmouse regularly consumed about 50% of eucosmid insect stocks (e.g., Ernamonia conicolani) in pine plantations, according to Gibb (1958, 1960, 1962). Tinbergen (1960) reported that 35% - 40% of lepidopteran larvae in pines were consumed by this species, while Betts (1955) found that about 20% of a female winter moth population was consumed by these birds.

Of the vireos occurring in pine-mixed hardwood habitats, the white-eyed vireo and red-eyed vireo are most common in the Sea Island Coastal Region. These two valuable insectivores are both common summer residents and nest in the forest trees. The white-eyed vireo eats a variety of insects, including moth and butterfly larvae and adults (Sprunt and Chamberlain 1970).

DOMINANT

Species			
Black vulture	C	PR	
Mourning dove	C	PR	
Screech owl	C	PR	
Chuck-will's-widow	C	SR	Mar.–Sept.
Ruby-throated hummingbird	C	SR	Mar.–Oct.
Common flicker	C	PR	
Pileated woodpecker	FC	PR	
Red-bellied woodpecker	C	PR	
Downy woodpecker	C	PR	
Great crested flycatcher	C	SR	April–Aug.
Blue jay	C	PR	
Common crow	C	PR	
Carolina chickadee	C	PR	
Tufted titmouse	C	PR	
House wren	C	WR	Sept.–May
Carolina wren	C	PR	
Catbird	C	PR	
Robin	C	WR	Oct.–April
Hermit thrush	C	WR	Oct.–April
Blue-gray gnatcatcher	C	PR	
Ruby-crowned kinglet	C	WR	Sept.–May
White-eyed vireo	C	PR	
Red-eyed vireo	C	SR	Mar.–Nov.
Northern warbler	C	SR	Mar.–Oct.

MODERATE

Species			
Turkey vulture	C	PR	
Sharp-shinned hawk	C	PR	
Red-tailed hawk	C	PR	
Red-shouldered hawk	C	PR	
American woodcock	FC	PR	
Yellow-billed cuckoo	C	SR	April–Nov.
Great horned owl	FC	PR	
Barred owl	C	PR	
Red-headed woodpecker	FC	PR	
Yellow-bellied sapsucker	FC	WR	Oct.–Mar.
Hairy woodpecker	FC	PR	
Eastern kingbird	C	SR	Mar.–Oct.
Eastern phoebe	C	WR	Aug.–April
Acadian flycatcher	C	SR	Mar.–Oct.
Eastern wood pewee	C	SR	Mar.–Oct.
Fish crow	C	PR	
Brown creeper	FC	WR	Oct.–April
Winter wren	FC	WR	Sept.–April
Brown thrasher	C	PR	
Wood thrush	C	SR	Mar.–Oct.
Cedar waxwing	C	WR	Aug.–June
Solitary vireo	FC	WR	Oct.–April
Black-and-white warbler	FC	WR	Sept.–May
Orange crowned warbler	FC	WR	Oct.–April

MINOR

Species			
Cooper's hawk	U	PR	
Broad-winged hawk	U	WR	Oct
Turkey	U	PR	
Black-billed cuckoo	FC	T	Apr / Aug
Whip-poor-will	FC	WR	Aug / Sep
White-breasted nuthatch	FC	PR	
Red-breasted nuthatch	U	WR	Sep
Veery	FC	T	Mar / Aug
Golden-crowned kinglet	FC	WR	Aug / Sep
Yellow-throated vireo	FC	SR	Mar
Magnolia warbler	U	T	Apr / Sep
Cape May warbler	FC	T	Apr / Sep
Black-throated blue warbler	C	T	Apr / Aug
Blackpoll warbler	FC	T	Apr / Sep
Fox sparrow	FC	WR	Nov

Table 6-14. Concluded

DOMINANT			MODERATE			MINOR
Yellow-rumped warbler	C	WR	Sept.-May	Ovenbird	C	T	April-May Aug.-Oct.
Yellow-throated warbler	C	PR		Northern waterthrush	C	T	Mar.-May July-Oct.
Hooded warbler	C	SR	Mar.-Oct.	Orchard oriole	C	SR	Mar.-Sept.
American redstart	C	T	April-May July-Oct.	Northern oriole	FC	WR	Sept.-May
Cardinal	C	PR		Summer tanager	C	SR	Mar.-Oct.
Rufous-sided towhee	C	PR		Purple finch	FC	WR	Oct.-April
Dark-eyed junco	C	WR	Oct.-May	American goldfinch	C	WR	Oct.-May
White-throated sparrow	C	WR	Oct.-May	Chipping sparrow	C	PR	
Song sparrow	C	WR	Sept.-April				

Note: Dominance indicates relative importance of the species as a group in the community. This concept is not based necessarily on taxonomic relationships but rather on numbers, size, and trophic dynamics.

KEY: C - Common, seen in good numbers
FC - Fairly common, moderate numbers
U - Uncommon, small numbers irregularly
PR - Permanent resident, present year around
WR - Winter resident
SR - Summer resident
T - Transient resident

470

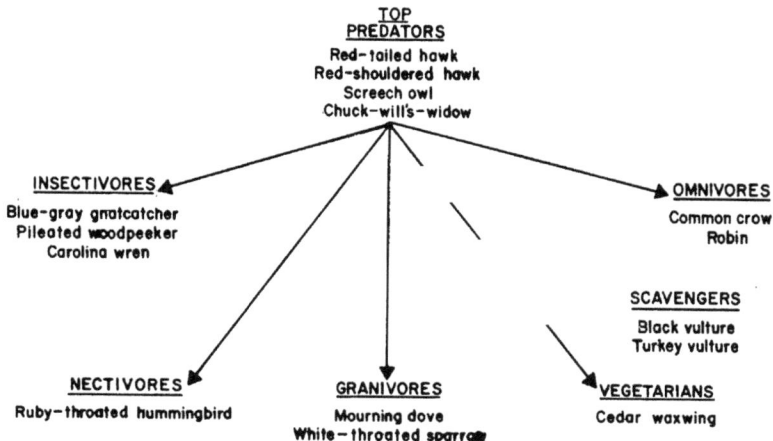

Figure 6-7. Generalized trophic relationships of representative birds of upland pine-mixed hardwood communities of the Sea Island Coastal Region.

The red-eyed vireo eats seeds of the bull bay (magnolia), in addition to insects. Wayne (in Sprunt and Chamberlain 1970) reported as many as 50 of these birds feeding on seeds of a single magnolia.

Habitat availability for dominant ground-dwelling species such as the Carolina wren, hermit thrush, robin, catbird, chuck-will's-widow, and eastern bobwhite varies with herbaceous and shrub vegetation density. Such density is a function of forest canopy closure. Since the pine-mixed hardwood forest has a well-developed understory and sub-canopy, habitat diversity usually is well-balanced and provides space for low to mid-level foragers and nesters. MacArthur and MacArthur (1961) found that foliage height profile determined bird diversity. The number of breeding birds was highest when the three horizontal layers corresponding to herbs, shrubs, and trees over 7.6 m (25 ft) tall had equal amounts of foliage. Bird diversity in a pine-mixed hardwood forest (basically three-layered) can thus be expected to be greater than in a young pine forest or old field.

Logging activities constitute the major influence of man on birds in this habitat. The grazing and browsing of livestock, and the rooting of swine, also impact on bird populations here.

The avifauna of mixed hardwood forest communities in the Sea Island Coastal Region is very similar to that of the pine-mixed hardwood forest discussed previously. Identification of key environmental features in the selection of one habitat over another is very difficult. Some species select on the basis of one or a few primary factors, while no single factor is known to be of primary importance for others. Presumably, the presence of nest holes, water, food sources, and gradients in vegetation densities affects the preference for one habitat over another. Willson (1974) concluded that the increase in bird species with the addition of shrub and tree layers may not be due to an increase in resource productivity, but rather to increased habitat patchiness in three dimensions, leading to new possibilities of space utilization by birds.

Vegetative structure appears to be a primary habitat characteristic controlling bird density and diversity. In the case of pine-mixed hardwood and mixed-hardwood communities, the dominant factor influencing avifauna would probably be understory and sub-canopy characteristics. The phyto-vertical distribution of birds in this habitat is based on three fundamental activities of bird behavior: 1) the height at which they feed, 2) the altitude of the nesting site, and 3) the elevation at which they seek refuge for protection (Dunlavy 1935).

In a classic paper dealing with plant succession in Georgia, Johnson and Odum (1956) found that maximum species diversity of birds was reached at about the 60th year, or during the pine-mixed hardwoods stage. The number of bird species then held steady through the oak-hickory climax stage. These authors recorded the dominant species (by density) for old pine forest with well-developed deciduous understory as pine warbler, Carolina wren, hooded warbler, and cardinal. In the oak-hickory climax, the red-eyed vireo, wood thrush, and cardinal were dominants.

3. Mammals

Many species of mammals cannot be assigned conveniently to a single terrestrial habitat. This is due in part to their great mobility, but perhaps more so to their rather generalized habitat requirements. Also, terrestrial forested habitats often occur as a mosaic of patches, including many transitional stages. For these reasons, the mammals of the upland system will be summarized into old field and forested communities.

a. Old Field Communities. The old field community occurs in all coastal counties of the Sea Island Coastal Region, but as a community type, it is not widespread on the barrier islands. This community is dominated in terms of its mammalian fauna by small mammals, particularly the rodents and lagomorphs.

Among the herbivorous mammals, the eastern cottontail rabbit is certainly one of the ecological dominants in this community and an important link in food chains of predacious mammals, a variety of raptorial birds, and some reptiles. Colley (1962) considered the old field community to be optimum habitat for cottontail rabbits because food and cover are abundant. Food habits of the eastern cottontail have been studied extensively in other areas, but not in the immediate Sea Island Coastal Region. Studies have shown that they consume a large variety of bark, twigs, leaves, seeds, and roots of various plants (Golley 1962). The eastern cottontail is generally the most abundant of the medium-sized mammals in this habitat, and probably the only one which may spend its entire life cycle within this community. Reproduction takes place over most of the year in the area, and the high reproductive capacity of the cottontail is well known. The gestation period is 30 days, and even in more northerly areas, three or more litters may be raised per year (Ecke 1955). The average litter size is four to six young. The grass-lined nests are usually well concealed within thickets and dense grass in the old field. The level of population density which can occur is suggested by Buele and Studholme (1942), who found an estimated 14 nests per acre in old field communities in Pennsylvania.

Eastern cottontail rabbits are generally more abundant on, or adjacent to, agricultural lands than in natural habitats, again because of the abundant food supply and numerous ecotonal situations. The cottontail does considerable damage to newly planted row crops and especially truck-garden crops if their numbers are not controlled either by predators or hunting pressure.

The high reproductive capacity of this species is matched by a high mortality rate, but neither fecundity nor mortality has been specifically measured in this area. Juvenile rabbits are common prey to other mammals, hawks, owls, and snakes. Domestic dogs and cats are effective predators on adult and juvenile rabbits, especially where old fields are close to human habitation. Hunting pressure on a year-round basis is often a minor mortality source.

The marsh rabbit is closely tied to aquatic communities, but this species enters the old field community for feeding if aquatic habitats are nearby. When in this community, its trophic role is essentially the same as the eastern cottontail rabbit.

A number of small rodents are commonly abundant in old field communities, including the marsh rice rat, harvest mouse, old-field mouse, cotton rat, pine mouse, and house mouse.

The distribution pattern of the old-field mouse is not clear in the study area. It occurs in several coastal areas of Georgia, but its presence in coastal South Carolina has not been documented. Cumberland Island is the only coastal island where its presence has been established. Even though the old field community is its preferred habitat, the old-field mouse is not significant in this area.

Colley (1962) suggested that the cotton rat may well be the most abundant mammal in Georgia. It is also abundant in South Carolina, where it finds optimum habitat in old field communities (Golley 1966), although it is seldom found in the earlier stages of re-vegetation. This species, unlike some mainland species, is present on most of the coastal islands. Johnson et al. (1974) recorded the species on Cockspur, Tybee, Blackbeard, Sapelo, St. Simons, Jekyll, Little Cumberland, and Cumberland islands in Georgia. In South Carolina, it is one of the more abundant small mammals of Kiawah Island (Pelton 1975). It is probably present in appropriate habitats on most South Carolina coastal islands, even though it has not been recorded from most.

472

According to Colley (1962), the food habits and food requirements of the cotton rat are well documented. Stomach analysis revealed that herbaceous plants such as lespedeza were most commonly eaten, with grasses such as purple top and broom sedge being of secondary importance. Like many rodents, this rat relies to a slight extent upon insects as a food source, but also grubs for roots and tubers.

This rat breeds over most of the year, with a resting period in mid-winter. The gestation period is 27 days, and an average litter consists of about five young.

The cotton rat is an important prey organism for hawks, owls, predacious mammals, and reptiles. For example, Johnson et al. (1974) reported that analysis of a large collection of owl pellets from Sapelo Island yielded 152 cotton rat skulls, compared with only 26 skulls of three other mammal species.

The opossum is generally the most abundant medium-sized mammalian omnivore to enter the old field community. Although it is basically a forest dweller, it may enter old fields to feed. Although opossums feed upon a wide variety of plant and animal materials, they rely heavily upon insects in the Southeast.

Among the strictly predacious mammals of old field communities are three species of shrews. The least shrew, which finds optimum habitat requirements in the old field community, is more abundant in the coastal plain than in the Piedmont. In Georgia, Johnson et al. (1974) cite its occurrence on Blackbeard, Sapelo, and Cumberland islands. In South Carolina, it is recorded in Charleston, Georgetown, and Jasper counties, although it probably occurs in all coastal counties of the Sea Island Coastal Region. Pelton (1975) found this species of shrew on Kiawah Island. The high metabolic rate and high level of food consumption of this and other shrews are well known. Insects, earthworms, centipedes, snails, and some plant material are consumed (Golley 1962).

The short-tailed shrew and the southeastern shrew can also be expected in the old field habitat of the Sea Island Coastal Region, although more moist habitats are considered more suitable for both species. The trophic positions of these two shrews are essentially the same as described for the least shrew.

The fossorial eastern mole is abundant in the old field community, but it is far from restricted in its habitat selection. The food habits of the mole have not been described for this area, but it probably depends largely on earthworms and soil insects as it does in other areas. Because of its fossorial habits, it is not highly subject to predation by raptorial birds or mammals. Snakes, which enter mole burrows, are probably the principal predators. The mole has a relatively low reproductive rate, only one litter of two to four young being produced per year (Golley 1962). A low mortality rate is often accompanied by low fecundity.

The larger predatory mammals which frequent the old field community can be treated as a group. This group consists of the gray and red foxes, the bobcat, the long-tailed weasel, and the striped skunk. None of these species is considered permanent residents of this habitat, but all are attracted to it by the abundance of small rodents and ground nesting birds.

Even without human intervention, the old field community is only a transitory habitat. If undisturbed and allowed to shift to forested land, the dense understory of grasses, vines, and thickets becomes dramatically altered, and its mammalian fauna shifts accordingly.

Human impacts are common, however, and two may be noted in particular. An old field community usually results when agricultural lands are abandoned or allowed to lie fallow for a period of years. Abandoned fields are not often unattended permanently, but are usually converted back into agricultural production. The clearing, plowing, and disking of old fields completely alters the habitat until such time as the field is abandoned again. Less dramatic and less permanent effects result from the frequent practice of controlled burning, preventing the old field community from progressing to more mature seral stages. While burning may have a negative impact on nesting mammals and juveniles, the long term effect may be favorable for some in stimulating subsequent growth of herbaceous plants and grasses.

b. Forest Communities. From a functional (trophic) standpoint, the forest mammals may be grouped into three categories: 1) herbivores, including browsers and grazers; 2) omnivores; and 3) predators. The largest of these units is the herbivore group, with species ranging in size from several of the smaller rodents to the white-tailed deer, and in abundance from many animals per acre (rodents) to many acres per animal (deer).

The white-tailed deer occurs in several habitats, but for convenience will be treated in greatest detail in this section. Several subspecies have been described from the coastal islands to the south, but the typical subspecies, Odocoileus virginanus virginanus, occurs on the mainland in the Sea Island Coastal Region. The several described subspecies are treated in the chapter on maritime forests (Chapter Three).

White-tailed deer are usually abundant in forested areas on both islands and the mainland. Much of the area is fairly isolated, and large acreages have been held in privately owned tracts on which hunting pressure has been tightly controlled. Of equal importance is the fact that deer production has been an important consideration in forest management practices. An excellent example of this can be found on Cat Island and South Island, South Carolina.

Because deer are predominantly browsers, feeding on leaves and green twigs of shrubs and trees, they are more at home in mixed hardwood forest than in open pine forest. Mixed hardwoods also provide better cover as well as mast from oaks during fall and winter. Deer also derive a significant portion of their nutrition from grazing. They often feed in more open pine woods, although predominantly under cover of darkness. Nutrient and energy requirements of deer have been fairly well studied, and it has been estimated that a 100-pound (45.5 kg) deer requires about 5 lb (2.3 kg) of fresh browse [2.5 lb (1.1 kg) air dried] daily (Colley 1962). Thorsland (1967) reported on the nutritional analysis of selected deer feeds in South Carolina.

Although deer are basically woodland mammals, they use both pastures and planted fields as nocturnal feeding grounds. Soybean fields in coastal Georgia and South Carolina are especially attractive to deer from adjacent forests. In many areas, increased deer populations pose a severe threat to soybean farmers. In some counties, depredation hunting permits are issued to farmers so they can protect their fields from deer. Newly planted cornfields can also be seriously damaged by grazing deer during night feeding forays.

The breeding season for white-tailed deer in the Sea Island Coastal Region extends from late August to mid-January, with peak breeding activity in mid-November. Based on a 196-day gestation period, fawning occurs from early March to early August, with a major peak in late May and early June. Yearling does generally produce a single fawn, although during the period from 1970 to 1975 almost one-third of the pregnant yearling does examined in coastal South Carolina were carrying two embryos. More than 50% of pregnant does examined carried twin embryos. Approximately 20% of the fawn does in some coastal areas were pregnant during spring. Fawns seldom give birth to more than a single offspring. The above cited data on reproduction are based on the studies by Moore (1976), as reported in a 6-year summary.

White-tailed deer have almost no predators, although some young fawns fall prey to bobcats and dogs. The principal sources of mortality, other than hunting, are disease and parasitism, both of which are enhanced by malnutrition. Many coastal forested areas are considered to be over-populated with deer at the present time.

Squirrels are probably the most conspicuous herbivorous mammals of upland forests. Three species occur in forest associations of the Sea Island Coastal Region, the gray squirrel, fox squirrel, and flying squirrel. All three occur in pine-mixed hardwood and mixed hardwood forests, but only the fox squirrel is likely to be seen in open pine forests. The gray squirrel, most abundant of the three, finds ideal habitat in mixed hardwood forests, especially those in which oak and hickories are abundant. Gray squirrels construct two types of nests in the hardwood forest. The so-called den nest is usually located within a hollow portion of a tree trunk, and is used almost exclusively for bearing and rearing young. The other type of nest, known as a leaf nest, is loosely constructed in tree branches. Such nests provide temporary shelter and are seldom used for reproduction. Two breeding seasons occur in coastal South Carolina and Georgia, the first starting in late January and the second in early June. Yearlings, however, which make up a large portion of the female population, breed only once a year (Golley 1962, 1966). The gestation period is 42 to 45 days, and small litters are the rule; generally, only two to three young are produced. The young remain in the nest for 6 to 8 weeks, and are generally weaned in 6 to 7 weeks.

A wide variety of nuts and seeds are consumed by the gray squirrel, with hickory nuts and acorns probably being the principal items when available. The green stage of the pine cone is also extensively utilized in season. The degree to which squirrels serve as prey is not as well known. No doubt, bobcats and long-tailed weasels take small numbers, but the larger hawks and owls are probably more serious predators. Where forest land is adjacent to human habitation, the house cat can be a significant predator. Hunting pressure is a major source of population control in many areas.

Many coastal farms in South Carolina and Georgia have at least a few pecan trees and, in some cases, rather extensive orchards. Wherever pecan or other nut-bearing trees are present, gray squirrels are abundant unless deliberately and continually eliminated.

The life history of the larger fox squirrel is similar to that of the gray squirrel. Like the latter, fox squirrels have winter and summer breeding seasons. Gestation period and litter size are also similar in the two species. Fox squirrels utilize den nests for reproduction and leaf nests for shelter, and food habits of the two squirrels are generally similar. The principal ecological differences between the two species are as follows: 1) fox squirrels seldom reach population densities comparable to those of gray squirrels, and 2) fox squirrels are much more tolerant of open land and open pine forest than are gray squirrels.

Along the coast, the small flying squirrel is primarily a resident of the hardwood forest. It is the least observed of all the squirrels in this area because of its nocturnal habits, and its abundance is generally underestimated. Under natural conditions, the flying squirrel utilizes tree cavities for dens, but they frequently occupy bird nest boxes when available. The flying squirrel is quite sociable; often four to six squirrels may be found using the same den or nest box (Colley 1966). While flying squirrels utilize nuts, seeds, and fruits as do the other two squirrels, they are more predacious. A variety of insects, as well as the eggs and young of birds, may be utilized (Colley 1962).

Several species of small rodents are common to abundant in various types of forest habitats, and are ecologically important both as herbivores and as prey species for mammals, reptiles, and raptorial birds.

The eastern wood rat may be the most typical rodent of forested habitats within the coastal area. Two subspecies occur in South Carolina, with Neotoma floridana floridana being the coastal form. Within the Sea Island Coastal Region, wood rats are more likely to be encountered in hardwood forest bottom lands than in high open pine forest. This species is neither arboreal nor fossorial, and constructs an unusually large ground nest for an animal its size. The house, built of a pile of sticks, leaves, and trash, may be several feet in diameter. The nest itself is placed well within the interior of this structure. Houses may be constructed in a hollow log or on the ground adjacent to a fallen log. Eastern wood rats develop food caches and toilet areas near the nest, and frequently add bits of glass, metal, paper, and other artifacts to the nest area. Breeding probably occurs year around in coastal areas of South Carolina and Georgia. The gestation period has been estimated as 32 days (Pearson 1952) or 42 days (Hamilton 1943). Litters are surprisingly small, consisting of two or three young. Rainey (1956) suggested that the age of sexual maturity may be as much as 1 year. The age of maturity and low litter size both suggest a low reproductive rate for a rodent. This rat is predominantly a herbivore, utilizing twigs, nuts,

fruit, and seeds (Colley 1966). Golley (1962) suggested that animal food is taken when available, but Lowery (1974) stated that the species is almost exclusively a vegetarian. The latter author also suggested that acorns are significant food items in the hardwood forest. This species is no doubt an important link in food chains of the forest. Since it is strongly nocturnal, it is probably not subject to significant predation by hawks, but the larger owls, nocturnal predacious mammals, and snakes feed heavily on the eastern wood rat (Lowery 1974).

The cotton mouse is a small but abundant rodent in lowland hardwood forests. This mouse often takes up residence in forest land adjacent to cotton fields and uses cotton to line the nest, whence its name. Although this species feeds upon seeds and other plant materials, it is far from an obligate herbivore. Animal material, principally insects, may constitute over half of its diet in summer.

The cotton mouse breeds throughout the year, but not uniformly by season, in the Southeast. Greatest breeding activity occurs in mid-winter, while the low ebb of reproductive activity occurs in mid-summer. This species possesses a high reproductive potential. Both sexes reach maturity in about 70 days. The gestation period ranges from 23 days in non-nursing females to 30 days for a nursing female. The mean number of young per litter is four. Population density may follow a regular cycle, with greatest numbers in winter and lowest numbers in mid-summer. This suggests a short life span and probably a high mortality rate due to predation. The same species that prey upon the eastern wood rat probably make even greater use of the cotton mouse as a food source.

Since agricultural lands are purposefully managed as plant-rich and seed-rich environments, rodents are abundant in such areas. Several, including the roof rat, Norway rat, and house mouse, are much more likely to be found in association with human habitation than in natural, unmodified environmental types. Barns and other farm storage buildings provide ideal habitat for such rodents. These mammals produce considerable losses to stored grains and, unless controlled, develop exceptionally high populations.

Other rodents such as the old-field mouse, cotton mouse, cotton rat, and eastern wood rat are more likely to be found in grassy fields, farm wood lots, or hedge rows surrounding row crop fields. Abundance of these rodents also serves to attract a variety of predators, including hawks and owls, reptiles, and other mammals.

Colonial pocket gophers occur through-out the extent of the Georgia portion of the Sea Island Coastal Region. Unless otherwise indicated, information on this group has been drawn from Colley (1962). Distribution of the several species is spotty and in some cases extremely limited within the Georgia coastal counties. All of these species are habitat-limited to pine forest and pine-mixed hardwoods. Ideal habitat appears to be the sandy soils of longleaf pine forests. The southeastern pocket gopher is the most widely distributed, being known over the full length of the Georgia coast from Camden to Chatham County. While it extends inland to the Blue Ridge Mountains, it is not known from any of the coastal islands. As pointed out earlier, this is a fossorial mammal limited to sandy soils, and major bodies of water apparently limit its spread. The Savannah River drainage system establishes its northern boundary. The colonial pocket gopher is limited to Camden County in the extreme southeastern corner of Georgia, but it is absent from nearby Cumberland Island. Sherman's pocket gopher is known only from a single colony located northwest of Savannah in Chatham County. We find no recent reference to this colony, and no information whether it has survived the recent expansion of metropolitan Savannah.

The history of the Cumberland Island pocket gopher, which is or was restricted to Cumberland Island, was reviewed by Johnson et al. (1974). When Bangs (1898) first collected and described the species, he indicated that its mounds were scattered through the pine woods for miles. E. B. Chamberlain (1978, The Charleston Museum, Charleston, South Carolina, pers. comm.) studied this species in the early 1930's, and described several substantial colonies in his notes. In 1970, Johnson et al. (1974) found only a single colony in an old field habitat. Several months after their observation, the field was plowed. Johnson et al. (1974) suggested that the species is endangered and should be added to the endangered species list. This may be too late, because Hillestad et al. (1975) suggested that the species may now be extinct.

The life histories of most of these pocket gopher species have not been studied, but presumably their mode of life is quite similar. The roots of herbaceous plants consitute their basic food supply.

Among omnivorous forest mammals, the opossum is probably the most abundant. This mammal is ubiquitous but finds its ideal habitat in hardwood forest, especially the dense woods along stream and river bottoms. The unusual reproductive habits of this marsupial are generally well known. The young are born in a very immature state after a short gestation period of only 12 to 13 days. Immediately after birth, the young make their way to the marsupium, where they seize and become temporarily attached to one of approximately 13 nipples. As many as 25 young may be born, but no more can be accommodated in the marsupium than there are nipples. The average number of successfully reared young is generally six or seven. In contrast to the short gestation period, the young may remain in the marsupium for several months and continue to nurse as much as 80 days after birth. In coastal South Carolina, there are two breeding seasons per year. The first period begins in January and continues into early March. The second breeding period is from mid-April into June. Females with young in the marsupium are almost never found in autumn.

One reason for the success of the opossum is its generalized food habits and requirements. Foods selected change with seasonal availability patterns. A wide variety of fruit, berries, and other plant matter may be consumed, although insects may constitute the bulk of its diet in the southern coastal plain. Small mammals and birds may also be consumed when they can be captured. When opossums occur near human dwellings, they are frequent scavengers of garbage.

With the high reproductive potential of this species, it is obvious that there must be a correspondingly high mortality rate if stable opossum populations are to be maintained. After leaving the marsupium, the small and inexperienced young fall easy prey to a number of predators, including foxes, weasels, bobcats, dogs, feral cats, the larger hawks and owls, and a variety of snakes. Aside from disease, which is largely an unknown factor, the automobile is probably one of the major sources of mortality for adults. The opossum appears to be particularly susceptible to being run over by automobiles, largely because of its nocturnal habits and slow movements.

The raccoon is another major omnivore of woodland habitats, particularly in hardwood forests where it finds an abundance of suitable den trees. In the forest, raccoons feed on a variety of nuts, seeds, and fruits, as well as insects, snails, small mammals, and birds. Where agricultural lands are adjacent to woodlands, raccoons may be quite destructive. Raccoons utilize a variety of fruits and vegetables, and are especially troublesome in cornfields where they frequently break down corn stalks to reach the ears of corn. The life cycle of the raccoon is described in Chapter Four of this volume, and will not be repeated here.

As noted for the opossum, the automobile may be the largest single source of mortality for adult raccoons. Where automobiles are not present, as on the isolated coastal islands, raccoon populations are quite large. Such is also the case in forested areas along the Santee River, as well as in isolated portions of the Francis Marion National Forest and major forested areas in Georgia. Hunting and trapping pressures are also very light. Where typical mortality sources are missing, endogenous density dependent factors come more into play. The raccoon seems especially susceptible to a distemper-like disease which, at times, causes dramatic population reductions. Rabies also appears to be widespread at times in raccoon populations of the Sea Island Coastal Region.

The striped skunk is a common omnivore of forested areas, though it appears to prefer more open land. Skunks usually remain in a burrow or den during daylight hours and forage for food at night. A variety of fruit and berries is consumed during summer and fall. Insects, small mammals, and carrion are utilized whenever available. Breeding occurs only once per year, during the spring. Litters, which average about four young, are produced in early summer after a gestation period of about 60 days (Colley 1962). Because of an effective defense mechanism, adult skunks are little subject to predation. However, the young do fall prey to owls, foxes, and other predators. The striped skunk is generally more abundant on agricultural lands than in unmodified forest lands, although it requires fence rows or wood lots for cover. Striped skunks frequently burrow under abandoned farm buildings to create den sites.

The nine-banded armadillo is a recent addition to the mammalian fauna of the Sea Island Coastal Region. Although not restricted by habitats, it is basically a forest animal and will be treated in this section. An early account of the biology of this mammal and its geographic dispersal was provided by Kalmbach (1943). In 1880, the nine-banded armadillo was restricted to the southern tip of Texas. By 1905, it had extended its range northward through much of central Texas. By 1943, the nine-banded armadillo had spread eastward, occupying most of Louisiana and crossing the Mississippi River; isolated specimens were found eastward of the area, but these were thought to be the result of accidental introduction rather than individuals contiguous with the main body of the species distribution. Colley (1962) reported that nine-banded armadillos were present in two southwestern counties in Georgia, as well as Bibb County in central Georgia. Colley speculated that these represented scattered introductions. At present, the nine-banded armadillo has occupied all of Florida and has spread at least halfway up the Georgia coast. E. B. Joseph (1979, South Carolina Marine Resources Division,

Charleston, unpub. data) reports seeing numerous road-killed animals in Camden and Glynn counties, and some as far north as Midway in McIntosh County. Considering the spread of this species over the past half-century, there is no reason to assume that its ultimate northern boundary has yet been reached. Apparently none of the coastal islands has yet been reached. Several isolated records for armadillo exist for South Carolina, but there is no evidence that any populations have become established in this State.

The motivation for Kalmbach's (1943) studies in the early forties was the fear that the armadillo might be a serious predator on quail nests. To clarify this issue, extensive food studies were conducted on Texas armadillos. On the basis of analyses conducted during each of the four seasons, insects and other invertebrates were found to constitute over 90% of its diet. Kalmbach concluded that moderate populations of armadillos constituted no threat to ground nesting birds. No similar studies have been conducted in this area, but there is no reason to assume that food habits would differ significantly.

The nine-banded armadillo has no important predators in our area, but they seem especially susceptible to being hit by automobiles. This is probably the major cause of mortality in our area.

The black bear, largest omnivorous mammal of the Eastern United States, occurs in the Francis Marion National Forest and perhaps in other portions of the study area. Although its numbers are presently so low that it has little ecological significance, it is still an item of popular interest. Black bears utilize almost any type of food available. Its diet includes berries, nuts, grasses, leaves, small mammals, insects, and fishes where available. The young of feral hogs, which are relatively abundant in the Francis Marion National Forest, may be a significant food source. The black bear has no predators here except man, who is no doubt the major element of population control for this species.

Entirely or predominantly carnivorous forest mammals range in size from the tiny southeastern shrew to the bobcat. The southeastern shrew, although apparently a rare species, is of historical interest in this characterization area. It was originally described by Bachman (1837) from the Santee Swamp, although Colley (1966) reported that the actual location was probably on Cat Island within the study area. Little is known of this predator except that it appears to prefer moist woods.

The most common forest shrew in the Sea Island Coastal Region is the short-tailed shrew. This shrew is most often encountered in moist woods where there is an abundance of litter and leaf mold, but it may be found in open pine forest. The short-tailed shrew feeds largely upon invertebrates; insects, snails, and earthworms are thought to constitute the bulk of its diet. When foraging on the surface at night, this shrew is subject to a variety of mammalian and avian predators. It is a relatively abundant species with a high reproductive capacity, and a high mortality rate is likely.

The long-tailed weasel is a highly predacious mammal in forested areas of the coast. Two subspecies occur in the Southeast, with Mustela frenata olivacea being the subspecies on the coast. This mammal is seldom seen and has not been investigated in the study area. Although its abundance is not known, there is much suitable habitat for this species in the Sea Island Coastal Region. Weasels feed largely upon small mammals; rats, mice, shrews, and young rabbits are all utilized. The weasel is a fast and efficient predator. Its slender body form allows it to enter small burrows in pursuit of prey. Almost nothing is known about the factors which control its population levels.

Two species of foxes, the gray and red, occur in the area, but the gray fox is the more common of the two on forested lands. The gray fox is also less strictly a predator than the red fox, as it feeds on fruits and nuts to a degree. A variety of small mammals, birds, amphibians, reptiles, and insects are preyed upon by both species. Only one litter per year is produced. Breeding occurs in late winter, and the young are born from March to May. Foxes are probably not subject to serious predation. Hunting, trapping, and automobile deaths are probably the principal mortality sources.

The gray fox is attracted to farms largely by the abundance of small mammals and birds, but it also consumes a variety of fruits and other plant materials. The modernization of agricultural practices has reduced the number of backyard or barnyard flocks of chickens, which were a consistent part of the family farm in times past. Still, many small family farms remain in coastal areas and unpenned flocks of poultry have not completely disappeared. The fox in the hen-house is still a food-chain link in this environment.

The bobcat remains a relatively abundant predator in coastal South Carolina and in portions of Georgia, especially in hardwood and bottomland forests. Their distribution on the coastal islands was discussed in Chapter Three, Maritime Ecosystem. These cats spend most of the day sleeping among the upper limbs of hardwood trees, from which they descend at night to hunt on the ground. Although largely nocturnal, they occasionally hunt during the day. The diet of this species is almost entirely meat. Where food studies have been conducted, rats and rabbits constitute the bulk of the diet. Almost any small mammal or bird that can be captured is likely to be consumed.

The bobcat is not seriously subjected to predation, and the major population controls are hunting, trapping, and automobile-caused mortalities. It is unlikely that food limits populations at their present levels, although this was probably the case before man-related mortality sources became important. Declines in availability of suitable habitat have undoubtedly been a factor in the reduction of bobcat populations to their relatively low current levels. The reproductive capacity is relatively high, and breeding can take place at any time of the year. Litters, which often consist of three to four young, are reared in dens usually located within a hollow log or under a fallen tree.

478

Appendix Table 1. List of vegetation identified in the Sea Island Characterization Study, arranged alphabetically by common name (Small 1933, Eyles and Robertson 1944, Bailey 1951, Mellinger and Mellinger 1962, Radford et al. 1968, Hotchkiss 1972, Bozeman 1975, Hosier 1975, McCollum and Ettman 1977, Teskey and Hinckley 1977, Tiner 1977, Porcher 1978, Wharton 1978, Rayner et al. 1979).

Alders	Alnus spp.
Alligator-weed	Alternanthera philoxeroides
Amaranth	Amaranthus spp.
American beech	Fagus grandifolia
American climbing fern	Lygodium palmatum
American elm	Ulmus americana
American holly	Ilex opaca
American hornbeam	Carpinus caroliniana
American three-square bulrush	Scirpus americanus
Annual salt marsh aster	Aster subulatus
Arrow-arum	Peltandra virginica
Arrowhead	Sagittaria graminea var. weatherbiana
Arrowheads	Sagittaria spp.
Arrowwood	Viburnum dentatum
Asiatic dayflower	Aneilema keisak
Asiatic panic grass	Panicum bisulcatum
Asiatic panicum	Panicum bisulcatum
Aster	Aster laevis var. concinnus
Aster	Aster laevis var. laevis
Aster	Aster praealtus
Aster	Aster puniceus
Aster	Aster simplex
Aster	Aster squarrosus
Aster	Aster tenuifolius
Asters	Aster spp.
Autumn coral-root	Corallorhiza odontorhiza
Baggy-knees	Sacciolepis striata
Bald cypress	Taxodium distichum
Bald rush	Psilocarya scirpoides
Baldwin's nutrush	Scleria baldwinii
Bamboo	Smilax laurifolia
Bamboo brier	Smilax auriculata
Banana water-lily	Nymphaea mexicana
Barbara's buttons	Marshallia graminifolia
Barley	Hordeum spp.
Bay starvine	Schisandra glabra
Beach elder	Iva imbricata
Beach grass	Panicum amarulum
Beach hogwort	Croton punctatus
Beach pea	Strophostyles helvola
Beach pennywort	Hydrocotyle bonariensis
Beak rush	Rhynchospora careyana
Beak rush	Rhynchospora corniculata
Beak rush	Rhynchospora decurrens
Beak rush	Rhynchospora glomerata
Beak rush	Rhynchospora macrostachya
Beak rush	Rhynchospora megalocarpa
Beak rush	Rhynchospora plumosa
Beak rushes	Rhynchospora spp.
Bearded grass-pink	Calopogon barbatus
Beard grass	Andropogon elliottii
Beard grass	Gymnopogon brevifolius
Beard grass	Andropogon sp.
Bedstraw	Galium circaezans
Beech	Fagus grandifolia
Beggar lice	Desmodium spp.
Beggar lice	Desmodium marilandicum
Beggar ticks	Desmodium spp.
Beggar ticks	Bidens spp.
Beggar ticks	Bidens pilosa
Beggar weeds	Desmodium spp.
Bermuda grass	Cynodon dactylon
Big duckweed	Spirodela polyrrhiza
Big floating heart	Nymphoides aquatica

Big primrose willow	Ludwigia peploides var. glabrescens
Big-rooted manroot	Ipomoea macrorhiza
Bird's eye	Veronica persica
Biscuit-flower	Sarracenia flava
Bitter gallberry	Ilex glabra
Blackberries	Rubus spp.
Black cherry	Prunus serotina
Black gum	Nyssa sylvatica
Black jack oak	Quercus marilandica
Black needlerush	Juncus roemerianus
Black oak	Quercus velutina
Black-root	Pterocaulon pycnostachym
Black rush	Juncus roemerianus
Black-stemmed spleenwort	Asplenium resiliens
Black titi	Cliftonia monophylla
Black willow	Salix nigra
Bladderwort	Utricularia inflata
Bladderwort	Utricularia inflata var. minor
Bladderwort	Utricularia olivacea
Bladderwort	Utricularia purpurea
Bladderwort	Utricularia subulata
Bladderwort	Utricularia vulgaris
Bladderworts	Utricularia spp.
Blazing star	Liatris tenuifolia
Blazing stars	Liatris spp.
Blue beach	Carpinus caroliniana
Blueberries	Vaccinium spp.
Blueberry	Vaccinium caesariense
Blueberry	Vaccinium myrsinites
Blue cat-tail	Typha glauca
Blue-eyed grass	Sisyrinchium mucronatum
Blue flag	Iris virginica
Bluegrass	Poa compressa
Bluejack oak	Quercus incana
Blue star	Amsonia ciliata
Bluestem	Andropogon elliottii
Bluestem	Andropogon gerardii
Bog buttons	Lachnocaulon beyrichianum
Bottlebrush three awn grass	Aristida spiciformis
Box elder	Acer negundo
Boykin's lobelia	Lobelia boykinii
Bracken fern	Pteridium aquilinum
Brazilian elodea	Egeria densa
Bristle-fruited spermolepsis	Spermolepsis echinata
Broadleaf waterplantain	Alisma plantago-aquatica
Broom sedge	Andropogon virginicus
Broom-straw	Andropogon sp.
Broom-straw	Andropogon elliottii
Broom-straw	Andropogon ternarius
Brown-top millet	Panicum ramosum
Buckwheat tree	Cliftonia monophylla
Bugleweed	Lycopus americanus
Bugleweed	Lycopus sessilifolius
Bull bay	Magnolia grandiflora
Bullgrass	Paspalum boscianum
Bullgrass	Paspalum dissectum
Bullgrass	Paspalum distichum
Bulrush	Scirpus americanus
Bulrush	Scirpus cyperinus
Bulrush	Scirpus etuberculatus
Bulrush	Scirpus robustus
Bulrush	Scirpus validus
Bulrushes	Scirpus spp.
Burmannia	Burmannia biflora
Bushy broom sedge	Andropogon virginicus
Bushy pondweed	Najas guadalupensis
Bushy pondweed	Najas minor
Bushy pondweeds	Najas spp.
Buttercup-leaved pennywort	Hydrocotyle ranunculoides
Butterfly-bush	Buddleja sp.
Butterfly pea	Centrosema virginianum

Butterfly pea	Clitoria mariana
Butter-print	Abutilon theophrastii
Butterweed	Senecio sp.
Butterwort	Pinguicula lutea
Butterworts	Pinguicula spp.
Button bush	Cephalanthus occidentalis
Buttonweed	Spermacoce glabra
Cabbage palmetto	Sabal palmetto
Cactus	Opuntia compressa
Calliopsis	Coreopsis tinctoria
Camphorweed	Heterotheca graminifolia
Camphorweed	Heterotheca subaxillaris
Camphorweed	Pluchea purpurascens
Canada bluegrass	Poa compressa
Cancer root	Conopholis americana
Cancer root	Orobanche uniflora
Cane	Arundinaria gigantea
Cape-weed	Lippia nodiflora
Cardinal flower	Lobelia cardinalis
Carolina cherry laurel	Prunus caroliniana
Carolina dog-hobble	Leucothoe populifolia
Carolina grass-of-parnassus	Parnassia caroliniana
Carolina spleenwort fern	Asplenium heteroresiliens
Carolina trillium	Trillium pusillum var. pusillum
Carpet grass	Reimarochloa oligostachya
Castor-bean	Ricinus communis
Castor oil plant	Ricinus communis
Catbrier	Smilax bona-nox
Catbriers	Smilax spp.
Cat-tail	Typha domingensis
Cat-tail	Typha glauca
Cat-tails	Typha spp.
Celery	Apium graveolens
Centipede grass	Eremochloa ophuroides
Chaff-seed	Schwalbea americana
Chapman oak	Quercus chapmanii
Chapman's sedge	Carex chapmanii
Cherrybark oak	Quercus falcata var. pagodaefloia
Cherry laurel	Prunus caroliniana
China-berry	Melia azedarach
Chinaman's shield	Centella asiatica
Chinquapin	Castanea pumila
Chinquapin oak	Quercus muehlenbergii
Chocolate-weed	Melochia corchorifolia
Chufa	Cyperus esculentus var. sativus
Cinnamon fern	Osmunda cinnamomea
Cinquefoil	Potentilla norvegica
Clearweed	Pilea pumila
Climbing fetterbush	Pieris phillyreifolia
Climbing hempweed	Mikania scandens
Close-flowered triple awn grass	Aristida condensata
Clovers	Trifolium spp.
Clubmosses	Lycopodium spp.
Coastal love grass	Eragrostis refracta
Coast bacopa	Bacopa monnieri
Coast pigweed	Amaranthus pumilus
Coffee-weed	Sesbania exaltata
Colic root	Aletris aurea
Colic root	Aletris farinosa
Colic root	Aletris lutea
Common bladderwort	Utricularia vulgaris
Common cat-tail	Typha latifolia
Common lespedezas	Lespedeza sp.
Common reed	Phragmites communis
Common sundew	Drosera rotundifolia
Common three-square	Scirpus americanus
Coontail	Ceratophyllum demersum
Coontail	Ceratophyllum echinatum
Coontails	Ceratophyllum spp.
Coral honeysuckle	Lonicera sempervirens

Cordgrass	Spartina alterniflora
Cordgrass	Spartina cynosuroides
Cordgrass	Spartina patens
Cordgrasses	Spartina spp.
Coreopsis	Coreopsis spp.
Coreopsis	Coreopsis falcata
Corn	Zea mays
Cotton rose	Filago germanica
Cottonweed	Froelichia floridana
Cottonwood	Populus deltoides
Cowpea	Vigna unguiculata
Crab grasses	Digitaria spp.
Cranberries	Vaccinium spp.
Crane-fly orchid	Tipularia discolor
Creeping cucumber	Melothria pendula
Creeping fig	Ficus pumila
Creeping rush	Juncus repens
Creeping spikerush	Eleocharis sp.
Creeping spurge	Euphorbia serpens
Creeping water plantain	Echinodorus cordifolius
Crinkled amaranth	Amaranthus crispus
Cross vine	Anisostichus capreolata
Croton	Croton punctatus
Crownbeard	Verbesina occidentalis
Crow-poison	Zigadenus densus
Cudweed	Gnaphalium purpureum
Cudweeds	Gnaphalium spp.
Curtiss' dropseed	Sporobolus curtissii
Cutgrass	Leersia hexandra
Cutgrass	Leersia lenticularis
Cutgrass	Leersia oryzoides
Cutgrass	Leersia virginica
Cypresses	Taxodium spp.
Dahoon	Ilex cassine
Dahoon	Ilex cassine var. myrtifolia
Daisy fleabane	Erigeron strigosus
Damask rose	Rosa damascena
Dangleberry	Gaylussacia frondosa
Dasheen	Colocasia esculentum
Dayflower	Commelina erecta
Dayflower	Commelina virginica
Deciduous holly	Ilex decidua
Delta duck potato	Sagittaria graminea
Dewberries	Rubus spp.
Diodia	Diodia teres
Dock	Rumex bucephalophorus
Dodder	Cuscuta sp.
Dodder	Cuscuta cephalanthii
Dodder	Cuscuta indecora
Dog fennel	Eupatorium capillifolium
Dog fennel	Eupatorium capillifolium var. leptophyllum
Dog-tongue	Eriogonum tomentosum
Dogwood	Cornus racemosa
Dotted smartweed	Polygonum punctatum
Downy rattlesnake plantain	Goodyera pubescens
Dropseed	Sporobolus teretifolius
Dropseed	Sporobolus virginicus
Dropwort	Oxypolis rigidior
Drummond's prickly pear	Opuntia drummondii
Duck potato	Sagittaria latifolia
Duckweed	Lemna minor
Duckweed	Lemna perpusilla
Duckweed	Lemna valdiviana
Duckweeds	Lemna spp.
Duckweeds	Spirodela spp.
Dwarf blueberry	Vaccinium myrsinites
Dwarf huckleberry	Gaylussacia dumosa
Dwarf laurel	Kalmia hirsuta
Dwarf palmetto	Sabal minor
Dwarf pawpaw	Asimina parviflora

Dwarf spikerush	Eleocharis parvula
Dwarf trillium	Trillium pusillum var. pusillum
Dwarf witch alder	Fothergilla gardenii
Eastern cottonwood	Populus deltoides
Eastern lilaeopsis	Lilaeopsis chinensis
Eastern red cedar	Juniperus virginiana
Eastern wolffiella	Wolffiella floridana
Eel grass	Zostera marina
Elderberry	Sambucus canadensis
Elderberry	Sambucus simpsonii
Elephant's foot	Elephantopus tomentosus
Elliot's blueberry	Vaccinium elliotii
English plantain	Plantago lanceolata
Eryngo	Eryngium integrifolium
Euphorbia	Euphorbia polygonifolia
Evening primrose	Oenothera humifusa
Evening primroses	Oenothera spp.
Everlasting	Gnaphalium obtusifolium
Everlastings	Gnaphalium spp.
Fall panic grass	Panicum dichotomiflorum
False asphodel	Tofieldia glabra
False asphodel	Tofieldia racemosa
False buckthorn	Bumelia lanuginosa
False indigo	Amorpha fruticosa
False nettle	Boehmeria cylindrica
False willow	Baccharis angustifolia
Fanwort	Cabomba caroliniana
Feathery bamboo	Bambusa vulgaris
Fern	Polypodium aureum
Fescue	Festuca myuros
Fescue	Festuca octoflora
Fescue	Festuca rubra
Fetter-bush	Leucothoe racemosa
Fetter-bush	Lyonia lucida
Fig	Ficus carica
Finger grass	Chloris petraea
Finger grass	Digitaria horizontalis
Fishweed	Potamogeton illinoiensis
Flag	Iris tridentata
Fleabane	Erigeron vernus
Floating heart	Nymphoides aquatica
Floating hearts	Nymphoides spp.
Floppy water milfoil	Myriophyllum laxum
Florida adder's mouth	Malaxis spicata
Florida bladderwort	Utricularia floridana
Florida dropseed	Sporobolus floridanus
Florida privet	Forestiera porulosa
Flowering dogwood	Cornus florida
Fly-catcher	Sarracenia flava
Fly-poison	Amianthium muscaetoxicum
Flytrap pitcher-plant	Sarracenia purpurea
Foxtail clubmoss	Lycopodium alopecuroides
Foxtail grass	Setaria geniculata
Foxtail grass	Setaria macrosperma
Foxtail grass	Setaria magna
French mulberry	Callicarpa americana
Fringed loosestrife	Lysimachia lanceolata
Fringe-leaved paspalum	Paspalum setaceum
Frog's bit	Limnobium spongia
Frost aster	Aster pilosus
Gaillardia	Gaillardia drummondii
Gamma grass	Tripsacum dactyloides
Gentians	Gentiana spp.
Georgia fever bark	Pickneya pubens
Georgia plume	Elliottia racemosa
Georgia's bulrush	Scirpus erismanae
Gerardia	Agalinis maritima
Giant cordgrass	Spartina cynosuroides
Giant cutgrass	Zizaniopsis miliacea

Giant foxtail grass	Setaria magna
Giant plume grass	Erianthus giganteus
Giant reed	Arundo donax
Giant-seeded beak rush	Rhynchospora megalocarpa
Giant spiral-orchid	Spiranthes longrilabris
Gladiolus	Gladiolus hortulana
Glasswort	Salicornia bigelovii
Glasswort	Salicornia europaea
Glasswort	Salicornia virginica
Glassworts	Salicornia spp.
Goat's rue	Tephrosia virginiana
Godfrey's sandwort	Arenaria godfreyi
Golden aster	Heterotheca floridana
Golden aster	Heterotheca graminifolia
Golden canna lily	Canna flaccida
Golden club	Orontium aquaticum
Goldenrod	Solidago chapmanii
Goldenrod	Solidago gymnospermoides
Goldenrod	Solidago sempervirens
Goldenrods	Solidago spp.
Gooseberries	Vaccinium spp.
Gopher apple	Chrysobalanus oblongifolius
Grain sorghum	Sorghum vulgare
Grapefruit	Citrus paradisi
Grass-leaved ladies' tresses	Spiranthes praecox
Grass-pinks	Calopogon spp.
Green ash	Fraxinus pennsylvanica
Greenbrier	Smilax auriculata
Greenbrier	Smilax boná-nox
Greenbrier	Smilax rotundifolia
Greenbrier	Smilax smallii
Greenbriers	Smilax spp.
Green fringed orchid	Habenaria lacera
Green fringeless orchid	Habenaria lacera
Ground cherry	Physalis pubescens var. grisea
Ground cherry	Physalis virginiana
Ground cherry	Physalis viscosa var. maritima
Groundnut	Apios americana
Gum	Nyssa sylvatica var. biflora
Hackberry	Celtis laevigata
Hair grass	Aira caryophyllea
Hairy wild-indigo	Baptisia arachnifera
Halberd-leaved marsh mallow	Hibiscus militaris
Hartwrightia	Hartwrightia floridana
Haws	Viburnum spp.
Hawthorn	Crataegus sp.
Hedge hyssop	Gratiola pilosa
Hercules' club	Aralia spinosa
Hercules' club	Zanthoxylum clava-herculis
Heterotheca	Heterotheca subaxillaris
Hickory	Carya spp.
Highbush blueberry	Vaccinium corymbosum
Hightide bushes	Baccharis spp.
Hollies	Ilex spp.
Hooded pitcher-plant	Sarracenia minor
Hop hornbeam	Ostrya virginiana
Horned bladderwort	Utricularia cornuta
Horned-pondweed	Zannichellia palustris
Hornwort	Ceratophyllum demersum
Hornwort	Ceratophyllum echinatum
Hornworts	Ceratophyllum spp.
Horse balm	Collinsonia canadensis
Horse sugar	Symplocos tinctoria
Horseweed	Erigeron canadensis
Huckleberries	Gaylussacia spp.
Huckleberries	Vaccinium spp.
Incised groovebur	Agrimonia incisa
Indian fig	Opuntia ficus-indica

Indian grass	Sorghastrum nutans
Indian grass	Sorghastrum secundum
Iris	Iris tridentata
Ironweed	Vernonia sp.
Ironweed	Vernonia altissima
Ironweed	Vernonia blodgettii
Ironweed	Vernonia harperi
Ironwood	Carpinus caroliniana
Italian rye grass	Lolium multiflorum
Japanese clover	Lespedeza striata
Jerusalem artichoke	Helianthus tuberosus
Jewel-weed	Impatiens capensis
Johnson grass	Sorghum halepense
Jointed spikerush	Eleocharis equisetoides
Jove's fruit	Lindera melissaefolium
June grass	Koeleria phleoides
Knawel	Scleranthus annuus
Knotweed	Polygonum lapathifolium
Knotweed	Polygonum persicaria
Knotweeds	Polygonum spp.
Lacegrass	Eragrostis capillaris
Lace-lip spiral orchid	Spiranthes laciniata
Ladies eardrops	Brunnichia cirrhosa
Lambkill	Kalmia angustifolia var. carolina
Lamb's quarters	Chenopodium album
Large-rooted morning glory	Ipomoea macrorhiza
Large-seed smartweed	Polygonum pensylvanicum
Laurel greenbrier	Smilax laurifolia
Laurel oak	Quercus laurifolia
Leafy pondweed	Potamogeton foliosus
Least adder's tongue	Ophioglossum nudicaule
Leather-flower	Clematis crispa
Leather-leaf	Cassandra calyculata
Lemon bacopa	Bacopa caroliniana
Lespedezas	Lespedeza bicolor
Leucothoe	Leucothoe axillaris
Leucothoe	Leucothoe populifolia
Lippia	Lippia nodiflora
Little bluestem	Andropogon scorparius
Little burhead	Echinodorus parvulus
Little floating heart	Nymphoides cordata
Live oak	Quercus virginiana
Lizard's tail	Saururus cernuus
Lobelia	Lobelia elongata
Loblolly bay	Gordonia lasianthus
Loblolly pine	Pinus taeda
Longleaf pine.	Pinus palustris
Long-styled smartweed	Polygonum longistylum
Loosestrife	Lythrum lineare
Loose water milfoil	Myriophyllum laxum
Lotus	Nelumbo lutea
Lotus	Nelumbo pentapetala
Love grass	Eragrostis capillaris
Love grass	Eragrostis pilosa
Love grass	Eragrostis refracta
Low millewort	Polygala nana
Low showy aster	Aster spectabilis
Lupine	Lupinus perennis
Macartney rose	Rosa bracteata
Maidencane	Panicum hemitomon
Male-berry	Lyonia ligustrina
Mangrove	Rhizophora mangle
Marsh cress	Rorippa islandica
Marsh cress	Rorippa sessiliflora
Marsh daisy	Boltonia asteroides
Marsh elder	Iva frutescens
Marsh eryngo	Eryngium aquaticum

Marsh fleabane	Pluchea purpurascens
Marsh fleabane	Pluchea rosea
Marsh fleabanes	Pluchea spp.
Marsh-gentian	Sabatia stellaris
Marsh hemp	Amaranthus cannabinus
Marsh pennywort	Hydrocotyle umbellata
Marsh-pink	Sabatia foliosa
Marsh purslane	Ludwigia natans
Meadow beauties	Rhexia spp.
Meadow beauty	Rhexia alifanus
Meadow beauty	Rhexia cubensis
Melonette	Melothria crassifolia
Memorial rose	Rosa wichuraniana
Milk pea	Galactia elliottii
Milk-vine	Cynanchum palustre
Milkwort	Polygala grandiflora
Millet	Pennisetum glaucum
Millets	Echinochloa spp.
Milo	Sorghum vulgare
Miterwort	Cynoctonum sessilifolium
Mock-bishopweed	Ptilimnium capillaceum
Mock-bishopweed	Ptilimnium costatum
Mockernut hickory	Carya tomentosa
Mosquito fern	Azolla caroliniana
Moundlily yucca	Yucca gloriosa
Muhlenberg's amphicarpum	Amphicarpum muhlenbergianum
Muhly grass	Muhlenbergia capillaris
Muhly grass	Muhlenbergia expansa
Mulberries	Morus spp.
Muscadine grape	Vitis rotundifolia
Muscle tree	Carpinus caroliniana
Muskgrasses	Chara spp.
Myrtle holly	Ilex cassine var. myrtifolia
Myrtle oak	Quercus myrtifolia
Myrtles	Myrica spp.
Narrow-leaved cat-tail	Typha angustifolia
Narrow-leaved pondweed	Potamogeton berchtoldii
Narrow-leaved rushfoil	Crotonopsis linearis
Needle palm	Rhapidophyllum hystrix
Netted chain fern	Woodwardia areolata
Nightshade	Solanum aculeatissimum
Nitella	Nitella sp.
Nodding smartweed	Polygonum lapathifolium
Nut grass	Cyperus esculentus
Nut rush	Scleria baldwinii
Nut rush	Scleria ciliata
Nut rush	Scleria trigolmerata
Nutmeg hickory	Carya myristicaeformis
Oatgrass	Arrhenatherum elatius
Oats	Avena sativa
Odorless wax myrtle	Myrica inodora
Ogeechee plum	Nyssa ogeche
Olive	Olea europaea
Olney's three-square bulrush	Scirpus olneyi
Orach	Atrixplex patula
Orchids	Habenaria spp.
Overcup oak	Quercus lyrata
Panic grass	Panicum amarum
Panic grass	Panicum leucothrix
Panic grass	Panicum virgatum
Panic grasses	Panicum spp.
Parrot-feather	Myriophyllum brasiliense
Parrot pitcher-plant	Sarracenia psittacina
Partridge berry	Mitchella repens
Partridge pea	Cassia fasciculata
Paspalum	Paspalum sp.
Passion-flower	Passiflora lutea
Pawpaw	Asimina incana
Pawpaw	Asimina pygmaea
Pawpaw	Asimina triloba
Peanut	Arachis hypogaea

Peat mosses	*Sphagnum* spp.
Pecan	*Carya illinoensis*
Pennywort	*Hydrocotyle ranunculoides*
Pennyworts	*Hydrocotyle* spp.
Pepper-vine	*Ampelopsis arborea*
Perennial glasswort	*Salicornia virginica*
Periwinkle	*Vinca major*
Persimmon	*Diospyros virginiana*
Petunia	*Petunia axillaris*
Pickerelweed	*Pontederia cordata*
Pigeon grape	*Vitis cinerea* var. *floridana*
Pigmy-pipes	*Monotropis odorata*
Pignut hickory	*Carya glabra*
Pigweed	*Amaranthus lividus*
Pigweed	*Amaranthus pumilus*
Pigweeds	*Amaranthus* spp.
Pineland agrimony	*Agrimonia incisa*
Pineweed	*Hypericum gentianoides*
Pin-weeds	*Lechea* spp.
Pipewort	*Eriocaulon compressum*
Pipewort	*Eriocaulon decangulare*
Pitcher-plants	*Sarracenia* spp.
Plume grass	*Erianthus giganteus*
Plume grasses	*Erianthus* spp.
Poison ivy	*Rhus radicans*
Poke weed	*Phytolacca rigida*
Polygala	*Polygala cymosa*
Polygala	*Polygala lutea*
Polygala	*Polygala sanguinea*
Polygalas	*Polygala* spp.
Polypremum	*Polypremum procumbens*
Pond cypress	*Taxodium ascendens*
Pond pine	*Pinus serotina*
Pond spice	*Litsea aestivalis*
Pondweed	*Potamogeton berchtoldii*
Pondweed	*Potamogeton foliosus*
Pondweed	*Potamogeton illinoiensis*
Pondweed	*Potamogeton nodosus*
Pondweed	*Potamogeton pectinatus*
Pondweeds	*Potamogeton* spp.
Poor-joe	*Diodia teres*
Poor-mans pepper	*Lepidium virginicum*
Popcorn tree	*Sapium sebiferum*
Poplar-leaved fetterbush	*Leucothoe populifolia*
Poplars	*Populus* spp.
Possum haw	*Ilex decidua*
Possum haw	*Viburnum nudum*
Post oak	*Quercus stellata*
Potato bean	*Apios americana*
Prickly mallow	*Sida spinosa*
Prickly pear	*Opuntia drummondii*
Princess-feather	*Polygonum orientale*
Privet	*Ligustrum japonicum*
Proliferating spikerush	*Eleocharis baldwinii*
Pumpkin ash	*Fraxinus tomentosa*
Purple bladderwort	*Utricularia purpurea*
Purple silkyscale	*Anthaenantia rufa*
Purple top	*Tridens flavus* var. *flavus*
Pussy-toes	*Antennaria* spp.
Queen's delight	*Stillingia sylvatica*
Rabbit tobacco	*Gnaphalium obtusifolium*
Ragweed	*Ambrosia artemisiifolia*
Ragweeds	*Ambrosia* spp.
Railroad vine	*Ipomoea pes-caprae*
Rain lily	*Zephyranthes simpsonii*
Rambler rose	*Rosa multiflora*
Raspberries	*Rubus* spp.
Rattanvine	*Berchemia scandens*
Rattlebox	*Crotalaria intermedia*

Rattlebox	_Crotalaria lanceolata_
Rattlebox	_Crotalaria retusa_
Rattlesnake fern	_Botrychium virginianum_
Ravenel's button snakeroot	_Eryngium aquaticum_ var. _ravenelii_
Ravenna-grass	_Erianthus ravennae_
Rayless goldenrod	_Chondrophora nudata_
Red ash	_Fraxinus pennsylvanica_
Red basil	_Satureja calamintha_
Red bay	_Persea borbonia_
Red buckeye	_Aesculus pavia_
Redbud	_Cercis canadensis_
Red cedar	_Juniperus virginiana_
Red chokeberry	_Sorbus arbutifolia_ var. _arbutifolia_
Reddish anthaenantia	_Anthaenantia rufa_
Red fescue	_Festuca rubra_
Red-hot poker	_Polygala lutea_
Red maple	_Acer rubrum_
Red oak	_Quercus rubra_
Redroot	_Lachnanthes caroliniana_
Redrooted nutgrass	_Cyperus erythrorhizos_
Redtop	_Agrostis stolonifera_
Red-veined dock	_Rumex sanguineus_
Reed	_Phragmites communis_
Reindeer lichen	_Cladonia_ spp.
Resurrection fern	_Polypodium polypodioides_
Rice cutgrass	_Leersia hexandra_
Rice cutgrass	_Leersia oryzoides_
Riverbank sandreed	_Calamovilfa brevipilis_ var. _brevipilis_
River birch	_Betula nigra_
Rosebud orchid	_Cleistes divaricata_
Rose dicerandra	_Dicerandra adoratissima_
Rose mallow	_Hisbiscus moscheutos_
Rose pogonia	_Pogonia ophioglossoides_
Roundleaf bacopa	_Bacopa rotundifolia_
Royal fern	_Osmunda regalis_
Running oak	_Quercus pumila_
Rush	_Juncus biflorus_
Rush	_Juncus nodatus_
Rush	_Juncus secundus_
Rush	_Juncus subcaudatus_
Rushes	_Juncus_ spp.
Russian thistle	_Salsola kali_
Rusty lyonia	_Lyonia ferruginea_
Rusty lyonia	_Rhynchospora megalocarpa_
Rye	_Secale cereale_
Rye grass	_Lolium_ sp.
Sabatia	_Sabatia brachiata_
Sabatia	_Sabatia dodecandra_
Sabatia	_Sabatia stellaris_
Sacciolepis	_Sacciolepis striata_
Sageretia	_Sageretia minutiflora_
Sago pondweed	_Potamogeton pectinatus_
Salt grass	_Distichlis spicata_
Salt marsh aster	_Aster tenuifolius_
Salt-marsh bulrush	_Scirpus robustus_
Salt marsh fimbristylis	_Fimbristylis spadicea_
Salt marsh millet	_Echinochloa walteri_
Saltmeadow cordgrass	_Spartina patens_
Saltwort	_Batis maritima_
Salvinia	_Salvinia rotundifolia_
Sand grass	_Triplasis purpurea_
Sand pine	_Pinus clausa_
Sand spurrey	_Spergularia marina_
Sandspur	_Cenchrus longispinus_
Sandspur	_Cenchrus tribuloides_
Sarvis holly	_Ilex amelanchier_
Sassafras	_Sassafras albidum_
Saw grass	_Cladium jamaicense_
Saw grass	_Mariscus jamaicense_
Saw palmetto	_Serenoa repens_

Scarlet oak	Quercus coccinea
Scarlet spiderling	Boerhaavia diffusa
Scrub oaks	Quercus spp.
Seabeach orach	Atriplex arenaria
Seabeach panic grass	Panicum amarum
Sea-blite	Suaeda linearis
Sea elder	Iva imbricata
Sea lavender	Limonium carolinianum
Sea lavender	Limonium nashii
Sea lavenders	Limonium spp.
Sea myrtle	Baccharis halimifolia
Sea myrtles	Baccharis spp.
Sea oats	Uniola paniculata
Sea ox-eye	Borrichia frutescens
Sea pink	Sabatia dodecandra
Sea pink	Sabatia foliosa
Sea purslane	Sesuvium maritimum
Sea purslane	Sesuvium portulacastrum
Sea purslanes	Sesuvium spp.
Sea rocket	Cakile harperi
Seashore mallow	Kosteleskya virginica
Seashore paspalum	Paspalum vaginatum
Seaside goldenrod	Solidago sempervirens
Seban	Sesbania macrocarpa
Sebastian bush	Sebastiania ligustrina
Sedge	Carex decomposita
Sedge	Carex joorii
Sedge	Carex shortiana
Sedge	Cyperus brevifolius
Sedge	Cyperus odoratus
Sedge	Cyperus polystachos
Sedge	Cyperus rivularis
Sedge	Cyperus rotundus
Sedge	Cyperus strigosus
Sedges	Carex spp.
Sedges	Cyperus spp.
Seed box	Ludwigia pilosa
Sensitive fern	Onoclea sensibilis
Sericea	Lespedza cuneata
Sesbania	Sesbania exaltata
Shepherd's purse	Capsella rubella
Short leaf pine	Pinus echinata
Showy aster	Aster spectabilis
Shrubby post oak	Quercus margaretta
Shrub oaks	Quercus spp.
Shumard oak	Quercus shumardii
Silver maple	Acer saccharinum
Single-flowered balduina	Balduina uniflora
Single-flowered cancer root	Orobanche uniflora
Sixweeks fescue	Festuca myuros
Skullcap	Scutellaria sp.
Slash pine	Pinus elliottii
Sleepy catchfly	Silene antirrhina
Slender spikerush	Eleocharis acicularis
Slippery elm	Ulmus rubra
Small-flowered buckthorn	Sageretia minutiflora
Small pondweed	Potamogeton berchtoldii
Smartweed	Polygonum hirsutum
Smartweeds	Polygonum spp.
Smooth aster	Aster laevis var. laevis
Smooth cordgrass	Spartina alterniflora
Smooth winterberry	Ilex laevigata
Sneeze-weed	Hélénium vernale
Snow-on-the-mountain	Euphorbia marginata
Soapberry	Sapindus marginatus
Soft-haired cornflower	Rudbeckia mollis
Soft rush	Juncus effusus
Soft-stem bulrush	Scirpus validus
Sour grass	Rumex acetosella
Sour orange	Citrus aurantium
Sourwood	Oxydendrum arboreum
Southern adder's tongue	Ophioglossum vulgatum var. pycnostichum

Southern bog buttons	Lachnocaulon beyrichianum
Southern bulrush	Scirpus californicus
Southern cat-tail	Typha domingensis
Southern elderberry	Sambucus simpsonii
Southern lady fern	Athyrium asplenioides
Southern lepuropetalon	Lepuropetalon spathulatum
Southern magnolia	Magnolia grandiflora
Southern naiad	Najas guadalupensis
Southern red cedar	Juniperus silicicola
Southern red oak	Quercus falcata
Southern rein orchid	Habenaria flava
Southern smartweed	Polygonum densiflorum
Southern smartweed	Polygonum portoricense
Southern spicebush	Lindera melissaefolium
Southern sugar maple	Acer saccharum floridanum
Southern wild rice	Zizaniopsis miliacea
Soybean	Glycine max
Spanish bayonet	Yucca aloifolia
Spanish moss	Tillandsia usneoides
Sparkleberries	Vaccinium spp.
Sparkleberry	Vaccinium arboreum
Spatter-dock	Nuphar advena
Sphagnum mosses	Sphagnum spp.
Spider-lily	Hymenocallis crassifolia
Spike-grass	Uniola latifolia
Spike-grass	Uniola laxa
Spike-grass	Uniola sessiliflora
Spikemoss	Selaginella arenicola
Spikerush	Eleocharis sp.
Spikerush	Eleocharis acicularis
Spikerush	Eleocharis albida
Spikerush	Eleocharis elongata
Spikerush	Eleocharis robbinsii
Spleenwort	Asplenium heteroresiliens
Sprangletop	Leptochloa sp.
Sprangletop	Leptochloa uninervia
Spreading pogonia	Cleistes divaricata
Spring coral-root	Corallorhiza wisteriana
Spring-flowered goldenrod	Solidago verna
Spruce pine	Pinus glabra
Square-stem spikerush	Eleocharis quadrangulata
Squaw-huckleberry	Vaccinium stamineum
Squaw-root	Conopholis americana
Stagger bush	Lyonia ferruginea
Stagger bush	Lyonia mariana
Star grass	Aletris lutea
Star grass	Hypoxis sessilis
Star-rush	Dichromena colorata
Starved aster	Aster lateriflorus
Sticky tofieldia	Tofieldia racemosa
Stillingia	Stillingia sylvatica
Stinging needle	Cnidoscolus stimulosus
St. John's-wort	Hypericum apocynifolium
St. John's-wort	Hypericum fasciculatum
St. John's-wort	Hypericum pseudomaculatum
St. John's-wort	Hypericum tubulosum
St. John's-wort	Hypericum virginicum
St. John's-wort	Hypericum walteri
St. John's-worts	Hypericum spp.
Storax	Styrax americana
Storax	Styrax grandifolia
Strawberry bush	Euonymus americanus
Sugarberry	Celtis laevigata
Sumac	Rhus spp.
Summer-farewell	Petalostemum pinnatum
Summer grape	Vitis aestivalis
Summer grape	Vitis aestivalis var. aestivalis
Sun-bonnets	Chaptalia tomentosa
Sundews	Drosera spp.
Sunflower	Helianthus angustifolius
Sunflower	Helianthus tuberosus
Sun-petaled meadow beauty	Rhexia aristosa
Swamp chestnut oak	Quercus michauxii

Swamp cottonwood	Populus heterophylla
Swamp dock	Rumex verticillatus
Swamp dogwood	Cornus stricta
Swamp holly	Ilex decidua
Swamp lily	Crinum americanum
Swamp milkweed	Asclepias incarnata sp. pulchra
Swamp privet	Forestiera acuminata
Swamp rose	Rosa palustris
Swamp smartweed	Polygonum hydropiperoides
Swamp smartweed	Polygonum setaceum
Swamp thistle	Carduus carolinianus
Swamp tupelo	Nyssa sylvatica var. biflora
Swamp willow	Salix caroliniana
Swaying bulrush	Scirpus subterminalis
Sweet bay	Magnolia virginiana
Sweetflag	Acorus calamus
Sweet gallberry	Ilex coriacea
Sweet grass	Muhlenbergia filipes
Sweet gum	Liquidambar styraciflua
Sweet leaf	Symplocos tinctoria
Sweet pepperbush	Clethra alnifolia
Sweet pignut hickory	Carya ovalis
Sweet pitcher-plant	Sarracenia rubra
Switchgrass	Panicum virgatum
Sycamore	Platanus occidentalis
Tag alder	Alnus serrulata
Tall oatgrass	Arrhenatherum elatius
Tansey-mustard	Descurainia pinnata
Tarflower	Befaria racemosa
Tearthumb	Polygonum arifolium
Tearthumb	Polygonum sagittatum
Thistle	Carduus carolinianus
Thoroughwort	Eupatorium album
Thoroughworts	Eupatorium spp.
Three awn grass	Aristida gyrans
Three awn grass	Aristida purpurascens
Three awn grass	Aristida spiciformis
Three awn grass	Aristida stricta
Three awn grasses	Aristida spp.
Three-birds orchid	Triphore trienthophora
Three-seeded mercury	Acalypha virginica
Three-way sedge	Dulichium arundinaceum
Thyme-leave speedwell	Veronica serpyllifolia
Tick trefoil	Desmodium sp.
Titi	Cyrilla racemiflora
Toad-flax	Linaria canadensis
Toad rush	Juncus bufonius
Toothache grass	Campulosus aromaticus
Toothache grass	Ctenium aromaticum
Toothache grasses	Ctenium spp.
Tough buckthorn	Bumelia tenax
Trailing lantana	Lantana montevidensis
Trianglestem spikerush	Eleocharis robbinsii
Trillium	Trillium pusillum
Tropical carpet grass	Axonopus compressus
Trumpet-plant	Sarracenia flava
Trumpet vine	Campsis radicans
Tulip tree	Liriodendron tulipifera
Turkey foot	Andropogon gerardii
Turkey oak	Quercus laevis
Turtle grass	Thalassia testudinum
Twig-rush	Cladium mariscoides
Umbrella tree	Magnolia macrophylla
Umbrella tree	Melia azedarach
Vanilla-plant	Trilisa odoratissima
Variable-leaved pondweed	Potamogeton diversifolius
Velvet-leaf	Abutilon theophrastii

Venus' fly .trap	Dionaea muscipula
Venus' looking-glass	Specularia perfoliata
Venus' looking-glasses	Specularia spp.
Vervain	Verbena officinalis
Violet	Viola cucullata
Violet	Viola papilionacea
Violet	Viola triloba
Virginia chain fern	Woodwardia virginica
Virginia creeper	Parthenocissus quinquefolia
Virginia willow	Itea virginica
Walter's sedge	Carex walteriana
Wampee	Eichhornia crassipes
Watches	Sarrencia flava
Water ash	Fraxinus caroliniana
Water elm	Planera aquatica
Water grass	Hydrochloa caroliniensis
Water hemlock	Cicuta maculata
Water hemp	Amaranthus cannabinus
Water hickory	Carya aquatica
Water-hoarhound	Lycopus virginicus
Water hyacinth	Eichhornia crassipes
Water hyssop	Bacopa monnieri
Water hyssop	Bacopa rotundifolia
Water-lily	Nymphaea bombycina
Water-lily	Nymphaea mexicana
Water-lily	Nymphaea odorata
Water locust	Gleditsia aquatica
Water loosestrife	Decodon verticillatus
Water-meal	Wolffia columbiana
Water milfoil	Myriophyllum heterophyllum
Water milfoil	Myriophyllum laxum
Water milfoil	Myriophyllum pinnatum
Water milfoils	Myriophyllum spp.
Water nymph	Najas gudalupensis
Water oak	Quercus nigra
Water parsnip	Sium suave
Water pimpernel	Samolus parviflorus
Water-primrose	Ludwigia peploides
Water-primroses	Ludwigia spp.
Water purslane	Ludwigia natans
Water-shield	Brasenia schreberi
Water spider orchid	Habenaria repens
Water spikerush	Eleocharis elongata
Water tupelo	Nyssa aquatica
Water-weed	Elodea canadensis
Water-weed	Egeria densa
Water-weeds	Elodea spp.
Wax myrtle	Myrica cerifera
Wedge grass	Sphenopholis intermedia
Weeping willow	Salix babylonica
Wheat	Triticum aestivum
White arrow-arum	Peltandra sagittaefolia
White ash	Fraxinus americana
White-bracted sedge	Dichromena latifolia
White colic root	Aletris obovata
White-fringed orchid	Habenaria blephariglottis var. integrilabia
White mulberry	Morus alba
White oak	Quercus alba
White water-lily	Nymphaea odorata
White wicky	Kalmia cuneata
Widgeon grass	Ruppia maritima
Wild azalea	Rhododendron canescens
Wild grapes	Vitis spp.
Wild licorice	Galium circaezans
Wild millet	Echinochloa crusgalli
Wild olive	Osmanthus americana
Wild plum	Prunus americana
Wild rice	Zizania aquatica

Wild rye	_Elymus villosus_
Wild rye grass	_Elymus villosus_
Wild rye grass	_Elymus virginicus_
Willow	_Salix caroliniana_
Willow oak	_Quercus phellos_
Willows	_Salix_ spp.
Winged elm	_Ulmus alata_
Wire grasses	_Aristida_ spp.
Wire-leaved dropseed	_Sporobolus teretifolia_
Witch alder	_Fothergilla gardenii_
Witch grass	_Leptoloma cognatum_
Witch grass	_Panicum capillare_
Witch-hazel	_Hamamelis virginiana_
Wood awn-grass	_Brachyelytrum erectum_
Wood fern	_Dryopteris dentata_
Wood grass	_Sorgastrum nutans_
Wood reed	_Cinna arundinacea_
Woolgrass bulrush	_Scirpus cyperinus_
Wreath aster	_Aster vimineus_
Yaupon holly	_Ilex vomitoria_
Yellow asphodel	_Narthacium americanum_
Yellow chestnut oak	_Quercus muehlenbergii_
Yellow cress	_Rorippa islandica_
Yellow cress	_Rorippa sessiliflora_
Yellow-eyed grass	_Xyris caroliniana_
Yellow-eyed grass	_Xyris elliottii_
Yellow-eyed grass	_Xyris smalliana_
Yellow-eyed grasses	_Xyris_ spp.
Yellow fringeless orchid	_Habenaria integra_
Yellow nelumbo	_Nelumbo lutea_
Yellow pitcher-plant	_Sarracenia flava_
Yellow pond-lily	_Nuphar luteum_
Yellow poplar	_Liriodendron tulipifera_
Yellow star grass	_Hypoxis sessilis_
Zenobia	_Zenobia pulverulenta_

Appendix Table   2.   List of vegetation identified in the Sea Island Characterization Study,
                      arranged alphabetically by scientific name (Small 1933, Eyles and
                      Robertson 1944, Bailey 1951, Mellinger and Mellinger 1962, Radford et al.
                      1968, Hotchkiss 1972, Bozeman 1975, Hosier 1975, McCollum and Ettman 1977,
                      Teskey and Hinckley 1977, Tiner 1977, Porcher 1978, Wharton 1978, Rayner
                      et al. 1979).

Abutilon theophrastii	Velvet-leaf/Butter-print
Acalypha virginica	Three-seeded mercury
Acer negundo	Box elder
Acer rubrum	Red maple
Acer saccharinum	Silver maple
Acer saccharum floridanum	Southern sugar maple
Acorus calamus	Sweetflag
Aesculus pavia	Red buckeye
Agalinis maritima	Gerardia
Agrimonia incisa	Incised groovebur/Pineland agrimony
Agrostis stolonifera	Redtop
Aira caryophyllea	Hair grass
Aletris aurea	Colic root
Aletris farinosa	Colic root
Aletris lutea	Stargrass/Colic root
Aletris obovata	White colic root
Alisma plantago-aquatica	Broadleaf waterplantain
Alnus spp.	Alders
Alnus serrulata	Tag alder
Alternanthera philoxeroides	Alligator-weed
Amaranthus spp.	Amaranth/Pigweeds
Amaranthus cannabinus	Water hemp/Marsh hemp
Amaranthus crispus	Crinkled amaranth
Amaranthus lividus	Pigweed
Amaranthus pumilus	Pigweed/Coast pigweed
Ambrosia spp.	Ragweeds
Ambrosia artemisiifolia	Ragweed
Amianthium muscaetoxicum	Fly-poison
Ammannia teres	--
Amorpha fruticosa	False indigo
Amorpha glabra	--
Amorpha herbacea	--
Ampelopsis arborea	Pepper-vine
Amphicarpum muhlenbergianum	Muhlenberg's amphicarpum
Amsonia ciliata	Blue star
Andropogon sp.	Broom-straw/Beard grass
Andropogon elliottii	Bluestem/Beard grass/Broom-straw
Andropogon gerardii	Bluestem/Turkeyfoot
Andropogon scorparius	Little bluestem
Andropogon ternarius	Broom-straw
Andropogon virginicus	Bushy broom sedge/ Broom sedge
Aneilema keisak	Asiatic dayflower
Anisostichus capreolata	Cross vine
Antennaria spp.	Pussy-toes
Anthaenantia rufa	Purple silkyscale/Reddish anthaenantia
Apios americana	Groundnut/Potato bean
Apium graveolens	Celery
Arachis hypogaea	Peanut
Aralia spinosa	Hercules' club
Arenaria godfreyi	Godfrey's sandwort
Aristida spp.	Three awn grasses/Wire grasses
Aristida condensata	Close-flowered triple crown grass
Aristida gyrans	Three awn grass
Aristida purpurascens	Three awn grass
Aristida spiciformis	Bottlebrush three awn grass/Three awn grass
Aristida stricta	Three awn grass
Arrhenatherum elatius	Tall oatgrass/Oatgrass
Arundinaria gigantea	Cane
Arundo donax	Giant reed
Asclepias incarnata ssp. pulchra	Swamp milkweed
Asclepias walteri	--
Asimina incana	Pawpaw

494

Asimina parviflora	Dwarf pawpaw
Asimina pygmaea	Pawpaw
Asimina triloba	Pawpaw
Asplenium heteroresiliens	Spleenwort/Carolina spleenwort fern
Asplenium resiliens	Black-stemmed spleenwort
Aster spp.	Asters
Aster laevis var. concinnus	Aster
Aster laevis var. laevis	Smooth aster/Aster
Aster lateriflorus	Starved aster
Aster pilosus	Frost aster
Aster praealtus	Aster
Aster puniceus	Aster
Aster simplex	Aster
Aster spectabilis	Showy aster/Low showy aster
Aster squarrosus	Aster
Aster subulatus	Annual salt marsh aster
Aster tenuifolius	Aster/Salt marsh aster
Aster vimineus	Wreath aster
Athyrium asplenioides	Southern lady fern
Atriplex arenaria	Seabeach orach
Atriplex patula	Orach
Avena sativa	Oats
Axonopus compressus	Tropical carpet grass
Azolla caroliniana	Mosquito fern
Baccharis spp.	Sea myrtles/High tide bushes
Baccharis angustifolia	False willow
Baccharis halimifolia	Sea myrtle
Bacopa caroliniana	Lemon bacopa
Bacopa monnieri	Water hyssop/Coast bacopa
Bacopa rotundifolia	Water hyssop/Roundleaf bacopa
Balduina angustifolia	--
Balduina uniflora	Single-flowered balduina
Bambusa vulgaris	Feathery bamboo
Baptisia arachnifera	Hairy wild-indigo
Batis maritima	Saltwort
Befaria racemosa	Tarflower
Berchemia scandens	Rattanvine
Betula nigra	River birch
Bidens spp.	Beggar ticks
Bidens pilosa	Beggar ticks
Boehmeria cylindrica	False nettle
Boerhaavia diffusa	Scarlet spiderling
Boltonia asteroides	Marsh daisy
Bonamia patens	--
Borrichia frutescens	Sea ox-eye
Botrychium virginianum	Rattlesnake fern
Brachyelytrum erectum	Wood awn-grass
Brasenia schreberi	Water-shield
Brunnichia cirrhosa	Ladies eardrops
Buddleja sp.	Butterfly-bush
Bumelia lanuginosa	False buckthorn
Bumelia tenax	Tough buckthorn
Burmannia biflora	Burmannia
Cabomba caroliniana	Fanwort
Cakile harperi	Sea rocket
Calamovilfa brevipilis var. brevipilis	Riverbank sandreed
Callicarpa americana	French mulberry
Calopogon spp.	Grass-pinks
Calopogon barbatus	Bearded grass-pink
Campsis radicans	Trumpet vine
Campulosus aromaticus	Toothache grass
Canna flaccida	Golden canna lily
Capsella rubella	Shepherd's purse
Carduus carolinianus	Thistle/Swamp thistle
Carex spp.	Sedges
Carex chapmanii	Chapman's sedge
Carex decomposita	Sedge
Carex joori	Sedge
Carex scoparia	--

Carex shortiana	Sedge
Carex stipata var. maxima	--
Carex typhina	--
Carex walteriana	Walter's sedge
Carpinus caroliniana	American hornbeam/Ironwood/Blue beach/Muscle tree
Carya spp.	Hickory
Carya aquatica	Water hickory
Carya glabra	Pignut hickory
Carya illinoensis	Pecan
Carya myristicaeformis	Nutmeg hickory
Carya ovalis	Sweet pignut hickory
Carya tomentosa	Mockernut hickory
Cassandra calyculata	Leather-leaf
Cassia fasciculata	Patridge pea
Castanea pumila	Chinquapin
Cayaponia boykinii	--
Celtis laevigata	Hackberry/Sugarberry
Cenchrus longispinus	Sandspurs
Cenchrus tribuloides	Sandspurs
Centella asiatica	Chinaman's shield
Centrosema virginianum	Butterfly pea
Cephalanthus occidentalis	Button bush
Ceratophyllum spp.	Coontails/Hornworts
Ceratophyllum demersum	Coontail/Hornwort
Ceratophyllum echinatum	Coontail/Hornwort
Cercis canadensis	Redbud
Chaptalia tomentosa	Sun-bonnets
Chara spp.	Muskgrasses
Chenopodium album	Lamb's quarters
Chloris petraea	Finger grass
Chondrophora nudata	Rayless goldenrod
Chrysobalanus oblongifolius	Gopher apples
Cicuta maculata	Water hemlock
Cinna arundinacea	Wood reed
Citrus aurantium	Sour orange
Citrus paradisi	Grapefruit
Cladium jamaicense	Saw grass
Cladium mariscoides	Twig-rush
Cladonia spp.	Reindeer lichen
Cleistes divaricata	Spreading pogonia/Rosebud orchid
Clematis crispa	Leather-flower
Cleome aldenella	--
Clethra alnifolia	Sweet pepperbush
Cliftonia monophylla	Black titi/Buckwheat tree
Clitoria mariana	Butterfly pea
Cnidoscolus stimulosus	Stinging needle
Collinsonia canadensis	Horse balm
Colocasia esculentum	Dasheen
Commelina erecta	Dayflower
Commelina virginica	Dayflower
Conopholis americana	Squaw-root/Cancer root
Corallorhiza odontorhiza	Autumn coral-root
Corallorhiza wisteriana	Spring coral-root
Coreopsis spp.	Coreopsis
Coreopsis falcata	Coreopsis
Coreopsis grandiflora	--
Coreopsis tinctoria	Calliopsis
Cornus florida	Flowering dogwood
Cornus racemosa	Dogwood
Cornus stricta	Swamp dogwood
Crataegus sp.	Hawthorn
Crinum americanum	Swamp lily
Crotalaria intermedia	Rattlebox
Crotalaria lanceolata	Rattlebox
Crotalaria retusa	Rattlebox
Croton punctatus	Croton/Beach hogwort
Crotonopsis linearis	Narrow-leaved rushfoil
Ctenium spp.	Toothache grasses
Ctenium aromaticum	Toothache grass

Cuscuta sp.	Dodder
Cuscuta cephalanthii	Dodder
Cuscuta indecora	Dodder
Cynanchum laeve	--
Cynanchum palustre	Milk-vine
Cynoctonum sessilifolium	Miterwort
Cynodon dactylon	Bermuda grass
Cyperus spp.	Sedges
Cyperus brevifolius	Sedge
Cyperus erythrorhizos	Redrooted nut grass
Cyperus esculentus	Nut grass
Cyperus esculentus var. *sativus*	Chufa
Cyperus odoratus	Sedge
Cyperus polystachos	Sedge
Cyperus rivularis	Sedge
Cyperus rotundus	Sedge
Cyperus strigosus	Sedge
Cyrilla racemiflora	Titi
Decodon verticillatus	Water loosestrife
Descurainia pinnata	Tansey-mustard
Desmodium sp.	Tick trefoil
Desmodium spp.	Beggar's ticks/Beggar lice/Beggar weeds
Desmodium marilandicum	Beggar lice
Dicerandra odoratissima	Rose dicerandra
Dichromena colorata	Star-rush
Dichromena latifolia	White-bracted sedge
Digitaria spp.	Crab grasses
Digitaria horizontalis	Finger grass
Diodia spp.	--
Diodia teres	Diodia/Poor-Joe
Dioda virginiana	--
Dionaea muscipula	Venus' fly trap
Diospyros virginiana	Persimmon
Distichlis spicata	Salt grass
Drosera spp.	Sundews
Drosera rotundifolia	Common sundew
Dryopteris dentata	Wood fern
Dulichium arundinaceum	Three-way sedge
Echinochloa spp.	Millets
Echinochloa crusgalli	Wild millet
Echinochloa polystachya	--
Echinochloa walteri	Salt marsh millet
Echinodorus cordifolius	Creeping water plantain
Echinodorus parvulus	Little burhead
Egeria densa	Water-weed/Brazilian elodea
Eichhornia crassipes	Water hyacinth/Wampee
Eleocharis sp.	Spikerush/Creeping spikerush
Eleocharis acicularis	Slender spikerush/Spikerush
Eleocharis albida	Spikerush
Eleocharis baldwinii	Proliferating spikerush
Eleocharis elongata	Spikerush/Water spikerush
Eleocharis equisetoides	Jointed spikerush
Eleocharis parvula	Dwarf spikerush
Eleocharis quadrangulata	Square-stem spikerush
Eleocharis robbinsii	Spikerush/Trianglestem spikerush
Elephantopus tomentosus	Elephant's-foot
Elliottia racemosa	Georgia plume
Elodea spp.	Water-weeds
Elodea canadensis	Water-weed
Elymus villosus	Wild rye/Wild rye grass
Elymus virginicus	Wild rye grass
Eragrostis capillaris	Lacegrass/Love grass
Eragrostis pilosa	Love grass
Eragrostis refracta	Coastal love grass/Love grass
Eremochloa ophiuroides	Centipede grass
Erianthus spp.	Plume grasses
Erianthus giganteus	Plume grass/Giant plume grass
Erianthus ravennae	Ravenna-grass
Erigeron canadensis	Horseweed
Erigeron strigosus	Daisy fleabane

Erigeron vernus	Fleabane
Eriocaulon compressum	Pipewort
Eriocaulon decangulare	Pipewort
Eriogonum tomentosum	Dog-tongue
Eryngium aquaticum	Marsh eryngo
Eryngium aquaticum var. ravenelii	Ravenel's button snakeroot
Eryngium integrifolium	Eryngo
Euonymus americanus	Strawberry bush
Eupatorium spp.	Thoroughworts
Eupatorium album	Thoroughwort
Eupatorium capillifolium	Dog fennel
Eupatorium capillifolium var. leptophyllum	Dog fennel
Euphorbia marginata	Snow-on-the-mountain
Euphorbia polygonifolia	Euphorbia
Euphorbia serpens	Creeping spurge
Fagus grandifolia	Beech/American beech
Festuca myuros	Sixweeks fescue/Fescue
Festuca octoflora	Fescue
Festuca rubra	Red fescue/Fescue
Ficus carica	Fig
Ficus pumila	Creeping fig
Filago germanica	Cotton rose
Fimbristylis harperi	--
Fimbristylis spadicea	Salt marsh fimbristylis
Forestiera acuminata	Swamp privet
Forestiera porulosa	Florida privet
Fothergilla gardenii	Dwarf witch alder/Witch alder
Fraxinus americana	White ash
Fraxinus caroliniana	Water ash
Fraxinus pennsylvanica	Red ash/Green ash
Fraxinus tomentosa	Pumpkin ash
Froelichia floridana	Cottonweed
Gaillardia drummondii	Gaillardia
Galactia elliottii	Milk pea
Galium circaezans	Wild licorice/Bedstraw
Gaura biennis	--
Gaylussacia spp.	Huckleberries
Gaylussacia dumosa	Dwarf huckleberry
Gaylussacia frondosa	Dangleberry
Gentiana spp.	Gentians
Gladiolus hortulana	Gladiolus
Gleditsia aquatica	Water locust
Glycine max	Soybean
Gnaphalium spp.	Cudweeds/Everlastings
Gnaphalium obtusifolium	Rabbit tobacco/Everlasting
Gnaphalium purpureum	Cudweed
Goodyera pubescens	Downy rattlesnake plantain
Gordonia lasianthus	Loblolly bay
Gratiola pilosa	Hedge hyssop
Gymnopogon brevifolius	Beard grass
Habenaria spp.	Orchids
Habenaria blephariglottis var. integrilabia	White-fringed orchid
Habenaria flava	Southern rein orchid
Habenaria integra	Yellow fringeless orchid
Habenaria lacera	Green fringed orchid/Green fringeless orchid
Habenaria repens	Water spider orchid
Halodule beaudettei	--
Hamamelis virginiana	Witch-hazel
Haplopappus divaricatus	--
Hartwrightia floridana	Hartwrightia
Helenium vernale	Sneeze-weed
Helianthus angustifolius	Sunflower
Helianthus tuberosus	Sunflower/Jerusalem artichoke
Heterotheca spp.	--
Heterotheca floridana	Golden aster
Heterotheca graminifolia	Camphor weed/Golden aster
Heterotheca nervosa	
Heterotheca subaxillaris	Heterotheca/Camphor weed

Hibiscus lasiocarpus	--
Hibiscus militaris	Halberd-leaved marsh mallow
Hibiscus moscheutos	Rose mallow
Hordeum spp.	Barley
Hydrochloa caroliniensis	Water grass
Hydrocotyle spp.	Pennyworts
Hydrocotyle bonariensis	Beach pennywort
Hydrocotyle ranunculoides	Pennywort/Buttercup-leaved pennywort
Hydrocotyle umbellata	Marsh pennywort
Hymenocallis crassifolia	Spider-lily
Hypericum spp.	St. John's-worts
Hypericum apocynifolium	St. John's-wort
Hypericum fasciculatum	St. John's-wort
Hypericum gentianoides	Pineweed
Hypericum pseudomaculatum	St. John's-wort
Hypericum tubulosum	St. John's-wort
Hypericum virginicum	St. John's-wort
Hypericum walteri	St. John's-wort
Hypoxis sessilis	Star grass/Yellow star grass
Ilex spp.	Hollies
Ilex amelanchier	Sarvis holly
Ilex cassine	Dahoon
Ilex cassine var. myrtifolia	Dahoon/Myrtle holly
Ilex coriacea	Sweet gallberry
Ilex decidua	Possum haw/Deciduous holly/Swamp holly
Ilex glabra	Bitter gallberry
Ilex laevigata	Smooth winterberry
Ilex opaca	American holly
Ilex vomitoria	Yaupon holly
Impatiens capensis	Jewel-weed
Ipomoea macrorhiza	Big-rooted manroot/Large-rooted morning glory
Ipomoea pes-caprae	Railroad vine
Iris tridentata	Flag/Iris
Iris virginica	Blue flag
Itea virginica	Virginia willow
Iva annua	--
Iva frutescens	Marsh elder
Iva imbricata	Beach elder/Sea elder
Juncus spp.	Rushés
Juncus biflorus	Rush
Juncus bufonius	Toad rush
Juncus effusus	Soft rush
Juncus nodatus	Rush
Juncus repens	Creeping rush
Juncus roemerianus	Black needlerush/Black rush
Juncus secundus	Rush
Juncus subcaudatus	Rush
Juniperus silicicola	Southern red cedar
Juniperus virginiana	Eastern red cedar/Red cedar
Justicia ovata	--
Kalmia angustifolia var. carolina	Lambkill
Kalmia cuneata	White wicky
Kalmia hirsuta	Dwarf laurel
Koeleria phleoides	June grass
Kosteletskya virginica	Seashore mallow
Lachnanthes caroliniana	Redroot
Lachnocaulon beyrichianum	Bog buttons/Southern bog buttons
Lantana montevidensis	Trailing lantana
Lechea spp.	Pin-weeds
Leersia hexandra	Rice cutgrass/Cutgrass
Leersia lenticularis	Cut grass
Leersia oryzoides	Rice cutgrass/Cutgrass
Leersia virginica	Cut grass
Lemna spp.	Duckweeds
Lemna minor	Duckweed
Lemna perpusilla	Duckweed
Lemna valdiviana	Duckweed

Lepidium virginicum	Poor-mans pepper
Leptóchloa sp.	Sprangletop
Leptochloa uninervia	Springletop
Leptoloma congnatum	Witch grass
Lepuropetalon spathulatum	Southern lepuropetalon
Lespedeza sp.	Common lespedezas
Lespedeza bicolor	Lespedezas
Lespedeza cuneata	Sericea
Lespedeza striata	Japanese clover
Leucothoe axillaris	Leucothoe
Leucothoe populifolia	Leucothoe/Carolina dog-hobble/Poplar-leaved fetterbush
Leucothoe racemosa	Fetter-bush
Liatris spp.	Blazing stars
Liatris tenuifolia	Blazing star
Ligustrum japonicum	Privet
Lilaeopsis chinensis	Eastern lilaeopsis
Limnobium spongia	Frog's-bit
Limonium spp.	Sea lavenders
Limonium carolinianum	Sea lavender
Limonium nashii	Sea lavender
Linaria canadensis	Toad-flax
Lindera melissaefolium	Jove's fruit/Southern spicebush
Lindernia crustacea	--
Lippia nodiflora	Cape-weed/Lippia
Liquidambar styraciflua	Sweet gum
Liriodendron tulipifera	Tulip tree/Yellow poplar
Litsea aestivalis	Pond spice
Lobelia boykinii	Boykin's lobelia
Lobelia cardinalis	Cardinal flower
Lobelia elongata	Lobelia
Lobelia floridana	--
Lolium sp.	Rye grass
Lolium multiflorum	Italian rye grass
Lonicera sempervirens	Coral honeysuckle
Ludwigia spp.	Water-primroses
Ludwigia bonariensis	--
Ludwigia natans	Water purslane/Marsh purslane
Ludwigia peploides	Water-primrose
Ludwigia peploides var. glabrescens	Big primrose willow
Ludwigia pilosa	Seed box
Lupinus perennis	Lupine
Lycopodium spp.	Clubmoss
Lycopodium alopecuroides	Foxtail clubmoss
Lycopus americanus	Bugleweed
Lycopus sessilifolius	Bugleweed
Lycopus virginicus	Water-hoarhound
Lygodium palmatum	American climbing fern
Lyonia ferruginea	Stagger bush/Rusty lyonia
Lyonia ligustrina	Male-berry
Lyonia lucida	Fetter-bush
Lyónia mariana	Stagger-bush
Lysimachia lanceolata	Fringed loosestrife
Lythrum lineare	Loosestrife
Magnolia grandiflora	Bull bay/Southern magnolia
Magnolia macrophylla	Umbrella tree
Magnolia virginiana	Sweet bay
Malaxis spicata	Florida adder's mouth
Mariscus jamaicense	Saw grass
Marshallia graminifolia	Barbara's buttons
Matelea gonocarpa	--
Melia azedarach	China-berry/Umbrella tree
Melochia corchorifolia	Chocolate-weed
Melothria crassifolia	Melonette
Melothria pendula	Creeping cucumber
Mikania scandens	Climbing hempweed
Mitchella repens	Partridge berry
Monotropsis odorata	Pigmy-pipes
Morus spp.	Mulberries

500

Morus alba	White mulberry
Muhlenbergia capillaris	Muhly grass
Muhlenbergia expansa	Muhly grass
Muhlenbergia filipes	Sweet grass
Myrica spp.	Myrtles
Myrica cerifera	Wax myrtle
Myrica inodora	Odorless wax myrtle
Myriophyllum spp.	Water milfoils
Myriophyllum brasiliense	Parrot-feather
Myriophyllum heterophyllum	Water milfoil
Myriophyllum laxum	Water milfoil/Floppy water milfoil/Loose watermilfoil
Myriophyllum pinnatum	Water milfoil
Najas spp.	Bushy pondweeds
Najas guadalupensis	Southern naiad/Bushy pondweed/Water nymph
Najas minor	Bushy pondweed
Narthecium americanum	Yellow asphodel
Nelumbo lutea	Lotus/Yellow nelumbo
Nelumbo pentapetela	Lotus
Nitella sp.	Nitella
Nuphar advena	Spatter-dock
Nuphar luteum	Yellow pond-lily
Nymphaea bombycina	Water-lily
Nymphaea mexicana	Banana water-lily/Water-lily
Nymphaea odorata	White water-lily/Water-lily
Nymphoides spp.	Floating hearts
Nymphoides aquatica	Floating heart/Big floating heart
Nymphoides cordata	Little floating heart
Nyssa aquatica	Water tupelo
Nyssa ogeche	Ogeechee plum
Nyssa sylvatica	Black gum
Nyssa sylvatica var. biflora	Swamp tupelo/Gum
Oenothera spp.	Evening primroses
Oenothera humifusa	Evening primrose
Olea europaea	Olive
Onoclea sensibilis	Sensitive fern
Ophioglossum nudicaule	Least adder's tongue
Ophioglossum vulgatum var. pycnostichum	Southern adder's tongue
Opuntia compressa	Cactus
Opuntia drummondii	Drummond's prickly pear/Prickly pear
Opuntia ficus-indica	Inidan fig
Orobanché uniflora	Single-flowered cancer root/Cancer root
Orontium aquaticum	Golden club
Osmanthus americana	Wild olive
Osmunda cinnamomea	Cinnamon fern
Osmunda regalis	Royal fern
Ostrya virginiana	Hop hornbeam
Oxydendrum arboreum	Sourwood
Oxypolis rigidior	Dropwort
Panicum agrostoides	--
Panicum spp.	Panic grasses
Panicum amarulum	Beach grass
Panicum amarum	Panic grass/Sea beach panic grass
Panicum bisulcatum	Asiatic panicum/Asiatic panic grass
Panicum capillare	Witch grass
Panicum dichotomiflorum	Fall panic grass
Panicum hemitomon	Maidencane
Panicum leucothrix	Panic grass
Panicum malacon	--
Panicum neuranthum	--
Panicum ovale	--
Panicum ramosum	Brown-top millet
Panicum virgatum	Switchgrass/Panic grass
Parnassia caroliniana	Carolina grass-of-parnassus
Paronychia fastigiata	--
Paronychia herniarioides	--
Parthenocissus quinquefolia	Virginia creeper

501

Paspalum sp.	Paspalum
Paspalum boscianum	Bullgrass
Paspalum dissectum	Bullgrass
Paspalum distichum	Bullgrass
Paspalum setaceum	Fringe-leaved paspalum
Paspalum vaginatum	Seashore paspalum
Passiflora lutea	Passion-flower
Peltandra sagittaefolia	White arrow-arum
Peltandra virginica	Arrow-arum
Pennisetum glaucum	Millet
Persea borbonia	Red bay
Petalostemum pinnatum	Summer-farewell
Petrocaulon pycnostachyum	Black-root
Petunia axillaris	Petunia
Phlox glaberrima	--
Phragmites communis	Common reed/Reed
Physalis pubescens var. grisea	Ground cherry
Physalis virginiana	Ground cherry
Physalis viscosa spp. maritima	Ground cherry
Phytolacca rigida	Poke weed
Pieris phillyreifolia	Climbing fetterbush
Pilea microphylla	--
Pilea pumila	Clearweed
Pinckneya pubens	Georgia fever bark
Pinguicula spp.	Butterworts
Pinguicula lutea	Butterwort
Pinus clausa	Sand pine
Pinus echinata	Short-leaf pine
Pinus elliottii	Slash pine
Pinus glabra	Spruce pine
Pinus palustris	Longleaf pine
Pinus serotina	Pond pine
Pinus taeda	Loblolly pine
Planera aquatica	Water elm
Plantago lanceolata	English plantain
Platanus occidentalis	Sycamore
Pluchea spp.	Marsh fleabanes
Pluchea purpurascens	Marsh fleabane/Camphorweed
Pluchea rosea	Marsh fleabane
Poa compressa	Canada bluegrass/Bluegrass
Pogonia ophioglossoides	Rose pogonia
Polygala spp.	Polygalas
Polygala cymosa	Polygala
Polygala grandiflora	Milkwort
Polygala lutea	Polygala/Red-hot poker
Polygala nana	Low millewort
Polygala sanguinea	Polygala
Polygonum spp.	Smartweeds/Knotweeds
Polygonum arifolium	Tearthumb
Polygonum densiflorum	Southern smartweed
Polygonum hirsutum	Smartweed
Polygonum hydropiperoides	Swamp smartweed
Polygonum lapathifolium	Nodding smartweed/Knotweed
Polygonum longistylum	Long-styled smartweed
Polygonum orientale	Princess-feather
Polygonum pensylvanicum	Large-seed smartweed
Polygonum persicaria	Knotweed
Polygonum portoricense	Southern smartweed
Polygonum punctatum	Dotted smartweed
Polygonum sagittatum	Tearthumb
Polygonum setaceum	Swamp smartweed
Polypodium aureum	Fern
Polypodium polypodioides	Resurrection fern
Polypremum procumbens	Polypremum
Pontederia cordata	Pickerelweed
Populus spp.	Poplars
Populus deltoides	Cottonwood/Eastern cottonwood
Populus heterophylla	Swamp cottonwood
Potamogeton spp.	Pondweeds
Potamogeton berchtoldii	Narrow-leaved pondweed/Pondweed/Small pondweed
Potamogeton diversifolius	Variable-leaved pondweed

Potamogeton foliosus	Leafy pondweed/Pondweed
Potamogeton illinoensis	Pondweed/Fishweed
Potamogeton nodosus	Pondweed
Potamogeton pectinatus	Sago pondweed/Pondweed
Potentilla norvegica	Cinquefoil
Prunus americana	Wild plum
Prunus caroliniana	Carolina cherry laurel/Cherry laurel
Prunus serotina	Black cherry
Psilocarya scirpoides	Bald rush
Pteridium aquilinum	Bracken fern
Pterocaulon pycnostachyum	Black-root
Ptilimnium capillaceum	Mock-bishopweed
Ptilimnium costatum	Mock-bishopweed
Quercus spp.	Scrub oaks/Shrub oaks
Quercus alba	White oak
Quercus chapmanii	Chapman oak
Quercus coccinea	Scarlet oak
Quercus falcata	Southern red oak
Quercus falcata var. pagodaefloia	Cherrybark oak
Quercus incana	Bluejack oak
Quercus laevis	Turkey oak
Quercus laurifolia	Laurel oak
Quercus lyrata	Overcup oak
Quercus margaretta	Shrubby post oak
Quercus marilandica	Black jack oak
Quercus michauxii	Swamp chestnut oak
Quercus muehlenbergii	Chinquapin oak/Yellow chestnut oak
Quercus myrtifolia	Myrtle oak
Quercus nigra	Water oak
Quercus phellos	Willow oak
Quercus pumila	Running oak
Quercus rubra	Red oak
Quercus shumardii	Shumard oak
Quercus stellata	Post oak
Quercus velutina	Black oak
Quercus virginiana	Live oak
Reimarochloa oligostachya	Carpet grass
Rhapidophyllum hystrix	Needle palm
Rhexia spp.	Meadow beauties
Rhexia alifanus	Meadow beauty
Rhexia aristosa	Sun-petaled meadow beauty
Rhexia cubensis	Meadow beauty
Rhizophora mangle	Mangrove
Rhododendron canescens	Wild azalea
Rhus spp.	Sumac
Rhus radicans	Poison ivy
Rhynchospora spp.	Beak rushes
Rhynchospora careyana	Beak rush
Rhynchospora corniculata	Beak rush
Rhynchospora decurrens	Beak rush
Rhynchospora glomerata	Beak rush
Rhynchospora macrostachya	Beak rush
Rhynchospora megalocarpa	Beak rush/Rusty lyonia/Giant-seeded beak rush
Rhynchospora plumosa	Beak rush
Ricinus communis	Castor oil plant/Castor-bean
Rorippa islandica	Marsh cress/Yellow cress
Rorippa sessiliflora	Marsh cress/Yellow cress
Rosa bracteata	Macartney rose
Rosa damascena	Damask rose
Rosa multiflora	Rambler rose
Rosa palustris	Swamp rose
Rosa wichuraiana	Memorial rose
Rubus spp.	Blackberries/Raspberries/Dewberries
Rudbeckia mollis	Soft-haired cornflower
Rumex acetosella	Sour-grass
Rumex bucephalophorus	Dock
Rumex sanguineus	Red-veined dock
Rumex verticillatus	Swamp dock
Ruppia maritima	Widgeon grass

Sabal minor	Dwarf palmetto
Sabal palmetto	Cabbage palmetto
Sabatia brachiata	Sabatia
Sabatia dodecandra	Sea pink/Sabatia
Sabatia foliosa	Marsh-pink/Sea pink
Sabatia stellaris	Marsh-gentian/Sabatia
Sacciolepis striata	Sacciolepis/Baggy-knees
Sageretia minutiflora	Sageretia/Small-flowered buckthorn
Sagittaria spp.	Arrowheads
Sagittaria graminea	Delta duck potato
Sagittaria graminea var. weatherbiana	Arrowhead
Sagittaria latifolia	Duck potato
Salicornia spp.	Glassworts
Salicornia bigelovii	Glasswort
Salicornia europaea	Glasswort
Salicornia virginica	Glasswort/Perennial glasswort
Salix spp.	Willows
Salix babylonica	Weeping willow
Salix caroliniana	Willow/Swamp willow
Salix nigra	Black willow
Salsola kali	Russian thistle
Salvinia rotundifolia	Salvinia
Sambucus canadensis	Elderberry
Sambucus simpsonii	Elderberry/Southern elderberry
Samolus parviflorus	Water pimpernel
Sapindus marginatus	Soapberry
Sapium sebiferum	Popcorn tree
Sarracenia spp.	Pitcher-plants
Sarracenia flava	Yellow pitcher-plant/Trumpet-plant/Watches/ Biscuit-flower/Fly-catcher
Sarracenia minor	Hooded pitcher-plant
Sarracenia psittacina	Parrot pitcher-plant
Sarracenia purpurea	Flytrap pitcher-plant
Sarracenia rubra	Sweet pitcher-plant
Sassafras albidum	Sassafras
Satureja calamintha	Red basil
Saururus cernuus	Lizard's tail
Schisandra glabra	Bay starvine
Schwalbea americana	Chaff-seed
Scirpus spp.	Bulrushes
Scirpus americanus	Bulrush/American three-square bulrush/ Common three-square
Scirpus californicus	Southern bulrush
Scirpus cyperinus	Wool grass bulrush
Scirpus erismanae	Georgia bulrush
Scirpus etuberculatus	Bulrush
Scirpus olneyi	Olney's threesquare bulrush
Scirpus robustus	Bulrush/Salt-marsh bulrush
Scirpus subterminalis	Swaying bullrush
Scirpus validus	Bulrush/Soft-stem bulrush
Scleranthus annuus	Knawel
Scleria baldwinii	Baldwin's nutrush/Nut rush
Scleria ciliata	Nut rush
Scleria trigolmerata	Nut rush
Scutellaria sp.	Skullcap
Sebastiania ligustrina	Sebastian bush
Secale cereale	Rye
Selaginella arenicola	Spikemoss
Senecio sp.	Butterweed
Serenoa repens	Saw palmetto
Sesbania exaltata	Coffee-weed/Sesbania
Sesbania macrocarpa	Seban
Sesuvium spp.	Sea purslanes
Sesuvium maritimum	Sea purslane
Sesuvium portulacastrum	Sea purslane
Setaria geniculata	Foxtail grass
Setaria macrosperma	Foxtail grass
Setaria magna	Foxtail grass/Giant foxtail grass
Sida spinosa	Prickly mallow
Silene antirrhina	Sleepy catchfly

Sisyrinchium mucronatum	Blue-eyed grass
Sium suave	Water parsnip
Smilax spp.	Catbriers/Greenbriers
Smilax auriculata	Bamboo brier/Greenbrier/Catbrier
Smilax bona-nox	Catbrier/Greenbrier
Smilax laurifolia	Laurel greenbrier/Bamboo
Smilax rotundifolia	Greenbrier
Smilax smallii	Greenbrier
Solanum aculeatissimum	Nightshade
Solidago spp.	Goldenrods
Solidago chapmanii	Goldenrod
Solidago gymnospermoides	Goldenrod
Solidago sempervirens	Seaside goldenrod/Goldenrod
Solidago verna	Spring-flowered goldenrod
Sorbus arbutifolia var. arbutifolia	Red chokeberry
Sorghastrum nutans	Indian grass/Woodgrass
Sorghastrum secundum	Indian grass
Sorghum halepense	Johnson grass
Sorghum vulgare	Milo/Grain sorghum
Spartina spp.	Cordgrasses
Spartina alterniflora	Smooth cordgrass/Cordgrass
Spartina cynosuroides	Cordgrass/Giant cordgrass
Spartina patens	Saltmeadow cordgrass/Cordgrass
Specularia spp.	Venus' looking-glass
Specularia perfoliata	Venus' looking-glass
Spergularia marina	Sand spurrey
Spermacoce glabra	Buttonweed
Spermolepsis echinata	Bristle-fruited spermolepsis
Sphagnum spp.	Peat mosses/Sphagnum mosses
Sphenopholis intermedia	Wedge grass
Spiranthes laciniata	Lace-lip spiral orchid/Lace-lip ladies tresses
Spiranthes longilabris	Giant spiral-orchid
Spiranthes praecox	Grass-leaved ladies' tresses
Spirodela spp.	Duckweeds
Spirodela polyrrhiza	Big duckweed
Sporobolus curtissii	Curtiss' dropseed
Sporobolus floridanus	Florida dropseed
Sporobolus teretifolius	Dropseed/Wire-leaved dropseed
Sporobolus virginicus	Dropseed
Stillingia sylvatica	Stillingia/Queen's delight
Stipulicida setacea	--
Strophostyles helvola	Beach pea
Styrax americana	Storax
Styrax grandifolia	Storax
Suaeda linearis	Sea-blite
Symplocos tinctoria	Horse sugar/Sweet leaf
Taxodium spp.	Cypresses
Taxodium ascendens	Pond cypress
Taxodium distichum	Bald cypress
Tephrosia florida	--
Tephrosia virginiana	Goat's rue
Thalassia testudinum	Turtle grass
Tillandsia usneoides	Spanish moss
Tipularia discolor	Crane-fly orchid
Tofieldia glabra	False asphodel
Tofieldia racemosa	False asphodel/Sticky tofieldia
Trachelospermum difforme	False asphodel
Tragia urens	--
Tridens flavus var. flavus	Purple top
Trifolium spp.	Clovers
Trilisa odoratissima	Vanilla-plant
Trilisa paniculata	--
Trillium pusillum	Trillium
Trillium pusillum var. pusillum	Dwarf trillium/Carolina trillium
Triphora trianthophora	Three-birds orchid
Triplasis purpurea	Sand grass
Tripsacum dactyloides	Gamma grass
Triticum aestivum	Wheat
Typha spp.	Cat-tails

Typha angustifolia	Narrow-leaved cat-tail
Typha domingensis	Southern cat-tail/Cat-tail
Typha glauca	Blue cat-tail/Cat-tail
Typha latifolia	Common cat-tail
Ulmus alata	Winged elm
Ulmus americana	American elm
Ulmus rubra	Slippery elm
Uniola latifolia	Spike-grass
Uniola laxa	Spike-grass
Uniola paniculata	Sea oats
Uniola sessiliflora	Spike grass
Utricularia spp.	Bladderworts
Utricularia cornuta	Horned bladderwort/Bladderwort
Utricularia floridana	Florida bladderwort
Utricularia inflata	Bladderwort
Utricularia inflata var. minor	Bladderwort
Utricularia olivacea	Bladderwort
Utricularia purpurea	Bladderwort/Purple bladderwort
Utricularia subulata	Bladderwort
Utricularia vulgaris	Common bladderwort/Bladderwort
Vaccinium spp.	Blueberries/Cranberries/Gooseberries/ Huckleberries/Sparkleberries
Vaccinium arboreum	Sparkleberry
Vaccinium caesariense	Blueberry
Vaccinium corymbosum	Highbush blueberry
Vaccinium elliottii	Elliot's blueberry
Vaccinium myrsinites	Dwarf blueberry/Blueberry
Vaccinium stamineum	Squaw-huckleberry
Verbena officinalis	Vervain
Verbesina occidentalis	Crownbeard
Vernonia sp.	Ironweed
Vernonia altissima	Ironweed
Vernonia blodgettii	Ironweed
Vernónia harperi	Ironweed
Veronica persica	Bird's eye
Veronica serpyllifolia	Thyme-leave speedwell
Viburnum spp.	Haws
Viburnum dentatum	Arrowwood
Viburnum nudum	Possum haw
Vigna unguiculata	Cowpea
Vinca major	Periwinkle
Viola cucullata	Violet
Viola papilionacea	Violet
Viola triloba	Violet
Vitis spp.	Wild grapes
Vitis aestivalis	Summer grape
Vitis aestivalis var. aestivalis	Summer grape
Vitis cinerea var. floridana	Pigeon grape
Vitis rotundifolia	Muscadine grape
Wahlenbergia gracilis	--
Wolffia columbiana	Water-meal
Wolffiella floridana	Eastern wolffiella
Woodwardia areolata	Netted chain fern
Woodwardia virginica	Virginia chain fern
Xyris spp.	Yellow-eyed grasses
Xyris caroliniana	Yellow-eyed grass
Xyris elliottii	Yellow-eyed grass
Xyris smalliana	Yellow-eyed grass
Yucca aloifolia	Spanish bayonet
Yucca gloriosa	Moundlily yucca
Zannichellia palustris	Horned-pondweed
Zanthoxylum clava-herculis	Hercules' club
Zea mays	Corn
Zenobia pulverulenta	Zenobia
Zephyranthes candida	--
Zephyranthes simpsonii	Rain lily
Zigadenus densus	Crow-poison
Zizania aquatica	Wild rice
Zizaniopsis miliacea	Southern wild rice/Giant cutgrass
Zostera marina	Eel grass

Alewife	Alosa pseudoharengus
Almaco jack	Seriola rivoliana
American eel	Anguilla rostrata
American shad	Alosa sapidissima
Angelfishes	Holacanthus spp.
Atlantic bonito	Sarda sarda
Atlantic bumper	Chloroscombrus chrysurus
Atlantic croaker	Micropogonias undulatus
Atlantic cutlassfish	Trichiurus lepturus
Atlantic guitarfish	Rhinobatos lentiginosus
Atlantic mackerel	Scomber scombrus
Atlantic menhaden	Brevoortia tyrannus
Atlantic midshipman	Porichthys porosissimus
Atlantic moonfish	Vomer setapinnis
Atlantic needlefish	Strongylura marina
Atlantic sharpnose shark	Rhizoprionodon terraenovae
Atlantic silverside	Menidia menidia
Atlantic spadefish	Chaetodipterus faber
Atlantic stingray	Dasyatis sabina
Atlantic sturgeon	Acipenser oxyrhynchus
Atlantic thread herring	Opisthonema oglinum
Balao	Hemiramphus balao
Banded darter	Etheostoma zonale
Banded drum	Larimus fasciatus
Banded pygmy sunfish	Elassoma zonatum
Banded sunfish	Enneacanthus obesus
Banded topminnow	Fundulus cingulatus
Bandtail puffer	Sphoeroides spengleri
Barracudas	Sphyraena spp.
Bay anchovy	Anchoa mitchilli
Bay whiff	Citharichthys spilopterus
Bighead searobin	Prionotus tribulus
Blackbanded darter	Percina nigrofasciata
Blackbanded sunfish	Enneacanthus chaetodon
Black bullhead	Ictalurus melas
Blackcheek tonguefish	Symphurus plagiusa
Black crappie	Pomoxis nigromaculatus
Black drum	Pogonias cromis
Blackfin snapper	Lutjanus buccanella
Blackline tilefish	Caulolatilus cyanops
Black jumprock	Moxostoma cervinum
Black madtom	Noturus funebris
Black sea bass	Centropristis striata
Blacktip shark	Carcharhinus limbatus
Blackwing searobin	Prionotus salmonicolor
Blennies	Chasmodes spp.
Blueback herring	Alosa aestivalis
Blue catfish	Ictalurus furcatus
Bluefin killifish	Lucania goodei
Bluefish	Pomatomus saltatrix
Bluegill	Lepomis macrochirus
Bluehead chub	Nocomis leptocephalus
Blue marlin	Makaira nigricans
Bluespotted sunfish	Enneacanthus gloriosus
Bluntnose stingray	Dasyatis sayi
Bonnethead	Sphyrna tiburo
Bowfin	Amia calva
Broad flounder	Paralichthys squamilentus
Brook silverside	Labidesthes sicculus
Brown bullhead	Ictalurus nebulosus
Burrfishes	Chilomycterus spp.
Butterfish	Peprilus triacanthus
Butterfishes	Peprilus spp.
Carolina hake	Urophycis earlli
Carp	Cyprinus carpio
Chain pickerel	Esox niger

507

Chain pipefish	Syngnathus louisianae
Channel catfish	Ictalurus punctatus
Christmas darter	Etheostoma hopkinsi
Clearnose skate	Raja eglanteria
Clown goby	Microgobius gulosus
Coastal shiner	Notropis petersoni
Cobia	Rachycentron canadum
Combtooth blennies	Hyposoblennius spp.
Conger eel	Conger oceanicus
Cownose ray	Rhinoptera bonasus
Crappies	Pomoxis spp.
Creek chubsucker	Erimyzon oblongus
Crested blenny	Hypleurochilus geminatus
Crevalle jack	Caranx hippos
Cypress minnow	Hybognathus hayi
Darter goby	Gobionellus boleosoma
Darters	Etheostoma spp.
Darters	Percina spp.
Dollar sunfish	Lepomis marginatus
Dolphins	Coryphaena spp.
Dusky anchovy	Anchoa lyolepis
Dusky pipefish	Syngnathus floridae
Dusky shark	Carcharhinus obscurus
Dusky shiner	Notropis cummingsae
Eastern mudminnow	Umbra pygmaea
Everglades pygmy sunfish	Elassoma evergladei
Fathead minnow	Pimephales promelas
Fat sleeper	Dormitator maculatus
Feather blenny	Hypsoblennius hentzi
Finetooth shark	Aprionodon isodon
Flat bullhead	Ictalurus platycephalus
Flathead catfish	Pylodictis olivaris
Flier	Centrarchus macropterus
Florida blenny	Chasmodes saburrae
Florida gar	Lepisosteus platyrhincus
Florida pompano	Trachinotus carolinus
Flounders	Paralichthys spp.
Freshwater drum	Aplodinotus grunniens
Freshwater goby	Gobionellus shufeldti
Fringed flounder	Etropus crossotus
Gafftopsail catfish	Bagre marinus
Gag grouper	Mycteroperca microlepis
Gars	Lepisosteus spp.
Gizzard shad	Dorosoma cepedianum
Glassy darter	Etheostoma vitreum
Gobies	Gobionellus spp.
Gobies	Gobiosoma spp.
Golden shiner	Notemigonus crysoleucas
Golden topminnow	Fundulus chrysotus
Goldfish	Carassius auratus
Gold tilefish	Lopholatilus chamaeleonticeps
Gray snapper	Lutjanus griseus
Great barracuda	Sphyraena barracuda
Greater amberjack	Seriola dumerili
Green goby	Microgobius thalassinus
Greenhead shiner	Notropis chlorocephalus
Green sunfish	Lepomis cyanellus
Groupers	Mycteroperca spp.
Guaguanche	Sphyraena guachancho
Gulf flounder	Paralichthys albigutta
Gulf kingfish	Menticirrhus littoralis
Hakes	Urophycis spp.
Halfbeak	Hyporhamphus unifasciatus
Hammerhead sharks	Sphyrna spp.
Harvestfish	Peprilus alepidotus
Herrings	Alosa spp.
Hickory shad	Alosa mediocris
Highfin carpsucker	Carpiodes velifer
Highfin goby	Gobionellus oceanicus

508

Highfin shiner	Notropis altipinnis
Hogchoker	Trinectes maculatus
Horse-eye jack	Caranx latus
Inshore lizardfish	Synodus foetens
Irish pompano	Diapterus olisthostomus
Ironcolor shiner	Notropis chalybaeus
Jacks	Caranx spp.
Killifishes	Fundulus spp.
Kingfishes	Menticirrhus spp.
King mackerel	Scomberomorus cavalla
King whitings	Menticirrhus spp.
Ladyfish	Elops saurus
Lake chubsucker	Erimyzon sucetta
Lancer stargazer	Kathetostoma albigutta
Largemouth bass	Micropterus salmoides
Least killifish	Heterandria formosa
Leatherjacket	Oligoplites saurus
Lefteye flounders	Citharichthys spp.
Lefteye flounders	Paralichthys spp.
Lemon shark	Negaprion brevirostris
Leopard searobin	Prionotus scitulus
Lined seahorse	Hippocampus erectus
Lined topminnow	Fundulus lineolatus
Logperch	Percina caprodes
Longear sunfish	Lepomis megalotis
Longnose gar	Lepisosteus osseus
Lookdown	Selene vomer
Lyre goby	Evorthodus lyricus
Madtoms	Noturus spp.
Margined madtom	Noturus insignis
Marked goby	Gobionellus stigmaticus
Marsh killifish	Fundulus confluentus
Menhaden	Brevoortia spp.
Mojarras	Eucinostomus spp.
Mosquitofish	Gambusia affinis
Mud sunfish	Acantharchus pomotis
Mullets	Mugil spp.
Mummichog	Fundulus heteroclitus
Naked goby	Gobiosoma bosci
Naked sole	Gymnachirus melas
Niangua darter	Etheostoma nianguae
Northern kingfish	Menticirrhus saxatilis
Northern pipefish	Syngnathus fuscus
Northern puffer	Sphoeroides maculatus
Northern searobin	Prionotus carolinus
Oceanic whitetip shark	Carcharhinus longimanus
Ocellated flounder	Ancylopsetta quadrocellata
Ocmulgee shiner	Notropis callisema
Ohoopee shiner	Notropis leedsi
Okefenokee pygmy sunfish	Elassoma okefenokee
Orange filefish	Aluterus schoepfi
Oyster toadfish	Opsanus tau
Palespotted eel	Ophichthus ocellatus
Palometa	Trachinotus goodei
Permit	Trachinotus falcatus
Piedmont darter	Percina crassa
Pigfish	Orthopristis chrysoptera
Pikes	Esox spp.
Pinfish	Lagodon rhomboides
Pirate perch	Aphredoderus sayanus
Planehead filefish	Monacanthus hispidus
Pompano	Trachinotus spp.
Porgies	Calamus spp.
Porgies	Stenotomus spp.
Puffers	Sphoeroides spp.

509

Pugnose minnow	Notropis emiliae
Pumpkinseed	Lepomis gibbosus
Pygmy filefish	Monacanthus setifer
Pygmy killifish	Leptolucania ommata
Pygmy sunfishes	Elassoma spp.
Rainwater killifish	Lucania parva
Redbreast sunfish	Lepomis auritus
Red drum	Sciaenops ocellata
Red porgy	Pagrus sedecim
Red snapper	Lutjanus campechanus
Redear sunfish	Lepomis microlophus
Redfin pickerel	Esox americanus americanus
Requiem sharks	Carcharhinus spp.
River carpsucker	Carpiodes carpio
Rock sea bass	Centropristis philadelphica
Rosefin shiner	Notropis ardens
Rosyface chub	Hybopsis rubrifrons
Rough silverside	Membras martinica
Sailfin-molly	Poecilia latipinna
Sailfin shiner	Notropis hypselopterus
Sailfish	Istiophorus platypterus
Sandbar shark	Carcharhinus milberti
Sand perch	Diplectrum formosum
Savannah darter	Etheostoma fricksium
Sawcheek darter	Etheostoma serriferum
Scaled sardine	Harengula pensacolae
Scalloped hammerhead	Sphyrna lewini
Scamp grouper	Mycteroperca phenax
Seaboard goby	Gobiosoma ginsburgi
Sea catfish	Arius felis
Searobins	Prionotus spp.
Seatrout	Cynoscion spp.
Sharksucker	Echeneis naucrates
Sharptail goby	Gobionellus hastatus
Sheepshead	Archosargus probatocephalus
Sheepshead minnow	Cyprinodon variegatus
Shield darter	Percina peltata
Shiners	Notropis spp.
Shorthead redhorse	Moxostoma macrolepidotum
Shortnose sturgeon	Acipenser brevirostrum
Shrimp eel	Ophicthus gomesi
Silk snapper	Lutjanus vivanus
Silver jenny	Eucinostomus gula
Silver perch	Bairdiella chrysura
Silver redhorse	Moxostoma anisurum
Silver seatrout	Cynoscion nothus
Silversides	Menidia spp.
Silvery minnow	Hybognathus nuchalis
Skilletfish	Gobiesox strumosus
Smallmouth bass	Micropterus dolomieui
Smooth butterfly ray	Gymnura micrura
Smooth dogfish	Mustelus canis
Smooth hammerhead	Sphyrna zygaena
Smooth puffer	Lagocephalus laevigatus
Snail bullhead	Ictalurus brunneus
Snappers	Lutjanus spp.
Snook	Centropomus undecimalis
Snowy grouper	Epinephelus niveatus
Southern flounder	Paralichthys lethostigma
Southern hake	Urophycis floridanus
Southern kingfish	Menticirrhus americanus
Southern stargazer	Astroscopus y-graecum
Southern stingray	Dasyatis americana
Spadefishes	Chaetodipterus spp.
Spanish mackerel	Scomberomorus maculatus
Spanish sardine	Sardinella anchovia
Speckled hind	Epinephelus drummondhayi
Speckled madtom	Noturus leptacanthus
Speckled worm eel	Myrophis punctatus
Spinycheek sleeper	Eleotris pisonis
Spiny dogfish	Squalus acanthias

Spot	Leiostomus xanthurus
Spotfin killifish	Fundulus luciae
Spotfin mojarra	Eucinostomus argenteus
Spottail shiner	Notropis hudsonius
Spotted eagle ray	Aetobatus narinari
Spotted hake	Urophycis regius
Spotted seatrout	Cynoscion nebulosus
Spotted sucker	Minytrema melanops
Spotted sunfish	Lepomis punctatus
Spotted whiff	Citharichthys macrops
Star drum	Stellifer lanceolatus
Starhead topminnow	Fundulus notti
Stingrays	Dasyatis spp.
Striped anchovy	Anchoa hepsetus
Striped bass	Morone saxatilis.
Striped blenny	Chasmodes bosquianus
Striped burrfish	Chilomycterus schoepfi
Striped cusk-eel	Rissola marginata
Striped killifish	Fundulus majalis
Striped mullet	Mugil cephalus
Striped searobin	Prionotus evolans
Suckermouth redhorse	Moxostoma pappillosum
Summer flounder	Paralichthys dentatus
Sunfishes	Enneacanthus spp.
Swallowtail shiner	Notropis procne
Swamp darter	Etheostoma fusiforme
Swampfish	Chologaster cornuta
Tadpole madtom	Noturus gyrinus
Taillight shiner	Notropis maculatus
Tarpon	Megalops atlantica
Temperate basses	Morone spp.
Tessellated darter	Etheostoma olmstedi
Threadfin shad	Dorosoma petenense
Thread herrings	Opisthonema spp.
Tidewater silverside	Menidia beryllina
Tilefish	Lopholatilus chamaeleonticeps
Tomtate	Haemulon aurolineatum
Tripletail	Lobotes surinamensis
Vermilion snapper	Rhomboplites aurorubens
Wahoo	Acanthocybium solanderi
Walleye	Stizostedion vitreum vitreum
Warmouth	Lepomis gulosus
Warsaw grouper	Epinephelus nigritus
Weakfish	Cynoscion regalis.
Whiffs	Citharichthys spp.
Whiffs	Etropus spp.
White bass	Morone chrysops
White catfish	Ictalurus catus
White crappie	Pomoxis annularis
Whitefin shiner	Notropis niveus
White hake	Urophycis tenuis
White marlin	Tetrapturus albidus
White mullet	Mugil curema
White perch	Morone americana
Windowpane	Scophthalmus aquosus
Yellow bullhead	Ictalurus natalis
Yellowedge grouper	Epinephelus flavolimbatus
Yellowfin menhaden	Brevoortia smithi
Yellowfin shiner	Notropis lutipinnis
Yellow perch	Perca flavescens

511

Appendix Table   4.   List of fishes identified in the Sea Island Characterization Study, arranged alphabetically by scientific name (Bailey et al. 1970).

Scientific name	Common name
Acantharchus pomotis	Mud sunfish
Acanthocybium solanderi	Wahoo
Acipenser brevirostrum	Shortnose sturgeon
Acipenser oxyrhynchus	Atlantic sturgeon
Aetobatus narinari	Spotted eagle ray
Alosa spp.	Herrings
Alosa aestivalis	Blueback herring
Alosa mediocris	Hickory shad
Alosa pseudoharengus	Alewife
Alosa sapidissima	American shad
Aluterus schoepfi	Orange filefish
Amia calva	Bowfin
Anchoa hepsetus	Striped anchovy
Anchoa lyolepis	Dusky anchovy
Anchoa mitchilli	Bay anchovy
Ancylopsetta quadrocellata	Ocellated flounder
Anguilla rostrata	American eel
Aphredoderus sayanus	Pirate perch
Aplodinotus grunniens	Freshwater drum
Aprionodon isodon	Finetooth shark
Archosargus probatocephalus	Sheepshead
Arius felis	Sea catfish
Astroscopus y-graecum	Southern stargazer
Bagre marinus	Gafftopsail catfish
Bairdiella chrysura	Silver perch
Brevoortia spp.	Menhadens
Brevoortia smithi	Yellowfin menhaden
Brevoortia tyrannus	Atlantic menhaden
Calamus spp.	Porgies
Caranx spp.	Jacks
Caranx hippos	Crevalle jack
Caranx latus	Horse-eye jack
Carassius auratus	Goldfish
Carcharhinus spp.	Requiem sharks
Carcharhinus limbatus	Blacktip shark
Carcharhinus longimanus	Oceanic whitetip shark
Carcharhinus milberti	Sandbar shark
Carcharhinus obscurus	Dusky shark
Carpiodes carpio	River carpsucker
Carpiodes velifer	Highfin carpsucker
Caulolatilus cyanops	Blackline tilefish
Centrarchus macropterus	Flier
Centropomus undecimalis	Snook
Centropristis philadelphica	Rock sea bass
Centropristis striata	Black sea bass
Chaetodipterus faber	Atlantic spadefish
Chasmodes spp.	Blennies
Chasmodes bosquianus	Striped blenny
Chasmodes saburrae	Florida blenny
Chilomycterus spp.	Burrfishes
Chilomycterus schoepfi	Striped burrfish
Chloroscombrus chrysurus	Atlantic bumper
Chologaster cornuta	Swampfish
Citharichthys spp.	Lefteye flounders/Whiffs
Citharichthys macrops	Spotted whiff
Citharichthys spilopterus	Bay whiff
Conger oceanicus	Conger eel
Coryphaena spp.	Dolphins
Cynoscion spp.	Seatrout
Cynoscion nebulosus	Spotted seatrout
Cynoscion nothus	Silver seatrout
Cynoscion regalis	Weakfish
Cyprinodon variegatus	Sheepshead minnow
Cyprinus carpio	Carp

Dasyatis spp.	Stingrays
Dasyatis americana	Southern stingray
Dasyatis sabina	Atlantic stingray
Dasyatis sayi	Bluntnose stingray
Diapterus olisthostomus	Irish pompano
Diplectrum formosum	Sand perch
Dormitator maculatus	Fat sleeper
Dorosoma cepedianum	Gizzard shad
Dorosoma petenense	Threadfin shad
Echeneis naucrates	Sharksucker
Elassoma spp.	Pygmy sunfishes
Elassoma evergladei	Everglades pygmy sunfish
Elassoma okefenokee	Okefenokee pygmy sunfish
Elassoma zonatum	Banded pygmy sunfish
Eleotris pisonis	Spinycheek sleeper
Elops saurus	Ladyfish
Enneacanthus spp.	Sunfishes
Enneacanthus chaetodon	Blackbanded sunfish
Enneacanthus gloriosus	Bluespotted sunfish
Enneacanthus obesus	Banded sunfish
Epinephelus drummondhayi	Speckled hind
Epinephelus flavolimbatus	Yellowedge grouper
Epinephelus nigritus	Warsaw grouper
Epinephelus niveatus	Snowy grouper
Erimyzon oblongus	Creek chubsucker
Erimyzon sucetta	Lake chubsucker
Esox spp.	Pikes
Esox americanus americanus	Redfin pickerel
Esox niger	Chain pickerel
Etheostoma spp.	Darters
Etheostoma fricksium	Savannah darter
Etheostoma fusiforme	Swamp darter
Etheostoma hopkinsi	Christmas darter
Etheostoma nianguae	Niangua darter
Etheostoma olmstedi	Tessellated darter
Etheostoma serriferum	Sawcheek darter
Etheostoma vitreum	Glassy darter
Etheostoma zonale	Banded darter
Etropus spp.	Left eye flounders
Etropus crossotus	Fringed flounder
Eucinostomus spp.	Mojarras
Eucinostomus argenteus	Spotfin mojarra
Eucinostomus gula	Silver jenny
Evorthodus lyricus	Lyre goby
Fundulus spp.	Killifishes
Fundulus chrysotus	Golden topminnow
Fundulus cingulatus	Banded topminnow
Fundulus confluentus	Marsh killifish
Fundulus heteroclitus	Mummichog
Fundulus lineolatus	Lined topminnow
Fundulus luciae	Spotfin killifish
Fundulus majalis	Striped killifish
Fundulus notti	Starhead topminnow
Gambusia affinis	Mosquitofish
Gobiesox strumosus	Skilletfish
Gobionellus spp.	Gobies
Gobionellus boleosoma	Darter goby
Gobionellus hastatus	Sharptail goby
Gobionellus oceanicus	Highfin goby
Gobionellus shufeldti	Freshwater goby
Gobionellus stigmaticus	Marked goby
Gobiosoma spp.	Gobies
Gobiosoma bosci	Naked goby
Gobiosoma ginsburgi	Seaboard goby
Gymnachirus melas	Naked sole
Gymnura micrura	Smooth butterfly ray

Haemulon aurolineatum	Tomtate
Harengula pensacolae	Scaled sardine
Hemiramphus balao	Balao
Heterandria formosa	Least killifish
Hippocampus erectus	Lined seahorse
Holacanthus spp.	Angelfishes
Hybognathus hayi	Cypress minnow
Hybognathus nuchalis	Silvery minnow
Hybopsis rubrifrons	Rosyface chub
Hypleurochilus geminatus	Crested blenny
Hyporhamphus unifasciatus	Halfbeak
Hypsoblennius spp.	Combtooth blennies
Hypsoblennius hentzi	Feather blenny
Ictalurus brunneus	Snail bullhead
Ictalurus catus	White catfish
Ictalurus furcatus	Blue catfish
Ictalurus melas	Black bullhead
Ictalurus natalis	Yellow bullhead
Ictalurus nebulosus	Brown bullhead
Ictalurus playtcephalus	Flat bullhead
Ictalurus punctatus	Channel catfish
Istiophorus platypterus	Sailfish
Kathetostoma albigutta	Lancer stargazer
Labidesthes sicculus	Brook silverside
Lagocephalus laevigatus	Smooth puffer
Lagodon rhomboides	Pinfish
Larimus fasciatus	Banded drum
Leiostomus xanthurus	Spot
Lepisosteus spp.	Gars
Lepisosteus platyrhincus	Florida gar
Lepisosteus osseus	Longnose gar
Lepomis auritus	Redbreast sunfish
Lepomis cyanellus	Green sunfish
Lepomis gibbosus	Pumpkinseed
Lepomis gulosus	Warmouth
Lepomis macrochirus	Bluegill
Lepomis marginatus	Dollar sunfish
Lepomis megalotis	Longear sunfish
Lepomis microlophus	Redear sunfish
Lepomis punctatus	Spotted sunfish
Leptolucania ommata	Pygmy killifish
Lobotes surinamensis	Tripletail
Lopholatilus chamaeleonticeps	Tilefish
Lucania goodei	Bluefin killifish
Lucania parva	Rainwater killifish
Lutjanus spp.	Snappers
Lutjanus buccanella	Blackfin snapper
Lutjanus campechanus	Red snapper
Lutjanus griseus	Gray snapper
Lutjanus vivanus	Silk snapper
Makaira nigricans	Blue marlin
Megalops atlantica	Tarpon
Membras martinica	Rough silverside
Menidia spp.	Silversides
Menidia beryllina	Tidewater silverside
Menidia menidia	Atlantic silverside
Menticirrhus spp.	King whitings/Kingfishes
Menticirrhus americanus	Southern kingfish
Menticirrhus littoralis	Gulf kingfish
Menticirrhus saxatilis	Northern kingfish
Microgobius gulosus	Clown goby
Microgobius thalassinus	Green goby
Micropogonias undulatus	Atlantic croaker
Micropterus dolomieui	Smallmouth bass
Micropterus salmoides	Largemouth bass
Minytrema melanops	Spotted sucker
Monacanthus hispidus	Planehead filefish
Monacanthus setifer	Pygmy filefish

514

Morone spp.	Temperate basses
Morone americana	White perch
Morone chrysops	White bass
Morone saxatilis	Striped bass
Moxostoma anisurum	Silver redhorse
Moxostoma cervinum	Black jumprock
Moxostoma macrolepidotum	Shorthead redhorse
Moxostoma pappillosum	Suckermouth redhorse
Mugil spp.	Mullets
Mugil cephalus	Striped mullet
Mugil curema	White mullet
Mustelus canis	Smooth dogfish
Mycteroperca spp.	Groupers
Mycteroperca microlepis	Gag grouper
Mycteroperca phenax	Scamp grouper
Myrophis punctatus	Speckled worm eel
Negaprion brevirostris	Lemon shark
Nocomis leptocephalus	Bluehead chub
Notemigonus crysoleucas	Golden shiner
Notropis spp.	Shiners
Notropis altipinnis	Highfin shiner
Notropis ardens	Rosefin shiner
Notropis callisema	Ocmulgee shiner
Notropis chalybaeus	Ironcolor shiner
Notropis chlorocephalus	Greenhead shiner
Notropis cummingsae	Dusky shiner
Notropis emiliae	Pugnose minnow
Notropis hudsonius	Spottail shiner
Notropis hypselopterus	Sailfin shiner
Notropis leedsi	Ohoopee shiner
Notropis lutipinnis	Yellowfin shiner
Notropis maculatus	Taillight shiner
Notropis niveus	Whitefin shiner
Notropis petersoni	Coastal shiner
Notropis procne	Swallowtail shiner
Noturus spp.	Madtoms
Noturus funebris	Black madtom
Noturus gyrinus	Tadpole madtom
Noturus insignis	Margined madtom
Noturus leptacanthus	Speckled madtom
Oligoplites saurus	Leatherjacket
Ophichthus ocellatus	Palespotted eel
Ophichthus gomesi	Shrimp eel
Opisthonema spp.	Thread herrings
Opisthonema oglinum	Atlantic thread herring
Opsanus tau	Oyster toadfish
Orthopristis chrysoptera	Pigfish
Pargus sedecim	Red porgy
Paralichthys spp.	Lefteye flounders/flounders
Paralichthys albigutta	Gulf flounder
Paralichthys dentatus	Summer flounder
Paralichthys lethostigma	Southern flounder
Paralichthys squamilentus	Broad flounder
Peprilus spp.	Butterfishes
Peprilus alepidotus	Harvestfish
Peprilus triacanthus	Butterfish
Perca flavescens	Yellow perch
Percina spp.	Darters
Percina caprodes	Logperch
Percina crassa	Piedmont darter
Percina nigrofasciata	Blackbanded darter
Percina peltata	Shield darter
Pimephales promelas	Fathead minnow
Poecilia latipinna	Sailfin molly
Pogonias cromis	Black drum
Pomatomus saltatrix	Bluefish
Pomoxis annularis	White crappie
Pomoxis nigromaculatus	Black crappie
Porichthys porosissimus	Atlantic midshipman

Prionotus spp.	Searobins
Prionotus carolinus	Northern searobin
Prionotus evolans	Striped searobin
Prionotus salmonicolor	Blackwing searobin
Prionotus scitulus	Leopard searobin
Prionotus tribulus	Bighead searobin
Pylodictis olivaris	Flathead catfish
Rachycentron canadum	Cobia
Raja eglanteria	Clearnose skate
Rhinobatos lentiginosus	Atlantic guitarfish
Rhinoptera bonasus	Cownose ray
Rhizoprionodon terraenovae	Atlantic sharpnose shark
Rhomboplites aurorubens	Vermilion snapper
Rissola marginata	Striped cusk-eel
Sarda sarda	Atlantic bonito
Sardinella anchovia	Spanish sardine
Sciaenops ocellata	Red drum
Scomber scombrus	Atlantic mackerel
Scomberomorus cavalla	King mackerel
Scomberomorus maculatus	Spanish mackerel
Scophthalmus aquosus	Windowpane
Selene vomer	Lookdown
Seriola dumerili	Greater amberjack
Seriola rivoliana	Almaco jack
Sphoeroides spp.	Puffers
Sphoeroides maculatus	Northern puffer
Sphoeroides spengleri	Bandtail puffer
Sphyraena spp.	Barracudas
Sphyraena barracuda	Great barracuda
Sphyraena guachancho	Guaguanche
Sphyrna spp.	Hammerhead sharks
Sphyrna lewini	Scalloped hammerhead
Sphyrna tiburo	Bonnethead
Sphyrna zygaena	Smooth hammerhead
Squalus acanthias	Spiny dogfish
Stellifer lanceolatus	Star drum
Stenotomus spp.	Porgies
Stizostedion vitreum vitreum	Walleye
Strongylura marina	Atlantic needlefish
Symphurus plagiusa	Blackcheek tonguefish
Syngnathus fuscus	Northern pipefish
Syngnathus louisianae	Chain pipefish
Syngnathus floridae	Dusky pipefish
Synodus foetens	Inshore lizardfish
Tetrapturus albidus	White marlin
Trachinotus spp.	Pompanos
Trachinotus carolinus	Florida pompano
Trachinotus falcatus	Permit
Trachinotus goodei	Palometa
Trichiurus lepturus	Atlantic cutlassfish
Trinectes maculatus	Hogchoker
Umbra pygmaea	Eastern mudminnow
Urophycis spp.	Hakes
Urophycis earll'	Carolina hake
Urophycis floridanus	Southern hake
Urophycis regius	Spotted hake
Urophycis tenuis	White hake
Vomer setapinnis	Atlantic moonfish

Appendix Table 5. List of amphibians and reptiles identified in the Sea Island Characterization Study, arranged alphabetically by common name (Conant 1975, Collins
et al. 1978).

American alligator   Alligator mississippiensis
American crocodile   Crocodylus acutus
Atlantic green turtle   Chelonia mydas mydas
Atlantic hawksbill turtle   Eretmochelys imbricata imbricata
Atlantic leatherback turtle   Dermochelys coriacea coriacea
Atlantic loggerhead turtle   Caretta caretta caretta
Atlantic ridley turtle   Lepidochelys kempi
Atlantic salt marsh snake   Nerodia fasciata taeniata

Banded water snake   Nerodia fasciata fasciata
Barking treefrog   Hyla gratiosa
Black swamp snake   Seminatrix pygaea
Box turtle   Terrapene carolina ssp.
Brimley's chorus frog   Pseudacris brimleyi
Broad-banded water snake   Nerodia fasciata confluens
Broadhead skink   Eumeces laticeps
Broad-striped dwarf siren   Pseudobranchus striatus striatus
Broken-striped newt   Notophthalmus viridescens dorsalis
Bronze frog   Rana clamitans clamitans
Brown snake   Storeria dekayi
Brown water snake   Nerodia taxispilota
Bullfrog   Rana catesbeiana

Canebrake rattlesnake   Crotalus horridus atricaudatus
Carolina crawfish frog   Rana areolata capito
Carolina diamondback terrapin   Malaclemys terrapin centrata
Carolina pigmy rattlesnake   Sistrurus miliarius miliarius
Carolina salt marsh snake   Nerodia sipedon williamengelsi
Carolina swamp snake   Seminatrix pygaea paludis
Carpenter frog   Rana virgatipes
Central newt   Notophthalmus viridescens louisianensis
Chicken turtle   Deirochelys reticularia ssp.
Common garter snake   Thamnophis sirtalis ssp.
Common snapping turtle   Chelydra serpentina serpentina
Cope's gray treefrog   Hyla chrysoscelis
Copperhead   Agkistrodon contortrix ssp.
Corn snake   Elaphe guttata ssp.
Corn snake   Elaphe guttata guttata
Cottonmouth   Agkistrodon piscivorus ssp.
Crawfish frog   Rana areolata ssp.

Dwarf salamander   Eurycea quadridigitata
Dwarf waterdog   Necturus punctatus

Earth snakes   Virginia spp.
Eastern bird-voiced treefrog   Hyla avivoca ogechiensis
Eastern box turtle   Terrapene carolina carolina
Eastern chicken turtle   Deirochelys reticularia reticularia
Eastern coachwhip   Masticophis flagellum flagellum
Eastern coral snake   Micrurus fulvius fulvius
Eastern cottonmouth   Agkistrodon piscivorus piscivorus
Eastern diamondback rattlesnake   Crotalus adamanteus
Eastern earth snake   Virginia valeriae valeriae
Eastern garter snake   Thamnophis sirtalis sirtalis
Eastern glass lizard   Ophisaurus ventralis
Eastern hognose snake   Heterodon platyrhinos
Eastern indigo snake   Drymarchon corais couperi
Eastern kingsnake   Lampropeltis getulus getulus
Eastern lesser siren   Siren intermedia intermedia
Eastern mud snake   Parancia abacura abacura
Eastern mud turtle   Kinosternon subrubrum subrubrum
Eastern narrowmouth toad   Gastrophryne carolinensis
Eastern ribbon snake   Thamnophis sauritus sauritus
Eastern river cooter   Chrysemys concinna concinna
Eastern slender glass lizard   Ophisaurus attenuatus longicaudus
Eastern spadefoot toad   Scaphiopus holbrooki
Eastern spadefoot toad   Scaphiopus holbrooki holbrooki
Eastern tiger salamander   Ambystoma tigrinum tigrinum
Eastern worm snake   Carphophis amoenus amoenus

517

Five-lined skink	Eumeces fasciatus
Flatwoods salamander	Ambystoma cingulatum
Florida cooter	Chrysemys floridana floridana
Florida cottonmouth	Agkistrodon piscivorus conanti
Florida crawfish frog	Rana areolata aesopus
Florida cricket frog	Acris gryllus dorsalis
Florida green water snake	Nerodia cyclopion floridana
Florida pine snake	Pituophis melanoleucus mugitus
Florida softshell	Trionyx ferox
Garter snakes	Thamnophis. spp.
Glossy crayfish snake	Rigina rigida rigida
Gopher tortoise	Gopherus polyphemus
Gray treefrog	Hyla versicolor
Greater siren	Siren lacertina
Green anole	Anolis carolinensis
Green sea turtle	Chelonia mydas
Green treefrog	Hyla cinerea
Green turtle	Chelonia mydas
Green water snake	Nerodia cyclopion ssp.
Ground skink	Scincella lateralis
Gulf Coast spiny softshell	Trionyx spiniferus asperus
Gulf salt marsh snake	Nerodia fasciata clarki
Hawksbill turtle	Eretmochelys imbricata
Indigo snake	Drymarchon corais
Island glass lizard	Ophisaurus compressus
Kemp's ridley turtle	Lepidochelys kempi
Leatherback turtle	Dermochelys coriacea
Lesser siren	Siren intermedia
Little grass frog	Limnaoedus ocularis
Loggerhead turtle	Caretta caretta
Longtail salamander	Eurycea longicauda longicauda
Mabee's salamander	Ambystoma mabeei
Many-lined salamander	Stereochilus marginatus
Marbled salamander	Ambystoma opacum
Mole kingsnake	Lampropeltis calligaster rhombomaculata
Mole salamander	Ambystoma talpoideum
Mole skink	Eumeces egregius
Mud salamander	Pseudotriton montanus ssp.
Mud salamander	Pseudotriton montanus
Mud snake	Farancia abacura ssp.
Mud turtle	Kinosternon subrubrum ssp.
Newts	Notophthalmus spp.
North Florida swamp snake	Seminatrix pygaea pygaea
Northern cricket frog	Acris crepitans crepitans
Northern diamondback terrapin	Malaclemys terrapin terrapin
Northern leopard frog	Rana pipiens
Northern mole skink	Eumeces egregius similis
Northern redbelly snake	Storeria occipitomaculata occipitomaculata
Northern scarlet snake	Cemophora coccinea copei
Northern spring peeper	Hyla crucifer crucifer
Oak toad	Bufo quercicus
Ornate chorus frog	Pseudacris ornata
Peninsula ribbon snake	Thamnophis sauritus sackeni
Pickerel frog	Rana palustris
Pig frog	Rana grylio
Pine snake	Pituophis melanoleucus ssp.
Pine woods snake	Rhadinaea flavilata
Pine woods treefrog	Hyla femoralis
Pigmy rattlesnake	Sistrurus miliarius ssp.
Rainbow snake	Farancia erytrogramma ssp.
Rainbow snake	Farancia erytrogramma erytrogramma
Ranid frogs	Rana spp.
Rat snake	Elaphe obsoleta ssp.

Redbelly snake	Storeria occipitomaculata
Redbelly water snake	Nerodia erythrogaster erythrogaster
Red salamander	Pseudotriton ruber
Red-spotted newt	Notophthalmus viridescens viridescens
Ribbon snake	Thamnophis sauritus ssp.
Ringneck snake	Diadophis punctatus
River cooter	Chrysemys concinna ssp.
River frog	Rana heckscheri
Rough earth snake	Virginia striatula
Rough green snake	Opheodrya aestivus
Scarlet kingsnake	Lampropeltis triangulum elapsoides
Six-lined racerunner	Cnemidophorus sexlineatus sexlineatus
Slender glass lizard	Ophisaurus attenuatus
Slimy salamander	Plethodon glutinosus glutinosus
Smooth earth snake	Virginia valeriae
Southeastern crowned snake	Tantilla coronata
Southeastern five-lined skink	Eumeces inexpectatus
Southern black racer	Coluber constrictor priapus
Southern chorus frog	Pseudacris nigrita
Southern chorus frog	Pseudacris nigrita nigrita
Southern copperhead	Agkistrodon contortrix contortrix
Southern cricket frog	Acris gryllus
Southern cricket frog	Acris gryllus gryllus
Southern dusky salamander	Desmognathus auriculatus
Southern fence lizard	Sceloporus undulatus undulatus
Southern hognose snake	Heterodon simus
Southern leopard frog	Rana sphenocephala
Southern red salamander	Pseudotriton ruber vioscai
Southern ringneck snake	Diadophis punctatus punctatus
Southern toad	Bufo terrestris
Southern two-lined salamander	Eurycea bislineata cirrigera
Spiny softshell	Trionyx spiniferus ssp.
Spotted salamander	Ambystoma maculatum
Spotted turtle	Clemmys guttata
Spring peeper	Hyla crucifer
Squirrel treefrog	Hyla squirella
Stinkpot	Sternotherus odoratus
Striped crayfish snake	Regina alleni
Striped mud turtle	Kinosternon bauri palmarum
Striped newt	Notophthalmus perstriatus
Texas horned lizards	Phrynosoma cornutum
Three-lined salamander	Eurycea longicauda guttolineata
Tiger salamander	Ambystoma tigrinum
Treefrogs	Hyla spp.
Two-lined salamander	Eurycea bislineata
Two-toed amphiuma	Amphiuma means
Upland chorus frog	Pseudacris triseriata feriarum
Worm snake	Carphophis amoenus ssp.
Yellowbelly slider	Chrysemys scripta scripta
Yellow rat snake	Elaphe obsoleta quadrivittata

Appendix Table 6. List of amphibians and reptiles identified in the Sea Island Characterization Study, arranged alphabetically by scientific name (Conant 1975,
Collins et al. 1978).

Acris crepitans crepitans   Northern cricket frog
Acris gryllus   Southern cricket frog
Acris gryllus dorsalis   Florida cricket frog
Acris gryllus gryllus   Southern cricket frog
Agkistrodon contortrix ssp.   Copperhead
Agkistrodon contortrix contortrix   Southern copperhead
Agkistrodon piscivorus ssp.   Cottonmouth
Agkistrodon piscivorus conanti   Florida cottonmouth
Agkistrodon piscivorus piscivorus   Eastern cottonmouth
Alligator mississippiensis   American alligator
Ambystoma cingulatum   Flatwoods salamander
Ambystoma mabeei   Mabee's salamander
Ambystoma maculatum   Spotted salamander
Ambystoma opacum   Marbled salamander
Ambystoma talpoideum   Mole salamander
Ambystoma tigrinum   Tiger salamander
Ambystoma tigrinum tigrinum   Eastern tiger salamander
Amphiuma means   Two-toed amphiuma
Anolis carolinensis   Green anole

Bufo quercicus   Oak toad
Bufo terrestris   Southern toad

Caretta caretta   Loggerhead turtle
Caretta caretta caretta   Atlantic loggerhead turtle
Carphophis amoenus ssp.   Worm snake
Carphophis amoenus amoenus   Eastern worm snake
Cemophora coccinea copei   Northern scarlet snake
Chelonia mydas   Green sea turtle/Green turtle
Chelonia mydas mydas   Atlantic green turtle
Chelydra serpentina serpentina   Common snapping turtle
Chrysemys concinna ssp.   River cooter
Chrysemys concinna concinna   Eastern river cooter
Chrysemys floridana floridana   Florida cooter
Chrysemys scripta scripta   Yellowbelly slider
Clemmys guttata   Spotted turtle
Cnemidophorus sexlineatus sexlineatus   Six-lined racerunner
Coluber constrictor priapus   Southern black racer
Crocodylus acutus   American crocodile
Crotalus adamanteus   Eastern diamondback rattlesnake
Crotalus horridus atricaudatus   Canebrake rattlesnake

Deirochelys reticularia ssp.   Chicken turtle
Deirochelys reticularia reticularia   Eastern chicken turtle
Dermochelys coriacea   Leatherback turtle
Dermochelys coriacea coriacea   Atlantic leatherback turtle
Desmognathus auriculatus   Southern dusky salamander
Diadophis punctatus   Ringneck snake
Diadophis punctatus punctatus   Southern ringneck snake
Drymarchon corais   Indigo snake
Drymarchon corais couperi   Eastern indigo snake

Elaphe guttata ssp.   Corn snake
Elaphe guttata guttata   Corn snake
Elaphe obsoleta ssp.   Rat snake
Elaphe obsoleta quadrivittata   Yellow rat snake
Eretmochelys imbricata   Hawksbill turtle
Eretmochelys imbricata imbricata   Atlantic hawksbill turtle
Eumeces egregius   Mole skink
Eumeces egregius similis   Northern mole skink
Eumeces fasciatus   Five-lined skink
Eumeces inexpectatus   Southeastern five-lined skink
Eumeces laticeps   Broadhead skink
Eurycea bislineata   Two-lined salamander
Eurycea bislineata cirrigera   Southern two-lined salamander
Eurycea longicauda guttolineata   Three-lined salamander

Eurycea longicauda longicauda	Longtail salamander
Eurycea quadridigitata	Dwarf salamander
Farancia abacura ssp.	Mud snake
Farancia abacura abacura	Eastern mud snake
Farancia erytrogramma ssp.	Rainbow snake
Farancia erytrogramma erytrogramma	Rainbow snake
Gastrophryne carolinensis	Eastern narrowmouth toad
Gopherus polyphemus	Gopher tortoise
Heterodon platyrhinos	Eastern hognose snake
Heterodon simus	Southern hognose snake
Hyla spp.	Treefrogs
Hyla avivoca ogechiensis	Eastern bird-voiced treefrog
Hyla chrysoscelis	Cope's gray treefrog
Hyla cinerea	Green treefrog
Hyla crucifer	Spring peeper
Hyla crucifer crucifer	Northern spring peeper
Hyla femoralis	Pine woods treefrog
Hyla gratiosa	Barking treefrog
Hyla squirella	Squirrel treefrog
Hyla versicolor	Gray treefrog
Kinosternon bauri palmarum	Striped mud turtle
Kinosternon subrubrum ssp.	Mud turtle
Kinosternon subrubrum subrubrum	Eastern mud turtle
Lampropeltis calligaster rhombomaculata	Mole kingsnake
Lampropeltis getulus getulus	Eastern kingsnake
Lampropeltis triangulum elapsoides	Scarlet kingsnake
Lepidochelys kempi	Atlantic ridley turtle/Kemp's ridley turtle
Limnaoedus ocularis	Little grass frog
Malaclemys terrapin centrata	Carolina diamondback terrapin
Malaclemys terrapin terrapin	Northern diamondback terrapin
Masticophis flagellum flagellum	Eastern coachwhip
Micrurus fulvius fulvius	Eastern coral snake
Necturus punctatus	Dwarf waterdog
Nerodia cyclopion ssp.	Green water snake
Nerodia cyclopion floridana	Florida green water snake
Nerodia erythrogaster erythrogaster	Redbelly water snake
Nerodia fasciata clarki	Gulf salt marsh snake
Nerodia fasciata confluens	Broad-banded water snake
Nerodia fasciata fasciata	Banded water snake
Nerodia fasciata taeniata	Atlantic salt marsh snake
Nerodia sipedon williamengelsi	Carolina salt marsh snake
Nerodia taxispilota	Brown water snake
Notophthalmus spp.	Newts
Notophthalmus perstriatus	Striped newt
Notophthalmus viridescens dorsalis	Broken-striped newt
Notophthalmus viridescens louisianensis	Central newt
Notophthalmus viridescens viridescens	Red-spotted newt
Opheodrys aestivus	Rough green snake
Ophisaurus attenuatus	Slender glass lizard
Ophisaurus attenuatus longicaudus	Eastern slender glass lizard
Ophisaurus compressus	Island glass lizard
Ophisaurus ventralis	Eastern glass lizard
Phrynosoma cornutum	Texas horned lizard
Pituophis melanoleucus ssp.	Pine snake
Pituophis melanoleucus mugitus	Florida pine snake
Plethodon glutinosus glutinosus	Slimy salamander
Pseudacris brimleyi	Brimley's chorus frog
Pseudacris nigrita	Southern chorus frog
Pseudacris nigrita nigrita	Southern chorus frog
Pseudacris ornata	Ornate chorus frog
Pseudacris triseriata feriarum	Upland chorus frog
Pseudobranchus striatus striatus	Broad-striped dwarf siren
Pseudotriton montanus ssp.	Mud salamander

521

Pseudotriton ruber	Red salamander
Pseudotriton ruber vioscai	Southern red salamander
Rana spp.	Ranid frogs
Rana areolata ssp.	Crawfish frog
Rana Areolata aesopus	Florida crawfish frog
Rana areolata capito	Carolina crawfish frog
Rana catesbeiana	Bullfrog
Rana clamitans clamitans	Bronze frog
Rana grylio	Pig frog
Rana heckscheri	River frog
Rana palustris	Pickerel frog
Rana pipiens	Northern leopard frog
Rana sphenocephala	Southern leopard frog
Rana virgatipes	Carpenter frog
Regina alleni	Striped crayfish snake
Regina rigida rigida	Glossy crayfish snake
Rhadinaea flavilata	Pine woods snake
Scaphiopus holbrooki	Eastern spadefoot toad
Scaphiopus holbrooki holbrooki	Eastern spadefoot toad
Sceloporus undulatus undulatus	Southern fence lizard
Scincella lateralis	Ground skink
Seminatrix pygaea	Black swamp snake
Seminatrix pygaea paludis	Carolina swamp snake
Seminatrix pygaea pygaea	Northern Florida swamp snake
Siren intermedia	Lesser siren
Siren intermedia intermedia	Eastern lesser siren
Siren lacertina	Greater siren
Sistrurus miliarius	Pygmy rattlesnake
Sistrurus miliarius miliarius	Carolina pygmy rattlesnake
Stereochilus marginatus	Many-lined salamander
Sternotherus odoratus	Stinkpot
Storeria dekayi	Brown snake
Storeria occipitomaculata	Redbelly snake
Storeria occipitomaculata occipitomaculata	Northern redbelly snake
Tantilla coronata	Southeastern crowned snake
Terrapene carolina ssp.	Box turtle
Terrapene carolina carolina	Eastern box turtle
Thamnophis spp.	Garter snakes
Thamnophis sauritus ssp.	Ribbon snake
Thamnophis sauritus sackeni	Peninsula ribbon snake
Thamnophis sauritus sauritus	Eastern ribbon snake
Thamnophis sirtalis ssp.	Common garter snake
Thamnophis sirtalis sirtalis	Eastern garter snake
Trionyx ferox	Flordia softshell
Trionyx spiniferus ssp.	Spiny softshell
Trionyx spiniferus asperus	Gulf Coast spiny softshell
Virginia spp.	Earth snakes
Virginia striatula	Rough earth snake
Virginia valeriae	Smooth earth snake
Virginia valeriae valeriae	Eastern earth snake

Appendix Table 7. List of birds identified in the Sea Island Characterization Study, arranged alphabetically by common name (Wetmore 1957, Sprunt and Chamberlain 1970, American Ornithologists' Union 1973, American Ornithologists' Union 1976).

Acadian flycatcher	*Empidonax virescens*
American avocet	*Recurvirostra americana*
American bittern	*Botaurus lentiginosus*
American coot	*Fulica americana*
American goldeneye	*Bucephala clangula americana*
American goldfinch	*Carduelis tristis*
American kestrel	*Falco sparverius*
American oystercatcher	*Haematopus palliatus palliatus*
American redstart	*Setophaga ruticilla ruticilla*
American wigeon	*Anas americana*
American woodcock	*Philohela minor*
Anhinga	*Anhinga anhinga*
Arctic peregrine falcon	*Falco peregrinus tundrius*
Audubon's shearwater	*Puffinus lherminieri*
Bachman's sparrow	*Aimophila aestivalis*
Bachman's warbler	*Vermivora bachmanii*
Bald eagle	*Haliaeetus leucocephalus*
Baldpate	*Anas americana*
Bank swallow	*Riparia riparia*
Barn owl	*Tyto alba*
Barn swallow	*Hirundo rustica erythrogaster*
Barred owl	*Strix varia*
Barrow's goldeneye	*Bucephala islandica*
Belted kingfisher	*Megaceryle alcyon*
Bewick's wren	*Thryomanes bewickii bewickii*
Black-and-white warbler	*Mniotilta varia*
Black-bellied plover	*Pluvialis squatarola*
Black-bellied whistling duck	*Dendrocygna autumnalis*
Black-billed cuckoo	*Coccyzus erythropthalmus*
Black-crowned night heron	*Nycticorax nycticorax*
Black duck	*Anas rubripes*
Black-headed gull	*Larus ridibundus*
Black-necked stilt	*Himantopus mexicanus mexicanus*
Blackpoll warbler	*Dendroica striata*
Black rail	*Laterallus jamaicensis*
Black scoter	*Melanitta nigra*
Black scoter	*Melanitta nigra americana*
Black skimmer	*Rynchops nigra*
Black tern	*Chlidonias niger*
Black-throated blue warbler	*Dendroica caerulescens*
Black-throated green warbler	*Dendroica virens waynei*
Black vulture	*Coragyps atratus*
Blue goose	*Chen caerulescens*
Blue-gray gnatcatcher	*Polioptila caerulea*
Blue grosbeak	*Guiraca caerulea*
Blue grosbeak	*Guiraca caerulea caerulea*
Blue-headed vireo	*Vireo solitarius solitarius*
Blue jay	*Cyanocitta cristata cristata*
Blue-winged teal	*Anas discors*
Blue-winged warbler	*Vermivora pinus*
Boat-tailed grackle	*Quiscalus major*
Bobolink	*Dolichonyx oryzivorus*
Bobwhite	*Colinus virginianus*
Bonaparte's gull	*Larus philadelphia*
Broad-winged hawk	*Buteo platypterus*
Brown creeper	*Certhia familiaris*
Brown-headed cowbird	*Molothrus ater*
Brown-headed nuthatch	*Sitta pusilla pusilla*
Brown pelican	*Pelecanus occidentalis*
Brown thrasher	*Toxostoma rufum rufum*
Bufflehead	*Bucephala albeola*
Cabot's tern	*Sterna sandvicensis acuflavidus*
Canada goose	*Branta canadensis*
Canvasback	*Aythya valisineria*
Cape May warbler	*Dendroica tigrina*

523

Cardinal	Richmondena cardinalis cardinalis
Carolina chickadee	Parus carolinensis
Carolina parakeet	Conuropsis carolinensis carolinensis
Carolina wren	Thryothorus ludovicianus
Caspian tern	Sterna caspia
Catbird	Dumetella carolinensis
Cattle egret	Bubulcus ibis ibis
Cedar waxwing	Bombycilla cedrorum
Chimney swift	Chaetura pelagica
Chipping sparrow	Spizella passerina passerina
Chuck-will's-widow	Caprimulgus carolinensis
Cinnamon teal	Anas cyanoptera
Clapper rail	Rallus longirostris
Cliff swallow	Petrochelidon pyrrhonota
Common crow	Corvus brachyrhynchos
Common eider	Somateria mollissima
Common flicker	Colaptes auratus
Common gallinule	Gallinula chloropus
Common goldeneye	Bucephala clangula
Common grackle	Quiscalus quiscula
Common loon	Gavia immer
Common merganser	Mergus merganser
Common snipe	Capella gallinago
Common tern	Sterna hirundo
Cooper's hawk	Accipiter cooperii
Dark-eyed junco	Junco hyemalis
Doubled-crested cormorant	Phalacrocorax auritus
Dowitchers	Limnodromus spp.
Downy woodpecker	Picoides pubescens
Dunlin	Calidris alpina
Dusky seaside sparrow	Ammospiza maritima nigrescens
Eastern bluebird	Sialia sialis
Eastern bobwhite	Colinus virginianus virginianus
Eastern brown pelican	Pelecanus occidentalis carolinensis
Eastern kingbird	Tyrannus tyrannus
Eastern meadowlark	Sturnella magna
Eastern phoebe	Sayornis phoebe
Eastern wood pewee	Contopus virens
Eskimo curlew	Numenius borealis
European wigeon	Anas penelope
Field sparrow	Spizella pusilla
Fish crow	Corvus ossifragus
Florida red-shouldered hawk	Buteo lineatus alleni
Forster's tern	Sterna forsteri
Fox sparrow	Passerella iliaca
Fulvous whistling duck	Dendrocygna bicolor
Gadwall	Anas strepera
Gannet	Morus bassanus
Glossy ibis	Plegadis falcinellus falcinellus
Golden-crowned kinglet	Regulus satrapa
Golden eagle	Aquila chrysaetos
Golden-winged warbler	Vermivora chrysoptera
Grasshopper sparrow	Ammodramus savannarum
Gray-cheeked thrush	Catharus minimus
Great black-backed gull	Larus marinus
Great blue heron	Ardea herodias
Great crested flycatcher	Myiarchus crinitus
Great egret	Casmerodius albus
Greater scaup	Aythya marila
Greater shearwater	Puffinus gravis
Greater yellowlegs	Tringa melanoleucus
Great horned owl	Bubo virginianus virginianus
Green heron	Butorides striatus
Green-winged teal	Anas crecca
Ground dove	Columbigallina passerina
Gull-billed tern	Gelochelidon nilotica

Hairy woodpecker	Picoides villosus
Harlequin duck	Histrionicus histrionicus
Henslow's sparrow	Passerherbulus henslowii
Hermit thrush	Catharus guttatus
Herring gull	Larus argentatus
Hooded merganser	Lophodytes cucullatus
Hooded warbler	Wilsonia citrina
Horned grebe	Podiceps auritus
House wren	Troglodytes aedon
Hudsonian curlew	Numenius phaeopus hudsonicus
Iceland gull	Larus glaucoides
Indigo bunting	Passerina cyanea
Ipswich sparrow	Passerculus sandwichensis princeps
Ivory-billed woodpecker	Campephilus principalis principalis
Kentucky warbler	Oporornis formosus
Kestrel	Falco tinnunculus
Killdeer	Charadrius vociferus
King eider	Somateria spectabilis
King rail	Rallus elegans
Kirtland's warbler	Dendroica kirtlandii
Knot	Calidris canutus rufa
Laughing gull	Larus atricilla
Least bittern	Ixobrychus exilis
Least sandpiper	Calidris minutilla
Least tern	Sterna albifrons
Lé Conte's sparrow	Passerherbulus caudacutus
Lesser scaup	Aythya affinis
Lesser yellowlegs	Tringa flavipes
Limpkin	Aramus guarauna
Little blue heron	Florida caerulea caerulea
Loggerhead shrike	Lanius ludovicianus
Long-billed curlew	Numenius americanus
Long-billed curlew	Numenius americanus americanus
Long-billed marsh wren	Cistothorus palustris
Long-billed marsh wren	Cistothorus palustris griseus
Louisiana heron	Hydranassa tricolor
Louisiana waterthrush	Seiurus motacilla
Macgillivray's seaside sparrow	Ammospiza maritima macgillivraii
Magnolia warbler	Dendroica magnolia
Mallard	Anas platyrhynchos
Marbled godwit	Limosa fedoa
Marsh hawk	Circus cyaneus
Marsh hen	Rallus longirostris
Masked duck	Oxyura dominica
Merlin	Falco columbarius
Mexican duck	Anas diazi
Mississippi kite	Ictinia mississippiensis
Mockingbird	Mimus polyglottos
Mockingbird	Mimus polyglottos polyglottos
Mottled duck	Anas fulvigula
Mourning dove	Zenaida macroura
Muscovy duck	Cairina moschata
Nighthawk	Chordeiles minor minor
Northern oriole	Icterus galbula
Northern parula	Parula americana
Northern phalarope	Lobipes lobatus
Northern shoveler	Anas clypeata
Northern waterthrush	Seiurus noveboracensis
Oldsquaw	Clangula hyemalis
Orange crowned warbler	Vermivora celata
Orchard oriole	Icterus spurius
Osprey	Pandion haliaetus
Osprey	Pandion haliaetus carolinensis
Ovenbird	Seiurus aurocapillus

Painted bunting	Passerina ciris ciris
Palm warbler	Dendroica palmarum
Peregrine falcon	Falco peregrinus anatum
Pied-billed grebe	Podilymbus podiceps
Pigeon hawk	Falco columbarius columbarius
Pileated woodpecker	Dryocopus pileatus pileatus
Pine siskin	Carduelis pinus
Pine warbler	Dendroica pinus
Pine warbler	Dendroica pinus pinus
Pintail	Anas acuta
Piping plover	Charadrius melodus
Prairie warbler	Dendroica discolor discolor
Prothonotary warbler	Protonotaria citrea
Purple finch	Carpodacus purpureus
Purple gallinule	Porphyrula martinica
Purple martin	Progne subis subis
Quail	Colinus virginianus
Red-bellied woodpecker	Melanerpes carolinus carolinus
Red-breasted merganser	Mergus serrator
Red-breasted nuthatch	Sitta canadensis
Red-cockaded woodpecker	Picoides borealis
Red-cockaded woodpecker	Picoides borealis borealis
Red-eyed vireo	Vireo olivaceus
Redhead	Aythya americana
Red-headed woodpecker	Melanerpes erythrocephalus erythrocephalus
Red-shouldered hawk	Buteo lineatus lineatus
Red-tailed hawk	Buteo jamaicensis borealis
Red-throated loon	Gavia stellata
Red-winged blackbird	Agelaius phoeniceus
Ring-billed gull	Larus delawarensis
Ring-necked duck	Aythya collaris
Robin	Turdus migratorius migratorius
Rough-winged swallow	Stelgidopteryx ruficollis serripennis
Royal tern	Sterna maxima
Ruby-crowned kinglet	Regulus calendula
Ruby-throated hummingbird	Archilochus colubris
Ruddy duck	Oxyura jamaicensis
Ruddy turnstone	Arenaria interpres
Ruffed grouse	Bonasa umbellus
Rufous-sided towhee	Pipilo erythrophthalmus
Rusty blackbird	Euphagus carolinus
Sanderling	Crocethia alba
Sandwich tern	Sterna sandvicensis
Savannah sparrow	Passerculus sandwichensis
Screech owl	Otus asio
Seaside sparrow	Ammospiza maritima
Semipalmated plover	Charadrius semipalmatus
Semipalmated sandpiper	Calidris pusillus
Sharp-shinned hawk	Accipiter striatus velox
Sharp-tailed sparrow	Ammospiza caudacuta
Short-billed dowitcher	Limnodromus griseus
Short-billed marsh wren	Cistothorus platensis
Short-eared owl	Asio flammeus
Shoveler	Anas clypeata
Snipe	Capella gallinago
Snow goose	Chen caerulescens
Snowy egret	Egretta thula thula
Solitary sandpiper	Tringa solitaria
Solitary vireo	Vireo solitarius
Song sparrow	Melospiza melodia melodia
Sora	Porzana carolina
Southern bald eagle	Haliaeetus leucocephalus leucocephalus
Southern crested flycatcher	Myiarchus crinitus crinitus
Southern crow	Corvus brachyrhynchos paulus
Southern downy woodpecker	Picoides pubescens pubescens
Southern screech owl	Otus asio asio
Spotted sandpiper	Actitis macularia
Starling	Sturnus vulgaris
Summer tanager	Piranga rubra rubra

Surf scoter	Melanitta perspicillata
Swainson's thrush	Catharus ustulatus
Swainson's warbler	Limnothlypis swainsonii
Swallow-tailed kite	Elanoides forficatus
Swamp sparrow	Melospiza georgiana
Swamp sparrow	Melospiza georgiana georgiana
Tennessee warbler	Vermivora peregrina
Tree swallow	Iridoprocne bicolor
Tufted titmouse	Parus bicolor
Turkey	Meleagris gallopavo
Turkey vulture	Cathartes aura aura
Veery	Catharus fuscescens
Vesper sparrow	Pooecetes gramineus
Virginia rail	Rallus limicola
Wayne's clapper rail	Rallus longirostris waynei
Western sandpiper	Calidris mauri
Whimbrel	Numenius phaeopus
Whip-poor-will	Caprimulgus vociferus
Whistling swan	Olor columbianus
White-breasted nuthatch	Sitta carolinensis
White-eyed vireo	Vireo griseus
White-fronted goose	Anser albifrons
White ibis	Eudocimus albus
White-throated sparrow	Zonotrichia albicollis
White-winged scoter	Melanitta deglandi
Willet	Catoptrophorus semipalmatus
Wilson's petrel	Oceanites oceanicus
Wilson's plover	Charadrius wilsonia
Wilson's snipe	Capella gallinago delicata
Winter wren	Troglodytes troglodytes
Woodcock	Philohela minor
Wood duck	Aix sponsa
Wood ibis	Mycteria americana
Wood stork	Mycteria americana
Wood thrush	Hylocichla mustelina
Worm-eating warbler	Helmitheros vermivorus
Yellow-bellied sapsucker	Sphyrapicus varius varius
Yellow-billed cuckoo	Coccyzus americanus americanus
Yellow-breasted chat	Icteria virens
Yellow-crowned night heron	Nyctanassa violacea
Yellow rail	Coturnicops noveboracensis
Yellow-rumped warbler	Dendroica coronata
Yellow-shafted flicker	Colaptes auratus auratus
Yellowthroat	Geothlypis trichas
Yellow-throated vireo	Vireo flavifrons
Yellow-throated warbler	Dendroica dominica dominica
Yellow warbler	Dendroica petechia

Appendix Table   8.   List of birds identified in the Sea Island Characterization Study,
                       arranged alphabetically by scientific name (Wetmore 1957, Sprunt and
                       Chamberlain 1970, American Ornithologists' Union 1973, American
                       Ornithologists' Union 1976).

Accipiter cooperii   Cooper's hawk
Accipiter striatus velox   Sharp-shinned hawk
Actitis macularia   Spotted sandpiper
Agelaius phoeniceus   Red-winged blackbird
Aimophila aestivalis   Bachman's sparrow
Aix sponsa   Wood duck
Ammodramus savannarum   Grasshopper sparrow
Ammospiza caudacuta   Sharp-tailed sparrow
Ammospiza maritima   Seaside sparrow
Ammospiza maritima macgillivraii   Macgillivray's seaside sparrow
Ammospiza maritima nigrescens   Dusky seaside sparrow
Anas acuta   Pintail
Anas americana   American wigeon/Baldpate
Anas clypeata   Shoveler/Northern shoveler
Anas crecca   Green-winged teal
Anas cyanoptera   Cinnamon teal
Anas diazi   Mexican duck
Anas discors   Blue-winged teal
Anas fulvigula   Mottled duck
Anas penelope   European wigeon
Anas platyrhynchos   Mallard
Anas rubripes   Black duck
Anas strepera   Gadwall
Anhinga anhinga   Anhinga
Anser albifrons   White-fronted goose
Aquila chrysaetos   Golden eagle
Aramus guarauna   Limpkin
Archilochus colubris   Ruby-throated hummingbird
Ardea herodias   Great blue heron
Arenaria interpres   Ruddy turnstone
Asio flammeus   Short-eared owl
Aythya affinis   Lesser scaup
Aythya americana   Redhead
Aythya collaris   Ring-necked duck
Aythya marila   Greater scaup
Aythya valisineria   Canvasback

Bombycilla cedrorum   Cedar waxwing
Bonasa umbellus   Ruffed grouse
Botaurus lentiginosus   American bittern
Branta canadensis   Canada goose
Bubo virginianus virginianus   Great horned owl
Bubulcus ibis ibis   Cattle egret
Bucephala albeola   Bufflehead
Bucephala clangula   Common goldeneye
Bucephala clangula americana   American goldeneye
Bucephala islandica   Barrow's goldeneye
Buteo jamaicensis borealis   Red-tailed hawk
Buteo lineatus alleni   Florida red-shouldered hawk
Buteo lineatus lineatus   Red-shouldered hawk
Buteo platypterus   Broad-winged hawk
Butorides striatus   Green heron

Cairina moschata   Muscovy duck
Calidris alpina   Dunlin
Calidris canutus rufa   Knot
Calidris mauri   Western sandpiper
Calidris minutilla   Least sandpiper
Calidris pusillus   Semipalmated sandpiper
Campephilus principalis principalis   Ivory-billed woodpecker
Capella gallinago   Common snipe/Snipe
Capella gallinago delicata   Wilson's snipe
Caprimulgus carolinensis   Chuck-will's-widow
Caprimulgus vociferus   Whip-poor-will
Carduelis pinus   Pine siskin
Carduelis tristis   American goldfinch

528

Carpodacus purpureus	Purple finch
Casmerodius albus	Great egret
Cathartes aura aura	Turkey vulture
Catharus fuscescens	Veery
Catharus guttatus	Hermit thrush
Catharus minimus	Gray-cheeked thrush
Catharus ustulatus	Swainson's thrush
Catoptrophorus semipalmatus	Willet
Certhia familiaris	Brown creeper
Chaetura pelagica	Chimney swift
Charadrius melodus	Piping plover
Charadrius semipalmatus	Semipalmated plover
Charadrius vociferus	Killdeer
Charadrius wilsonia	Wilson's plover
Chen caerulescens	Blue goose/Snow goose
Chlidonias niger	Black tern
Chordeiles minor minor	Nighthawk
Circus cyaneus	Marsh hawk
Cistothorus palustris	Long-billed marsh wren
Cistothorus palustris griseus	Long-billed marsh wren
Cistothorus platensis	Short-billed marsh wren
Clangula hyemalis	Oldsquaw
Coccyzus americanus americanus	Yellow-billed cuckoo
Coccyzus erythrophthalmus	Black-billed cuckoo
Colaptes auratus	Common flicker
Colaptes auratus auratus	Yellow-shafted flicker
Colinus virginianus	Bobwhite/Quail
Colinus virginianus virginianus	Eastern bobwhite
Columbigallina passerina	Ground dove
Contopus virens	Eastern wood pewee
Conuropsis carolinensis carolinensis	Carolina parakeet
Coragyps atratus	Black vulture
Corvus brachyrhynchos	Common crow
Corvus brachyrhynchos paulus	Southern crow
Corvus ossifragus	Fish crow
Coturnicops noveboracensis	Yellow rail
Crocethia alba	Sanderling
Cyanocitta cristata cristata	Blue jay
Dendrocygna autumnalis	Black-bellied whistling duck
Dendrocygna bicolor	Fulvous whistling duck
Dendroica caerulescens	Black-throated blue warbler
Dendroica coronata	Yellow-rumped warbler
Dendroica discolor discolor	Prairie warbler
Dendroica dominica dominica	Yellow-throated warbler
Dendroica kirtlandii	Kirtland's warbler
Dendroica magnolia	Magnolia warbler
Dendroica palmarum	Palm warbler
Dendroica petechia	Yellow warbler
Dendroica pinus	Pine warbler
Dendroica pinus pinus	Pine warbler
Dendroica striata	Blackpoll warbler
Dendroica tigrina	Cape May warbler
Dendroica virens waynei	Black-throated green warbler
Dolichonyx oryzivorus	Bobolink
Dryocopus pileatus pileatus	Pileated woodpecker
Dumetella carolinensis	Catbird
Egretta thula thula	Snowy egret
Elanoides forficatus	Swallow-tailed kite
Empidonax virescens	Acadian flycatcher
Eudocimus albus	White ibis
Euphagus carolinus	Rusty blackbird
Falco columbarius	Merlin
Falco columbarius columbarius	Pigeon hawk
Falco peregrinus anatum	Peregrine falcon
Falco peregrinus tundrius	Arctic peregrine falcon
Falco sparverius	American kestrel
Falco tinnunculus	Kestrel
Florida caerulea caerulea	Little blue heron
Fulica americana	American coot

Gallinula chloropus	Common gallinule
Gavia immer	Common loon
Gavia stellata	Red-throated loon
Gelochelidon nilotica	Gull-billed tern
Geothlypis trichas	Yellowthroat
Guiraca caerulea	Blue grosbeak
Cuiraca caerulea caerulea	Blue grosbeak
Haematopus palliatus palliatus	American oystercatcher
Haliaeetus leucocephalus	Bald eagle
Haliaeetus leucocephalus leucocephalus	Southern bald eagle
Helmitheros vermivorus	Worm-eating warbler
Himantopus mexicanus mexicanus	Black-necked stilt
Hirundo rustica erythrogaster	Barn swallow
Histrionicus histrionicus	Harlequin duck
Hydranassa tricolor	Louisiana heron
Hylocichla mustelina	Wood thrush
Icteria virens	Yellow-breasted chat
Icterus galbula	Northern oriole
Icterus spurius	Orchard oriole
Ictinia mississippiensis	Mississippi kite
Iridoprocne bicolor	Tree swallow
Ixobrychus exilis	Least bittern
Junco hyemalis	Dark-eyed junco
Lanius ludovicianus	Loggerhead shrike
Larus argentatus	Herring gull
Larus atricilla	Laughing gull
Larus delawarensis	Ring-billed gull
Larus glaucoides	Iceland gull
Larus marinus	Great black-backed gull
Larus philadelphia	Bonaparte's gull
Larus ridibundus	Black-headed gull
Laterallus jamaicensis	Black rail
Limnodromus spp.	Dowitchers
Limnodromus griseus	Short-billed dowitcher
Limnothlypis swainsonii	Swainson's warbler
Limosa fedoa	Marbled godwit
Lobipes lobatus	Northern phalarope
Lophodytes cucullatus	Hooded merganser
Megaceryle alcyon	Belted kingfisher
Melanerpes carolinus carolinus	Red-bellied woodpecker
Melanerpes erythrocephalus erythrocephalus	Red-headed woodpecker
Melanitta deglandi	White-winged scoter
Melanitta nigra	Black scoter
Melanitta nigra americana	Black scoter
Melanitta perspicillata	Surf scoter
Meleagris gallopavo	Turkey
Melospiza georgiana	Swamp sparrow
Melospiza georgiana georgiana	Swamp sparrow
Melospiza melodia atlantica	Song sparrow
Mergus merganser	Common merganser
Mergus serrator	Red-breasted merganser
Mimus polygolottos	Mockingbird
Mimus polyglottos polyglottos	Mockingbird
Mniotilta varia	Black-and-white warbler
Molothrus aster	Brown-headed cowbird
Morus bassanus	Gannet
Mycteria americana	Wood ibis/Wood stork
Myiarchus crinitus	Great crested flycatcher
Myiarchus crinitus crinitus	Southern crested flycatcher
Numenius americanus	Long-billed curlew
Numenius americanus americanus	Long-billed curlew
Numenius borealis	Eskimo curlew
Numenius phaeopus	Whimbrel
Numenius phaeopus hudsonicus	Hudsonian curlew
Nyctanassa violacea	Yellow-crowned night heron
Nycticorax nycticorax	Black-crowned night heron

Oceanites oceanicus	Wilson's petrel
Olor columbianus	Whistling swan
Oporornis formosus	Kentucky warbler
Otus asio	Screech owl
Otus asio asio	Southern screech owl
Oxyura dominica	Masked duck
Oxyura jamaicensis	Ruddy duck
Pandion haliaetus	Osprey
Pandion haliaetus carolinensis	Osprey
Parula americana	Northern parula
Parus bicolor	Tufted titmouse
Parus carolinensis	Carolina chickadee
Passerculus sandwichensis	Savannah sparrow
Passerculus sandwichensis princeps	Ipswich sparrow
Passerella iliaca	Fox sparrow
Passerherbulus caudacutus	Le Conte's sparrow
Passerherbulus henslowii	Henslow's sparrow
Passerina ciris ciris	Painted bunting
Passerina cyanea	Indigo bunting
Pelecanus occidentalis	Brown pelican
Pelecanus occidentalis carolinensis	Eastern brown pelican
Petrochelidon pyrrhonota	Cliff swallow
Phalacrocorax auritus	Double-crested cormorant
Philohela minor	American woodcock/Woodcock
Picoides borealis	Red-cockaded woodpecker
Picoides borealis borealis	Red-cockaded woodpecker
Picoides pubescens	Downy woodpecker
Picoides pubescens pubescens	Southern downy woodpecker
Picoides villosus	Hairy woodpecker
Pipilo erythrophthalmus	Rufous-sided towhee
Piranga rubra rubra	Summer tanager
Plegadis falcinellus falcinellus	Glossy ibis
Pluvialis squatarola	Black-bellied plover
Podiceps auritus	Horned grebe
Podilymbus podiceps	Pied-billed grebe
Polioptila caerulea	Blue-gray gnatcatcher
Pooecetes gramineus	Vesper sparrow
Porphyrula martinica	Purple gallinule
Porzana carolina	Sora
Progne subis subis	Purple martin
Protonotaria citrea	Prothonotary warbler
Puffinus gravis	Greater shearwater
Puffinus lherminieri	Audubon's shearwater
Quiscalus major	Boat-tailed grackle
Quiscalus quiscula	Common grackle
Rallus elegans	King rail
Rallus limicola	Virginia rail
Rallus longirostris	Clapper rail/Marsh hen
Rallus longirostris waynei	Wayne's clapper rail
Recurvirostra americana	American avocet
Regulus calendula	Ruby-crowned kinglet
Regulus satrapa	Golden-crowned kinglet
Richmondena cardinalis cardinalis	Cardinal
Riparia riparia	Bank swallow
Rynchops nigra	Black skimmer
Sayornis phoebe	Eastern phoebe
Seiurus aurocapillus	Ovenbird
Seiurus motacilla	Louisiana waterthrush
Seiurus noveboracensis	Northern waterthrush
Setophaga ruticilla ruticilla	American redstart
Sialia sialis	Eastern bluebird
Sitta canadensis	Red-breasted nuthatch
Sitta carolinensis	White-breasted nuthatch
Sitta pusilla pusilla	Brown-headed nuthatch
Somateria mollissima	Common eider
Somateria spectabilis	King eider
Sphyrapicus varius varius	Yellow-bellied sapsucker
Spizella passerina passerina	Chipping sparrow

Spizella pusilla	Field sparrow
Stelgidopteryx ruficollis serripennis	Rough-winged swallow
Sterna albifrons	Least tern
Sterna caspia	Caspian tern
Sterna forsteri	Forster's tern
Sterna hirundo	Common tern
Sterna maxima	Royal tern
Sterna sandvicensis	Sandwich tern
Sterna sandvicensis acuflavidus	Cabot's tern
Strix varia	Barred owl
Sturnella magna	Eastern meadowlark
Sturnus vulgaris	Starling
Thryomanes bewickii bewickii	Bewick's wren
Thryothorus ludovicianus	Carolina wren
Toxostoma rufum rufum	Brown thrasher
Tringa flavipes	Lesser yellowlegs
Tringa melanoleucus	Greater yellowlegs
Tringa solitaria	Solitary sandpiper
Troglodytes aedon	House wren
Troglodytes troglodytes	Winter wren
Turdus migratorius migratorius	Robin
Tyrannus tyrannus	Eastern kingbird
Tyto alba	Barn owl
Vermivora bachmanii	Bachman's warbler
Vermivora celata	Orange crowned warbler
Vermivora chrysoptera	Golden-winged warbler
Vermivora peregrina	Tennessee warbler
Vermivora pinus	Blue-winged warbler
Vireo flavifrons	Yellow-throated vireo
Vireo griseus	White-eyed vireo
Vireo olivaceus	Red-eyed vireo
Vireo solitarius	Solitary vireo
Vireo solitarius solitarius	Blue-headed vireo
Wilsonia citrina	Hooded warbler
Zenaida macroura	Mourning dove
Zonotrichia albicollis	White-throated sparrow

532

Appendix Table   9.   List of mammals identified in the Sea Island Characterization Study,
arranged alphabetically by common name (Hall and Kelson 1959; Colley
1962, 1966; Ridgeway 1972; Johnson et al. 1974; Lowery 1974; Zingmark
1978).

Antillean beaked whale   Mesoplodon europaeus
Atlantic beaked whale   Mesoplodon densirostris
Atlantic bottle-nosed dolphin   Tursiops truncatus
Atlantic right whale   Eubalaena glacialis

Beaver   Castor canadensis
Big brown bat   Eptesicus fuscus fuscus
Black bear   Ursus americanus
Blue whale   Sibbaldus musculus
Bobcat   Lynx rufus
Bottle-nosed dolphin   Tursiops truncatus
Bowhead whale   Balaena mysticetus
Brazilian free-tailed bat   Tadarida brasiliensis cynocephala
Bridled dolphin   Stenella frontalis
Bryde's whale   Balaenoptera edeni

California sea lion   Zalophus californianus
Colonial pocket gopher   Geomys colonus
Common dolphin   Delphinus delphis
Common porpoise   Phocoena phocoena
Cotton mouse   Peromyscus gossypinus/Peromyscus gossypinus
    anastasae
Cotton rat   Sigmodon hispidus
Cow   Bos taurus
Cumberland Island pocket gopher   Geomys cumberlandius

Dolphin   Coryphaena hippurus
Domestic hog   Sus scrofa domesticus
Dwarf sperm whale   Kogia simus

Eastern cottontail   Sylvilagus floridanus
Eastern cougar   Felis concolor cougar
Eastern mole   Scalopus aquaticus/Scalopus aquaticus howelli
Eastern pipistrelle   Pipistrellus subflavus subflavus
Eastern wood rat   Neotoma floridana/Neotoma floridana floridana
European fallow deer   Dama dama
European wild hog   Sus scrofa cristatus
Evening bat   Nycticeius humeralis humeralis

False killer whale   Pseudorca crassidens
Feral hog   Sus scrofa
Finback whale   Balaenoptera physalus
Florida manatee   Trichechus manatus latirostris
Florida panther   Felis concolor coryi
Flying squirrel   Glaucomys volans saturatus
Fox squirrel   Sciurus Niger/Sciurus niger rufiventer

Goat   Capra hircus
Goose-beaked whale   Ziphius cavirostris
Grampus   Grampus griseus
Gray fox   Urocyon cinereoargenteus
Gray squirrel   Sciurus carolinensis
Gray wolf   Canis lupus

Harbor porpoise   Phocoena phocoena
Harbor seal   Phoca vitulina concolor
Harvest mouse   Reithrodontomys humulis
Hoary bat   Lasiurus cinereus cinereus
Horse   Equus caballus
House mouse   Mus musculus
Humpback whale   Megaptera novaeangliae

Killer whale   Orcinus orca

Least shrew   Cryptotis parva
Little brown myotis   Myotis lucifugus lucifugus

Long-beaked dolphin	*Stenella longirostris*
Long-beaked porpoise	*Stenella longirostris*
Long-tailed weasel	*Mustela frenata olivacea*
Man	*Homo sapiens*
Marsh rabbit	*Sylvilagus palustris*
Marsh rice rat	*Oryzomys palustris*
Meadow vole	*Microtus pennsylvanicus pennsylvanicus*
Mink	*Mustela vison*
Minke whale	*Balaenoptera acutorostrata*
Muskrat	*Ondatra zibethicus*
Nine-banded armadillo	*Dasypus novemcinctus/Dasypus novemcinctus mexicanus*
Northern yellow bat	*Lasiurus intermedius floridanus*
Norway rat	*Rattus norvegicus*
Nutria	*Myocastor coypus*
Old-field mouse	*Peromyscus polionotus*
Opossum	*Didelphis marsupialis*
Pine mouse	*Pitymys pinetorum*
Pygym sperm whale	*Kogia breviceps*
Raccoon	*Procyon lotor*
Rafinesque's big-eared bat	*Plecotus rafinesquii macrotis*
Red bat	*Lasiurus borealis borealis*
Red deer	*Cervus elaphus*
Red fox	*Vulpes fulva*
River otter	*Lutra canadensis*
Roof rat	*Rattus rattus*
Rough-toothed dolphin	*Steno bredanensis*
Rough-tooth porpoise	*Steno bredanensis*
Sei whale	*Balaenoptera borealis*
Seminole bat	*Lasiurus seminolus*
Sheep	*Ovis aries*
Sherman's pocker gopher	*Geomys fontanelus*
Short-finned blackfish	*Globicephala macrorhyncha*
Short-finned pilot whale	*Globicephala macrorhyncha*
Short-tailed shrew	*Blarina brevicauda*
Silver-haired bat	*Lasionycteris noctivagans*
Southeastern myotis	*Myotis austroriparius*
Southeastern pocket gopher	*Geomys pinetis*
Southeastern shrew	*Sorex longirostris/Sorex longirostris longirostris*
Southern flying squirrel	*Glaucomys volans saturatus*
Sperm whale	*Physeter catodon*
Spotted dolphin	*Stenella plagiodon*
Spotted porpoise	*Stenella plagiodon*
Star-nosed mole	*Condylura cristata/Condylura cristata parva*
Striped dolphin	*Stenella coeruleoalba*
Striped skunk	*Mephitis mephitis*
Swamp rabbit	*Sylvilagus aquaticus*
True's beaked whale	*Mesoplodon mirus*
Virginia opossum	*Didelphis virginiana*
West Indian manatee	*Trichechus manatus latirostris*
White-tailed deer	*Odocoileus virginianus*
White-tailed deer	*Odocoileus virginianus hiltonensis*
White-tailed deer	*Odocoileus virginianus nigribarbis*
White-tailed deer	*Odocoileus virginianus taurinsulae*
White-tailed deer	*Odocoileus virginianus virginianus*

Appendix Table 10. List of mammals identified in the Sea Island Characterization
                     study, arranged alphabetically by scientific name (Hall and
                     Kelson 1959; Golley 1962, 1966; Ridgeway 1972; Johnson el al.
                     1974; Lower 1974; Zingmark 1978.

Balaena mysticetus   Bowhead whale
Balaenoptera acutorostrata   Minke whale
Balaenoptera borealis   Sei whale
Balaenoptera edeni   Bryde's whale
Balaenoptera physalus   Finback whale
Blarina brevicauda   Short-tailed shrew
Bos taurus   Cow

Canis lupus   Gray wolf
Capra hircus   Goat
Castor canadensis   Beaver
Cervus elaphus   Red deer
Condylura cristata   Star-nosed mole
Condylura cristata parva   Star-nosed mole
Coryphaena hippurus   Dolphin
Cryptotis parva   Least shrew

Dama dama   European fallow deer
Dasypus novemcinctus   Nine-banded armadillo
Dasypus novemcinctus mexicanus   Nine-banded armadillo
Delphinus delphis   Common dolphin
Didelphis marsupialis   Opossum
Didelphis virginiana   Virginia opossum

Eptesicus fuscus fuscus   Big brown bat
Equus caballus   Horse
Eubalaena glacialis   Atlantic right whale

Felis concolor coryi   Florida panther
Felis concolor cougar   Eastern cougar

Geomys colonus   Colonial pocket gopher
Geomys cumberlandius   Cumberland Island pocket gopher
Geomys fontanelus   Sherman's pocket gopher
Geomys pinetis   Southeastern pocket gopher
Glaucomys volans saturatus   Flying squirrel/Southern flying squirrel
Globicephala macrorhyncha   Short-finned pilot whale/Short-finned blackfish
Grampus griseus   Grampus

Homo sapiens   Man

Kogia breviceps   Pygmy sperm whale
Kogia simus   Dwarf sperm whale

Lasionycteris noctivagans   Silver-haired bat
Lasiurus borealis borealis   Red bat
Lasiurus cinereus cinereus   Hoary bat
Lasiurus intermedius floridanus   Northern yellow bat
Lasiurus seminolus   Seminole bat
Lutra canadensis   River otter
Lynx rufus   Bobcat

Megaptera novaeangliae   Humpback whale
Mephitis mephitis   Striped skunk
Mesoplodon densirostris   Atlantic beaked whale
Mesoplodon europaeus   Antillean beaked whale
Mesoplodon mirus   True's beaked whale
Microtus pennsylvanicus pennsylvanicus   Meadow vole
Mus musculus   House mouse
Mustela frenata olivacea   Long-tailed weasel
Mustela vison   Mink
Myocastor coypus   Nutria
Myotis austroriparius   Southeastern myotis
Myotis lucifugus lucifugus   Little brown myotis

Neotoma floridana   Eastern wood rat
Neotoma floridana floridana   Eastern wood rat
Nycticeius humeralis humeralis   Evening bat

535

Odocoileus virginianus	White-tailed deer
Odocoileus virginianus hiltonensis	White-tailed deer
Odocoileus virginianus nigribarbis	White-tailed deer
Odocoileus virginianus taurinsulae	White-tailed deer
Odocoileus virginianus virginianus	White-tailed deer
Ondatra zibethicus	Muskrat
Orcinus orca	Killer whale
Oryzomys palustris	Marsh rice rat
Ovis aries	Sheep
Peromyscus gossypinus	Cotton mouse
Peromyscus gossypinus anastasae	Cotton mouse
Peromyscus polionotus	Old-field mouse
Phoca vitulina concolor	Harbor seal
Phocoena phocoena	Harbor porpoise/Common porpoise
Physeter catodon	Sperm whale
Pipistrellus subflavus subflavus	Eastern pipistrelle
Pitymys pinetorum	Pine mouse
Plecotus rafinesquii macrotis	Rafinesque's big-eared bat
Procyon lotor	Raccoon
Pseudorca crassidens	False killer whale
Rattus norvegicus	Norway rat
Rattus rattus	Roof rat
Reithrodontomys humulis	Harvest mouse
Scalopus aquaticus	Eastern mole
Scalopus aquaticus howelli	Eastern mole
Sciurus carolinensis	Gray squirrel
Sciurus niger	Fox squirrel
Sciurus niger rufiventer	Fox squirrel
Sibbaldus musculus	Blue whale
Sigmodon hispidus	Cotton rat
Sorex longirostris	Southeastern shrew
Sorex longirostris longirostris	Southeastern shrew
Stenella coeruleoalba	Striped dolphin
Stenella frontalis	Bridled dolphin
Stenella longirostris	Long-beaked dolphin/Long-beaked porpoise
Stenella plagiodon	Spotted dolphin/Spotted porpoise
Steno bredanensis	Rough-toothed porpoise/Rough-toothed dolphin
Sus scrofa	Feral hog
Sus scrofa cristatus	European wild hog
Sus scrofa domesticus	Domestic hog
Sylvilagus aquaticus	Swamp rabbit
Sylvilagus floridanus	Eastern cottontail
Sylvilagus palustris	Marsh rabbit
Tadarida brasiliensis cynocephala	Brazilian free-tailed bat
Trichechus manatus latirostris	Florida manatee/West Indian manatee
Tursiops truncatus	Atlantic bottle-nosed dolphin
Tursiops truncatus	Bottle-nosed dolphin
Urocyon cinereoargenteus	Gray fox
Ursus americanus	Black bear
Vulpes fulva	Red fox
Zalophus californianus	California sea lion
Ziphius cavirostris	Goose-beaked whale

REFERENCES CITED

Abbas, L. E. 1977. "To eel or not to eel." An economic analysis of a part-time eel fishing enterprise. Sea Grant Publication UNC-SG-77-02. 8 pp.

Academy of Natural Sciences of Philadelphia. 1974. Cooper River survey for 1973 for the E. I. DuPont de Nemours & Company. Philadelphia, Pa. 150 pp.

Adams, D. A. 1963. Factors influencing vascular plant zonation in North Carolina salt marshes. Ecology 44 (3):445-456.

Adams, J. G. 1970. Clupeids in the Altamaha River, Georgia. Ga. Game Fish Comm., Coastal Fish. Div. Contrib. Ser. No. 20. Brunswick. 27 pp.

Ager, L. A. 1975. Monthly food habits of various size groups of black crappie in Lake Okeechobee. Proc. Annu. Conf. Southeast. Assoc. Game Fish Comm. 29:336-341.

Ahearn, D. G., S. A. Crow, and W. L. Cook. 1977. Microbial interactions with pesticides in estuarine surface slicks. EPA Ecological Research Series, EPA-600/3-77-050.

Ahlstrom, E. H. 1938. Plankton rotatoria from North Carolina. J. Elisha Mitchell Soc. 54:88-110.

Allen, E. R. and W. T. Neill. 1949. Increasing abundance of the alligator in the eastern portion of its range. Herpetologica 5:109-112.

Allen, E. R. and W. T. Neill. 1952. Know your reptiles: the indigo snake. Fla. Wildl. Aug.:44-47.

Allen, E. R. and D. Swindell. 1948. Cottonmouth moccasin of Florida. Herpetologica 4(Supp. 1):1-15.

Allen, M. J. 1932. A survey of the amphibians and reptiles of Harrison County, Mississippi. Am. Mus. Novit. 542. 20 pp.

Allen, P. F. 1950. Ecological bases for land use planning in gulf coast marsh-lans. J. Soil Water Conserv. 5(2): 57-62.

Allen, P. H. 1958. A tidewater swamp forest and succession after clear-cutting. M.A. Thesis. Duke Univ., Durham, N.C. 48 pp.

Allen, R. M. and N. M. Scarbough. 1960. Growth of slash pine and pond pine on wet sites. U.S. Dep. Agric. Forest Serv. Res. Note SO-125.

Allwein, J. 1967. North American hydro-medusae from Beaufort, North Carolina. Vidensk. Medd. Dan. Naturhist. Foren. 130:117-136.

Almeida, J. L. 1969. A floristic and edaphic comparison of burned and un-burned pineland plots in Georgetown, South Carolina. M.S. Thesis. Univ. S.C., Columbia. 62 pp.

American Alligator Recovery Team. 1979. American alligator recovery plan. Technical draft. U.S. Dep. Inter., Fish Wildl. Serv. 28 pp.

American Birds. 1971. Seventy-first Christmas bird count. Am. Birds 25 (2):268-277.

American Birds. 1972. Seventy-second Christmas bird count. Am. Birds 26 (2):287-295.

American Birds. 1973. Seventy-third Christmas bird count. Am. Birds 27 (2):286-295.

American Birds. 1974. Seventy-fourth Christmas bird count. Am. Birds 28 (2):297-306.

American Birds. 1975. Seventy-fifth Christmas bird count. Am. Birds 29 (2):324-332.

American Birds. 1976. Seventy-sixth Christmas bird count. Am. Birds 30 (2):334-344.

American Birds. 1977. Seventy-seventh Christmas bird count. Am. Birds 31 (4):584-596.

American Chemical Society. 1969. Cleaning our environment: the chemical basis for action. Am. Chem. Soc., Washington, D.C.

American Ornithologists' Union. 1973. Thirty-second supplement to The American Ornithologists' Union check-list of North American birds. Auk 90 (2):411-419.

American Ornithologists' Union. 1976. Thirty-third supplement to the American Ornithologists' Union check-list of North American Birds. Auk 93(4):875-879.

Ancellin, J., M. Avargues, P. Bovard, P. Gueguéniat, and A. Vilquin. 1973. Biological and physicochemical aspects of the radioactive contamination of marine organisms and marine sediments, pp. 225-243. In: Symposium on the interaction of radioactive contaminants with the constituents of the marine environment. July 1972. Seattle, Wash. Int. Atomic Energy Agency, Vienna.

Anderson, J. D., S. L. Anderson, K. A.
Hawthorne, and S. C. Litwin. 1976.
Amphibians and reptiles from Jekyll
Island, Glynn County, Georgia.
Herpetol. Rev. 7(4):179-180.

Anderson, N. M. and J. T. Polhemus. 1976.
Water-striders (Hemiptera: Gerridae,
Veliidae, etc.), pp. 187-224. In:
L. Cheng, ed. Marine insects. American
Elsevier Publ. Co., New York.

Anderson, R. S. 1970. Predator-prey relationships and predation rates for
crustacean zooplankters from some
lakes in Western Canada. Can. J.
Zool. 48:1229-1240.

Anderson, W. D. 1973. A preliminary
study of the oyster parasite,
Labyrinthomyxa marina in the Wando
River, Charleston, South Carolina.
S.C. Mar. Resour. Cent., Charleston.
43 pp. (Unpubl.)

Anderson, W. D. 1976. A comparative
study of a salt water impoundment
with its adjacent tidal creek
pertinent to culture of Crassostrea
virginica (Gmelin). M.S. Thesis.
Old Dominion Univ., Norfolk, Va.
65 pp.

Anderson, W. D., Jr. 1964a. Fishes of
some South Carolina coastal plain
streams. Q. J. Fla. Acad. Sci.
27(3):31-54.

Anderson, W. D., Jr. 1964b. New records
of fishes for South Carolina, and
the occurrences of Chaetodipterus
Faber in fresh water. Copeia 1964:
243-244.

Anderson, W. D., Jr., J., K. Dias, R. K.
Dias, D. M. Cupka, and N. A.
Chamberlain. 1977. The macrofauna
of the surf zone off Folly Beach,
South Carolina. NOAA Tech. Rep.
NMFS SSRF 704. 23 pp.

Anderson, W. W. 1957. Early development,
spawning, growth, and occurrence of
silver mullet (Mugil curema) along
South Atlantic coast of United
States. Fish. Bull. 57:397-414.

Anderson, W. W. 1958. Larval development,
growth, and spawning of striped
mullet (Mugil cephalus) along the
South Atlantic coast of the United
States. Fish. Bull. 58:501-519.

Anderson, W. W. and J. W. Gehringer. 1957a.
Physical oceanographic, biological,
and chemical data--South Atlantic
coast of the United States, M/V
Theodore N. Gill Cruise 3. U.S. Fish
Wildl. Serv. Spec. Sci. Rep. Fish.
No. 210. 208 pp.

Anderson, W. W. and J. W. Gehringer. 1957b.
Physical oceanographic, biological,
and chemical data--South Atlantic
coast of the United States, M/V
Theodore N. Gill Cruise 4. U.S. Fish
Wildl. Serv. Spec. Sci. Rep. Fish.
No. 234. 192 pp.

Anderson, W. W. and J. W. Gehringer. 1958a.
Physical oceanographic, biological,
and chemical data--South Atlantic
coast of the United States, M/V
Theodore N. Gill Cruise 5. U.S. Fish
Wildl. Serv. Spec. Sci. Rep. Fish.
No. 248. 220 pp.

Anderson, W. W. and J. W. Gehringer. 1958b.
Physical oceanographic, biological,
and chemical data--South Atlantic
coast of the United States, M/V
Theodore N. Gill Cruise 6. U.S. Fish
Wildl. Serv. Spec. Sci. Rep. Fish.
No. 265. 99 pp.

Anderson, W. W. and J. W. Gehringer. 1959a.
Physical oceanographic, biological,
and chemical data--South Atlantic
coast of the United States, M/V
Theodore N. Gill Cruise 7. U.S. Fish
Wildl. Serv. Spec. Sci. Rep. Fish.
No. 278. 277 pp.

Anderson, W. W. and J. W. Gehringer. 1959b.
Physical oceanographic, biological,
and chemical data--South Atlantic
coast of the United States, M/V
Theodore N. Gill Cruise 8. U.S. Fish
Wildl. Serv. Spec. Sci. Rep. Fish.
No. 303. 227 pp.

Anderson, W. W. and J. W. Gehringer. 1959c.
Physical oceanographic, biological,
and chemical data--South Atlantic
coast of the United States, M/V
Theodore N. Gill Cruise 9. U.S. Fish
Wildl. Serv. Spec. Sci. Rep. Fish.
No. 313. 226 pp.

Anderson, W. W., J. W. Gehringer, and E.
Cohen. 1956a. Physical oceanographic,
biological and chemical data--South
Atlantic coast of the United States,
M/V Theodore N. Gill Cruise 1. U.S.
Fish Wildl. Serv. Spec. Sci. Rep.
Fish. No. 178. 160 pp.

Anderson, W. W., J. W. Gehringer, and E.
Cohen. 1956b. Physical oceanographic,
biological and chemical data--South
Atlantic coast of the United States,
M/V Theodore N. Gill Cruise 2. U.S.
Fish Wildl. Serv. Spec. Sci. Rep.
Fish. No. 198. 270 pp.

Andrews, E. F. 1917. Agency of fire in
the propagation of longleaf pine.
Bot. Gaz. 64:497-508.

Arbib, R. 1975. The Blue List for 1976.
American Birds 29(6):1067-1072.

538

Ardo, P. 1957. Studies in the marine shore dune ecosystem with special reference to the dipterous fauna. Opusc. Entomol. Suppl. 14:1-255.

Arndt, C. H. 1914. Some insects of the between tides zone. Proc. Indiana Acad. Sci. 1914:323-336.

Arner, D. H., H. R. Robinette, J. E. Frasier, and M. H. Gray. 1976. Effects of channelization of the Lunapolila River on fish, aquatic invertebrates, water quality and furbearers. U.S. Fish Wildl. Serv., Biol. Serv. Program. FWS 76/08.

Arthur, S. C. 1928. The fur animals of Louisiana. La. Dep. Conserv. Bull. 18. 433 pp.

Art, H. W. 1976. Ecological studies of the Sunken Forest, Fire Island National Seashore, New York. Natl. Park Serv. Sci. Monograph Ser. No. 7. U.S. Gov. Printing Off., Washington, D.C. 237 pp.

Art, H. W., F. H. Bormann, G. K. Voight, and G. M. Woodwell. 1974. Barrier island forest ecosystem: role of meterologic nutrient inputs. Science 184(4132):60-62.

Ashmole, N. P. 1971. Sea bird ecology and the marine environment, pp. 223-286. In: D. S. Farmer and J. R. King, eds. Avian biology. Vol. 1. Academic Press, New York.

Ashmole, N. P. and M. J. Ashmole. 1967. Comparative feeding ecology of sea birds of a tropical oceanic island. Peabody Mus. Nat. Hist. Yale Univ. Bull. 24:1-131.

Atkinson, L. P. 1975. Oceanographic observations in the Georgia Bight: data report for R/V Eastward cruises E-13-73 (4-11 September) and E-19-73 (8-9 December 1973). Ga. Mar. Sci. Cent. Tech. Rep. Ser. No. 75-6. Skidaway Island. 156 pp.

Audubon Field Notes. 1967. Sixty-seventh Christmas bird count. Audubon Field Notes 21(2):184-191.

Audubon Field Notes. 1968. Sixty-eighth Christmas bird count. Audubon Field Notes 22(2):203-209.

Audubon Field Notes. 1969. Sixty-ninth Christmas bird count. Audubon Field Notes 23(2):229-235.

Audubon Field Notes. 1970. Seventieth Christmas bird count. Audubon Field Notes 24(2):228-236.

Audubon, J. J. 1840. The birds of America. New York.

Axelrad, D. M., K. A. Moore, and M. E. Bender. 1976. Nitrogen, phosphorus, and carbon flux in Chesapeake Bay marshes. Va. Polytech. Inst. Water Resour. Res. Cent. Bull. 79. Blacksburg. 182 pp.

Axtell, R. C., ed. 1974. Training manual for mosquito and biting fly control in coastal areas. Univ. N.C. Sea Grant Publ. UNC-SG-74-08. N.C. State Univ., Raleigh. 250 pp.

Axtell, R. C. 1976. Horse flies and deer flies (Diptera: Tabanidae), pp. 415-446. In: L. Cheng, ed. Marine insects. American Elsevier Publ. Co., New York.

Bachman, J. 1837. Some remarks on the genus Sorex, with a monograph of the North American species. J. Acad. Nat. Sci. Phila. 7(2):362-402.

Baden, J. III, W. T. Batson, and R. Stalter. 1975. Factors affecting the distribution of vegetation of abandoned rice fields, Georgetown County, South Carolina. Castanea 40(3):171-184.

Badger, A. G. 1968. Oyster research in South Carolina, pp. 67-69. In: Proc. oyster culture workshop. Ga. Game Fish Comm. Contrib. Ser. 6. Atlanta.

Bahr, L. M., Jr. 1976. Energetic aspects of the intertidal oyster reef community at Sapelo Island, Georgia (U.S.A.). Ecology 57:121-131.

Bailey, J. W. 1851. Microscopical observations made in South Carolina, Georgia and Florida. Smithson. Contrib. Knowledge 2(8):1-48.

Bailey, L. H. 1951. Manual of cultivated plants.

Bailey, R. M., J. E. Fitch, E. S. Herald, E. Z. Lachner, C. C. Lindsey, C. R. Robins, and W. B. Scott. 1970. A list of common and scientific names of fishes from the United States and Canada. Am. Fish. Soc. Spec. Publ. 6. 150 pp. (First publ. 1960).

Bailey, R. M. and H. M. Harrison. 1948. Food habits of the southern channel catfish Ictalurus lacustris punctatus in the Des Moines River, Iowa. Trans. Am. Fish. Soc. 75:11-138.

Baker, J. M. 1970. The effects of oils on plants. Environ. Pollut. 1:27-44.

Baldwin, W. P. 1950. Recent advances in managing coastal plain impoundments for waterfowl. Presented Southeast. Assoc. Game & Fish Comm. Conf., Richmond, Va. 12 pp. (unpubl.)

Baldwin, W. P. 1956. Food supply key to attracting ducks. S.C. Wildl. 3(1):5-12.

Baldwin, W. P. 1968. Impoundments for waterfowl on South Atlantic and Gulf coastal marshes, pp. 127-133. In: J. D. Newsom, ed. Proceedings of the marsh and estuary management symposium. La. State Univ., Baton Rouge.

Balfour-Brown, F. 1958. British water beetles. Vol. 3. Ray Society, London.

Ballard, G. S. 1975a. A consideration of oyster culture in old rice fields, pp. 13-17. In: J. M. Dean, ed. The potential use of South Carolina rice fields for aquaculture. S.C. State Dev. Board, Columbia.

Ballard, G. S. 1975b. The ecology of the rice field, pp. 1-12. In: J. M. Dean, ed. The potential use of South Carolina rice fields for aquaculture. S.C. State Dev. Board, Columbia.

Bangs, O. 1898. The land mammals of peninsular Florida and the coast region of Georgia. Proc. Boston Soc. Nat. Hist. 28(7):157-235.

Banner, A. H. 1953. On a new genus and species of mysid from Southern Florida (Crustaceans, Malacostraca). Tulane Stud. Zool. 1:3-8.

Bara, M. O. 1971. Alligator research project. Annu. progress rep. Aug. 1970 - Dec. 1971. S.C. Wildl. Mar. Resour. Dep., Columbia. 39 pp.

Bara, M. O. 1976. American alligator investigations. Final study rep. Aug. 1970 - Dec. 1975. S.C. Wildl. Mar. Resour. Dep., Columbia. 39 pp.

Barans, C. A. and V. G. Burrell, Jr. 1976. Preliminary findings of trawling on the continental shelf off the Southeastern United States during four seasons (1973-1975). S.C. Mar. Resour. Cent. Tech. Rep. No. 13. 16 pp.

Barbour, R. W. 1956. A study of the cottonmouth Agkistrodon piscivorus leucostoma Troost, in Kentucky. Trans. Ky. Acad. Sci. 17:33-41.

Barbour, R. W. and W. H. Davis. 1969. Bats of America. Univ. Press of Kentucky. 286 pp.

Barclay, L. A. 1978. An ecological characterization of the sea islands and coastal plain of South Carolina and Georgia, pp. 711-720. In: Coastal zone '78· Proc. symposium on technical, environmental, socio-economic and regulatory aspects of coastal zone management, 14-16 March 1978, San Francisco. Am. Soc. Civil Eng., New York.

Barnes, R. D. 1953. The ecological distribution of spiders in non-forest maritime communities at Beaufort, North Carolina. Ecol. Monogr. 23: 315-337.

Barnes, R. D. 1963. Invertebrate zoology. W. B. Saunders Co., Philadelphia, Pa. 632 pp.

Barry, J. M. 1968. A survey of the native vascular plants of the Baruch Plantation. M.S. Thesis. Univ. S.C., Columbia. 197 pp.

Barry, J. M. and W. T. Batson. 1969. The vegetation of the Baruch Plantation, Georgetown, South Carolina, in relation to soil types. Castanea 34:71-77.

Bass, D. G., Jr. and V. G. Hitt. 1974. Ecological aspects of the redbreast sunfish, Lepomis auritus, in Florida. Proc. Annu. Conf. Southeast. Assoc. Game Fish Comm. 28:296-307.

Batson, W. T. and B. C. Blackwelder. 1974. Vertical distribution of epiphytic algae on Spartina alterniflora from transects along the Cooper and Wando rivers, pp. 22-35. In: F. P. Nelson, ed. The Cooper River environmental study. S.C. Water Resources Comm. State Water Plan Rep. No. 117. Columbia.

Batson, W. T., W. R. Kelley, L. F. Swails, and F. F. Welbourne, Jr. 1957. The vegetation of a mature beech-magnolia forest within the Gantt Tract. Univ. S.C. Publ. Biol. Ser. III, 2(2): 65-71.

Battle, J. D. 1892. An investigation of the coast waters of South Carolina with reference to oyster culture. Fish. Bull. 10:303-330.

Baughman, J. L. 1948. An annotated bibliography of oysters with pertinent material on mussels and other shellfish and an appendix on pollution. Tex. Agric. Mining Res. Found., Rockport Pilot College Stn. Tex. A&M Univ., College Stn. 794 pp.

Bayless, J. D. 1968. Survey and classification of the Edisto River and tributaries, South Carolina. S.C. Wildl. Mar. Resour. Dep. Rep., Columbia. 40 pp.

Bayless, J. D. and H. A. Loyacano, Jr. 1979. Coastal plains eel study completion report. Coastal Plains Regional Commission and S.C. Wildl. Mar. Resour. Dep., Charleston. 78 pp.

Baynard, O. E. 1912. Food of herons and ibises. Wilson Bull. 24(4):167-169.

Bearden, C. M. 1961. Notes on postlarvae of commercial shrimp (Penaeus) in South Carolina. Contrib. Bears Bluff Lab. No. 33. Wadmalaw Island, S.C. 8 pp.

Bearden, C. M. 1963. A contribution to the biology of the king whitings, genus Menticirrhus, of South Carolina. Contrib. Bears Bluff Lab. No. 38. Wadmalaw Island, S.C. 27 pp.

Bearden, C. M. 1964. Distribution and abundance of Atlantic croaker, Micropogon undulatus, in South Carolina. Contrib. Bears Bluff Lab. No. 40. Wadmalaw Island, S.C. 23 pp.

Bearden, C. M. 1965. Elasmobranch fishes of South Carolina. Contrib. Bears Bluff Lab. No. 42. Wadmalaw Island, S.C. 22 pp.

Bearden, C. M. 1967. Salt-water impoundments for game fish in South Carolina. Prog. Fish-Cult. 29(3):123-128.

Bearden, C. M. and C. H. Farmer III. 1972. Fishery resources of Port Royal Sound estuary, pp. 203-212. In: Port Royal Sound Environmental Study. S.C. Water Resources Comm., Columbia.

Bearden, C. M. and M. D. McKenzie. 1971. An investigation of the offshore demersal fish resources of South Carolina. S.C. Mar. Resour. Cent. Tech. Rep. No. 2. 19 pp.

Bears Bluff Laboratories, Inc. 1956. Annual report 1954-1955. Contrib. Bears Bluff Lab. No. 18. Wadmalaw Island, S.C. 15 pp.

Bears Bluff Laboratories, Inc. 1964. Biological studies of Charleston Harbor, S.C., and the Santee River. Wadmalaw Island, S.C. Submitted to U.S. Fish Wildl. Serv., Bur. Sport Fish. Wildl. Contract No. 14-16-0004-1024. 106 pp.

Bears Bluff Laboratories, Inc. 1965. Biological studies of Price Inlet area from Charleston Harbor to Bulls Bay, South Carolina. Wadmalaw Island, S.C. 54 pp. (Unpubl.)

Beckett, T. A. 1966. Deveaux Bank - 1964 and 1965. Chat 30(4): 93-100.

Beckett, T. A. 1975. Ornithological study of the areas proposed for cutting in the Santee-Cooper Basin, pp. 96-99. In: F. P. Nelson, ed. Interim report of the taskforce under authority of the legislative committee created by concurrent Resolution 217 to evaluate the harvesting of timber in the Santee Swamp. S.C. Water Resources Comm., Columbia.

Bell, R. and J. I. Richardson. 1978. An analysis of tag recoveries from loggerhead sea turtles (Caretta caretta) nesting on Little Cumberland Island, Georgia, pp. 20-24. In: G. E. Henderson, ed. Proceedings of the Florida and interregional conference on sea turtles, 24-25 July 1976, Jensen Beach, Florida. Fla. Mar. Res. Publ. 33.

Bell, S. S. 1979. Short- and long-term variation in high marsh meiofauna community. Estuarine Coastal Mar. Sci. 9:331-350.

Bell, S. S. and B. C. Coull. 1978. Field evidence that shrimp predation regulates meiofauna. Oecologia, Berl. 35:141-148.

Bell, S. S., M. C. Watzin, and B. C. Coull. 1978. Biogenic structure and its effect on the spatial heterogeneity of meiofauna in a salt marsh. J. Exp. Mar. Biol. Ecol. 35:141-148.

Bellis, V. 1974. Medium salinity plankton systems, pp. 358-396. In: H. T. Odum, B. J. Copeland, E. A. McMahan, eds. Coastal ecological systems of the United States. Vol. 2. The Conservation Foundation, Washington, D.C.

Benijts-Claus, C. and F. Benijts. 1975. The effect of low lead and zinc concentrations on the larval development of the mud-crab Rhithropanopeus harrisii Gould, pp. 43-52. In: J. H. Koeman and J. J. T. W. A. Strik, eds. Sublethal effects of toxic chemicals on aquatic animals. Elsevier Sci. Publ. Co., New York.

Benke, A. C. 1972. An experimental field study on the ecology of co-existing larval odonates. Ph.D. Dissertation. Univ. Ga., Athens. 112 pp.

Benke, A. C. 1976. Dragonfly production and prey turnover. Ecology 57:915-927.

Benke, A. C. and S. S. Benke. 1975. Comparative dynamics and life histories of co-existing dragonfly populations. Ecology 56:302-317.

541

Bennett, D. H., J. W. Gibbons, and J. C. Franson. 1970. Terrestrial activity in aquatic turtles. Ecology 51:738-740.

Bennett, G. W. 1962. Management of artificial lakes and ponds. Reinhold Publ. Corp., New York. 283 pp.

Bent, A. C. 1961. Life histories of North American birds of prey. 2 pts. Dover Publ., Inc., New York. (First publ. 1937, 1938.)

Bent, A. C. 1962a. Life histories of North American shore birds. 2 pts. Dover Publ., Inc., New York. (First publ. 1927, 1929.)

Bent, A. C. 1962b. Life histories of North American wild fowl. 2 pts. Dover Publ., Inc., New York. 314 pp. (First publ. 1923.)

Bent, A. C. 1963a. Life histories of North American gulls and terns. Order Longipennes. Dover Publ., Inc., New York. 337 pp. (First publ. 1921.)

Bent, A. C. 1963b. Life histories of North American wood warblers. 2 pts. Dover Publ., Inc., New York. 734 pp. (First publ. 1953.)

Bent, A. C. 1963c. Life histories of North American marsh birds. Dover Publ., Inc., New York. 392 pp. (First publ. 1926.)

Bent, A. C. 1964. Life histories of North American woodpeckers. Dover Publ., Inc., New York. 334 pp. (First publ. 1939.)

Bent, A. C. 1965. Life histories of North American blackbirds, orioles, tanagers, and allies. Dover Publ., Inc., New York. 549 pp. (First publ. 1958.)

Bent, A. C. 1968. Life histories of North American cardinals, grosbeaks, buntings, towhees, finches, sparrows, and allies. 3 pts. Dover Publ., Inc., New York. 1889 pp.

Berg, C. O., D. J. Karlin, and J. C. Mackiewicz. 1955. Fly larvae that kill snails. Am. Malacol. Union Inc. Annu. Rep. 1955:9-10.

Berry, F. H. 1959. Young jack crevalles (Caranx sp.) off the Southeastern Atlantic Coast of the United States. Fish. Bull. 59(152):417-535.

Betts, M. M. 1955. The food of titmice in oak woodland. J. Anim. Ecol. 24:282-323.

Bigelow, H. B. and W. W. Welsh. 1925. Fishes of the Gulf of Maine. Bull. U.S. Bur. Fish., 40(1):1-567.

Binmore, T. V. 1964. Oyster seed farming proves good venture. Natl. Fishermen - Maine Coast 45(9):24-28.

Bishop, J. M. and J. V. Miglarese. 1978. Carnivorous feeding in adult striped mullet. Copeia 1978(4):705-707.

Biswell, H. H., W. O. Shepherd, B. L. Southwell, and T. S. Boggess, Jr. 1943. Native forage plants of cutover forest lands in the coastal plain of Georgia. Ga. Coastal Plain Exp. Stn. (Tifton) Bull. 37. 43 pp.

Blackwelder, B. C. 1972. Algae of the marshlands, pp. 265-269. In: Port Royal Sound environmental study. S.C. Water Resources Comm., Columbia.

Blaney, R. M. 1971. An annotated checklist and biogeographic analysis of the insular herpetofauna of the Appalachicola region, Florida. Herpetologica 27:406-430.

Blus, L. J. 1977. Status of colonial nesters in South Carolina. (Unpubl.)

Blus, L. J., A. A. Belisle, and R. M. Prouty. 1974a. Relations of the brown pelican to certain environmental pollutants. Pestic. Monit. J. 7(3/4):181-194.

Blus, L. J., T. G. Lamont, and B. S. Neely, Jr. 1979. Effects of organochlorine residues on eggshell thickness, reproduction and population status of brown pelicans (Pelecanus occidentalis) in South Carolina and Forida, 1969-76. Pestic. Monit. J. 12(4):172-184.

Blus, L. J., B. S. Neely, Jr., A. A. Belisle, and R. M. Prouty. 1974b. Organochlorine residues in brown pelican eggs: relation to reproductive success. Environ. Pollut. 7:81-91.

Blus, L. J., B. S. Neeley, Jr., T. G. Lamont, and B. Mulhern. 1977. Residues of organochlorines and heavy metals in tissues and eggs of brown pelicans, 1969-73. Pestic. Monit. J. 11(1):40-53.

Boesch, D. F. 1972. Species diversity of marine macrobenthos in the Virginia area. Chesapeake Sci. 13:206-211.

Boesch, D. F. 1973. Classification and community structure of macrobenthos in the Hampton Roads area, Virginia. Mar. Biol. 21:226-244.

Boesch, D. F. 1974. Diversity, stability and response to human disturbance in estuarine ecosystems. Proc. Int. Congr. Zool., The Hague. 1:109-114.

Boesch, D. F., R. J. Diaz, and R. W. Virnstein. 1976b. Effects of tropical storm Agnes on soft-bottom macrobenthic communities of the James and York estuaries and the Lower Chesapeake Bay. Chesapeake Sci. 17:246-259.

Boesch, D. F., C. H. Hershner, and J. H. Milgram. 1974. Oil spills and the marine environment. Ballinger Publ. Co., Cambridge, Mass. 114 pp.

Boesch, D. F., M. L. Wass, and R. W. Virnstein. 1976a. The dynamics of estuarine benthic communities, pp. 177-196. In: M. L. Wiley, Adaptation to the estuary. Academic Press, Inc., New York.

Boney, A. D. 1970. Toxicity studies with an oil-spill emulsifier and the green alga Prasinocladus marinus. J. Mar. Biol. Assoc. U.K. 50:461-473.

Boney, A. D. 1975. Phytoplankton. Inst. Biol. Stud. Biol. No. 52. Edward Arnold Publ. Ltd., London. 116 pp.

Bookhout, C. G., J. D. Costlow, Jr., and R. Monroe. 1976. Effects of methoxychlor on larval development of mud-crab and blue crab. Water Air Soil Pollut. 5:349-365.

Bookhout, C. G., J. D. Costlow, Jr., and R. Monroe. 1979. Kepone R effects on development of Callinectes sapidus and Rhithropanopeus harrisii. EPA Ecol. Res. Ser. EPA-600/3-79-104.

Bookhout, C. G. and R. J. Monroe. 1977. Effects of malathion on the development of crabs, pp. 3-19. In: F. J. Vernberg, A. Calabrese, F. P. Thunberg, and W. B. Vernberg, eds. Physiological responses of marine biota to pollutants. Academic Press, New York.

Bookhout, C. G., A. J. Wilson, Jr., T. W. Duke, and J. I. Lowe. 1972. Effects of mirex on the larval development of two crabs. Water Air Soil Pollut. 1:165-180.

Bormann, F. H. 1956. Ecological implications of changes in photosynthetic response of Pinus taeda seedlings during ontogeny. Ecology 37:70-74.

Borror, D. J. and D. M. Delong. 1964. An introduction to the study of insects. Holt, Rinehart, and Winston, Inc., New York. 819 pp.

Borthwick, P. W., T. W. Duke, A. J. Wilson, Jr., J. I. Lowe, J. M. Patrick, Jr., and J. C. Oberheu. 1973. Accumulation and movement of mirex in selected estuaries of South Carolina, 1969-71. Pestic. Monit. J. 7(1):6-26.

Bouchon-Brandely, M. 1882. De la sexualite chez l'huitre ordinaire (O. edulis) et chez l'huitre Portugaise (O. angulata). Fecondation artificielle de l'huitre Portugaise. C. R. Hebd. Seances Acad. Sci. 95(5):256-259. Translated by J. A. Ryder. Bull. U.S. Fish Comm. 1882(2):339-341.

Bourdeau, P. F. and D. A. Adams. 1956. Factors in vegetational zonation of salt marshes near Southport, N.C. Bull. Ecol. Soc. Am. 37(3):68.

Bourdeau, P. F. and H. J. Oosting. 1959. The maritime live oak forest in North Carolina. Ecology 40:148-152.

Bourn, W. S. and C. Cottam. 1939. The effect of lowering water levels on marsh wildlife. Trans. N. Am. Wildl. Conf. 4:343-350.

Bourn, W. S. and C. Cottam. 1950. Some biological effects of ditching tidewater marshes. U.S. Fish Wildl. Serv. Res. Rep. 19. 30 pp.

Bourne, W. R. 1968. Observation of an encounter between birds and floating oil. Nature 219(5154):632.

Bousfield, E. L. 1955. Ecological control of the occurrence of barnacles in the Miramichi Estuary. Natl. Mus. Can. Bull. No. 137. Biol. Ser. 46:1-69.

Bowden, J. and C. G. Johnson. 1976. Migrating and other terrestrial insects at sea, pp. 97-117. In: L. Cheng, ed. Marine insects. American Elsevier Publ. Co., New York.

Bowman, T. E. 1971. The distribution of calanoid copepods off the Southeastern United States between Cape Hatteras and Southern Florida. Smithson. Contrib. Zool. 96. 58 pp.

Bowman, T. E. and H. E. Gruner. 1973. The families and genera of Hyperiidea (Crustacea, Amphipoda). Smithson. Contrib. Zool. 146:1-64.

Bowman, T. E. and J. C. McCain. 1967. Distribution of the planktonic shrimp, _Lucifer_, in the Western North Atlantic. Bull. Mar. Sci. 17:660-671.

Boyce, S. G. 1954. The salt spray community. Ecol. Monogr. 24:29-67.

Boyd, C. E. 1970. Production, mineral accumulation and pigment concentrations in _Typha latifolia_ and _Scirpus americanus_. Ecology 51(2): 288-290.

Boyd, C. E. 1971. The limnological role of aquatic macrophytes in their relationship to reservoir management, pp. 153-166. In: G. E. Hall, ed. Reservoir fisheries and limnology. Am. Fish. Soc. Spec. Publ. No. 8. Washington, D.C.

Boyd, C. E. and L. W. Hess. 1970. Factors influencing shoot production and mineral nutrient levels in _Typha latifolia_. Ecology 51:296-300.

Bozeman, J. R. 1965. Floristic and edaphic studies of the Altamaha River sand ridge, Georgia. (Notes from the herbarium of Georgia Southern College.) Elliottia 3:1-74.

Bozeman, J. R. 1971. A sociologic and geographic study of the sand ridge vegetation in the coastal plain of Georgia. Ph.D. Dissertation. Univ. N.C., Chapel Hill. 272 pp.

Bozeman, J. R. 1975. Vegetation, pp. 63-117. In: H. O. Hillestad, J. R. Bozeman, A. S. Johnson, C. W. Berisford, and J. I. Richardson. The ecology of the Cumberland Island National Seashore, Camden County, Georgia. Ga. Mar. Sci. Cent. Tech. Rep. Ser. No. 75-5. Skidaway Island.

Bozeman, J. R. and J. R. Darrell. 1975. The river swamp ecosystem and related vegetation. A study of Georgia's coastal area. Ga. Dep. Nat. Resour., Off. Planning Res., Atlanta. 37 pp.

Bradbury, H. M. 1938. Mosquito control operations on tide marshes in Massachusetts and their effect on shore birds and waterfowl. J. Wildl. Manage. 2(2):49-52.

Braun, E. L. 1950. The southeastern evergreen forest, pp. 280-304. In: Deciduous forests of Eastern North America. The Blarision Co., Philadelphia.

Briese, L. A. and M. H. Smith. 1974. Seasonal abundance and movement of nine species of small mammals. J. Mammal. 55(3):615-629.

Brinkhurst, R. O., R. E. Boltt, M. G. Johnson, S. Mozley, and A. V. Tyler. 1974. The benthos of lakes. St. Martin's Press, New York. 190 pp.

Brinson, M. M. 1975. A study of nutrient cycling in a riverine swamp forest ecosystem in North Carolina. Final Rep. to N.C. Sci. Tech. Comm. East Carolina Univ., Greenville, N.C. 65 pp.

Briscoe, C. B. 1961. Germination of cherrybark and nuttall oak acorns following flooding. Ecology 42(2): 430-431.

Broadfoot, W. M. 1967. Shallow-water impoundment increases soil moisture and growth of hardwoods. Soil Sci. Soc. Am. Proc. 31:562-564.

Broadfoot, W. M. 1973. Raised water tables affect southern hardwood growth. U.S. Dep. Agric. Forest Serv. Res. Note SO-168.

Broadfoot, W. M. and H. L. Williston. 1973. Flooding effects on southern forests. J. For. 71(9): 584-587.

Brongersma, L. D. 1972. European Atlantic turtles. Zool. Verh. (Leiden) 121:245-247.

Brooks, R. H., P. L. Brezonik, H. D. Putnam, and M. A. Keirn. 1971. Nitrogen fixation in an estuarine environment: The Wacasassa on the Florida gulf coast. Limnol. Oceanogr. 16:701-710.

Brown, C. A. 1959. Vegetation of the Outer Banks of North Carolina. La. State Univ. Stud. Coastal Stud. Ser. No. 4. 179 pp.

Brown, E. B. 1955. Some current British soil pest problems, pp. 256-267. In: K. M. Kevan, ed. Soil zoology. Academic Press, Inc., New York.

Brown, H. J. 1930. The desmids of the southeastern coastal plain region of United States. Trans. Am. Microsc. Soc. 49(2):97-137.

Brown, S. G. 1975. Relation between stranding mortality and population abundance of smaller cetacea in the Northeast Atlantic Ocean. J. Fish. Res. Board Can. 32:1095-1099.

Brown, T. W. 1963. The ecology of cypress heads in North Central Florida. M.S. Thesis. Univ. Fla., Gainesville. 59 pp.

Buckley, P. A. and F. G. Buckley. 1976. Guidelines for the protection and management of colonially nesting waterbirds. U.S. Dep. Inter., Natl. Park Serv., North Atlantic Regional Off., Boston, Mass.

Buckley, P. A. and F. G. Buckley. 1977. Hexagonal packing of royal tern nests. Auk 94:36-43.

Buele, J. D. and A. T. Studholme. 1942. Cottontail rabbit nests and nestlings. J. Wildl. Manage. 6: 133-140.

Buell, M. F. 1939. Peat formation in the Carolina Bays. Bull. Torrey Bot. Club 66:483-487.

Buell, M. F. 1946. Jerome Bog, a peat-filled "Carolina Bay." Bull. Torrey Bot. Club 73(1):24-33.

Bulak, J. S. and T. A. Curtis. 1978. Santee Cooper rediversion project. Annu. progress rep. Project SCR-1-1. 1 Nov. 1976 - 31 Dec. 1977. S.C. Wildl. Mar. Resour. Dep., Columbia. 70 pp.

Bullis, H. R., Jr. and S. B. Drummond. 1978. Sea turtle captures off the Southeastern United States by exploratory fishing vessels 1950-1976, pp. 45-50. In: G. E. Henderson, ed. Proceedings of the Florida and interregional conference on sea turtles. Jensen Beach, Fla., July 1976. Fla. Mar. Res. Publ. 33.

Buntz, J. 1966. Stomach analysis of chain pickerel, *Esox niger*, of South Central Florida. Proc. Annu. Conf. Southeast. Assoc. Game Fish Comm. 20:315-318.

Burchard, R. P. 1972. Chesapeake Bay bacteria able to cycle carbon, nitrogen, sulfur, and phosphorus. Chesapeake Sci. 13(2):179-180.

Burger, J. 1976. Behavior of hatchling diamondback terrapins (*Malaclemys terrapin*) in the field. Copeia 4: 742-748.

Burger, J. 1978. The pattern and mechanism of nesting in mixed-species heronries, pp. 45-58. In: A. Sprunt, J. C. Ogden, and S. Winckler, eds. Wading birds. Natl. Audubon Soc. Res. Rep. 7. New York.

Burger, W. L., P. W. Smith, and H. M. Smith. 1949. Notable records of reptiles and amphibians in Oklahoma, Arkansas, and Texas. J. Tenn. Acad. Sci. 24:130-134.

Burges, A. and F. Rawe. 1967. Soil biology. Academic Press, New York. 532 pp.

Burleigh, T. D. 1958. Georgia birds. Univ. Okla. Press, Norman. 746 pp.

Burns, R. W. 1974. Species and diversity of larval fishes in a high marsh tidal creek. M.S. Thesis. Univ. S.C., Columbia.

Burrell, V. G. 1968. The ecological significance of a ctenophore, *Mnemiopsis leidyi* (A. Agassiz), in a fish nursery ground. M.S. Thesis. Coll. William and Mary, Williamsburg, Va. 62 pp.

Burrell, V. G. 1972. Distribution and abundance of calanoid copepods in the York River Estuary, Virginia 1968 and 1969. Ph.D. Dissertation. Coll. William and Mary, Williamsburg, Va. 234 pp.

Burrell, V. G., Jr. 1975. The relationship of proposed offshore nuclear power plants to marine fisheries of the South Atlantic region of the United States, pp. 491-495. In: Proc. Ocean '75 Conf., San Diego, Calif.

Burrell, V. G., Jr. 1977. Mortalities of oysters and hard clams associated with heavy runoff in the Santee River system, South Carolina, in the spring of '75. Proc. Natl. Shellfish. Assoc. 67: 35-43.

Burrell, V. G., Jr. and W. A. Van Engel. 1976. Predation by and distribution of a ctenophore, *Mnemiopsis leidyi* A. Agassiz, in the York River Estuary. Estuarine Coastal Mar. Sci. 4:235-242.

Burton, E. M. 1970. Supplement, pp. 573-642. In: A. Sprunt, Jr. and E. B. Chamberlain. South Carolina bird life. Univ. S.C. Press, Columbia. (First publ. 1949.)

Burton, J. D. 1972. Prolonged flooding inhibits growth of loblolly pine seedlings. U.S. Dep. Agric. Forest Serv. Res. Note SO-124.

Butler, P. A. 1966. Pesticides in the marine environment. J. Appl. Ecol. 3(Suppl.):253-259.

Butler, P. A. 1968a. Pesticides in the estuary, pp. 120-124. In: J. D. Newsom, ed. Proceedings of the marsh and estuary management symposium. La. State Univ., Baton Rouge.

Butler, P. A. 1968b. Pesticide residues in estuarine mollusks, pp. 107-121. In: Proc. natl. symposium estuarine pollution. Stanford Univ., Stanford, Calif. 1967.

Butler, P. A. 1969a. Monitoring pesticide pollution. Bioscience 19(10):889-891.

Butler, P. A. 1969b. Significance of DDT residues in estuarine fauna, pp. 205-220. In: Chemical fall-out. Charles C. Thomas, Springfield, Ill.

Butler, P. A. 1973. Residues in fish, wildlife, and estuaries: organochlorine residues in estuarine mollusks, 1965-72. National Pesticides Monitoring Program. Pestic. Monit. J. 6(4):

Butler, R. D. 1968. Report of the survey of county agents on the imported fire ant. U.S. Dep. Agric., ARS.

Cadee, G. C. and J. Hegeman. 1974. Primary production of the benthic microflora living on tidal flats in the Dutch Wadden Sea. Neth. J. Sea Res. 8:260-291.

Cadieu, C. R. and J. D. Bayless. 1968. Anadromous fish survey of the Santee and Cooper rivers, South Carolina. Job completion rep. Project AFS-2-1. S.C. Wildl. Mar. Resour. Dep., Columbia. 92 pp. (Unpubl.)

Cagle, F. R. 1952. A Louisiana terrapin population (Malaclemys), Copeia 1952:74-76.

Cain, R. L. 1973. The annual occurrence, abundance and diversity of fishes in an intertidal creek. M.S. Thesis. Univ. S.C., Columbia. 80 pp.

Cain, R. L. and J. M. Dean. 1976. Annual occurrence, abundance and diversity of fish in a South Carolina intertidal creek. Mar. Biol. 36:369-379.

Cain, T. D. 1972. Additional epifauna of a reef off North Carolina. J. Elisha Mitchell Sci. Soc. 88:79-82.

Cairns, J., Jr. 1968. A comparison of the toxicity of some common industrial waste components tested individually and combined. Prog. Fish-Cult. 30(1):3-8.

Calabrese, A., J. R. MacInnes, D. A. Nelson, and J. E. Miller. 1977. Survival and growth of bivalve larvae under heavy-metal stress. Mar. Biol. 41:179-184.

Calder, D. R. 1976. The zonation of hydroids along salinity gradients in South Carolina estuaries, pp. 165-174. In: G. O. Mackie, ed. Coelenterate ecology and behavior. Plenum Press, New York.

Calder, D. R., C. M. Bearden, and B. B. Boothe, Jr. 1976. Environmental inventory of a small neutral embayment: Murrells Inlet, South Carolina. S.C. Mar. Resour. Cent. Tech. Rep. No. 10. 58 pp.

Calder, D. R., C. M. Bearden, B. B. Boothe, Jr., and R. W. Tiner, Jr. 1977a. A reconnaissance of the macrobenthic communities, wetlands, and shellfish resources of Little River Inlet, North Carolina and South Carolina. S.C. Mar. Resour. Cent. Tech. Rep. No. 17. 58 pp.

Calder, D. R. and B. B. Boothe. 1977. Data from some subtidal quantitative benthic samples taken in estuaries of South Carolina. S.C. Mar. Resour. Cent. Data Rep. No. 3, Charleston. 35 pp.

Calder, D. R., B. B. Boothe, Jr., and M. S. Maclin. 1977b. A preliminary report on estuarine macrobenthos of the Edisto and Santee river systems, South Carolina. S.C. Mar. Resour. Cent. Tech. Rep. No. 22. 50 pp.

Calder, D. R. and V. G. Burrell, Jr. 1978. Phylum Ctenophora, pp. 94-96. In: R. G. Zingmark, ed. An annotated checklist of the biota of the coastal zone of South Carolina. Univ. S.C. Press, Columbia.

Calder, D. R., P. J. Eldridge, and E. B. Joseph, eds. 1974a. The shrimp fishery of the Southeastern United States: a management planning profile. S.C. Mar. Resour. Cent. Tech. Rep. No. 5. 229 pp.

Calder, D. R., P. J. Eldridge, and M. H. Shealy, Jr. 1974b. Description of resource. S.C. Mar. Resour. Cent. Tech. Rep. No. 5:4-38.

Calder, D. R. and B. S. Hester. 1978. Phylum Cnidaria, pp. 87-93. In: R. G. Zingmark, ed. An annotated checklist of the biota of the coastal zone of South Carolina. Univ. S.C. Press, Columbia.

Caldwell, D. K., ed. 1959. The loggerhead turtles of Cape Romain, South Carolina. Bull. Fla. State Mus. Biol. Sci. 4(10):319-348.

Caldwell, D. K. 1962. Comments on the nesting behavior of Atlantic loggerhead sea turtles, based primarily on tagging returns. Q. J. Fla. Acad. Sci. 25(4):287-302.

Caldwell, D. K. 1968. Baby loggerhead turtles associated with sargassum weed. Q. J. Fla. Acad. Sci. 31(4): 271-272.

Caldwell, D. K. and M C. Caldwell. 1974. Marine mammals from the Southeastern United States coast: Cape Hatteras to Cape Canaveral, pp. 704-772. In: M. H. Roberts, Jr. A socio-economic environmental baseline summary for the South Atlantic region between Cape Hatteras, North Carolina and Cape Canaveral, Florida. Vol. 3. Chemical and biological oceanography. Va. Inst. Mar. Sci.

Caldwell, D. K., H. N. Neuhauser, M. C. Caldwell, and H. W. Collidge. 1971. Recent records of marine mammals from the coasts of Georgia and South Carolina. Cetology 5: 1-12.

Caldwell, J. A. 1963. An investigation of raccoons in North-central Florida. M.S. Thesis. Univ. Fla., Gainesville. 108 pp.

Camburn, K. E., R. L. Lowe, and D. L. Stoneburner. 1978. The hapto-benthic diatom flora of Longbranch Creek, South Carolina. Nova Hedwigia, Band 30:149-279.

Cameron, G. N. 1972. Analysis of insect trophic diversity in two salt marsh communities. Ecology 53:58-73.

Cammen, L. M. 1976. Macroinvertebrate colonization of Spartina marshes artificially established in dredge spoil. Estuarine Coastal Mar. Sci. 4:357-372.

Campbell, P. H. 1973. Studies on brackish water phytoplankton. Univ. N.C. Sea Grant, Chapel Hill. UNC-SG-73. 395 pp.

Campbell, W. M., J. M. Dean, and W. D. Chamberlain. 1975. Environmental inventory of Kiawah Island. Environmental Research Center, Inc. Columbia, S.C. 703 pp.

Carlander, K. D. 1969. Handbook of freshwater fishery biology. Vol. 1. Iowa State Univ. Press, Ames. 752 pp.

Carpenter, E. J., P. D. Van Raalte, and T. Valiela. 1978. Nitrogen fixation by algae in a Massachusetts salt marsh. Limnol. Oceanogr. 23(2):318-327.

Carr, A. 1940. A contribution to the herpetology of Florida. Univ. Fla. Publ. Biol. Sci. Ser. 3(1):1-118.

Carr, A. 1967. So excellente a fishe. Natural History Press, Ithaca, N.Y. 212 pp.

Carr, A. F. and C. J. Goin. 1942. Rehabilitation of Natrix sipedon taeniaia Cope. Proc. N. Engl. Zool. Club 21:47-54.

Carriker, M. R. 1951. Ecological observations on the distribution of oyster larvae in New Jersey estuaries. Ecol. Monogr. 21(1): 19-38.

Carriker, M. R. 1956. Biology and propagation of young hard clams, Mercenaria mercenaria. J. Elisha Mitchell Sci. Soc. 72:57-60.

Carriker, M. R. 1959. The role of physical and biological factors in the culture of Crassostrea and Mercenaria in a salt-water pond. Ecol. Monogr. 29(3):219-266.

Carriker, M. R. 1967. Ecology of estuarine benthic invertebrates: a perspective, pp. 442-487. In: G. H. Lauff, ed. Estuaries. Am. Assoc. Adv. Sci. Publ. No. 83. Washington, D.C.

Carter, M. R., L. A. Burns, T. R. Cavinder, K. R. Dugger, P. L. Fore, D. B. Hicks, H. L. Revells, and T. W. Schmidt. 1973. Ecosystems analysis of the Big Cypress Swamp and estuaries. South Florida Ecological Study. U.S. Environ. Prot. Agency, Athens.

Carter, N. 1932. A comparative study of the algal flora of two salt marshes. Part 1. J. Ecol. 20: 341-370.

Carter, N. 1933. A comparative study of the algal flora of two salt marshes. Parts 2 and 3. J. Ecol. 21:128-208; 385-403.

Caswell, W. B. 1977. Ground water guidebook for the State of Maine. Maine Geol. Surv. Open File Rep., Augusta. 202 pp.

Catesby, M. 1771. The natural history of Carolina, Florida and the Bahama Islands. 2 vols. London. (First publ. 1754.)

Catts, E. P., Jr. 1957. Mosquito prevalence on impounded and ditched salt marshes, Assawoman Wildlife Area, Delaware. M.S. Thesis, Univ. Del., Newark.

547

Cerame-Vivas, M. J. and I. E. Gray.
1966. The distributional pattern
of benthic invertebrates of the
continental shelf off North
Carolina. Ecology 47:260-270.

Chabreck, R. H. 1960. Coastal marsh
impoundments for ducks in
Louisiana. Proc. Annu. Conf.
Southeast. Assoc. Game Fish Comm.
14:24-29.

Chabreck, R. H. 1966. The movement of
alligators in Louisiana. Proc.
Annu. Conf. Southeast. Assoc. Game
Fish Comm. 19:102-110.

Chabreck, R. H. 1968. The relation of
cattle and cattle grazing to marsh
wildlife and plants in Louisiana.
Proc. Annu. Conf. Southeast. Assoc.
Game Fish Comm. 22:55-58.

Chabreck, R. H. 1972. The foods and
feeding habits of alligators from
fresh and saline environments in
Louisiana. Proc. Annu. Conf.
Southeast. Assoc. Game Fish Comm.
25:117-124.

Chamberlain, E. B., ed. 1962. Southern
Atlantic coast region. Audubon
Field Notes 16(5):467.

Chamberlain, E. B. 1968. Birds of the
Francis Marion National Forest.
U.S. Dep. Agric., Forest Serv.,
Atlanta. 32 pp.

Chamberlain, E. B. 1974. Rare and
endangered birds of the southern
national forests. U.S. Dep.
Agric., Forest Serv., Atlanta, Ga.

Chamberlain, W. D., Jr. [1979]. A
survey of the flora and fauna unique
to the barrier islands of North
Carolina, South Carolina, and
Georgia. Contract with Natl. and
Fla. Audubon Soc. (Unpubl.)

Chamberlain, W. D., Jr. and E. B.
Chamberlain. 1975. Avifauna of
Kiawah Island, pp. AV1-AV107. In:
W. M. Campbell, J. M. Dean, and
W. D. Chamberlain, eds. Environ-
mental inventory of Kiawah Island.
Environmental Research Center, Inc.,
Columbia, S.C.

Chan, G. 1972. A study of the effects
of San Francisco oil spill on marine
organisms. Part I. College of
Marin, Kentfield, Calif. 78 pp.

Chapman, H. H. 1926. Factors determining
natural reproduction of longleaf
pine on cut-over lands in La Salle
Parish, Louisiana. Yale Univ. School
For. Bull. 16.

Chapman, H. H. 1932a. Longleaf pine a
climax? Ecology 13:328-334.

Chapman, H. H. 1932b. Some further relations of fire to longleaf pine.
J. For. 30:602-604.

Chapman, H. H. 1936. Effects of fire in
preparation of seedbed for longleaf
pine seedlings. J. For. 34:852-854.

Chapman, R. L. 1971. The macroscopic
marine algae of Sapelo Island and
other sites on the Georgia coast.
Bull. Ga. Acad. Sci. 29:77-89.

Chapman, V. J. 1938. Studies in salt
marsh ecology: Sections I to III.
J. Ecol. 26(1):144-179.

Chapman, V. J. 1960. The plant ecology
of Scolt Head Island, pp. 85-163.
In: J. A. Steers, ed. Scolt Head
Island. W. Heffer and Sons, Ltd.,
Cambridge.

Cheek, R. P. 1968. The American shad.
U.S. Fish Wildl. Serv. Fish. Leafl.
No. 614. 13 pp.

Cheng, L. 1973. The ocean strider
Halobates (Heperoptera, Gerridae)
in the Atlantic Ocean. Okeanologia
13:683-690.

Cheng, L. 1974. Notes on the ecology
of the oceanic insect Halobates.
Mar. Fish. Rev. 36:1-7.

Cheng, L. 1975. Marine pleuston -
animals at the sea-air interface.
Annu. Rev. Oceanogr. Mar. Biol.
13:181-212.

Cheng, L. 1976. Insects in marine
environments, pp. 1-4. In: L.
Cheng, ed. Marine insects. American
Elsevier Publ. Co., New York.

Cherry, R. D., J. W. Higgo, and S. W.
Fowler. 1978. Zooplankton fecal
pellets and element residence time
in the ocean. Nature 274:246-248.

Christensen, N. L. 1976. The role of
carnivory in Sarracenia flava
L. with regard to specific nutrient
deficiencies. J. Elisha Mitchell
Sci. Soc. 92:144-147.

Christian, R. R., K. Bancroft, and W. J.
Wiebe. 1978. Resistance of the
microbial community within salt
marsh soils to selected perturbations.
Ecology 59(6):1200-1210.

Christian, R. R. and W. J. Wiebe. 1978.
Anaerobic microbial community
metabolism in Spartina alterniflora
soils. Limnol. Oceanogr. 23(2):328-
336.

Christiansen, M. E., J. D. Costlow, Jr., and R. J. Monroe. 1977a. Effects of the juvenile hormone mimic ZR-515 (Altosid [R]) on larval development of the mud-crab Rhithropanopeus harrisii in various salinities and cyclic temperatures. Mar. Biol. 39: 169-179.

Christiansen, M. E., J. D. Costlow, Jr., and R. J. Monroe. 1977b. Effects of the juvenile hormone mimic ZR-512 (Altozar [R]) on larval development of the mud-crab Rhithropanopeus harrisii at various cyclic temperatures. Mar. Biol. 39:281-288.

Chubb, J. C. 1954. Observations on oiled birds, 1951-1953. Northwest Nat. 2:460-461.

Churchill, M. A. 1958. Effects of storage impoundments on water quality. Trans. Am. Soc. Civil Eng. 123:419-464.

Clark, J. 1974. Coastal ecosystems – ecological considerations for management of the coastal zone. The Conservation Foundation, Washington, D.C. In cooperation with U.S. Dep. Commer., NOAA. 178 pp.

Clark, R. B. 1969. Oil pollution and the conservation of sea birds, pp. 76-112. In: Proceedings international conference on oil pollution of the sea. Rome.

Clark, R. B. 1971. Oil pollution and its biological consequences. A review of current scientific literature. Prepared for Great Barrier Reef Petroleum Drilling Royal Commissions. 11 pp.

Coker, R. E. 1906. The natural history and cultivation of the diamondback terrapin. N.C. Geol. Surv. Bull. No. 14. 69 pp.

Coker, R. E. 1920. The diamond-back terrapin: past, present and future. Sci. Mon. N.Y. 11:171-186.

Coker, R. E. 1926. Plankton collections in Lake James, North Carolina--copepods and cladocera. J. Elisha Mitchell Sci Soc. 41(3-4):228-258.

Coker, R. E. 1938. Notes on the peculiar crustacean fauna of White Lake, North Carolina. Arch. Hydrobiol. 34:130-133.

Coker, R. E. 1951. The diamond-back terrapin in North Carolina, pp. 219-230. In: H. F. Taylor, ed. Survey of marine fisheries of North Carolina. Univ. N.C. Press, Chapel Hill.

Coker, W. C. 1905. Observation on the flora of the Isle of Palms, Charleston, South Carolina. Torreya 5(8):135-145.

Cole, G. A. 1975. Textbook of limnology. C. V. Mosby Co., St. Louis, Mo. 283 pp.

Collins, H. L., J. R. Davis, and G. P. Markin. 1973. Residues of mirex in channel catfish and other aquatic organisms. Bull. Environ. Contam. Toxicol. 10(2):73-77.

Collins, J. T., J. E. Huheey, J. L. Knight, and H. M. Smith. 1978. Standard common and current scientific names for North American amphibians and reptiles. Society for the Study of Amphibians and Reptiles, Misc. Publ. Herpetol. Circ. No. 7. 36 pp.

Colquhoun, D. J. and W. Pierce. 1971. Pleistocene transgressive-regressive sequences on the Atlantic Coastal Plain. Quaternaria 15:35-50.

Colson, C. B. 1888. History of the mill pond oyster and cause of its disappearance. Elliott Soc. Proc. 5 pp.

Colwell, R. R. and J. Kaper. 1978. Distribution, survival, and significance of pathogenic bacteria and viruses in estuaries, pp. 443-457. In: M. L. Wiley, ed. Estuarine interactions. Academic Press, New York.

Conant, R. 1975. A field guide to reptiles and amphibians of Eastern and Central North America. Houghton-Mifflin Co., Boston, Mass. 429 pp. (First publ. 1958.)

Conant, R. and J. D. Lazell, Jr. 1973. The Carolina salt marsh snake: a distinct form of Natrix sipedon. Breviora 400:1-13.

Connell, W. A. 1940. Tidal inundation as a factor limiting the distribution of Aedes spp. on a Delaware marsh. Proc. N. J. Mosq. Exterm. Assoc. 27:166-177.

Conner, W. H. and J. W. Day, Jr. 1976. Productivity and composition of a bald cypress-water tupelo site and a bottomland hardwood site in a Louisiana swamp. Am. J. Bot. 63(10):1354-1364.

Conover, R. J. 1971. Some relations between zooplankton and Bunker C oil in Chaclabructo Bay following the wreck of the tanker Arrow. J. Fish. Res. Board Can. 28:1327-1330.

Conrad, W. B., Jr. 1966. A food habit study of ducks wintering on the lower Pee Dee and Waccamaw rivers. Georgetown, South Carolina. Proc. Annu. Conf. Southeast. Assoc. Game Fish Comm. 19:93-98.

Cook, T. M. and C. K. Goldman. 1976. Bacteriology of Chesapeake Bay surface waters. Chesapeake Sci. 17(1):40-49.

Cooley, N. R., J. M. Keltner, Jr. , and J. Forester. 1972. Mirex and Aroclor 1254: effect on an accumulation by Tetrahymena pyriformis strain W. J. Protozool. 19(4)636-638.

Coomer, C. E., Jr., D. R. Holder, and C. D. Swanson. 1978. A comparison of the diets of redbreast sunfish and spotted sucker in a coastal plain stream. Proc. Annu. Conf. Southeast. Assoc. Fish Wildl. Agencies 32. (In press.)

Cooper, A. W. 1974. Salt marshes, pp. 55-98. In: H. T. Odum, B. J. Copeland, and E. A. McMahan, eds. Coastal ecological systems of the United States. Vol. II. The Conservation Foundation, Washington, D.C.

Copeland, B. J. 1970. Estuarine classification and responses to disturbances. Trans. Am. Fish. Soc. 99(4):826-235.

Copeland, B. J. 1974. Impoundment systems, pp. 168-179. In: H. T. Odum, B. J. Copeland, and E. A. McMahan, eds. Coastal ecological systems of the United States. Vol. III. The Conservation Foundation, Washington, D.C.

Copeland, B. J., H. T. Odum, and F. N. Moseley. 1974b. Migrating subsystems, pp. 422-453. In: H. T. Odum, B. J. Copeland, E. A. McMahan, eds. Coastal ecological systems of the United States. Vol. III. The Conservation Foundation, Washington, D.C.

Copeland, B. J., K. R. Tenore, and D. B. Horton. 1974a. Oligohaline regime, pp. 315-357. In: H. T. Odum, B. J. Copeland, and E. A. McMahan, eds. Coastal ecological systems of the United States. Vol. II. The Conservation Foundation, Washington, D.C.

Corbin, F. 1951. Notes on the blue-green algae of South Carolina with some additions to the flora of the State. Bios 22:129.

Costlow, J. D., Jr. 1977. The effect of juvenile hormone mimics on development of the mud crab, Rhithropanopeus harrisii (Gould), pp. 439-457. In: F. J. Vernberg, A. Calabrese, F. P. Thunberg, and W. B. Vernberg, eds. Physiological responses of marine biota to pollutants. Academic Press, New York.

Cottam, C. 1939. Food habits of North American diving ducks. U.S. Dep. Agric. Tech. Bull. 643. 139 pp. 10 pl.

Coull, B. C. 1973. Estuarine meiofauna: a review: trophic relationships and microbial interactions, pp. 499-511. In: L. H. Stevenson and R. R. Colwell, eds. Estuarine microbial ecology. Univ. S.C. Press, Columbia.

Coull, B. C. 1975. Macrobenthic communities of Kiawah Island, pp. MB1-MB12. In: W. M. Campbell, J. M. Dean, and W. D. Chamberlain, eds. Environmental inventory of Kiawah Island. Environmental Research Center, Inc. Columbia, S.C.

Coull, B. C., S. S. Bell, A. M. Savory, and B. W. Dudley. 1979. Zonation of meiobenthic copepods in a Southeastern United States salt marsh. Estuarine Coastal Mar. Sci. 9:181-188.

Coull, B. C. and J. W. Fleeger. 1977. Long-term temporal variation and community dynamics of meiobenthic copepods. Ecology 58:1136-1143.

Coull, B. C. and W. B. Vernberg. 1975. Reproductive periodicity of meiobenthic copepods: seasonal or continuous? Mar. Biol. 32:289-293.

Cowardin, L. M., V. Carter, F. C. Golet, and E. T. Laroe. 1977. Classification of wetlands and deep-water habitats of the United States. An operational draft. U.S. Fish Wildl. Serv., Off. Biol. Serv., Washington, D.C. 100 pp.

Cox, J. L. 1971. DDT residues in coastal marine phytoplankton and their transfer in pelagic food chains. Ph.D. Dissertation. Stanford Univ., Stanford, Calif. 149 pp.

Craighead, F. C. 1950. Insect enemies of eastern forests. U.S. Dep. Agric. Misc. Publ. No. 657. Washington, D.C. 679 pp.

Crane, J. 1940. Eastern Pacific expeditions of the New York Zoological Society. On the post-embryonic development of brachyuran crabs of the genus Ocypode. Zoologica (N.Y.) 25:65-82.

Crenshaw, J. W., Jr. 1955. The life history of the southern spiny lizard, Sceloporus undulatus undulatus Latreille. Am. Midl. Nat. 54(2):257-298.

Crochet, D. W. [1976]. Commercial anadromous fishery, Waccamaw and Pee Dee rivers. Annu. progress rep. Project AFC-5-2. July 1, 1974 - June 30, 1975. S.C. Wildl. Mar. Resour. Dept., Div. Game Freshwater Fish. Columbia. 36 pp.

Crochet, D. W. 1977. Commercial anadromous fishery, Waccamaw and Pee Dee rivers. Job completion rep. AFC-5. S.C. Wildl. Mar. Resour. Dep., Div. Game Freshwater Fish., Columbia.

Crochet, D. W., D. E. Allen, and M. 'L. Hornberger. 1976. Commercial anadromous fishery, Waccamaw and Pee Dee rivers. Job completion rep. Project AFC-5. Oct. 1, 1973 - Sept. 1, 1976. S.C. Wildl. Mar. Resour. Dep., Div. Game Freshwater Fish. Columbia. 114 pp. (Unpubl.)

Crocker, A. D., J. Cronshaw, and W. N. Homes. 1974. The effect of crude oil on interstitial intestinal absorption in ducklings (Anas platyrhychos). Environ. Pollut. 7:165-177.

Croker, R. A. 1967. Niche diversity in five sympatric species of intertidal amphipods (Crustacea: Haustoriidae). Ecol. Monogr. 37(3):173-200.

Croker, R. A. 1968. Distribution and abundance of some intertidal sand beach amphipods accompanying the passage of two hurricanes. Chesapeake Sci. 9(3):157-162.

Crossley, D. A., Jr. and K. K. Bohnsack. 1960. Long-term ecological study in the Oak Ridge area. III. The oribatid mite fauna in pine litter. Ecology 41(4):628-638.

Cross, W. H. 1955. Anisopteran Odonata of the Savannah River Plant, South Carolina. J. Elisha Mitchell Sci. Soc. 71:9-17.

Cummins, K. W. 1975. The importance of different energy sources in freshwater ecosystems, pp. 50-54. In: productivity of world ecosystems. National Academy of Sciences. Washington, D.C.

Cupka, D. M. 1972. A survey of the ichthyofauna of the surf zone in South Carolina. S.C. Mar. Resour. Cent. Tech. Rep. No. 4. 19 pp.

Curtis, T. A. 1970a. Anadromous fish survey of the Ashley River watershed. Job completion rep. Project AFS-2-2. July 1, 1969 - June 30, 1970. S.C. Wildl. Mar. Resour. Dep., Columbia. 91 pp. (Unpubl.)

Curtis, T. A. 1970b. Anadromous fish survey of the Combahee River watershed. Job completion rep. Project AFS-2-4. July 1, 1969 - June 30, 1970. S.C. Wildl. Mar. Resour. Dep., Columbia. 130 pp. (Unpubl.)

Curtis, T. A. 1971. Macrobenthos studies, pp. 10-13. In: T. A. Curtis. Anadromous fish survey of the Santee and Cooper system. Job progress rep. Project AFS 3-1. S.C. Wildl. Mar. Resour. Dep., Columbia.

Curtis, T. A. 1972. Anadromous fish survey of the Santee and Cooper rivers. Job progress rep. Project AFS-3-2. July 1, 1971 - June 30, 1972. S.C. Wildl. Mar. Resour. Dep., Columbia. 40 pp.

Curtis, T. A. 1973. Anadromous fish survey of the Santee and Cooper river system. Annu. progress rep. Project AFS-3-3. July 1, 1972 - June 30, 1973. S.C. Wildl. Mar. Resour. Dep., Div. Game Freshwater Fish, Columbia. 44 pp.

Curtis, T. A. 1974. Anadromous fish survey of the Santee and Cooper river system. Annu. progress rep. Project AFS-3-4. July 1, 1973 - June 30, 1974. S.C. Wildl. Mar. Resour. Dep., Div. Game Freshwater Fish, Columbia. 54 pp.

Curtis, T. A. 1975. Anadromous fish survey of the Santee and Cooper river system. Annu. progress rep. Project AFS-3-5. July 1, 1974 - June 30, 1975. S.C. Wildl. Mar. Resour. Dep., Div. Game Freshwater Fish, Columbia. 31 pp.

Curtis, T. A. 1976. Anadromous fish survey of the Santee and Cooper river system. Annu. progress rep. Project AFS-3-6. July 1, 1975 - June 30, 1976. S.C. Wildl. Mar. Resour. Dep., Div. Game Freshwater Fish, Columbia. 28 pp.

Custer, T. W. and R. G. Osborn. 1977. Wading birds as biological indicators: 1975 colony survey. U.S. Fish Wildl. Serv. Spec. Sci. Rep.--Wildl.

Cutts, E. 1955. Depredation at a breeding colony. Chat 19:70.

Cypert, E. 1961. The effects of fires in the Okefenokee Swamp in 1954 and 1955. Am. Midl. Nat. 66(2):485-503.

Dahlberg, M. D. 1972. An ecological study of Georgia coastal fishes. Fish. Bull. 70(2):323-355.

Dahlberg, M. D. 1978. Applying survival curves to assessment of fish larval entrainment impact, pp. 39-48. In: J. H. Thorp and J. W. Gibbons, eds. Energy and environmental stress in aquatic systems. Tech. Inf. Center, U.S. Dep. Energy, CONF-771114.

Dahlberg, M. D. and J. C. Conyers. 1973. An ecological study of *Gobiosoma bosci* and *G. ginsburgi* (Family Gobiidae) on the Georgia coast. Fish. Bull. 71(1): 279-287.

Dahlberg, M. D. and E. P. Odum. 1970. Annual cycles of species occurrence, abundance, and diversity in Georgia estuarine fish populations. Am. Midl. Nat. 83(2):382-392.

Dahlberg, M. D. and D. C. Scott. 1971. The freshwater fishes of Georgia. Bull. Ga. Acad. Sci. 29:1-64.

Daiber, F. C. 1974. Salt marsh plants and future coastal salt marshes in relation to animals, pp. 475-508. In: R. J. Reimold and W. H. Queen, eds. Ecology of halophytes. Academic Press, New York.

Dame, R. F. 1972a. Variations in the caloric value of the American oyster *Crassostrea virginica*. Proc. Natl. Shellfish. Assoc. 62:86-88.

Dame, R. F. 1972b. Comparison of various allometric relationships in intertidal and subtidal American oysters. Fish. Bull. 70:1121-1126.

Dame, R. F. 1974. The ecological energies of growth, respiration and assimilation in the intertidal American oyster, *Crassostrea virginica*. Mar. Biol. 17:243-250.

Dame, R. F. 1976. Energy flow in an intertidal oyster population. Estuarine Coastal Mar. Sci. 4:243-253.

Dames and Moore Associates. 1975. Environmental assessment report proposed chemical plant, Berkeley County, South Carolina for Amoco Chemicals Corporation. Park Ridge, Ill.

Darnell, R. M. 1958. Food habits of fishes and larger invertebrates of Lake Ponchartrain, Louisiana, an estuarine community. Publ. Inst. Mar. Sci. Univ. Tex. 5:353-416.

Darnell, R. M. 1967. Organic detritus in relation to the estuarine ecosystem, pp. 376-382. In: G. H. Lauff, ed. Estuaries. Am. Assoc. Adv. Sci. Publ. No. 83. Washington, D.C.

Darsie, R. F., Jr. and P. F. Springer. 1957. Three-year investigation of mosquito breeding in natural and impounded tidal marshes in Delaware. Univ. Delaware Agric. Exp. Stn. Bull. 1957:320-365.

David, P. M. 1965. The surface fauna of the ocean. Endeavour 24(92):95-100.

Davis, C. C. 1955. The marine and freshwater plankton. Mich. State Univ., East Lansing. 562 pp.

Davis, D. E. 1960. The spread of the cattle egret in the United States. Auk 77:421-424.

Davis, G. E. and M. C. Whiting. 1977. Loggerhead sea turtle nesting in Everglades National Park, Florida, U.S.A. Herpetologica 33(1):18-28.

Davis, J. 1974. The fire ant wars. S.C. Wildl. 21(1):4-8.

Davis, J. H. 1946. The peat deposits of Florida. Fla. Geol. Surv. Bull. 30.

Davis, L. V. 1978. Class Insecta, pp. 186-220. In: R. G. Zingmark, ed. An annotated checklist of the biota of the coastal zone of South Carolina. Univ. S.C. Press, Columbia.

Davis, L. V. and I. E. Gray. 1966. Zonal and seasonal distribution of insects in North Carolina salt marshes. Ecol. Monogr. 36:275-295.

Davis, W. H. and C. L. Rippy. 1968. Distribution of *Myotis lucifugus* and *Myotis austroriparius* in the Southeastern United States. J. Mammal. 49(1):113-117.

Davis, W. S. 1957. Ova production of American shad in Atlantic Coast rivers. U.S. Fish Wildl. Serv. Res. Rep. No. 49. 5 pp.

Dawson, C. E. 1958. A study of the biology and life history of the spot, *Leiostomus xanthurus lacepede*, with special reference to South Carolina. Contrib. Bears Bluff Lab. No. 28. Wadmalaw Island, S.C. 48 pp.

Dawson, R. H. [1977]. Partial review of the literature of marsh macrophyte production. S.C. Mar. Resour. Cent., Charleston. 23 pp. (Unpubl.)

Day, J. H. 1951. The ecology of South African estuaries. Part 1. A review of estuarine conditions in general. Trans. R. Soc. S. Afr. 33:53-91.

Day, J. H. 1964. The origin and distribution of estuarine animals in South Africa. Monogr. Biol. 14:159-173.

Day, J. H., H. G. Field, and M. P. Montgomery. 1971. The use of numerical methods to determine the distribution of the benthic fauna across the continental shelf of North Carolina. J. Anim. Ecol. 40:93-125.

Dean, B. 1892. The physical and biological characteristics of the natural oyster-grounds of South Carolina. Fish. Bull. 10:335-361.

Dean, B. 1893. Report on the European methods of oyster-culture. Bull. U.S. Fish. Comm. 11(1891):357-406.

Dean, J. M. 1975. The potential use of South Carolina rice fields for aquaculture. S.C. State Dev. Board, Columbia. 96 pp.

Dean, J. M. and O. R. Talbert. 1975. The loggerhead turtles of Kiawah Island, pp. T1-T19. In: W. M. Campbell, J. M. Dean, and W. D. Chamberlain, eds. Environmental inventory of Kiawah Island. Environmental Research Center, Inc., Columbia, S.C.

Debell, D. S. and I. D. Auld. 1971. Establishment of swamp tupelo seedlings after regeneration cuts. U.S. Dep. Agric. Forest Serv. Res. Note SE-164.

DeCoursey, P. J. and W. B. Vernberg. 1972. Effect of mercury on survival, metabolism, and behavior of larval Uca pugilator (Brachyura). Oikos 23: 241-247.

DeCoursey, P. J. and W. B. Vernberg. 1975. The effect of dredging in a polluted estuary on the physiology of larval zooplankton. Water Res. 9:149-154.

Deevey, G. B. 1948. The zooplankton of Tisbury Great Pond. Bull. Bingham Oceanogr. Collect. Yale Univ. 12(1): 1-44.

de la Cruz, A. A. 1973. The role of tidal marshes in the productivity of coastal waters. Assoc. Southeast. Biol. Bull. 20(4):147-156.

Demaree, D. 1932. Submerging experiments with taxodium. Ecology 13:258-262.

Dendy, J. S. and R. H. Stroud. 1949. The dominating influence of Fontana Reservoir on temperature and dissolved oxygen in the Little Tennessee River and its impoundments. J. Tenn. Acad. Sci. 24(1):41-51.

Dennis, J. V. 1966. Preliminary list of birds of the Congaree Swamp in South Carolina. Belle W. Baruch Foundation, Georgetown, S.C. (Unpubl.)

Dennis, J. V. 1975a. Carolina Bays of the Santee Coastal Reserve. 19 pp. (Unpubl.)

Dennis, J. V. 1975b. Blake's Reserve. 12 pp. (Unpubl.)

Dennis, W. M. 1973. A synecological study of the Santee Swamp, Sumter County, S.C. M.S. Thesis. Univ. S.C., Columbia. 61 pp.

Dennis, W. M. and W. T. Batson. 1974. Floating log and stump communities in the Santee Swamp in South Carolina. Castanea 39:166-170.

Denton, F. J. 1977. The recent nesting osprey population in Georgia. Oriole 42(3):41-45.

Denton, J. F., E. B. Chamberlain, and B. R. Chamberlain. 1950. South Atlantic coast region. Audubon Field Notes 4:270.

Desola, C. R. and F. Abrams. 1933. Testudinata from South-eastern Georgia, including the Okefenokee Swamp. Copeia 1933:10-12.

de Sylva, D. P. 1975. Nektonic food webs in estuaries, pp. 420-447. In: L. E. Cronin, ed. Estuarine research. Vol. 1. Academic Press, New York.

Dexter, D. M. 1967. Distribution and niche diversity of haustoriid amphipods in North Carolina. Chesapeake Sci. 8(3) 187-192.

Dexter, D. M. 1969. Structure of an intertidal sandy-beach community in North Carolina. Chesapeake Sci. 10:93-98.

Dexter, R. W. 1943. Anurida maritima: an important sea-shore scavenger. J. Econ. Entomol. 36:797.

Diaz, H. and J. D. Costlow. 1972. Larval development of Ocypode quadrata (Brachyura: Crustacea) under laboratory conditions. Mar. Biol. 15:120-131.

Dickson, R. E., J. F. Hosner, and N. W. Hosley. 1965. The effects of four water regimes upon the growth of four bottomland tree species. For. Sci. 11(3):299-305.

Dillard, G. E. 1966. A floristic and ecological study of benthic algae in two North Carolina streams. Ph.D. Dissertation. N.C. State Univ., Raleigh.

Dillard, G. E. 1967. The freshwater algae of South Carolina. I. Previous work and recent additions. J. Elisha Mitchell Sci. Soc. 83:128-131.

Dingle, E. S. 1926. Red-cockaded woodpeckers in cornfields. Bird-lore 28:124-125.

Dobson, T. 1976. Seaweed flies (Diptera: Coelopidae, etc.), pp. 447-464. In: L. Cheng, ed. Marine insects. American Elsevier Publ. Co., New York.

Dopson, C. W., Jr. and J. I. Richardson. 1968. Late summer and fall observations from the Georgia coast. Oriole 33(4):41-43.

Dörjes, J. 1972. Georgia coastal region, Sapelo Island, U.S.A.: sedimentology and biology. VII. Distribution and zonation of macrobenthic animals. Senckenb. Marit. 4:183-216.

Dörjes, J. 1977. Marine macrobenthic communities of the Sapelo Island, Georgia region, pp. 399-421. In: B. C. Coull, ed. Ecology of marine benthos. Univ. S.C. Press, Columbia.

Dörjes, J. and J. D. Howard. 1975. Estuaries of the Georgia coast, U.S.A.: sedimentology and biology. IV. Fluvial-marine transition indicators in an estuarine environment, Ogeechee River, Ossabaw Sound. Senckenb. Marit. 7:137-179.

Doutt, J. K. 1941. Wind pruning and salt spray as factors in ecology. Ecology 22(2):195-196.

Drake, J. C. 1891. On the sounds and estuaries of Georgia with reference to oyster culture. U.S. Coast and Geodetic Surv. Bull. 19:179-209.

Dubarry, A. P. 1963. Germination of bottomland tree seed while immersed in water. J. For. 61(3):225-226.

Dudley, R. G., A. W. Mullis, and J. W. Terrell. 1977. Movements of striped bass (Morone saxatilis) in the Savannah River, Georgia. Trans. Am. Fish. Soc. 106(4):314-322.

Duncan, R. E. 1975. Wando River aerial imagery and marsh productivity study. S.C. Water Resources Comm. Spec. Stud. Rep. No. 120. Columbia. 28 pp.

Duncan, W. M. 1944. A new species of Baptisia. Rhodora 46:29-31.

Duncan, W. H. 1955. Woody vegetation of Sapelo Island, Georgia. Assoc. Southeast. Biol. Bull. 2:5. (Abstr.)

Dunlavy, J. C. 1935. Studies on the phyto-vertical distribution of birds. Auk 52:425-431.

Dunstan, W. M. and L. P. Atkinson. 1976. Sources of new nitrogen for the South Atlantic Bight, pp. 69-78. In: M. Wiley, ed. Estuarine processes. Vol. 1. Uses, stresses, and adaptation to the estuary. Academic Press, New York.

Dunstan, W. M. and J. Hosford. 1977. The distribution of planktonic blue-green algae related to the hydrography of the Georgia Bight. Bull. Mar. Sci. 27(4):824-829.

Dunstan, W. M., G. L. McIntire, and H. L. Windom. 1975a. Revegetation on dredge spoil in the SE Spartina marshes. J. Waterways Harbors Coastal Eng. Div., Proc. Am. Soc. Civ. Eng. 101(WW3) (Pap. 11530):269-276.

Dunstan, W. M., H. L. Windom, and G. L. McIntire. 1975b. The role of Spartina alterniflora in the flow of lead, cadmium, and copper through the salt-marsh ecosystem, pp. 250-256. In: F. G. Howell, J. B. Gentry, and M. H. Smith, eds. Mineral cycling in southeastern ecosystems. U.S. Energy Res. Dev. Adm. Symp. Ser. Conf.-740513. Springfield, Va.

Durant, J. E. [1970]. The effects of temperature and salinity upon the gonadal cycle of Crassostrea virginica (Gmelin) in Georgia waters, pp. 2-13. In: T. L. Linton, ed. Feasibility study of methods for improving oyster production in Georgia. Completion rep. 2-10-R. Univ. Ga. Mar. Inst., Sapelo Island and Ga. Game Fish Comm., Atlanta.

Dusi, J. L. and R. T. Dusi. 1968. Ecological factors contributing to nesting failure in a heron colony. Wilson Bull. 80:458-466.

Dusi, J. L. and R. T. Dusi. 1970. Nesting success and mortality of nestlings in a cattle egret colony. Wilson Bull. 82:458-460.

Earle, L. C. and H. J. Humm. 1964. Intertidal zonation of algae in Beaufort Harbor. J. Elisha Mitchell Sci. Soc. 80(2):78-82.

Eastern Peregrine Falcon Recovery Team. 1976. Eastern peregrine falcon recovery plan. Technial draft. U.S. Dep. Inter., Fish Wildl. Serv. 130 pp. (Unpubl.)

Ecke, D. H. 1955. The reproductive cycle of the Mearns cottontail in Illinois. Am. Midl. Nat. 53:294-311.

Eddy, J. E., V. J. Henry, J. H. Hoyt, and E. Bradley. 1967. Description and use of an underwater television system in the Atlantic Continental Shelf. U.S. Geol. Surv. Prof. Pap. 575-C: 72-76.

Eddy, S. 1932. The plankton of the Sangamon River in the summer of 1929. Ill. Nat. Hist. Surv. Bull. 19:469-486.

Edmisten, J. A. 1963. The ecology of the Florida pine flatwoods. Ph.D. Dissertation. Univ. Fla., Gainesville. 108 pp.

Edmondson, W. T. 1959. Fresh-water biology. John Wiley & Sons, New York. 1248 pp.

Edwards, C. A., D. E. Reichle, and D. A. Crossley, Jr. 1973. The role of soil invertebrates in turnover of organic matter and nutrients, pp. 147-172. In: D. E. Reichle, ed. Analysis of temperate forest ecosystems. Springer-Verlag, New York.

Edwards, M. G. 1978. Recovery efforts for the peregrine falcon in the Southeast, pp. 85-87. In: R. O. Odom and L. Landers, eds. Proceedings of the rare and endangered wildlife symposium, 3-4 Aug. 1978, Athens, Ga. Ga. Dep. Nat. Resour., Game Fish Div., Tech. Bull. WL 4.

Eldridge, P. J., A. G. Eversole, and J. M. Whetstone. 1979. Comparative survival and growth rates of hard clams, Mercenaria mercenaria, planted in trays subtidally and intertidally at varying densities in a South Carolina estuary. Proc. Natl. Shellfish Assoc. 69. (In press.)

Eldridge, P. J., W. Waltz, and H. Mills. 1976. Relative abundance of Mercenaria mercenaria notata in estuaries in South Carolina. Veliger 18(4):396-397.

Eleuterius, L. N. 1975. The life history of the salt marsh rush, Juncus roemerianus. Bull. Torrey Bot. Club 102(3):135-140.

Ellis, M. M. 1937. Detection and measurement of stream pollution. Fish. Bull. 48(22):365-437.

Elton, C. 1968. Animal ecology. Assoc. Book Publ., Ltd., London. 207 pp.

Emery, K. O. and E. Uchupi. 1972. Western North Atlantic Ocean: topography, rocks, structure, water, life and sediments. Am. Assoc. Petrol. Geol. Mem. 17. 532 pp.

Energy Resources Co., Inc. 1978. An ecological characterization of Maine's coast north and east of Cape Elizabeth. Draft Test Characterization. 7 vols. (Unpubl.)

Engelmann, M. D. 1961. The role of soil arthropods in the energetics of an old field community. Ecol. Monogr. 31:221-238.

Engels, W. L. 1942. Vertebrate fauna of North Carolina coastal islands, a study in the dynamics of animal distribution. I. Ocracoke Island. Am. Midl. Nat. 28(2):273-304.

Engels, W. L. 1952. Vertebrate fauna of North Carolina coastal islands. II. Shackleford Banks. Am. Midl. Nat. 47(3):702-742.

Environmental Research Center, Inc. 1976. Environmental inventory of Lake Murray, South Carolina. Columbia, S.C. 289 pp.

Enwright Associates, Inc. 1977. Arthur M. Williams Station 316 (A) and (B) demonstrations. Greenville, S.C. S.C. Electric and Gas Co., Columbia. 184 pp.

Erickson, R. C. 1963. Oil pollution and migratory birds. Atl. Nat. 18(1):5-14.

Erkenbrecher, C. W. and L. H. Stevenson. 1975. The influence of tidal flux on microbial biomass in salt marsh creeks. Limnol. Oceanogr. 20(4):618-625.

Erkenbrecher, C. W. and L. H. Stevenson. 1977. Factors related to the distribution of microbial biomass in salt-marsh-creeks. Mar. Biol. 40:121-125.

Evans, P. D., C. N. E. Ruscoe, and J. E. Treherne. 1971. Observations on the biology and submergence behavior of some littoral beetles. J. Mar. Biol. Assoc. U.K. 51:375-386.

Ewel, K. C. 1976. Cypress ponds, pp. 14-49. In: C. H. Wharton, H. T. Odum, K. Ewel, M. Duever, A. Lugo, R. Boyt, J. Bartholomew, E. Debellevae, S. Brown, M. Brown and L. Duever, eds. Forested wetlands of Florida - their management and use. Univ. Fla., Cent. Wetlands, Phelps Lab., Gainesville.

Ewel, K. C. and W. J. Mitsch. 1975. The effect of fire on the species composition of trees in two cypress domes, pp. 215-222. In: H. T. Odum, K. C. Ewel, J. W. Ordway, and M. K. Johnston. Cypress wetlands for water management, recycling and conservation. Second annu. rep. to Natl. Sci. Found. and Rockefeller Found. Univ. Fla., Cent. Wetlands, Phelps Lab., Gainesville.

Eyles, D. E. 1939. Studies on the vegetation of certain coastal islands in the mouth of the Savannah River, Georgia. M.S. Thesis. Emory Univ., Atlanta. 225 pp.

555

Eyles, D. E. 1941. A phytosociological study of the Castalia-Myriophyllum community of Georgia coastal plain boggy ponds. Am. Midl. Nat. 26:421-438.

Eyles, D. E. and J. L. Robertson. 1944. A guide and key to the aquatic plants of the Southeastern United States. U.S. Public Health Serv. Bull. 286. 151 pp.

Faircloth, W. 1971. The vascular flora of central south Georgia. Ph.D. Dissertation. Univ. Ga., Athens. 217 pp.

Farmer, C. H. III and J. D. Whitaker. 1978. Overwintering white shrimp: a report to the fishermen. S.C. Wildl. Mar. Resour. Dep., Charleston. 32 pp. (Map.)

Farnworth, E. G. and F. B. Golley. 1973. Fragile ecosystems: evaluation of research and applications in the neotropics. A report of The Institute of Ecology (TIE). Springer-Verlag, New York. 258 pp.

Fassett, N. C. 1957. A manual of aquatic plants. Univ. Wis. Press, Madison. 405 pp.

Fedorenko, A. Y. 1975. Feeding characteristics and predation impact of Chaoborus (Diptera, Chaoboridae) larvae in a small lake. Limnol. Oceanogr. 20(2):250-258.

Felfoldy, L. J. M. 1961. On the chlorophyll content and biological productivity of periphytic diatom communities on the stony shores of Lake Balaton. Ann. Biol. Tihany 28:99-104.

Ferrigno, F. 1961. Variations in mosquito-wildlife associations on coastal marshes. Proc. N.J. Mosq. Exterm. Assoc. 48:193-203.

Fields, H. M. 1962. Pompanos (Trachinotus spp.) of South Atlantic coast of the United States. Fish. Bull. 62(207): 189-222.

Fisher, T. R., Jr. 1975. The quantity of estuarine seston as food for suspension feeders. 38th Annu. Meeting A.S.L.O., Dalhousie Univ., Halifax, Nova Scotia.

Fletemeyer, J. R. 1978. Underwater tracking evidence of neonate loggerhead sea turtles seeking shelter in drifting Sargassum. Copeia 1978(1):148-149.

Florida Panther Recovery Team. 1978. Florida panther recovery plan. Technical draft. U.S. Dep. Inter., Fish Wildl. Serv. 14 pp.

Florschutz, O., Jr. 1959. Mosquito production and wildlife usage in natural, ditched, and unditched tidal marshes at Assawomen Wildlife Area, Delaware. Proc. N.J. Mosq. Exterm. Assoc. 46:103-111.

Folk, G. E., Jr. 1939. Easter on Wassaw Island. Oriole 4:21-23.

Forsythe, D. M. 1973. Gull populations at Charleston, S.C., June 1971 to June 1972. Chat 37(3):57-62.

Forsythe, D. M. 1978. Birds, pp. 277-295. In: R. G. Zingmark, ed. An annotated checklist of the biota of the coastal zone of South Carolina. Univ. S.C. Press, Columbia.

Forward, R. B., Jr. and J. D. Costlow, Jr. 1976. Crustacean larval behavior as an indicator of sublethal effects of an insect juvenile hormone mimic, pp. 279-289. In: M. Wiley, ed. Estuarine processes, Vol. I. Uses, stresses, and adaptations to the estuary. Academic Press, New York.

Foster, W. A. 1968. Studies on the distribution and growth of Juncus roemerianus in southeastern Brunswick County, North Carolina. M.S. Thesis. N.C. State Univ., Raleigh. 72 pp.

Foster, W. A. and J. E. Treherne. 1975. The distribution of an intertidal aphid, Pemphigus trehernei Foster, on marine salt marshes. Oecologia, Berl. 21:141-155.

Foster, W. A. and J. E. Treherne. 1976. Insects of marine salt marshes: problems and adaptations, pp. 5-42. In: L. Cheng, ed. Marine insects. American Elsevier Publ. Co., New York.

Fowler, H. W. 1935. Notes on South Carolina fresh-water fishes. Charleston Mus. Contrib. No. 7. 28 pp.

Fowler, H. W. 1945. A study of the fishes of the Southern Piedmont and coastal plain. Proc. Acad. Nat. Sci. Monogr. No. 7. 408 pp.

Fowler, S. W., M. Heyrand, and R. D. Cherry. 1976a. Accumulation and retention of plutonium by marine zooplankton, pp. 42-50. In: Activities of the international lab., mar. radioactivity.

Fowler, S. W., M. Heyrand, and J. la Rosa. 1976b. Mercury kinetics in marine zooplankton. Activities of the international lab., mar. radioactivity rep. 1976. 20 pp.

Fowler, S. W., B. Oregioni, and J. la Rosa. 1976c. Trace metals in pelagic organisms from the Mediterranean Sea, pp. 110-122. In: Activities of the international lab., mar. radioactivity.

Fox, R. S. 1978. Order Amphipoda, pp. 161-166. In: R. G. Zingmark, ed. An annotated checklist of the biota of the coastal zone of South Carolina. Univ. S.C. Press, Columbia.

Francisco, W., L. T. Jenkins, Jr., M. McWilliams, and J. Rathmann. 1970. Wassaw Island study. Univ. Ga., Environmental Design Ser. No. 1. Athens. 129 pp.

Frankenberg, D. 1965. Variability in marine benthic communities off Georgia. Trans. Joint Conf. Ocean Sci. Ocean Eng. 2:1111.

Frankenberg, D. 1971. The dynamics of benthic communities off Georgia, USA. Thalassia Jugosl. 7(1):49-55.

Frankenberg, D. and A. S. Leiper. 1977. Seasonal cycles in benthic communities of the Georgia continental shelf, pp. 383-397. In: B. C. Coull, ed. Ecology of marine benthos. Univ. S.C. Press, Columbia.

Frey, D. G. 1948. A biological survey of Lake Waccamaw. Wildl. In N.C. 1948 (July):4-6, 23.

Frey, D. G. 1949. Morphometry and hydrography of some natural lakes of the North Carolina coastal plain: the bay lake as a morphometric type. J. Elisha Mitchell Sci. Soc. 65:1-37.

Frey, D. G. 1951. The fishes of North Carolina's bay lakes and their intraspecific variation. J. Elisha Mitchell Sci. Soc. 67:1-44.

Frick, J. 1976. Orientation and behaviour of hatchling green turtles (Chelonia mydas) in the sea. Anim. Behav. 24(4): 849-857.

Fritz, G. 1975. Measurement of the standing crop biomass. Part V. pp. 37-42. In: J. M. Dean, ed. The potential use of South Carolina rice fields for aquaculture. S.C. State Development Board, Columbia.

Frohne, W. C. 1942. Notes on Phymatodocis norostedtiana Wolle (Desmidiaceae) from South Carolina and its var. minor Borgesen from Georgia. Trans. Am. Microsc. Soc. 61:438-441.

Fryer, G. 1957. The food of some freshwater cyclopoid copepods and its ecological significance. J. Anim. Ecol. 26:263-286.

Fuller, S. 1974. Macroinvertebrates, pp. 49-61. In: Cooper River survey 1973 for the E.I. Dupont de Nemours & Company. Acad. Nat. Sci. Phila., Philadelphia, Pa.

Funderburg, J. B., Jr. 1956. An ecological study of the summer maritime birds of southeastern North Carolina. M.S. Thesis. N.C. State Univ., Raleigh.

Funderburg, J. B., Jr. and T. L. Quay. 1959. Summer maritime birds of southeastern North Carolina. J. Elisha Mitchell Soc. 75:13-18.

Furukawa, A. and T. L. Linton. 1970. Oyster spat setting patterns in a medium salinity sound of Georgia, pp. 1-9. In: T. L. Linton, ed. Feasibility study of methods for improving oyster production in Georgia. Completion rep. 2-10-R. Ga. Game Fish Comm., Atlanta.

Gaarder, T. and R. Sparck. 1932. Hydrographisch-biochemische untersuchungen in Norwegischen Austern-pollen. Bergeus Mus. Arab. Naturv. Rek. 1:1-44.

Gabrielson, F. C. 1968. The effects of shade, litter and root competition on old field vegetation in Aiken County, South Carolina. Ph.D. Dissertation. Univ. Ga., Athens. 56 pp.

Gaddy, L. L. 1977. A physiographic and vegetational study of brown pelican nesting habitat on Deveaux Bank, S.C. S.C. Wildl. Mar. Resour. Dep., Div. Wildl. Freshwater Fish., Columbia. 26 pp. (Unpubl.)

Gaddy, L. L., T. S. Kohlsaat, E. A. Laurent, and K. B. Stansell. 1975. A vegetation analysis of preserve alternatives involving the Beidler Tract of the Congaree Swamp. S.C. Wildl. Mar. Resour. Dep., Div. Nat. Area Acquisition Res. Planning, Columbia. 99 pp. (Unpubl.)

Gallagher, J. L. 1975. The significance of the surface film in salt marsh plankton metabolism. Limnol. Oceanogr. 20:120-123.

Gallagher, J. L., R. A. Linthurst, and W. J. Pfeiffer. 1979. Aerial production, mortality and mineral accumulation dynamics in Spartina alterniflora and Juncus roemerianus in a Georgia salt marsh. Ecology (In press.)

Gallagher, J. L., R. A. Linthurst, and R. J. Reimold. 1975. Marshes of McIntosh County, Georgia. Univ. Ga. Mar. Inst., Sapelo Island. 8 pp.+ Map. (Unpubl.)

Gallagher, J. L. and R. J. Reimold. 1973. Tidal marsh plant distribution and productivity patterns from the sea to fresh water--a challenge in resolution and discrimination, pp. 166-183. In: Proc. fourth biennial workshop on color aerial photography. Sponsored by Am. Soc. Photogram.

Gallagher, R. M., M. L. Hollinger, R. M. Ingle, and C. R. Futch. 1972. Marine turtle nesting on Hutchinson Island, Florida, in 1971. Fla. Dep. Nat. Resour. Mar. Res. Lab. Spec. Sci. Rep. No. 37. 11 pp.

Galtsoff, P. S. 1964. The American oyster Crassostrea virginica Gmelin. Fish. Bull. 64. 480 pp.

Galtsoff, P. S. and R. H. Luce. 1930. Oyster investigations in Georgia. U.S. Dep. Commer., Bur. Fish. Doc. No. 1077:62-100.

Gannon, J. E. and R. S. Stemberger. 1978. Zooplankton (especially crustaceans and rotifers) as indicators of water quality. Trans. Am. Microsc. Soc. 97(1):16-35.

Gano, L. 1917. A study of physiographic ecology in northern Florida. Bot. Gaz. 63:337-372.

Gardner, L. R. 1975. Runoff from an intertidal marsh during tidal exposure--recession curves and chemical characteristics. Limnol. Oceanogr. 20 (1):81-89.

Gardner, L. R. and W. Kitchens. 1978. Sediment and chemical exchanges between salt marshes and coastal waters-a review, pp. 191-207. In: B. J. Kjerfve, ed. Estuarine processes. Univ. S.C. Press, Columbia.

Garren, K. H. 1943. Effects of fire on the vegetation of the Southeastern United States. Bot. Rev. 9(9):617-654.

Garrick, L. D. and J. W. Lang. 1977. The alligator revealed. Nat. Hist. 86 (6):54-61.

Gauthreaux, S. A., Jr., J. P. Holt, F. M. Probst, T. A. Beckett III, and R. N. McFarlane. 1979. Status report - the birds, pp. 82-87. In: D. M. Forsythe and W. B. Ezell, eds. Proceedings of the first South Carolina endangered species symposium. Nov. 1976. S.C. Wildl. Mar. Resour. Dep. and the Citadel. Charleston, S.C.

Gehringer, J. W. 1959. Early development and metamorphosis of the ten-pounder Elops saurus Linnaeus. Fish. Bull. 59(155):619-647.

Georgia Department of Natural Resources. 1976. The environmental impact of freshwater wetland alterations on coastal estuaries. Tech. Planning Seminar. Resour. Planning Sec., Off. Planning Res. Atlanta. In cooperation with Coastal Plains Center for Marine Development Services, Wilmington, N.C. 85 pp.

Georgia Department of Natural Resources. 1978. The Altamaha: a scenic and recreational river proposal for the Great Altamaha Swamp. Off. Planning Res., Resour. Planning Sec., River Planning Unit. 162 pp.

Georgia Water Quality Control Board. 1971. Water quality study of Chatham County recreational waters. Atlanta. In cooperation with Chatham County Health Dep., Savannah, Ga. 62 pp.

Georgia Water Quality Control Board. 1972. Inventory of water pollution control facilities. 4 vols. Atlanta.

Gerba, C. P. and J. S. McLeod. 1976. Effect of sediments on the survival of Escherichia coli in marine waters. Appl. Environ. Microbiol. 32:114-120.

Gerlach, S. A. 1971. On the importance of marine meiofauna for benthos communities. Oecologia (Berlin) 6:176-190.

Germann, J. F., L. E. McSwain, D. R. Holder, and C. D. Swanson. 1975. Life history of warmouth in the Suwannee River and Okefenokee Swamp, Georgia. Proc. Annu. Conf. Southeast. Assoc. Game Fish Comm. 28:259-278.

Giam, C. S., M. K. Wong, A. R. Hanks, W. M. Sackett, and R. L. Richardson. 1973. Chlorinated hydrocarbons in plankton from the Gulf of Mexico and Northern Caribbean. Bull. Environ. Contam. Toxicol. 9(6):376-382.

Gibb, J. A. 1958. Predation by tits and squirrels on the eucosmid Ernarmonia conicolana. J. Anim. Ecol. 27:375-396.

Gibb, J. A. 1960. Populations of tits and goldcrests and their food supply in pine plantations. Ibis 102:163-208.

Gibb, J. A. 1962. L. Tinbergen's hypothesis of the role of specific search images. Ibis 104:106-111.

Gibbons, J. W. 1978. Reptiles, pp. 270-276. In: R. G. Zingmark, ed. An annotated checklist of the biota of the coastal zone of South Carolina. Univ. S.C. Press, Columbia.

Gibbons, J. W. and J. W. Coker. 1978. Herpetofaunal colonization patterns of Atlantic Coast barrier islands. Am. Midl. Nat. 99(1):219-233.

Gibbons, J. W. and J. R. Harrison. 1975. Reptiles and amphibians of Kiawah Island, pp. R1-R26. In: W. M. Campbell, J. M. Dean, and W. D. Chamberlain, eds. Environmental Inventory of Kiawah Island. Environmental Research Center, Inc., Columbia, S.C.

Gibson, V. R. and G. D. Grice. 1977.
Response of macrozooplankton populations to copper. Controlled Ecosystem
Pollution Experiment. Bull. Mar. Sci.
27(1):85-91.

Godfrey, H. 1951. A report on the eel
investigations for 1950-1951. Fish.
Res. Board Can. MS Rep. Biol. Sta.
439.

Godfrey, P. J. and M. M. Godfrey. 1976.
Barrier island ecology of Cape Lookout
National Seashore and vicinity, North
Carolina. Natl. Park Serv. Sci.
Monogr. Ser. No. 9. 160 pp.

Godwin, W. F. 1968. The distribution and
density of the hard clam, Mercenaria
mercenaria, on the Georgia coast. Ga.
Game Fish Comm., Mar. Fish. Div.
Contrib. Ser. No. 10. Brunswick.
30 pp.

Godwin, W. F. and J. G. Adams. 1969.
Young clupeids of the Altamaha River,
Georgia. Ga. Game Fish Comm., Mar.
Fish. Div. Contrib. Ser. No. 15.
Brunswick. 30 pp.

Godwin, W. F. and L. G. McBay. 1967. Preliminary studies of the shad fishery
of the Altamaha River, Georgia. Ga.
Game Fish Comm., Mar. Fish. Div.
Contrib. Ser. No. 2. Brunswick.
24 pp.

Goethe, F. 1968. The effects of oil pollution on populations of marine and
coastal birds. Helgol. Wiss.
Meeresunters. 17:370-374.

Goldman, E. A. and R. Kellogg. 1940. Ten
new white-tailed deer from North and
Middle America. Proc. Biol. Soc.
Wash. 53:81-89.

Goldstein, A. K. and J. J. Manzi. 1976.
Additions to the freshwater algae of
South Carolina. J. Elisha Mitchell
Sci. Soc. 92(1):9-13.

Golley, F. B. 1962. Mammals of Georgia.
Univ. Ga. Press, Athens. 218 pp.

Colley, F. B. 1965. Structure and
function of an old-field broomsedge
community. Ecol. Monogr. 35:113-137.

Golley, F. B. 1966. South Carolina mammals. Charleston Mus., Charleston,
S.C. 181 pp.

Goodman, D. 1975. The theory of diversity
stability relationships in ecology.
Quarterly Rev. Biol. 50(3):237-266.

Gosner, K. L. 1971. Guide to identification of marine and estuarine invertebrates. John Wiley and Sons, Inc.,
New York. 693 pp.

Gosselink, J. [1978]. Mississippi deltaic
plain ecological characterization.
Planning Model. 25 pp. (Unpubl.)

Gosselink, J. G. and C. J. Kirby. 1974.
Decomposition of salt marsh grass,
Spartina alterniflora Loisel.
Limnol. Oceanogr. 19(5):825-832.

Goyal, S. M., C. P. Gerba, and J. L.
Melnick. 1977. Occurrence and distribution of bacterial indicators
and pathogens in canal communities
along the Texas coast. Appl. Environ.
Microbiol. 34:139-149.

Goyal, S. M., C. P. Gerba, and J. L.
Melnick. 1979. R+ bacteria in estuarine sediments. Mar. Pollut. Bull.
10:25-27.

Gran, H. H. and T. Braarud. 1935. A
quantitative study of the phytoplankton in the Bay of Fundy and the Gulf
of Maine. J. Biol. Board Can. 1:279-
467.

Grant, R. R., Jr. 1974. Algae, pp. 16-30.
In: Cooper River survey 1973 for the
E. I. DuPont de Nemours and Company.
Acad. Nat. Sci. Philadelphia.

Gray, I. E. 1974. Worm and clam flats,
pp. 204-243. In: H. T. Odum, B. J.
Copeland, and E. A. McMahan, eds.
Coastal ecological systems of the
United States. Vol. 2. The Conserva-
tion Foundation, Washington, D.C.

Green, H. B. 1967. The imported fire ant
in Mississippi. Miss. State Univ.
Agric. Exp. Stn. Bull. 737:4-23.

Green, J. 1968a. The biology of estuarine
animals. Univ. Wash., Seattle. 401 pp.

Green, R. H. 1968b. Mortality and stability in a low diversity subtropical
community. Ecology 49(5):848-854.

Green, W. E. 1947. Effect of water impoundment on tree mortality and growth.
J. For. 45:118-120.

Grice, G. D. 1951. Observations on
Polydora (mudworm) in South Carolina
oysters. Contrib. Bears Bluff Lab.
No. 11. Wadmalaw Island, S.C. 8 pp.

Grice, G. D. and A. D. Hart. 1962. The
abundance, seasonal occurrence and
distribution of the epizooplankton
between New York and Bermuda. Ecol.
Monogr. 32(4):287-309.

Grice, G. D., M. R. Reeve, P. Koeller,
and D. W. Menzel. 1977. The use of
large volume, transparent, enclosed
sea-surface water columns in the
study of stress on plankton ecosystems.
Helgol. Wiss. Meeresunters. 30:118-
133.

Grice, G. D., P. H. Wiebe, and E. Hoagland. 1973. Acid-iron waste as a factor affecting the distribution and abundance of zooplankton in the New York Bight. No. 1. Laboratory studies on the effects of acid waste on copepods. Estuarine Coastal Mar. Sci. 1:45–50.

Griffith, R. E. 1940. Waterfowl management of Atlantic Coast refuges. Trans. N. Am. Wildl. Conf. 5:373–377.

Grontved, J. 1962. Preliminary report on the productivity of microbenthos and phytoplankton in the Danish Wadden Sea. Medd. Dan. Fisk.--Havunders. N.S. 3:347–372.

Gunter, G. 1950. Seasonal population changes and distributions as related to salinity of certain invertebrates on the Texas coast, including the commercial shrimp. Publ. Inst. Mar. Sci. Univ. Tex. 1:7–51.

Gunter, G., J. Y. Christmas, and R. Kilabrew. 1964. Some relations of salinity to population distributions of motile estuarine organisms with special reference to penaeid shrimp. Ecology 45(1):181–185.

Haines, E. B. 1974. Processes affecting production in Georgia coastal waters. Ph.D. Dissertation. Duke Univ., Durham, N.C.

Haines, E. B. 1976. Nitrogen content and acidity of rain on the Georgia coast. Water Resour. Bull. 12:1223–1231.

Haines, E. B. and W. M. Dunstan. 1975. The distribution and relation of particulate organic material and primary productivity in the Georgia Bight, 1973–1974. Estuarine Coastal Mar. Sci. 3:431–441.

Haley, S. R. 1969. Relative growth and sexual maturity of the Texas ghost crab, Ocypode quadrata (Fabr.) (Brachyura, Ocypodidae). Crustaceana 17(3):285–297.

Haley, S. R. 1972. Reproductive cycling in the ghost crab, Ocypode quadrata (Fabr.) (Brachyura, Ocypodidae). Crustaceana 23(1):1–11.

Hall, E. R. and K. R. Kelson. 1959. The mammals of North America. Ronald Press Co., New York, N.Y. 1083 pp.

Hall, J. M. 1960. A gathering of shorebirds. Devin-Adair Co., New York. 242 pp.

Halls, L. K., B. L. Southwell, and F. E. Knox. 1952. Burning and grazing in coastal plain forests. Ga. Coastal Plain Exp. Stn. (Tifton) Bull. 51. 33 pp.

Hall, T. F. and G. E. Smith. 1955. Effects of flooding on woody plants, West Sandy Dewatering Project, Kentucky Reservoir. J. For. 53:281–285.

Hamel, P. B. 1977. The wood stork in South Carolina, a review. Chat 41(2):24–26.

Hamel, P. B. and R. G. Hooper. 1978. Status of Bachman's warbler, a progress report, pp. 112–121. In: R. O. Odom and L. Landers, eds. Proceedings of the rare and endangered wildlife symposium, 3–4 Aug. 1978, Athens, Ga. Ga. Dep. Nat. Resour., Game Fish Div., Tech. Bull. WL 4.

Hamilton, W. J., Jr. 1943. The mammals of Eastern United States. Comstock Publ. Co., Inc., Ithaca, N.Y.

Hamilton, W. J., Jr. and J. A. Pollack. 1955. The food of some crotalid snakes from Ft. Benning, Georgia. Nat. Hist. Misc. (Chicago) 140:1–4.

Hamilton, W. J., Jr. and J. A. Pollack. 1956. The food of some colubrid snakes from Fort Benning, Georgia. Ecology 37:519–526.

Hamilton, W. J., Jr. and J. A. Pollack. 1961. The food of some lizards from Fort Benning, Georgia. Herpetologica 17(2):99–106.

Hammer, Siler, and George Associates. 1975. Regional assessment study of the Chattahoochee-Flint-Apalachicola Basin. Draft phase II report. Part II: data, methodologies and detailed analysis. Beak Consultants, Inc., Harmon Engineers, The Research Group, Inc. and Water Resources Engineers. 255 pp.

Hammond, D. L. and D. M. Cupka. 1977. An economic and biological evaluation of the South Carolina pier fishery. S.C. Mar. Resour. Cent. Tech. Rep. No. 20. 14 pp.

Hannan, P. J. and C. Patouillet. 1972a. Effect of mercury on algal growth rates. Biotechnol. Bioeng. 14:93–101.

Hannan, P. J. and C. Patouillet. 1972b. Effects of pollutants on growth of algae. U.S. Naval Res. Lab., Feb. 1972 progress rep. Washington, D.C.

Hannan, P. J. and C. Patouillet. 1972c. Nutrient and pollutant concentrations as determinants in algal growth rates, pp. 340–342. In: M. Ruivo, ed. Marine pollution and sea life. Fishing News Books Ltd., Surrey, England.

Hansen, J. L. 1975. Status of the wood stork and the wood stork feeding project at Corkscrew Swamp Sanctuary. Natl. Audubon Soc. Meeting, New Orleans, 1975.

Hanson, R. P. and L. Karstad. 1959. Feral swine in the Southeastern United States. J. Wildl. Manage. 23(1):64-74.

Hardy, J. D., Jr. 1953. Notes on the distribution of *Mycrohyla carolinensis* in southern Maryland. Herpetologica 8:162-166.

Harms, W. R. 1973. Some effects of soil type and water regime on growth of tupelo seedlings. Ecology 54(1):188-193.

Harper, F. 1927. The mammals of the Okefinokee (sic) Swamp region of Georgia. Proc. Boston Soc. Nat. Hist. 38(7):191-396.

Harper, F. 1932. A dweller in the piney woods. Sci. Month. 32:176-181.

Harrell, R. M. 1977. Age, growth, and sex ratio of the American eel, *Anguilla rostrata*, in the Cooper River, South Carolina. M.S. Thesis. Clemson Univ., Clemson, S.C.

Harris, M. D. 1975. Effects of flooding on forest vegetation at two Oklahoma lakes. J. Soil Water Conserv. 30(6):294-295.

Harrison, J. R. 1970. The northern cricket frog, *Acris crepitans* Baird, in the lower coastal plain of South Carolina. Bull. S.C. Acad. Sci. 32:39. (Abstr.)

Harrison, J. R. 1978. Herpetofauna: amphibians, pp. 260-269. In: R. G. Zingmark, ed. An annotated checklist of the biota of the coastal zone of South Carolina. Univ. S.C. Press, Columbia.

Harrison, J. R., J. W. Gibbons, D. H. Nelson, and C. L. Abercrombie III. 1979. Status report: amphibians, pp. 73-78. In: D. M. Forsythe and W. B. Ezell, eds. Proceedings of the first South Caroina endangered species symposium. Nov. 1976. S.C. Wildl. Mar. Resour. Dep. and the Citadel. Charleston, S.C.

Hart, C. W., Jr. and S. L. H. Fuller. 1974. Pollution ecology of freshwater invertebrates. Academic Press, New York. 389 pp.

Hartman, D. S. 1969. Florida's manatee, mermaids in peril. Natl. Geogr. Mag. 136:342-353.

Hartman, D. S. 1971. Behavior and ecology of the Florida manatee, *Trichechus manatus latirostus* (Horlan), at Crystal River, Citrus County. Ph.D. Dissertation. Cornell Univ., Ithaca, N.Y. 285 pp.

Hartshorn, G. S. 1972. Vegetation and soil relationships in southern Beaufort County, North Carolina. J. Elisha Mitchell Sci. Soc. 88:226-240.

Hartung, R. 1967. Energy metabolism in oil-covered ducks. J. Wildl. Manage. 31(4):798-804.

Harvey, W. H. 1853. *Nereis boreali-americana*. II. Rhodospermeae. Smithson. Contrib. Knowledge 5(5):1-128.

Hasler, A. D. 1947. Eutrophication of lakes by domestic drainage. Ecology 28:383-395.

Hatch, J. J. 1970. Predation and piracy by gulls at a ternery in Maine. Auk 87:244-254.

Havel, J. F. 1976. Vascular plant survey of the lower Santee River floodplain, pp. 23-32. In: F. P. Nelson, ed. Lower Santee River environmental quality study. S.C. Water Resources Comm. Rep. No. 122. Columbia.

Hayes, M. O. and T. W. Kana, eds. 1976. Terrigenous clastic depositional environments. Univ. S.C. Dep. Geol., Coastal Mar. Div. Tech. Rep. No. 11-CRD. 302 pp.

Hayes, M. O., S. J. Wilson, D. M. Fitzgerald, L. J. Hulmes, and D. K. Hubbard. 1975. Coastal processes and geomorphology, pp. G1-G165. In: W. M. Campbell, J. M. Dean, and W. D. Chamberlain. Environmental inventory of Kiawah Island. Environmental Research Center, Inc., Columbia, S.C.

Hay, W. P. 1917. Artificial propagation of the diamond-back terrapin. U.S. Bur. Fish. Econ. Cir. No. 5. 27 pp.

Heard, R. W. 1975. Feeding habits of white catfish from a Georgia estuary. Fla. Sci. 38(1):20-28.

Heard, R. W. and E. J. Heard. 1971. Invertebrate fauna of the North and South Newport rivers and adjacent waters, pp. 122-246. In: An ecological survey of the North and South Newport rivers and adjacent waters with respect to possible effects of treated Kraft Mill effluent. Univ. Ga. Mar. Inst., Sapelo Island. Final rep. to Ga. Water Quality Control Board. UGA #D2422-122.

Hebard, F. V. 1941. Winter birds of the Okefenokee and Coleraine. Bull. 3. Ga. Soc. of Naturalists. 84 pp.

Hebard, F. V. 1950. Glossy ibis breeding in Georgia. Oriole 15:9-10.

Heerdt, P. F. van and M. F. Morzer Bruyns.
1960. A biocoenological investigation in the Yellow Dune region of
Terschelling. Tijdschr. Entomol.
103:225-275.

Heidt, A. R. and R. J. Gilbert. 1978.
The shortnose sturgeon in the Altamaha
River drainage, Georgia, pp. 54-60.
In: R. O. Odom and L. Landers, eds.
Proceedings of the rare and endangered
wildlife symposium, 3-4 Aug. 1978,
Athens, Ga. Ga. Dep. Nat. Resour.,
Game Fish Div., Tech. Bull. WL4.

Heinle, D. R. 1969. Temperature and zooplankton. Chesapeake Sci. 10(3&4):
186-209.

Heinle, D. R. and D. A. Flemer. 1976.
Flows of material between poorly
flooded tidal marshes and an estuary.
Mar. Biol. 35:359-373.

Henry, C. J. 1972. An analysis of the
population dynamics of selected avian
species--with special reference to
changes during the modern pesticide
era. U.S. Fish Wildl. Serv. Wildl.
Res. Rep. No. 1. 99 pp.

Henry, C. J. and A. P. Noltemeier. 1975.
Osprey nesting populations in the
coastal Carolinas. Am. Birds 29:
1073-1079.

Hepburn, J. S., E. Q. St. John, and F. M.
Jones. 1920. The absorption of
nutrients and allied phenomena in the
pitchers of the Sarraceniaceae.
J. Franklin Inst. 189:147-184.

Herke, W. H. 1971. Use of natural and
semi-impounded Louisiana tidal marshes
as nurseries for fishes and crustaceans.
Ph.D. Dissertation. La. State Univ.,
Baton Rouge. 242 pp.

Herrick, F. H. 1968. Audubon the
naturalist: a history of his life
and time. 2 vols. Dover Publ., Inc.,
New York. 951 pp.

Herricks, E. E. and A. S. Buikema. 1977.
Effects of pollution on freshwater
invertebrates. J. Water Pollut.
Control Fed. 120:1492-1506.

Hester, B. S. 1976. Distribution and
seasonality of Hydromedusae in South
Carolina estuaries. M.S. Thesis.
Coll. Charleston, Charleston, S.C.
88 pp.

Hester, F. E. and J. Dermid. 1973. The
world of the wood duck. J. B.
Lippincott Co., Philadelphia. 155 pp.

Heyward, F. 1939. The relation of fire
to stand composition of long-leaf pine
forest. Ecology 20:287-304.

Hicks, D. B. 1972. Seasonal distribution
and relative abundance of fishes in
the channel reaches and shore areas,
pp. 193-201. In: Port Royal Sound
environmental study. S.C. Water Resources Comm., Columbia.

Hicks, T. 1977. Beaver and their control
in Georgia. Ga. Dep. Nat. Resour.,
Game Fish Div. Tech. Bull. WL 2.
Atlanta. 23 pp.

Hildebrand, S. F. 1932. Growth of diamondback terrapins, size attained, sex
ratio and longevity. Zoologica 9(15):
551-563.

Hildebrand, S. F. and L. E. Cable. 1938.
Further notes on the development and
life history of some teleosts at
Beaufort, N.C. Fish. Bull. 24:505-642.

Hillestad, H. O., J. R. Bozeman, A. S.
Johnson, C. W. Berisford, and J. I.
Richardson. 1975. The ecology of the
Cumberland Island National Seashore,
Camden County, Georgia. Ga. Mar. Sci.
Cent. Tech. Rep. Ser. No. 75-5.
Skidaway Island. 299 pp.

Hillestad, H. O., J. I. Richardson, and
G. K. Williamson. 1977. Incidental
capture of sea turtles by shrimp
trawlermen in Georgia. Rep. to U.S.
Natl. Mar. Fish. Serv. Contract No.
03-7-042-35129. Southeastern Wildl.
Serv., Inc., Athens, Ga. 104 pp.
(Unpubl.)

Hillestad, H. O., J. I. Richardson, and
G. K. Williamson. 1978. Incidental
capture of sea turtles by shrimp
trawlermen in Georgia. Proc. Annu.
Conf. Southeast. Assoc. Fish Wildl.
Agencies 32:167-178.

Hinde, H. P. 1954. The vertical distribution of salt marsh phanerogams in
relation to tide levels. Ecol.
Monogr. 24(2):209-225.

Hobbie, J. E. 1976. Nutrients in estuaries. Oceanus 19(5):41-47.

Hobbie, J. E., O. Holm-Hansen, T. T.
Packard, L. R. Pomeroy, R. W. Sheldon,
J. P. Thomas, and W. J. Wiebe. 1972.
A study of the distribution and
activity of microorganisms in ocean
water. Limnol. Oceanogr. 17(4):544-
555.

Hoese, H. D. 1970. Studies on the parasitic oyster fungus, Labyrinthomyxa
sp., in Georgia salt waters, pp. 1-
19. In: T. L. Linton, ed. Feasibility study of methods for improving
oyster production in Georgia. Completion Rep. 2-10-R. Ga. Game Fish
Comm., Atlanta.

Hoese, H. D. 1973. A trawl study of near-shore fishes and invertebrates of the Georgia coast. Contrib. Mar. Sci. 17:63-98.

Hoese, H. D. and J. E. Durant. 1970. Notes on the boring sponges of Georgia, pp. 1-7. In: T. L. Linton, ed. Feasibility study of methods, for improving oyster production in Georgia. Completion Rep. 2-10-R. Ga. Game Fish Comm., Atlanta.

Holder, D. R. 1970. A study of fish movements from the Okefenokee Swamp into the Suwannee River. Ga. Dep. Nat. Resour., Game Fish Div., Sports Fish. Div., Atlanta.

Holder, D. R. 1971. Life history of stream fish from the Suwannee River. Study VI. Annu. Progress Rep. Project F-21-3. Ga. Dep. Nat. Resour., Game Fish Div., Atlanta.

Holder, D. R. 1973. Age and growth, reproduction and food habits of bowfin in the Okefenokee Swamp and the Suwanee River. Annu. Progress Rep. Project F-21-4. Ga. Dep. Nat. Resour., Game Fish Div., Atlanta.

Holland, A. F. 1974. A study of the intertidal macrofaunal communities inhabiting sand and mud bars of the North Inlet area near Georgetown, S.C., U.S.A. Ph.D. Dissertation. Univ. S.C., Columbia. (Libr. Congr. Card No. Mic. 75-16, 483). 196 pp. Univ. Microfilms, Ann Arbor, Mich.

Holland, A. F. and J. M. Dean. 1977a. The community biology of intertidal macrofauna inhabiting sandbars in the North Inlet Area, South Carolina, pp. 423-438. In: B. C. Coull, ed. Ecology of marine benthos. Univ. S.C. Press, Columbia.

Holland, A. F. and J. M. Dean. 1977b. The biology of the stout razor clam Tagelus plebeius: I. Animal-sediment relationships, feeding mechanisms, and community biology. Chesapeake Sci. 18:58-66.

Holland, A. F. and T. T. Polgar. 1976. Seasonal changes in the structure of an intertidal community. Mar. Biol. 37:341-348.

Holland, A. F., R. G. Zingmark, and J. M. Dean. 1974. Quantitative evidence concerning the stabilization of sediments by marine benthic diatoms. Mar. Biol. 27:191-196.

Hollis, E. H. 1948. The homing tendency of shad. Science 108:332-333.

Hook, D. D. and C. L. Brown. 1973. Root adaptations and relative flood tolerance of five hardwood species. For. Sci. 19(3):225-229.

Hook, D. D., C. L. Brown, and P. P. Kormanik. 1970. Lenticel and water root development of swamp tupelo under various flooding conditions. Bot. Gaz. 131: 217-224.

Hook, D. D., O. G. Langdon, and W. A. Hamilton. 1973. The swamp and its water nymph. Am. For. May 1973.

Hopkins, M. L. and T. E. Lynn. 1971. Some characteristics of the red-cockaded woodpecker cavity trees and management implications in South Carolina, pp. 140-169. In: R. L. Thompson, ed. Proceedings symposium on the ecology and management of the red-cockaded woodpecker. Okefenokee Natl. Wildl. Refuge, 26-27 May 1971, Folkston, Ga. U.S. Dep. Inter., Fish Wildl. Serv.

Hopkins, S. H. 1956. The boring sponges which attack South Carolina oysters, with notes on some associated organisms. Contrib. Bears Bluff Lab. No. 23. Wadmalaw Island, S.C. 30 pp.

Hopkins, S. R., T. M. Murphy, K. B. Stansell, and P. M. Wilkinson. 1978. Biotic and abiotic factors affecting nest mortality in the Atlantic loggerhead turtle. Proc. Annu. Conf. Southeast. Assoc. Fish Wildl. Agencies 32:213-223.

Horak, G. J. 1970. A comparative study of the foods of the sora and Virginia rail. Wilson Bull. 82(2):206-213.

Horch, K. W. and M. Salmon. 1969. Production, perception and reception of acoustic stimuli by semiterrestrial crabs (Genus Ocypode and Uca, Family Ocypodidae). Forma Functio 1:1-25.

Hornberger, M. L. 1979. Synopsis of data on the American eel. S.C. Wildl. Mar. Resour. Dep., Wildl. Freshwater Fish. Div. 14 pp.

Hosier, P. E. 1975. Dunes and marsh vegetation, pp. D1-D96. In: W. M. Campbell, J. M. Dean, and W. D. Chamberlain, eds. Environmental inventory of Kiawah Island. Environmental Research Center, Inc. Columbia, S.C.

Hosner, J. F. 1957. Effects of water upon the seed germination of bottomland trees. For. Sci. 3(1):67-70.

Hosner, J. F. 1958. The effects of complete inundation upon seedlings of six bottomland tree species. Ecology 39(2):371-373.

Hosner, J. F. 1959. Survival, root, and shoot growth of six bottomland tree species following flooding. J. For. 57:927-928.

Hosner, J. F. 1960. Relative tolerance to complete inundation of fourteen bottomland tree species. For. Sci. 6(3):246-251.

Hosner, J. F. and S. G. Boyce. 1962. Tolerance to water saturated soils of various bottomland hardwoods. For. Sci. 8(2):180-186.

Hoss, D. E., L. C. Clements, and D. R. Colby. 1977. Synergistic effects of exposure to temperature and chlorine on survival of young-of-the-year estuarine fishes, pp. 345-355. In: F. J. Vernberg, A. Calabrese, F. P. Thurberg, W. B. Vernberg, eds. Physiological responses of marine biota to pollutants. Academic Press, New York.

Hoss, D. E., L. C. Coston, J. P. Baptist, and B. W. Engel. 1975. Effects of temperature, copper and chlorine on fish during simulated entrainment in power plant condenser cooling systems. Environmental effects of cooling systems at nuclear plants. Int. Atomic Energy Agency SM-187-19:519-527.

Hoss, D. E., L. C. Coston, and W. E. Schaaf. 1973. Effects of dredged harbor sediments on larval estuarine fish common to Charleston Harbor, South Carolina. U.S. Dep. Commer., Natl. Mar. Fish. Serv., Atlantic Estuarine Fish. Cent. 27 pp.

Hoss, D. E., L. C. Coston, and W. E. Schaaf. 1974. Effects of sea water extracts of sediments from Charleston Harbor, South Carolina on larval estuarine fishes. Estuarine Coastal Mar. Sci. 2:323-328.

Hotchkiss, N. 1972. Common marsh, underwater and floating-leaved plants of the United States and Canada. Dover Publ., Inc., New York. 124 pp.

Hotchkiss, N. and R. E. Stewart. 1947. Vegetation of the Patuxent Research Refuge, Maryland. Am. Midl. Nat. 38(1):1-75.

Howard, J. D. and J. Dörjes. 1972. Animal-sediment relationships in two beach-related tidal flats: Sapelo Island, Georgia. J. Sediment. Petrol. 42(3):608-623.

Howard, J. D. and R. W. Frey. 1975. Estuaries of the Georgia coast, U.S.A.: sedimentology and biology. II. Regional animal-sediment characteristics of Georgia estuaries. Senckenb. Marit. 7:33-103.

Howard, J. D., R. W. Frey, and H. E. Reineck. 1972. Georgia coastal region, Sapelo Island, U.S.A.: sedimentology and biology. I. Introduction. Senckenb. Marit. 4:3-14.

Hoyt, J. H. 1968. Geology of the Golden Isles and lower Georgia coastal plain, pp. 18-34. In: D. S. Maney, F. C. Marland, and C. B. West, eds. Conference on the future of the marshlands and sea islands of Georgia. Oct. 13-14, 1968. Ga. Nat. Areas Council and Coastal Area Planning Dev. Comm.

Hudson, P. L. 1975. 1975 Annual report southeastern reservoir investigation. U.S. Fish Wildl. Serv., Clemson, S.C. 52 pp.

Huff, J. A. 1975. Life history of Gulf of Mexico sturgeon, Acipenser oxyrhynchus desotoi in Suwannee River, Florida. Fla. Mar. Res. Publ. No. 16. 30 pp.

Hulburt, E. M. 1963a. The diversity of phytoplanktonic populations in oceanic, coastal, and estuarine regions. J. Mar. Res. 21(2):81-93.

Hulburt, E. M. 1963b. The occurrence of Skeletonema costatum (Bacillariophyceae) in the Gulf Stream and Sargasso Sea. Bull. Mar. Sci. Gulf Caribb. 13:219-223.

Hulburt, E. M. 1967. Some notes on the phytoplankton off the southeastern coast of the United States. Bull. Mar. Sci. 17:330-337.

Hulburt, E. M. and R. S. MacKenzie. 1971. Distribution of phytoplankton species at the western margin of the North Atlantic Ocean. Bull. Mar. Sci. 21:603-612.

Hulburt, E. M. and J. Rodman. 1963. Distribution of phytoplankton species with respect to salinity between the coast of Southern New England and Bermuda. Limnol. Oceanogr. 8:263-269.

Hunding, C. 1971. Production of benthic microalgae in the littoral zone of a eutrophic lake. Oikos 22:389-397.

Hunt, F. M. 1951. Effects of flooded soil on growth of pine seedlings. Plant Physiol. 26(2):363-368.

Hunt, J. L., Jr. 1974. The geology and origin of Gray's Reef, Georgia Continental Shelf. M.S. Thesis. Univ. Ga., Athens. 83 pp.

Hunt, K. W. 1943. Floating mats on a southeastern coastal plain reservoir. Bull. Torrey Bot. Club 70(5):481-488.

Huntsman, G. R. and I. G. MacIntyre. 1971. Tropical coral patches in Onslow Bay. Underwater Nat. 7:32-34.

Hustedt, F. 1955. Marine littoral diatoms of Beaufort, North Carolina. Duke Univ. Marine Station Bull. No. 6. 67 pp.

Huston, M. 1979. A general hypothesis of species diversity. Am. Naturalist 113(1):81-101.

Hutchinson, G. E. 1957. A treatise on Limnology. Vol. 1. Geography, physics, and chemistry. John Wiley & Sons, New York. 1015 pp.

Hutchinson, G. E. 1967. A treatise on limnology. Vol. II. Introduction to lake biology and the limnoplankton. John Wiley & Sons, New York. 1115 pp.

Hynes, H. B. N. 1970. The ecology of running waters. Univ. Toronto Press, Toronto. 555 pp.

Ineson, F. A. and I. F. Eldredge. 1938. Forest resources of northeastern Florida. U.S. Dep. Agric. Misc. Publ. 313.

Ingle, R. M. 1953. Studies on the effect of dredging operations upon fish and shellfish. Gulf Caribb. Fish. Inst. Proc. 5:106.

Ivie, G. W., H. W. Dorough, and E. G. Alley. 1974. Photodecomposition of mirex on silica gel chromatoplates exposed to natural and artificial light. J. Agric. Food Chem. 22(6):933-935.

Jackson, J. A. 1978. Analysis of the distribution and population status of the red-cockaded woodpecker, pp. 101-111. In: R. O. Odom and L. Landers, eds. Proceedings of the rare and endangered wildlife symposium, 3-4 Aug. 1978. Athens, Ga. Ga. Dep. Nat. Resour., Game Fish Div., Tech. Bull. WL 4.

Jacobs, J. E. 1968. A preliminary check list of fresh-water algae in South Carolina. J. Elisha Mitchell Sci. Soc. 84(4):454-457.

Jacobs, J. E. 1971. A preliminary taxonomic survey of the freshwater algae of the Belle W. Baruch Plantation in Georgetown County, South Carolina. J. Elisha Mitchell Sci. Soc. 87:26-30.

Jeffries, R. L. 1972. Aspects of salt marsh ecology with particular reference to inorganic plant nutrition, pp. 61-85. In: R. S. K. Barnes and J. Green, eds. The estuarine environment. Applied Science Publ. Ltd., London.

Jensen, A. S. 1937. Remarks on the Greenland eel, its occurrence and reference to Anguilla rostrata. Medd. Grønland 118(9):1-8.

Jervis, R. A. 1964. Primary production in a freshwater marsh ecosystem. Ph.D. Dissertation. Rutgers Univ., New Brunswick, N.J. (Libr. Congr. Card No. Mic. 64-10,924.) 58 pp. Univ. Microfilms, Ann Arbor, Mich.

Jester, D. B. 1974. Life history, ecology and management of the carp Cyprinus carpio Linnaeus in Elephant Butte Lake. New Mexico State Univ. Agric. Exp. Stn. Res. Rep. 273. 80 pp.

Jetter, W. and L. D. Harris. 1976. The effects of perturbation on cypress dome animal communities, pp. 577-653. In: H. T. Odum, E. C. Ewel, J. W. Ordway, and M. K. Johnston, eds. Cypress wetlands for water management, recycling and conservation. Univ. Fla. Phelps Lab., Center for Wetlands Third Annu. Rep., Gainesville.

Joanen, T. 1969. Nesting ecology of alligators in Louisiana. Proc. Annu. Conf. Southeast. Assoc. Game Fish Comm. 23: 141-151.

Joanen, T. 1974. Population status and distribution of alligators in the Southeastern United States. Presented at Southeast. Regional Endangered Species Workshop. Sept. 1974. Tallahassee, Fla. 44 pp. (Unpubl.)

Joanen, T. and L. McNease. 1970. A telemetric study of nesting female alligators on Rockefeller Refuge, Louisiana. Proc. Annu. Conf. Southeast. Assoc. Game Fish Comm. 24:175-193.

Joanen, T. and L. McNease. 1973. A telemetric study of adult male alligators on Rockerfeller Refuge, Louisiana. Proc. Annu. Conf. Southeast. Assoc. Game Fish Comm. 26:252-275.

Joanen, T., L. McNease, H. DuPuie, and W. G. Perry. 1972. Salinity tolerance of hatchling alligators. La. Wildl. Fish. Comm., 1970-71 Biennial Rep. 14:56. New Orleans.

Jobson, H. G. M. 1940. Reptiles and amphibians from Georgetown County, South Carolina. Herpetologica 2:39-43.

Johnsgard, P. A. 1975. Waterfowl of North America. Indiana Univ. Press, Bloomington. 571 pp.

Johnson, A. S. 1970. Biology of the raccoon (Procyon lotor varius) Nelson and Goldman in Alabama. Auburn Univ. Agric. Exp. Stn. Bull. 402. Auburn. 148 pp.

Johnson, A. S., H. O. Hillestad, S. F. Shanholtzer, and G. F. Shanholtzer. 1974. An ecological survey of the coastal region of Georgia. Natl. Park Serv. Sci. Monogr. Ser. No. 3. Washington, D.C. 233 pp.

Johnson, D. S. and H. H. York. 1915. Relation of plants to tide levels. Carnegie Inst. Wash. Publ. No. 206. Washington, D.C. 162 pp.

Johnson, D. W. and E. P. Odum. 1956. Breeding bird populations in relation to plant succession on the piedmont of Georgia. Ecology 37:50-62.

Johnson, T. L. and L. E. McSwain. 1973. Age and growth, reproduction, food habits and distribution of spotted sucker in the Flint River. Statewide Fisheries Investigations. Annu. Rep. F-21-4. Vol. 2. Ga. Dep. Nat. Resour., Game Fish Div., Atlanta.

Johnson, T. L. and L. E. McSwain. 1974. Age and growth, reproduction, food habits and distribution of spotted sucker in the Flint River. Final rep. Project F-21-5. Ga. Dep. Nat. Resour., Game Fish Div., Atlanta.

Johnson, T. W., Jr. 1967. The estuarine mycoflora, pp. 303-305. In: G. H. Lauff, ed. Estuaries. Am. Assoc. Adv. Sci. Publ. No. 83. Washington, D.C.

Johnson, T. W., Jr. and F. K. Sparrow, Jr. 1961. Fungi in oceans and estuaries. J. Cramer, Weinheim, Germany. 668 pp.

Johnston, J. B. 1978. Ecological characterization--an overview, pp. 692-710. In: Coastal zone '78. Proc. symposium on technical, environmental, socioeconomic and regulatory aspects of coastal zone management, 14-16 March 1978, San Francisco. Am. Soc. Civil Eng., New York.

Jordan, D. S. 1877. Contributions to North American ichthyology II. Synopsis of the Silvridae of the freshwaters of North America. U.S. Natl. Mus. Bull. 10:69-120.

Jordan, D. S. 1878. A catalogue of the fishes of the freshwaters of North America. U.S. Geol. Geogr. Surv. Terr. 4:407-442.

Jordan, D. S. 1885. A catalogue of fishes known to inhabit the waters of North America, north of the Tropic of Cancer, with notes on the species discovered in 1883 and 1884. Annu. Rep. U.S. Comm. Fish and Fish. 1884:787-973.

Joyce, E. A. 1972. A partial bibliography of oysters, with annotations. Fla. Dep. Nat. Resour. Mar. Res. Lab. Spec. Sci. Rep. No. 34. 846 pp.

Kahl, M. P. 1964. Food ecology of the wood stork (Mycteria americana) in Florida. Ecol. Monogr. 34:97-117.

Kaiser, K. L. E. 1978. The rise and fall of mirex. Environ. Sci. Technol. 12(5): 520-528.

Kale, H. W. II. 1965. Ecology and bioenergetics of the long-billed marsh wren. Telmatodytes palustris griseus, in a salt marsh ecosystem. Nuttall Ornithol. Club Publ. No. 5. 142 pp.

Kale, H. W. II. 1967. Recoveries of black skimmers banded on Little Egg Island, Georgia. Oriole 32(2):13-16.

Kale, H. W. II, G. W. Sciple, and I. R. Tomkins. 1965. The royal tern colony of Little Egg Island, Georgia. Bird-banding 36(1):21-27.

Kale, H. W. II and J. M. Teal. 1958. Royal tern nesting on Little Egg Island. Oriole 23(3):36-37.

Kale, H. W. II and L. A. Webber. 1972. Breeding bird populations in a native coastal hammock and golf course subdivision. Paper presented at the American Ornithologists' Union Meeting, August.

Kalmbach, E. R. 1943. The armadillo: its relation to agriculture and game. Tex. Game Fish Oyster Comm., Austin. 61 pp.

Kaneko, T. and R. R. Colwell. 1973. Ecology of Vibrio parahaemolyticus in Chesapeake Bay. J. Bacteriol. 113: 24-32.

Kapoor, I. P., R. L. Metcalf, A. S. Hirwe, P. Y. Lu, and J. R. Coats. 1972. Comparative metabolism of DDT, methylchlor and ethoxychlor in a model ecosystem. J. Agric. Food Chem. 20(1):1-6.

Karr, J. R. and R. R. Roth. 1971. Vegetation structure and avian diversity in several New World areas. Am. Nat. 105(945):423-435.

Keefe, C. W. 1972. Marsh production: a summary of the literature. Contrib. Mar. Sci. 16:163-181.

Keiser, R. K., Jr. 1976. Species composition, magnitude and utilization of the incidental catch of the South Carolina shrimp fishery. S.C. Mar. Resour. Cent. Tech. Rep. No. 16. 94 pp.

Keith, J. O. 1968. Insecticide residues in fish-eating birds and their environment. Aves 5(1):28-41.

Keith, W. J. and H. S. Cochran, Jr. 1968. Charting of subtidal oyster beds and experimental planting of seed oysters in South Carolina. Contrib. Bears Bluff Lab. No. 48. Wadmalaw Island, S.C. 19 pp.

Keith, W. J. and R. C. Gracy. 1972. History of the South Carolina oyster. S.C. Mar. Resour. Cent. Educ. Rep. No. 1. Charleston. 19 pp.

Kelley, B. J., Jr. 1978. Order Mysidacea, pp. 170. In: R. G. Zingmark, ed. An annotated checklist of the biota of the coastal zone of South Carolina. Univ. S.C. Press, Columbia, S.C.

Kemp, H. T., R. L. Little, V. L. Holoman, and R. L. Darby. 1973. Water quality criteria data book. Vol. 5: effects of chemicals on aquatic life. U.S. Environ. Prot. Agency, Natl. Water Quality Lab., Duluth, Minn. 511 pp.

Kennedy, H. E. 1970. Growth of newly planted water tupelo seedlings after flooding and siltation. For. Sci. 16:250-256.

Kennedy, H. E. and R. M. Kinard. 1974. 1973 Mississippi River flood's impact on natural hardwood forests and plantations. U.S. Dep. Agric. Forest Serv. Res. Note SO-177.

Kerwin, J. A. 1966. Classification and structure of the tidal marshes of the Poropotank River, Virginia. Assoc. Southeast. Biol. Bull. 13:40. (Abstr.)

Kerwin, J. A. and L. G. Webb. 1972. Foods of ducks wintering in coastal South Carolina, 1965-1967. Proc. Annu. Conf. Southeast. Assoc. Game Fish Comm. 25:223-245.

Ketchum, B. H. 1967. Phytoplankton nutrients in estuaries, pp. 329-335. In: G. H. Lauff, ed. Estuaries. Am. Assoc. Adv. Sci. Publ. No. 83. Washington, D.C.

Kevan, D. K. M., ed. 1955. Soil zoology. Academic Press, Inc., New York. 512 pp.

Kevan, D. K. M. 1968. Soil animals. H. F. & G. Witherby Ltd., London. 244 pp. (First publ. 1962.)

Kight, J. 1962. An ecological study of the bobcat, Lynx rufus (Schreber), in west-central South Carolina. M.S. Thesis. Univ. Ga., Athens. 52 pp.

King, D. L. and R. C. Ball. 1966. A qualitative and quantitative measure of Aufwuchs production. Trans. Am. Microsc. Soc. 85:232-240.

Kirby-Smith, W. W. 1976. The detritus problem and the feeding and digestion of estuarine organisms, pp. 469-479. In: M. Wiley, ed. Estuarine processes. Vol. I. Uses, stresses, and adaptation to the estuary. Academic Press, New York.

Kirtland's Warbler Recovery Team. 1976. Kirtland's warbler recovery plan. Technical draft. U.S. Dep. Inter., Fish Wildl. Serv. (Unpubl.)

Kitchens, W. M., Jr., J. M. Dean, L. H. Stevenson, and J. H. Cooper. 1975. The santee swamp as a nutrient sink, pp. 349-366. In: F. G. Howell, J. B. Gentry and M. H. Smith, eds. Mineral cycling in southeastern ecosystems. U.S. Energy Res. Dev. Adm. Symp. Ser. Conf-740513.

Kjelson, M. A. and G. N. Johnson. 1976. Further observations of the feeding ecology of postlarval pinfish, Lagodon rhomboides and spot Leiostomus xanthurus. Fish. Bull. 74(2):423-432.

Kjelson, M. A., D. S. Peters, G. W. Thayer, and G. N. Johnson. 1975. The general feeding ecology of postlarval fishes in the Newport River estuary. Fish. Bull. 73(1):137-144.

Klawitter, R. A. 1962. Sweetgum, swamp tupelo, and water tupelo sites in a South Carolina bottomland forest. Ph.D. Dissertation. Duke Univ. School of Forestry, Durham, N.C. 175 pp.

Knepton, J. C., Jr. 1954. A note on the burrowing habits of the salamander, Amphiuma means means. Copeia 1954:68.

Knepton, J. C., Jr. 1956. County records of Testudinata collected in Georgia. J. Tenn. Acad. Sci. 31(4):322-324.

Knight, F. B. 1958. The effects of woodpeckers on populations of the Englemann spruce beetle. J. Econ. Entomol. 51: 603-607.

Knott, D. M. 1980. The zooplankton of the North Edisto River and two artificial saltwater impoundments. M.S. Thesis, College of Charleston.

Knowlton, C. J. 1972. Fishes taken
during commercial shrimp fishing in
Georgia's close inshore ocean
waters. Ga. Game Fish Comm.,
Coastal Fish. Off., Contrib. Ser. 21.
Brunswick. 42 pp.

Kohlmeyer, J. 1978. Marine fungi, pp.
37-38. In: R. G. Zingmark, ed.
An annotated checklist of the biota
of the coastal zone of South Carolina.
Univ. S.C. Press, Columbia.

Kohlsaat, T. 1974. Elements of diversity:
the natural communities of South
Carolina. Nature Conservancy,
Washington, D.C. 109 pp.

Komarek, E. V. 1969. Fire and animal
behavior. Proc. Annu. Tall Timbers
Fire Ecol. Conf. 9:161-207.

Korringa, P. and H. Postuma. 1957. Investigations into the fertility of
the Gulf of Naples and adjacent
salt water lakes with special reference to shellfish cultivation.
Pubbl. Stn. Zool. Napoli 29:229-284.

Kowal, N. E. 1969. Ingestion rate of a
pine-mor oribatid mite. Am. Midl.
Nat. 81:595-598.

Kowal, N. E. and D. A. Crossley, Jr.
1971. The ingestion rates of microarthropods in pine-mor, estimated
with radioactive calcium. Ecology
52:444-452.

Kraeuter, J. N. 1976. Biodeposition by
salt-marsh invertebrates. Mar.
Biol. 35:215-223.

Kraeuter, J. N. and E. Setzler. 1975.
The seasonal cycle of Scyphozoa and
Cubozoa in Georgia estuaries. Bull.
Mar. Sci. 25:66-74.

Kraeuter, J. N. and P. L. Wolf. 1974.
The relationships of marine macroinvertebrates to salt marsh plants,
pp. 449-462. In: R. J. Reimold
and W. H. Queen, eds. Ecology of
Halophytes. Academic Press, New
York.

Krenkel, J. H. and J. C. Burdick III.
1976. Dredging and its environmental
effects. Proc. of special conference
on dredging. Am. Soc. Civil Eng.
1035 pp.

Kroger, R. L. and J. F. Guthrie. 1972.
Incidence of the parasitic isopod
(Olencira praegustator), in juvenile
Atlantic menhaden. Copeia 1972 (2):
370-374.

Kubiena, W. L. 1955. Animal activity as
a decisive factor in establishment
of humus forms, pp. 73-80. In: D. K.
M. Kevan, ed. Soil zoology. Academic
Press, Inc., New York.

Kuenzler, E. J. 1961. Structure and
energy flow of a mussel population in
a Georgia salt marsh. Limnol.
Oceanogr. 6(2):191-204.

Kühnelt, W. 1955. A brief introduction to
the major groups of soil animals and
their biology, pp. 29-43. In: D. K.
M. Kevan, ed. Soil zoology. Academic
Press, Inc., New York.

Kühnelt, W. 1963. Soil-inhabiting arthropoda.
Annu. Rev. Entomol. 8:115-136.

Kurz, H. and D. Demaree. 1934. Cypress
buttresses and knees in relation to
water and air. Ecology 15:36-41.

Kurz, H. and K. Wagner. 1957. Tidal marshes
of the Gulf and Atlantic coasts of
northern Florida and Charleston, South
Carolina. Fla. State Univ. Stud. No.
24. Tallahassee. 168 pp.

Laessle, A. M. 1942. The plant communities
of the Welaka area. Univ. Fla. Press,
Biol. Sci. Ser. 9(1):1-143.

Laessle, A. M. 1958. The origin and
successional relationships of sandhill
vegetation and sand-pine scrub. Ecol.
Monogr. 28:361-387.

Laessle, A. M. and C. D. Monk. 1961. Some
live oak forests of northeastern
Florida. Q. J. Fla. Acad. Sci. 24(1):
39-55.

Lagler, K. F. 1956. Freshwater fishery
biology. Wm. C. Brown. Co., Dubuque,
Iowa. 421 pp.

Landers, J. L., A. S. Johnson, P. H.
Morgan, and W. P. Baldwin. 1976. Duck
foods in managed tidal impoundments in
South Carolina. J. Wildl. Manage.
40(4):721-728.

Land, L. S. and J. H. Hoyt. 1966. Sedimentation in a meandering estuary.
Sedimentology 6(3):191-207.

Larsen, H. 1962. Halophilism, pp. 297-342.
In: I. C. Gunsalus and R. Y. Stanier,
eds. The bacteria. Vol. 4: growth.
Academic Press, New York.

Larsen, H. S. 1963. Effects of soaking
in water on acorn germination of
four southern oaks. For. Sci. 9(2):
236-241.

Laurie, P. 1978. Capers Island. S.C. Mar.
Resour. Cent. Publ., Charleston. 7 pp.

LaVal, R. K. 1967. Records of bats from
the Southeastern United States. J.
Mammal. 48(4):645-648.

Lawson, J. 1937. The history of North
Carolina; containing the exact
description and natural history of the
country. Garrett and Massie, Richmond,
Va. 258 pp. (First publ. 1714.)

Lay, D. W. and D. N. Russell. 1970. Notes on the red-cockaded woodpecker in Texas. Auk 87:781–786.

Leach, G. C. 1925. Artificial propagation of shad. Rep. U.S. Comm. Fish. 1924: 459–486.

Leach, J. H. 1970. Epibenthic algal production in an intertidal mudflat. Limnol. Oceanogr. 15:514–521.

Leatherwood, S. D., D. K. Caldwell, and H. E. Winn. 1976. Whales, dolphins, and porpoises of the western North Atlantic; a guide to their identification. NOAA Tech. Rep., Natl. Mar. Fish. Serv. Circ. 306. 176 pp.

Lebuff, C. R., Jr. 1974. Unusual nesting relocation in the loggerhead turtle, Caretta caretta. Herpetologica 30(1):29–31.

Lee, R. F. and L. Cheng. 1974. A comparative study of the lipids of water-striders from marine, estuarine and freshwater environments: Halobates, Rheumatobates, Gerris (Heteroptera: Gerridae). Limnol. Oceanogr. 19(6): 958–965.

Lee, W. Y. and J. C. C. Nicol. 1977. The effects of the water soluble fractions of No. 2 fuel oil on the survival and behavior of coastal and oceanic zooplankton. Environ. Pollut. 12:279–291.

Leffler, J. W. 1978. Ecosystem response to stress in aquatic microcosms, pp. 102–119. In: J. H. Thorp and J. W. Gibbons, eds. Energy and environmental stress in aquatic systems. U.S. Dep. Energy, Tech. Inf. Center Symp. Ser. 48.

Lehrbas, M. M. and I. F. Eldredge. 1941. Forest resources of south Georgia. U.S. Dep. Agric. Misc. Publ. No. 390. 50 pp.

Leim, A. H. 1924. The life history of the shad (Alosa sapidissima) (Wilson) with special reference to the factors limiting its abundance. Biol. Board Can. Contrib. Can. Biol. 2(11):163–284.

Leiper, A. S. 1973. Seasonal change in the structure of three sublittoral marine benthic communities off Sapelo Island, Georgia. Ph.D. Dissertation. Univ. Ga., Athens. 296 pp.

Lemon, P. C. 1949. Successional responses of herbs in the longleaf-slash pine forest after fire. Ecology 30:135–145.

Lepple, F. K. 1973. Mercury in the environment; a global review including recent studies in the Delaware Bay region. Univ. Del., Newark. DEL-SG-8-73. 75 pp.

Lesser, F. H. 1965. Some environmental considerations of impounded tidal marshes on mosquito and waterbird prevalence, Little Creek Wildlife Area, Delaware, M.S. Thesis. Univ. Delaware, Newark.

Levin, A. A., T. J. Birch, R. E. Hillman, and G. E. Raines. 1972. Thermal discharges: ecological effects. Environ. Sci. Technol. 6(3):224–230.

Ligon, J. D. 1970. Behavior and breeding biology of the red-cockaded woodpecker. Auk 87:255–278.

Likens, G. E. 1972. Eutrophication and aquatic ecosystems, pp. 3–13. In: G. E. Likens, ed. Nutrients and eutrophication: the limiting-nutrient controversy. Am. Soc. Limnol. Oceanogr. Spec. Symp. Vol. 1. Lawrence, Kans.

Likens, G. E. 1975. Primary production of inland aquatic ecosystems. In: H. Lieth and R. H. Whittaker, eds. The primary productivity of the biosphere. Spring-Verlag, New York.

Likens, G. E., F. H. Bormann, R. S. Pierce, J. S. Eaton, and N. M. Johnson. Biogeochemistry of a forested ecosystem. Spring-Verlag, New York. 146 pp.

Lind, C. T. and G. Cottam. 1969. The submerged aquatics of University Bay: a study in eutrophication. Am. Midl. Nat. 81:353–369.

Lindeman, R. L. 1942. Seasonal food-cycle dynamics in a senescent lake. Am. Midl. Nat. 26:636–673.

Lindner, M. J. and W. W. Anderson. 1956. Growth, migrations, spawning and size distribution of shrimp Penaeus setiferus. Fish. Bull. 56(106): 556–645.

Liner, E. A. 1954. The herpetofauna of Lafayette, Terrebonne and Vermilion parishes, Louisiana. Proc. La. Acad. Sci. 17:65–85.

Linley, J. R. 1976. Biting midges of mangrove swamps and salt marshes (Diptera: Ceratopogonidae), pp. 335–376. In: L. Cheng, ed. Marine Insects. American Elsevier Publ. Co., New York.

Linton, T. L., ed. [1970]. Feasibility study of methods for improving oyster production in Georgia. Completion rep. 2-10-R. Univ. Ga. Mar. Inst., Sapelo Island, and Ga. Game Fish Comm., Atlanta. 172 pp.

Linton, T. L. and W. L. Rickards. 1965. Young common snook on the coast of Georgia. Q. J. Fla. Acad. Sci. 28(2):185-189.

Longwell, A. C. 1976. Chromosome mutagenesis in developing mackerel eggs sampled from the New York Bight. Tech. Memo. Rep. No. NOAA-TM-ERL-MESA 7. 68 pp.

Lonsdale, D. J. and B. C. Coull. 1977. Composition and seasonality of zoo-plankton of North Inlet, South Carolina. Chesapeake Sci. 18(3):272-283.

Loosanoff, V. L. 1964. New shellfish farming. Trans. N. Am. Wildl. Nat. Resour. Conf. 29:332-336.

Loosanoff, V. L. 1965. The American or eastern oyster. U.S. Fish. Wildl. Serv. Circ. 205. 36 pp.

Loosanoff, V. L. and H. C. Davis. 1963. Rearing of bivalve mollusks. Adv. Mar. Biol. 1:1-136.

Loucks, W. L. and R. A. Keen. 1973. Submersion tolerance of selected seedling trees. J. For. 71:496-497.

Lowe, J. I., P. R Parrish, A. J. Wilson, Jr., P. D. Wilson, and T. W. Duke. 1971. Effects of mirex on selected estuarine organisms. Trans. N. Am. Wildl. Nat. Resour. Conf. 36:171-186.

Lowery, G. H., Jr. 1974. The mammals of Louisiana and its adjacent waters. La. State Univ. Press, Baton Rouge. 565 pp.

Loyacano, H. A., Jr. 1975. A list of freshwater fishes of South Carolina. Clemson Univ., S.C. Agric. Exp. Stn. Bull. 580. Clemson, S.C. 12 pp.

Lunz, G. R. 1935. Oyster pest control investigations 1935. A preliminary report of the survey of coastal waters of South Carolina. 5 pp. (Unpubl.)

Lunz, G. R. 1938a. Oyster culture with reference to dredging operations in South Carolina. (Part I). U.S. Army Corps Eng., Charleston District, S.C. 135 pp.

Lunz, G. R. 1938b. The effects of the flooding of the Santee River in April 1936 on oysters in the Cape Romain area of South Carolina. Part 2. U.S. Army Corps Eng., Charleston District, S.C. 33 pp.

Lunz, G. R. 1940. The annelid worm, Polydora, as an oyster pest. Science 72:310.

Lunz, G. R. 1941. Polydora, a pest in South Carolina oysters. J. Elisha Mitchell Sci. Soc. 57:273-283.

Lunz, G. R. 1943. The yield of certain oyster lands in South Carolina. Am. Midl. Nat. 30:806-808.

Lunz, G. R. 1951. A salt water fish pond. Contrib. Bears Bluff Lab. No. 12. Wadmalaw Island, S.C. 12 pp.

Lunz, G. R. 1952. Oysters in South Carolina grow above low tide level. Atl. Fisherman 18:42-43.

Lunz, G. R. 1955. Cultivation of oysters in ponds at Bears Bluff Laboratories. Proc. Natl. Shellfish. Assoc. 46:83-87.

Lunz, G. R. 1956. Harvest from an experimental one acre salt-water pond at Bears Bluff Laboratories, South Carolina. Prog. Fish-Cult. 18(2):92-94.

Lunz, G. R. 1957. Pond cultivation of shrimp in South Carolina. Gulf Caribb. Fish. Inst. Proc. 10:44-48.

Lunz, G. R. 1958. Salt ponds can raise shrimp cheaper than sea trawlers can catch them. Quick Frozen Foods 1958:111-112, 114.

Lunz, G. R. 1968. Farming the salt marshes, pp. 172-177. In: J. D. Newsom, ed. Proceedings of the marsh and estuary management symposium. La. State Univ., Baton Rouge.

Lunz, G. R. and C. M. Bearden. 1963. Control of predaceous fishes in shrimp farming in South Carolina. Contrib. Bears Bluff Lab. No. 36. Wadmalaw Island, S.C. 9 pp.

Lunz, G. R. and F. J. Schwartz. [1970]. Analysis of eighteen year trawl catches of seatrout (Cynoscion sp: Sciaenidae) from South Carolina. Contrib. Bears Bluff Lab. No. 53. Wadmalaw Island, S.C. 29 pp.

MacArthur, R. H. and J. W. MacArthur. 1961. On bird species diversity. Ecology 42:594-598.

MacArthur, R. H. and E. R. Pianka. 1966. On optimal use of a patchy environment. Am. Nat. 100(916):603-607.

Macfie, M. E. and L. F. Swails, Jr. 1957. The algae. A new distribution record of a rare variety of Micrasterias. Univ. S.C. Publ., Biol. Ser. III, 2(2):61-62.

MacGinitie, G. E. and N. MacGinitie. 1968. Natural history of marine animals. McGraw-Hill, New York. 523 pp.

MacGregor, D. S. 1970. Studies of pond culture of oysters, pp. 1-36. In: T. L. Linton, ed. Feasibility study of methods for improving oyster production in Georgia. Completion rep. 2-10-R. Ga. Game Fish Comm., Mar. Fish. Div. and Univ. Ga.

MacIntyre, I. G. 1970. New data on the occurrence of tropical reef corals on the North Carolina continental shelf. J. Elisha Mitchell Sci. Soc. 86(4):178 (Abstr.)

MacIntyre, I. G. and J. D. Milliman. 1970. Physiographic features on the outer shelf and upper slope, Atlantic Continental Margin, Southeastern United States. Geol. Soc. Am. Bull. 81:2577-2598.

MacIntyre, I. G. and O. H. Pilkey. 1969. Tropical reef corals: tolerance of low temperatures on the Carolina continental shelf. Science 166:374-375.

Mackenthun, K. M. 1965. Nitrogen and phosphorus in water: an annotated selected bibliography of their biological effects. U.S. Dep. Health, Education and Welfare, Public Health Serv., Div. Water Supply and Pollution Control. 111 pp.

Mackenthun, K. M. and W. M. Ingram. 1967. Biological associated problems in freshwater environments, their identification, investigation and control. U.S. Dep. Inter., Fed. Water Pollut. Control Admin., Washington, D.C. 287 pp.

Mackin, J. G. 1961. Canal dredging and silting in Louisiana bays. Publ. Inst. Mar. Sci. Univ. Texas 7: 262-314.

MacLeod, D. A. 1967. The morphology and genesis of salt marsh soils and the changes taking place in them on reclamation. Ph.D. Dissertation. Univ. Cambridge, U.K.

MacNamara, R. W. 1949. Salt marsh development of Tuckahoe, New Jersey. Trans. N. Am. Wildl. Conf. 14:100-117.

Mahan, W. E., M. O. Bara, T. Strange, Jr., M. G. White, L. W. Gahan, D. W. York, and W. J. Wiebe. 1975. Santee Swamp wildlife resources report. S.C. Wildl. Mar. Resour. Dep., Div. Game Freshwater Fish. 106 pp.

Mahood, R. K. 1974. Seatrout of the genus Cynoscion in coastal waters of Georgia. Ga. Dep. Nat. Resour., Game Fish Div., Coastal Fish. Off. Contrib. Ser. No. 26. Brunswick. 36 pp.

Mahood, R. K., C. D. Harris, J. L. Music, Jr., and B. A. Palmer. 1974a. Survey of the fisheries resources in Georgia's estuarine and inshore ocean waters. Part I: Southern section, St. Andrews Sound and St. Simons Sound estuaries. Ga. Dep. Nat. Resour., Game Fish Div., Coastal Fish. Off. Contrib. Ser. No. 22. Brunswick. 104 pp.

Mahood, R. K., C. D. Harris, J. L. Music, Jr., and B. A. Palmer. 1974b. Survey of the fisheries resources in Georgia's estuarine and inshore ocean waters. Part II: Central section, Doboy Sound and Sapelo Sound estuaries. Ga. Dep. Nat. Resour., Game Fish Div., Coastal Fish. Off. Contrib. Ser. No. 23. Brunswick. 99 pp.

Mahood, R. K., C. D. Harris, J. L. Music, Jr., and B. A. Palmer. 1974c. Survey of the fisheries resources in Georgia's estuarine and inshore ocean waters. Part III: Northern section, Ossabaw Sound and Wassaw Sound estuaries. Ga. Dep. Nat. Resour., Game Fish Div., Coastal Fish. Off. Contrib. Ser. No. 24. Brunswick. 100 pp.

Mahood, R. K., C. D. Harris, J. L. Music, Jr., and B. A. Palmer. 1974d. Survey of the fisheries resources in Georgia's estuarine and inshore ocean waters. Part IV: Coastal Georgia--southern, central and northern sections. Ga. Dep. Nat. Resour., Game Fish Div., Coastal Fish. Off. Contrib. Ser. No. 25. Brunswick. 201 pp.

Mangold, R. E. 1962. The role of low-level dike salt impoundments in mosquito control and wildlife utilization. Proc. N.J. Mosq. Exterm. Assoc. 49:117-120.

Mann, T. M. 1977. Impact of developed coastline on nesting and hatchling sea turtles in Southeastern Florida. M.S. Thesis. Fla. Atlantic Univ., Boca Raton. 100 pp.

Mann, T. M. 1978. Impact of developed coastline on nesting and hatchling sea turtles in Southeastern Florida, pp. 53-55. In: G. E. Henderson, ed. Proceedings of the Florida and inter-regional conference on sea turtles, 24-25 July 1976, Jensen Beach, Florida. Fla. Mar. Res. Publ. 33.

Mansueti, R. J. 1962. Eggs, larvae, and young of the hickory shad, _Alosa mediocris_, with comments on its ecology in the estuary. Chesapeake Sci. 3(3):173-205.

Manzi, J. J., V. G. Burrell, and W. Z. Carson. 1977b. A comparison of growth and survival of subtidal _Crassostrea virginica_ (Gmelin) in South Carolina salt marsh impoundments. Aquaculture 12:293-310.

Manzi, J. J., P. E. Stofan, and J. L. Dupuy. 1977a. Spatial heterogeneity of phytoplankton populations in estuarine surface microlayers. Mar. Biol. 41:29-38.

Manzi, J. J. and R. G. Zingmark. 1978. Phytoplankton, pp. 2-22. In: R. G. Zingmark, ed. An annotated checklist of the biota of the coastal zone of. South Carolina. Univ. S.C. Press, Columbia.

Marcy, B. C., P. M. Jacobson, and R. L. Nankee. 1972. Observations on the reactions of young American shad to a heated effluent. Trans. Am. Fish. Soc. 101(4):740-743.

Markin, G. P., H. Collins, and J. Diller. 1972. Colony founding by queens of the red imported fire ant _Solenopsis invicta_. J. Econ. Entomol. 65(5): 1053-1058.

Markin, G. P., J. C. Hawthorne, H. C. Collins, and J. H. Ford. 1974. Levels of mirex and some other organochlorine residues in seafood from Atlantic and Gulf Coast states. Pestic. Monit. J. 7(3):139-143.

Marples, T. G. 1966. A radionuclide tracer study of arthropod food chains in a _Spartina_ salt marsh ecosystem. Ecology 47(2):270-277.

Marples, T. G. and E. P. Odum. 1964. A radionuclide tracer study of arthropod food chains in a _Spartina alterniflora_ salt marsh. Ecol. Soc. Am. Bull. 45:81. (Abstr.)

Marshall, H. G. 1966. Observations on the vertical distribution of cocolithophores in the northwestern Sargasso Sea. Limnol. Oceanogr. 11:432-435.

Marshall, H. G. 1968. Cocolithophores in the northwest Sargasso Sea. Limnol. Oceanogr. 13:370-376.

Marshall, H. G. 1969a. Observations on the spatial concentrations of phytoplankton. Castanea 34:217-222.

Marshall, H. G. 1969b. Phytoplankton distribution off the North Carolina coast. Am. Midl. Nat. 82(1):241-257.

Marshall, H. G. 1971. Composition of phytoplankton off the Southeastern coast of the United States. Bull. Mar. Sci. 21(4):806-825.

Marshall, H. G. 1976. Phytoplankton distribution along the Eastern coast of the USA. I. Phytoplankton composition. Mar. Biol. 38:81-89.

Marshall, H. G. 1978. Phytoplankton distribution along the Eastern coast of the U.S.A. Part II: seasonal assemblages north of Cape Hatteras, North Carolina. Mar. Biol. 45:203-208.

Marshall, J. T. 1963. Fire and birds in the mountains of Southern Arizona. Proc. Annu. Tall Timbers Fire Ecol. Conf. 2:135-141.

Marshall, N., C. A. Oviatt, and D. M. Skauen. 1971. Productivity of the benthic microflora of shoal estuarine environments in Southern New England. Int. Rev. Gesamten Hydrobiol. 56:947-955.

Martin, A. C. 1935. Quail-food plants of the Southeastern United States. U.S. Dep. Agric. Circ. 348. 16 pp.

Martin, A. C., H. S. Zim, and A. L. Nelson. 1951. American wildlife and plants. Dover Publ., Inc., New York. 500 pp.

Martin, B. J. 1980. Effects of petroleum compounds on estuarine fishes. Environ. Res. Lab., Gulf Breeze. EPA-600/3-80-19. 31 pp.

Martof, B. S. 1956. Amphibians and reptiles of Georgia, a guide. Univ. Ga. Press, Athens. 94 pp.

Martof, B. S. 1963. Some observations on the herpetofauna of Sapelo Island, Georgia. Herpetologica 19(1):70-72.

Mathews, T. D. and O. Pashuk. 1977. A description of oceanographic conditions off the Southeastern United States during 1973. S.C. Mar. Resour. Cent. Tech. Rep. No. 19. 105 pp.

Mathur, D. 1970. Food habits and feeding chronology of channel catfish _Ictalurus punctatus_ (Rafinesque) in Conowingo Reservoir. Proc. Annu. Conf. Southeast. Assoc. Game Fish Comm. 24:377-386.

Mattoon, W. R. 1915. The southern cypress. U.S. Dep. Agric. Bull No. 272. 74 pp.

May, E. B. 1969. Feasibility of off-bottom culture in Alabama. Ala. Mar. Resour. Bull. 3:1-14.

Maynard, C. J. 1896. The birds of eastern North America. C. J. Maynard, Newtonville, Mass. 721 pp. (First publ. 1879.)

Mayou, T. V. and J. D. Howard. 1975. Estuaries of the Georgia coast, U.S.A.: sedimentology and biology. VI. Animal-sediment relationships of a salt marsh estuary - Doboy Sound. Senckenb. Marit. 7:205-236.

McAlpine, R. B. 1961. Yellow-poplar seedlings intolerant to flooding. J. For. 59:566-568.

McCloskey, L. R. 1970. The dynamics of the community associated with a marine scleractinian coral. Int. Rev. Gesamten Hydrobiol. 55:13-81.

McClurkin, D. C. 1965. Diameter growth and phenology of trees on sites with high water tables. U.S. Dep. Agric. Forest Serv. Res. Note SO-22.

McCollum, J. L. and D. R. Ettman. 1977. Georgia's protected plants. Ga. Dep. Nat. Resour., Resour. Planning Sec., Endangered Plant Program, Atlanta. 64 pp.

McConkey, E. H. 1954. A systematic study of the North American lizards of the genus Ophisaurus. Am. Midl. Nat. 51:133-171.

McCord, J. W. 1977. Food habits and elver migration of American eel, Anguilla rostrata, in Cooper River, South Carolina. M.S. Thesis. Clemson Univ., Clemson, S.C.

McDermott, R. E. 1954. Effects of saturated soil on seedling growth of some bottomland hardwood species. Ecology 35:36-41.

McEwan, L. C. 1976. Patterns of species diversity of plants growing on floating logs in a cypress dome, pp. 228-236. In: H. T. Odum, K. C. Ewel, J. W. Ordway, and M. K. Johnston. Cypress wetlands for water management, recycling and conservation. Third annu. rep. to Natl. Sci. Found. and Rockefeller Found. Nov. 1, 1975 - Dec. 15, 1976. Univ. Fla., Cent. Wetlands, Phelps Lab., Gainesville.

McFarlane, R. W. 1963. Disorientation of loggerhead hatchlings by artificial road lighting. Copeia 1963(1):153.

McFarlane, R. W. 1976a. Fish diversity in adjacent ambient, thermal, and post-thermal freshwater streams, pp. 268-271. In: G. W. Esch and R. W. McFarlane, eds. Thermal ecology II. U.S. Energy Res. Dev. Adm. Symp. Ser. Conf-750425.

McFarlane, R. W. 1976b. Birds as agents of biological control. Biologists 58(4):123-140.

McIlhenny, E. A. 1934. Notes on incubation and growth of alligators. Copeia 1934:80-88.

McIlhenny, E. A. 1935. The alligator's life history. The Christopher Publishing House, Boston. 117 pp.

McIntire, G. L. and W. M. Dunstan. 1975. The seasonal cycle of growth and production in three salt marshes adjacent to the Savannah River. Ga. Mar. Sci. Cent. Tech. Rep. Ser. No. 75-2. Skidaway Island. 13 pp.

McKenzie, M. D. 1970. Fluctuations in abundance of the blue crab and factors affecting mortalities. S.C. Mar. Resour. Cent. Tech. Rep. No. 1. 45 pp.

McKenzie, M. D. and A. C. Badger. 1969. A systematic survey of intertidal oysters in the Savannah River basin area of South Carolina. Contrib. Bears Bluff Lab. 50. Wadmalaw Island, S.C. 15 pp.

McMinn, J. W. and W. H. McNab. 1971. Early growth and development of slash pine under drought and flooding. U.S. Dep. Agric. Forest Serv. Res. Pap. SE-89.

McNulty, J. K. 1953. Seasonal and vertical patterns of oyster setting off Wadmalaw Island, S.C. Contrib. Bears Bluff. Lab. No. 15. Wadmalaw Island, S.C. 17 pp.

McSwain, L. E. 1971. Life history of stream fish - spotted sucker Minytrema melanops. Study No. XII. Annu. progress rep. Statewide fisheries investigations F-21-3, Vol. 2. July 1, 1970 - June 30, 1971. Ga. Dep. Nat. Resour., Game Fish Div., Atlanta.

Meanley, B. 1972. Swamps, river bottoms, and canebrakes. Barre Publ., Barre, Mass. 142 pp.

Mellinger, E. O. and M. B. Mellinger. [1962]. Plants of the Savannah National Wildlife Refuge. Savannah Natl. Wildl. Refuge, Bur. Sport Fish. Wildl., Hardeeville, S.C. 61 pp.

Mehninick, E. F. 1962. Comparison of invertebrate populations of soil and litter of mowed grasslands in areas treated and untreated with pesticides. Ecology 43:556-561.

Menzies, R. J., O. H. Pilkey, B. W. Blackwelder, D. Dexter, P. Huling, L. McCloskey, and W. J. Wiebe. 1966. A submerged reef off North Carolina. Int. Rev. Gesamten Hydrobiol. 51:393-431.

Merrimen, D. 1941. Studies of the striped bass (Roccus saxitilis) of the Atlantic coast. Fish. Bull. 50(35):1-77.

Metcalf, I. S. H. 1947. Preliminary note on algae and protozoa found in a fresh water pond. Contrib. Bears Bluff Lab. No. 3. Wadmalaw Island, S.C. 5 pp.

Meyerriecks, A. J. 1965. Ringed-billed gulls gorge on fiddler crab. Wilson Bull. 77(4):402.

Middaugh, D. P. and G. Floyd. 1978. The effect of prehatch and posthatch exposure to cadmium on salinity tolerance of larval grass shrimp, Palaemonetes pugio. Estuaries 1(2): 123-125.

Miller, G. L. and J. C. Jorgenson. 1969. Seasonal abundance and length frequency distribution of some marine fishes in coastal Georgia. U.S Fish Wildl. Serv. Data Rep. 35. 102 pp.

Miller, R. R. 1963. Genus Dorsoma Rafinesque 1820. Gizzard shads, threadfin shads, pp. 443-451. In: H. B. Bigelow, ed. Fishes of the Western North Atlantic. Part 3. Sears Found. Mar. Res. Mem. No. 1. Yale Univ., New Haven, Conn.

Miller, W. R. and F. E. Egler. 1950. Vegetation of the Wequetequock-Pawcatuck tidal marshes, Connecticut. Ecol. Monogr. 20:143-172.

Milliman, J. D., O. H. Pilkey, and D. A. Ross. 1972. Sediments of the continental margin of the eastern United States. Geol. Soc. Am. Bull. 83:1315-1334.

Milne, L. J. and M. J. Milne. 1946. Notes on the behavior of the ghost crab. Am Nat. 80:362-380.

Mitsch, W. J. 1975. Systems analysis of nutrient disposal in cypress wetlands and lake ecosystems in Florida. Ph.D. Dissertation. Univ. Fla., Gainesville. 421 pp.

Mitsch, W. J. and K. C. Ewel. 1976. Net primary productivity in cypress ecosystems, pp. 237-254. In: H. T. Odum, K. C. Ewel, J. W. Ordway, and M. K. Johnston. Cypress wetlands for water management, recycling and conservation. Second annu. rep. to Natl. Sci. Found. and Rockefeller Found. Nov. 1, 1974 - Oct. 31, 1975. Univ. Fla., Cent. Wetlands, Phelps Lab., Gainesville.

Moll, R. A. 1975. Production and consumption of phytoplankton in a salt marsh ecosystem. 38th annu. meeting, A.S.L.O., Dalhousie Univ., Halifax, Nova Scotia.

Monk, C. D. 1960. A preliminary study of the relationship between the vegetation of a mesic hammock community and a sandhill community. Q. J. Fla. Acad. Sci. 23(1):1-12.

Monk, C. D. 1965. Southern mixed hardwood forest of North Central Florida. Ecol. Monogr. 35(4):335-354.

Monk, C. D. 1966. Ecological significance of evergreenness. Ecology 47(3):504-505.

Monk, C. D. 1968. Successional and environmental relationships of the forest vegetation of North Central Florida. Am. Midl. Nat. 79(2):441-457.

Monk, C. D. and T. W. Brown. 1965. Ecological consideration of cypress heads in North Central Florida. Am. Wildl. Nat. 74:126-140.

Moore, W. G. 1955. The life history of the spiny-tailed fairy shrimp in Louisiana. Ecology 36:176-184.

Moore, W. G. 1976. White-tailed deer investigations. A six-year summary. Oct. 1970 - June 1976. S.C. Wildl. Mar. Resour. Dep., Div. Wildl. Freshwater Fish., Columbia. 32 pp.

Morgan, P. H. 1974. A study of tidelands and impoundments within a three-river delta system--the South Edisto, Ashepoo, and Combahee rivers of South Carolina. M.S. Thesis. Univ. Ga., Athens. 92 pp.

Morgan, P. H., A. S. Johnson, W. P. Baldwin, and J. L. Landers. 1975. Characteristics and management of tidal impoundments for wildlife in a South Carolina estuary. Proc. Annu. Conf. Southeast. Assoc. Game Fish Comm. 29:526-539.

574

Morrow, J. E. 1972. Economic evaluation of wildlife and wildlife oriented resources in the South-eastern United States. A guide to the data collecting and processing system. Ga. State Univ., Atlanta. Environ. Res. Group. 166 pp.

Moss, S. A. 1970. The responses of young American shad to rapid temperature changes. Trans. Am. Fish. Soc. 99(2):381-384.

Mount, R. H. 1975. The reptiles and amphibians of Alabama. Auburn Univ. Agric. Exp. Stn., Auburn, Ala. 347 pp.

Mount, R. H. 1976. Amphibians and reptiles, pp. 66-79. In: H. Boschung, ed. Endangered and threatened plants and animals of Alabama. Bull. Ala. Mus. Nat. Hist. No. 2.

Murawski, S. A. and A. L. Pacheco. 1977. Biological and fisheries data on Atlantic sturgeon, Acipenser oxyrhynchus (Mitchill). U.S. Natl. Mar. Fish. Serv. Tech. Ser. Rep. No. 10. 69 pp.

Murchelano, R. A. and C. Brown. 1968. Bacteriological study of the natural flora of the eastern oyster, Crassostrea virginica. J. Invertebr. Pathol. 11(3):520-521.

Murchelano, R. A. and C. Brown. 1970. Heterotrophic bacteria in Long Island Sound. Mar. Biol. 7(1):1-6.

Murphy, P. W. 1955. Ecology of the fauna of forest soils, pp. 99-123. In: D. K. M. Kevan, ed. Soil Zoology. Academic Press, New York.

Murphy, P. W., ed. 1962. Progress in soil zoology. Butterworth and Co., London. 398 pp.

Murphy, R. C. 1967. Serial atlas of the marine environment. Folio 14: distribution of North Atlantic pelagic birds. Am. Geogr. Soc., New York. 4 pp. + 8 pl.

Murphy, T. A. 1971. Environmental effects of oil pollution. J. Sanit. Eng. Div., Proc. Am. Soc. Civ. Eng. 97(SA3):361-371.

Murphy, T. M., Jr. 1977. Movement, distribution, and population dynamics of an alligator population inhabiting a nuclear reactor cooling reservoir. M.S. Thesis. Univ. Ga., School of Forest Resour., Athens.

Murphy, T. M., Jr. and J. W. Coker. 1978. The status of the bald eagle in South Carolina, pp. 89-93. In: R. O. Odom and L. Landers, eds. Proceedings of the rare and endangered wildlife symposium, 3-4 Aug. 1978, Athens, Ga. Ga. Dep. Nat. Resour., Game Div., Tech. Bull. WL 4.

Music, J. L., Jr. 1974. Observations on the spot (Leiostomus xanthurus) in Georgia's estuarine and close inshore ocean waters. Ga. Dep. Nat. Resour., Game Fish Div., Coastal Fish. Off. Contrib. Ser. No. 28. Brunswick. 29 pp.

National Wetlands Inventory. 1978. Field notes from coastal South Carolina. U.S. Fish Wildl. Serv. Off., St. Petersburg, Fla. (Unpubl.)

Neely, W. W. 1960. Managing Scirpus robustus for ducks. Proc. Annu. Conf. Southeast. Assoc. Game Fish Comm. 14:30-34.

Neely, W. W. 1962. Saline soils and brackish waters in management of wildlife, fish, and shrimp. Trans. N. Am. Wildl. Nat. Resour. Conf. 27:321-335.

Negus, N.C., E. Gould, and R. K. Chipman. 1961. Ecology of the rice rat, Oryzomys palustris (Harlan), on Breton Island, Gulf of Mexico, with a critique of the social stress theory. Tulane Stud. Zool. 8(4):93-123.

Neill, W. T. 1948. The lizards of Georgia. Herpetologica 4(5):153-158.

Neill, W. T. 1951. Notes on the natural history of certain North American snakes. Ross Allen's Reptile Inst., Publ. Res. Div. 1(5):47-60.

Neill, W. T. 1958. The occurrence of amphibians and reptiles in saltwater areas, and a bibliography. Bull. Mar. Sci. Gulf Caribb. 8(1):1-97.

Neill, W. T. 1971. The last of the ruling reptiles; alligators, crocodiles, and their kin. Columbia Univ. Press, New York. 486 pp.

Nelson, A. L. 1934. Two mammal records for South Carolina. J. Mammal. 15:253-254.

Nelson, D. J. and D. C. Scott. 1962. Role of detritus in the productivity of a rock-outcrop community in a piedmont stream. Limnol. Oceanogr. 7:396-413.

Nelson, F. P., ed. 1974. The Cooper River environmental study. S.C. Water Resour. Comm. State Water Plan Rep. No. 117. Columbia. 164 pp.

Nelson-Smith, A. 1977. Estuaries, pp. 123-146. In: R. S. K. Barnes, ed. The coastline. John Wiley and Sons, New York.

Neuhauser, H. N. and W. W. Baker. 1974. Annotated list of mammals of the coastal islands of Georgia, pp. 197-209. In: A. S. Shanholtzer. An ecological survey of the coastal region of Georgia. Natl. Park Serv. Sci. Monogr. No. 3. Washington, D.C.

Neuhauser, H. N. and C. Ruckdeschel. 1978. Whales (Cetacea) of Georgia, pp. 38-53. In: R. R. Odom and L. Landers, eds. Proceedings of the rare and endangered wildlife symposium. Ga. Dep. Nat. Resour. Tech. Bull. WL 4.

Nichols, J. D., L. Viehman, R. H. Chabreck, and B. Fenderson. 1976. Simulation of a commercially harvested alligator population in Louisiana. La. State Univ. Agric. Exp. Stn., Baton Rouge. 59 pp.

Nichols, P. R. and P. M. Keney. 1963. Crab larvae (Callinectes), in plankton collections from cruises of M/V Theodore N. Gill. South Coast of the United States, 1953-54. U.S. Fish Wildl. Serv. Spec. Sci. Rep. Fish. No. 448. 14 pp.

Nicol, E. A. T. 1935. The ecology of a salt-marsh. J. Mar. Biol. Assoc. U.K. 20:203-261.

Nimmo, D. R. and L. H. Bahner. 1976. Metals, pesticides and PCB's: toxicities to shrimp singly and on combination, pp. 523-532. In: M. Wiley, ed. Estuarine processes. Vol. I. Uses, stresses, and adaptation to the estuary. Academic Press, New York

Nimmo, D. R., J. Forrester, P. T. Heitmuller, and G. H. Cook. 1974. Accumulation of Aroclor 1254 in grass shrimp, Palaemonetes pugio, in laboratory and field exposures. Bull. Environ. Contam. Toxicol. 11: 303-308.

Nimmo, D. W. R., D. V. Lightner, and L. H. Bahner. 1977. Effects of cadmium on the shrimps, Penaeus duorarum, Palaemonetes pugio, and Palaemonetes vulgaris, pp. 131-183. In: F. J. Vernberg, A. Calabrese, F. P. Thunberg, and W. B. Vernberg, eds. Physiological responses of marine biota to pollutants. Academic Press, New York.

Noble, R. E. 1974. Forestry - a means to better wildlife habitat. Wildl. in N.C. 38(2):20-23.

Noble, R. E. and R. B. Hamilton. 1976. Bird populations in even-aged loblolly pine forests of Southeastern Louisiana. Proc. Annu. Conf. Southeast. Assoc. Game Fish Comm. 29:441-450.

Noe, F. 1977. Gill nets: A survey of the literature and a preliminary analysis of gill net catches from an intertidal area of Charleston Harbor. S.C. Mar. Resour. Cent., Charleston. 45 pp. (Unpubl.)

Norris, R. A. 1963. Birds of the A.E.C. Savannah River Plant area. Charleston Mus. Contrib. No. 14. Charleston, S.C. 78 pp.

Nowak, R. T. and E. D. Brodie, Jr. 1977. Palatability of selected salamanders to shrews, Blarina. Paper presented 57th annu. meeting Am. Soc. Ichthyologists and Herpetologists. Gainesville, Fla. 1 p. (Abstr.)

Obrecht, G. B. 1946. Notes on South Carolina reptiles and amphibians. Copeia 1946(2):71-74.

Odell, D. K. 1975. Status and aspects of the life history of the bottlenose dolphin, Tursiops truncatus, in Florida. J. Fish. Res. Board Can. 32(7):1055-1058.

Odom, R. R. 1976. 1975 heronry survey of the Georgia coast. Oriole 41 (2/3):19-35.

Odom, R. R., J. L. McCollum, M. A. Neville, and D. R. Ettman. 1977. Georgia's protected wildlife. Ga. Dep. Nat. Resour., Game Fish Div., Endangered Wildl. Program, Social Circle. 51 pp.

Odum, E. P. 1947. Young southern loblolly-shortleaf pine. 11th breeding bird census. Audubon Field Notes 1:197-198.

Odum, E. P. 1950. Bird populations of the Highlands (North Carolina) Plateau in relation to plant succession and avian invasion. Ecology 31(4):587-605.

Odum, E. P. 1959. Fundamentals of ecology. Prepared in collaboration with H. T. Odum. W. B. Saunders Co., Philadelphia. 546 pp. (First publ. 1953.)

Odum, E. P. 1960. Organic production and turnover in old-field succession. Ecology 41:34-49.

Odum, E. P. 1963. Primary and secondary energy flow in relation to ecosystem structure. Proc. 16th Int. Congr. Zool. 4:336-338.

Odum, E. P. 1964. New ecology. Bioscience 14(7):14-16.

Odum, E. P. 1969. The strategy of ecosystem development. Science 164:262-270.

Odum, E. P. 1971. Fundamentals of ecology. W. B. Saunders Co., Philadelphia. 574 pp. (First publ. 1953.)

Odum, E. P. 1975. Ecology: the link between the natural and the social sciences. Holt, Rinehart and Winston, New York.

Odum, E. P. and A. A. de la Cruz. 1967. Particulate organic detritus in a Georgia salt marsh-estuarine ecosystem, pp. 383-388. In: G. H. Lauff, ed. Estuaries. Am. Assoc. Adv. Sci. Publ. No. 83. Washington, D.C.

Odum, E. P. and M. E. Fanning. 1973. Comparison of the productivity of Spartina alterniflora and Spartina cynosuroides in Georgia coastal marshes. Bull. Ga. Acad. Sci. 31:1-12.

Odum, E. P. and A. E. Smalley. 1959. Comparison of population energy flow of a herbivorous and a deposit-feeding invertebrate in a salt marsh ecosystem. Proc. Natl. Acad. Sci. 45(4):617-622.

Odum, H. T. 1967. Energetics of world food production, pp. 55-94. In: The world food problem. Vol. III. U.S. Gov. Print. Off., Washington, D.C.

Odum, H. T., B. J. Copeland, and E. A. McMahan. 1974. Coastal ecological systems of the United States. Vols. I-IV. Conservation Foundation, Washington, D.C.

Odum, H. T., K. C. Ewel, J. W. Ordway, and M. K. Johnston. 1975. Cypress wetlands for water management, recycling and conservation. Second annu. rep. to Natl. Sci. Found. and Rockefeller Found. Nov..1, 1974 - Oct. 31, 1975. Univ. Fla., Cent. Wetlands, Phelps Lab., Gainesville. 817 pp.

Odum, H. T., K. C. Ewel, J. W. Ordway, and M. K. Johnston. 1976. Cypress wetlands for water management, recycling and conservation. Third annu. rep. to Natl. Sci. Found. and Rockefeller Found. Nov. 1, 1975 - Dec. 15, 1976. Univ. Fla. Cent. Wetlands, Phelps Lab., Gainesville. 879 pp.

Odum, W. E. 1968. Mullet grazing on a dinoflagellate bloom. Chesapeake Sci. 9(3):202-204.

Odum, W. E. 1970a. Utilization of the direct grazing and plant detritus food chains by the striped mullet Mugil cephalus, pp. 222-240. In: J. Steele, ed. Marine food chains, a symposium. Univ. Calif. Press, Berkeley.

Odum, W. E. 1970b. Pathways of energy flow in a south Florida estuary. Ph.D. Dissertation. Univ. Miami, Coral Gables, Fla. 163 pp.

Odum, W. E. 1978. The importance of tidal freshwater wetlands in coastal zone management, pp. 1196-1203. In: Coastal zone '78. Proc. symposium on technical, environmental, socioeconomic and regulatory aspects of coastal zone management, March 1978, San Francisco. Am. Soc. Civil Eng., New York.

Oertel, G. F. and M. Larsen. 1976. Developmental sequences in Georgia coastal dunes and distribution of dune plants. Bull. Ga. Acad. Sci. 34:35-48.

Ogden, J. C. 1978. Recent population trends of colonial wading birds on the Atlantic and Gulf coastal plains, pp. 137-153. In: A. Sprunt, J. C. Ogden, and S. Winckler, eds. Wading birds. Natl. Audubon Soc. Res. Rep. 7. New York.

Ohlendorf, H. M., E. E. Klaas, and T. E. Kaiser. 1978. Environmental pollutants and eggshell thinning in the black-crowned night heron, pp. 63-82. In: A. Sprunt, J. C. Ogden, and S. Winckler, eds. Wading Birds. Natl. Audubon Soc. Res. Rep. 7. New York.

Olmsted, L. and R. V. Kilambi. 1970. Stomach content analysis of white bass, Roccus chrysops, in Beaver Reservoir, Arkansas. Proc. Annu. Conf. Southeast. Assoc. Game Fish Comm. 23:244-250.

O'Meara, G. F. 1976. Saltmarsh mosquitoes (Diptera: Culicidae), pp. 303-334. In: L. Cheng, ed. Marine insects. American Elsevier Publ. Co., New York.

Oosting, H. J. 1945. Tolerance to salt spray of plants of coastal dunes. Ecology 26(1):85-89.

Oosting, H. J. 1954. Ecological processes and vegetation of the maritime strand in the Southeastern U.S. Bot. Rev. 20:226-262.

Oosting, H. J. 1956. The study of plant communities. W. H. Freeman & Co., San Francisco, Calif. 440 pp. (First publ. 1948.)

Oosting, H. J. and W. D. Billings. 1942. Factors effecting vegetational zonation on coastal dunes. Ecology 23: 131-142.

Osborn, R. G. and T. W. Custer. 1978. Herons and their allies; atlas of Atlantic coast colonies, 1975 and 1976. Patuxent Wildl. Res. Cent., Off. Biol. Serv., U.S. Fish Wildl. Serv. 211 pp.

Owre, H. B. and M. Foyo. 1967. Copepods of the Florida Current. Fauna Caribaea No. 1. Crustacea, Part 1: Copepoda. Univ. Miami Inst. Mar. Sci., Miami. 137 pp.

Pacheco, A. L., ed. 1968. Alewife, blueback herring and American shad. Proceedings of workshop on egg, larval, and juvenile stages of fish in Atlantic coast estuaries. Natl. Mar. Fish. Serv. Tech. Publ. No. 1.

Paffenhofer, G. A. and S. C. Knowles. 1978. Laboratory experiments on feeding, growth and fecundity of effects of cadmium on _Pseudodiaptomus_. Bull. Mar. Sci. 28(3):574.

Palmisano, A. W. 1978. Ecosystem characterization - an approach to coastal planning and management, pp. 4-9. In: J. B. Johnston and L. A. Barclay, eds. Contributed papers on coastal ecological characterization studies, presented at the fourth biennial int. estuarine res. conf., 2-5 Oct. 1977, Mt. Pocono, Pa. U.S. Fish Wildl. Serv., Off. Biol. Serv. FWS/OBS-77/37.

Pamatmat, M. M. 1968. Ecology and metabolism of a benthic community on an intertidal sand flat. Int. Rev. Gesamten Hydrobiol. 53:211-298.

Parnell, J. F. and D. A. Adams. 1971. Smith Island: a resource capability study. Interim report. Univ. N.C., Dep. Biol., Wilmington, N.C. 83 pp.

Parrish, L. P. 1972. Seasonal abundance and distribution of the benthic invertebrates community, pp. 241-247. In: Port Royal Sound environmental study. S.C. Water Resources Comm., Columbia.

Parrish, W. F. 1960. The status of the beaver in Georgia. M.S. Thesis. Univ. Ga., Athens.

Pasch, R. W. 1973. Populations studies - reservoir - Lake Blackshear. Study XVI, Job 1. Annu. progress rep. Project F-21-4. Vol. 4. Ga. Dep. Nat. Resour., Game Fish Div., Atlanta.

Pasch, R. W. 1974. Some relationships between food habits and growth of largemouth bass in Lake Blackshear, Georgia. Proc. Annu. Conf. Southeast. Assoc. Game Fish Comm. 28:307-321.

Pate, P. P., Jr. 1972. Life history aspects of the hickory shad, _Alosa mediocris_ (Mitchill), in the Neuse River, North Carolina. M.S. Thesis. N.C. State Univ., Raleigh. 66 pp.

Patrick, R., J. Cairns, Jr., and S. S. Roback. 1967. An ecosystematic study of the fauna and flora of the Savannah River. Proc. Acad. Nat. Sci. Phil. 118(5):109-407.

Patten, B. C., ed. 1971. Systems analysis and simulation in ecology. Vol. 1. Academic Press, New York. 607 pp.

Payne, K. 1972. A survey of the _Spartina_-feeding insects in Poole Harbour, Dorset. Entomol. Mon. Mag. 108:66-79.

Pearse, A. S. 1936. Estuarine animals at Beaufort, North Carolina. J. Elisha Mitchell Sci. Soc. 52:174-222.

Pearse, A. S. 1943. Effects of burning over and raking off litter on certain soil animals in the Duke Forest. Am. Midl. Nat. 29:406-424.

Pearse, A. S. 1946. Observations on the microfauna of the Duke Forest. Ecol. Monogr. 16:127-150.

Pearse, A. S. 1953. Some observations on the land snails in the Duke Forest. J. Elisha Mitchell Sci. Soc. 69:60-62.

Pearse, A. S., H. J. Humm, and G. W. Wharton. 1942. Ecology of sand beaches at Beaufort, N.C. Ecol. Monogr. 12(2): 136-190.

Pearse, A. S. and L. G. Williams. 1951. The biota of the reefs off the Carolinas. J. Elisha Mitchell Sci. Soc. 67(1):133-161.

Pearson, P. G. 1952. Observations concerning the life history and ecology of the wood rat, _Neotoma floridana floridana_ (Ord). J. Mammal. 33:459-463.

Peet, R. K. 1975. Relative diversity indices. Ecology 56:496-498.

578

Pelton, M. R. 1975. The mammals of Kiawah Island, pp. M1-M45. In: W. M. Campbell, J. M. Dean, and W. D. Chamberlain, eds. Environmental inventory of Kiawah Island. Environmental Research Center, Inc., Columbia, S.C.

Penfound, W. T. 1952. Southern swamps and marshes. Bot. Rev. 18(6):413-446.

Penfound, W. T. and T. T. Earle. 1948. The biology of the water hyacinth. Ecol. Monogr. 18:447-472.

Penfound, W. T. and E. S. Hathaway. 1938. Plant communities in the marshlands of Southeastern Louisiana. Ecol. Monogr. 8(1):1-56.

Pennak, R. W. 1953. Fresh-water invertebrates of the United States. Ronald Press Co., New York. 769 pp.

Percival, E. 1929. A report on the fauna of the estuaries of the River Tamar and the River Lynher. J. Mar. Biol. Assoc. U.K. 16:81-108.

Percival, H. F. 1968. Ecological study of selected waterfowl food plants. Job completion rep. Project W-38-R-5, Job XI-A. S.C. Wildl. Mar. Resour. Dep., S.C. Agric. Exp. Stn., and Belle W. Baruch Found. Columbia. 109 pp.

Perkins, E. F. 1974. The biology of estuaries and coastal water. Academic Press, New York. 678 pp.

Pessin, L. J. 1933. Forest associations in the uplands of the lower Gulf Coastal Plain. Ecology 14:1-14.

Peters, B. G. 1955. Soil inhabiting nematodes, pp. 44-54. In: D. K. M. Kevan, ed. Soil zoology. Academic Press, Inc., New York.

Pettus, D. 1958. Water relationships in Natrix sipedon. Copeia 1958:207-211.

Pfeiffer, W. J. 1974. Cattle egrets feeding in salt marsh. Oriole 39(4):44-45.

Pfitzer, D. W. 1954. Investigations of waters below storage reservoirs in Tennessee. Trans. N. Am. Wildl. Conf. 19:271-282.

Philson, P. J. 1939. Freshwater algae of North and South Carolina. Part I. Cyanophyceae. J. Elisha Mitchell Sci. Soc. 55:83-116.

Philson, P. J. 1940. Species of Oedogonium new to South Carolina. J. Elisha Mitchell Sci. Soc. 56:106-110.

Pierce, E. L. and M. L. Wass. 1962. Chaetognatha from the Florida Current and coastal water of the Southeastern United States. Bull. Mar. Sci. Gulf. Caribb. 12(3):403-431.

Pinder, J. E. III. 1975. Effects of species removal on an old-field plant community. Ecology 56:747-751.

Pinson, J. N., Jr. 1973. A floristic analysis of open dunes in South Carolina. Ph.D. Dissertation. Univ. S.C., Columbia. 82 pp.

Plummer, G. 1963. Soils of the pitcher plant habitats in the Georgia coastal plain. Ecology 44(4):727-734.

Plummer, G. L. and J. B. Kethley. 1964. Foliar absorption of amino acids, peptides and other nutrients by the pitcher plant Sarracenia flava. Bot. Gaz. 125:245-260.

Poer, L. D., Jr. 1967. A herpetological survey of the Isle of Palms, a South Carolina coastal island. M.S. Thesis. Univ. S.C., Columbia. 43 pp.

Polhemus, J. T. 1976. Shore bugs (Hemiptera: Saldidae, etc.), pp. 225-262. In: L. Cheng, ed. Marine insects. American Elsevier Publ. Co., New York.

Pomeroy, L. R. 1959. Algal productivity in salt marshes of Georgia. Limnol. Oceanogr. 4(4):386-398.

Pomeroy, L. R., R. E. Johannes, E. P. Odum, and B. Roffman. 1969. The phosphorus and zinc cycles and productivity of a salt marsh, pp. 412-419. In: D. J. Nelson and F. C. Evans, eds. Symposium on radioecology. U.S. Atomic Energy Comm., Washington, D.C.

Pomeroy, L. R., L. R. Shenton, R. D. H. Jones, and R. J. Reimold. 1972. Nutrient flux in estuaries, pp. 274-291. In: G. E. Likens, ed. Nutrients and eutrophication: the limiting-nutrient controversy. Am. Soc. Limnol. Oceanogr. Spec. Sym. Vol. 1. Lawrence, Kans.

Poole, K. T. 1978. Fish, pp. 236-259. In: R. G. Zingmark, ed. An annotated checklist of the biota of the coastal zone of South Carolina. Univ. S.C. Press, Columbia.

Porcher, R. D., Jr. 1966. A floristic study of the vascular plants in nine selected Carolina Bays in Berkeley County, South Carolina. M.S. Thesis. Univ. S.C., Columbia. 123 pp.

Porcher, R. D., Jr. 1974. A study of plant communities on Belmont Plantation, Hampton County, South Carolina. Ph.D. Dissertation. Univ. S.C., Columbia. 113 pp.

Porcher, R. D., Jr. 1978. Vascular plants, pp. 39-82. In: R. G. Zingmark, ed. An annotated checklist of the biota of the coastal zone of South Carolina. Univ. S.C. Press, Columbia.

Porter, H. J. and C. E. Jenner. 1967. Notes on some mollusca off the coast of North Carolina. Am. Malacol. Union, Inc. Annu. Rep. 1967:23-24.

Post, E. 1936. Systematische und pflanzengeographische notizen sur Bostrychia-Caloglossa assoziation. Rev. Algol. 9:1-84.

Post, W. 1970. Range expansion of the cattle egret into interior South Carolina. Chat 34(2):31-33.

Powell, J. H , Jr. 1971. The status of crocodilians in the United States, Mexico, Central America, and the West Indies. Proc. working meeting of crocodile specialists, IUCN Suppl. Pap. 32:72-82.

Powles, H. and B. W. Stender. 1976. Observations on composition, seasonality and distribution of ichthyoplankton from MARMAP cruises in the South Atlantic Bight, 1973. S.C. Mar. Resour. Cent. Tech. Rep. No. 11. 47 pp.

Powles, H. and B. W. Stender. 1978. Taxonomic data on the early life history stages of Sciaenidae of the South Atlantic Bight of the United States. S.C. Mar. Resour. Cent. Tech. Rep. No. 31. 64 pp.

Pratt, D. 1974. Salt requirements for growth and function of marine bacteria, pp. 3-13. In: R. R. Colwell and R. Y. Morita, eds. Effect of the ocean environment on microbial activities. Univ. Park Press, Baltimore, Md.

Pratt, D. and S. Tedder. 1974. Variations in the _alt requirements for optimum growth rate of marine bacteria, pp. 38-45. In: R. R. Colwell and R. Y. Morita, eds. Effect of the ocean environment on microbial activites. Univ. Park Press, Baltimore, Md.

Pritchard, D. W. 1955. Estuarine circulation patterns. Proc. Am. Soc. Civ. Eng. 81(717):1-11.

Provost, M. W. 1968. Managing impounded salt marshes for mosquito control and estuarine resource conservation, pp. 163-171. In: J. D. Newsom, ed. Proceedings of the marsh and estuary management symposium. La. State Univ., Baton Rouge.

Provost, M. W. 1969. Ecological control of salt marsh mosquitoes with side benefits to birds. Proc. Tall Timbers Conf. Ecol. Anim. Control Habitat Manage. 1:193-206.

Quarterman, E. and C. Keever. 1962. Southern mixed hardwood forest: climax in the southeastern coastal plain, U.S.A. Ecol. Monogr. 32(2): 167-185.

Quay, T. L. 1947. Winter birds of upland plant communities. Auk 64: 382-388.

Quay, T. L. 1959. The birds, mammals, reptiles, and amphibians of Cape Hatteras National Seashore Recreation Area. Proj. completion rep., Cape Hatteras National Seashore Recreation Area, Manteo, N.C.

Radford, A. E. 1976. Vegetation – habitats – floras natural areas in the Southeastern United States: field data and information. Univ. N.C. Student Stores, Chapel Hill. 289 pp.

Radford, A. E., H. E. Ahles, and C. R. Bell. 1968. Manual of the vascular flora of the Carolinas. Univ. N.C. Press, Chapel Hill. 1183 pp.

Ragotzkie, R. A. 1959a. Mortality of loggerhead turtle eggs from excessive rainfall. Ecology 40:303-305.

Ragotzkie, R. A. 1959b. Plankton productivity in estuarine waters of Georgia. Publ. Inst. Mar. Sci. Univ. Tex. 6:146-158.

Rainey, D. G. 1956. Eastern wood rat, Neotoma floridana: life history and ecology. Univ. Kansas, Mus. Nat. Hist. Publ. 8:535-646.

Ralph, R. D. 1975. Blue-green algae of the shores and marshes of southern Delaware. Univ. Del., Newark. DEL-SG-20-75. 139 pp.

Ralph, R. D. 1977. The Myxophyceae of the marshes of southern Delaware. Chesapeake Sci. 18:208-221.

Raney, E. C. 1952. The life history of the striped bass (Roccus saxatilis) (Walbaum). Bull. Bingham Oceanogr. Collect. Yale Univ. 14(1):5-97.

Ranwell, D. S. 1972. Ecology of salt marshes and sand dunes. Chapman and Hall, London, England. 258 pp.

Rawson, D. S. 1939. Some physical and chemical factors in the metabolism of lakes, pp. 9-26. In: E. P. Moulton, ed. Problems of lake biology. Am. Assoc. Adv. Sci. Publ. No. 10. Washington, D.C.

Raymont, J. E. G. 1963. Plankton and productivity in the oceans. Pergamon Press, New York. 668 pp.

Rayner, D. A. 1974. An analysis of maritime closed dunes vegetation in South Carolina. M.S. Thesis. Univ. S.C., Columbia. 128 pp.

Rayner, D. A., C. L. Rodgers, and R. C. Clark. 1979. Native vascular plants, endangered, threatened, or otherwise in jeopardy in South Carolina. S.C. Committee on Endangered, Threatened, and Rare Plants, Columbia. S.C. Mus. Comm. Bull. No. 4. 22 pp.

Red-cockaded Woodpecker Recovery Team. 1979. Red-cockaded woodpecker recovery plan. Technical draft. U.S. Dep. Inter., Fish Wildl. Serv. 38 pp.

Redmann, G. 1973. Studies on the toxicity of mirex to the estuarine grass shrimp, Palaemonetes pugio. Gulf Res. Rep. 4(2):272-277.

Rees, R. A. 1968. A survey of the aquatic organism population of the Savannah and Ogeechee river estuaries and their relationship in the diet of the striped bass Morone saxatilis (Walbaum). M.S. Thesis. Univ. Ga., Athens.

Reeve, M. R. 1975. The ecological significance of the zooplankton in the shallow subtropical waters of South Florida, pp. 352-371. In: L. E. Cronin, ed. Estuarine research. Vol. 1. Academic Press, New York.

Reeve, M. R., G. D. Grice, V. R. Gibson, M. A. Walker, K. Darcy, T. Ikeda, L. G. Maurer, W. J. Wiebe, R. G. Wiegert, and R. L. Wetzel. 1976. A controlled environmental pollution experiment (CEPEX) and its usefulness in the study of larger marine zooplankton under toxic stress, pp. 145-162. In: A. P. M. Lockwood, ed. Effects of pollutants on aquatic organisms. Soc. Exp. Biol. Sem. Ser. 2. Vol. 2. Cambridge Univ. Press, New York.

Reimer, C. W. 1966. Consideration of fifteen diatom taxa (Bacillariophyta) from the Savannah River, including seven described as new. Not. Nat. (Phila.) 397. 15 pp.

Reimold. R. J., P. C. Adams, and C. J. Durant. 1973. Effects of toxaphene contamination on estuarine ecology. Ga. Mar. Sci. Cent. Tech. Rep. Ser. No. 73-8. Skidaway Island. 100 pp.

Reimold, R. J. and F. C. Daiber. 1967. Eutrophication of estuarine areas by rainwater. Chesapeake Sci. 8 (2):132-133.

Reimold, R. J. and F. C. Daiber. 1970. Dissolved phosphorus concentrations in a natural salt-marsh of Delaware. Hydrobiologia 36(3-4):361-371.

Reimold, R. J. and R. A. Linthurst. 1977. Primary productivity of minor marsh plants in Delaware, Georgia, and Maine. Dredged Material Res. Program Tech. Rep. D-77-36. U.S. Army Eng. Waterways Exp. Stn., Vicksburg. 104 pp.

Reimold, R. J. and M. H. Shealy, Jr. 1976. Chlorinated hydrocarbon pesticides and mercury in coastal young-of-the-year finfish, South Carolina and Georgia--1972-1974. Pestic. Monit. J. 9(4):170-175.

Reish, D. J., T. J. Kauwling, A. J. Mearns, P. S. Oshida, S. S. Ross, F. G. Wilkes, and M. J. Ray. 1978. Marine and estuarine pollution. J. Water Pollut. Control Fed. 50:1424-1469.

Remane, A. 1934. Die brackwasserfauna. Verh. Dtsch. Zool. Ges. 36:34-74.

Remane, A. 1971. Ecology of brackish water, pp. 1-210. In: A. Remane and C. Schlieper. Biology of brackish water. Wiley Interscience, New York.

Remane, A. and C. Schlieper. 1971. Biology of brackish water. Wiley Interscience, New York. 372 pp.

Rheinheimer, G. 1967. Okologische untersuchungen zur intrifikation in nord und ostsee. Helgol. Wiss. Meeresunters. 15:243-252.

Rheinheimer, G. 1970. Uber das vórkómmen von brackwasserbakterien in der ostsee. Vie Milieu 22 (Suppl.):281-291.

Rheinheimer, G. 1974. Aquatic microbiology. John Wiley & Sons, New York. 184 pp.

Rice, T. R. and R. L. Ferguson. 1975. Response of estuarine phytoplankton to environmental conditions, pp. 1-44. In: F. J. Vernberg, ed. Physiological ecology of estuarine organisms. Univ. S.C. Press, Columbia.

Richards, B. N. 1974. Introduction to the soil ecosystem. Longman, Inc., New York. 266 pp.

Richardson, J. 1974. Insects, pp. 35-48. In: Cooper River survey for 1973 for the E. I. DuPont de Nemours and Company. Acad. Nat. Sci. of Philadelphia.

Richardson, J. I. 1978. Results of a hatchery for incubating loggerhead sea turtle (Carretta caretta Linne) eggs on Little Cumberland Island, Georgia, p. 15. In: G. E. Henderson, ed. Proceedings of the Florida and interregional conference on sea turtles, 24-25 July 1976, Jensen Beach, Florida. Fla. Mar. Res. Publ. 33.

Richardson, J. I. and H. D. Hillestad. 1978. Ecology of a loggerhead sea turtle population in Georgia, pp. 22-37. In: R. R. Odom and L. Landers, eds. Proceedings of the rare and endangered wildlife symposium. Ga. Dep. Nat. Resour. Tech. Bull. WL 4.

Richardson, J. I., T. H. Richardson, and M. W. Dix. 1978. Population estimates for nesting female loggerhead sea turtles (Caretta caretta) in the St. Andrews Sound area of southeastern Georgia, U.S.A., pp. 34-38. In: G. E. Henderson, ed. Proceedings of the Florida and interregional conference on sea turtles, 24-25 July 1976, Jensen Beach, Florida. Fla. Mar. Res. Publ. 33.

Richardson, J. I. and J. Worthington. 1975. Value and vulnerability of Georgia's coastal island system, pp. 37-109. In: The value and vulnerability of coastal resources--background papers for review and discussion. Ga. Dep. Nat. Resour., Planning Section, Atlanta. 291 pp.

Richardson, T. H., J. I. Richardson, C. Ruckdeschel, and M. W. Dix. 1978. Remigration patterns of loggerhead sea turtles (Caretta caretta) nesting on Little Cumberland and Cumberland islands, Georgia, pp. 39-44. In: G. E. Henderson, ed. Proceedings Florida and interregional conference on sea turtles. Jensen Beach, Fla., July 1976. Fla. Mar. Res. Publ. 33.

Richmond, N. D. 1945. The habits of the rainbow snake in Virginia. Copeia 1945:28-30.

Ridgeway, S. H., ed. 1972. Mammals of the sea, biology and medicine. Charles C. Thomas, Springfield, Ill. 812 pp.

Riedl, R. and E. A. McMahan. 1974. High energy beaches, pp. 180-251. In: H. T. Odum, B. J. Copeland and E. A. McMahan, eds. Coastal ecological systems of the United States. Vol. 1. The Conservation Foundation, Washington, D.C.

Ringler, S. B. 1977. The herpetofauna of Ossabaw Island, Chatham County, Georgia. Herpetol. Rev. 8(2):39.

Riznyk, R. Z., J. L. Edens, and R. C. Libby. 1978. Production of epibenthic diatoms in a southern California impounded estuary. J. Phycol. 14:273-279.

Roberts, A. 1963. Breeding bird census of two pine forests with special reference to the pine warbler. Oriole 28(4):63-71.

Roberts, M. H., Jr. 1968. Functional morphology of mouth parts of the hermit crabs, Pagurus longicarpus and Pagurus pollicaris. Chesapeake Sci. 9(1):9-20.

Roberts, M. H., Jr. 1974a. Biology of benthic fauna, pp. 156-327. In: M. H. Roberts, Jr., ed. A socioeconomic environmental baseline summary for the South Atlantic region between Cape Hatteras, North Carolina and Cape Canaveral, Florida. Vol. 3. Chemical and biological oceanography. Va. Inst. Mar. Sci., Gloucester Point.

Roberts, M. H., Jr., ed. 1974b. A socioeconomic environmental baseline summary for the South Atlantic region between Cape Hatteras, North Carolina and Cape Canaveral, Florida. Vol. 3. Chemical and biological oceanography. Va. Inst. Mar. Sci., Gloucester Point. 722 pp.

Roberts, M. H., Jr., C. E. Laird, and J. E. Illowsky. 1979. Effects of chlorinated seawater on decapod crustacean and Mulinia larvae. EPA Ecological Research Series, EPA-600/3-79-031.

Robertson, A. 1972. Calanoid copepods: new records from Oklahoma. Southwest. Nat. 17(2):201-203.

Rodgers, C. L. and R. C. Clark. 1977. Native vascular plants endangered, threatened, or otherwise in jeopardy in South Carolina. S.C. Advisory Committee on Endangered, Threatened and Rare Plants. S.C. Wildl. Mar. Resour. Dep., Heritage Trust Program, Columbia. 18 pp. (Unpubl.)

Rounsefell, G. A. 1972. Ecological effects of offshore construction. J. Mar. Sci. 2(1). 208 pp.

Ruibal, R. 1959. The ecology of a brackish water population of _Rana pipiens_. Copeia 1959:315-322.

Russell-Hunter, W. D. 1970. Aquatic productivity: an introduction to some basic aspects of biological oceanography and limnology. Macmillan, New York. 306 pp.

Ryder, J. A. 1890. The sturgeon and sturgeon industries of the eastern coast of the United States, with an account of experiments bearing upon sturgeon culture. U.S. Fish. Comm. Bull. (1888) 8:231-238.

Ryther, J. H. 1969. Photosynthesis and fish production in the sea. Science 166:72-76.

Salt, G., F. S. J. Hollick, F. Raw, and M. V. Brian. 1948. The arthropod population of pasture soil. J. Anim. Ecol. 17:139-150.

Salt, G. W. 1953. An ecological analysis of three California avifaunas. Condor 55:258-273.

Sanders, A. E. 1978. Mammals, pp. 296-308. In: R. G. Zingmark, ed. An annotated checklist of the biota of the coastal zone of South Carolina. Univ. S.C. Press, Columbia.

Sanders, H. L., E. M. Goudsmit, E. L. Mills, and G. E. Hampson. 1962. A study of the intertidal fauna of Barnstable Harbor, Massachusetts. Limnol. Oceanogr. 7:63-79.

Sandifer, P. A. 1975. The role of pelagic larvae in recruitment to populations of adult decapod crustaceans in the York River Estuary and adjacent Lower Chesapeake Bay, Virginia. Estuarine Coastal Mar. Sci. 3:269-279.

Sandifer, P. A. and P. J. Eldridge. 1976. A study of the distribution and abundance of _Penaeus_ larvae off the Southeastern United States in an attempt to identify spawning grounds of the brown shrimp. Final Rep., State-Federal Shrimp Manage. Program, Contract No. 03-5-042-16. S.C. Mar. Resour. Cent., Charleston. (Unpubl.)

Sandow, J. T., Jr. 1973. Stream population studies, Satilla River. State-wide fisheries investigations. Annu. progress rep. F-21-4. Ga. Dep. Nat. Resour., Game Fish Div., Atlanta.

Sandow, J. T., Jr. 1974. Stream population studies, Satilla River. Study 16, Job 2. Annu. progress rep. Project F-21-5. Vol. 3. Ga. Dep. Nat. Resour., Game Fish Div., Atlanta.

Sandow, J. T., Jr., D. R. Holder, and L. E. McSwain. 1974. Life history of the redbreast sunfish in the Satilla River, Georgia. Proc. Annu. Conf. Southeast. Assoc. Game Fish Comm. 28:279-295.

Sass, H. R. 1926. Adventures in green places. G. P. Putnam's Sons, New York. 297 pp.

Schelske, C. L. and E. P. Odum. 1962. Mechanisms maintaining high productivity in Georgia estuaries. Gulf Caribb. Fish. Inst. Proc. 14:75-80.

Schlesinger, W. H. and P. L. Marks. 1975. Okefenokee cypress swamp: forest biomass, production and phytosociology. Bull. Ecol. Soc. Am. 56:28.

Schmidt, J. 1922. The breeding places of the eel. Phil. Trans., Series B, 211:179-208.

Schneider, C. W. 1976. Spatial and temporal distributions of benthic marine algae on the continental shelf of the Carolinas. Bull. Mar. Sci. 26(2):133-151.

Schneider, C. W. and R. B. Searles. 1978. Is North Carolina a transitional benthic algal phytogeographic zone in the western Atlantic? J. Phycol. 14:25. (Abstr.)

Schramm, W. 1973. Investigations on the influence of oil pollutions on marine algae. Mar. Biol. 14(3):189-198.

Schreiber, R. W. and R. W. Risebrough. 1972. Studies of the brown pelican. 1. Status of brown pelican populations in the United States. Wilson Bull. 84(2):119-135.

Schubel, J. R. 1974. Effects of exposure to time-excess temperature histories typically experienced at power plants on the hatching success of fish eggs. Estuarine Coastal Mar. Sci. 2:105-116.

Schwartz, A. 1956. Geographic variation in the chicken turtle _Deirochelys reticulartia latreille_. Fieldiana Zool. 34(41):461-503.

Schwartz, A. 1957. Chorus frogs (_Pseudacris nigrita_ Leconte) in South Carolina. Am. Mus. Novit. 1838:1-12.

Sciple, G. W. 1963. Changing status of occurrence of the glossy ibis. Oriole 28(2):23-25.

Scott, D. C. 1959. The nymph of the mayfly genus _Tortopus_ (Ephemeroptera: Polymitarcidae). Ann. Entomol. Soc. Am. 52:205-213.

Scully, E. P. 1978. Utilization of surface foam as a food source by the hermit crab, Pagurus longicarpus Say, 1817. Mar. Behav. Physiol. 5:159–162.

Scura, E. D. and G. H. Theilacker. 1977. Transfer of chlorinated hydrocarbon PCB in a laboratory food chain. Mar. Biol. 40:317–325.

Seehorn, M. E. 1975. Fishes of southeastern national forests. Proc. Annu. Conf. Southeast. Assoc. Game Fish Comm. 29:10–27.

Segrè, A., J. P. Hailman, and C. G. Beer. 1968. Complex interactions between clapper rails and laughing gulls. Wilson Bull. 80(2):213–219.

Sekavec, G. B. and G. R. Huntsman. 1973. Reef fishing in the Carolina shelf, pp. 78–86. In: Proc. 15th Annu. Int. Game Fish Res. Conf., Miami Beach.

Sellner, K. G. 1973. Primary production and the release of dissolved organic matter from natural communities of estuarine phytoplankton. M.S. Thesis. Univ. S.C., Columbia. 45 pp.

Sellner, K. G., R. G. Zingmark, and T. G. Miller. 1976. Interpretations of the C14 method of measuring the total annual production of phytoplankton in a South Carolina estuary. Bot. Mar. 19:119–125.

Shanholtzer, G. F. 1970. Breeding records and distribution of the glossy ibis on the Georgia coast. Oriole 35:37–39.

Shanholtzer, G. F. 1974a. Locations and descriptions of wading bird rookeries on the Georgia coast, pp. 191–196. In: A. S. Johnson, H. O. Hillestad, S. F. Shanholtzer, and G. F. Shanholtzer. An ecological survey of the coastal region of Georgia. Natl. Park Serv. Sci. Monogr. Ser. No. 3. Washington, D.C.

Shanholtzer, G. F. 1974b. Checklist of birds occurring in the coastal region of Georgia: species, abundance and habitat, pp. 171–190. In: A. S. Johnson, H. O. Hillestad, S. F. Shanholtzer, and G. F. Shanholtzer. An ecological survey of the coastal region of Georgia. Natl. Park Serv. Sci. Monogr. Ser. No. 3. Washington, D.C.

Shanholtzer, G. F. 1974c. Relationships of vertebrates to salt marsh plants, pp. 463–474. In: R. J. Reimold and W. H. Queen, eds. Ecology of halophytes. Academic Press, Inc., New York.

Sharitz, R. R. 1975. Forest communities of Kiawah Island, pp. F1–F39. In: W. M. Campbell, J. M. Dean, and W. D. Chamberlain, eds. Environmental inventory of Kiawah Island. Environmental Research Center, Inc., Columbia, S.C.

Sharitz, R. R., J. E. Irwin, and E. J. Christy. 1974. Vegetation of swamps receiving reactor effluents. Oikos 25:7–13.

Sharp, H. F., Jr. 1962. Trophic relationships of the rice rat, Oryzomys palustris palustris (Harlan), living in a Georgia salt marsh. M.S. Thesis. Univ. Ga., Athens.

Shaw, W. N. 1965. Pond culture of oysters – past, present, and future. Trans. N. Am. Wildl. Nat. Resour. Conf. 30:114–120.

Shaw, W. N. 1969. Oysters, pp. 469–476. In: F. E. Firth, ed. The encyclopedia of marine resources. Van Nostrand Co., New York.

Sheals, J. G. 1955. The effects of DDT and BHC on soil Collembola and Acarina, pp. 241–250. In: D. K. M. Kevan, ed. Soil zoology. Academic Press, Inc., New York. 512 pp.

Shealy, M. H., Jr. 1974. Bottom trawl data from South Carolina estuarine survey cruises, 1973. S.C. Mar. Resour. Cent. Data Rep. No. 1. 113 pp.

Shealy, M. H., Jr. 1975. Midwater trawl data from South Carolina estuarine survey cruises (North Edisto, South Edisto, and Cooper rivers), 1973 and 1974. S.C. Mar. Resour. Cent. Data Rep. No. 2. 64 pp.

Shealy, M. H., Jr., B. B. Boothe, and C. M. Bearden. 1975. A survey of the benthic macrofauna of Fripp Inlet and Hunting Island, South Carolina, prior to beach nourishment. S.C. Mar. Resour. Cent. Tech. Rep. No. 7. 30 pp.

Shealy, M. H., Jr., J. V. Miglarese, and E. B. Joseph. 1974. Bottom fishes of South Carolina estuaries – relative abundance, seasonal distribution and length-frequency relationships. S.C. Mar. Resour. Cent. Tech. Rep. No. 6. 189 pp.

Shealy, M. H., Jr. and P. A. Sandifer. 1975. Effects of mercury on survival and development of the larval grass shrimp Palaemonetes vulgaris. Mar. Biol. 33:7–16.

Sheldon, W. W. 1974. Elvers in Maine: techniques of locating, catching, and holding. Maine Dep. Mar. Resour., Augusta. 25 pp.

Shenker, J. M. and J. M. Dean. 1979.
The utilization of an intertidal
salt marsh creek by larval and
juvenile fishes: abundance,
diversity, and temporal variation.
Estuaries 2(3):154-163.

Sherberger, S. R. and A. L. Buikema, Jr.
1976. Effects of light intensity
wave lengths and photoperiods on the
toxicity of chromium to Daphnia pulex
(Cladocera). Assoc. Southeast Biol.
Bull. 23(2):95. (Abstr.)

Shoemaker, A. H. 1972. Reef mollusks of
South Carolina. Nautilus 85:114-120.

Shoemaker, W. E. 1964. A biological
control for Aedes sollicitans are
the resulting effect upon wildlife.
Proc. N.J. Mosq. Exterm. Assoc.
51:93-97.

Shriner, D. 1972. Vascular plant inventory of selected marsh transects
in Port Royal Sound, pp. 249-259.
In: Port Royal Sound environmental
study. S.C. Water Resources Comm.,
Columbia.

Shuler, J. 1977. Bachman's warbler
habitat. Chat 41(2):19-20.

Shunk, I. V. 1939. Oxygen requirements
for germination of seeds of Nyssa
aquatica, tupelo gum. Science 90:
565-566.

Sieburth, J. M. 1976. Bacterial substrates and productivity in marine
ecosystems. Annu. Rev. Ecol.
Syst. 7:259-285.

Sikora, W. B., R. H. Heard, and M. D.
Dahlberg. 1972. The occurrence
and food habits of two species of
hake, Urophycis regius and U.
floridanus in Georgia estuaries.
Trans. Am. Fish. Soc. 101(3):513-
525.

Silver, M. W., A. L. Shanhs, and J. D.
Trent. 1978. Marine snow: micro-
plankton habitat and source of
small-scale patchiness in pelagic
populations. Science 201:371-373.

Simons, J. 1974. Vaucheria compacta:
a euryhaline estuarine algal species.
Acta Bot. Neerl. 23:613-626.

Simpson, K. W. 1976. Shore flies and
brine flies (Diptera: Ephydridae),
pp. 465-496. In: L. Cheng, ed.
Marine insects. American Elsevier
Publ. Co., New York.

Sizemore, R. K., L. H. Stevenson, and
B. H. Hebeler. 1973. Distribution
and activity of proteolytic bacteria
in estuarine sediments, pp. 133-143.
In: L. H. Stevenson and R. R. Colwell,
eds. Estuarine microbial ecology.
Univ. S.C. Press, Columbia.

Small, J. K. 1933. Manual of the southeastern flora. Univ. N.C. Press,
Chapel Hill. 1554 pp.

Smalley, A. E. 1959. The growth cycle of
Spartina and its relation to the
insect populations in the marsh,
pp. 96-100. In: R. A. Ragotzkie,
L. R. Pomeroy, J. M. Teal, and D. C.
Scott, eds. Proceedings salt marsh
conference. Univ. Ga. Mar. Inst.
March 25-28, 1958. Sapelo Island.

Smalley, A. E. 1960. Energy flow of a
salt marsh grasshopper population.
Ecology 41:785-790.

Smith, H. M. 1907. The fishes of North
Carolina. N.C. Geol. Econ. Surv.
Vol. II. 453 pp.

Smith, K. L. 1971. Structural and functional
aspects of a sublittoral community.
Ph.D. Dissertation. Univ. Ga.,
Athens. 160 pp.

Smith, K. L. 1973. Respiration of a sublittoral community. Ecology 54:
1065-1075.

Smith, L. D. 1968. Notes on the distribution, relative abundance and
growth of juvenile anadromous fish
in the Altamaha River system,
Georgia, with specific reference to
striped bass, Roccus saxatilis
(Walbaum). Ga. Game Fish Comm.,
Sport Fish Div. Contrib. Ser. No. 1.
Atlanta. 22 pp.

Smith, L. D. 1970. Life history studies
of striped bass. Final rep.
project AFS-2. 1 Jan. 1967 - 30
June 1970. Ga. Dep. Nat. Resour.,
Game Fish Comm., Sport Fish Div.,
Atlanta. 134 pp.

Smith, M. W. and J. W. Saunders. 1955.
The American eel in certin freshwaters
of the maritime provinces of Canada.
J. Fish. Res. Board Can. 12(2):
238-269.

Smith, P. W. and J. C. List. 1955. Notes
on Mississippi amphibians and reptiles.
Am. Midl. Nat. 53:115-125.

Smith, R. L. 1976. Ecological genesis of
endangered species: the philosophy
of preservation. Annu. Rev. Ecol.
Syst. 7:33-55.

Smith, R. O. 1949. Summary of oyster
farming experiments in S.C. 1939-1940.
U.S. Fish Wildl. Serv. Spec. Sci.
Rep. No. 63. 20 pp.

Smith, T. I. J., P. A. Sandifer, and W. C.
Trimble. 1976. Pond culture of
the Malaysian prawn, Macrobrachium
rosenbergii (de Man) in South
Carolina. Proc. World Mariculture
Soc. 7:625-645.

585

Smith, W. G. 1970. _Spartina_ "die-back" in Louisiana marshlands. La. State Univ. Stud. Coastal Stud. Bull. 5: 89-95.

Snyder, N. F. R. and H. A. Snyder. 1969. A comparative study of mollusc predation by limpkins, Everglade kites, and boat-tailed grackles. Living Bird 8:177-223.

Soloman, C. H. and S. P. Naughton. 1977. Effect of Hurricane Eloise on the benthic fauna of Panama City Beach, Florida, U.S.A. Mar. Biol. 42:357-363.

Solomon, J. D. 1969. Woodpecker predation on insect borers in living hardwoods. Ann. Entomol. Soc. Am. 62(5):1214-1215.

Somers, G. F. 1975. Biology of blue-green algae of tide marshes. Univ. Del., Coll. Mar. Stud., Newark. Sea Grant Publ. DEL-SG-14-75. 62 pp.

Somers, G. F. and M. Brown. 1978. The affinity of trichomes of blue-green algae for calcium ions. Estuaries 1(1):17-28.

Somes, H. A., Jr. and T. R. Ashbaugh. 1973. Vegetation of St. Catherine's Island, Georgia. Jack McCormick and Associates, Devon, Pa. Prepared for Am. Mus. Nat. Hist., New York. 47 pp.

Soots, R. J. and J. F. Parnell. 1975. Ecological succession of breeding birds in relation to plant succession on dredge islands in North Carolina. Univ. N.C. Sea Grant Publ. UNC-SG-75-27. Raleigh. 91 pp.

Sorokin, J. I. 1971. On the role of bacteria in the productivity of tropical oceanic waters. Int. Rev. Gesamten Hydrobiol. 56:1-48.

South Carolina Legislative Council of the General Assembly. 1977. Nongame and endangered species, pp. 626-627. In: Code of laws of South Carolina 1976. Vol. 26. Rules and regulations. The Lawyers Cooperative Publ. Co., Rochester, N.Y.

South Carolina Wildlife and Marine Resources Department. 1974a. Report on Big Opening (Berkeley County). Div. Natural Area Acquisition Resour. Planning, Columbia. 3 pp.

South Carolina Wildlife and Marine Resources Department. 1974b. Report on Socastee Savannah. Div. Natural Area Acquisition Resour. Planning, Columbia. 5 pp. (Unpubl.)

South Carolina Wildlife and Marine Resources Department. 1974c. Report on Wambaw Creek. Div. Natural Area Acquistion Resour. Planning, Columbia. 4 pp. (Unpubl.)

South Carolina Wildlfie and Marine Resources Department. 1974d. Report on Honey Hill lime sinks. Div. Natural Area Acquisition Resour. Planning, Columbia. 3 pp. (Unpubl.)

South Carolina Wildlife and Marine Resources Department. 1974e. Report on Tillman Sand Ridge. Div. Natural Area Acquisition Resour. Planning, Columbia.

South Carolina Wildlife and Marine Resources Department. 1974f. Report on Huger mixed mesophytic woods. Div. Natural Area Acquisition Resour. Planning, Columbia. 10 pp.

South Carolina Wildlife and Marine Resources Department. 1975. Report on Little Wambaw Swamp. Div. Natural Area Acquisition Resour. Planning, Columbia. 3 pp. (Unpubl.)

Speake, D. W., J. A. McGlincy, and T. R. Colvin. 1978. Ecology and management of the eastern indigo snake in Georgia: a progress report, pp. 64-73. In: R. O. Odum and L. Landers, eds. Proceedings of the rare and endangered wildlife symposium, 3-4 Aug. 1978, Athens, Ga. Ga. Dep. Nat. Resour., Game Fish Div., Tech. Bull. WL 4.

Speake, D. W. and R. H. Mount. 1974. Some possible ecological effects of "rattlenake roundups" in the southeastern coastal plain. Proc. Annu. Conf. Southeast. Assoc. Game Fish Comm. 27:267-277.

Spinner, G. P. 1969. Serial atlas of the marine environment. Folio 18. The wildlife wetlands and shellfish areas of the Atlantic Coastal Zone. Am. Geogr. Soc., New York. 4 pp. + 12 pl.

Sprague, J. W. 1961. Report of fisheries investigations during the seventh year of impoundment of Fort Randall Reservoir, South Dakota, 1959. South Dakota D-J Fed. Aid Project F-1-R-9. Rep. 42-43.

Springer, P. F. 1961. The effects on wildlife of applications of DDT and other insecticides for larval mosquito control in tidal marshes of the Eastern United States. Ph.D. Dissertation. Cornell Univ., Ithaca, N.Y. (Libr. Congr. Card No. Mic. 61-5173.) 85 pp. Univ. Microfilms, Ann Arbor, Mich.

Springer, P. F. and J. R. Webster. 1951. Biological effects of DDT application on tidal salt marshes. Mosquito News 11(2):67-74.

Springer, S. 1938. On the size of _Rana spenocephala_. Copeia 1938:49.

Sprunt, A., Jr. 1922. Discovery of the breeding of the white ibis in South Carolina. Oologist 39(10):142-144.

Sprunt, A., Jr. 1924. Breeding of MacGillivary's seaside sparrow in South Carolina. Auk 41:482-484.

Sprunt, A., Jr. 1925. An avian city of the South Carolina coast. Auk 42: 311-319.

Sprunt, A., Jr. 1936. The roseate spoonbill in Georgia. Auk. 52:203-204.

Sprunt, A., Jr. 1954. Florida bird life. Coward-McCann, New York.

Sprunt, A., Jr. and E. B. Chamberlain. 1949. South Carolina birdlife. Univ. S.C. Press, Columbia. 567 pp.

Sprunt, A., Jr. and E. B. Chamberlain. 1970. South Carolina bird life. Univ. S.C. Press, Columbia. 655 pp. (First publ. 1949.)

Sprunt, A., IV, J. C. Ogden, and S. Winckler. 1978. Wading birds. Natl. Audubon Soc. Res. Rep. No. 7. New York. 381 pp.

Sroka, E. E. 1975. A vegetational survey on cross sectional transects in Austin Cary Forest cypress dome, pp. 223-236. In: H. T. Odum, K. C. Ewel, J. W. Ordway, and M. K. Johnston. Cypress wetlands for water management, recycling and conservation. Second annu. rep. to Natl. Sci. Found. and Rockefeller Found. Univ. Fla. Cent. Wetlands, Phelps Lab., Gainesville.

Stalter, R. 1968. An ecological study of a South Carolina salt marsh. Ph.D. Dissertation. Univ. S.C., Columbia. 62 pp.

Stalter, R. 1971. The summer and fall flora of Huntington Beach State Park, Georgetown County, South Carolina. Castanea 36:167-174.

Stalter, R. 1972a. The flora of the Outer Otter Island, Colleton County, South Carolina. Castanea 37:298-300

Stalter, R. 1972b. The summer and fall flora of Brookgreen Gardens, Murrells Inlet, South Carolina. Castanea 37:214-227.

Stalter, R. 1973a. The flora of Turtle Island, Jasper County, South Carolina. Castanea 38:35-37.

Stalter, R. 1973b. Factors affecting vegetation distribution in the Cooper River estuary, South Carolina. Castanea 38:18-24.

Stalter, R. 1974a. Vegetation in coastal dunes of South Carolina. Castanea 39:95-103.

Stalter, R. 1974b. The vegetation of the Cooper River estuary, pp. 41-45. In: F. P. Nelson, ed. The Cooper River environmental study. S.C. Water Resources Comm. State Water Plan Rep. No. 117. Columbia.

Stalter, R. 1974c. The evergreen maritime forest of South Carolina's barrier islands. Am. J. Bot. 54:66. (Abstr.)

Stalter, R. 1974d. A synecological study of the evergreen maritime forest of three South Carolina barrier islands. Assoc. Southeast. Biol. Bull. 21:86. (Abstr.)

Stalter, R. 1974e. A synecological study of the evergreen maritime forest of Hunting Island, South Carolina. Bull. S.C. Acad. Sci. 36 91. (Abstr.)

Stalter, R. 1975a. The flora of the Isle of Palms, South Carolina. Castanea 40:4-13.

Stalter, R. 1975b. Factors effecting vegetational zonation on coastal dunes, Georgetown County, South Carolina. Bull. S.C. Acad. Sci. 37:75. (Abstr.)

Stalter, R. 1975c. A floristic study of South Carolina's barrier islands. Bull. S.C. Acad. Sci. 37:74. (Abstr.)

Stalter, R. and W. T. Batson. 1969. Transplantation of salt marsh vegetation, Georgetown, South Carolina. Ecology 50:1087-1089.

Steele, J. H. 1974. The structure of marine ecosystems. Harvard Univ. Press, Cambridge, Mass. 128 pp.

Steele, J. H. and I. E. Baird. 1968. Production ecology of a sandy beach. Limnol. Oceanogr. 13:14-25.

Steidinger. K. A. and J. F. Van Breedveld. 1971. Benthic marine algae from waters adjacent to the Crystal River Electric Power Plant. Fla. Dep. Nat. Resour. Mar. Res. Lab. Prof. Pap. Ser. No. 16. 46 pp.

Stephenson, T. A. and A. Stephenson. 1952. Life between tide-marks in North America. II: Northern Florida and the Carolinas. J. Ecol. 40(1): 1-49.

Stevenson, L. H., C. E. Millwood, and
B. H. Hebeler. 1974. Aerobic,
heterotrophic bacterial populations
in estuarine water and sediments,
pp. 268-285. In: R. R. Colwell
and R. Y. Morita, eds. Effects of
the ocean environment on microbial
activities. Univ. Park Press,
Baltimore, Md.

Stevens, R. E. 1959. The white and
channel catfishes of the Santee-
Cooper Reservoir and Tailrace
Sanctuary. Proc. Annu. Conf.
Southeast. Assoc. Game Fish Comm.
13:203-219.

Stewart, P. A. 1975. Cases of birds
reducing or eliminating infestations
of tabacco insects. Wilson Bull.
87(1):107-109.

Stewart, R. E. 1951. Clapper rail populations of the Middle Atlantic States.
Trans. N. Am. Wildl. Conf. 16:421-
430.

Stickel, L. F. and M. P. Dieten. 1979.
Ecological and physiological/
toxicological effects of petroleum
on aquatic birds. U.S. Fish Wildl.
Serv., Biol. Serv. Program. FWS/OBS-
79/23. 14 pp.

Stickney, R. R. 1972. Effects of intracoastal waterway dredging on
ichthyofauna and benthic macroinvertebrates. Ga. Mar. Sci. Cent.
Tech. Rep. Ser. No. 72-4. Skidaway
Island. 26 pp.

Stickney, R. R. 1976. Food habits of
Georgia estuarine fishes. II.
Symphurus plagiusa (Pleuronectiformes:
Cynoglossidae). Trans. Am. Fish.
Soc. 105:202-207.

Stickney, R. R. and S. C. Knowles. 1975.
Summer zooplankton distribution in
Georgia estuary. Mar. Biol. 33:147-
154.

Stickney, R. R. and S. C. Knowles. 1976.
Seasonal zooplankton patterns in a
Georgia estuary. Bull. Ga. Acad.
Sci. 34:121-128.

Stickney, R. R. and D. Perlmutter. 1975.
Impact of intracoastal waterway
maintenance dredging on a mud bottom
benthos community. Biol. Conserv.
7(3):211-226.

Stickney, R. R., G. L. Taylor, and R. H.
Heard III. 1974. Food habits of
Georgia estuarine fishes. I. Four
species of flounders (Pleuronectiformes: Bothidae). Fish. Bull. 72:
515-525.

Stickney, R. R., G. L. Taylor, and B. B.
White. 1975. Food habits of five
species of young Southeastern United
States estuarine Sciaenidae.
Chesapeake Sci. 16(2):104-114.

Stoddard, H. L. 1931. The bobwhite
quail: its habits, preservation
and increase. Charles Scribners
Sons, New York. 559 pp.

Stoddard, H. L., Sr. 1963. Bird habitat
and fire. Proc. Tall Timbers Fire
Ecol. Conf. 2:163-175.

Straughan, D. 1972. The influence of the
Santa Barbara oil spill (January-
February, 1969) on the intertidal
distribution of marine organisms.
Rep. to Western Oil and Gas Assoc.
59 pp.

Street, M. W. 1969. Fecundity of the
hickory shad in the Altamaha River,
Georgia. Ga. Game Fish Comm.,
Mar. Fish. Div., Contrib. Ser. No.
14. Brunswick. 11 pp.

Street, M. W. 1970. Some aspects of
the life histories of hickory shad,
Alosa mediocris (Mitchill), and
blueback herring, Alosa aestivalis
(Mitchill) in the Altamaha River,
Georgia. M.S. Thesis. Univ. Ga.,
Athens. 89 pp.

Street, M. W. and J. G. Adams. 1969.
Aging of hickory shad and blueback
herring in Georgia by the scale
method. Ga. Game Fish Comm., Mar.
Fish. Div., Contrib. Ser. No. 18.
Brunswick. 13 pp.

Struhsaker, P. 1969. Demersal fish
resources: composition, distribution and commercial potential of the
continental shelf stocks off Southeastern United States. Fish. Ind.
Res. 4(7):261-300.

Sullivan, M. J. 1975. Diatom communities
from a Delaware salt marsh. J.
Phycol. 11:384-390.

Sullivan, M. J. 1976. Long-term effects
of manipulating light intensity and
nutrient enrichment of the structure
of a salt marsh diatom community. J.
Phycol. 12:205-210.

Sunda, W. G., D. W. Engel, and R. M.
Thuotte. 1978. Effect of chemical
speciation on toxicity of cadmium to
grass shrimp, Palaemonetes pugio:
importance of free cadmium ion.
Environ. Sci. Technol. 12:409-413.

Sutcliffe, D. W. 1961a. Studies on salt and water balance in caddis larvae (Trichoptera): I. Osmotic and ionic regulation of body fluids in *Limnephilus affinis* Curtis. J. Exp. Biol. 38:501-519.

Sutcliffe, D. W. 1961b. Salinity fluctuations and the fauna in a salt marsh with special reference to aquatic insects. Trans. Nat. Hist. Soc. Northumberl. 14:37-56.

Sutcliffe, W. H. 1950. A qualitative and quantitative study of the surface zooplankton at Beaufort, North Carolina. Ph.D. Dissertation. Duke Univ., Durham, N.C. 137 pp.

Sverdrup, H. U. 1953. On the conditions for the vernal blooming of phytoplankton. Rapp. P-V Reun. Cons. Int. Explor. Mer 18:287-295.

Swails, L. F., Jr., F. F. Welbourne, Jr., and W. E. Hoy. 1957. The flora of the bottom lands of the Savannah River Swamp. Part VII. Univ. S.C. Publ., Biol. Ser. III, 2(2):72-77. Columbia.

Swingle, H. S. 1950. Relationships and dynamics of balanced and unbalanced fish populations. Ala. Polytech. Inst. Agric. Exp. Stn. Bull. No. 274. 73 pp.

Swingle, H. S. 1966. Fish kills caused by phytoplankton blooms and their prevention. Proc. world symposium on warm water pond fish culture. Rep. No. 44.

Symposium on the Classification of Brackish Waters. 1958. The Venice system for the classification of marine waters according to salinity. Oikos 9:311-312.

Tabb, D. C. 1966. The estuary as a habitat for spotted seatrout, *Cynoscion nebulosus*, pp. 59-67. In: R. F. Smith, A. H. Swartz and W. H. Massman, eds. A symposium on estuarine fisheries. Am. Fish. Soc. Spec. Publ. No. 3. Washington, D. C.

Tagatz, M. E. 1961. Tolerance of striped bass and American shad to changes of temperature and salinity. U.S. Fish Wildl. Serv. Spec. Sci. Rep. Fish. No. 388. 8 pp.

Tagatz, M. E., P. W. Borthwick, J. M. Ivey, and J. Knight. 1976. Effects of leached mirex on experimental communities of estuarine animals. Arch. Environ. Contam. Toxicol. 4(4):435-442.

Tagatz, M. E., J. M. Ivey, J. C. Moore, and M. Tobia. 1977. Effects of pentachlorophenol on the development of estuarine communities. J. Toxicol. Environ. Health 5:501-506.

Tagatz, M. E., J. M. Ivey, and M. Tobia. 1978. Effects of Dowicide R G-ST on development of experimental estuarine macrobenthic communities, pp. 157-163. In: K. Ranga Rao, ed. Pentachlorophenol. Plenum Publ. Corp., New York.

Tait, H. D. 1977. Ecological characterization--an aid to decision-making. Progress report presented at Environmental Protection Agency Program Review, 2 March 1977, Newport, R.I.

Tanner, J. T. 1942. The ivory-billed woodpecker. Natl. Audubon Soc., Res. Rep. No. 1. New York. 111 pp.

Tanner, W. F. 1960. Florida coastal classification. Trans. Gulf Coast Assoc. Geol. Soc. 10:259-266.

Tarplee, W. H., Jr., D. E. Louder, and A. J. Weber. 1971. Evaluation of the effects of channelization on fish populations in North Carolina's coastal plain streams. N.C. Wildl. Resour. Comm., Raleigh. 22 pp.

Tatem, H. E. 1975. The toxicity and physiological effects of oil and petroleum hydrocarbons on estuarine grass shrimp *Palaemonetes pugio* Holthuis. Ph.D. Dissertation. Texas A&M Univ., College Station. 133 pp.

Taylor, W. R. 1969. Marine algae of the Northeastern coast of North America. Univ. Mich. Press, Ann Arbor. 509 pp. (First publ. 1957.)

Teal, J. M. 1958a. Additions to the breeding birds of Sapelo Island. Oriole 23:8.

Teal, J. M. 1958b. Distribution of fiddler crabs in Georgia salt marshes. Ecology 39(2):185-193.

Teal, J. M. 1959a. Birds of Sapelo Island and vicinity. Oriole 24:1-14, 17-20.

Teal, J. M. 1959b. Energy flow in the salt marsh ecosystem, pp. 101-107. In: R. A. Ragotzkie, L. R. Pomeroy, J. M. Teal, and D. C. Scott, eds. Proceedings salt marsh conference, Univ. Ga. Mar. Inst. March 25-28, 1958. Sapelo Island.

Teal, J. M. 1962. Energy flow in the salt marsh ecosystem of Georgia. Ecology 43(4):614-624.

Teal, M, and J, Teal, 1964, Portrait of an island. Atheneum, New York. 167 pp.

Teal, J. M. and W. Weiser. 1966. The distribution and ecology of nematodes in a Georgia salt marsh. Limnol. Oceanogr. 11:217-222.

Tebo, L. B. and W. W. Hassler. 1961. Seasonal abundance of aquatic insects in western North Carolina trout streams. J. Elisha Mitchell Sci. Soc. 77:249-259.

Tenore, K. R., C. F. Chamberlain, W. M. Dunstan, and R. B. Hanson. 1978. Possible effects of Gulf Stream instrusions and coastal runoff on the benthos of the continental shelf of the Georgia Bight, pp. 577-598. In: M. C. Wiley, ed. Estuarine interactions. Academic Press, New York.

Teskey, R. O. and T. M. Hinckley. 1977. Impact of water level changes on woody riparian and wetland communities. Vol. 2: Southern forest region. U.S. Fish Wildl. Serv., Off. Biol. Serv., Natl. Stream Alteration Team, Columbia, Mo. 46 pp.

Texas Instruments, Inc. 1978. South Atlantic benchmark program, outer continental shelf (OCS) environmental studies. Vol. 3. Results of studies of Georgia Bight of North Atlantic Ocean. Draft final rep., Contract No. AA550-CT7-2. 439 pp.

Thayer, G. W. 1971. Phytoplankton production and the distribution of nutrients in a shallow unstratified estuarine system near Beaufort, N.C. Chesapeake Sci. 12(4):240-253.

Thayer, G. W. 1974. Identity and regulation of nutrients limiting phytoplankton production in the shallow estuaries near Beaufort, N.C. Oecologia 14:75-92.

Theiling, D. L. and H. A. Loyacano, Jr. 1976. Age and growth of red drum from a saltwater marsh impoundment in South Carolina. Trans. Am. Fish. Soc. 105(1):41-44.

Thompson, J. R. 1973. Ecological effects of offshore dredging and beach nourishment: a review. U.S. Army Corps Eng., Coastal Eng. Res. Cent. Misc. Pap. No. 1-73. Fort Belvoir, Va. 39 pp.

Thorsland, O. A. 1967. Nutritional analyses of selected deer foods in South Carolina. Proc. Annu. Conf. Southeast. Assoc. Game Fish Comm. 20:84-104.

Tinbergen, L. 1960. The dynamics of insects and bird populations in pine woods. Arch. Neerl. Zool. 13(3):259-472.

Tindall, E. E. 1961. A two-year study of mosquito breeding and wildlife usage in Little Creek impounded salt marsh, Little Creek Wildlife Area, Delaware, 1950-60. Proc. N.J. Mosq. Exterm. Assoc. 48:100-105.

Tiner, R. W., Jr. 1977. An inventory of South Carolina's coastal marshes. S.C. Mar. Resour. Cent. Tech. Rep. No. 23. 33 pp.

Tiner, R. W., Jr., R. H. Dunlap, and J. P. Deveaux. 1976. Interim report on an inventory of South Carolina non-forested tidal wetlands. S.C. Mar. Resour. Cent., Charleston. (Unpubl.)

Tinkham, E. R. 1955. A note on longevity of soil larvaecides, pp. 253-255. In: D. K. M. Kevan, ed. Soil Zoology. Academic Press, Inc., New York.

Tomkins, I. R. 1934. Notes from Chatham County, Georgia. Auk 51(2):252-253.

Tomkins, I. R. 1935. The marsh rabbit: an incomplete life history. J. Mammal. 16(3):201-205.

Tomkins, I. R. 1936. The stilt sandpiper again on the lower Savannah River. Auk 53:329.

Tomkins, I. R. 1937. Wayne's clapper rail carries its young. Wilson Bull. 49:296-297.

Tomkins, I. R. 1947. The oyster-catcher of the Atlantic coast of North America and its relation to oysters. Wilson Bull. 59(4):204-208.

Tomkins, I. R. 1954. Life history notes on the American oyster-catcher. Oriole 19(4):37-45.

Tomkins, I. R. 1955. The distribution of the marsh rabbit in Georgia. J. Mammal. 36:144-145.

Tomkins, I. R. 1956. The manatee along the Georgia coast. J. Mammal. 37(2):288-289.

Tomkins, I. R. 1958. A Georgia specimen of the manatee. J. Mammal. 39:154.

Tomkins, I. R. 1959. Life history notes of the least tern. Wilson Bull. 71:313-322.

Tomkins, I. R. 1965a. Absence of the blue jay on some of Georgia's coastal islands. Oriole 30(2):77-79.

Tomkins, I. R. 1965b. Mammalian field notes. Ga. Historical Soc., Savannah. 8 pp. (Unpubl.)

Trent, L. and W. W. Hassler. 1968. Gill net selection, migrations, size, and age composition, sex ratio, harvest efficiency amd management of striped bass in the Roanoke River, North Carolina. Chesapeake Sci. 9(4):217-232.

Tsuda, R. T. and D. A. Grosenbaugh. 1977. Agat sewage treatment plant: impact of secondary treated effluent on Guam coastal waters. Univ. Guam Mar. Lab. Tech. Rep. 3. 43 pp.

Tuck, L. M. 1960. Department of Northern Affairs and National Research National Parks Branch, Canadian Wildlife Service 1. 260 pp.

Tucker, W. H. 1972. Food habits, growth, and length-weight relationships of young-of-the-year black crappie and largemouth bass in ponds. Proc. Annu. Conf. Southeast. Assoc. Game Fish Comm. 26:565-576.

Turner, C. H. 1910. Ecological notes on the cladocera and copepoda of Augusta, Georgia, with description of new or little known species. Trans. Acad. Sci. St. Louis 19(10): 151-176.

Turner, H. J. 1951. Shellfish culture in salt water ponds, pp. 9-10. In: Fourth report investigations of shellfisheries. Mass. Dep. Nat. Resour., Div. Mar. Fish., Boston.

Turner, H. J., Jr. and D. L. Belding. 1957. The tidal migrations of Donax variabilis (Say). Limnol. Oceanogr. 2:120-124.

Turner, R. E. 1976. Geographic variations in salt marsh macrophyte production: a review. Contrib. Mar. Sci. 20: 47-68.

Turner, R. E., S. W. Woo, and H. R. Jitts. 1979. Estuarine influences on a continental shelf plankton community. Science 206:218-220.

Turner, W. R. and G. N. Johnson. 1972. Standing crops of aquatic organisms in five South Carolina tidal streams, pp. 179-191. In: Port Royal Sound environmental study. S.C. Water Resources Comm., Columbia.

Turner, W. R. and G. N. Johnson. 1973. Distribution and relative abundance of fishes in Newport River, North Carolina. NOAA Tech. Rep. NMFS SSRF 666. 23 pp.

Turner, W. R. and G. N. Johnson. 1974. Standing crops of aquatic organisms in tidal streams of the lower Cooper River system, South Carolina, pp. 13-20. In: F. P. Nelson, ed. The Cooper River environmental study. S.C. Water Resources Comm. State Water Plan Rep. No. 117. Columbia.

Tyler-Schroeder, D. B. 1976. Effects of two polychlorinated biphenyls, Aroclor R 1016 and 1242, on the grass shrimp, Palaemonetes pugio. M.S. Thesis. Univ. West Florida, Pensacola. 128 pp.

Tyler-Schroeder, D. B. 1979. Use of the grass shrimp (Palaemonetes pugio) in a life-cycle toxicity test, pp. 159-170. In: L. L. Marking and R. A. Kimerle, eds. Aquatic toxicology. Am. Soc. Test. Mater., Spec. Tech. Publ. 667.

Ulrich, G. F. 1978. Incidental catch of loggerhead turtles by South Carolina commerical fisheries. S.C. Mar. Resour. Cent., Charleston. Rep. to U.S. Natl. Mar. Fish. Serv. Contract Nos. 03-7-042-35151 and 03-7-042-35121. 36 pp. (Unpubl.)

University of Georgia, Institute of Ecology. 1978. Microbial processes and biomass on the Southeastern Continental Shelf: Cruise Report 1975-1976. Ga. Mar. Sci. Cent. Tech. Rep. Ser. No. 78-4. Skidaway Island. 65 pp.

Univeristy of Georgia, Marine Institute. 1971. An ecological survey of the North and South Newport rivers and adjacent waters with respect to possible effects of treated Kraft Mill effluent. Final rep. to Ga. Water Quality Control Board. UGA #D2422-122. 280 pp.

Univeristy of South Carolina. 1973. Bioassay studies, Charleston Harbor, South Carolina. The effects of dredging harbor sediments on plankton. Final rep. to U.S. Army Corps Eng., Charleston District. Belle W. Baruch Coastal Res. Inst., Columbia. 54 pp.

Urner, A. 1935. Relation of mosquito control in New Jersey to bird life of salt marshes. Proc. N.J. Mosq. Exterm. Assoc. 22:130-136.

591

Ursin, M. J. 1975. Life in and around freshwater wetlands. Thomas Y. Crowell Co., New York. 116 pp.

U.S. Army Corps of Engineers. 1975. Cooper River Rediversion Project, Charleston Harbor, South Carolina. Final environ. statement. U.S. Army Eng. District, Charleston. 201 pp.

U.S. Army Corps of Engineers. 1977. Water resources development by the U.S. Army Corps of Engineers in Georgia. South Atlantic Div., Atlanta. 111 pp.

U.S. Department of Agriculture. 1973. Environmental statement, imported fire ant cooperative Federal-State control and regulatory program. Animal and Plant Health Inspection Serv., Washington, D.C. 49 pp.

U.S. Department of Agriculture. 1975. Control of water pollution from cropland. Vol. 1. A manual for guideline development. In cooperation with the U.S. Environ. Prot. Agency, Washington, D.C. Rep. No. EPA-600/2-75-026A. 111 pp.

U.S. Department of Agriculture, Forest Service. 1972. Insects and diseases of trees in the South. Southeast. Area Forest Pest Manage. Group, Atlanta. 81 pp.

U.S. Department of Agriculture, Forest Service. 1975. Proceedings of the symposium on management of forest and range habitats for nongame birds. May 6-9, 1975, Tuscon, Ariz. For. Serv. General Tech. Rep. WD-1. Washington, D.C. 343 pp.

U.S. Department of Health, Education and Welfare. 1965. Ashley River pollution study. Summary rep. 6 pp.

U.S. Department of Interior, Bureau of Land Management. 1977. Proposed 1978 outer continental shelf oil and gas lease sale, South Atlantic OCS Sale No. 43. Final environ. impact statement. 3 vols. OCS Off., New Orleans, La. 899 pp. + 15 maps.

U.S. Department of Interior, Fish and Wildlife Service. 1962. Effects of pesticides on fish and wildlife in 1960. U.S. Fish Wildl. Serv. Circ. 143. 52 pp.

U.S. Department of Interior, Fish and Wildlife Service. 1963. Pesticide-wildlife studies: a review of Fish and Wildlife Service investigations during 1961 and 1962. U.S. Fish Wildl. Serv. Circ. 167. 109 pp.

U.S. Department of Interior, Fish and Wildlife Service. 1964. Pesticide-wildlife studies 1963: a review of Fish and Wildlife Service investigations during the calendar year. U.S. Fish Wildl. Serv. Circ. 199. 129 pp.

U.S. Department of Interior, Fish and Wildlife Service. 1965. The effects of pesticides on fish and wildlife. U.S. Fish Wildl. Serv. Cir. 226. 77 pp.

U.S. Department of Interior, Fish and Wildlife Service. 1971. Birds of the Cape Romain National Wildlife Refuge. 8 pp.

U.S. Department of Interior, Fish and Wildlife Service. 1973. Threatened wildlife of the United States. Off. Endangered Species and Int. Act. Washington, D.C.

U.S. Department of Interior, Fish and Wildlife Service. 1976. Mosquito control procedures and practices and their effect on the environment, a literature search and annotated bibliography. Off. Biol. Serv. FWS/OBS/76-22. Washington, D.C. 78 pp.

U.S. Department of Interior, Fish and Wildlife Service. 1979a. List of endangered and threatened wildlife and plants. Federal Register 44(12): 3636-3654.

U.S. Department of Interior, Fish and Wildlife Service. 1979b. Endangered and threatened species recovery planning guidelines. 27 pp. (Unpubl.)

U.S. Environmental Protection Agency. 1971. Report on pollution of the St. Marys and Amelia estuaries, Georgia - Florida. Southeast Water Lab., Athens, Ga. NTIS #PB-213-394. 52 pp.

U.S. Naval Facilities Engineering Command. 1977. Draft environmental impact statement for preferred alternative location for a fleet ballistic missile (FBM) submarine support base Kings Bay, Georgia. Appendix D: Existing biological environment. 613 pp.

Vaccaro, R. F., G. D. Grice, G. T. Rowe, and P. H. Wiebe. 1972. Acid iron waste disposal and the summer distribution of standing crops in the New York Bight. Water Res. 6:231-256.

Valentine, J. M., J. R. Walther, K. M. McCartney, and L. M. Ivy. 1972. Alligator diets on the Sabine National Wildlife Refuge, Louisiana. J. Wildl. Manage. 36:809-815.

Valiela, I., J. M. Teal, S. Volkmann, D. Shafer, and E. J. Carpenter. 1978. Nutrient and particulate fluxes in a salt marsh ecosystem: tidal exchanges and inputs by precipitation and groundwater. Limnol. Oceanogr. 23(4):788-812.

Van Dolah, R. F., D. R. Calder, D. M. Knott, and M. S. Maclin. 1979. Effects of dredging and unconfined disposal of dredged material on macrobenthic communities in Sewee Bay, South Carolina. S.C. Mar. Resour. Cent. Tech. Rep. No. 39. 54 pp.

Van Engle, W. A. and E. B. Joseph. 1968. The characterization of coastal and estuarine fish nursery grounds as natural communities. Va. Inst. Mar. Sci. Final Rep. Nov. 1965 - Aug. 1967. Contract No. 14-17-0007-531. Comm. Fish. Res. Dev. Act. 43 pp. (Unpubl.)

Van Pelt, A. F. 1966' Activity and density of old-field ants of the Savannah River Plant, South Carolina. J. Elisha Mitchell Sci. Soc. 82:35-43.

Van Tyne, J. and A. J. Berger. 1959. Fundamentals of ornithology, Dover Publ., Inc., New York. 254 pp.

Van Valkenburg, S. D. and D. A. Flemer. 1974. The distribution and productivity of nannoplankton in a temperate estuarine area. Estuarine Coastal Mar. Sci. 2:311-322.

Vaughn, T. L. 1967. Fecundity of the American shad in the Altamaha River system. Ga. Game Fish Comm., Mar. Fish. Div. Contrib. Ser. No. 3, Brunswick. 9 pp.

Vernberg, F. J., M. S. Guram, and A. Savory. 1977. Survival of larval and adult fiddler crabs exposed to Aroclor[R] 1016 and 1254 and different temperature-salinity combinations, pp. 37-50. In: F. J. Vernberg, A. Calabrese, F. P. Thunberg, and W. B. Vernberg, eds. Physiological responses of marine biota to pollutants. Academic Press, New York.

Vernberg, F. J. and C. E. Sansbury. 1972. Studies on salt marsh invertebrates of Port Royal Sound, pp. 271-277. In: Port Royal Sound environmental study. S.C. Water Resources Comm., Columbia.

Vernberg, W. B. and B. C. Coull. 1975. Multiple factor effects of environmental parameters on the physiology, ecology, and distribution of some marine meiofauna. Cah. Biol. Mar. 16:721-732.

Vernberg, W. B., P. J. DeCoursey, and W. J. Padgett. 1973. Synergistic effects of environmental variables in larvae of Uca pugilator. Mar. Biol. 22:307-312.

Vigerstad, J. J. and L. J. Tilly. 1980. Effects of submerged macrophytes on heleoplanktonic cladocera. Hydrobiologia. (In press.)

Viosca, P., Jr. 1923. An ecological study of the cold blooded vertebrates of southeastern Louisiana. Copeia 1923:35-44.

Viosca, P., Jr. 1924. A terrestrial form of siren Lacertina. Copeia 1924:102-104.

Vladykov, V. D. 1966. Remarks on the American eel (Anguilla rostrata LeSueur): sizes of elvers entering streams; the relative abundance of adult males and females; and present economic importance of eels in North America. Verb. Int. Ver. Limnol. 16:1007-1017.

Vladykov, V. D. and J. R. Greeley. 1963. Order Acipenseroidei, pp. 24-60. In: H. B. Bigelow, ed. Fishes of the Western North Atlantic. Sears Found. Mar. Res., Mem. No. 1. Part 3. Yale Univ.

Vogl, R. J. 1969. One hundred and thirty years of plant succession in a southeastern Wisconsin lowland. Ecology 50:248-255.

Wade, C. W. 1971. Commercial anadromous fishery, Edisto River, South Carolina. Annu. progress rep. Project AFC-4-1. S.C. Wildl. Mar. Resour. Dep., Div. Game Freshwater Fish., Columbia. 79 pp. (Unpubl.)

Wagner, R. H. 1964. The ecology of Uniola paniculata L. in the dune-strand habitat of North Carolina. Ecol. Monogr. 34(1):79-96.

Wahlenberg, W. G., ed. 1965. A guide to loblolly and slash pine plantation management in Southeastern USA. Ga. Forest Res. Council Rep. No. 14. Macon. 360 pp.

Walburg, C. H. 1956. Commercial and sport shad fisheries of the Edisto River, South Carolina, 1955. U.S. Fish Wildl. Serv. Spec. Sci. Rep. Fish. No. 187. 9 pp.

Walburg, C. H. and P. R. Nichols. 1967. Biology and management of the American shad and status of the fisheries, Atlantic coast of the United States, 1960. U.S. Fish Wildl. Serv. Spec. Sci. Rep. Fish. No. 550. 105 pp.

Wallwork, J. A. 1970. Ecology of soil animals. McGraw-Hill, New York. 283 pp.

Wass, M. L. 1974. Birds of the coastal zone, pp. 512-703. In: M. H. Roberts, Jr. A socio-economic environmental baseline summary for the South Atlantic region between Cape Hatteras, North Carolina and Cape Canaveral, Florida. Vol. 3: Chemical and biological oceanography. Va. Inst. Mar. Sci., Gloucester Point.

Watling, L., J. Lindsay, R. Smith, and D. Maurer. 1974. The distribution of isopoda in the Delaware Bay region. Int. Rev. Gesamten Hydrobiol. 59: 343-351.

Watson, S. W. 1963. Autotrophic nitrification in the ocean, pp. 73-84. In: C. H. Oppenheimer. Symposium on marine microbiology. Thomas, Springfield, Ill.

Wayne, A. T. 1902. The Ipswich sparrow (Ammodramus princeps) on the mainland of South Carolina. Auk 19:203.

Wayne, A. T. 1910. Birds of South Carolina. Charleston Mus. Contrib. 1. Charleston, S.C. 254 pp.

Welch, E. B. 1967. Factors initiating phytoplankton blooms and resulting effects on dissolved oxygen in an enriched estuary. Ph.D. Dissertation. Univ. Wash., Seattle. 102 pp.

Weller, W. F. 1977. Migration routes of the salamanders Ambystoma jeffersonianum and A. platineum to and from a spring breeding pond. Presented 57th annual meeting Am. Soc. Ichthyologists and Herpetologists. Gainesville, Fla. 1 p. (Abstr.)

Wells, B. W. 1928. Plant communities of the coastal plain of North Carolina and their successional relations. Ecology 9(2):230-242.

Wells, B. W. 1932. The natural gardens of North Carolina with keys and descriptions of the herbaceous wild flowers found therein. Univ. N.C. Press, Chapel Hill. 458 pp.

Wells, B. W. 1939. A new forest climax: the salt spray climax of Smith Island, N.C. Bull. Torrey Bot. Club 66:629-634.

Wells, B. W. 1942. Ecological problems of the Southeastern United States coastal plain. Bot. Rev. 8:533-561.

Wells, B. W. 1946. Vegetation of Holly Shelter Wildlife Management Area. N.C. Dep. Conserv. Bull. 2. 40 pp.

Wells, B. W. and I. V. Shunk. 1928. A southern upland grass-sedge bog: an ecological study. N.C. Agric. Exp. Stn. Tech. Bull. 32. Raleigh.

Wells, B. W. and I. V. Shunk. 1931. The vegetation and habitat factors of the coarser sands of the North Carolina coastal plain: an ecological study. Ecol. Monogr. 1(4):465-520.

Wells, B. W. and I. V. Shunk. 1937. Seaside shrubs: windforms vs. spray forms. Science 85(2212):499.

Wells, B. W. and I. V. Shunk. 1938. Salt spray: an important factor in coastal ecology. Bull. Torrey Bot. Club 85:485-492.

Welsh, W. M. 1916. Notes on the fishes of the Pee Dee River basin, North and South Carolina. Copeia 1916: 54-56.

Wenner, C. A. and J. A. Musick. 1974. Fecundity and gonad observations of the American eel, Anguilla rostrata, migrating from Chesapeake Bay, Virginia. J. Fish. Res. Board Can. 31:1387-1391.

West Indian (Florida) Manatee Recovery Team. 1979. West Indian (Florida) manatee recovery plan. Technical draft. U.S. Dep. Inter., Fish Wildl. Serv. 35 pp.

Westvaco Corporation. 1972. A summary of biological studies of the Cooper River and adjacent waters in the vicinity of Charleston, South Carolina 1967-1971. Westvaco Kraft Div., Charleston, S.C. 72 pp.

Wetmore, A. 1957. Check-list of North American birds. Committee on Classification and Nomenclature. Am. Ornithologists' Union, New York.

Wetzel, R. G. 1975. Limnology. W. B. Saunders Co., Philadelphia, Pa. 743 pp.

Wharton, C. H. 1969. The cottonmouth moccasin on Sea Horse Key, Florida. Bull. Fla. State Mus. 14(3):227-272.

Wharton, C. H. 1970. The Southern river swamp--a multiple-use environment. Ga. State Univ., School Business Adm., Atlanta. 48 pp.

Wharton, C. H. 1978. The natural environments of Georgia. Ga. Dep. Nat. Resour., Off. Planning Res. and Geologic and Water Resour., Atlanta. 227 pp.

Wharton, C. H., H. T. Odum, K. Ewel, M. Duever, A. Lugo, R. Boyt, J. Bartholomew, E. Debellevue, S. Brown, M. Brown, and L. Duever. 1976. Forested wetlands of Florida - their management and use. Final rep. to Fla. Div. State Planning. Univ. Fla., Cent. Wetlands, Phelps Lab., Gainesville. 421 pp.

Wheeler, R. J. 1939. Food habits of the opossum in Sumter County, Alabama. M.S. Thesis. Alabama Polytechnic Inst., Auburn. 56 pp.

Wherry, E. T. 1920. Plant distribution around salt marshes. Ecology 1: 42–48.

Whetstone, J. M. and A. G. Eversole. 1978. Predation on hard clams, Mercenaria mercenaria, by mud crabs, Panopeus herbstii. Proc. Natl. Shellfish. Assoc. 68:42–48.

Whigham, D. F., R. McCormick, R. E. Goud, and R. L. Simpson. 1978. Biomass and primary production in freshwater tidal wetlands of the Middle Atlantic coast. In: R. E. Goud, D. F. Whigham, and R. L. Simpson, eds. Freshwater wetlands: production processes and management potential. Academic Press, New York.

Whitaker, J. D. 1978. A contribution to the biology of Loligo pealei and Logigo plei (Cephalopoda, Myopsida) off the Southeastern coast of the United States. M.S. Thesis, College of Charleston, Charleston, S.C. 164 pp.

White, M. G. 1969. Anadromous fish survey of the Edisto and Coosawhatchie rivers - South Carolina. Job completion rep. Project AFS-2-2. S.C. Wildl. Mar. Resour. Dep., Columbia. 94 pp. (Unpubl.)

White, M. G. 1970. Anadromous fish survey of the Savannah and Ashepoo rivers. Job completion rep. Project AFS-2-3. S.C. Wildl. Mar. Resour. Dep., Columbia. 94 pp. (Unpubl.)

White, M. G. 1971. Fisheries studies in District V. S.C. Wildl. Mar. Resour. Dep., Columbia. 29 pp.

White, M. G. 1972. Fisheries studies in District V. Annu. progress rep. Project F-16-2. July 1, 1971 - June 30, 1972. S.C. Wildl. Mar. Resour. Dep., Columbia. 34 pp.

White, M. G. 1973. Fisheries studies in District V. Annu. progress rep. Project F-16-3. July 1, 1972 - June 30, 1973. S.C. Wildl. Mar. Resour. Dep., Div. Game Freshwater Fish., Columbia. 37 pp.

White, M. G. 1974. Fisheries studies in District V. Annu. progress rep. Project F-16-4. July 1, 1973 - June 30, 1974. S.C. Wildl. Mar. Resour. Dep., Div. Game Freshwater Fish. Columbia. 19 pp.

White, M. G. 1975. Fisheries studies in District V. Annu. progress rep. Project F-16-5. July 1, 1974 - June 30, 1975. S.C. Wildl. Mar. Resour. Dep., Div. Game Freshwater Fish. Columbia. 15 pp.

White, M. G. 1976. Fisheries studies in District V. Annu. progress rep. Project F-16-5. July 1, 1975 - June 30, 1976. S.C. Wildl. Mar. Resour. Dep., Div. Game Freshwater Fish. Columbia. 14 pp.

White, M. G. and T. A. Curtis. 1969. Anadromous fish survey of the Black and Pee Dee river watersheds. Project AFS 2-4. Job completion rep. 1 July 1968 - 30 June 1969. S.C. Wildl. Mar. Resour. Dep., Columbia. 73 pp.

Whitney, D. E., G. M. Woodwell, and R. W. Howarth. 1975. Nitrogen fixation in Flax Pond: a Long Island salt marsh. Limnol. Oceanogr. 20:640–643.

Wiebe, W. 1975. Microorganism-detritus relationships in estuarine waters. Symposium on detritus and its biological role in aquatic environments. Pallanzo, Italy.

Wiebe, W. J. and C. W. Hendricks. 1974. Distribution of heterotrophic bacteria in a transect of the Atlantic Ocean, pp. 524-535. In: R. R. Colwell and R. Y. Morita, eds. Effect of the ocean environment on microbial activities. Univ. Park Press, Baltimore.

Wiegert, R. G. 1974. Litterbag studies of microarthropod populations in three South Carolina old fields. Ecology 55:94–102.

Wiese, J. H. 1978. Heron nest-site selection and its ecological effects, pp. 27-34. In: A. Sprunt, J. C. Ogden, and S. Winckler, eds. Wading birds. National Audubon Society Res. Rep. 7. New York.

Wilbur, R. L. 1969. The redear sunfish in Florida. Fla. Game Freshwater Fish. Comm. Fish. Bull. No. 5. 64 pp.

Wilkinson, P. M. 1970. Vegetative succession in newly controlled marshes. Job completion rep. Project W-38-6. April 1, 1967 - June 30, 1970. S.C. Wildl. Mar. Resour. Dep., Columbia. 37 pp.

Williams, A. B. 1965. Marine decapod crustaceans of the Carolinas. Fish. Bull. 65(1):1-298.

Williams, L. E., Jr., ed. 1978. Recovery plan, eastern brown pelican (Pelecanus occidentalis carolinensis). U.S. Dep. Inter., Fish Wildl. Serv. 45 pp.

Williams, L. G. 1951. Algae of the Black Rocks, pp. 149-161. In: A. S. Pearse and L. G. Williams. The biota of the reefs off the Carolinas. J. Elisha Mitchell Sci. Soc. 67:133-161.

Williams, L. G. 1966. Dominant planktonic rotifers of the major waterways of the United States. Limnol. Oceanogr. 11(1):83-91.

Williams, R. B. 1962. The ecology of diatom populations in a Georgia salt marsh. Ph.D. Dissertation. Harvard Univ., Cambridge, Mass. 146 pp.

Williams, R. B. and A. V. Holden. 1973. Organochlorine residues from plankton. Mar. Pollut. Bull. 4(7): 109-111.

Williams, R. B. and M. B. Murdoch. 1966. Annual production of Spartina and Juncus in North Carolina salt marshes. Presented Am. Inst. Biol. Sci. Meeting. Aug. 15, 1966. College Park, Md.

Williamson, G. K. and R. A. Moulis. 1979. Distribution of Georgia amphibians and reptiles in the Savannah Science Museum Collection. Savannah Science Museum, Inc., Spec. Publ. No. 1. Savannah, Ga.

Willson, M. F. 1974. Avian community organization and habitat structure. Ecology 55:1017-1029.

Wilson, K. A. 1954. Role of mink and otter as muskrat predators in northeastern North Carolina. J. Wildl. Manage. 18(2):199-207

Wilson, K. A. 1968. Fur production on southeastern coastal marshes, pp. 149-162. In: J. D. Newsom, ed. Proceedings of the marsh and estuary management symposium. La. State Univ., Baton Rouge.

Winchester, B. H., R. S. Delotelle, J. R. Newman, and J. T. McClave. 1978. Ecology and management of the colonial pocket gopher: a progress report, pp. 173-184. In: R. O. Odom and L. Landers, eds. Proceedings of the rare and endangered wildlife symposium, 3-4 Aug. 1978, Athens, Ga. Ga. Dep. Nat. Resour., Game Fish Div., Tech. Bull. WL 4.

Windom, H. L. 1975. Heavy metal fluxes through saltmarsh estuaries, pp. 137-152. In: L. E. Cronin, ed. Estuarine research. Vol. 1. Chemistry, biology, and the estuarine system. Academic Press, Inc., New York.

Windom, H. L. 1976. Environmental aspects of dredging in the coastal zone. CRC Critical Rev. Environ. Control. 1976:91-109.

Windom, H. L., W. M. Dunstan, and W. S. Gardner. 1975. River input of inorganic phosphorus and nitrogen to the southeastern salt-marsh estuarine environment, pp. 309-313. In: F. G. Howell, J. B. Gentry, anad M. H. Smith, eds. Mineral cycling in southeastern ecosystems. U.S. Energy Res. Dev. Adm. Symp. Ser. Conf-740513. Springfield, Va.

Winner, J. M. 1975. Zooplankton, pp. 155-169. In: B. A. Whitman, ed. River ecology. Univ. Calif. Press, Berkeley.

Wisely, B. and R. A. P. Blick. 1967. Mortality of marine invertebrate larvae in mercury, copper and zinc solutions. Australian J. Mar. Freshwater Res. 18:63-72.

Wiseman, D. R. 1978. Benthic marine algae, pp. 23-36. In: R. G. Zingmark, ed. An annotated checklist of the biota of the coastal zone of South Carolina. Univ. S.C. Press, Columbia.

Wiseman, D. R. and C. W. Schneider. 1976. Investigations of the marine algae of South Carolina. I. New records of rhodophyta. Rhodora 78(815):516-524.

Wishart, M. A. and H. A. Loyacano. 1974. A survey of edible crawfish for the coastal plain of South Carolina. Completion report for the Coastal Plains Regional Commission. Dep. Entomology and Economic Zoology, Clemson Univ., Clemson.

Whitham, P. R. 1974. Neonate sea turtles from the stomach of a pelagic fish. Copeia 1974(2):548.

Whitham, P. R. 1976. Evidence for ocean-current mediated dispersal in young green turtles, _Chelonia mydas_ (Linnaeus). M.S. Thesis. Univ. Okla., Norman. 48 pp.

Wolcott, T. G. 1976a. Uptake of soil capillary water by ghost crabs. Nature 264:756-757.

Wolcott, T. G. 1976b. Uptake of soil capillary water by ghost crabs. Am. Zool. 16(2):240. (Abstr.)

Wolcott, T. G. 1978. Ecological role of ghost crabs, _Ocypode quadrata_ (Fabricius) on an ocean beach: scavengers or predators? J. Exp. Mar. Biol. Ecol. 31:67-82.

Wolf, P. L., S. F. Shanholtzer, and R. J. Reimold. 1975. Population estimates for _Uca pugnax_ (Smith, 1870) on the Duplin estuary marsh, Georgia, U.S.A. (Decapoda Brachyura, Ocypodidae). Crustaceana 29(1): 79-91.

Woodall, W. R., J. G. Adams, and J. Heise. 1975. Invertebrates eaten by Altamaha River fish. Paper presented at 39th annu. meeting Ga. Entomol. Soc., March 19-21, St. Simons Island, Ga.

Wood, E. J. F. 1965. Marine microbial ecology. Reinhold Publ. Corp, New York. 243 pp.

Wood, E. J. F. 1967. Microbiology of oceans and estuaries. Elsevier Publ. Co., Amsterdam. 319 pp.

Wood, H. C. 1874. A contribution to the history of fresh-water algae of the United States. Smithson. Contrib. Knowledge 19:1-202.

Woodwell, G. M. 1956. Phytosociology of coastal plain wetlands of the Carolinas. M.A. Thesis. Duke Univ., Durham, N.C. 51 pp.

Woodwell, G. M. 1958. Factors controlling growth of pond pine seedlings in organic soil of the Carolinas. Ecol. Monogr. 28:219-236.

Woodwell, G. M., C. A. Hall, D. Whitney, and R. A. Houghton. 1979. The Flax Pond ecosystems study: the annual metabolism and nutrient budgets of a salt marsh. In: R. Jeffries and A. Davy, eds. Ecological processes in coastal environments. British Ecol. Soc. Symposium, Blackwell. (In press.)

Woodwell, G. M., D. E. Whitney, C. A. Hall, and R. A. Houghton. 1977. The Flax Pond ecosystem study: exchanges of carbon in water between a salt marsh and Long Island Sound. Limnol. Oceanogr. 22:833-838.

Worthington, J. S. 1972. An evaluation of environmental imapct: Little Cumberland Island, Georgia. M.S. Thesis. Univ. Mass., Amherst. 161 pp.

Worthington, W. W. 1890. South Carolina notes. Auk 7:82.

Wright, A. H. and A. A. Wright. 1932. The habitats and composition of the vegetation of Okefenokee Swamp, Georgia. Ecol. Monogr. 2:109-232.

Wright, A. H. and A. A. Wright. 1949. Handbook of frogs and toads of the United States and Canada. Comstock Publ. Co., Ithaca, N.Y. 640 pp.

Wright, R. T. and J. E. Hobbie. 1966. Use of glucose and acetate by bacteria and algae in aquatic systems. Ecology. 47:447-464.

Wyatt, H. and D. R. Holder. 1969. Life history studies. Annu. Rep. Study II, Job 3. Project F-19-R-3. Ga. Dep. Nat. Resour., Game Fish Comm., Atlanta.

Yeager, L. E. 1949. Effect of permanent flooding in a river-bottom timber area. Ill. Nat. Hist. Surv. Bull. 25(2):33-65.

Yentsch, C. S. 1971. The harvest-primary production, pp. 150-164. In: P. J. Herring and M. R. Clarke, eds. Deep oceans. Arthur Baker Ltd., London.

Yentsch, C. S. 1977. Plankton production. Mar. ecosystems analysis (MESA) N.Y. Bight Atlas Monograph 12. State Univ. N.Y., Sea Grant Inst., Albany. NYSSGP-AM-77-007. 25 pp.

Yonge, C. M. 1960. Oysters. Collins, London. 209 pp.

Young, A. M. 1978. Superorder Eucarda: order Decapoda, pp. 171-185. In: R. G. Zingmark, ed. An annotated checklist of the biota of the coastal zone of South Carolina. Univ. S.C. Press, Columbia.

Yount, J. L. 1966. South Atlantic states, pp. 269-286. In: D. G. Frey, ed. Limnology in North America. Univ. Wis. Press, Madison.

Zingmark, R. G. 1975. The phytoplankton of Kiawah Island, pp. P1-P38. In: W. M. Campbell, J. M. Dean and W. D. Chamberlain, eds. Environmental inventory of Kiawah Island. Environmental Research Center, Inc., Columbia, S.C.

Zingmark, R. G. 1977. Studies on the
phytoplankton and microbenthic
algae in the North Inlet estuary,
pp. 35-39. In: F. J. Vernberg,
R. Bonnell, B. Coull, R. Dame, Jr.,
P. DeCoursey, W. Kitchens, Jr.,
B. Kjerfve, H. Stevenson, W.
Vernberg, R. Zingmark. The dynamics
of an estuary as a natural eco-
system. U.S. Environ. Prot. Agency
EPA-600/3-77-016. Gulf Breeze,
Fla.

Zingmark, R. G. 1978. An annotated
checklist of the biota of the
coastal zone of South Carolina. Univ.
S.C. Press, Columbia. 364 pp.

Zingmark, R. G. and T. G. Miller. 1975.
The effects of mercury on the
photosynthesis and growth of
estuarine and oceanic phytoplankton,
pp. 45-57. In: F. J. Vernberg,
ed. Physiological ecology of
estuarine organisms. Univ. S.C. Press.
Columbia.

Zobell, C. E. 1946. Marine microbiology.
Chronica Botanica Press, Waltham,
Mass. 240 pp.

Zobell, C. E. 1963. The occurrence,
effects and fate of oil polluting
the sea. Int. J. Water Pollut.
7:173-197.

Zobell, C. E. 1973. Microbial and
environmental transitions in
estuaries, pp. 9-31. In: L. H.
Stevenson and R. R. Colwell, eds.
Estuarine microbial ecology. Univ.
S.C. Press, Columbia.

Zweifel, R. G. and C. J. Cole. 1974. An
annotated checklist of the
amphibians and reptiles of St.
Catherines Island, Georgia. Am.
Mus. Nat. Hist., New York. 32 pp.

(These abbreviations, taken from the References Cited, follow the BIOSIS list of Serials, 1976, BioScience Information Service of Biological Abstracts.)

Abstr., Abstract; Abstracts
Acad., Academia; Academie; Academy
Act., Activities
Adm., Administration; Administrative
Adv., Advancement; Advances
AEC, Atomic Energy Commission
Afr., Africa
Agric., Agricultural; Agriculture
Agron., Agronomy
Akad., Akademie
Ala., Alabama
Algol., Algologique
Am., American
Amphib., Amphibians
An., Anais
Anal., Analysis; Analytical
Anat., Anatomical
Anim., Animal
Ann., Annals
Annu., Annual
Anthropol., Anthropologist; Anthropology
App., Appendix
Appl., Applicata; Applied
Aquat., Aquatic
Arch., Archiv; Archives
Archeol., Archeological; Archeologist
Ariz., Arizona
Ark., Arkansas
Assoc., Association
Atl., Atlantic
Atmos., Atmospheric
Aud., Audubon
Aug., August
Aust., Australia
Auth., Authority

Bacteriol., Bacteriology
Behav., Behavior
Berl., Berlin
Bibliogr., Bibliographic
Biochem., Biochemical; Biochemistry
Bioeng., Bioengineering
Biol., Biologicae; Biological; Biologie; Biology
Biomed., Biomedical
Biometeorol., Biometeorology
Biosci., Biosciences
Biotechnol., Biotechnology
Bot., Botanica; Botanical; Botany
Br., Britain; British
Bras., Brasileira
Bros., Brothers
Bull., Bulletin
Bur., Bureau
Bus., Business

Cah., Cahiers
Calif., California
Camb., Cambridge
Can., Canada; Canadien
Caribb., Caribbean
Cat., Catalogue

Cent., Center
C.F.S., Commercial Fishery Statistics
Chem., Chemical; Chemicko; Chemistry
Chromatogr., Chromatography
Cienc., Ciencies
Circ., Circular
Civ., Civil
Co., Company; County
Coll., College
Collect., Collection
Colo., Colorado
Comm., Commission
Commer., Commerce; Commercial
Comp., Comparative
Conf., Conference
Congr., Congress
Conn., Connecticut
Cons., Conseil
Conserv., Conservation; Conservationist
Const., Constitution; Constitutional
Contam., Contamination
Contrib., Contribution; Contributions
Coop., Cooperative
Corp., Corporation
Cosmochim., Cosmochimica
C.R., Comptes Rendus
Cult., Culturist
Curr., Current

Dan., Danmarks; Dansk
D.C., District of Columbia
Dec., December
Del., Delaware
Dep., Department
Dev., Development; Developmental
Dig., Digest
Dis., Disease; Diseases
Div., Division
D.J., Dingell Johnson
DMRP, Dredged Material Research Program
Doc., Document; Documentation
Dokl., Doklady
Dtsch., Deutschen

Ecol., Ecological; Ecology
Econ., Economic; Economics
Ed., Editor
Eds., Editors
EDS, Environmental Data Service
Educ., Educational
Ekol., Ekologia
Encycl., Encyclopedia
Eng., Engineering; Engineers
Engl., England
Entomol., Entomological; Entomologie; Entomologist; Entomology
Environ., Environment
EPA, Environmental Protection Agency
ERL, Environmental Research Laboratory
ERDA, Environmental Research & Development Administration
ESSA, Environmental Satellite Services Administration
Establ., Establishment
Ethol., Ethology
Evol., Evolutionary
Exp., Experiment; Experimental
Explor., Exploration

599

Ext., Extension
Exterm., Extermination

Fac., Faculties; Faculty
FAO, Food and Agriculture Organization
Fed., Federal; Federation
Fenn., Fennici
Fish., Fisheries; Fishing
Fisk., Fiskeri
Fla., Florida
For., Forestry; Forests
Foreni, Forening
Found., Foundation
FWS, Fish and Wildlife Service

G., Giornale
Ga., Georgia
G.B., Great Britain
Gard., Garden
Gaz., Gazette
Gen., General
Geochim., Geochimica
Geogr., Geographic; Geographical; Geography
Geol., Geological; Geologie; Geology
Geophys., Geophysical
Ges., Gesellschaft
Gidrobiol., Gidrobiologlcheskogo
Gov., Government

Harv., Harvard
Havunders., Havundersogelser
Hebd., Hebdomadaires
Helgol., Hogolander
Helminthol., Helminthological
Herb., Herbarium
Herpetol., Herpetological
Hist., Historical; History
Hortic., Horticulture
Hwy., Highway
Hydraul., Hydraulic
Hydrobiol., Hydrobiologie; Hydrobiology
Hydrosci., Hydroscience

IBP, International Biological Program
Ichthyol., Ichthyological
Idrobiol., Idrobiologia
Ill., Illinois
Immunol., Immunology
Inc., Incorporated
Ind., Indiana
Ind., Industrial; Industry
Inf., Information
Insp., Inspection
Inst., Institute; Instituts
Int., International; Internationalen
Inter., Interior
Intern., Internal
Invertebr., Invertebrate
Inv., Inventory
Invest., Investigacion; Investigation
Ist., lstituto
Ital., Italiano
IUCN, International Union for Conservation
    of Nature

J., Journal
Jan., January
Jpn., Japanese
Jugosl., Jugoslavica

K., Koninklijke
Kans., Kansas

Ky., Kentucky

La., Louisiana
Lab., Laboratory
Leafl., Leaflet
Lett., Letters
Libr., Library
Limnol., Limnologie; Limnology
Linn., Linnaean; Linnaeus
Lond., London
Los Ang., Los Angeles
Ltd., Limited

M.A., Master of Arts
Mag., Magazine
Malacol., Malacological
Mammal., Mammalogy
Manage., Management
Manuf., Manufacturer
Mar., Marina; Marine
Maricult., Mariculture
Marit., Maritime
Market., Marketing
Mass., Massachusetts
Mater., Materials
MD., Maryland
Meas., Measurement
Mech., Mechanic
Med., Medical; Medicine
Medd., Meddeleser
Meeresforsch., Meeresforschung
Meeresunters., Meeresuntersuchungen
Mem., Memoir; Memoirs; Memorial; Memorie
Memo., Memorandum
MESA, Marine Ecosystems Analysis
Meteorol., Meteorology
Methodol., Methodology
Mic., Microfilm
Mich., Michigan
Microbiol., Microbiologica; Microbiology
Microsc., Microscopical
Midl., Midland
Mimeogra., Mimeograph
Minist., Ministry
Minn., Minnesota
Misc., Miscellaneous
Miss., Mississippi
Mitt., Mittelilungen
Mo., Missouri
Model., Modelling
Mon., Monthly
Monit., Monitoring
Monogr., Monograph; Monographiae
Mont., Montana
Morphol., Morphology
Mosq., Mosquito
M.S., Master of Science
Mus., Museum
Mycol., Mycological

N., North
Nat., Naturae; Natural; Naturalist; Naturaliste; Nature
Natl., National
Naturhist., Naturhistorisk
N.C., North Carolina
Ned., Nederlandsa
Neerl., Neerlandaises
N. Engl., New England
Neth., Netherlands
Newsl., Newsletter

N.H., New Hampshire
N.J., New Jersey
NMFS, National Marine Fisheries Service
No., Number
NOAA, National Oceanic and Atmospheric
  Administration
North., Northern
Northeast., Northeastern
Northumberl., Northumberland
Not., Notulae
Nov., November
Novit., Novitates
N.S., New Series
N.S.W., New South Wales
N.T.I.S., National Technical Informa-
  tion Service
Nucl., Nuclear
N.Y., New York
N.Z., New Zealand

OBS, Office of Biological Services
OCS, Outer Continental Shelf
Occas., Occasional
Oceanogr., Oceanographic; Oceanography
Oceanol., Oceanology
Off., Office
Okla., Oklahoma
Ool., Ooology
Opusc., Opuscula
Oreg., Oregon
Organ., Organization
Ornithol., Ornithological
O-Va., Obshchestva
OWRR., Office of Water Research

Pa., Pennsylvania
Pac., Pacific
Palaeoecol., Palaeoecology
Palaeoclimatol., Palaeoclimatology
Palaeogeogr., Palaeogeography
Paleontol., Paleontology
Pamphl., Pamphlet
Pap., Paper; Papers
Parasitol., Parasitology
Pathol., Pathology
Penn., Pennsylvania
Pesq., Pesquera
Pestic., Pesticides
Petrol., Petroleum; Petrology
Phila., Philadelphia
Philos., Philosophical
Photogram., Photogrammetry
Phycol., Phycology
Phys., Physical; Physics
Physiol., Physiologia; Physiological; Physiology
Phytol., Phytologist
Plant., Plantarum
Pls., Plates
Pol., Polish; Polonica; Polska
Pollut., Pollution
Polytech., Polytechnica; Polytechnical
PP., Pages
Prefect., Prefectural
Preserv., Preservation
Proc., Proceedings
Prof., Professional
Prog., Progress; Progressive
Prot., Protection
Protozool., Protozoology
Pts., Parts
Pubbl., Pubblicazioni

Publ., Publication; Published; Publishers
  Publishing
Purif., Purification
P-V., Proces-Verbeaux

Q., Quarterly
Qual , Quality

R., Royal
Radiobiol., Radiobiological
Rapp., Rapports
Rec., Records
Reg., Regional
Rep., Report
Repos., Repository
Res., Research
Resour., Resources
Reun., Reunions
Rev., Review; Revue

S., South
Saf., Safety
Sanit., Sanitary
Sat., Saturday
Sb., Sbornik
S.C., South Carolina
Sch., School
Sci., Science; Sciences; Scientific; Scientist
Sec., Section
Sediment., Sedimentary
Sem., Seminar
Senckenb., Senckenbergiana
Ser., Series
Serv., Service
Sess., Session
Shellfish., Shellfisheries
Sk., Skoly
Smithson., Smithsonian
Soc., Society
South., Southern
Southeast., Southeastern
Southwest., Southwestern
Spec., Special
SSRF, Special Scientific Report Fisheries
Stand., Standards
Stat., Statistical; Statistics
Stn., Station; Stazione
Stns., Stations
Stud., Studies; Study
Suppl., Supplement; Supplementary
Surg., Surgical
Surv., Survey
Symp., Symposium
Synop., Synopsis
Syst., Systematics; Systems

Teach., Teacher
Tech., Technical
Technol., Technologicke; Technologie; Technology
TM, Technical Memorandum
Tenn., Tennessee
Terr., Territory
Test., Testing
Tex., Texas
Theor., Theoretical
Theriol., Theriologica
Tijdschr., Tijdschrift
Toxicol., Toxicology
Trans., Transactions
Trop., Tropical
Tr., Trudy

U.K., United Kingdom
U.N., United Nations
Univ., University
Unpubl., Unpublished
U.S., United States

Va., Virginia
Ver., Vereins
Verb., Verhandlungen
Veroeff., Veroeffentlichungen
Vet., Veterinary
Vidensk., Videnskabelige
Vol., Volume
Vses., Vsesoyuznogo
Vys., Vysoke

Wash., Washington
Wet., Wetenschappen
Wildl., Wildlife
Wis., Wisconsin
Wiss., Wissenschaftliche
W.Va., West Virginia

Z., Zeitschrift
Zool , Zoologica; Zoological; Zoologie;
  Zoologischen; Zoologist; Zoology

Estuarine Ecosystem, 158-294; Lacustrine Ecosystem, 382-396; Marine Ecosystem, 30-107; Maritime Ecosystem, 108-157; Palustrine Ecosystem, 295-382; Riverine Ecosystem, 396-433; Upland Ecosystem, 434-478.

Lake Murray macrobenthos, 388
Lakes
  benthic invertebrates, 387-388
  eutrophic, 386
  fertility, 305
  fishes, 392-394,411
  food web, 388
  insects, 389
  nutrient cycling, 386-387
  oligotrophic, 386
  oxbow, 383,386
  trophy, 305
Largemouth bass, 392-393,411
Larvae
  fishes, 47-48,196,198,240,414
  insects, 234,335,404-407
  penaeid, 45-46
Laughing gull, 71-72,137,203
  breeding success, 138
  effects of toxicity on, 76
Law of Conservation of Matter, 1
Leadenwah Creek algae, 209,211-212
Least shrew, 156,473
Least tern, 26,138,142,145,270
Leatherback turtle (See Atlantic leather-
    back turtle)
Liberty County
  alligators, 289
  rookery, 365
Liebig's Law of the Minimum, 2
Limnetic subsystem, definition, 295,384
Limnetic zone, 159
Litter fauna, 109,112
Little blue heron, 368
  breeding success, 138
  food habits, 293
Little Cumberland Island
  Atlantic loggerhead turtle, 96,99
  cotton rat, 153,472
  herpetofauna, 128-134
  squirrels, 155
Little Egg Island
  rookery, 135
  royal tern nesting, 205
Little River algae, 40
Little River Inlet macroinvertebrates,
    176-177
Little St. Simons Island
  deer, 155
  loggerhead turtle, 99
Little Tybee Island
  loggerhead turtle, 99
  rookery, 367
Little Wambaw Swamp, 328
Little Wassaw Island, loggerhead turtle,
    99
Littoral subsystem, definition, 295,
    382-383
Live bottom, 36,52 (See also reefs)
Lizards, 127-128,131,133-134,457-459 (See
    also Herpetofauna)
Loggerhead turtle (See Atlantic logger-
    head turtle)
Long Bay algae, 40
Long-billed curlew, 26,257
Long-billed marsh wren, 257,370
Long Branch Creek haptobenthic diatoms, 302

Long County vegetation, 445,447
Longshore currents, 31
Long-tailed weasel, 478
Louisiana heron, 138,268,293
Low marsh, 212-213

Macgillivray's seaside sparrow, 257
Macroinvertebrates
  Annandale Plantation, 281-282
  distribution, 264
  diversity, 177,265
  dredging effects, 178
  intertidal flats, 263-264
  Kiawah Island, 176-177
  Lake Murray, 388
  Little River Inlet, 176-177
  man's impact on, 178
  Murrells Inlet, 176
  North Edisto River, 176-177
  North Inlet, 265
  North Newport River, 177
  North Santee River, 176
  Ogeechee River, 177
  Ossabaw Sound, 179
  pesticide impacts, 178
  Port Royal Sound, 176-177,227,229
  salinity effects, 176
  saltmarsh, 228
  Sapelo Island, 227,229
  Sapelo Sound, 177
  Sewee Bay, 178
  South Edisto River, 176,405
  South Newport River, 177
  South Santee River, 176
  species composition, 176
  St. Catherines Sound, 177
Magnolia Gardens rookery, 362
Malaysian prawn aquaculture, 281
Mammals
  dredging effects, 261
  dune, 153
  endangered species, 18-19,23
  estuarine, 259-262
  fire, effects of, 473
  Kiawah Island, 106
  man's impact, 106,157,261,473,476-477
  marine, 23,76-80
    stranding network, 80
    strandings, 77-79
Manatee (See West Indian manatee)
Man's impact
  biocides, 174-175
  birds (See Birds)
  channelization, 423,433
  clearcutting, 424,459,468
  damming, 432
  development and construction, 106,
    153-154,304
  drainage, wetlands, 423
  dredging and filling (See Dredging and
    filling)
  fishes, 50,65,95,199-200,424,432
  heavy metals, 49
  herpetofauna, 459
  impoundments, 258
  insects, 238-239
  invertebrates, 49,178,230

Estuarine Ecosystem, 158-294; Lacustrine Ecosystem, 382-396; Marine Ecosystem, 30-107; Maritime
Ecosystem, 108-157; Palustrine Ecosystem, 295-382; Riverine Ecosystem, 396-433; Upland Ecosystem,
434-478.

White ibis, 253
  breeding habits, 368
  food habits, 375
  nesting habits, 380
White shrimp, 55 (See also Penaeid shrimp)
White shrimp, mortalities, 50
White-tailed deer, 381,473-474
  control, 154-155
  distribution, 260
  food habits, 154,259,474
  Jekyll Island, 155
  Little St. Simons Island, 155
  population density, 154
  reproduction, 474
  subspecies, 154
White water-lily, 383
Whooping Crane Pond rookery, 363
Wild rice, 309
Willet, 138,293-294
Wilmington River smooth cordgrass, 220-221
Wilson's petrel, 69
Wilson's plover, 105,138
Windthrow, 318
Winyah Bay
  Atlantic sturgeon, 201
  bald eagle, 374
  eels, 418
  zooplankton, 171
Wolf Island loggerhead turtle, 99
Wood duck, 380,420
Wood ibis, 253
Wood stork, 26,380

Xeric slacks, 116

Yellowbelly slider, 355
Yellow breasted chat, 147
Yellow-crowned night heron, 368
Yellow perch, 411
Yellow rail, 374
Yellow rat snakes, 252
Yellow-rumped warbler, 153

Zooplankton, 344
  abundance, 43-44,172
  Altamaha River, 401
  Ashley River, 171
  Cooper River, 171
  Cumberland Sound, 171
  current effects, 49
  dispersion, 171
  Edisto River, 171
  effects of toxicity on, 49-50
  entrainment, 175
  food source, 82
  Francis Marion National Forest, 335
  impoundments, 280-282
  Kings Bay, 42
  lake, 387
  man's impact, 172,174-175
  metabolic waste, 42
  North Edisto River, 282
  North Inlet, 171-172
  North Santee River, 172
  Ochlockonee River, 401
  oil spill, impacts, 49

  origin, 120
  perturbations, 432-433
  pH effects, 335
  pond, 335
  population structure, 400-401
  predators, 44,387
  Price Inlet, 171
  reproduction, 171
  Riceboro Creek, 402
  river, 399
  salinity variations, 170
  Santee River, 171
  Savannah River, 400-402
  seasonality, 43,172-173
  Skidaway River, 171
  species composition, 42,387
  St. Marys River, 42-44, 171
  thermal shock, 175
  transport, 171
  Wando River, 171
  Winyah Bay, 171

Estuarine Ecosystem, 158-294; Lacustrine Ecosystem, 382-396; Marine Ecosystem, 30-107; Maritime Ecosystem, 108-157; Palustrine Ecosystem, 295-382; Riverine Ecosystem, 396-433; Upland Ecosystem, 434-478.

620

*U.S. GOVERNMENT PRINTING OFFICE: 1981--772-154

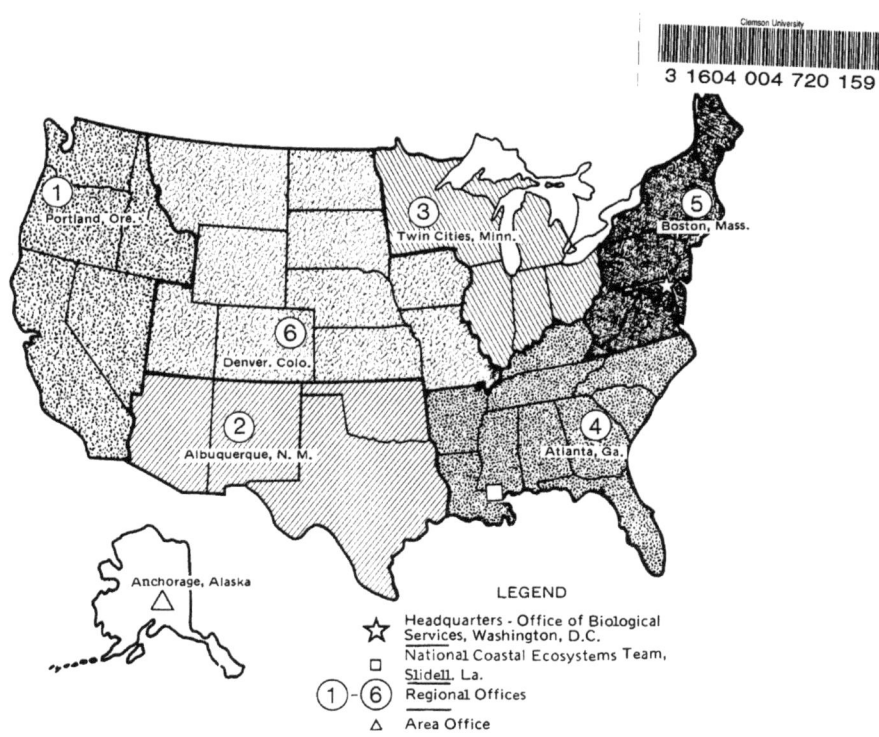

①
Portland, Ore.

③
Twin Cities, Minn.

⑤
Boston, Mass.

⑥
Denver, Colo.

②
Albuquerque, N. M.

④
Atlanta, Ga.

△
Anchorage, Alaska

LEGEND

☆ Headquarters - Office of Biological
Services, Washington, D.C.

□ National Coastal Ecosystems Team,
Slidell, La.

①-⑥ Regional Offices

△ Area Office

Lightning Source UK Ltd.
Milton Keynes UK
UKHW010135300119
336364UK00007B/518/P